Analytical Instrumentation

R. E. Sherman, Editor

L. Rhodes, Associate Editor

Analytical Instrumentation

R. E. Sherman, Editor
L. Rhodes, Associate Editor

Practical Guides for Measurement and Control

ISA

Printed in the United States of America

ISA
67 Alexander Drive
P.O. Box 12277
Research Triangle Park, NC 27709
U.S.A.

Library of Congress Cataloging-in-Publication Data

Analytical instrumentation : practical guides for measurement and control / R. Sherman, editor.
p. cm.
Includes bibliographical references.
ISBN 1-55617-581-7
1. Chemical process control--Instruments. I. Sherman, R. (Robert) II. Rhodes, L. (Larry)
TP155.75.A535 1996 95-51364
660'.283—dc20 CIP

About This Series

This volume is part of the Practical Guides Series developed and published by the ISA, the International Society for Measurement and Control.

The Practical Guides were conceived because of a shortage of published material in the field of measurement and control that bridges the gap between theory and actual industrial practice. Many books in the field have catered to the needs of technical students, who need to be oriented to basic control theory and concepts, or college-level readers, who are interested in engineering mainly from a classroom perspective. There are handbooks for practicing engineers that cover measurement and control, but these handbooks often devote only a chapter or two to topics that merit more attention. Within the Practical Guides Series, separate volumes address each of the important topics and give them comprehensive, book-length treatments. Each book in the series can be understood and used by technical students, sales engineers, sales personnel, and managers, and relied upon by those who have "real-life" industrial concerns such as correct application, safety, installation, and maintenance.

Another unique feature of the Practical Guides is the stress placed on the actual experience of measurement and control practitioners. The Practical Guides are overseen by various Volume Editors and a Series Technical Editor, who have extensive experience in measurement and control. The Volume Editors have been selected for their specific expertise in the volume topics, and bring together numerous Contributing Writers with even more specialized knowledge.

The Series Technical Editor, who is responsible for general technical consistency within each volume and across all volumes, helps guide the Volume Editors.

The Practical Guides capture the hard-earned experience of the writers and, by employing examples and recording anecdotal observations, make that experience as applicable for the reader as possible. Case studies, either hypothetical or based on real case histories, are used to illustrate typical situations and show how good planning and practical applications made the difference between success and failure. Some of this information has never been documented before.

This volume is designed to be at home in a library, in a classroom, or on the plant floor. The comfortable reading style, large pages, and frequent illustrations will contribute to ease of use. The page design uses graphics to "call out" some of the major points of the text, such as crucial safety checks and important examples. Each Practical Guide gathers widely scattered information in a single text, with bibliographies directing the reader to other sources.

Call for Participation

The major purpose of this series is to collect in one volume most of the existing practical knowledge about specific measurement and control subjects.

Additional material for inclusion in subsequent updates of this volume is most welcome.

If you wish to contribute any helpful hints, case studies, or other material, please contact:

Manager, Acquisition and Development
Training & Publication Services
P.O. Box 12277
Research Triangle Park, NC 27709
Phone: (919) 549-8411
Fax: (919) 549-8288

Table of Contents

Preface

Your volume editor and the many process-experienced authors of this volume have all struggled with the question "What should be written for a volume titled *Practical Guide Series: Analytical Instrumentation Volume*? What will really constitute 'Practical Guidance?'

Most of the chapter authors have more than 20 years of experience, either applying analyzers in process plants or applying analyzers for process manufacturers. I personally knew most of them before they contributed to this volume. I am now more knowledgeable for working with those I did not know prior to this volume. There are many more of you out there that could teach me and the Process Analyzer community about your application and expertise. Feel free to contact myself or ISA with your ideas and submissions. We realize that there are many topics not covered by this initial effort.

To a person, the current authors' problem was "How do I put much of what I know on paper? What small tip or sketch will make the difference between 'another' Process Analyzer installation and 'another successful' Process Analyzer installation?"

I hope that you will find this volume equally important and useful to your personal success in Process Analyzer applications. It is divided into two parts for your convenience:

Part I — *Implementation of Process Analyzers*

This part provides the infrastructure that is required for all process analyzer applications — Justifications, Interface details, Sample Conditioning, Calibration/Validation, SPC/SQC, Specification/Purchase details, Maintenance, and Training.

Part II — *Analyzer Application Fundamentals*

This part addresses Specific Types of process analyzers. It provides a bit of theory and a lot of Tips, Traps, Hints, and Hazards on the Process Application of many specific process analyzers grouped as:

- Physical Property Analyzers
- Electrochemical Analyzers
- Spectrophotometric Analyzers
- Compositional Analyzers

If you do not find your process analyzer application in here, contact us and help us write the Second Edition!

In the following pages you will find a bit of theory and a lot of practical knowledge. For more theory and specific analyzer reference (current to 1986), I cannot recommend any volume more highly than Clevett's *Process Analyzer Technology* (Wiley Interscience, 1986 — available through ISA). The only major current Process Analyzer missing from this volume is the mass spectrometer and that is treated on one excellent and comprehensive chapter in Nichols' *On-Line Process Analyzers* (Wiley Interscience, 1988). I personally have three copies of Clevett's text — one at the office, one at home (to clarify "night thoughts"), and one in my process safety equipment travel bag for use at jobsites. The book is truly that useful.

I would like to thank My Wife for Her many nights and weekends of sacrificing me to the basement office for this volume, Larry Rhodes for recruiting the authors of the first nine (9) chapters and graciously trusting me to do the honors for the last twenty-eight (28), and Leo F. Hohnstedt, PhD, for his emphasis on practicality throughout my undergraduate and graduate science education.

About the Editors

About the Volume Editor

Robert E. Sherman has more than 25 years of experience in the design and commissioning of process analyzer and continuous emission monitor systems (CEMS). These systems are monitoring refining, petrochemical, electric power generation, cement production, steel production and many other industrial process plants. Specific expertise includes design of sample probes, sample transfer lines, sample conditioning systems, CEMS for regulatory reporting, analyzer system cabinets and walk-in climate-controlled analyzer shelters. He has managed process analyzer projects including specification, engineering, procurement, fabrication, acceptance testing, field commissioning, documentation, operations and maintenance training.

Bob has been on the Technical Advisory Boards of DuPont (photometric and moisture analyzers — now a division of Ametek), Applied Automation, Inc. (gas chromatographs — now a division of Hartmann & Braun who is owned by ABB), and Precision Scientific, Inc. (physical property analyzers — now a division of Varlen).

Bob is an Invited Speaker to Pittcon, has presented many training sessions to industry technical personnel, and has published/presented more than 30 technical articles and papers relevant to process analysis, process control and process industry management issues. He is currently authoring a monograph, "Process Sampling Technology," to be published in June 2000.

His employment has included three years with Beckman Instruments, 12 years with Amoco Oil Company, five years with Fluor Daniel, three years with Applied Automation, and six years operating his own design/fabrication/training and consulting firm, Consulting Technologists Inc./Process Analyzer System Technologists (CTI/PAST). Bob is currently employed with Dearborn Professional Services Inc. as Manager - Advanced Solutions Group. He can be reached via Internet at roberteshermans@compuserve.com or at res@dps-hq.com.

About the Associate Editor

Larry Rhodes is a Senior Process Analyzer Engineer in the Control Engineering Directorate of Shell Oil Products Company. After receiving his Ph.D. in Analytical Chemistry from LSU in 1979, he taught Chemistry prior to joining Monsanto Chemical Company as a Process Analyzer Chemist. He joined Shell in 1983 and has been primarily involved with providing technical support to multiple facilities, and implementing process analyzers on projects. He is currently a member of the Analyzer Standards Committee of the API.

About the Series Technical Editor

John W. Bernard has worked on the leading edge of industrial automation systems for 35 years. He has extensive experience in process management and research. Among his many achievements is installing the first digital computers for direct process control. Bernard has written numerous papers on computer control and systems engineering and has been deeply involved in standards activities.

Contributors

Biographies of each of the following contributors are included at the end of their respective chapter(s).

Ed Baldwin
Moisture Systems Corp.

Jim Converse
Sterling Chemical

Rick Cooley
Eli Lilly

Bill Dailey
Process Analyzer Resources

Conard Donnell
Sun Oil (Retired)

Bob Farmer
Applied Automation, Inc.

Ken Fletcher
Foxboro Analytical (Siebe)

Buzz Goins
Consultant

John Groetsch, Jr.
K-Patents Inc.

Bill Hardy
Shell Oil Products Co.

Milton Jackson
Analytical Detectors Corp.

Dieter Kita
COSA Instrument Corp.

Jane Konieczny-Burdick
Foxboro Analytical (Siebe)

John Konopacki
Analytical Detectors Corp.

Jerry McCleary
Milton Roy (Retired)

Bruce McIntosh
KVB/Analect

Terry McMahon
McMahon Technology Associates

Patrick Moore
Houston Atlas Inc.

Gary Nichols
Sterling Chemical

Byrd Parmalee
American Air Liquide (Retired)

Larry Rhodes
Shell Oil Products Co.

Bob Saltzman
Saltzman Associates

Robert E. Sherman
Fluor Daniel

Jim Tatera
Dow Corning Corp.

Ian Verhappen
Syncrude Canada Ltd.

Fusheng Wang
Shell Oil Products Co.

Edward L. Wright
Edward L. Wright & Associates, Inc.

Dennis Yavorsky
Scientific Pittsburgh

Introduction

Modern industrial process analytical instrumentation was conceived in the late 1930s in the research laboratory of I. G. Farbenindustrie at Ludwigshafen, Germany, by Dr. K. F. Luft and Dr. E. Lehrer, who developed the infrared absorption recorder and the magnetic oxygen recorder, respectively. These instruments were born in the instrumentation workshop at the I. G. Plant in Oppau. More than 50 years later, nondispersive infrared absorption using the Luft acoustic detector and oxygen measurement by paramagnetic susceptibility are still widely used in industrial applications.

Although the development of instrumental techniques for qualitative and quantitative chemical analysis began earlier in the 20th century, this activity was confined to scientific research laboratories. Early analytical instruments were designed and fabricated by the researchers themselves. The commercial analytical instrument industry began to form in the 1930s. Beckman Instruments' commercial pH meter (patent application filed October 12, 1934) and later the ultraviolet spectrophotometer (Model DU-1940) were among the first such developments. Applied Research Laboratories' optical emission spectrometer (QUANTOMETER) for metal/mineral assays was another early instrument manufacturing venture.

Mine Safety Appliances (MSA), an experienced hand at toxic vapor detection, had applied a Hopkalite carbon monoxide sensor to control the ventilating apparatus in Boston's Sumner Tunnel as early as 1929. Other early analytical sensors and instruments were used outside the laboratory environment, but none could be considered precursors of present day on-stream process analyzers as were those at I. G. Farbenindustrie.

In the spring of 1946, a four-man team from the British Intelligence Operations Subcommittee (BIOS, Trip #2081) interviewed personnel at 23 chemical plants and 18 instrument firms in compiling their report, "Instrumentation and Control in the German Chemical Industry" (BIOS Final Report No. 1007, undated). The stated purpose of this investigation was "to study the application of industrial instruments in the German chemical industry; to examine the designs and to investigate the manufacture of standard and special instruments; to obtain information on recent research and development of industrial instruments."

The BIOS team determined that several hundred infrared absorption recorders and up to 15 magnetic oxygen recorders were in use during World War II. The investigators noted that newer I. G. plants such as those at Oppau and Huels were extensively instrumented, while older facilities such as those at Mainkur and Leverkusen had comparatively few modern instruments. Two of I. G.'s most extensively instrumented plants, Leuna and Schkopau, were in the Soviet Zone and consequently could not be surveyed.

The investigating team — team leader C. H. Gregory, H. B. Appleton, A. P. Lowes, and F. C. Whalen — recognized the importance of the new analysis instruments (italics added):

> *The development in recent years of the infrared absorption meter and the magnetic oxygen recorder by the I. G. Farbenindustrie represents a great advance*. The former instrument will find wide applications in the future. As far as could be ascertained, most of the new developments on industrial analysis instruments appear to originate from the research laboratories of the I. G. Farbenindustrie. In spite of the above progress, *analysis instruments in general still require much more care and attention than flow, pressure, temperature and level instruments and few, if any, are sufficiently reliable for automatic control purposes.*
>
> With certain important exceptions it is concluded that there are available in this country [Great Britain] instruments as good as any in Germany for flow, pressure, temperature and liquid level measurements. *Regarding analysis instruments, it would appear that there is something to learn about the development of such meters.*

Among instruments "evacuated" to Great Britain from I. G.'s plant at Huels were:

- One magnetic oxygen recorder
  (Range 0-2% O2 in N2, complete with all accessories)
- One infrared absorption recorder
  (Range 0-0.5% CO, complete with 6-month supply of charts)

Hartmann & Braun (Frankfurt am Main) was the original licensee for the I. G. infrared absorption recorder (URAS-Ultrarotabsorption schreiber).

> In general the instruments made by the I. G. Farbenindustrie appear to be heavier in design than is necessary, their finish leaves much to be desired and they give the impression of having been designed by laboratory physicists or engineers unfamiliar with modern technique. The instruments supplied by instrument firms are certainly better finished than the I. G. ones but on such points as standardization of parts, use of die castings and suitability of material for instrument cases, these firms lag behind practices in this country and America.

Dr. Engel, the chief instrument specialist at Huels, strongly favored instruments made by specialist firms. Dr. Ernst, head of the Instrument Department at Ludwigshafen, strongly favored those built at the Oppau workshop.

Instrument safety in "flameproof areas" was another critical factor:

> The design of safe circuits for many instruments such as the infrared absorption meter, the magnetic oxygen recorder, electrical resistance meters, gas thermal conductivity meters, the Hartmann & Braun photo-cell recorder, has not been considered possible up to the present time. If possible, these instruments are housed outside of the danger area. If this is not possible, the Safety Officer has to approve the use of these instruments if they are essential to the process.

The Safety Officer in I. G. plants had the final authority for determining whether the risk of an uncertified instrument in a flameproof area was justified.

By the late 1940s, process analyzers began to be applied in U.S. refineries and chemical plants. Initially, infrared absorption photometers using the Luft detector as well as fixed interference filters, paramagnetic oxygen analyzers for combustion control, differential thermal conductivity analyzers for hydrocarbon gases, and distillation-type analyzers for refinery streams predominated.

By the late 1950s, gas chromatography had become the preferred analytical technology in the hydrocarbon processing industry because of its ability to identify quantitatively individual chemical species within a complex mixture of components with similar spectroscopic characteristics. Union Carbide's plant in South Charleston, West Virginia, was a leader in developing on-stream applications for gas chromatographs. On-stream process gas chromatography grew to become the most important class of process analysis instrumentation by the late 1960s.

It was also in the late 1950s that the DuPont Company developed an on-stream application for fixed-filter ultraviolet-visible photometry in the context of a new process for titanium dioxide pigment production. DuPont later organized an independent business unit to exploit this technology (process UV) commercially and pursued it for about 30 years before divesting the unit in 1990 to AMETEK. In the early 1980s, DuPont enhanced its process UV photometer with photodiode-array detector technology to give this instrument multicomponent analysis capability in the context of work with the Department of Energy's Savannah River Nuclear Plant.

By 1960, process analyzer technology was being applied routinely at large petroleum refineries and petrochemical plants. Except for combustion control applications, on-stream analyzers were not used extensively outside these industries. The analyzer complement at Standard Oil Company of New Jersey's large Baton Rouge refinery in shown in Table I-1. Suppliers of this instrumentation are listed in Table I-2.

Table I-1. Baton Rouge Refinery: Analyzer Complement (1960)

Available Analyzers	Installed Cost, dollars	Suppliers*
Gas chromatograph (GC)	10,000-15,000	2, 3, 4, 10
Colorimeter	8,000	2
Densitometer	8,000	12
Final boiling point (FBP)	9,000	5, 11, 13
Flash point	10,000	11
Hydrogen sulfide (Pb acetate tape)	11,000	9
Ionization chamber (trace gases, ppb)	10,000	8
Initial boiling point (IBP)	5,000-7,250	5, 13
Infrared (NDIR)	8,000-12,000	2, 6, 8
Moisture (electrolytic)	7,000	2, 3, 7
Moisture (heat of adsorption)	14,000	8
Differential refractometer	6,000-10,000	3, 4
Reid vapor pressure (RVP)	7,000	11
Viscometer (VIS)	8,000	5
Ultraviolet (UV)	10,000	1

* See Table I-2.

The emergence of on-line process control computers in the 1960s accelerated the application of on-stream analyzers as process analytical data became more crucial to process control strategies. These on-line computers also expedited the adoption of on-stream analyzers by the pulp and paper, primary metals, cement/glass, and other industries. These industries had not used analog feedback control extensively because of their process characteristics, but were able to use computer-based strategies with calculated performance parameters. Analytical data were often crucial to developing these calculated parameters.

As digital process control systems (voracious consumers of process data) proliferated in the 1970s, the need for fast multicomponent on-stream analyzers became apparent. Dow Chemical pioneered the use of on-stream mass spectrometers in an ethylene oxide process at Freeport, Texas, in 1975. This technology spread to other applications with relatively simple chemistries that allowed molecular-ion collection to be used (ammonia synthesis and fermentation off-gas analysis). Other techniques for fast multicomponent analysis including Fourier transform infrared (FTIR) and Raman spectroscopy remain largely experimental despite extensive development efforts over the last 10 to 15 years.

Table I-2. Baton Rouge Refinery: Analyzer Suppliers

In 1960	Known as (in 1996)
1. Analytic Systems Co.	1. Unknown
2. Beckman Instrument Co.	2. Rosemount
3. Consolidated Electrodynamics Corp.	3. Unknown
4. Greenbriar Instruments, Inc.	4. ABB Process Analytics
5. Hallikainen Instruments, Inc.	5. Precision Scientific Petroleum Instruments
6. Liston-Becker	6. Unknown
7. Manufacturers Engineering & Equipment Corp.	7. MEECO
8. Mine Safety Appliances Co.	8. MSA
9. Minneapolis-Honeywell Rubicon Instruments	9. Honeywell
10. Perkin-Elmer Corp.	10. Perkin-Elmer Real-Time Instruments
11. Precision Scientific Co.	11. Precision Scientific Petroleum Instruments
12. Precision Thermometer & Instrument Co. (Densitrol)	12. PRINCO
13. Technical Oil Tool Co. (TOTCO)	13. Precision Scientific Petroleum Instruments

The range of analytical techniques used in on-stream applications continues to expand. Ion chromatography has proved effective in high-purity water monitoring, in electronics manufacturing and nuclear power generation. Supercritical fluid chromatography has been used with high-boiling streams in petroleum refining applications. Microwave spectroscopy is being used to control power plant stack-gas stripping processes. Amoco Chemical and others are exploring on-stream process nuclear magnetic resonance (NMR) applications. Emerging biochemical processes will present continuing opportunities to apply on-stream fluorescence and other spectrophotometric methods. On-stream high-performance liquid chromatography (HPLC), despite years of effort by highly skilled firms, has not yet been applied to a broad class of applications.

This treatment of process analytical technology, by a distinguished array of experts, chronicles more than 50 years of process analyzer development — from its origin in the research laboratory at Ludwigshafen in the late 1930s to a dynamic worldwide technology in the mid-1990s. It is ISA's hope that this body of knowledge will be helpful as a launching platform to those who will lead the development of process analytical technology in its second half-century.

About the Author

Terry McMahon was a member of IBM's original market development group for computer-based process control in the early 1960s. He made IBM's 100% Club in 1964-65 selling the IBM 1710 and 1800 process control computers. As marketing manager of Realtime Systems Inc. in 1966-68, he participated in early applications of computer control to multiple process GC applications at Sinclair Refining and other sites. An independent consultant since 1968 (McMahon Technology Associates), he has implemented hundreds of market and technology studies, many dealing with process analyzer markets and new analytical technologies for process use. Terry has a bachelor's degree from M.I.T. as well as master and doctorate degrees from Yale in chemical engineering. His "Around the Loop" column is published monthly in *Control* magazine.

Edward L. Wright has been involved in process analyzer systems design since 1956 and is currently president of Edward L. Wright and Associates Incorporated, which provides analyzer systems consulting.

An ISA Fellow since 1990 and member since 1954, he was cited for advancements in the application of online process analyzers through innovation in sample handling, systems integration and maintenance practice. Mr. Wright is a member of the Editorial Board of Process Control Quality and an annuitant from Exxon Research and Engineering where he was an engineering associate specializing in process analyzer systems. His career started at the Bayway Refinery of the Esso Oil Company in 1954 in process control and switched over to process analyzers in 1956.

Mr. Wright started a commercial prepackaged analyzer system operation in 1962 at Customline Control Products. Other employers include Comsip Customline Division, Xertex United Controls Division, and Henderson Industries. He holds a BSME from Howard University in Washington, D.C.

Part I

Implementation of Process Analyzer Systems

The application of analyzers involves a great deal more than the selection of the most appropriate analysis technique. Many companies today are looking at the total "cost of ownership" of an analyzer that may have an operational life of 10 to 20 years. The initial capital cost of an analyzer is a small portion of this total expenditure. It is not uncommon for the total installation (analyzer, shelter, sample system, sample lines, etc.) and startup costs to exceed 3 times the cost of the analyzer alone. Long-term maintenance costs include manpower, utilities (power, instrument air, nitrogen, carrier gases, steam, etc.), spare parts, calibration standards, etc. The total cost of owning and operating an analyzer over its lifetime may well exceed 10 times the initial cost of the analyzer alone. Decisions made early in the application process can have a significant impact on the total lifetime costs. The first part of this volume will deal with issues common to the application of all analyzers.

There are four basic justifications for the installation of analyzers: legal, environmental, safety-related, and economic. The application of profitability-based instrumentation is undergoing increased scrutiny, and analyzers are no exception. Chapter 1, "Typical Analyzer Application Justifications," provides assistance and methods for justifying analyzers. Chapter 2, "Interfacing with Analyzer System," describes the various ways analyzers transfer the analysis data to an information or control system.

It is a widely-held opinion among the experienced practitioners of analyzer technology that the success or failure of an analyzer application is more dependent upon the proper design and construction of the analyzer sample system than on the choice of analysis method among several manufacturers. Chapter 3, "Sample Conditioning Systems," provides guidance on the proper design and engineering of sample preparation systems.

No less important than proper sample conditioning is the choice and design of the calibration system or methodology for an analyzer. Chapter 4, "Principles of Calibration," describes the many aspects involved in the proper calibration of an analyzer. Mathematical treatments of calibration and process data are becoming common practice as tools employed to optimize calibration frequency or even predict when an analyzer is going to fail (so that it may be taken down and overhauled during a scheduled downtime of the process unit). Chapter 5, "SPC/SQC for Analyzers," describes these techniques, and various methods of mathematically treating process and calibration data.

Process analyzers are often compared to off-line results obtained by operators using laboratory equipment. Chapter 6, "Validation of Process Analyzers," describes standard methods for validating process analyzers (number of samples required, frequency of calibration, etc.). Note that validation relates to calibration as precision relates to accuracy in the big picture.

One an analyzer has been justified, the analysis technique chose, the sample conditioning techniques considered, along with necessary validation testing to off-line analyzers, it is then desirable to purchase the analyzer system in the most cost-effective manner. Chapter 7, "Specification and Purchasing of Analyzers," provides guidance in the commercial aspects of purchasing an analyzer system. This often includes preparing a specification which can be distributed for bid submittals, preparing a bid analysis after all bids are received, resolving technical ("... or equal") issues, and obtaining clarification on all aspects of the bids received. The specification package for an analyzer system often includes information on required shelters and field installation details.

Chapter 8, "Training Aspects," addresses one of the biggest problems facing personnel that will be responsible for the long-term maintenance of the analyzer — proper training. Whether it is provided by a "system's integrator" or the analyzer manufacturer, proper training should include troubleshooting tips and techniques, and "over-the-shoulder' type (hands-on) training for the routine maintenance tasks that will be required.

It is clear from periodic surveys of industries that maintenance groups may be organized in many different ways, and still provide effective maintenance practices. Chapter 9, "Personnel and Organization Issues," compares various maintenance organizations and how personnel in these organizations are trained and developed.

The issues dealt with in Part I are common to all analyzer applications, and may be summarized as follows:

Introduction to Process Analyzers

Chapter 1	Typical Process Analyzer Application Justifications
Chapter 2	Interfacing with Process Analyzer Systems
Chapter 3	Process Sample Conditioning Systems
Chapter 4	Principles of Process Analyzer Calibration
Chapter 5	SPC/SQC for Process Analyzer Systems
Chapter 6	Validation of Process Analyzers
Chapter 7	Specification and Purchase of Process Analyzer Systems
Chapter 8	Process Analyzer System Training Considerations
Chapter 9	Process Analyzer System Maintenance: Personnel and Organizational Issues

Each Chapter in Part II of this volume will provide detailed information on individual analysis techniques commonly employed in industry.

1

Typical Process Analyzer Application Justifications

This chapter addresses the economic aspects of process analyzer installations. Discussed first are simple need-to-install situations, followed by "almost-need-to-install" cases based on safety and loss prevention. Next is the more common situation of trying to sell a system based on how much extra revenue its operation will return back to the company after it is installed and commissioned. This section also contains all the formulas needed to calculate what those numbers might be.

Included in this chapter are a number of examples of various working installations that you can either implement at your facility or use as a catalyst to develop unique applications of your own. With the rapidly changing technology and increased power of the analytical techniques available today, the applications out there waiting to be identified and developed to profitable advantage are many.

Compliance with Government Regulations

Environmental regulations are in a continuous state of flux and can be quite different from country to country, state to state (province to province), and even county to county. It is therefore imperative that the information in this section be reviewed with your present regulatory agency to confirm its correctness.

Continuous Emissions Monitoring Systems (CEMS)

In today's environmentally-conscious society, the products being emitted from your facility must not only be monitored but often controlled as well. Continuous emission monitoring systems (CEMS) perform this task around the clock (continuous!). In addition to providing a monitoring function in many cases, these devices also provide a monetary return, either as an input on a closed-loop control system or as verification of the actual "production" of a controlled substance which can be traded on the stock exchange (SO_2 credits). CEMS are packaged in a variety of formats, and a preponderance of technology is available with which to make the necessary measurements.

In the United States and many other industrialized countries, "clean air" legislation has mandated that facilities continuously monitor emissions. In the U.S., the two main pieces of legislation are 40 CFR 60, 40 CFR 75 (and soon 40 CFR 64 — Compliance Assurance Monitoring System [CAMS]), which define how and to what level of accuracy emissions must be measured.

STACK GAS MONITORS

Stack gas monitoring systems required by State or Federal regulation are commonly referred to as CEMS. As the name implies, these units obtain their samples from a stack effluent stream which is normally emitted to the atmosphere. These units are tightly regulated with respect to their accuracy, repeatability, and availability.

Many difficulties are associated with CEMS, not the least of which is obtaining a representative sample and then maintaining it in a single-phase condition until it is ready for analysis. Many papers have been published in the past 5-7 years on the installation, operation, and design of CEMS units.

In any event, CEMS are a legislated requirement. If you are diligent, you can use the output signal for your profit as well. This can be done by using the CEMS signals as feedback signals to upstream processing units, being mindful of the lag time, in order to fine-tune operations as close as you feel comfortable to your facility's operational limits. Typically, facility production is directly linked to emissions level. Therefore, the closer you can operate to the emission limit, the more you can produce. A second profitmaking venture for your CEMS is in the stock market, where you can trade certain defined allowances for dollars per ton. This assumes, of course, that you have "capacity" to spare.

The objective with CEMS is to obtain a highly reliable system of the highest possible accuracy (operating as close to the permit limit as possible) for the least effort.

AREA MONITORS

Area monitors measure the level of a certain constituent in ambient air at ground conditions. Whereas stack monitors measure how much of what is coming from where, these units determine exactly how much of what is collecting where.

In some cases, area monitors are batch operations, in which samplers are installed in the field for a fixed period of time and then are collected and brought back to the laboratory for analysis. Another form of area monitoring now coming into more common use is the perimeter monitor system, which usually consists of a transceiver and reflector and a light beam (infrared or ultraviolet) along a facility boundary. The absorption of compounds of interest is detected in levels of concentration/distance. A high concentration over a very short distance will read the same as a low concentration over a proportionately longer distance.

Point monitors, such as gas detectors, are another form of real-time emission monitors, but they have the disadvantage of a very limited detection radius. Therefore, unless the gas of interest is concentrated in the same vicinity as the detector, it will pass unnoticed.

These are monitor devices are often a legislated requirement. The only time that they can provide some direct real-time benefit is if you are using perimeter monitors that can detect a leak or emission and report an alarm. Passive devices such as samplers will be unable to do so due to their batch processing.

SPECIAL CALIBRATION REQUIREMENTS

All CEMS must be calibrated on a regular basis, usually daily, to confirm that the system is maintaining reliable and accurate readings. Unique to the calibration procedure is the fact that the calibration material must flow through the entire system from the probe to the detector. Often, the material used for the calibration must also be independently certified, which makes the entire procedure very expensive.

Since many of these calibration materials are unstable in storage or have a legislated storage life, it is very important to investigate your calibration fluid needs early in the project for availability and delivery.

Since many of these calibration materials are unstable in storage or have a legislated storage life, it is very important to investigate your calibration material needs early in the project for availability and delivery.

DATA REPORTING REQUIREMENTS

Results of all monitoring systems and their quality control programs must be submitted. Each jurisdiction and industry has its own reporting requirements to the appropriate authorities.

The frequency of certain testing procedures is often a legislated function of how well the systems operate. Therefore, a company's quality assurance or control procedures can have a significant impact on the cost of compliance and operation of a monitoring program.

Remember, only chemicals that contribute to ozone formation will be defined as volatile organic compounds (VOCs) by the Environmental Protection Agency (EPA) for the purposes of the smog provision of the Clean Air Act Amendment - 1991.

Wastewater Monitoring

Most plants emit liquid waste streams as part of their operation. The most common liquid effluent is the universal solvent, water. In some cases, facilities operate their own treatment plant to remove impurities and assure that water leaving the plant is of the same or better quality than the original supply water. It is thus necessary to monitor not only the quality of the water entering the plant to be sure it will not be harmful to your process but to also monitor the quality of the water leaving the facility to assure it meets legislated outfall requirements.

BIOTREATERS

One of the most common ways to treat wastewater is with a biotreater. A large number of parameters in a biotreater must be controlled to assure optimum operation and to protect the essential bacteria in the vessel. For this reason, many companies are now using either a BOD (biochemical oxygen demand) or TOD (total oxygen demand) analyzer to continuously monitor the respiratory rate and hence the health of these cultures. The payback for this monitoring is greater throughput and longer run lengths for the same energy input, while also assuring that environmental regulations are met.

EFFLUENT DISCHARGE MONITORS

In addition to monitoring the oxygen demand requirements of effluent, it is also necessary to control such parameters as ion concentrations (especially for metals), pH, and color. As stated before, a sudden change in one of these parameters not only can result in a production loss and violation of your operating license but also can serve as an indication of an upstream operating problem. Don't forget to consider these extra "hidden" benefits when installing an effluent or other discharge monitor.

ODOR CONTROL SYSTEMS

pH and oxidation reduction potential (ORP) sensors are critical to the operation of wastewater treatment facilities to control the formation of chlorine offgassing. The pH sensor controls the NaOH level in the scrubber, while the ORP sensor controls the amount of NaOCl required. The pH must be controlled to a caustic state, thus keeping the chlorine in solution as hypochlorite ions. Obviously, the chlorine ions are required to effectively control the bacteria and other levels in the plant effluent [Ref. 1].

Ground-Water Monitoring

The earth has a large body of water deep beneath its surface. This ground-water table is the ultimate repository for all the moisture that falls to earth. Hence, by monitoring this water source it is possible to detect many "hidden" sources of potentially serious environmental damage.

DEEP-WELL MONITORING

The most reliable means of monitoring ground water is by deep-well monitoring, which, as its name implies, consists of drilling a well or bore into the ground-water table and installing the sensing devices or sample-removal devices at this level. With today's technology, it is often possible to monitor ground water using a small-diameter pipe driven into the ground and a fiber-optic cable/sensor in the pipe to obtain spectra in real time. These spectra are then used to provide a measure of the level of contaminants in the upstream area. A series of additional wells or bore sites are then used in conjunction with geological knowledge to determine the likely source of the emission.

Ground-water monitoring is becoming increasingly important. Legislation can place the responsibility for a site cleanup on the present property owner rather than on the previous owner(s).

UNDERGROUND STORAGE TANK (UST) MONITORING

All storage tanks eventually leak. It is therefore imperative that some form of monitoring program be in place to assure that when a leak occurs, it is detected quickly and repaired to minimize the damage to the surrounding environment and hence the associated cleanup costs.

The most common method of underground storage tank monitoring is to install the tank as a double-wall vessel and then monitor the condensed water in this space for the presence of contamination. If contamination is detected, the vessel can be emptied and repaired or replaced before any damage is actually done to the environment, thus saving the owner substantial cleanup and disposal expenses.

Environmental Controls

This section lists examples of environmental controls and how analytical systems are contributing to their success.

RECYCLING OPERATIONS

A Fourier transform infrared (FTIR) analyzer has been used to differentiate between the six most common types of recyclable plastics. The analyzer took five scans per sample and about 10 s per item to obtain a 95% success rate. Completion of the project will minimize the requirement for manual sorting of plastics at major recycling facilities.

POWER PLANT OPERATIONS

As a result of new limitations on SO_2 emissions by the EPA, many coal-fired facilities are using coal-blending to meet the new targets. By blending a low-sulfur coal with cheaper high-sulfur coal, the producer is able to generate power at a reduced cost. In one plant, an on-line x-ray analyzer was used to determine the sulfur levels in the coals as they were blended. As a result, the facility was able to use as little as 7% low-sulfur coal rather than the 30% that would have been needed if the traditional off-line blending method had been used [Ref. 2].

Safety Considerations

Safety is of paramount importance in any facility. Process analyzers or sensing devices are frequently used to monitor the levels of contaminants in a process stream, the environment or the ambient air. In many processing operations, the addition of even a small amount of a certain contaminant to the process stream can result in a catastrophic explosion. Even something as simple as carbon monoxide can result in death; hence the recent increase in sales of these sensors for installation in private residences.

No facility will reject a project proposal if it will increase the safety of plant operation. The difficulty is in quantifying and balancing the risk versus the benefit. Although it does not make sense to invest $1,000,000,000 to prevent an incident that will impact production to the tune of $10,000 once every 10 years, it may make sense in order to prevent a $10,000,000 occurrence every other year or potential fatalities. It is up to you as a professional to determine where the risks lie and how to minimize them. The life you save may be your own.

Personnel Protection

No one can quantify the value of a human life. Therefore, all workers in a facility must be protected. Many types of analyzers are installed to prevent unsafe situations from arising.

With today's trend toward miniaturization of electronics and sensors, it is not uncommon to have a multiple-compound sensor packaged in a lightweight unit the size of a cellular phone to provide 24-h coverage and alarming capability. In addition to such sensors, permanent fixed-compound sensing systems are often necessary to monitor an area before it is entered.

Government Regulations

Government regulation is a common justification for analyzer and safety-related equipment installations. The federal Occupational Safety and Health Administration (OSHA) imposes heavy penalties for noncompliance. One means for assuring compliance with regulations as they pertain to exposure levels is to continuously monitor those levels. Many new devices and unique systems are being designed to meet the challenge of continuous measurement of exotic compounds at very low levels. These levels can vary from jurisdiction to jurisdiction, so it is important to have a copy of your local regulations on hand. Check these regulations when purchasing a new system.

These levels can vary from jurisdiction to jurisdiction, so it is important to have a copy of your local regulations on hand. Check these regulations when purchasing a new system.

Material Safety Data Sheets (MSDS) are just one example of the new OSHA requirements. Every chemical on site must have an MSDS, including calibration materials, reagents, carrier gases, and maintenance fluids such as electrolytes or special cleaning solutions used on probes or sensors.

One of the simplest yet most vital analyzer applications is for monitoring the oxygen content in breathing air. These analyzers, generally in the form of portable personnel monitors, are typically found in airtight enclosures and vessels, climate-controlled lockers, vessels that are normally blanketed with inert gas, and breathing-air lines for workers on cylinder or compressed-air supplies.

Equipment Protection

Many processes operate under tight specifications and could easily fail by an explosion if contaminated by certain substances. Process analyzers are often the only means of assuring that a contaminant does not manifest itself and that, if it does, it is managed to a level below its hazardous limit.

A good guideline is to use 25% of the explosive limit, hazardous limit, or safety limit as the low-level alarming point and 50% of that limit as the point at which automatic shutdown or increased alarming actions occur.

Automatic Shutdown Systems

In automatic shutdown systems, it is imperative when an alarm occurs it be a true alarm and not the result of an equipment malfunction. Standards are being developed to identify means of minimizing the risk of false alarms while at the same time providing the level of protection required. A common method in use is a two-out-of-three "voting" process, in which two of three devices must register the same reading in order for an alarm to occur.

Automatic shutdown systems, as their name implies, initiate some method of isolating the equipment or material presenting the hazard, thus preventing it from increasing or spreading further. The alarm limits at which the actions occur are site and situation-dependent and must be determined by risk assessment.

Economic Justifications

There are likely as many analyzer justifications and ways to provide for their continued operation as there are analyzers and facilities. This section is intended to make you aware of the basis on which to make the justifications expected of you.

Always perform economic justifications in terms of quantity of product saved. Convert it to dollars at the very end. This allows you to update benefit calculations as commodity prices change without having to redo the entire calculation.

Always perform economic justifications in terms of quantity of product saved. Convert it to dollars at the very end. This allows you to update benefit calculations as commodity prices change without having to redo the entire calculation.

Remember, justifications must be made not only on the purchase of the analyzer but on its continued operation as well. In today's lean and mean operating environment, everyone is expected to do more with less. So if you expect an extra analyzer technician to be assigned to your analyzer once it is commissioned, that request had better be in the business plan submitted with the proposal. The analyzer that is not maintained is the analyzer that will not be operating, and without the analyzer operating your net benefit as a result of its installation will be negative.

Benefits

In advanced control projects that require process analyzer systems, it is not unusual for analyzer results to provide up to 70% of the entire program benefits. This breaks out to 40% as a result of the analyzer signals and an additional 10 to 30% which can be replaced by inferential models [Ref. 3].

The negative attitude of many users toward process analyzers has sometimes resulted in replacing analyzers with more complex computer control systems performing ever more complicated plant balance, modeling, and optimization routines. The difficulty here is that it is nearly impossible to model all the potential upset conditions and respond to them in real time. Thus, a successful operating analyzer system is the only solution.

Historically, it has been shown that if an analyzer has an availability of less than 95% for monitoring applications, 98% for closed loop control applications, or 100% for safety and environmental applications, it is not trusted by operations. It is for this reason and the fact that many process operators still do not trust analyzers, which at many installations are used for indication only, with the plant being run off of more "reliable" but dated laboratory samples.

Availability and utilization are defined as:

$$Availability = \frac{Operating\&Calibrated\ Analyzer}{ProcessOperatingtime}$$

$$Utilization = \frac{Analyzer-in-Control}{ProcessOperatingtime}$$

Remember, the benefits of an analyzer system are only being received when the analyzer and the plant are operating.

Remember, the benefits of an analyzer system are only being received when the analyzer and the plant are operating. Therefore, all benefits discussed below should be discounted by the operating days and utilization factor each year. For example, if the plant operated 350 days a year at 98% utilization, calculated benefits should be discounted by

$$\left(\frac{350}{365}\right) \times \left(\frac{98}{100}\right) = 94 = 94\%$$

However, since most analyzers are calibrated against the laboratory, there are always deviations. These deviations can come from some or all of the following sources: analyzer repeatability (statistical error), laboratory repeatability (statistical error), analyzer systematic error, laboratory systematic error, sampling errors at the analyzer, sample transport errors, and manual handling errors [Ref. 4].

If a process analyzer is used for indication only, seriously consider not installing it since it will not receive adequate maintenance and support needed to maintain a high availability (on-line) factor. This will perpetuate the self-fulfilling prophecy that "the analyzer doesn't work."

IMPROVED PROCESS CONTROL

Objectively, composition is not as difficult to control as flow, for example, but the specifications placed on product quality are so stringent that ordinary performance is seldom acceptable. The impurity of a product stream leaving a fractionator, for example, may be specified at 1.0 ± 0.2%. It is virtually impossible to regulate flow within ±1% in the unsteady state, yet the composition control is asked to perform five times as well. This is perhaps the greatest single reason why composition control has the distinction of being a problem area. Because quality can be measured to 0.1% is apparently reason enough to expect composition to be controlled to the same tolerance [Ref. 5].

A control loop is only as good as its weakest element. If the final control element is accurate to only ±1%, you will never control to ±0.2%.

A control loop is only as good as its weakest element. If the final control element is accurate to only ±1%, you will never control to ±0.2%.

Economic incentive for improved control is most likely to be found where consistent quality of a valuable product is important. Many unit operations are conducted at reduced capacity to assure that quality will surpass specifications, even with poor control. Naturally, the rate of payout from these sources will vary directly with the production rate of the plant. Large plants therefore encounter reduced risks [Ref. 6].

The normal procedure for justifying investments in improved process control is to estimate the economic improvement by moving the average closer to the specification limit. This assumes that the measurement is normally distributed and that by reducing the variability in the process one is able to move the average closer to the set point. It is this second step of moving the set point that provides the economic benefit.

As just stated, the basis for estimating benefits of a control system is to determine how much the mean (x) can be moved toward the set point. Assuming that the profit increases with the resultant move, the relationship is given by: $\Delta x = x_1 - x_0$, where x_1 is the new condition and x_0 is the original mean. The five rules that follow can be used to determine Δx [Ref. 7]. The following table is used for the first three of these rules, which assumes a normal distribution:

% Data Outside Limit	Probability, $P(m)$	Confidence Interval, %
25.0	0.675	75.0
15.86	1.000	84.14
5.00	1.650	95.0
2.27	2.000	97.7
0.14	3.000	99.86

Thus, for a 95% confidence interval where only 5% of the data would fall outside the control limit if the mean were moved, the probability is 1.65.

Same Limit Rule

In this case, the after-control data and the before-limit data are assumed to violate the set limit by the same percentage. The amount the process variable can be moved is therefore given by:

$$\Delta x = P(m)(s_0 - s_1)$$

It is typically assumed that $s_1 = s_0 / 2$, in which case:

$$\Delta x = P(m)(s_0 / 2)$$

where:

s_0 = Original standard deviation

s_1 = Standard deviation after closed loop control

Same Percentage Rule

In the event that it is possible to have an acceptable percentage of the data fall outside the control limit before and after implementation of the control strategy, the amount that the process variable can be moved is calculated by the same percentage rule:

$$\Delta x = (x_1 - x_0)\left(1 - \frac{s_1}{s_0}\right)$$

Once again, if it is assumed that $s_1 = s_0 / 2$, this can be simplified to:

$$\Delta x = \left(\frac{x_1 - x_0}{2}\right)$$

where x_1 represents the control limit.

Final Percentage Rule

If the main concern is the percentage of after-control data that exceeds the control limit, the following formula applies:

$$\Delta x = x_1 - (x_0 + P(m)\,s_1)$$

Once again, if it is assumed that $s_1 = s_0 / 2$, this can be simplified to:

$$\Delta x = x_1 - \left(x_0 + P(m)\frac{s}{2}\right)$$

Curvature Rules

Curvature rules are based on the assumption that the majority of the process profit can be fit by a polynomial function in terms of the process variable [Ref. 8]. For example:

$$profit = A + Bx + Cx^2 \quad \text{or} \quad profit = A + Bx + Cx^2 + Dx^3$$

which for the quadratic equation is given by:

$$\Delta profit = C(s_1^2 - s_0^2) \text{ for } C < 0$$

If $s_1 = s_0 / 2$, this can be simplified to:

$$\Delta profit = -3C\frac{s_0^2}{4}$$

For a cubic equation the equations are:

$$\Delta profit = (c + 3Dx)\,(s_1^2 - s_0^2) \quad \text{for } (C + 3Dx) < 0$$

If $s_1 = s_0 / 2$, this can be simplified to:

$$\Delta profit = -3\,(C + 3Dx)\,\frac{s_0^2}{4}$$

Achievable Operation Rule

This final rule is based on the assumption that a significant percentage of operation at or beyond a certain achievable limit implies that after the control change the performance will be allowed to continue at that limit of excursions, ignoring the statistical spread of the data. In this case, the process change can be calculated by $\Delta x = x_1 - x_0$.

ELIMINATION OF OFF-SPEC PRODUCT

Off-specification means that a product is outside the specified tolerances. To prevent this from occurring or at least to minimize its likelihood, facility operators run at a point a certain distance away from the target or specification limit. This difference or offset costs money, and it is through the minimization of this offset that financial gains are to be made. The above series of formulas present a number of ways to quantify these gains, but it is up to you as an engineer to work with laboratory, production accounting, and statistical personnel to determine how much you can realistically move toward the specification limit and still meet your customers' needs.

Costs

This is the one area that project managers and accountants really watch, so it is your job to assure that you are getting full value for your money. If you want to buy something other than the low bid, you will have to present solid technical reasons to justify why supplier H should be chosen over supplier L. Costs include life cycle costs which include not only the price of the equipment but also the cost of the peripherals, lifelong operation and maintenance, disposable materials such as reagents and calibration material, personnel training, and the salvage value of the equipment at the end of its useful life.

Costs include not only the price of the equipment but also the cost of peripherals, lifelong operation and maintenance, disposable materials such as reagents and calibration material, personnel training, and the salvage value of the equipment at the end of its useful life.

ENGINEERING/DESIGN

Many factors must be considered as part of the analyzer selection process, including selection, sampling system design, utilities, design practices/guidelines/ standards, and the overall environment in which the system is to be installed.

EQUIPMENT

The lion's share of the cost of a project is associated with the equipment and its installation. A rough rule of thumb for estimating the installed cost of a system is to obtain a price for the equipment and then double it to cover the cost of such items as power, control signals, utilities, and site preparation.

A rough rule of thumb for estimating the installed cost of a system is to obtain a price for the equipment and double it to cover the cost of such items as power, control signals, utilities, and site preparation.

A separate section of this chapter covers how to choose between two options on systems, but it does not really capture some of the intangibles that affect the total project cost, such as spare parts availability, training and familiarity (the necessary learning curve), and the quality and availability of local support. Communication between yourself and your supplier or representative is critical. Once an analyzer system is installed, it will typically last for 10 years. The two of you will have to work together during that period to assure that the system keeps working.

FIELD INSTALLATION

Part of the reason analyzer systems sometimes develop a "bad" reputation is that their environment was not considered when the installation was designed. Availability, calibration needs, maintenance, equipment configuration, associated laboratory procedures, and analyzer performance monitoring all affect the quality of the field installation.

DOCUMENTATION

You may have the greatest system in the world, but without proper system documentation no one will know how to keep it working. This may make you feel indispensable, but repeated middle-of-the-night telephone calls are not too appealing. A good documentation system will make it easier and possible for everyone involved to build what you want, how you want, and to keep it running after the fact. Training is also essential to assure your equipment is kept running.

Documentation includes user manuals, troubleshooting manuals, sample system drawings, spare parts lists, a theory-of-system explanation, and lastly, your home phone number for those rare occasions when all else fails.

TRAINING

Training those who will keep the analyzer system working is an important part of the project. The person who will be responsible for the actual operation of the system once it has been turned over from the project group should become involved in the project as early as possible. This person is also a key member of the factory acceptance test team. During the acceptance process, he or she will have an opportunity to talk to the people who know the equipment inside out and will thus be able to learn some important tips and tricks. Involvement will also give the maintenance person a sense of ownership and belonging as well as an opportunity to express concerns early on in the project.

Training of maintenance personnel should also be considered as part of the project costs, because the equipment must be kept running in order to obtain the claimed benefits. Plant engineers and process operators must also learn how the new analyzer system and its associated controls are going to affect future plant operation.

STARTUP

The big day is here, and you're ready to start up the new analyzer system. It is important at this stage that you involve everyone who will be responsible for the system. However, it is not necessary to let the whole world know what the system is doing. It is a good idea to operate the system to a local data acquisition device or to the background on your distributed control system for 2 to 4 weeks to let everything stabilize and work the bugs out.

It is a good idea to operate the system to a local data acquisition device or to the background on your distributed control system for 2 to 4 weeks to let everything stabilize and work the bugs out.

Remember, you can only make one first impression. If plant personnel see you tweaking things every day or so, the impression will be that the analyzer system will never work or that its results are unreliable. Once you have the system all lined out for a period of time, then have your console specialist add the signals and control loops to the local displays in the control room. Now all that the workers will see is a wonderful system giving them good, reliable numbers.

ROUTINE MAINTENANCE

For an analyzer to continue to work properly, trained craftsmen dedicated to its maintenance and support are essential. It is not unusual for some analyzers to require 10% of their original purchase price per year for maintenance [Ref. 3]. The use of an automated process analyzer maintenance monitoring program containing the following data and reports as a minimum should be considered: individual analyzer performance, preventive maintenance performed, preventive maintenance scheduling, device breakdown causes, control charts, availability, tag number reports, and standard deviation between calibrations for statistical quality control (SQC) charts [Ref. 9].

Routine maintenance must include not only preventive and predictive maintenance using SQC techniques but also data such as calibration frequencies, laboratory sample validations, and associated calibration standards. The laboratory calibration methods and associated analytical techniques can affect the process analyzer, especially since these measurements will be compared. It is thus important that good communication be maintained between laboratory and field personnel to assure that the same analytical method is being used for both measurements, that both results are being reported in the same units, and that laboratory (as well as process analyzer) standards are consistent.

It has been the author's experience that although measurements may appear similar and repeat well, as they should coming from the same source, the analytical method can often be markedly different, especially for a high-purity compound. Process analyzers will often measure the actual compound concentration as 95%, while the laboratory method will take the alternate route of solving for the compound by difference.

Normalization of results for a process analyzer is strongly discouraged because this can easily mask symptoms of an operating problem.

Another common area of difference is the normalization of results. Normalization of results for a process analyzer is strongly discouraged because this can easily mask symptoms of an operating problem. Since normalization always makes everything add up to 100%, and shares the error equally among all components according to their portion of the total, a sum of the measurements could be at 75% of the actual reading but still be transmitted "correctly" since the component ratios relative to each other are still correct.

It is far better to provide for a margin of error in your sum and then initiate an alarm than to always transmit a result that reads 100%.

Good maintenance practices include the following: a separate and devoted analyzer maintenance organization with close ties to the engineering group, good cooperation with the site laboratory, maintenance training, breakdown maintenance repair facilities, accurate logging of all analyzer performance data (with today's computer technology, this can easily become a highly automated procedure), breakdown and preventive maintenance data, coding of all maintenance actions in a consistent manner (one technician may call a procedure a calibration, while another will see it as part of the preventive maintenance program), and the use of control charts for performance monitoring.

The control charts should trend the introduction of calibration materials to the analyzer system as reference materials. A change in the calibration factors is made only if the control chart violates one of the set alarm or action limits; then the cause of the error should be corrected and the analyzer system recalibrated. If this procedure is not followed, introducing a new calibration factor every time will merely worsen control. You will be forever hunting around the true mean of the calibration sample by trying to correct for inherent system variability.

SPARE PARTS

It is inevitable that at some point a critical component will fail. When preparing quotations, most vendors will make available a list of recommended spare parts for service of varying lengths.

Before ordering these spares, however, ask your supplier and local representative about the extent of the inventory they carry. If they have what you need and can get it to you on short notice, there may be no reason for spare parts to collect dust on your warehouse shelf. Such items should be considered when performing your project risk analysis to determine just how critical the new analyzer system and its associated applications will be once the system is up and running. The key to making a good parts inventory system work is communication between yourself, your field people, and your supplier's local representative.

Investment Criteria

What may seem like a bad investment opportunity is often merely a bad timing situation.

Each company uses different criteria to determine whether an investment will proceed. These criteria will fluctuate, depending not only on the cost of borrowing money but also on the financial well-being of the company as a whole. Therefore, if your project is not justified today, don't discard all your hard work. In a different operating environment (i.e., a change in commodity prices or interest rates), the project may be given the green light. What may seem like a bad investment opportunity is often merely a bad timing situation.

RETURN ON INVESTMENT (ROI)

The rate of return on investment is ordinarily expressed on an annual percentage basis. The yearly profit divided by the total investment necessary represents the fractional return, and this faction times 100 is the standard percent return on investment. To account for the time value of money and the fact that if the money was left on deposit somewhere you would obtain a return on investment, the concept of discounted cash flow is normally used to determine a project's rate of return. This is called the rate of return based on discounted cash flow and is based on the amount of the investment that is unreturned at the end of each year during the estimated life of the project. An iterative trial-and-error approach is used to establish an interest rate that can be applied to yearly cash flow so that the original investment is reduced to zero after salvage and land values are excluded during the projected project life.

The following formula is used to calculate the interest (i) earned or applied during n years:

$$(1 + i)^n$$

For a discounted cash flow, the discount factor is calculated as:

$$d_n = \frac{1}{(1 + i)^n}$$

Thus, for example, to calculate the discounted cash flow rate of return for an item with a five-year life earning $30,000 in year 1, $31,000 in year 2, $36,000 in year 3, $40,000 in year 4, and $43,000 in year 5 and costing $100,000 fixed capital and $10,000 working capital with a salvage value of $10,000, the following two equations would be used and balanced. For the cash flow:

$$S = 30{,}000(1 + i)^4 + 31{,}000(1 + i)^3 + 36{,}000(1 + i)^2 + 40{,}000(1 + i) + 43{,}000$$

Against the future worth of the initial investment:

$$S = (100{,}000 + 10{,}000)(1 + i)^5 - 10{,}000 - 10{,}000$$

where the first term represents the capital invested and the two terms being subtracted are for the working capital employed and the salvage value. Solving these two equations by trial and error, we find that the interest rate necessary to set them equal is 20.7%.

The following table summarizes the output of an economic evaluation Excel™ spreadsheet that was developed by Thomas Ward of Clarkson University. Using this spreadsheet for which the listing is included in the appendix will allow for rapid estimation of the economic returns of a particular project while also serving as a "checklist" of items to be considered as part of the evaluation.

INPUT PARAMETERS

DISCOUNT RATE	0.1	PROJECT START	1993
DEBT/TOTAL CAPITAL	0	PROJECT END:	2002
FED INC TAX RATE	0.34	DEPRECIATION: STRAIGHT LINE	
INVEST TAX CREDIT	0	SALVAGE: NEGLIGIBLE	
INTEREST RATE	0	END-OF-YEAR FLOWS	
PRESENT TIME:	1993	LINE-ITEM INTEREST NEGLECTED	

LINE	END OF	1993	1994	1995	1996	1997	1998	1999	2000	2001	2002
	OPERATING YEAR	-2	-1	0	1	2	3	4	5	6	7
	DISCOUNT YEAR n	0	1	2	3	4	5	6	7	8	9
1	FIXED CAPITAL		-2000	-2000							
2	LAND		-200								200
3	WORKING CAPITAL			-400							400
4	TOTAL CAPITAL		-2200	-2400							
5	SALES REVENUE				4000	5500	6000	6000	6000	6000	6000
6	PROD COST [LESS DEPR & INT]				2400	2000	2000	2000	2000	2000	2000
7	CAPITALIZED FIXED CAPITAL			-4200							
8	CAPITALIZED TOTAL CAPITAL			-4820							
9	OPERATING INCOME				1600	3500	4000	4000	4000	4000	3800
10	INTEREST ON DEBT				0	0	0	0	0	0	0
11	DEPRECIATION [TAX BASIS]				600	600	600	600	600	600	600
12	NET TAXABLE INCOME				1000	2900	3400	3400	3400	3400	3200
13	FEDERAL INCOME TAX				340	986	1156	1156	1156	1156	1088
14	INVESTMENT TAX CREDIT				0	0	0	0	0	0	0
15	NET INCOME				660	1914	2244	2244	2244	2244	2112
16	CASH FLOW	0	-2200	-2400	1260	2514	2844	2844	2844	2844	3312
17	CASH FLOW + INTEREST	0	-2200	-2400	1260	2514	2844	2844	2844	2844	3312
18	DISCOUNTED FLOW										
		0									
		-2000									
		-1983									
		947									
		1717									
		1766									
		1605									
		1459									
		1327									
		1405									

PROFITABILITY PARAMETERS

DISCOUNTED TOTAL CAP	3983
NET PRESENT VALUE	6242
NET RETURN RATE [%/YR]	17.4
NET PAYOUT TIME [YEAR]	2.75
NET PAYOUT FRACTION	0.392
VENTURE LIFE [YEARS]	9
OPERATING LIFE [YEARS]	7

PAYBACK PERIOD

The payback period is defined as the minimum length of time theoretically necessary to recover the original capital investment in the form of cash flow to a project on the basis of total income minus all costs except depreciation. Generally, this means the original depreciable fixed capital investment only, with interest effects neglected. This is expressed mathematically as:

$$\text{Payback} = \frac{(\text{depreciable capital investment})}{\left(\dfrac{\text{average profit}}{\text{year}} + \dfrac{\text{average depreciation}}{\text{year}}\right)}$$

The following formulas can be used to compare and possibly justify the installation of a process analyzer based on savings versus manual laboratory analysis. In each case, you must be certain to compare the actual costs in each category: number of analyses, labor, amortized capital, equipment maintenance, and laboratory cost. It is then possible to perform the following two calculations [Ref. 10]:

$$\text{Cost/analysis} = \frac{\text{Total operating cost}}{\#\ \text{Analysis}}$$

The payback can be calculated from:

$$\frac{\text{Installed cost}}{(\text{Lab cost} - \text{Analyzer cost})\,/\,\text{yr}} = \text{years}$$

Several factors affect the payback ratio (P_r), including maximum payback (P_a in k\$/yr), where it costs nothing to operate the equipment and it is available 100% of the time; the analyzer range (A_m); the combined analyzer and laboratory repeatabilities (R); the systematic error (C); the systematic calibration error (S); the analyzer availability (A in %); the cost of maintaining the equipment (C_m, also in k\$/yr); and the actual system investment (A_i in k\$/yr). The following equation [Ref. 4] summarizes all these factors as:

$$P_r = \frac{\left\{P_a - \left(\frac{KP_a}{A_m}\right)\left[\frac{(R+C+S)^2}{A_m}\right]\right\}\left[\frac{(A-95)}{5}\right] - C_m}{A_i}$$

where K is a constant. This relationship can be used for all processes except for projects based on environmental compliance or safety. The term $(A - 95)/5$ represents the required minimum level of availability of 95% also mentioned earlier. Table 1-1 [Ref. 4] represents typical ranges of payback rates that can be expected from various types of analyzers.

Table 1-1. Typical Payback Rates

Analyzer	Application	Payback (k$/yr)	
		1979	1985
Chromatograph	FCCU light ends	50-300	180-1080
	Refinery light-ends section	30-50	108-180
Cloud-point analyzer	Blender - gas oil	65-70	234-252
Distillation analyzer	Gasoline	17-100	51.2-360
Flash-point analyzer	Kerosene	12-50	43.2-190
	Blender	25-40	90-144
Gas density analyzer	Fuel gas	10-54	39-194
Moisture analyzer	Reformer cycle gas	40-50	144-180
Octane engine (MON)	Gasoline	100	360
Oxygen analyzer	Stack gas	16-90	24-135
	Reformer regeneration gas	20-40	20-40
	FCCU regenerator stack gas	40	60
pH Meter	Overhead sour water	15-40	22.5-60
Reid vapor pressure analyzer	Gasoline blender	40-150	144-540
Sulfur analyzer	Gas oil	40	144
Viscosity analyzer	High-sulfur fuel oil	40-70	144-252

PRIORITIES IN CAPITAL EXPENDITURES

The way to determine if one alternative within a project is economically more attractive than another is to use the method of capitalized costs, where capitalized costs is defined as the original cost of the equipment plus the present value of investing the same money for the same time period. The basic equation for capitalized cost is:

$$K = C_v + \frac{C_r}{(1+i)^n - 1} = \frac{C_r(1+i)^n}{(1+i)^n - 1} + V_s$$

where:

K = capitalized cost, \$

C_v = original cost of equipment, \$

C_r = replacement cost, \$

V_s = salvage value at end of useful life, \$

i = interest rate, %

The term $\frac{(1+i)^n}{(1+i)^n - 1}$ represents the capital cost factor.

An example of where this may be applied is in determining whether it is better to purchase an analyzer system with shelter for \$100,000 with a salvage value of \$30,000 after 10 years versus a \$50,000 system without a shelter having no salvage value after 8 years. The replacement cost in both cases will be \$50,000 while the cost of capital (i.e., interest rate) is 10%.

For the system with shelter:

$$K = 100,000 + \frac{50,000}{(1+0.10)^{10} - 1} \qquad K = \frac{50,000 \times (1+0.10)^{10}}{(1+0.10)^{10} - 1} + 30,000$$

$$K = 100,000 + \frac{50,000}{2.594 - 1} = 131372.70 \qquad K = 81,372.70 + 30,000 = 111,372.70$$

For the system alone:

$$K = 50,000 + \frac{50,000}{(1+0.10)^{8} - 1}$$

$$K = 50,000 + \frac{50,000}{2.144 - 1} = 93722.01$$

Therefore, the capitalized cost of the second option is lower and should be selected if economics were the only concern in the equipment selection.

Capital versus Expense Considerations

There are a multitude of investment and tax-related reasons for installing analyzers as either part of a capital or an expense item. This situation is unique to each operating facility and its associated tax regime.

Generally, any new grassroots installation will be ruled capital since it increases the asset value of the facility to which it is added. Conversely, replacement of an analyzer in kind is normally considered an expense project. It is, however, possible to install replacement equipment as a capital project if its value is above a certain percentage of the total facility/unit or if the upgrade is required for some other reason legislated or imposed on the operator. In the event that old equipment is being replaced by a new capitalized piece of equipment, it is often necessary to prepare two requests for funding: one to demolish the existing system, thus removing it from the account, and a second for installation of the new system.

Legal Requirements

Many of the legal requirements for process analyzers are based on some form of legislation for the protection of personnel or the environment. Both of these topics have been covered earlier, but it should be emphasized that when an analyzer system is installed to meet a legal requirement, such as a contract or custody transfer, care should be taken to assure that it meets or exceeds all specifications.

Process Quality Monitoring

Improved quality control is the most common reason for most analyzer installations. However, improved quality control can be obtained only when the analyzer is operating and is placed in closed-loop control. An analyzer used only for indication is like a vehicle equipped only with a light that says too fast or too slow but not by how much. A control system will tell you "how much" and actually maintain the offset at a predetermined differential.

With today's high-powered computer processors, most facilities now have some form of statistical quality control program in place. In fact, all major distributed control system (DCS) and control system vendors offer real-time SQC as an option. The expense to the user when implementing an SQC system includes not just the hardware but also the larger expense of training personnel in the use of these techniques and developing the historical data to set realistic operating ranges, midpoints, and alarm ranges. SQC is simply elementary statistics, and in return you will receive a higher quality product with less variation. Remember, though, that your control is only as good as the information it receives, and time is critical. Control theory states that to assure you have a sample that represents the actual process change, the measurement should be taken at least six times for every time-cycle of the process. For example, if a distillation tower has a response time of 12 min, the analysis on the overhead should be updated every 2 min to obtain optimal control.

To ensure you have a sample that represents the actual process change, the measurement should be taken at least six times for every time-cycle of the process.

Statistical analysis of near-infrared (NIR) methods has revealed that they can be as accurate as the traditional method on which they are based and have better precision.

Common Applications by Industry

This final section lists by industry some typical analyzer applications gleaned from the literature and, where possible, their estimated payback. To obtain more information on particular applications, consult the original article or book listed in the bibliography and contact the author directly.

Steel Industry

A laser-produced plasma sensor directs a laser at molten steel inside the furnace, causing it to vaporize and emit radiation. The unit then analyzes the emission spectrum and calculates the composition of a steel alloy that may contain 10 to 15 different elements. Typical analysis time is less than a minute, and the output can be used to assure that blending of exotic metals is within specification [Ref. 11].

An on-line coal analyzer is available to measure sulfur content of the raw and blended coal being supplied to a coal-fired power plant. Use of the on-line analyzer helps avoid overblending of large quantities of low-sulfur coal to ensure that the facility stays within its EPA-regulated sulfur emission limits. Use of statistical methods would require approximately 30% of the blended coal be low sulfur, but with installation of the analyzer as little as 7% low-sulfur coal would have to be used. Blending is done on either percent sulfur content, ash content, or SO_2/MMBTU (Million British Thermal Units) content [Ref. 12].

An on-line coal analyzer is available to measure sulfur content of the raw and blended coal being supplied to a coal-fired power plant.

Another analysis that can be performed on coal-fired facilities is to monitor the product off-gases produced from the manufacture of metallurgical coke for their energy content or calorific value. By mixing these off-gases in the proper proportion, the plant heat balance can be optimized [Ref. 13].

Monitor the product off-gases produced from the manufacture of metallurgical coke for their energy content or calorific value.

Off-gas from blast furnaces is measured and controlled for oxygen content to monitor the carbon reduction in the charge. An increase in the oxygen content indicates completion of the reduction reaction. The vacuum oxygen decarburization (VOD) process also requires measurement of CO and CO_2 above the liquid steel. In particular, the CO level declines rapidly when argon is introduced to the pool to flush out any remaining oxygen.

Another energy-efficiency calculation for blast furnaces is measurement of the energy index, which represents the energy consumed in the current month as a percentage of the energy consumed in a historical base period under similar operating conditions. An index of less than 100% as calculated by the following equation is desirable:

$$Index = -0.997G + 1.725M + 1.917P + 0.448L + 0.565S - 0.077H - 0.752N + 191$$

where:

- G = blast furnace gas heating value in BTU/ft^3
- M = mill delays in percent of operating time
- P = pit delays in percent of operating time
- L = lost heat factor as percent of ingots charged cold
- S = product mix factor in percent stainless and hot top ingots
- H = pit use performance as tons/hr/ft^2 hearth versus normal
- N = fuel mix in percent natural gas fired

The blast furnace gas heating value can be measured by a calorimeter and the fuel delivery can be measured with a flow transmitter, thus providing a real-time indication of furnace performance [Ref. 14].

The blast furnace gas heating value can be measured by a calorimeter and the fuel delivery can be measured with a flow transmitter, thus providing a real-time indication of furnace performance [Ref. 14].

To assure that an effective cleaning job is achieved on a finished metal product, it must be thoroughly cleaned and prepared for the coating process. A conductivity analyzer is therefore required in the cleaning tank to continuously monitor the caustic level and to control the caustic addition pump.

Pulp and Paper Industry

Alkalinity analyzers are frequently used in the kraft process to control white liquor flow to the digester and also reduce kappa number variability. A second application of this type of analyzer is used on the weak wash liquor to control the water flow to the lime mud filters, control the green liquid density, and indicate lime mud separation process malfunctions that could lead to excess total reduced sulfur emissions.

Biochemical oxygen demand (BOD) analyzers with a 3-min cycle time are used to monitor mill wastewater and thus allow the operators to predict 5-day BOD values. The benefits to be captured from this application include evaluating final effluent quality, monitoring of waste sources that flow into the wastewater treatment stream, and providing feedback within the primary and secondary treatment systems so that operation of these systems can be set at optimal efficiency for reducing 5-day BOD and toxicity. As a result of these benefits, the mill is able to stay in production during periods of highly restrictive water allocations, no National Pollution Discharge Elimination System (NPDES) permit limit excursions were reported as being attributed to the mill. Water quality below the mill equals or exceeds the environmental standards for the stream [Ref. 15].

A method has been commercialized to control the starch used to bond virgin kraft in a corrugated converter plant.

A method has been commercialized to control the starch used to bond virgin kraft in a corrugated converter plant. In a batch system, a carrier starch is mixed proportionally with water and chemicals, and is then cooked to a temperature within the range of 130 to 140°F (55 to 60°C). Quench water is then added to reduce the temperature to 100°F (40°C) when raw starch and biocide are mixed in. The final 20 to 24% solids mixture is then transferred to storage tanks. The analyzer controls the viscosity and hence the starch level in these tanks. As a result of this control, personnel safety and environmental safety have increased due to reduced production of foul and noxious odors. Additional benefits include reduced biocide costs and increased product quality [Ref. 16].

Starch solutions are also controlled by monitoring the viscosity of the solutions at 1.4 to 10 bars (14×10^4 to 10^6 Pa) and 185°F (85°C) in range of 1,000 to 1,500 cP (1 to 1.5 Pa · s). Similarly, vibrating tube density meters are being used to measure the consistency of paper slurry by detecting air bubbles.

Petrochemical Industry

A process gas chromatograph has been used to monitor a process manufacturing carbon disulfide by the hydrolyzing of hydrogen sulfide. The thermal conductivity detector base chromatograph was used to monitor air, CO_2, H_2S, COS, methane, water, and CS_2 on a 5-min cycle time. The return on investment was 1.14 years, based on the formula used above comparing lab analysis costs to process analyzer operational costs [Ref. 10].

Monitoring of hydroxyl number is critical to the successful operation of polyester, urethane, epoxy resin, and organic ester facilities. NIR analysis is a good fit for this measurement due to the large characteristic signature around the 20,000 to 21,500 Å (2,000 to 2,150 nm) wavelength range. Gains of 5% or more efficiency have been achieved as a result of using this information to continuously control the process [Ref. 17].

Caprolactum is used to manufacture nylon 6, and if made by oxidizing cyclohexane to cyclohexanol, control of the oxygen content in the reactor inlet vapor is critical. Cyclohexane at its boiling point is the background vapor. As a result, this measurement must be made using a paramagnetic oxygen analyzer.

As a result of this correlation, it is now possible to perform the analysis in 5 min versus the 5 hr required by the former method.

An NIR analyzer and associated calibration set have been developed by a leading adhesive manufacturer to control the surface moisture of its adhesive pellets. As a result of this correlation, it is now possible to perform the analysis in 5 min versus the 5 hr required by the former method. Installation of this rapid-response analyzer has drastically reduced plant recycle, since the batch polyamide resin process can now be closely monitored.

Photometric analyzers are frequently used at the following points in a methanol production facility: before the compressor to monitor the hydrocarbon feed composition/purity; at the compressor to monitor the industrial gas feed; before the reactor to monitor the feedstock purity, carbon monoxide, and steam feed; and after the condenser to monitor the reactor crudes and the moisture content [Ref. 18].

Pharmaceutical Industry

Oxygen measurement is one means of monitoring the efficiency of fermentation processes used in the manufacture of pharmaceuticals and other biotechnology products. Accurate monitoring of the oxygen level in the vessel assures that the biological processes are operating in the proper balance of nutrients. The rate of oxygen consumption also serves to indicate the level of enzyme activity. In addition to oxygen, nitrogen, argon, and carbon dioxide are measured to calculate the respiratory quotient — a critical indicator of the health of fermentation processes. This multiple gas analysis is most often made with a process mass spectrometer.

Refinery Applications

The Ryan/Holmes process is a lean oil process used to separate paraffin-series hydrocarbons, carbon dioxide, and hydrogen sulfide. The CO_2 is recovered as one product stream, while the hydrocarbons and H_2S are removed as the other sour (contains sulfur) natural gas liquid (NGL) stream. Two analyzers are used as part of a multivariate control algorithm to optimize recovery of NGLs and CO_2. These analyzers are installed on the gas feed inlet and the overhead (vapor stream) of the ethane recovery column. The feed-gas composition directly controls the lean gas oil feed additive rate, while the ethane overhead analysis is used to control the ethane column recycle. Upon completion of the multivariate control system and analyzers, ethane recovery improved by 8% and the reboiler heat consumption was lowered by 12% [Ref. 19].

Upon completion of the multivariate control system and analyzers, ethane recovery improved by 8% and the reboiler heat consumption was lowered by 12% [Ref. 19].

Catalyst regeneration is a common occurrence in the refinery industry where the carbon layer on the catalyst is burned off. Oxygen analysis indicates the status of the process. Should the catalyst be regenerated too quickly, the catalyst could be destroyed. If, however, the regeneration is too slow, the cost is lost production. In addition, a sudden breakthrough in oxygen indicates the end point of the regeneration process.

Table 1-2 summarizes typical applications for physical property analysis used in a refinery operation [Ref. 20]. Many of these analyses are now also being accomplished with NIR analyzers.

Table 1-2. Typical Applications for Physical Property Analysis in a Refinery

Measurement	ASTM Method	Common Application
Viscosity		
Capillary	D446	Asphalts
Saybolt	D88	Oils
Distillation		
Atmospheric	D86	Diesel fuel and lighter
Reduced pressure	D1160	Lube oils
Flash		
Tag closed cup (TCC)	D56	Gasoline
Cleveland open cup (COC)	D92	Gasoline
Pensky-Martens closed cup (P-M)	D93	Fuel oils
Vapor pressure		
Reid	D323	Gasoline
LPG	D1267	LPG
Absolute	D2551	Heavy oils
Pour point	D97	Fuel oils
Freeze point	D1659	Jet fuels
Cloud point	D2500	Clear distillates

Table 1-3 [Ref. 20] lists additional common on-line analyzer applications.

The analysis of hydrogen sulfide is critical in many areas of modern refineries. Typical examples include amine treaters for H_2S removal, synthesis/fuel gas, reforming, and product/intermediate liquids. The first two of these services assure that the vapor streams to various parts of the plant are within specification to minimize the emission of pollutants and corrosion of piping. The other two examples will affect the plant run length and efficiency due to the effect that H_2S can have on catalysts or on the price obtained for products.

Table 1-3. Typical On-line Analyzer Applications

Technique	Principle of Operation	Sample Type	Comments
Conductivity	Measurement of the total amount of charge carriers in an aqueous solution	Best for one ionizable component in water or as a nonselective detector of dissolved ionized species	Widely used; available in electrodeless models for dirty samples
Ion-selective electrodes	Respond to a particular ion in solution with an electrical signal proportional to the logarithm of the concentration present	Species to be measured *must* be ionized	Primarily used in power and wastewater applications
Gas-sensing electrodes	Operate similar to ion-selective electrodes	Oxygen can be done in gas or liquid phases, ammonia and carbon dioxide in dissolved water only	Oxygen most commonly used
Total organic carbon	Converts organics to CO_2 for infrared measurement	Water samples	Low- and high-temperature techniques available
Gas chromatography	Converts a mixture of gases into a sequence of individual constituents in a flowing gas stream	Any mixture of gases of low-boiling-point liquids. Number of components affects cycle time	Widely used, but can be relatively slow (2-10 min per analysis)
Liquid chromatography	Converts a mixture of solutes into a sequence of individual constituents in a flowing liquid stream	Any liquid mixture	Rarely used due to expense and complexity
Supercritical fluid chromatography (SFC)	Converts a mixture of individual constituents in a flowing supercritical fluid stream (normally CO2)	High-molecular-weight compounds	Similar to gas chromatography, but more complex
Nondispersive infrared (NDIR)	Measures the infrared energy absorbed by the component of interest at its specific wavelength versus a pure material	Limited number of components per sample	Widely used for CO, CO_2, methane, and water; not suitable for diatomic molecules such as oxygen
Infrared (NIR, FTIR)	Measures the infrared energy absorbed by the component of interest at its specific wavelength compared to the total energy flux at that wavelength	Both liquids and gases with more components than NDIR	Greater selectivity than NDIR; no diatomics
Oxygen	Measures by various techniques: zirconia, galvanic, paramagnetic, etc.	Gas samples	Widely used for combustion control
Mass spectrometry	Differentiates components by molecular fragments, measure this quantity, then computes the original quantity	Gases or low-boiling-point liquids	Limited constituent range; high-speed and wide dynamic range

Gasoline blending is another significant benefit application in refineries. A typical blend application controls the following qualities: specific gravity at 60°F (15°C), research octane number, motor octane number, Reid vapor pressure (RVP), volume percent distilled at 160°F (70°C), volume percent distilled at 212°F (100°C), volume percent distilled at 260°F (125°C), volume percent distilled at 355°F (180°C), and lead content. Economic calculations on this type of project should be based not only on the reduced giveaway but should also include the additional credits due to reduction of reblends, optimal use of blend components, and possible laboratory sample reductions. Typical shifts in product usage that can be obtained are more butane in the product (RVP), and use of light and pyrolyzed heavy naphtha as blend constituents [Ref. 21]. All but the last of the above measurements are readily available through either stand-alone analyzer systems or more recently with the use of NIR instrumentation.

The following relationships can be used as a screening process to determine if further work should be done on a proposed analyzer installation. The equations are based on a series of application reports prepared for Precision Scientific by Turner, Mason & Associates in 1980/81. The relationships are based on the differential price in cents per unit production (i.e., ¢/1000 Bbl) between two product streams and the facility production rate (i.e., 1000 Bbl/day). Hence, they are still relevant today.

Reid vapor pressure as a function of unleaded gasoline versus butane prices:

$$\$ \times 1,000/yr = \frac{(c - 0.0756)}{1.79632} \times throughput$$

Pour point on price spread of kerosene jet and No. 2 fuel:

$$\$ \times 1,000/yr = \frac{(c + 0.02076)}{0.13826} \times throughput$$

Cloud point for the same fuels as immediately above:

$$\$ \times 1,000/yr = \frac{(c + 0.0544)}{0.13627} \times throughput$$

Viscosity between No. 2 fuel oil and 3% sulfur No. 6 fuel oil:

$$\$ \times 1,000/yr = \frac{(c - 0.0767)}{7.42571} \times throughput$$

In all of these equations, c is the price differential in cents between the two products, and throughput is the unit capacity in MBBL per calendar day.

Reid vapor pressure can also be calculated using a gas chromatograph and the following information:

Component	Vapor Pressure of Pure Component at 100°F (40°C), psia
x_i	P_i^0
iC_4	72.2
nC_4	51.6
iC_5	20.4
nC_5	15.6
nC_6	5.0
nC_7	1.6
nC_8	0.5
nC_9	0.2
nC_{10}	0.1

The partial pressure is determined by summing over all the components using $p_{RVP} = \sum \gamma_i C P_i^0 x_i$, and C is given by $C = \frac{n_0}{n_0 + k(\gamma_i P_i^0 - P)}$, where n_0 is the number of moles material in the hydrocarbon liquid phase and k = 1.346 mmol/psia. The values of γ_i are taken from the American Petroleum Institute (API) handbook.

To related Reid vapor pressure (RVP) to true vapor pressure (TVP) in psia for motor fuels, the following equation can be used:

$$RVP = 1.078TVP - 0.737$$

Motor octane analyzers are installed and justified on the basis that they make possible savings on fuel additives such as MTBE and TAME while achieving longer catalyst life because of lower operating temperature in the distillation process. To assure that the product octane is never less than the specification minimum, refiners often operate with a positive margin (octane giveaway) of " 0.3 to 0.5 octane number, which can be reduced to 0.1 to 0.3 octane number if closed-loop control is implemented [Ref. 18].

It is better to use the 5% or 10% boiling point rather than the initial boiling point (IBP = 0%) because it is more repeatable and thus provides greater meaning. Similarly, the 90% or 95% boiling point should be used rather than the final boiling point (FBP = 100%) for determining loss.

When boiling-point analyzers are used for process control, it is better to use the 5% or 10% boiling point rather than the initial boiling point (IBP = 0%) because it is more repeatable and thus provides greater meaning. Similarly, the 90% or 95% boiling point should be used rather than the final boiling point (FBP = 100%) for determining loss. Use of the lower boiling point at the upper end of the range will also improve instrument reliability through reduced coke formation in the analyzer. Table 1-4 summarizes some typical boiling-point ranges for various petroleum products.

Pour-point analyzers are used for testing unblended light distillate fuels prior to the addition of pour-point depressants. Once pour-point depressants have been added, the results are significantly less reliable. On very viscous fluids at ambient temperature, the transition from fluid to the no-flow solids-condition is difficult to detect. In these circumstances, the wax-point or cloud-point analyzer is more commonly used. The pour-point analyzer has nearly been replaced by the viscosity analyzer, which gives a better indication of pumpability. An oil will move into a suction line at a temperature of 40 to 45°F (5 to 8°C) because shear forces in a vessel are typically higher than those in the analyzer [Ref. 22].

Table 1-4. Typical Boiling-Point Ranges for Petroleum Products

Product	Range	% Recovered Points, °F (°C)		
	(IBP-FBP), °F (°C)	10%	50%	90%
Unstabilized gasoline	—	120 (50)	230 (110)	285 (140)
Gasoline	85-430 (30-220)	140 (60)	230 (110)	300 (150)
Aviation gasoline	100-340 (40-170)	150 (65)	212 (100)	250 (120)
Benzene	250-300 (120-150)	—	—	30 (140)
Naphtha	—	260 (125)	310 (155)	355 (180)
White mineral spirits	320-390 (160-200)	—	—	—
Light gas oil	—	430 (220)	490 (255)	610 (320)
Kerosene	300-570 (150-300)	30 (140)	390 (200)	430 (220)

Process gas chromatographs are frequently used to monitor the performance of separation processes by providing near-real-time measurement of stream composition. Closed-loop control using a real-time signal as from a chromatograph can:

- Maintain satisfactory product quality
- Enhance the yields of more valuable products
- Reduce energy consumption
- Minimize waste
- Increase unit throughput

Figure 1-1 shows a typical control scheme on a distillation tower. Overfractionation, or making product purity higher than contractually required, is a common practice in distillation processes where only basic regulatory controls are used. This margin reduces profitability in the forms of reduced throughput, excess energy consumption, and higher value product giveaway in lower valued products.

Overfractionation, or making product purity higher than contractually required, is a common practice in distillation processes where only basic regulatory controls are used.

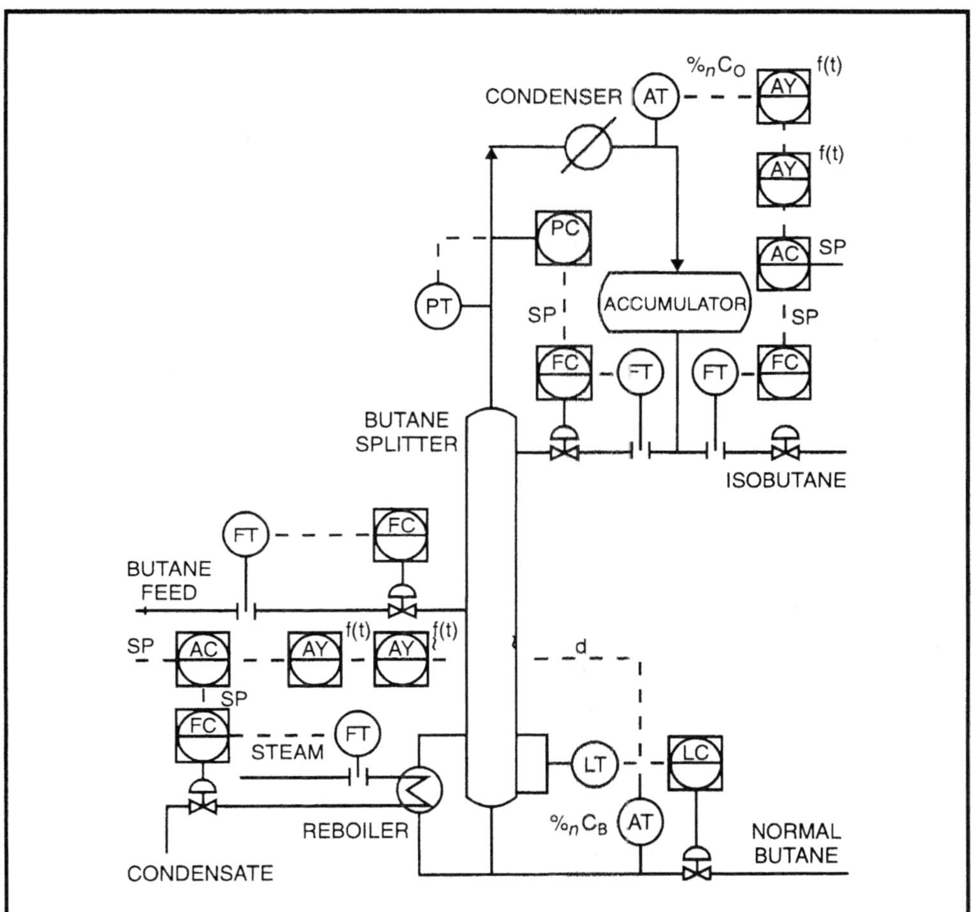

Figure 1-1. Typical Control Scheme on a Distillation Tower

In applications such as these, the benefits are normally calculated from energy savings (approximately 5% of the total) and reduced giveaway. It is important to remember that you cannot claim a credit for the energy savings if the energy saved cannot be used elsewhere in the plant [Ref. 23].

The analyzer on the product heavy key, in the above figure isobutane, should be located between the condenser and the accumulator, where the vapors are totally condensed, to minimize response time. Likewise, the bottoms product key, normal butane in Figure 1-1, is sampled from the product line rather than the reboiler to avoid the lag time associated with the reboiler volume. A material balance on the vessel is necessary to develop the control algorithm.

In applications such as these, the benefits are normally calculated from energy savings (approximately 5% of the total) and reduced giveaway. It is important to remember that you cannot claim a credit for the energy savings if the energy saved cannot be used elsewhere in the plant [Ref. 23].

Food and Beverage Industry

Infrared analysis of raw milk entering a high-temperature short-time pasteurizer is used in the dairy industry to standardize protein levels. The analyzer output controls the protein-to-fat ratio by blending an ultrafiltered milk with the raw milk to provide a consistent year-round protein-to-fat-level ratio.

Infrared absorption has been used to measure the moisture content of cottage cheese curd. It is possible to use this on-line analytical technique to control moisture in the 8 to 85% range on a wet weight basis in a low fat content (less than 1%) product.

NIR analysis has been used by the food industry for many years. With the advances in the design of these analyzers, their probes, and the associated chemometrics software, it is not uncommon to monitor the operation of a fermenter and its broth constituents with a remote vessel-mounted probe.

Another common NIR application is the on-line measurement of flour proteins and moisture using transferable and hence almost universal calibrations to control and supplement as necessary the gluten content of a flour product.

Conductivity cells are commonly used to detect the interface for clean-in-place food-processing systems where caustic solutions are used for cleaning process equipment. After the equipment is washed with caustic, it is then flushed with water and the caustic diluted to an acceptable discharge value. This is where the conductivity meter is used to indicate when one operation has ended and the next has begun. Caustic can then be diverted to a recovery system until it is dilute enough to be added to the plant sewer or waste streams.

Electronics Industry

A mass spectrometer can be used to monitor semiconductor processes, providing that dynamic sampling pressures range from atmospheric to 10^{-3} mbar. This yields good sensitivity to semiconductor gases, including dopant gases. As a result, the semiconductors have fewer defects and product yields are dramatically increased.

Another mass spectrometer application is to monitor impurities in sputtering (ion or film deposition) system process gases. A positive correlation has been found between the presence of oxidizing impurities and failure-causing effects in submicron geometry and ultralarge-scale integrated (ULSI) circuits, as well as application of extremely low loss optical coatings. A good example of the effect of impurities is that oxygen and water at parts per million concentrations can cause electromigration of sputtered metal interconnects.

Component washing is an extremely high-purity process in the semiconductor manufacturing industry. For this reason, a series of resistivity measurements are made on the wafer wash water before it is used to clean the semiconductor wafers to assure optimal processing with minimal water usage.

Cement Industry

Measurement of the combustion process in cement kilns focuses particularly on oxygen and carbon monoxide as the two standard parameters for control. Recently, however, it has become feasible and necessary to monitor NOx and SO2 as well — not only for environmental reasons but also to more reliably control the more sophisticated and complicated processes in a closed-loop system. The return on the investment is very rapid, with estimated savings of more than $2.00/tonne of product. This correlates to a combined 3% reduction in material and operation costs [Ref. 24].

Mining Applications

A color line-scan camera is being used in a limestone mine in Finland to differentiate and sort red pegmatite amphibolite and gray granite from white limestones by analyzing the reflection of light from the stones as they pass the illumination area at 4 m/s. Half a meter after passing the camera, air-pressure valves are used to "blow" the rejects off the main conveyor [Ref. 25].

Paint Applications

The measurement and control of pH is a critical step in the manufacture of paint pigments. The measurements required can range from 0.5 to 6 pH at the start of each batch, with a target range of 4 to 7 pH ±0.5 to 1.5 pH. Due to the aggressive nature of this slurry service, a retractable probe with a polymerized reference electrode was selected to eliminate the need for a porous reference electrode and thus minimize plugging problems. By inserting the probe into the vessel approximately 6 in. at a 15° angle, a reliable system is possible and has been in service for several years [Ref. 26].

Flue Gas Analysis and Burner Control

Since flue gas analysis is common to many industries, it is being considered as a separate application.

When a zirconia-based analyzer is used to monitor the oxygen level in the presence of sulfur dioxide, the sulfur dioxide will cause the oxygen analyzer signal output to read false on the low side by as little as 10% or as much as 50%. Be sure to check with your analyzer manufacturer as to how their sensor is affected by this problem.

When a zirconia-based analyzer is used to monitor the oxygen level in the presence of sulfur dioxide, the sulfur dioxide will cause the oxygen analyzer signal output to read false on the low side by as little as 10% or as much as 50%. Be sure to check with your analyzer manufacturer as to how their sensor is affected by this problem.

Most modern flue gas analysis systems measure not only oxygen but also carbon monoxide. The typical control range for CO is 200 ppm. This is 0.6% excess oxygen for natural-gas-fired furnaces, 1.3% for oil-fired furnaces, and 5.2% for coal-fired furnaces.

The CO measurement is independent of air leakage except for dilution. A 10% air leakage will affect a CO reading of 200 ppm at 1% excess oxygen by reducing it to 180 ppm, but it will increase the excess oxygen reading by 2% to give a 3% excess oxygen signal.

Monitoring the oxygen content of furnaces provides savings inasmuch as that each percent reduction in excess air correlates to roughly a similar savings in the amount of fuel consumed (i.e., 1% less excess oxygen = 1% less fuel to burners). The benefits of the additional CO reading include: faulty burner detection by a sudden increase in CO, minimization of flue gas exit temperature, reduction of fan motor horsepower (induced draft furnaces), detection of air leakage, and reduction of pollutants as a result of more optimal firing.

The amount of excess air depends on the type of fuel being used. Natural gas efficiently burns at 8 to 10% excess air (1.6 to 2% excess oxygen), oil at 10 to 15% excess air, and coal at 18 to 25% excess air, with the reasons for these

Safety dictates certain operating precautions for fuel-air controls to assure that a deficiency of air does not allow an excess of fuel to accumulate in the furnace.

differences being due to the differences in the state of the fuel as well as the amount of noncombustibles present.

Safety dictates certain operating precautions for fuel-air controls to assure that a deficiency of air does not allow an excess of fuel to accumulate in the furnace. Fuel and air flow can both be set from a master firing-rate controller, but this feature requires automatic selection. As shown in Figure 1-2, the fuel-air ratio is adjusted through the manipulation of the span of the air measurement by the oxygen controller. Normally the set point would be adjusted, but in order for the selection system to operate, the two controllers must have the same values. If air flow is lost, its flow is preferentially selected by the fuel controller for the set point. Alternately, if fuel flow is higher than necessary, its measurement is automatically selected to test the air flow. This protects the furnace from blower failure (loss of air supply) and from lags of the set point in either loop.

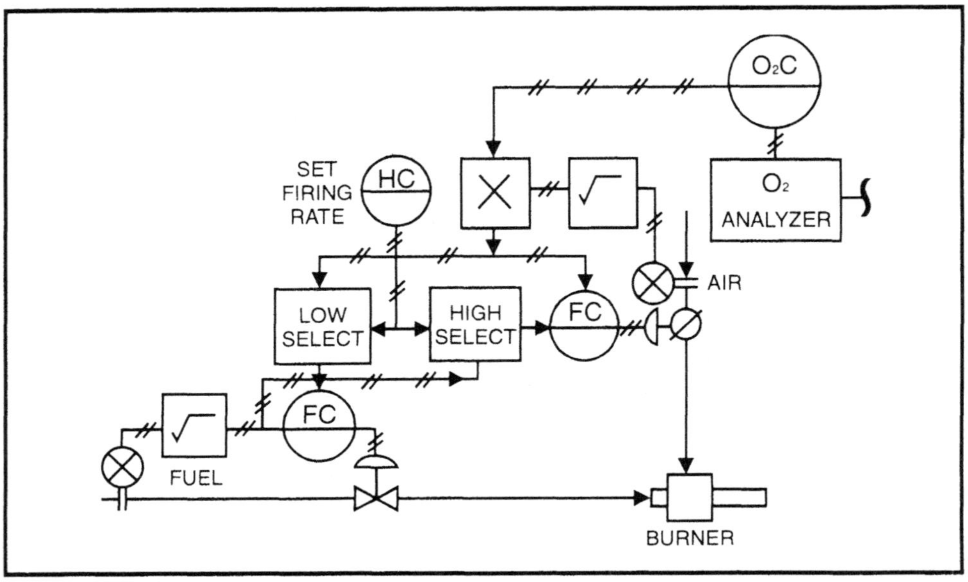

Figure 1-2. Furnace Lead-Lag Control Logic

A calorimeter and flow orifice can be used with a feedforward control loop to assure a constant heat flow to furnace burners. Combining the heat value and flow signal results in a heat flow-rate signal. When considering this application, you must be mindful of the process lag time, the analyzer system lag time (including sample system), and the control element response time. Due to the large difference between process time (very fast) and the control system lag time, a large sudden variation in one measured variable will not be controlled. However, gradual changes in the signals can be reliably minimized with such a system. A typical control diagram is shown in Figure 1-3.

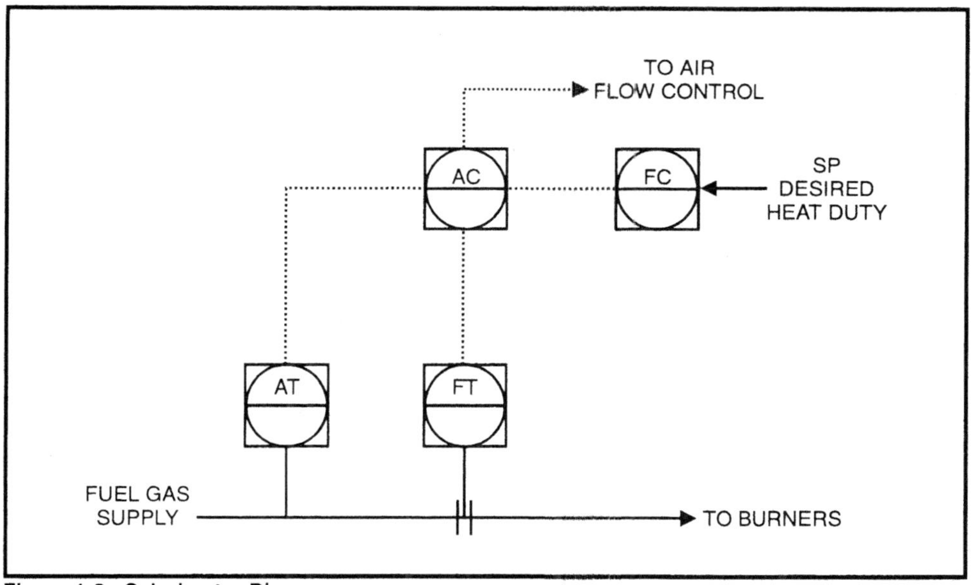

Figure 1-3. Calorimeter Diagram

The continuous monitoring of boiler feedwater for silica can be accomplished by a combination of wet chemistry and photometric color intensity detection. The chemistry involved in the monitoring reaction is:

$$\text{Total silica} + \text{molybdate}^{3+} + H^+ \rightarrow \text{silicomolybdate}^{4-} + \text{ammonium sulfate} + \text{water}$$

The silicomolybdate is yellow-colored in proportion to the amount of silica present in the original sample. Since silica is often one of the first contaminants to break out of an ion-exchange bed, this is an excellent method for monitoring ion-exchange bed efficiency.

Another important parameter in boiler feedwater control to prevent corrosion and hydrogen embrittlement of the tubes is pH. pH must be contained within a limited range to assure that the chemical treatment added to the boiler for protection against these problems remains active and in solution. Sodium-sensitive specific-ion electrodes are also often used on boiler feedwater systems since sodium ions also contribute heavily to the corrosion in boiler water systems.

Miscellaneous Applications

X-ray analyzers are being used in sawmills to image feed timber to find knots and other imperfections in the logs in real time to maximize the optimum use of each log.

The accurate control of the caustic saturator solution is one of the most important single factors in the efficient operation of the bleach range used in the textile industry. If the caustic concentration is too low, the scouring is incomplete; too high, and the cotton fibers are damaged. The caustic signal, as measured by a conductivity meter, is used in a closed-loop system to control the strong caustic addition to the saturator.

References

1. Waarvick, C., Vigila, R., and Titus, R. J., "Automated Odor Control Systems," *INTECH* 40(8): 28-31, August 1993.

2. Reynolds, M., "On-line Coal Analyzer Helps Plant Meet EPA Standards," *Power Engineering*, p. 57, February 1993.

3. Freidman, Y. Z., "Avoid Advanced Control Project Mistakes," *Hydrocarbon Processing*, 71(10): 115-120, October 1992.
4. Reeves, P., "The Status of Process Analytical Instrumentation: Recommended Practices and Future Trends," *Analytica Chemica Acta*, 190: 45-65, 1986.
5. Shinskey, F. G., *Process Control Systems Application/Design/Adjustment*, 2nd ed. New York: McGraw Hill, 1979, p. 78-79.
6. Shinskey, F. G., *Process Control Systems Application/Design/Adjustment*, 2nd ed. New York: McGraw Hill, 1979, p. 191.
7. Cline, R. P., Martin, G., and Turpin L., "Estimating Control Function Benefits," *Hydrocarbon Processing*, 70(6): 68-73, June 1991.
8. Stout, T. M., and Cline, R. P., "Control System Justification," *Instrumentation Technology*, 23(9): 51-58, September 1976.
9. Dailey, W. V., "Improve Process Analyzer Availability," *Hydrocarbon Processing*, 72(1): 69-71, January 1993.
10. Karr, J. J., Mcleod, N., and Feiler, R., "Analysis of H_2S/CS_2 Process Streams," *Process Control and Quality* 5: 179-186, 1993.
11. *ISA Mining & Metallurgy Industries Newsletter*, 26(1): February 1992.
12. *Power Engineering*, p. 57, February 1992.
13. Nicholas, P., "Comprehensive Gas Analysis for Processing," *Control & Instrumentation*, p. 35-37, March 1994.
14. Bridges, W. H., and Marolf, P., "Reduced Fuel Costs by Enriching Blast Furnace Gas with Natural Gas," *Iron and Steel Engineer*, p. 33-36, September 1982.
15. *Chilton's I&CS (Instrumentation & Control Systems)*, 65(2): 78-81, February 1992.
16. *Canadian Process Equipment & Control News*, p. 68, February 1992.
17. "N.I.R. Monitoring of Hydroxyl Number," *Chemical Equipment*, p. 6-10, September 1992.
18. Nichols, G. D., *On-line Process Analyzers*, New York: John Wiley & Sons, 1988.
19. Berkowitz, P. N., Papdopoulos, M. N., Gamez, J. P., and Clay, R. M., "On-line Optimal Control Improves Gas Processing," *Hydrocarbon Processing*, 71(4): 61-63, April 1992.
20. Sherman, R. E., "Process Analyzers: Vital Control System Components," *Chemical Processing*, p. 70-78, June 1988.
21. Serpemen, Y., Wenzel, F. W., and Hubel, A., "Blending Technology Key to Making New Gasolines," *Oil & Gas Journal*, p. 62-74, May 18, 1991.
22. Huskins, D. J., *Quality Measuring Instruments in On-line Process Analysis*, New York: John Wiley & Sons, 1982.
23. Hall, G. F., "GC-Directed Control Improves Refining," *Hydrocarbon Processing*, p. 47-50, February 1991.
24. Parvin, T., and Henderson, W., "Gas Analysis for Profit," *Control & Instrumentation*, p. 87-89, March 1994.

25. Tulloch, M. H., "Optical Sorter Boosts Limestone-Mine Profits in Finland," *Photonics Spectra*, p. 20, 1993.

26. Paul, B. O., and Pruzan, T., "pH Sensors and Retractable Housings Slash Batch Processing Duration," *Technology for Process Control*, p. 55-58, August 1993.

About the Author

Ian Verhappen is a professional engineer in the province of Alberta where he works as an engineering associate for Syncrude Canada Ltd. He is responsible for the engineering support of the company's analyzer specification and installations. He has worked on a variety of projects in all stages from specification to procurement, installation, commissioning, troubleshooting, and decommissioning. Mr. Verhappen has authored or co-authored over a dozen papers on process analysis including topics such as project execution, applications, and justification.

Mr. Verhappen graduated from the University of Alberta in 1982 with a B.Sc. in Chemical Engineering. He had previously received a B.Sc. in Environmental Science in 1979. In addition to being a professional engineer, Mr. Verhappen is one of the original Certified Specialists in Analytical Technology (CSAT), a member of ISA where he presently serves as both the Analysis Division Director and newsletter editor, a member of the American Institute of Chemical Engineers, and was recipient of the 1994 ISA Analysis Division Member of the Year award.

2

Interfacing With Process Analyzer Systems

The increasing use of process analyzers throughout the world has occurred for one reason: they provide valuable information to their users. In many cases, the data provided validates some property of the material being manufactured, or assures suitability for sale, or use as a feedstock for another part of the plant. Analyzers would be of little value if they were simply black boxes that performed analyses all day but had no means for sharing their data with the outside world. Figure 2-1 illustrates the importance of analyzer communication.

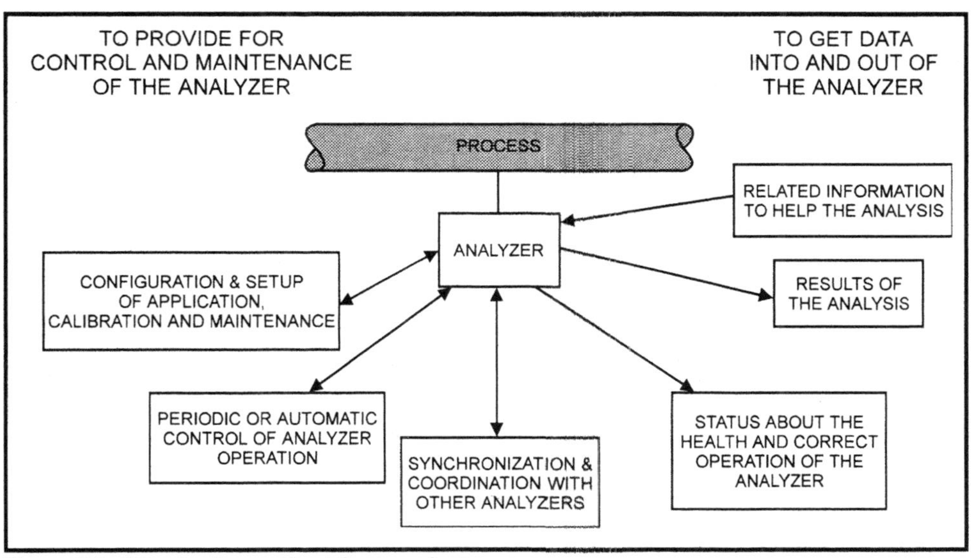

Figure 2-1. Why Analyzers Must Communicate

Most modern analytical equipment can provide a great deal of information, both about the analytical results and about the equipment. Of primary interest is the value or values of the variable measured by the instrument and presented to the user as results. These results can be communicated to the user in various ways. The oldest (and still prevalent) method is the local display of the results on an analog or, as is more commonly used today, a digital meter. These results may represent some percent of a scaled amount or they may be converted into relevant engineering units (parts per million, mole percent, etc.). Analyzer results are also commonly provided as scaled analog signals to remote process control systems, displays, or chart recorders. Gaining tremendous popularity is a means for transferring data by direct digital transfer of the analytical results to a process control system or personal computer.

Gaining tremendous popularity is a means for transferring data by direct digital transfer of the analytical results to a process control system or personal computer.

Analyzers also have the ability to provide information about their operating condition. The unit constantly checks itself for malfunctions and reports on its status. Many analyzers are programmable so that the level or severity of the various malfunctions reported can be set by the user. Also, special user-defined alarm points may be added to the analyzer's normal internal checks. These checks may be hardware or software-integrity-related and in some cases may inform the user of problems related not to the analyzer itself but to the process it is monitoring.

Computer control systems often have the ability to activate some event in the analyzer through various forms of communication.

Computer control systems often have the ability to activate some event in the analyzer through various forms of communication. A common example of this would be an operator requesting an analyzer to switch from measuring the process to performing a calibration check. Another type of action may be changing which process stream the analyzer is monitoring by sending such a command from the process operator's control system workstation. In some cases, an operator may need to take an analyzer completely off-line from monitoring the process.

Another area of communication with analytical equipment is the user interface to the analyzer itself. Many analyzers today have some type of keyboard/display interface, which may allow a number of functions. The analyzers may require some type of configuration information, such as measurement range or calibration frequency. The user may need to configure the analyzer to meet the specific requirements of the measurement application or to set up the analyzer's hardware options for their best usage in the system in which it is being installed.

This same interface may also be used to initiate some operational actions in the analyzer. The same type of actions mentioned earlier, like performing calibrations or process stream switching, may also be actuated through the analyzer interface. In most cases, the user has much more control of the analyzer with this local interface than is allowed through serial interfaces from the analyzer to the process control system. All data types provided within the analyzer are available with the keyboard/display interface.

Maintenance personnel can review how well their analyzers are functioning and make changes in the analyzer configurations from within their maintenance shop or other central plant location.

One other area of communication that is gaining importance in analyzer systems is remote control and remote maintenance access. Modern analytical systems may provide the same type of access from a remote interface that is available with local keyboards and display interfaces. These remote interfaces may be based on personal computer systems, or they may be a slightly modified version of the local interface. Some analytical systems even provide this type of remote access from multiple locations in a plant to all of the analyzers installed in the plant through an analyzer network communications system. Maintenance personnel can review how well their analyzers are functioning and make changes in the analyzer configurations from within their maintenance shop or other central plant location. Some systems even allow the analyzer configurations to be uploaded and stored in a computer over the communications network.

These types of communications to and from analyzers are essential in today's process analytical systems. They provide the required information for improved analyzer maintenance and the valuable data needed to better control the process plants in which they are installed.

Communication Techniques

Most people who work with analyzers have some familiarity with electronic systems. Furthermore, because of the proliferation of personal computers, most people have at least some awareness of communications that involve computers. As a result, "communication" often is associated with elaborate or sophisticated schemes where computers talk to one another. It is important that those interested in the practical aspects of analyzer communication do not become biased by these preconceptions toward excessive (and potentially expensive) complexity. In fact, communication involving analyzers takes on many forms, and some of the most rudimentary are completely adequate for many applications.

Obviously, simple systems are not always adequate. Still, when an analyzer system designer is considering the requirements, it is often helpful to view the needs in simple terms first. Then grow into more sophisticated systems which, of themselves, have "grown up" from the fundamentals. Additionally, for any person using or maintaining a complex analyzer system, it is usually helpful to consider the fundamental components of the system.

In fact, communication involving analyzers takes on many forms, and some of the most rudimentary are completely adequate for many applications.

The methods available for analyzer communication can be divided into three basic groups:

- Analog techniques
- Discrete digital techniques
- Serial digital techniques

The following sections will describe variations on these basic techniques. A later section of this chapter will discuss the considerations that should lead to choices among these techniques in any particular system application.

Analog Communication Techniques

Analog communication techniques are perhaps the most inherently obvious form of communication. Since the work is often analog (continuously variable) in nature, the measurement of events that occur in the world are often analog in form. For example, the concentration of a component in a process occurs between 0 and 100%, and the pressure of a fluid in a vessel is rarely exactly at one point or another but is usually at any possible pressure between the low and high physical limits of the vessel.

As described above, analyzer systems that measure physical processes must deliver outputs that represent the physical state being measured. Since the physical processes are continuous, it is convenient and it seems natural that the output be capable of being continuous.

CONTINUOUS ANALOG OUTPUTS

In the simplest case, this is accomplished by causing the output of an electronic amplifier to mirror the process variable being measured. If this mirroring is done continuously, the output of the amplifier can be plotted on paper. The resulting line is called a "trend" recording because it demonstrates visually the "trend" of the process variable.

For instance, a voltage amplifier can be committed to the task of representing the process variable. The lowest and highest outputs of the amplifier can be made to correspond to the lowest and highest possible conditions of the process variable.

An infrared analyzer might be used to measure the concentration of carbon dioxide in a stack effluent. The analyzer system designer intends to let the output be represented by a voltage amplifier that can deliver a signal from 0 to 10 VDC. If the possible concentrations of the CO_2 might range from a low of 0% (no CO_2 present at all) to a high of 20%, then the output of the amplifier is allowed to represent the CO_2 as follows:

0 VDC means CO_2 at 0%

10 VDC means CO_2 at 20%

Values in between are represented by voltages between the high and low outputs; for instance, 5 VDC means CO_2 at 10%.

In the previous example, a voltage amplifier was used to represent the output. It is actually more common that an amplifier with a current output is used.

A chromatograph is used to measure the concentrations of two components (methane and ethane) in a natural gas stream. Two current amplifiers (mADC) are dedicated to the task of transmitting the results of the measurement (one component on each amplifier). Both current amplifiers are capable of delivering a current no lower than 4 mADC and no higher than 20 mADC. If the methane ranges from 70 to 90% and the ethane ranges from 5 to 20%, then the outputs (Figure 2-2) of the current amplifiers represent the actual measurements as follows:

Amplifier 1
4 mADC means methane at 70%
12 mADC means methane at 80%
20 mADC means methane at 90%

Amplifier 2
4 mADC means ethane at 5%
12 mADC means ethane at 12.5%
20 mADC means ethane at 20%

Note that 12 mADC is the mid-range of the current amplifiers (that is, halfway between 4 and 20 mADC) and is therefore used to represent the halfway point of the process variables being measured.
See Figure 2-2.

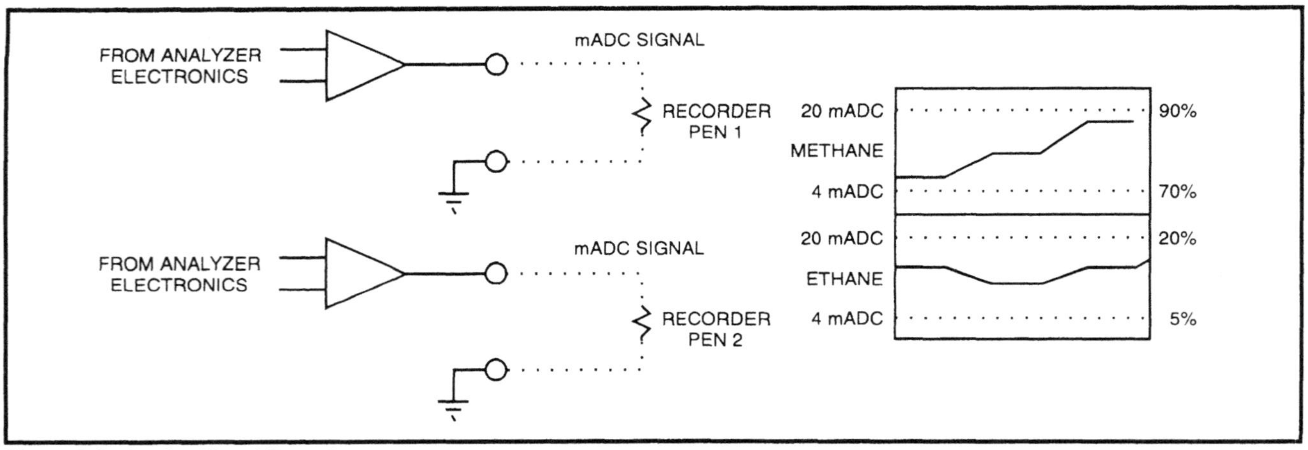

Figure 2-2. Analog Trend Outputs

The previous example illustrates one of the common variations in the use of analog output techniques for analyzer communication—that is, variations in the electrical hardware used. The most common variations in use today are:

Voltage-based analog outputs
0 to 10 VDC
2 to 10 VDC (live zero basis)

Current-based analog outputs
0 to 20 mADC (live zero basis)
4 to 20 mADC
10 to 50 mADC

A large number of other configurations are also used, including negatives of the ranges listed above and schemes based on 24 VDC sources and 110/220 VAC sources. It is common in the United States to use analog sources that have an elevated minimum level (such as 4 mADC), whereas it tends to be more common in other countries to use sources that drop to an actual zero output level (example, 0 mADC). Occasionally, when an elevated zero is used in normal operation, the occurrence of a true zero can be used to indicate an equipment error or malfunction.

When an elevated zero is used in normal operation, the occurrence of a true zero can be used to indicate an equipment error or malfunction.

However, despite their apparent simplicity, analog-output schemes have two significant deficiencies, which are illustrated by the examples above. First is the necessity to assume a scale for the process variable being measured. It is, in fact, *usually* true that a process variable being measured has a normal low and normal high level in real operation. However, the use of the analog-output communication scheme enforces this assumption. As in the example above, if CO_2 exceeds 20% in real operation, the electronic amplifier has no way to indicate that this over-range situation has occurred. Additionally, the assumption of a range for the process becomes a source of confusion for the human user. As in the example above, it is not self-evident that "12 mADC means ethane at 12.5%" while "20 mADC means ethane at 20%." This assignment of physical value to electronic signal level must be understood by the user as a matter of training about the system he is designing or trying to use.

The second most significant deficiency is the heavy use of hardware. As illustrated by the second example above, it is necessary to dedicate an electronic amplifier to each process variable being measured. This fact may not be significant in a small analyzer system performing only a few measurements. However, in larger systems where more data must be collected, the fact becomes more important.

TIME-MULTIPLEXED ANALOG OUTPUTS

To overcome this second deficiency, a specialized analog communications scheme has been used in process analyzer systems. This scheme involves time-sharing the same analog-output amplifier for multiple process-variable outputs. The most classic representation of this scheme is the generation of a bar graph. The output here is quite different from the "trend" output described previously. In this method of communicating from the analyzer to the human being, each time a particular measurement is made, its value is sent to an electronic amplifier for only a short time—typically 1 to 10 s. When this short time has expired, the amplifier is turned off.

When the amplifier is off, it is then "available" to be used to output a value corresponding to a different process variable. Therefore, when measurement of the second variable is completed, this second value is sent to the amplifier, again for just a short time. Using this scheme, several different process variables can be transmitted sequentially to the same electronic amplifier. If the output of the amplifier is plotted on paper over time, the signal looks like a series of vertical bars. Figure 2-3 shows the output of an analyzer that measures two different components.

This type of output is often referred to as a "bar chart."

While this output scheme has the advantage of reducing the cost of output hardware, it also has several disadvantages. The first disadvantage is fairly obvious: It adds complexity to a system that began as something simple. It is now necessary for a person or another machine using the information to know which bar on a bar-graph chart represents what process variable being measured.

The first disadvantage is fairly obvious: It adds complexity to a system that began as something simple.

This problem can be solved by using "synchronizing" methods to indicate the start of a series of bars. This can be done by having the electronics draw special bars that are extra wide or extra tall to indicate the first of a series of outputs. Additionally, a relay or other signal might be turned on coincident with the first bar of a series of outputs. This scheme — sometimes called "data-ready/cycle-sync" (Figure 2-4) — provides a method to overcome the problem. However, the solution again adds complexity.

The second disadvantage of time-sharing analog hardware is that the output is no longer continuous.

The second disadvantage of time-sharing analog hardware is that the output is no longer continuous. When a process variable being measured changes after the time its bar is drawn, the change cannot be sent to the output until it is allowed to draw a new bar. This situation is not a problem for chromatographs or more complex "batch-type" analyzers, but it can be a significant limitation on spectrophotometers or other "continuous" analyzers.

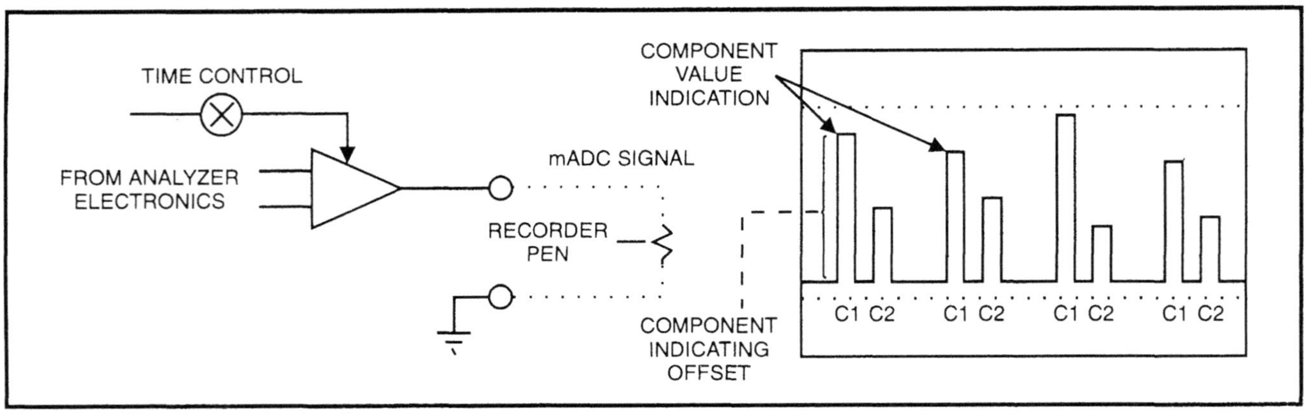

Figure 2-3. Time-Shared Analog Output

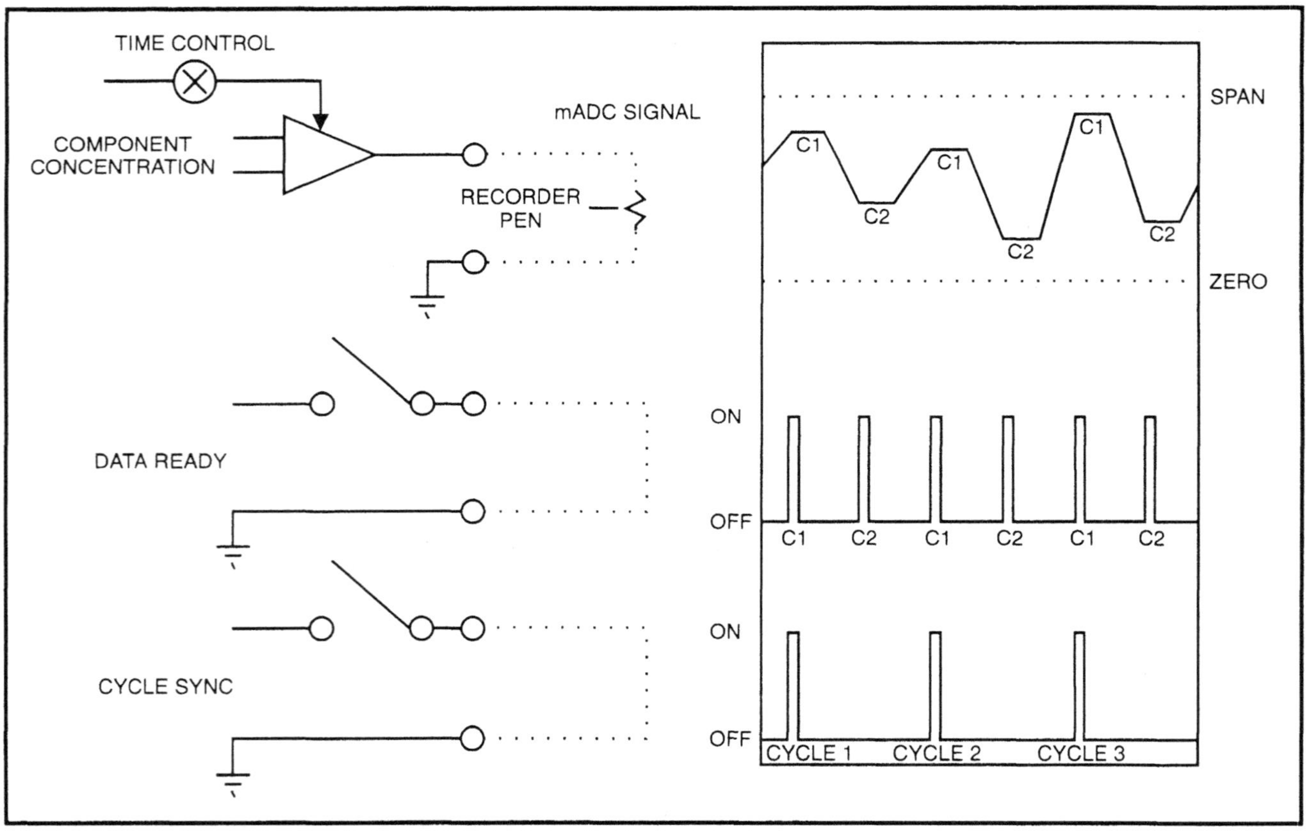

Figure 2-4. Time-Shared Analog Output with Data-Ready/Cycle-Sync Controls

ANALOG COMMUNICATION NETWORKS

Through the use of time-shared analog communication techniques, entire communication "networks" can be built up. In general, this practice is common only in one specific form. When multiple analyzers are in use in a single facility, it is possible to connect analog outputs from several different analyzers onto a single cable by making the connection through switches or relays (Figure 2-5). This permits multiple analyzers to share a common output device — such as a recorder — and common field wiring, thereby saving costs. Because this system is not useful in continuous operation, its most common application is in connecting information that is normally used intermittently, such as for maintenance.

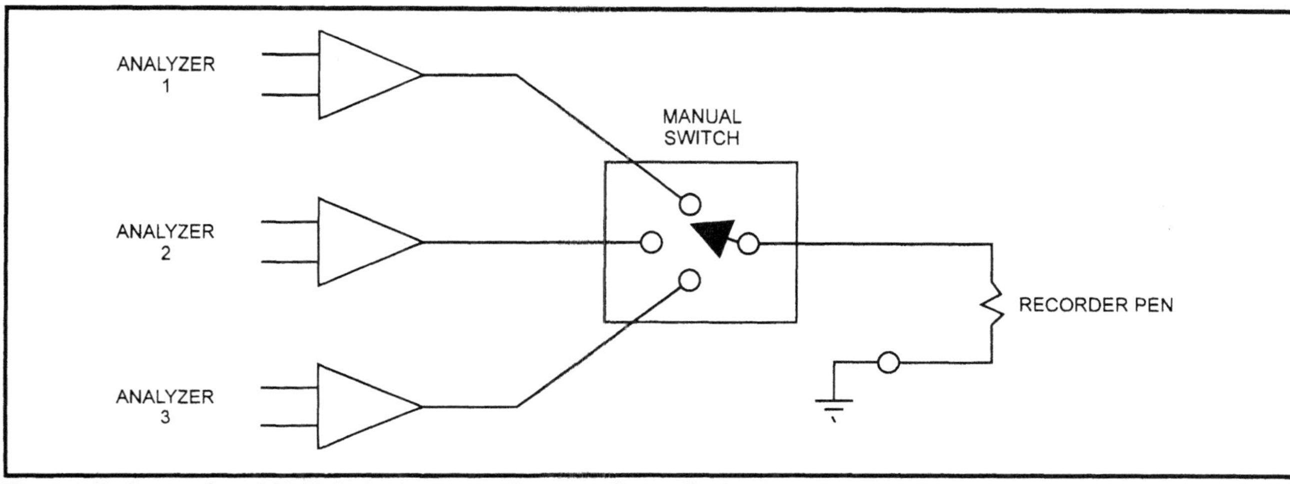

Figure 2-5. Connection of Multiple Analyzers to a Common Output Device

ANALOG TECHNIQUES USED IN ANALYZER INPUTS

As noted previously, analyzer communication systems must provide methods for communicating into the analyzer as well as out from it. In current practice, the main reason for sending information to an analyzer with analog methods is to provide process-related information to the analyzer so that it can better perform its analysis.

For example, an analyzer might be used to measure the components present in a natural gas pipeline and to produce an output scaled to represent the heating value of the gas (measured in BTU or kilocalories). In this application, the analyzer might be set up to make use of a flow measurement—provided by another analyzer or a transmitter—to convert the simple measure of heating value into an integrated value for volume of heating value per unit time.

Obviously, this information might be entered into the analyzer by hand, such as by setting a potentiometer. Or this information might be automatically entered by use of an analog connection. In this case, the analyzer would use analog input electronics to receive the input signal. In either case, information is communicated to the analyzer in analog form.

Analog systems are susceptible to error introduced by electronic noise.

Analog systems depend on absolute electrical values.

If the ground reference point is different in the receiving and the transmitting equipment, this can result in a significant error.

Voltage-signal hardware should only be used for very short signal runs.

PRACTICAL CONSIDERATIONS IN ANALOG COMMUNICATION TECHNIQUES

Later sections of this chapter will discuss reasons for choosing one communication technique over another. However, if a decision has been made to use analog communication, certain practical considerations must also be made.

First, analog systems are susceptible to error introduced by electronic noise. Such noise can come from many sources, including lightning, motors or other electrical equipment, radios, and walkie-talkies. This usually means that all analog signals must be connected using shielded wire, and the shield must be attached to electrical ground to minimize noise interference.

Failure to isolate system signal from noise can introduce significant errors in the system signal.

Second, analog systems depend on absolute electrical values. For example, if an analog range of 4 to 20 mADC represents a component concentration of 0 to 50%, then very small deviations in current from the electronic amplifier can represent significant error in the component concentration.

Such error usually comes from three sources:

1. The electronic amplifier can be out of proper calibration. Electronic zero and span adjustments are normally provided and should be checked as a matter of routine maintenance at least once per month on all critical signal paths.
2. Corrosion, oxidation, dirt, or other aging can cause electrical resistance to develop in analog-signal wiring. Care should be taken in the installation of critical wiring to protect connections and tie points. Plug-in connectors used in the installation should be inspected to assure that fit is snug. Additionally, when electronic calibration is performed, measurements should be taken at the end of field-wiring points — not directly at the output of the amplifier. Visual cabling inspection should be performed annually as a part of routine maintenance on critical signals.
3. Error can be introduced as a result of "ground loops" (Figure 2-6). Most electronic systems use a common physical location as the electronically-referenced ground. Sometimes this location is the mounting chassis of the equipment, although in most new equipment the common reference is kept isolated from the chassis. Numerous input and output voltages and currents are measured relative to this common reference. If the ground reference point is different in the receiving and the transmitting equipment, this can result in a significant error. In some cases, designers have attempted to use the earth itself as a common reference. Unfortunately, the earth can have very large voltage differences even in the narrow confines of a single process plant.

To prevent ground loop error, both the transmitting and receiving devices should share the same electrical reference point. If this is not physically possible, then ground-isolated analog-electronic hardware should be used.

Third, analog hardware should be chosen with consideration for the final installation. Current-based signal hardware is commonly used because it is less susceptible to electronic noise. Additionally, the effect of wire resistance on a current signal is compensated for by the amplifier itself. Voltage-signal hardware should only be used for very short signal runs.

In all cases, the user must consider the analog load imposed by the receiving device. Generally, in modern electronic systems, a single analog source can provide its signal to nearly any single analog receiver. However, if multiple receivers are to be used — such as multiple recorders placed on a single measurement output — then the load imposed by the recorders (measured in ohms) must be checked against the maximum capability of the source (measured in volts or milliamps). Ohm's Law can be applied to assure that the maximum source voltage available divided by the maximum receiver impedance applied does not fall below the required maximum current signal required. In this case, "total receiver impedance" must include the resistance of signal wiring.

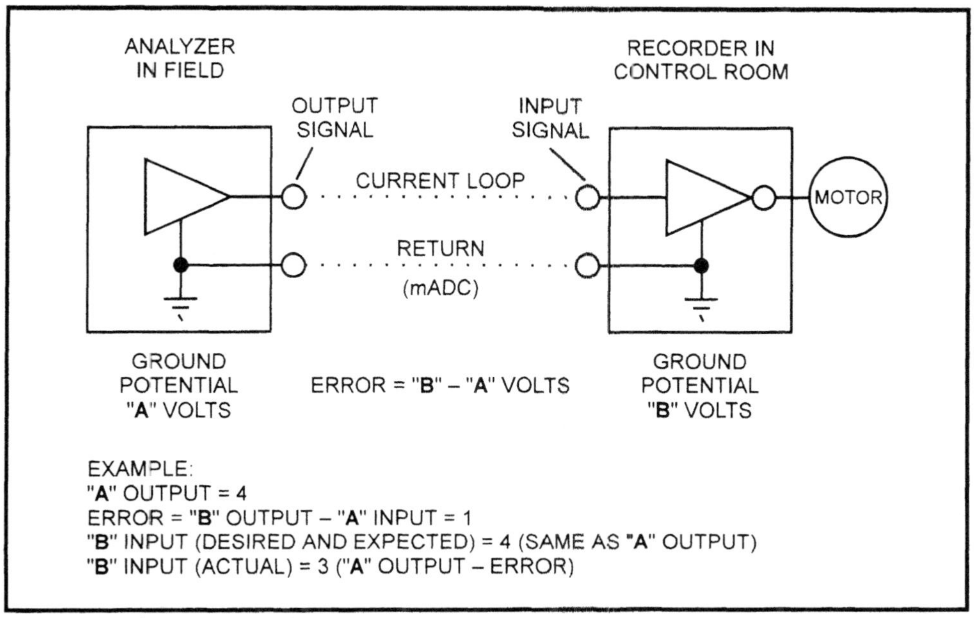

Figure 2-6. "Ground-Loop" Errors

Discrete Digital Communication Techniques

The use of discrete digital hardware enables the simplest form of communication. Most analyzer systems have a need to communicate to the human operator or to other electronic equipment items of information that are binary in nature. Examples of this include:

- Operating status: analyzer is running/stopped
- Physical condition: oven temperature is OK/too hot
- Measurement information: pressure is OK/too low
- Warning information: compressed calibration gases are OK/nearly gone

Obviously, a large number of conditions might be listed that can be communicated with simple relays that are switched on or off (binary information).

Other items of information pertain more directly to the measurement itself. Examples include:

- Component concentration: above/below set point
- Variable measured: equal to/not equal to zero
- Fire-hazard limit: has/has not been reached

Discrete digital communications are implemented in two ways. The first is by using contacts that open or close. This is normally performed by a relay that is used to switch alarm lights or bells off and on. The second is through the use of electronic logic circuitry. In this case, the presence or absence of a voltage (often 5 VDC) indicates the binary information being transmitted.

There are some practical considerations in such communication systems. The first is the electronic capacity of the switching device. Relays usually are very limited in the levels of voltage or current they can switch. A relay may be used which can switch 110 VAC to activate a large alarm in the United States. However, the same relay may be unable to handle the 220 VAC used in other parts of the world. Likewise, that relay may be unable to cleanly turn off lower voltages such as 5 or 10 or 24 VDC. This is often true if the "relay" is really a solid-state device such as a triac. By the same token, a small-capacity switching device may turn off or on without any electronic noise or "bounce." However, that device may burn out if higher voltages are applied.

A second consideration is output isolation. A relay may be used as an output. But if one of its contacts is tied permanently to the analyzer's chassis or ground reference, then this can impose limitations on the ability of a receiving device to use the relay. This is particularly true if logic-level binary-voltage devices are in use; ground loops can cause error in readings. Ground loops are discussed in more detail in the previous section on analog communication techniques.

A single relay on an analyzer can be used to collectively indicate all possible error conditions in the analyzer.

A third consideration is hardware capacity for a large number of output signals. A single relay on an analyzer can be used to collectively indicate all possible error conditions in the analyzer. However, if it is desired to provide diagnostic indications, then many different digital outputs will be required.

Serial Digital Communication Techniques

The availability of computer electronics has made serial digital communication a popular and rapidly growing communication technique. This methodology provides a means to overcome many of the limitations in discrete digital and analog communication systems. Furthermore, serial communication permits an analyzer to both send and receive data.

It is this last capability — combining input and output with a user interface into one system — that has brought on the extensive use of communication networks. This topic is discussed at the end of the chapter. This section will deal only with the various primary techniques used to implement serial communication and some of the practical aspects of using such systems.

BASICS OF SERIAL COMMUNICATION METHODS

The essence of serial communication hardware is a single discrete digital electronics device.

The essence of serial communication hardware is a single discrete digital electronics device. In principle, the device could be a relay; in practice, electronic voltage or current amplifiers are used. The point is that the electronics involved are capable of assuming only two conditions: off/on or lower/higher voltage level.

Such electronic hardware can normally indicate only two items of information (see the previous section about discrete digital techniques). However, by introducing a time element into the picture, the same electronic hardware can be used to express an unlimited amount of information, either into or out of an analyzer system.

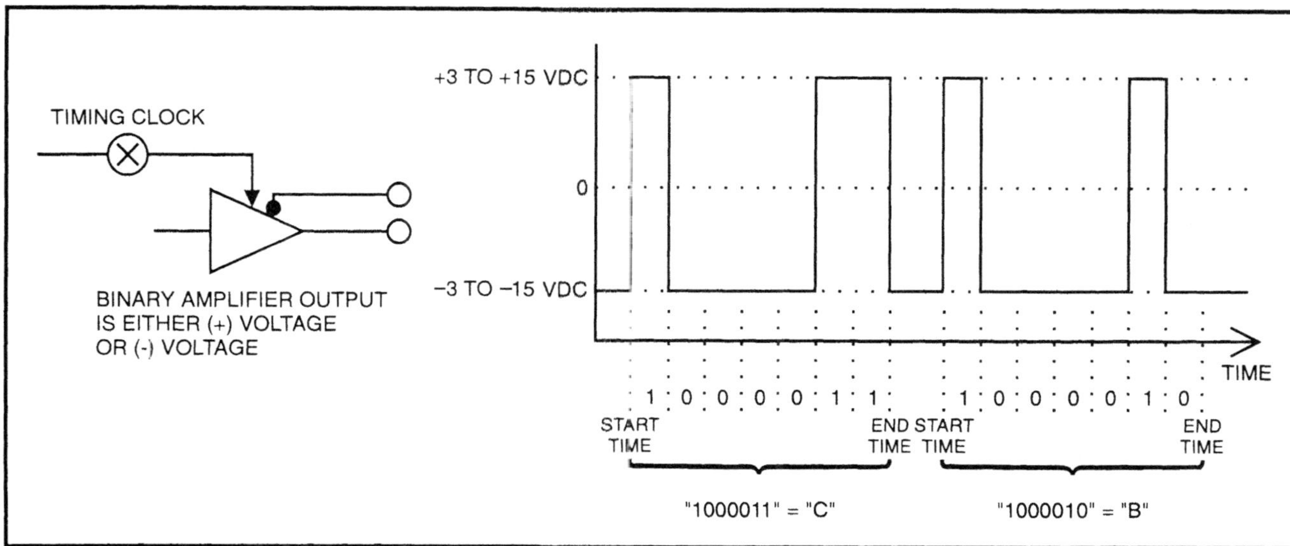

Figure 2-7. Basic Serial Representation

Figure 2-7 illustrates this technique. At a single point in time, the electronic hardware is in only one condition: high or low. But if a timer is started and the pattern of high/low voltages coming from the electronic hardware is watched, then the pattern itself can be used to represent an item of complex information. For example, as shown in Figure 2-7, in the ASCII standard code interpretation, a pattern of seven states is used to represent the alphabetic letter "C." Therefore, by starting a timer that synchronizes the signal source (such as the analyzer) and the listener (such as a printer) and by observing the pattern of signals that occurs during a specified time interval, it is possible for the analyzer to send the letter "C" to the printer. Similarly, if the timer is then restarted and in the next time interval a different pattern is created by the analyzer, it is possible for the analyzer to send yet another letter to the printer. As indicated in the figure, this could be the letter "B." Over some length of time, it is possible for the analyzer to send patterns representing the entire alphabet to the printer. Or, more practically, the analyzer can send written information.

The key elements are: first, that the analyzer and the printer "agree" on the binary patterns that will represent the various letters; second, that the printer and the analyzer use a consistent and identical time standard; third, that enough time is allowed to pass for significant information to be transmitted from one device to the other.

HARDWARE VARIATIONS

In the next section, some practical considerations of serial communication will be discussed. However, it is first appropriate to consider some of the variations in hardware that are used for this purpose.

Several systems are in common use. The older of these are:

- RS-232-C voltage hardware
- 0/20 mADC current loop hardware

Most newer systems are readily understood as variations of these two.

The RS-232-C system is a convention that was specified for military use and is generally accepted as a standard hardware technique throughout industry.

RS-232-C Convention

The RS-232-C system is a convention that was specified for military use and is generally accepted as a standard hardware technique throughout industry. It uses voltage amplifiers to create the two voltage signal levels needed to represent the two binary conditions. The hardware has, at its most basic level, three signal wires: which are transmit, receive, and ground. The transmit line sources voltages relative to the ground line and the receive line observes voltages relative to the same ground line.

The voltages in use are:

–3 to –15 VDC	means binary 1, or set, or on, or high
+3 to +15 VDC	means binary 0, or reset, or off, or low
>15 VDC	may cause hardware damage
–3 to +3 VDC	is indeterminate and may cause error

The binary conditions are often expressed in three ways as indicated. It is significant to note that the high/low binary condition is the opposite of the actual voltage signal.

The RS-232-C convention includes several other optional signals besides transmit and receive. These include:

- RTS: Request To Send
- CTS: Clear To Send
- DSR: Data Set Ready
- DTR: Data Terminal Ready

These other signals are also defined and are used for control of communication devices such as modems. However, a more complete discussion is beyond the scope of this chapter.

Physical wiring is also often standardized under the RS-232-C convention. A "D"-shape 25-pin cable connector is used. In this system, pin 2 is the transmit line, pin 3 is the receive line, and pin 7 is the common. The other pins are also defined in a standard manner. In general, a voltmeter or an oscilloscope can be used to observe RS-232-C signal activity.

This technique was originally established as a means of operating mechanical teletype (TTY) machines.

Current Loop Convention

This technique was originally established as a means of operating mechanical teletype (TTY) machines. As long as a signal current was flowing, moving components in a teletype machine would remain in one physical location. If the current in the signal line was interrupted, mechanical components would release. At rather slow speeds, it was possible to cause these mechanical fingers to activate appropriate keys on the mechanical typewriter of the teletype.

The convention establishes two current levels as the operating levels that represent binary states. These are:

(1) 20 milliamps	means binary 0 or space or reset
(2) 0 milliamps	means binary 1 or mark or set

While the technique has almost no present use in teletype activation, it has been retained as a de facto industry standard for several reasons.

First, there was a large installed base of teletype equipment in the world, and it was suitable for equipment manufacturers to make new equipment that was compatible with the old.

Second, the current loop offers higher signal noise immunity than the newer RS-232-C convention. In general, there is wide latitude around the nominal 0 and 20 mADC setpoints of the current loop convention.

Third, current loops can be powered by current amplifiers that can operate into longer transmission lines. The voltage amplifiers of the RS-232-C are limited to voltage losses that occur in the resistive wiring. The RS-232-C specification itself limits line length to 50 ft (18 m).

Fourth, multiple receiving devices can usually be installed in series on a current loop. Multiple devices installed on a voltage line like RS-232-C can overload the voltage drive capabilities of the electronic amplifiers.

Fifth, current loop systems do not have to be referenced to a common voltage reference such as ground. Therefore, current loops are more likely to be immune to ground-loop problems.

Current-loop systems also pose some disadvantages. Such systems do not have well-defined control signals to permit more complex device control, such as that needed for modems. And because of electronic rise-and-fall times, current-loop systems may be constrained to operate at slower speeds.

Other Hardware Variations

Most modern communication systems support RS-232-C and current-loop conventions in standard configurations. Most also implement newer conventions. These include a later military specification, RS-422, and others.

These newer conventions are based on voltage amplifiers that operate with differential line drivers (see Figure 2-8). Such systems usually depend on a 0 VDC/5 VDC signal differential that switches polarity between two signal wires. A ground line may be provided as a noise shield, but this line is not part of the transmission loop. Therefore, ground loops do not occur.

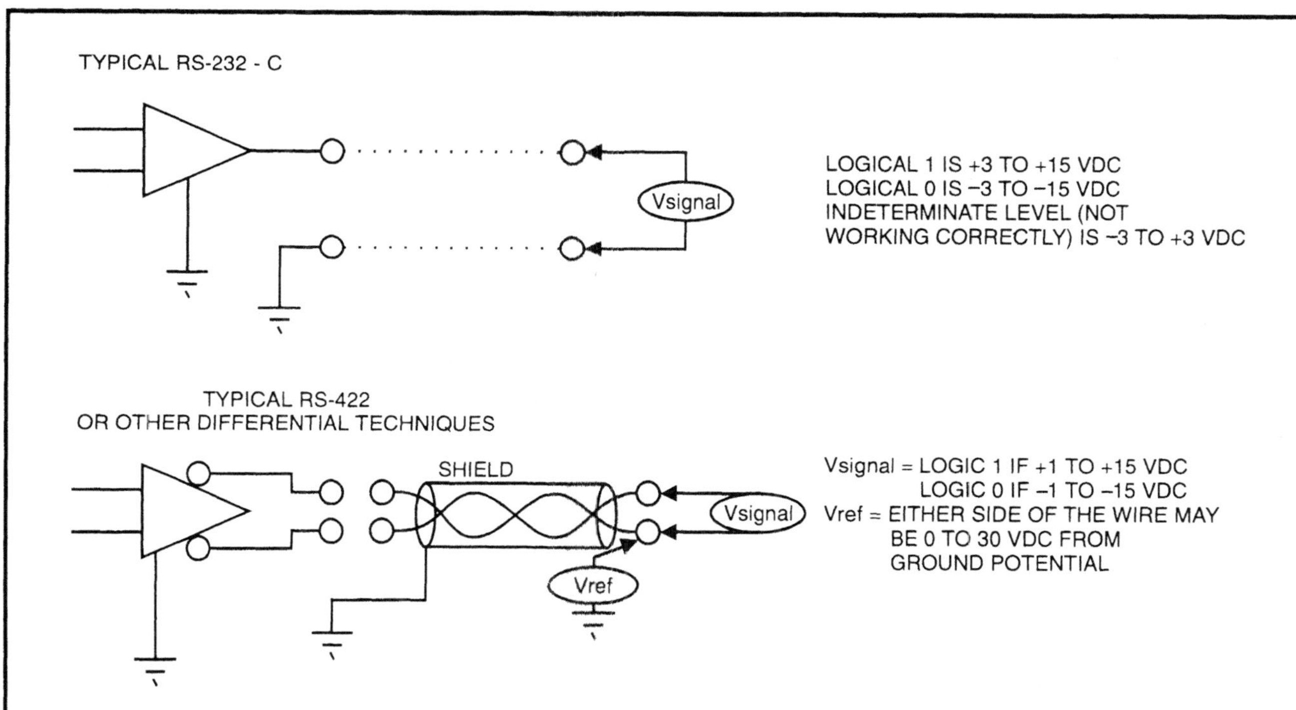

Figure 2-8. Voltage Amplification for Serial Communication

Additionally, modern transmission line drive circuitry supports a "third" state besides the binary high/low. This third state is "off," and in this condition, a transmitter on a transmission line is apparently "disconnected." This operating mode permits several different devices to be connected to a single communication line in a "multidrop" configuration (Figure 2-9).

A final transmission mode used in serial communication systems is based on a carrier signal or "tone" that carries the signal information.

A final transmission mode used in serial communication systems is based on a carrier signal or "tone" that carries the signal information. This carrier signal generally operates at a very high frequency. Actually, the carrier itself is driven by hardware very similar to the multidrop transmission line drivers just described. However, the on/off states are operated continuously and at a very high frequency. Subsequently, the data signal is superimposed on the carrier "tone" at a somewhat lower frequency. This results in a modulation similar to radio operation. Although the fundamental techniques are the same as slower systems, this method permits very-high-speed communications.

PRACTICAL CONSIDERATIONS OF SERIAL COMMUNICATION TECHNIQUES

Several considerations must be made in dealing with serial communication systems. Such systems are in common use because of the advantages they offer. These include:

- The capability to combine both input and output functions into a system that is common with maintenance and control functions.
- The capability to transmit and receive large amounts of information with very little hardware cost.
- The ability to use standard hardware to connect two different systems from two different manufacturers.

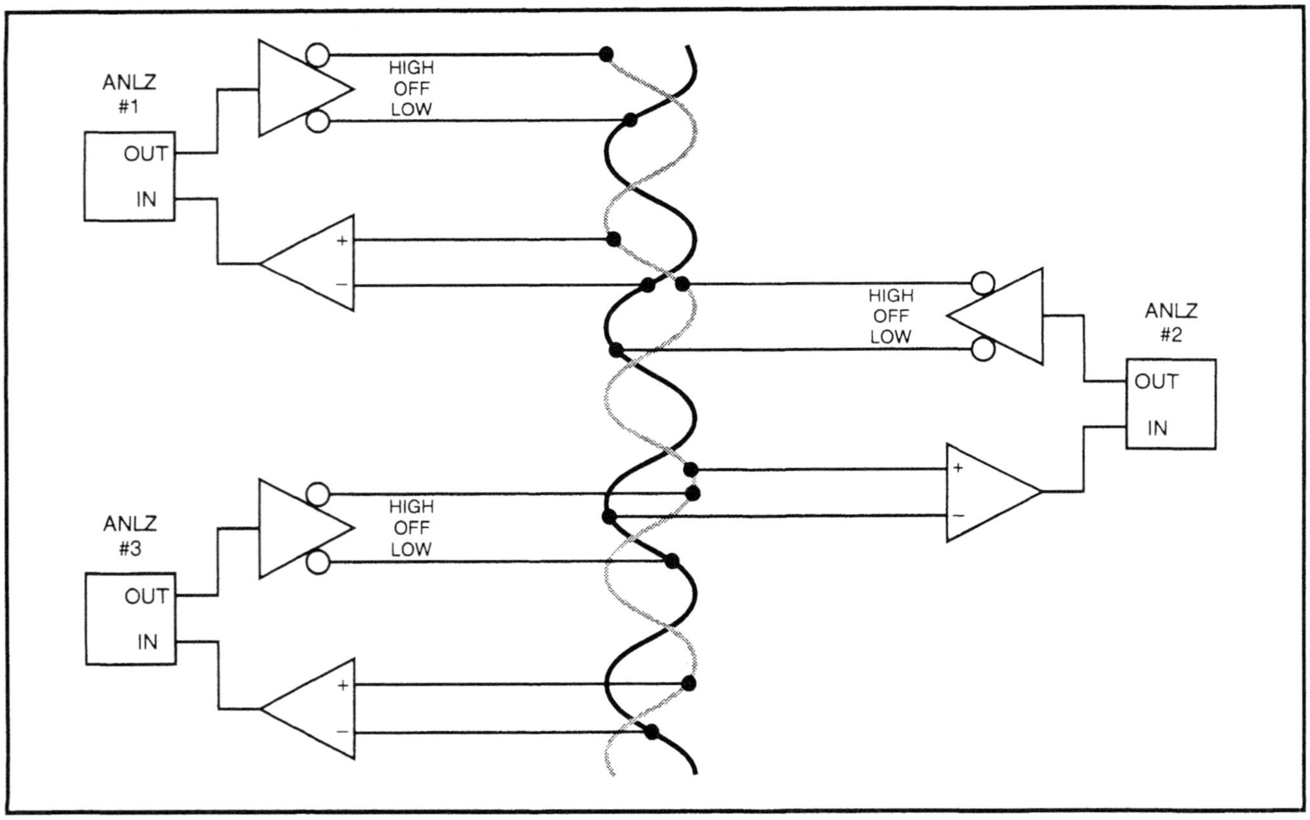

Figure 2-9. Multidrop Serial Connections

However, for each of these advantages, there is a corresponding disadvantage. These are:

- The difficulty of training human operators to understand the function that is being performed at any one time.
- The requirement for very specialized hardware and potentially complex software, both of which lead to relatively high expense.
- The lack of standardization at any level above hardware, including software protocols and application software.

In general, a person will not question the use of a serial communication system to drive a printer. Such systems are in common use and are very well defined and inexpensive. In systems in which an analyzer is to communicate with another computer, care should be taken and analog or discrete digital options should be considered.

In systems in which an analyzer is to communicate with another computer, care should be taken and analog or discrete digital options should be considered.

The following sections detail two of the most common problems that occur in a real-world environment.

Software and Protocol

As stated previously, successful serial communication requires agreement between the sending and receiving devices on the code patterns that will be used to represent information and the timing and time intervals through which this information will be sent. The time intervals are the easiest to agree on. In most systems, standard time intervals are defined and accepted in the industry. Expressed as "baud rate," these frequencies specify the number of binary bits of information per second that are sent from one device to another. Standard baud rates are 300, 1200, 2400, 4800, 9600, and 19,200 baud. Some other variations are occasionally used.

Timing is often controlled by a bit-synchronization sequence. For instance, if an alphabetic character requires seven binary bits to represent it, then these seven bits are preceded by a "start" bit that is used by the sending device to "wake up" the receiving device. Following the seven data bits is usually an error control bit, called parity, and one or two end bits. The start and end bits allow a known time interval for both transmitting and receiving devices to synchronize.

Protocol agreement is much more difficult to establish. At the most rudimentary level, certain industry standards have been adopted which allow representation of alphabetic characters in a "pre-agreed" form. The most commonly accepted standard is ASCII (pronounced "as-key").

However, raw computer information is often transmitted as an image of the computer's memory — that is, in its direct binary (machine language) form. Such information may be "received" by a receiving device and still be of no value if the receiver does not "understand" the meaning (sequence) of the data. Such understanding is utilized by complementary software in both the sending and receiving devices.

Additionally, there must be rules as to which device may "talk" and which must "listen" at any given point in time. And there must be agreement about "how long" one device may talk before it must listen and what it may talk about.

Protocol software is usually complex and usually created on a custom basis for each application because very few industrial standards exist in this area.

This understanding and these rules are implemented by complementary software in the sending and receiving devices. This software is referred to as communications "protocol." Protocol software is usually complex and usually created on a custom basis for each application because very few industrial standards exist in this area. Fortunately, the analyzer industry has made some progress toward mutual agreement on protocol standardization. One of the more commonly accepted protocols was originally published by Gould, Inc. as part of the Modicon line of programmable logic controllers. The original protocol and various adaptations of it are referred to as the Modbus[1] protocol.

Hardware

Although hardware conventions are well established and standardized, practical considerations of dealing with communications hardware are still numerous. In general, this is because specialized equipment is required to test or troubleshoot such systems.

It was mentioned earlier that "time is required" to pass information from one device to another in serial fashion. This is one of the biggest disadvantages of serial communication systems; that is, information transfer from one point is delayed while transfer from another point completes. Transfer is not continuous or "instantaneous."

To overcome this problem, manufacturers have improved electronics to allow very-high-speed transfers of data. This has reduced problems of delays in communication systems but it has not eliminated them. For example, if many analyzers have current analysis information and want to send their information to output devices at the same time, one analyzer must go first, another second, and so on. Ultimately, the last analyzer to be permitted to use the communications system may have to wait a long time.

If a communications system is made faster, this delay is reduced — but it is never eliminated. In no case is such output truly "continuous."

Although an oscilloscope might be used to observe voltage changes at high speed, an oscilloscope does not have the software to recognize and understand binary patterns.

However, in the process of making the system faster, manufacturers have also made the hardware more difficult to maintain. It is not possible to use a voltmeter to observe communication occurring at a high rate of transmission. The best that can be done is to use the meter to troubleshoot simple short circuits and open lines in the communications wire. Additionally, although an oscilloscope might be used to observe voltage changes at high speed, an oscilloscope does not have the software to recognize and understand binary patterns.

Specialized equipment, such as a serial data analyzer may be used for this purpose. This makes troubleshooting possible, but introduces a new problem. Now, maintenance personnel must understand the software protocol that the machines are using.

Now, maintenance personnel must understand the software protocol that the machines are using.

Despite these difficulties, the advantages are strong and development continues on serial digital communication techniques. Problems are constantly being resolved and these systems are being made easier to use.

Implementing Analyzer Communication

Selection of a communication technique for handling analyzer information involves many considerations. Each of these considerations is important, but typically the selection will be based on those issues which are most important for the particular installation.

1. Modbus is a registered trademark of Gould, Inc.

Installed Cost Per Point

The installed cost per point will vary depending upon the conditions existing at each user's site. Items that affect installed cost include:

- Manpower costs: labor rates for running cable
- Area electrical-hazard classification for cable route: NEC/NFPA 496 Class, Group, and Division or general-purpose safety requirements
- Local codes: corporate or governmental guidelines which mandate special installation conformance
- Material costs: cable, conduit, seals, junction boxes, etc.
- Existing cable available for use: extra signal wiring pairs previously installed

Figure 2-10 illustrates relative installed cost information for various approaches as the quantity of I/O points installed is increased. This figure shows the cost trends for various communication techniques. Exact costs will vary depending on the variables given in the preceding paragraph.

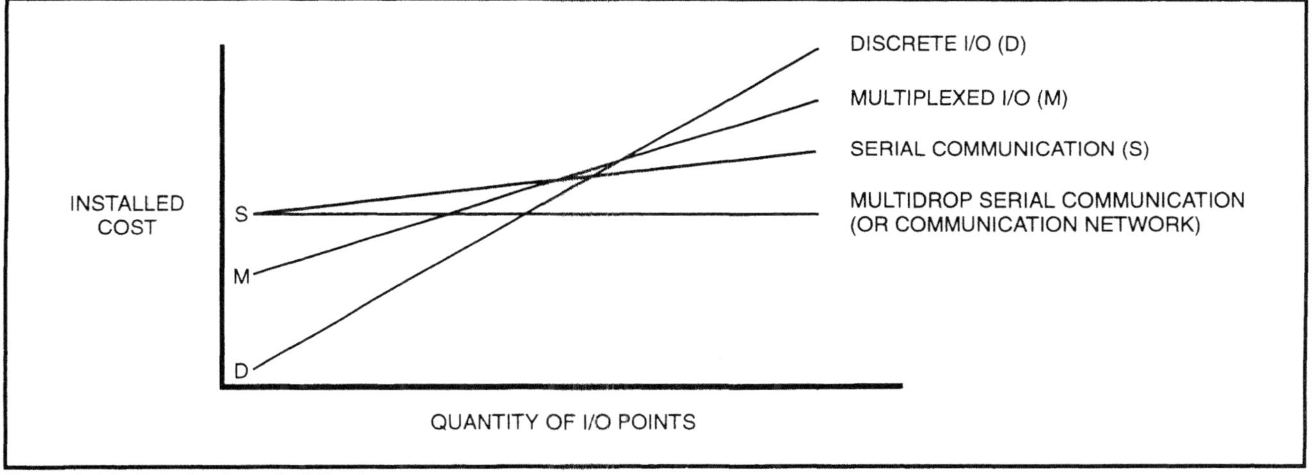

Figure 2-10. Relative Installed Cost

Data Verification and Security

Serial communication techniques have a distinct advantage over analog techniques in the area of data verification. Each serial data transmission normally contains some form of message integrity information transmitted at the end of the message. Common security methods used include checksums of the message and cyclic redundancy checking (CRC) information added at the end of the message. If an invalid checksum or CRC value is detected in a message, the message is typically retransmitted until it is acknowledged successfully or a retry limit is reached.

Security can be further enhanced for serial communications by installing backup or "redundant" communication links where data can be transmitted over multiple paths. These installations typically route two cables in separate cable trays so that if one cable is damaged, the other cable will be able to continue to maintain communications.

Security can be further enhanced for serial communications by installing backup or "redundant" communication links where data can be transmitted over multiple paths.

Signal security of analog signals is higher if signals are wired individually in that each data point is independent from the other analog signals. A failure in one point does not affect other data points from that same analyzer. Signal security of analog signals is lower if signals are multiplexed and therefore dependent on the multiplexer and the common output cable used. Data verification relies upon periodic calibration since no data validation is transmitted with the analog signals. Any induced noise or signal degradation will not be detected under normal use.

Data Rangeability and Resolution

The serial data are typically transmitted in engineering units with resolution limited only by the analyzer itself.

Serial communications techniques have a distinct advantage over analog techniques in the area of data rangeability and resolution. The serial data are typically transmitted in engineering units with resolution limited only by the analyzer itself. For those serial systems which do not transmit values in engineering units, range information is typically transmitted so that a conversion can be made at the device receiving the data.

Analog signals are typically limited to a resolution of one part in one thousand or one-tenth of one percent of full-scale value. The range of engineering units which is represented by the current or voltage presented must be established during design for the analyzer and the device collecting the data. This necessitates that all range changes be manually made on both ends since this information is not transmitted with the data.

UNIDIRECTIONAL VERSUS BIDIRECTIONAL COMMUNICATION

Serial communications usually have the ability to transmit data in both directions (full-duplex). This mode transfers data from the analyzer to another device, but also allows information to be transmitted back to the analyzer. The capability of sending information to the analyzer allows the analyzer to be controlled remotely. Remote requests can perform a wide variety of functions limited only by the analyzer and the device used to transmit instructions. Examples of requests that can be sent back to the analyzer through remote serial communications are:

- Analysis stream selection
- Analysis stream skip
- Calibration
- Report log request
- Remote data monitoring
- Remote start-up

Analog signals are unidirectional in that information is transmitted from the analyzer to a receiving device. This technique does not have the ability to transmit information back to the analyzer.

TRANSMISSION DISTANCE

Maximum transmission distances vary widely with respect to the communications technique. Each of these methods can have its distance extended with the use of signal extenders (repeaters). Some general rules of thumb for transmission distances without extenders are as follows:

Signal Type	Drive Type	Maximum Length, ft (m)
Serial*	Single ended (RS-232-C)	50 (15)
	Differential voltage	4,000 (1,220)
	Current loop	4,000 (1,220)
	Network	5,000 (1,525)
	Modem	Limited only by phone service
Analog	4-20 mADC	4,000 (1,220)
	2-10 VDC	1,000 (300)

* Depending on transmission speed and type of cable used.

Detailed information should be readily available from the component device manufacturer.

Practical Considerations When Implementing Analyzer Interfaces

The successful installation of an analyzer interface can best be assured by following a few simple guidelines.

INSTALLATION CONSIDERATIONS

Low-impedance cables are often used to allow the maximum available line length and to minimize signal degradation. Typically, the analyzer manufacturer will specify cables that maximize the signal performance for each of the various communication techniques.

Low-impedance cables are often used to allow the maximum available line length and to minimize signal degradation.

Shielding is important for each of the communication types. This provides protection against "crosstalk" between cables as well as protection from noise induced by electromagnetic interference (EMI). Shielding is most effective when signal lines are twisted-pair and are individually shielded. To prevent ground loops, the cable shields should be connected on one end only and to an instrument-quality ground.

Maximum transmission distances vary widely with respect to the communication technique. Each of these methods can have its distance extended with the use of signal extenders (repeaters) or by reducing baud rates. Some general rules of thumb for maximum distance without extenders are given above in the section "Transmission Distance."

Shielding is most effective when signal lines are twisted-pair and are individually shielded.

Analog signal cabling and signal level matching are relatively easy to install and to verify. Each signal has only a "plus" and a "minus" polarity and a shield wire for a total of two connections plus shield per point.

Serial cabling may be more complex to install since multiple connections can be required. The device receiving analyzer data must be wired in a compatible fashion to the analyzer and must match the analyzer's protocol and format.

Interfacing serial signals is most difficult when the device receiving the data is not supplied by the analyzer manufacturer. Custom software drivers must usually be written for the device that receives the data unless that device is supplied by the analyzer manufacturer.

USER FLEXIBILITY

User flexibility for adding data points is maximized with serial communications. The serial approach handles multiple points of data over a single serial communication cable. Additional data points can be added by reprogramming the analyzer to send the additional data. The device receiving the analysis data can be reprogrammed to accept the new data. This also accounts for the relatively flat curve when calculating the cost per input shown in Figure 2-10.

The analog signal approach is the least flexible for the following reasons:

- New cable connections must be made.
- Additional analog output cards will be required for the analyzer unless spare output channels are available on existing cards.
- Additional analog input cards will be required for the device collecting analyzer data unless spares inputs are available on existing cards.
- New cable will need to be installed if extra signal lines are not available in cables which extend from the analyzer to the device receiving the data.

PROBLEM AVOIDANCE

Signal isolation is important so that any noise picked up through the analyzer electronics is not passed into the signal lines. It is also used so that any noise induced on the signal lines will not be passed into the analyzer electronics. Many instrument installations require that all instruments have isolated signals in order to maintain the high-quality electrical ground.

All electrical termination connections must be tight, or signal degradation will occur. Insulation must not be pinched in the connection, and the bare wire should extend into the terminal for the full length of the exposed surface. In general, all appropriate safety codes must be followed for each installation.

All twisted-pair connections should be checked to make sure that positive and negative connections are not reversed.

All signal grounds must be instrument-quality. Shields must not be grounded at both ends or ground loops will be introduced. All twisted-pair connections should be checked to make sure that positive and negative connections are not reversed.

The Use of Analyzer Networks

A new communications option available to process analyzer systems is the use of analyzer communication networks (Figure 2-11). While networking has been available to distributed control systems (DCSs) for many years, its use in process analyzer communications has only been available since the mid-1980s. The use of networks, or data highways as they are sometimes called, is meant to satisfy three emerging requirements in the process analyzer market:

- Reduce analyzer I/O quantity and complexity
- Simplify analyzer maintenance and control
- Allow sophisticated data archiving and retrieval

Reducing analyzer I/O quantity and complexity is in response to the dramatic growth in the use of process analyzers. As the number of process analyzers grew at a plant site, so did the number of I/O points that needed to be installed.

Analyzer networks allow the data from all connected analyzers to be gathered via a common set of hardware and cables.

Analyzer networks allow the data from all connected analyzers to be gathered via a common set of hardware and cables. The network cable can then be brought to central control rooms, maintenance shops, and engineering offices. In these locations, the data can be dispersed into the desired format — for example, printer logs, trend recorders, and alarm panels. While the initial costs for establishing a network in a plant are higher than traditional techniques, the savings in hardware and cabling can be tremendous as the plant's process analyzer count increases.

Another user requirement increasing the need for analyzer networks is the desire to simplify analyzer maintenance and control. With the increased use of process analyzers comes the increased need for analyzer field maintenance. Networks provide the ability to centralize engineering and maintenance functions by using service panels and PC workstations connected to the network. Through the use of these devices, a technician or engineer can review and adjust an analyzer without having to physically visit each analyzer in the plant.

An extension of the desire to make analyzer maintenance more efficient is the move toward more sophisticated data archiving and retrieval. Although process analyzer data are often being stored in distributed control systems, it is not always readily available to engineering and maintenance staffs. The ability to connect personal computers to the analyzer network allows direct collection of the information generated by the analyzers.

Once this information is stored on the data system, it can be used for generating preventive maintenance schedules, development of statistical quality control and statistical process control programs, and spotting trends in analyzer performance. Often the data are stored in a format compatible with commercial, off-the-shelf software packages.

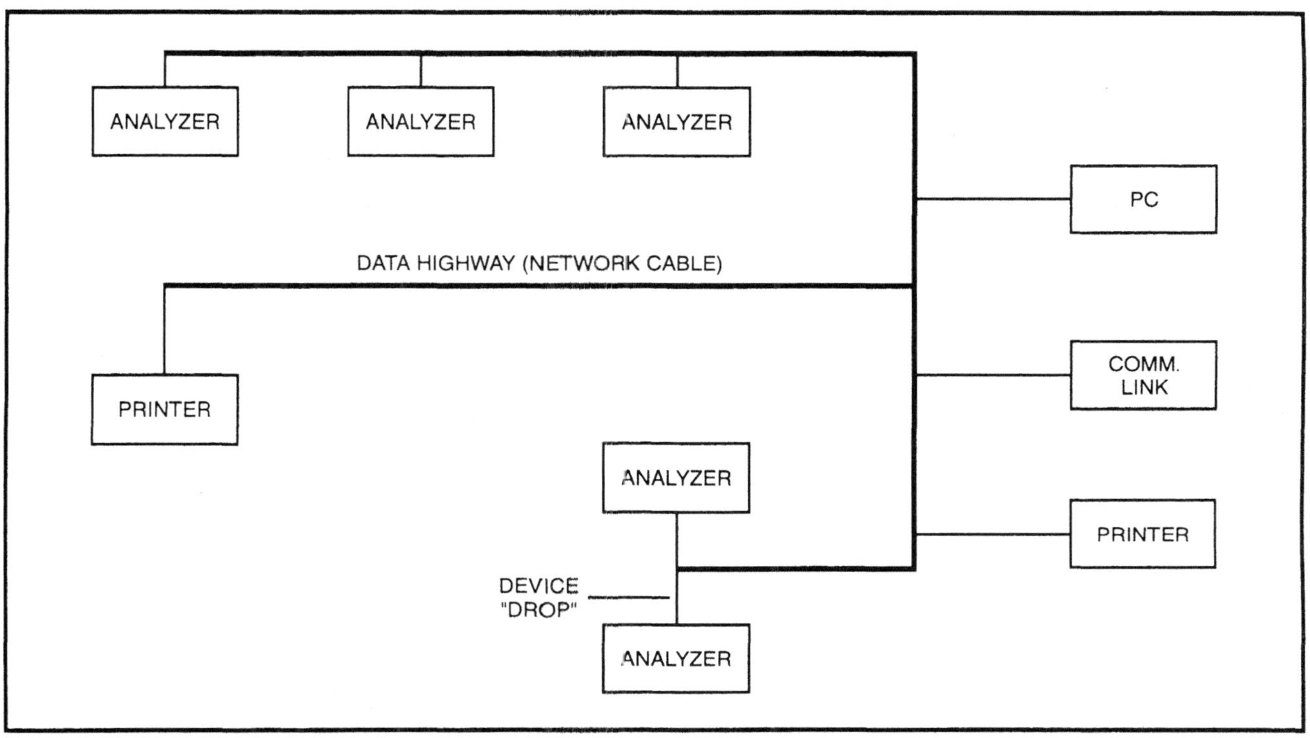

Figure 2-11. Typical Analyzer Network

Networking Techniques Available

Either centralized control is used (as in the master/slave designs) or distributed control is used (as in the masterless designs).

Analyzer networks tend to fall into two categories: master/slave designs and masterless designs. These two types also go by a wide variety of other names, such as "central manager-based," "polling architecture," and "peer-to-peer." But they all boil down to two basic approaches on how the system regulates the access of connected equipment needing to communicate on the network. Either centralized control is used (as in the master/slave designs) or distributed control is used (as in the masterless designs).

MASTER / SLAVE NETWORKS

The master/slave design uses a centralized communications controller to regulate data flow and access onto the network (Figure 2-12). The centralized controller is typically a personal computer and is often called a network master or data manager since it dictates to all other hardware the availability of the network when data transfer is needed. All other devices on the network (analyzers, printers, service panels, PC workstations) are slaves and can only communicate when given permission by the master device. In some designs, the master device will "poll" each of the attached analyzers to collect data and then pass it on to the destination device. The network master also typically acts as the central I/O device for connecting serial links to DCS systems, printers, and personal computers.

The master/slave design is well suited for applications requiring lower hardware costs. The electronics and software to implement this technique are readily available. Modification of the process analyzer is minimized since it plays a passive role in the network communications. This has resulted in the master/slave design being the most common with the largest selection of designs.

The master/slave technique suffers from two serious shortcomings: lack of full redundancy and lack of system configuration flexibility.

The master/slave technique suffers from two serious shortcomings: lack of full redundancy and lack of system configuration flexibility. In the master/slave network design, there can only be one active network master. If it fails, all communication on the network is stopped. Also lost is all I/O originating from the network master. Some designs allow a second network master to be connected in a standby mode, but this often results in complicated wiring and switchover procedures. There is also the possibility of loss of data during the switch from one network master to the other.

The master/slave design also limits network configuration flexibility by restricting the use of personal computers, printers and central I/O. Most designs require these devices to be connected directly to the network master or even use the network master itself in the dual role of a PC workstation. This limits the number and location of these devices. However, for smaller installations, this may not be a problem.

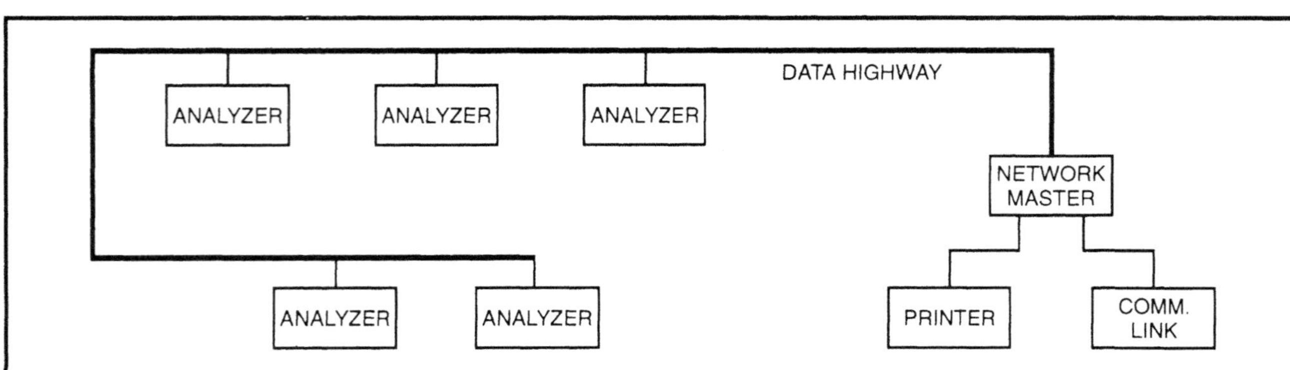

Figure 2-12. Typical Master/Slave Network

MASTERLESS NETWORKS

In the masterless network, each piece of equipment that connects onto the network (analyzers, printers, service panels, personal computers, etc.) contains its own electronics and software to regulate communication onto the network (Figure 2-13). No network master is required; each device cooperates with another on network access through a defined "access contention" procedure. The access contention procedure varies from supplier to supplier, but the primary purpose is the same: to keep two devices from trying to talk on the network at the same time. When an analyzer has data it wants to transmit onto the network, it follows the access contention procedure to assure that the network is available for use.

In the masterless network, each piece of equipment that connects onto the network (analyzers, printers, service panels, personal computers, etc.) contains its own electronics and software to regulate communication onto the network.

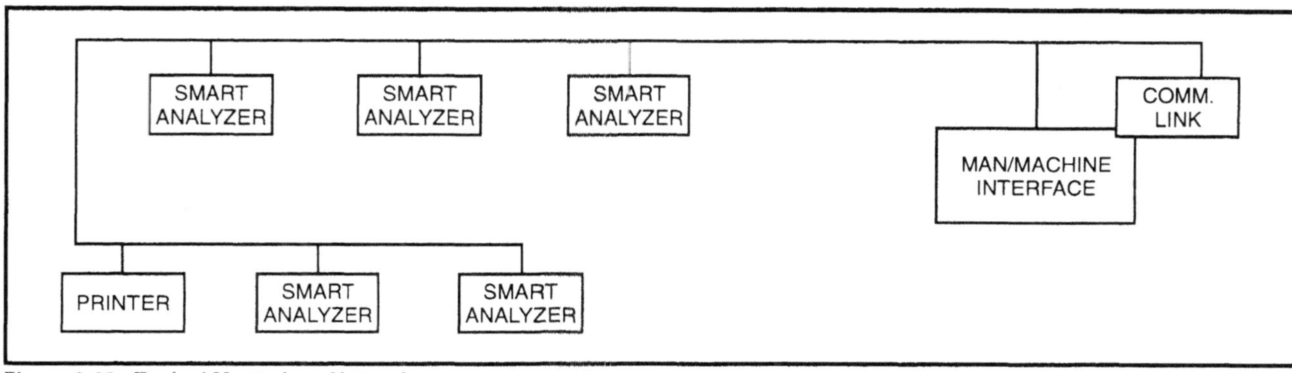

Figure 2-13. Typical Masterless Network

The primary advantage of the masterless design is its inherently redundant nature. Since there is no "single point of failure" that will stop all network communication, this design provides the stability and redundancy needed for larger systems and where important data are being handled.

The primary advantage of the masterless design is its inherently redundant nature.

The masterless design also allows more flexibility in how the network is configured. Since there is no network master controlling communication, there is no problem with having system printers, personal computers, and I/O in numerous locations on the network. This is extremely beneficial to larger facilities and those locations with maintenance, engineering, and operations departments in different locations.

The disadvantages of the masterless design are that its highway sophistication has resulted in more expensive hardware, leading to somewhat higher installation costs, and it has also resulted in limited availability. It can also be more difficult to troubleshoot the masterless network for communications problems because there is no single device you can go to for monitoring and interrogation of the network.

It can also be more difficult to troubleshoot the masterless network for communications problems because there is no single device you can go to for monitoring and interrogation of the network.

Network Integrity and Redundancy

One of the most important points a user needs to consider when evaluating an analyzer network is its ability to handle network failures without the loss of data. In the process environment there is always the possibility of hardware failing, cable being accidentally cut or shorted, and a host of other potential problems.

A well-designed analyzer network needs to be able to handle system failures without loss of data. This can be accomplished by a number of techniques, such as redundant electronics, dual network cable sets, and automatic switchover on loss of communication.

Some of the important questions to ask when evaluating any analyzer network are:

- Is there any single piece of hardware needed for the network to operate?
- How is this single point of failure prevented from happening?
- What happens when the network cable is accidentally cut or shorted?
- How does the network handle devices that "lock" onto the network?

Network Topology

Due to the higher initial installation costs of analyzer networks over traditional techniques, it is important that the network be able to handle the plant's requirements for many years. Therefore, the ability and ease of expanding the network to handle future demands need to be considered. The issues involved are maximum network size, both physical and system capacity, ease of expansion, cabling restrictions, and the ability to add multiple types of analyzers onto the network.

Most plants will never put 100% of their process analyzers onto the analyzer network, but, as discussed earlier, there are certainly tangible benefits to putting on as many as practical.

Most plants will never put 100% of their process analyzers onto the analyzer network, but, as discussed earlier, there are certainly tangible benefits to putting on as many as practical. However, the sheer length of cable needed to connect everything can be intimidating. Fortunately, most networks support 1 to 5 miles of cable, with even greater lengths available as options.

It is important to keep any cabling restrictions in mind when estimating cable requirements. For example, many networks require everything to be connected one after another (daisy-chained) such that everything is connected in a single long cable path. Some designs require that the cable radiate out from a central point, often the network master, into a star configuration. The third and most flexible approach are networks that allow "branching." Branching often results in the minimum amount of cable used to connect everything together. Figure 2-14 illustrates these three configurations.

System capacity, also called unit-loading, refers to the maximum number of devices and I/O that can be handled by the network. All networks have a maximum number of analyzers and devices that can be connected together. Fortunately, many of the newer designs allow 100 or more analyzers to be connected into one system. However, an important question to ask is, does the addition of future analyzers require changes in existing equipment. This is often the case with master/slave designs, where the network master needs to be modified as new equipment is added. Also, new or replacement equipment may be needed to get the network up to maximum capacity.

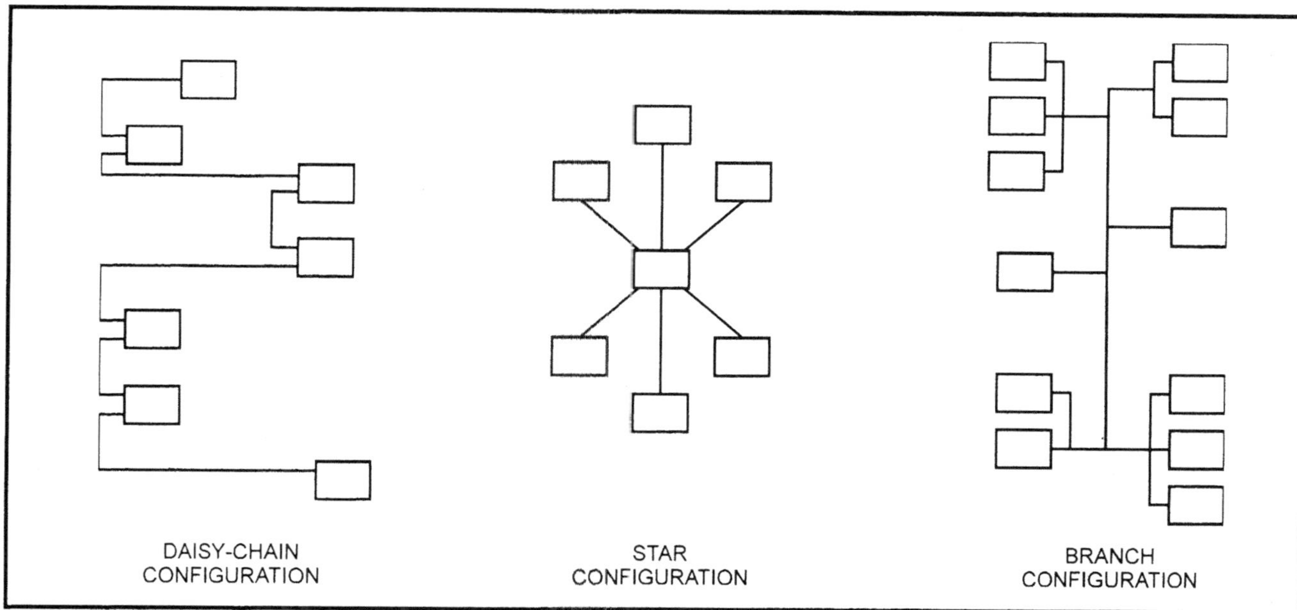

Figure 2-14. Network Cabling Configurations

Multianalyzer Integration

The integration of multiple types of analyzers onto one network (data highway) can be a problem, because all process analyzer networks currently available are developed by individual process analyzer manufacturers as part of their analyzer system. Therefore, the user is often limited in choices based on which type of process analyzer is currently being used. Fortunately, most designs allow the connection of other analyzers onto the network through the use of an electronics interface device (EID).

Fortunately, most designs allow the connection of other analyzers onto the network through the use of an electronics interface device (EID).

Typically, the EID would be found near the analyzer and would receive the analog and status signal generated by the process analyzer (Figure 2-15). Once this information is in the EID, it is available for transmission via the network (data highway) to a printer, trend recorder, or any other type of I/O device. The EIDs are also capable of performing calculations as part of their data handling. It is also frequently possible to program the EIDs to initiate calibration of the attached analyzers through the use of relay contacts.

The level of access available over the network for the analyzers connected by EIDs is seldom as great as analyzers designed to connect directly to the network. It is not possible to get all of the operational information via the EID, such as operating temperature, power supply voltage, baseline offset, calibration factors, and so on. This may or may not be a problem, depending on the complexity of the analyzer.

It is not possible to get all of the operational information via the EID, such as operating temperature, power supply voltage, baseline offset, calibration factors, and so on.

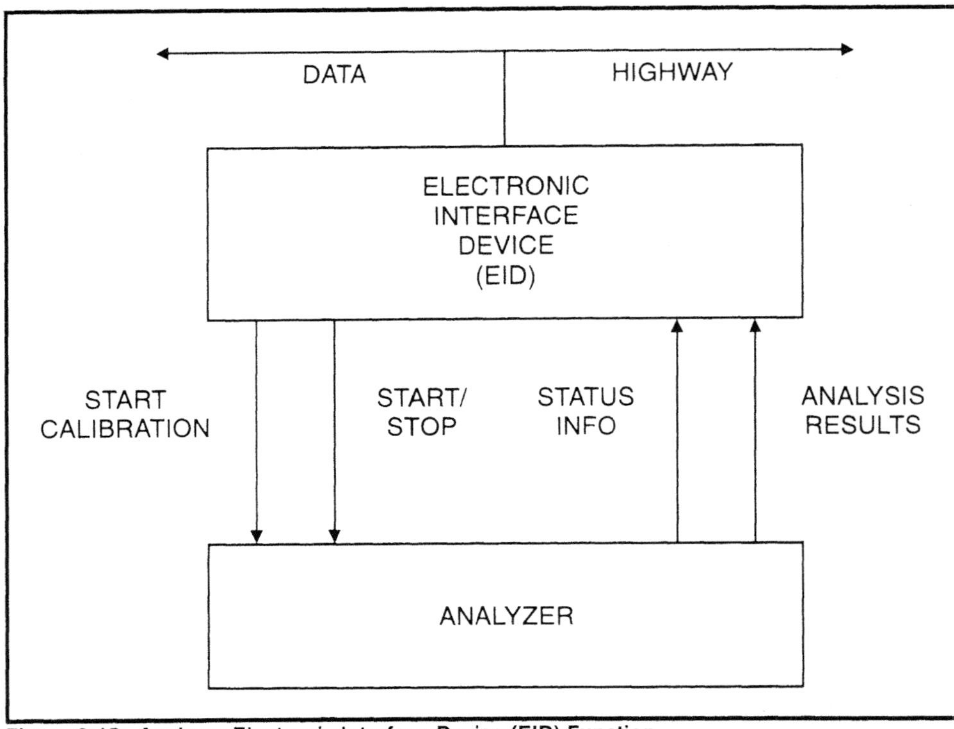

Figure 2-15. Analyzer Electronic Interface Device (EID) Function

About the Author

Bob Farmer graduated with a B.S. in Electrical Engineering from Kansas State University in 1974 with specialization in electronic design and communication systems. Following graduation, he went to work for Applied Automation, Bartlesville, OK, and has since acquired over 20 years of experience in process analyzer system design, application, and marketing. His work credits include original system design on serial communication systems for process analyzer systems, and he holds a U.S. patent for a balanced serial transmission system which has been used throughout the world. Mr. Farmer has extensive international field application experience in analyzer communication systems. He has spoken to numerous users and has taught classes concerning communication techniques, and he has spoken at ISA and other trade group technical seminars. He is also a member of various committees and subcommittees pertaining to Fieldbus and other process analyzer communication topics. Mr. Farmer, a Registered Professional Engineer, has a previous ISA paper published concerning use of personal computers in analyzer systems. At the present time, he is the Product Manager for Process Analytical Systems at Elsag Bailey / Applied Automation.

3

Process Sample Conditioning Systems

A "decision tree" would be one trunk and many branches.

This chapter will be more of a "decision 'nursery' "; three small trunks, each with many small branches. The "trunks" will be Process, System and Component characteristics. The "many small branches" will be the details in each category basic to the design of a sample conditioning system (SCS).

Immediately you will notice the term *system*. It cannot be emphasized strongly enough that the process, the sample conditioning and the analyzer are a *system* and it will be understood before the reading of this chapter is complete that sample conditioning itself must truly be a *system*.

Each application must be worked through in a logical sequence. That sequence must be to understand the process characteristics, the system characteristics and, finally, the component characteristics which are necessary to provide sample conditioning of the process sample for this specific process analyzer application.

Process Characteristics

Process characteristics encompasses the pertinent physical characteristics of the process to be sampled and the sample to be conditioned. The major categories are gas, liquid or solid. This document will mainly address gaseous samples with specific details for liquid samples in certain applications. The field of sampling solids is a complete different technology and will not be addressed in this document. ISA, the editor or any author in this volume may be contacted directly for assistance in addressing solids sampling. Within gas or liquid sampling, one must address the aspects of pressure, temperature and flow for each sample conditioning system. Following are details on these initial aspects of sample conditioning system design.

Sample *phase* will be one of the following:

- **Gas** — molecules are quite active, relatively far away from each other and take the shape of the vessel in which they are contained.
- **Liquid** — molecules are less active than in a gas, are closer together than in a gas yet still take the shape of the vessel in which they are contained.
- **Solid** — molecules are less active than in a liquid, are not necessarily closer together than in a liquid but are held in rigid arrangements and establish their own physical boundary without the absolute need of a vessel for containment.

A successful process analyzer application ***must*** *have its sample presented as* ***one and only one*** *sample phase. No process analyzer can handle a mixed phase sample.*

Sample *pressure* will be one of the following:

- **Positive** — at a pressure greater than that of the atmosphere

> 14.7 psig	> 100 kPa
> 29.9 inHg	> 0.0 psia
> 407 inWC	> 760 mmHg
> 33.9 ftWC	> 1.00 bar

Recall that PSIG = PSIA + 14.69

- **Atmospheric** — at a pressure equal to that of the atmosphere

= 14.7 psig	= 100 kPa
= 29.9 inHg	= 0.0 psia
= 407 inWC	= 760 mmHg
= 33.9 ftWC	= 1.00 bar

- **Negative** — at a pressure less than that of the atmosphere (vacuum)

< 14.7 psig	< 100 kPa
< 29.9 inHg	< 0.0 psia
< 407 inWC	< 760 mmHg
< 33.9 ftWC	<1.00 bar

Sample *temperature* will be one of the following:

- **High** — this is a very relative term but a reasonable definition is >400°F (204°C) with the reason being that this is a temperature above which several common elastomers (VITON™ and TEFLON™) experience thermal breakdown.
- **Working Range** — this is again a very relative term but a reasonable definition is 400°F (204°C) to –40°F (–40°C) with the reason being that this is a temperature range within which several common elastomers will satisfactorily function.
- **Cryogenic** — this is a term commonly used to indicate a temperature below –40°F (–40°C) all the way down to absolute zero –459.69°F (–273.16°C), a temperature at which all molecular movement ceases and all gases and all liquids both become solids.

Sample *flow* will be one of the following:

- **Hypersonic** — very seldom encountered in process analyzer sample conditioning systems but defined as five times the speed of sound (which is 741 MPH at sea level) or about 5,434 FT/SEC (FPS) at atmospheric pressure.
- **Sonic** — defined as 1,087 FPS at atmospheric pressure.
- **Turbulent** — this condition is defined as flow which is erratic in magnitude and direction, is dependent upon the Reynolds Number (R_e) of the flow condition and is defined as above an R_e of 4,000.

- **Critical** — this flow condition is defined as having an R_e of between 2,000 and 4,000 and flow characteristics of the flowing media are unpredictable.
- **Laminar** — this condition is defined as flow which is streamlined (velocity anywhere in the flowing media is constant and conforming to a solid boundary), is dependent upon the Reynolds Number (R_e) of the flow condition and is defined as below an R_e of 2,000.
- **Low** — this condition is a very relative term but for process analyzer sample conditioning systems it would typically indicate flows in the 2,500 cc/min (5.3 SCFH) to 50 cc/min (0.1 SCFH) range.
- **No flow** — this condition is also a very relative term but for process analyzer sample conditioning systems it would typically indicate a flow rate which would result in no more than one bubble per minute from 0.125 in. (1/8 in.) O.D. tube into water whose surface is open to the atmosphere. This is the typical leak check bottle arrangement for leakage verification across critical valves in (especially double-block-and-bleed) sample stream switching applications.

System Characteristics

To explain the system characteristics of the process, sample conditioning, analyzer relationship the following diagram will provide a helpful reference.

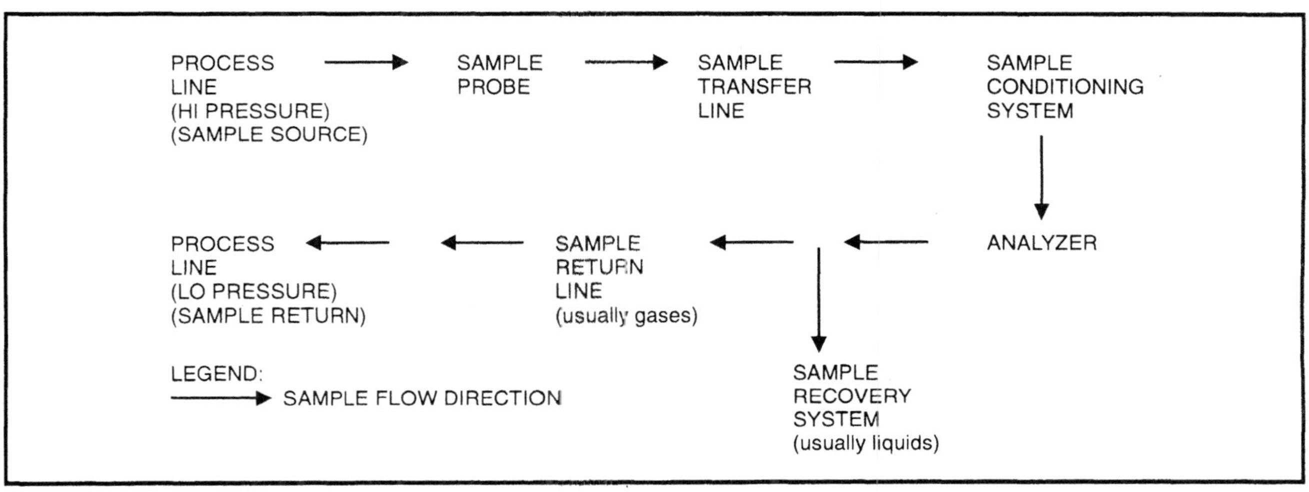

In order to establish flow through the analyzer there must be a pressure differential or pressure drop between the sample source and the sample return (across the analyzer).

This is noted above as flow from HI pressure process line to LO pressure process line. Typically, an absolute minimum pressure difference of 15 psig is necessary to establish flow through the system (sample probe, sample transfer line, sample conditioning system, analyzer, sample return/recovery). A range of 25 - 50 psig is useful while pressure drops over 50 psig often require special treatment (pressure-reducing regulators, heated pressure-reducing regulators, etc.). The terms HI and LO above are "relative" terms, i.e. they could be HI = 35 psig and LO = 20 psig or they could be HI = 235 psig and LO = 220 psig. The pertinent fact here is the "differential" pressure or "pressure drop" between the HI and LO process pressures to establish flow through the analyzer system. If sufficient differential pressure is not available from sample source and sample return in the process, then a sample pump would have to be used. This sample pump may be a diaphragm-type (e.g., KNF Neuberger-609.890.8600 or ADI-305.428.7333) "pushing" sample through the analyzer. This sample pump could also be a jet (aspirator)-type (Penberthy-616.396.5900) "pulling" sample through the analyzer. Contact these manufacturers for application of their pumps.

Sample probes are a technology to themselves. Each application has characteristics which require design considerations for its sample probe. Two treatments of sample probe types and their applications are discussed in References 1 and 2. Several helpful considerations are given in the "Tips, Traps, Hints and Hazards" section of this chapter. References 3 through 6 are excellent practical texts and booklets.

Sample transfer line and *sample return line* considerations are similar and are treated collectively in the section "Sample Transfer Lines."

A *sample conditioning system* is a calculated and orderly arrangement of components, many of which are treated individually under "Component Characteristics" in this chapter. An excellent source of suppliers of sampling systems can be found in the ISA Directory (published annually) in the Products Sections under the headings: Sampling Systems, Gas and Sampling Systems, Liquid.

Analyzers are the subject of Part II of this Volume and are treated under the following categories:

- Physical Property Analyzers
- Electrochemical Analyzers
- Spectrophotometric Analyzers
- Compositional Analyzers

This coverage is representative rather than comprehensive due to the large number of process analyzer types and manufacturers.

Sample recovery systems are utilized for liquid samples where the analyzer alters the chemical or physical characteristics of the sample, liquid samples where multiple streams are sampled and sample return would be difficult to coordinate return to the proper stream or samples where return must be to a higher pressure than the sample point. Again, this consideration is highly specialized and contacting a manufacturer of this specialized equipment would be recommended (Liquid Sample Recovery Systems: Precision Scientific Petroleum Instruments Company-708.493.0100).

Sample Transfer Lines

Sample transfer lines can be heated (heat-traced-and-insulated [HT&I]) or unheated (insulated to retain heat or totally uninsulated).

Unheated lines are usually used in the case of mild climates or the transmission of dry gases, where moisture freezing or product crystallization/viscosity is not a factor.

Heated sample transfer lines can utilize steam or electricity for the heat source.

Steam-heated sample transfer lines can be trapped (controlled steam flow) or untrapped (constant flow of steam). Steam heated lines are typically controlled only to a ±25°F temperature range, since the temperature control device is usually a thermally-actuated bucket trap. Steam pressure and superheat, the ambient temperature, and the initial product temperature determine the trap distance from the steam supply at which the steam trap may be located in order to maintain the desired temperature in the process transfer tube. Steam is often assumed to be a lower cost alternative to electricity for heat in electrical-hazard-rated (explosionproof) areas. This reasoning may be considered correct where freeze-protection is the goal or where an elevated temperature is desired with little control of the temperature to be maintained. The difficulty of properly installing the steam tracer line and properly insulating and protecting the insulation from moisture intrusion and weather-related decay usually provide sufficient incentive to use electric-traced /integrally-electric-heat-traced-and-insulated products. Steam tracer leaks can be easy to tolerate and expensive to ignore whereas electric trace/heat (properly designed initially) either maintains the proper temperature or it definitively does not work, providing an immediate incentive for its replacement. Also, steam leaks can be very dangerous to personnel and destructive to transfer lines and analytical equipment.

Electric heat-traced-and-insulated sample transfer lines can be of several types:

- Heated hose
- Parallel resistance self-regulating ("twin-lead")
- Parallel resistance constant wattage ("zoned cable")
- Parallel resistance power-limiting

Heated Hose

Heated hose is usually a Teflon™ sample transfer tube with stainless steel overbraid and factory-installed (often "crimped-on") end fittings, although quality tube fittings can be factory or field installed (see Figure 3-1).

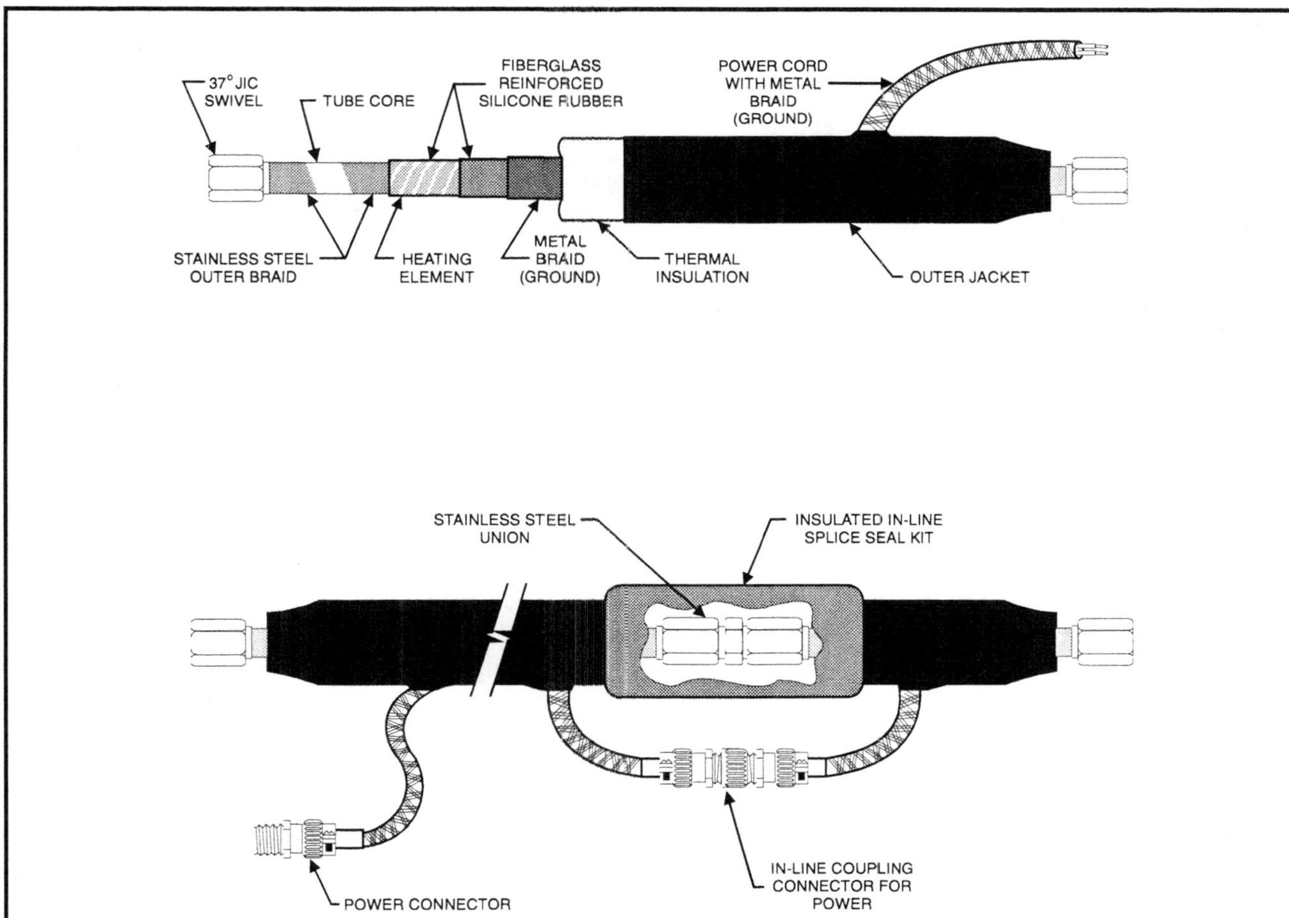

Figure 3-1. Heated Hose: Typical Construction Detail and Typical Hose Section Coupling Detail
(Courtesy of Cellex Manufacturing Inc.)

The sample transfer tube can also be stainless steel. The heating element is often a resistance wire (note: series resistance) spiral-wrapped over electrical insulation applied outside of the stainless steel overbraid on the Teflon sample transfer tube. Finally, this heater/hose assembly is covered with thermal insulation and the entire assembly is protected with a flexible plastic outer covering. Series resistance means that the heating element changes in resistance as the length changes. As the resistance changes, there is a corresponding change in the watts-per-foot. Thus, each hose must be custom-designed and the length cannot be field modified. This presents the problem that you cannot stretch a too-short hose and it is difficult to aesthetically and mechanically utilize excess length hose in field applications. The tendency for inexperienced field personnel to utilize excess hose length in certain ways results in specific problems (see Table 3-1).

Table 3-1. Utilization Problems

Utilization	Problem
loop it	Looping can result in sample tube blockage from solid material "falling out" of the sample stream or "pressure-pulsing" where liquid condenses out of a gaseous sample stream, collects at the bottom of a loop and the gaseous sample continues to "bubble" through the liquid pocket. Today's sensitive gas analyzers can see these pressure pulsations as packing more molecules into the sample cell, resulting in repetitive abnormally-high "spiking" values for the variable of interest. Sample transfer lines should have a continuous downward slope of (nominal) 0.5 inch per 1 linear foot. The manner to transfer excess length to proper installation practice is to "snake" the line down a vertical surface or down a vertical plane (with proper physical support).
coil it	Coiling (multiple loops) can result in sample hose meltdown or, in extreme cases, ignition of the combustible components of the hose (outer jacket and Teflon sample transfer tube). This destruction of the sample hose results from the fact that its heater is series resistance and is designed to be referenced to ambient temperature at all times. Coiling results in heat transfer to heating media and subsequent meltdown or autoignition.

The main advantage of the heated hose is that it is very flexible and can be utilized where the sample transfer line is subject to constant or timed repetitive motion. The main disadvantages of the heated hose is the inability to field modify the length of the hose and the fact that they are usually handmade and thus a long delivery item.

Parallel-Resistance Self-Regulating ("Twin-Lead")

Parallel-resistance self-regulating ("twin-lead") heating cables utilize a continuous extrusion of a temperature-sensitive semiconductive polymer matrix between two current-carrying buss wires and have no preset "zone" length (see Figure 3-2). Therefore, self-regulating cables can be cut to virtually any length without regard for the heating "zone" length or where the heating "zones" are connected to the buss wires. Self-regulating cables require careful preparation for the power supply connection to avoid having heat-producing cable inside the electrical -power junction box. Self-regulating lines are typically more forgiving in situations of poor design or poor installation practices.

In the event that a self-regulating cable is accidently installed inside of the thermal insulation rather than attached directly to the sample transfer line, the cable will operate at a higher than normal temperature and reduce its power without overheating. Self-regulating cables have their watt densities (heat generation capacities) referenced to a 50°F sample transfer tube temperature. This published watt density rating at 50°F is the minimum power output of the cable that can be expected in the field installation. Most self-regulating cables are damaged when exposed to elevated temperatures (process excursions or sample transfer line "steam-out") and tend to deteriorate after extended periods of extreme thermal cycling. There are two types of self-regulating cable available. The energized maximum operating temperatures are 150°F and 250°F with a corresponding deenergized maximum exposure temperature of 185°F and 375°F, respectively. The 150°F rated cable is used primarily for low heat process applications and freeze-protection. The 250°F rated cable is used for high heat process applications and where elevated "power-on" current levels are unacceptable. These cables are typically rated to 277 VAC maximum operating voltage (not to be confused with the DC-electric rating of the cable insulation itself which is typically 600 VAC). Another problem with self-regulating cables is cold-start power requirement. Their resistance is designed to decrease with a decrease in temperature and increase with an increase in temperature. This leads to the situation of maximum operating current draw at initial power-up, regardless of the initial sample transfer line temperature. Compensation for this fact must be included in the design of the power source (sizing of the circuit breaker, wiring, etc.). This makes the installation of the simple "self-regulating" heater more technology-intensive than the installation of a parallel-resistance constant-wattage ("zoned cable" series) temperature-controlled integrally-heat-traced-and-insulated sample transfer line system.

The typical application of this self-regulating ("twin-lead") integrally heat-traced-and-insulated sample transfer line (see Figure 3-3) is in freeze-protection applications and in applications where the temperature of the transferred sample is not critical but is only desired to be kept above some defined minimum value. This self-regulating cable can be "clamped" at an upper temperature limit by the application of a thermostat but it is still going to have heating characteristics where it will start heating at some approximate minimum temperature and float on top of any ambient temperature differential from this approximate minimum to its maximum "clamped" value.

In summary, with self-regulating ("twin-lead") cable you typically do not know at what temperature your sample is maintained; you only know that it will be maintained between some approximate defined minimum and maximum temperature values.

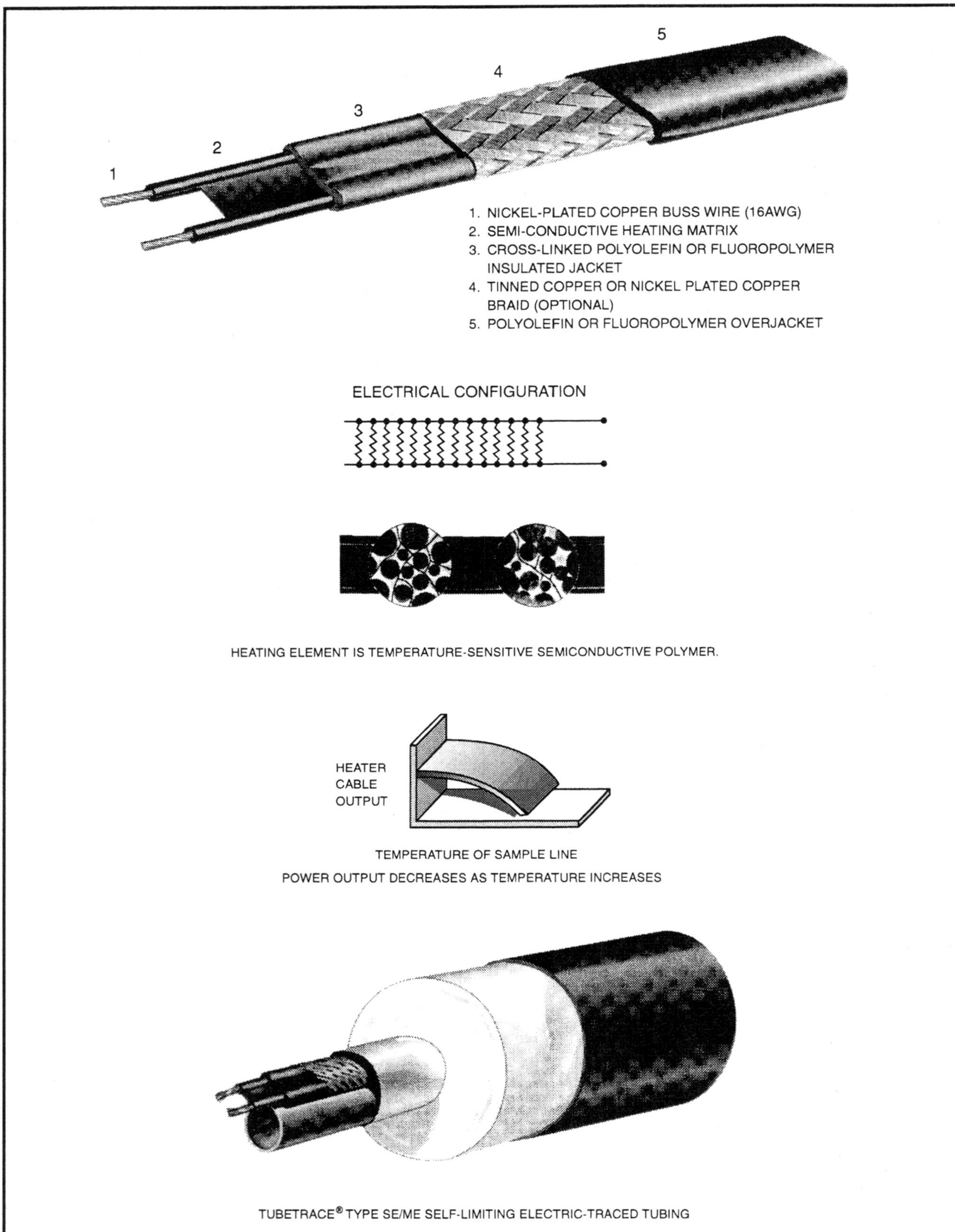

Figure 3-2. Self-Limiting Cables ("Twin-Lead")
(Courtesy of Cellex Manufacturing Inc.)

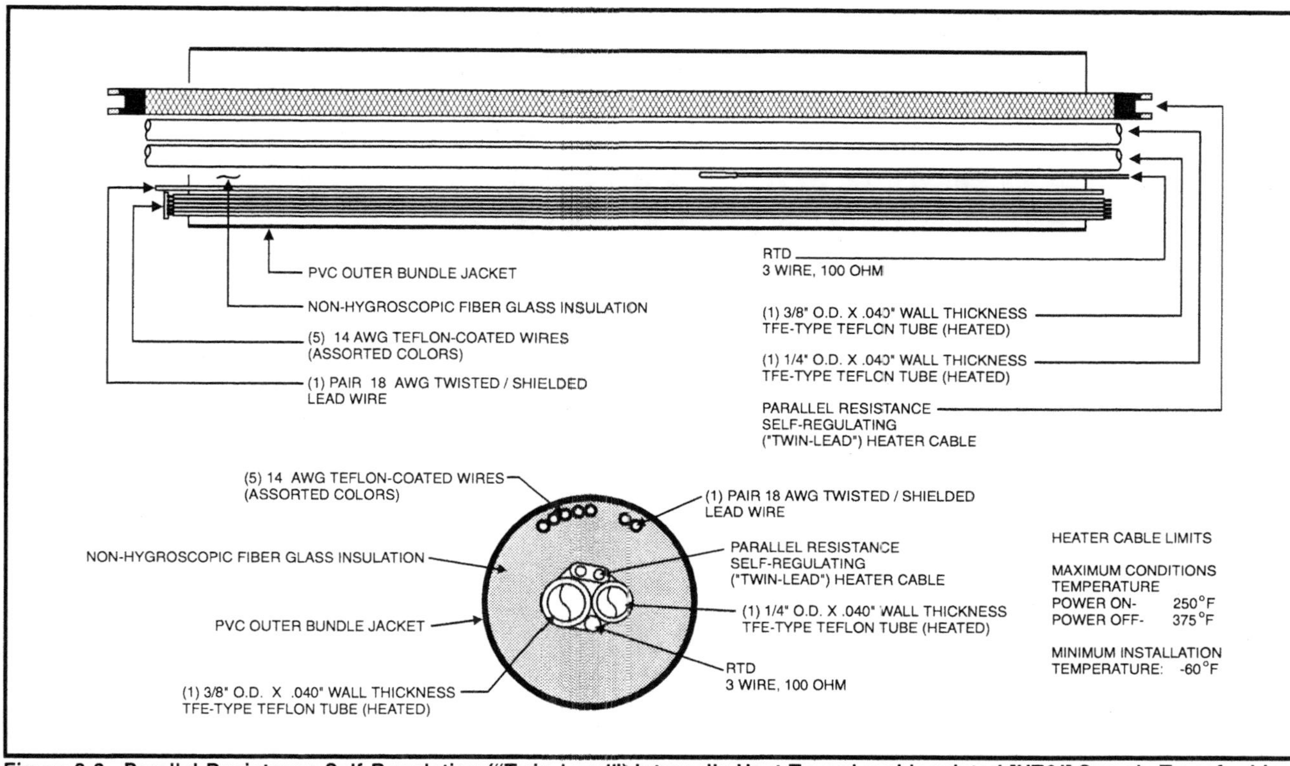

Figure 3-3. Parallel-Resistance Self-Regulating ("Twin-Lead") Integrally Heat-Traced-and-Insulated [HT&I] Sample Transfer Line (Typical Line Fabrication Detail)
(Courtesy of Cellex Manufacturing Inc.)

Parallel-Resistance Constant-Wattage ("Zoned Cable")

Parallel-resistance constant-wattage heat trace cable is often referred to as "zoned cable" because the heating "zones" (usually 2-3 feet in length) are physically-discernible nichrome heating elements connected in parallel with prior and following heating "zones" (see Figure 3-4).

At the power connection, "zoned cable" can be fabricated to have a built-in, non-heat-producing "cold lead." This can easily be accomplished by cutting the cable between the buss connection in the field to have a non-heat-producing "zone" for power lead junction box terminations. Although constant-wattage ("zoned cable") can be crossed during installation without serious problems when low watt densities or low process temperatures are involved, at higher watt densities and higher process temperatures there are possibilities of serious line overheating where the line crosses itself. Historically, the big advantages of parallel-resistance constant-wattage ("zoned cable") has been lower cost and maintaining a known sample transfer temperature. Parallel-resistance constant-wattage ("zoned cable") can be rated for much higher temperatures than parallel-resistance self-regulating ("twin-lead") cables. Zoned lines can be rated for 400°F (power on) and 500°F (power off) without regard to total service time or thermal cycling conditions. This is particularly useful on sample transfer lines which will experience elevated temperature during process excursions or during sample transfer line "steam out."

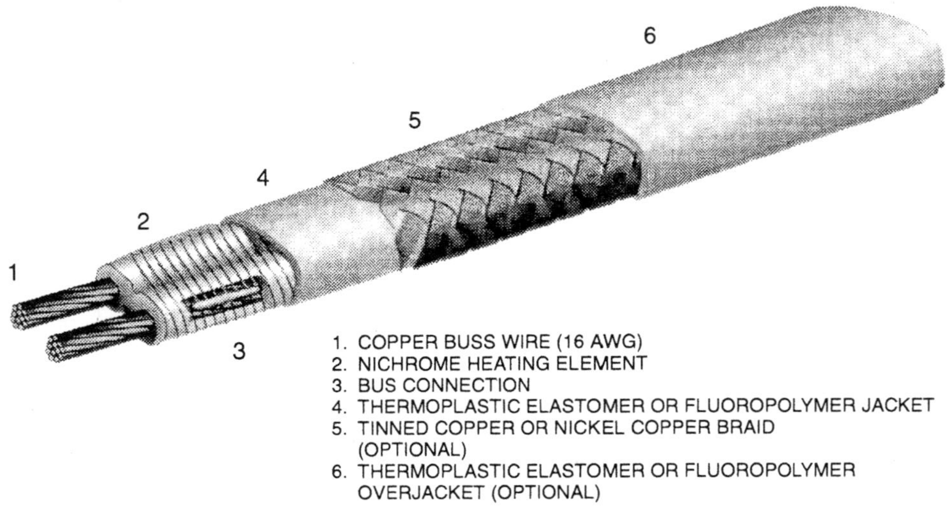

ELECTRICAL CONFIGURATION

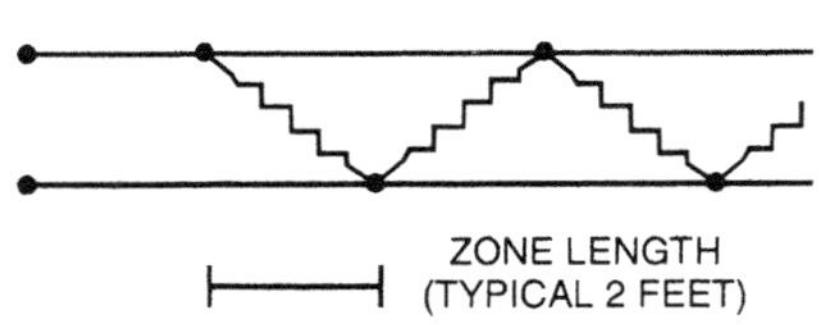

HEATING ELEMENT IS STABLE NICHROME WIRE ELECTRICALLY CONNECTED TO BUSS CONDUCTORS AT BUSS CONNECTION. EACH HEATING "ZONE" OPERATES INDEPENDENTLY OF EVERY OTHER ZONE, AND ALLOWS CABLE TO BE "CUT-TO-LENGTH" IN THE FIELD.

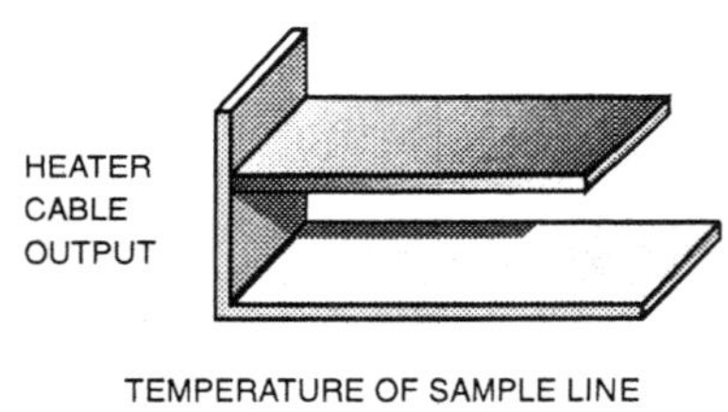

POWER OUTPUT REMAINS CONSTANT OVER TEMPERATURE RANGE

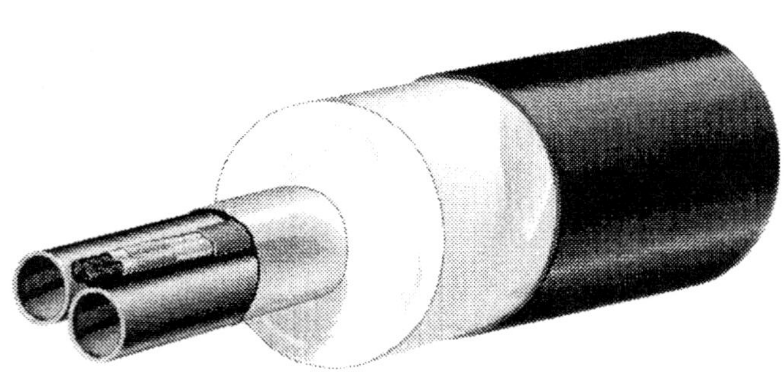

TUBETRACE® TYPE SE/ME CONSTANT-WATT ELECTRIC-TRACED TUBING

Figure 3-4. Parallel-Resistance Constant-Wattage Cables ("Zoned Cable")
(Courtesy of Cellex Manufacturing Inc.)

Some zoned cables are capable of directly being connected to 480 VAC. Higher voltages allow longer sample transfer line length per power point. Although many of the methods of monitoring an integrally heat-traced-and-insulated sample transfer lines' performance are common to both parallel-resistance self-regulating ("twin-lead") and parallel-resistance constant-wattage ("zoned cable"), a monitoring method that stands out for "zoned cable" is heating element electrical current monitoring. This can be performed through visual-indicating locally-installed ammeters or with microprocessor-based systems having the capability of programming operating current limits, above and below which a remote alarm is transmitted. A major advantage of "zoned cable" systems is controlled temperature from three-mode-controlled initial heat-up through ±2F° temperature control without regard to varying process sample or varying ambient temperatures. The use of single-mode control (bimetallic switch) temperature controllers can result in premature failure of "zoned cable" systems.

The failure mode is as follows:

The integrally heat-traced-and-insulated sample transfer line is designed to maintain a given elevated temperature during low ambient and/or low process sample temperature conditions and there is some safety factor built-in to maintain this desired temperature (typically a 20% safety factor). It is apparent that there is much more heat available than is needed during typical ambient and process conditions. As the system loses heat and the sample temperature falls through the setpoint of the controller, this action turns on power to the line's heating element. Without three-mode control: the heater immediately generates maximum power (heat) output, rapidly heats up the sample transfer tube(s) and overshoots the temperature setpoint as thermal inertia carries the temperature rapidly upward., turning off the simple temperature controller and allowing the temperature to again gather thermal inertia but in the opposite direction as the temperature falls through the temperature setpoint and we repeat this "rapid and deep" thermal cycle. Including three-mode control: the controller gradually feeds in power to the heater as the falling temperature reaches the setpoint and gradually withdraws power to the heater as the rising temperature approaches the setpoint, reducing both the frequency and the magnitude of the "no-longer-rapid-and-deep" thermal cycle.

"Rapid-and-deep" thermal cycling will cause heater cable failure in "zoned cables" that are not designed to withstand this action. Most "zoned cable" heaters are manufactured by insulating the electrical conductors with Teflon, adding a small-diameter nichrome resistance wire and then containing this fabricated heating element inside of a Teflon casing molded to the heater element. One of the disadvantages of Teflon is that it expands and contracts with the temperature changes. Expansion and contraction of the Teflon-encased heater will cause the linear nichrome element to be stressed (compressed and extended) which will eventually cause the heater element's failure.

Parallel-Resistance Power-Limiting

Parallel-resistance power-limiting heaters are a major improvement over "zoned cable." The manufacturers of power-limiting cable will spiral-wrap a positive temperature coefficient (PTC) alloy resistor wire around a core fiber. This spiral resistance element is then spiral-wrapped around the heating buss wires. As this system expands and contracts, the buss wire/resistance-element-heater expands and contracts smoothly as a coil spring, thus extending the life of the heater essentially indefinitely. Another benefit of power-limiting cable is its ability to turn down in power output as it gets hotter. The heating element is a "positive temperature coefficient (PTC)" alloy that increases resistance with increasing temperature; this minimizes the chance of overheating.

If a temperature controller is desired or needed, it is recommended that the user purchase the temperature feedback element as a factory-installed item.

One final note about "zoned cable" and power-limiting integrally heat-traced-and-insulated sample transfer lines (see Figure 3-5). If a temperature controller is desired or needed (and it is highly recommended by this author, utilizing a three-mode control unit), it is recommended that the user purchase the temperature feedback element (RTD, thermocouple, thermistor) as a factory-installed item. "Zoned cable" sample transfer lines are very efficient and can maintain ±2F° control, as previously noted. Disruptions in the line's insulation are read by the controller as a reduced temperature area. A disruption in the insulation is defined especially as a place where the bundle has been cut to field-install a temperature sensor. Also, field-installed temperature sensors are often not placed in direct and intimate contact with the sample transfer tube. This is particularly noticeable in line applications controlling above 200°F. Manufacturer-installed feedback sensors can read the sample transfer tube temperature very accurately and can be installed at any distance from the powered end of the sample transfer line. The cut-to-length feature of the "zoned cable" or power-limiting integrally heat-traced-and-insulated sample transfer line is not compromised because the line length can be field-altered at the opposite end from the powered end (the end that has the power and temperature feedback device wires protruding). This also allows the powered end of this line to be factory-finished, eliminating one more labor-intensive field operation when the line is installed at the job site.

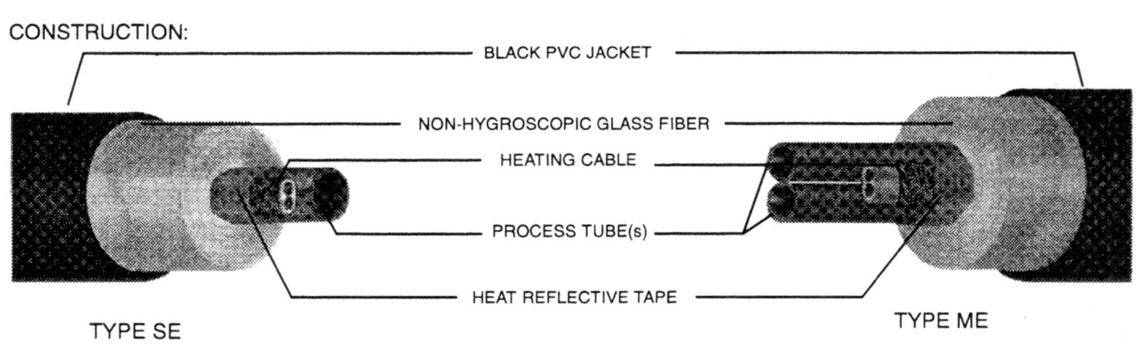

RATINGS (TYPICAL)			
MAXIMUM OUTERJACKET EXPOSURE TEMPERATURE	220°F	THERMAL CONDUCTIVITY OF THE INSULATION	.42 BTU/hr-ft-°F
MAXIMUM OPERATING JACKET SURFACE TEMPERATURE (BASED ON 80°F AMBIENT AND 400°F TRACER	140°F	SUPPORT DISTANCE: HORIZONTAL RUNS: 5 - 6 ft VERTICAL RUNS: 10 - 16 ft	
MINIMUM INSTALLATION TEMPERATURE (STD. PVC JACKET)	+15°F		

Figure 3-5. TubeTrace® Type SE/ME Self-Regulating Electric-Traced Tubing (Typical) (Parallel-Resistance Constant-Wattage ["Zoned Cable"] Integrally-Heat-Traced-and-Insulated Sample Transfer Line (Typical Line Fabrication Detail)
(Courtesy of Cellex Manufacturing Inc.)

Filter Types and Filtration

The following material addresses only the treatment of gaseous samples. The treatment of liquid samples is an entirely different but not necessarily parallel consideration. For a typical liquid bypass filter, see Figure 3-6.

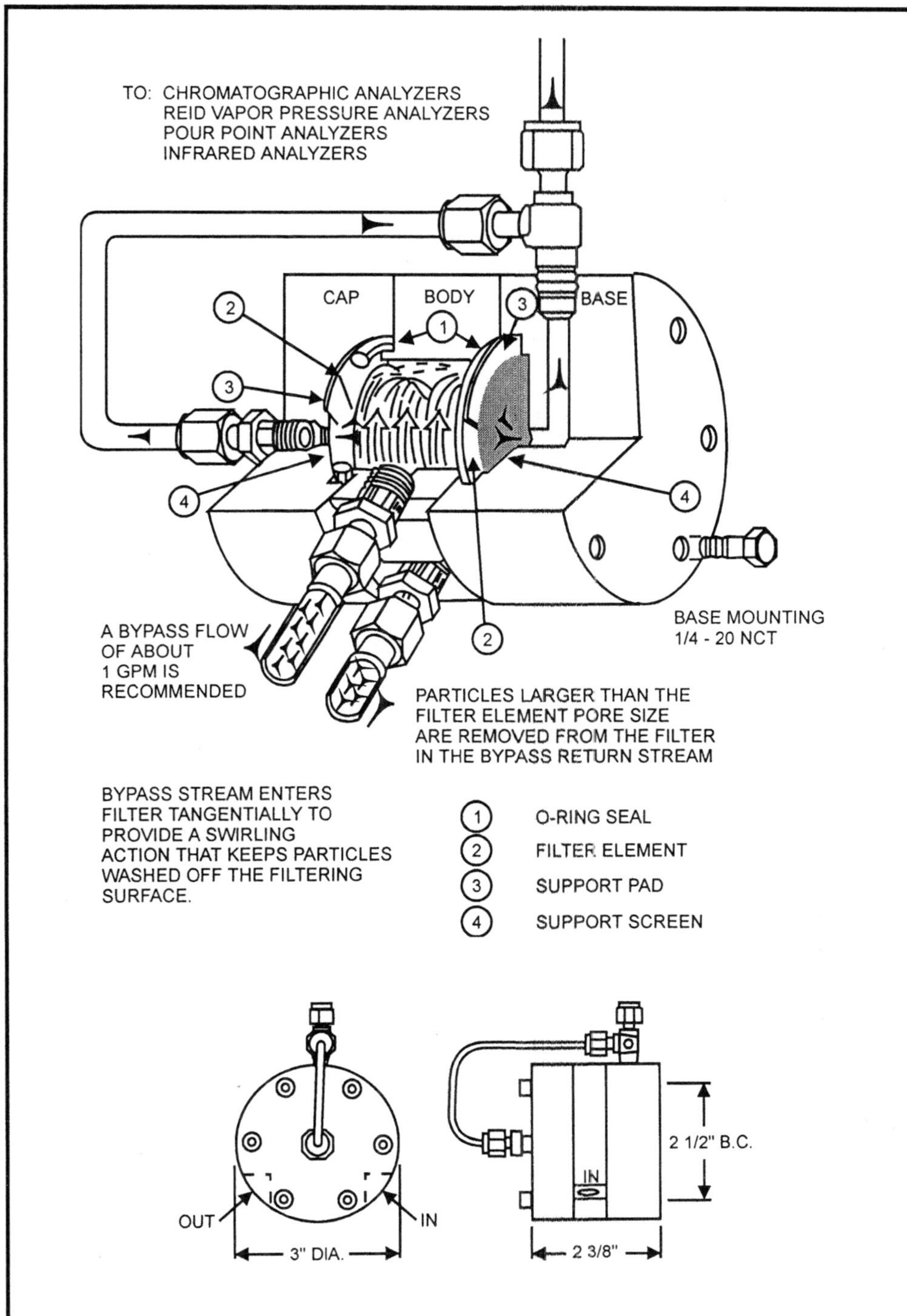

Figure 3-6. The Collins Swirlklean Bypass Filter
(Courtesy of Collins Products Company)

Filter types are typically divided into two general categories: particulate and coalescing. The particulate filters can be inertial, bypass (slipstream) or depth. The coalescing filters can be depth or semipermeable membrane.

Particulate filters do just as their name implies-they filter particulate! In performing this task you must take into account the size and density of the particulate which you are filtering. If the particulate is large enough and of sufficient density, it can be separated from the gaseous sample stream by *inertial filtration*, also often known as air- or hydro-cyclone (See Figure 3-7) (if the separation is rotational) or Mott™ (if the separation is linear). This separation principle is such that heavier particles go to the outside of a cyclone or pass directly through a Mott™ while the clean gas sample passes up through the center of the cyclone or out 90° to the linear flow for the Mott™. If the particulate is smaller and lighter, it can be separated from the gaseous sample stream by *bypass (slipstream) filtration* (see Figure 3-8). This separation method utilizes a filter body enclosing a filter element with the sample gas flow being from the filter element's outer surface through to the filter element's inner surface while the bypass (slipstream) gas flow passes out of the filter body without passing through the filter element. This method of filtration allows only as much gaseous sample as is required by the analyzer to be filtered while the bypass (slipstream) gas flow passes out of the filter body without filtration of the particulate from the sample gas. Once again, the size and density of the particulate relative to the gaseous sample must be considered.

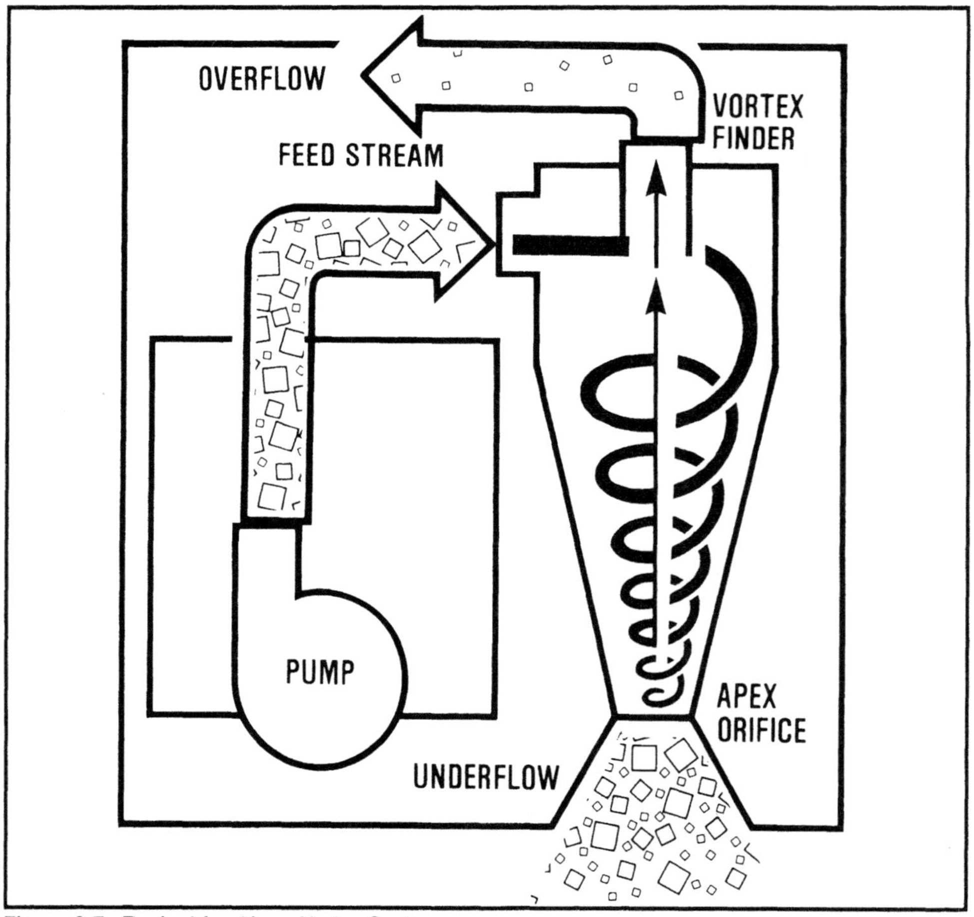

Figure 3-7. Typical for Air-or Hydro-Cyclone
(Courtesy of Krebs Engineers)

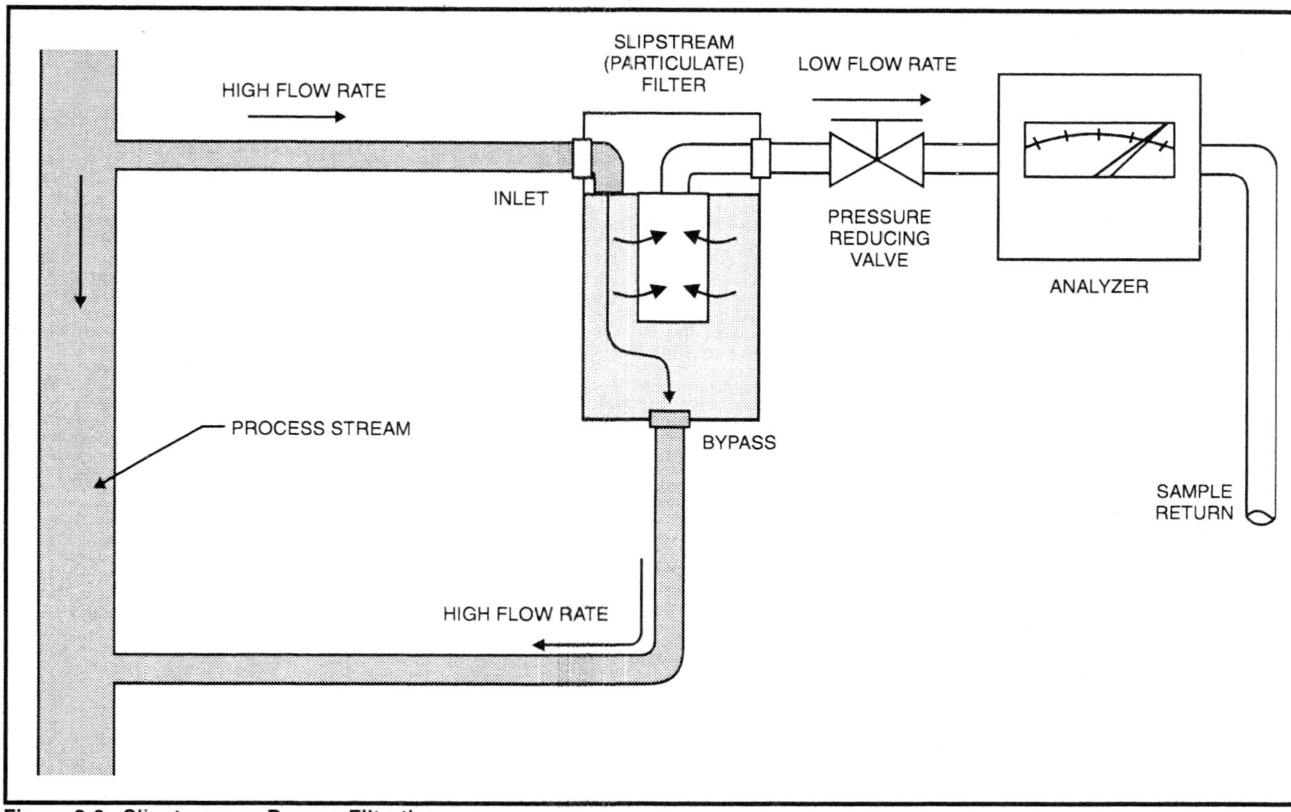

Figure 3-8. Slipstream or Bypass Filtration
(Courtesy of Balston/Whatman, Inc.)

If the filter element is too fine [too small of a porosity (micron) rating], the element will plug prematurely; if the filter element is too coarse [too large of a porosity (micron) rating], particulate will pass through the filter element and either prematurely clog the final depth filter or foul the analyzer sample cell.

With membrane filter elements, particles larger than the micron ratings do not enter the pores. Good turbulent flow can remove particles from the surface. The smaller the pore size the better. A very turbulent condition is a performance requirement.

A *depth filter* (Figure 3-12) is a total flow-through filter and most often it is the final filter in an analyzer system, installed just prior to the analyzer sample entry port to protect the analyzer's sample cell from particulate contamination. This filter element is most usually a small swept-volume filter body with a borosilicate-glass fiber or sintered stainless steel element of very fine (0.1, 0.5, 1, 3, 5, 10, 15 microns) porosity. Sintered stainless steel is fine particles of stainless steel which have been partially fused together to give a filter element of a known porosity.

Coalescing filters perform the function of creating large droplets from small droplets by depth filter element filtration of gaseous sample flow from the inside of the coalescing filter element, through the filter element, and off of the outside of the coalescing filter element. At this point we have large droplets of coalesced liquid dripping off of the outside surface of the coalescing filter element. In order to continue operation of this filter, this coalesced liquid must be removed on a continuous basis. Thus, the coalescing filter element must be installed in a bypass (slipstream) filter housing. See Figures 3-9 to 3-11 and Examples 3-1 and 3-2 for types of coalescing filters. The continuous removal of this coalesced liquid (condensate) will be addressed later in this chapter under "Condensate Removal and Condensate Entrainment Protection."

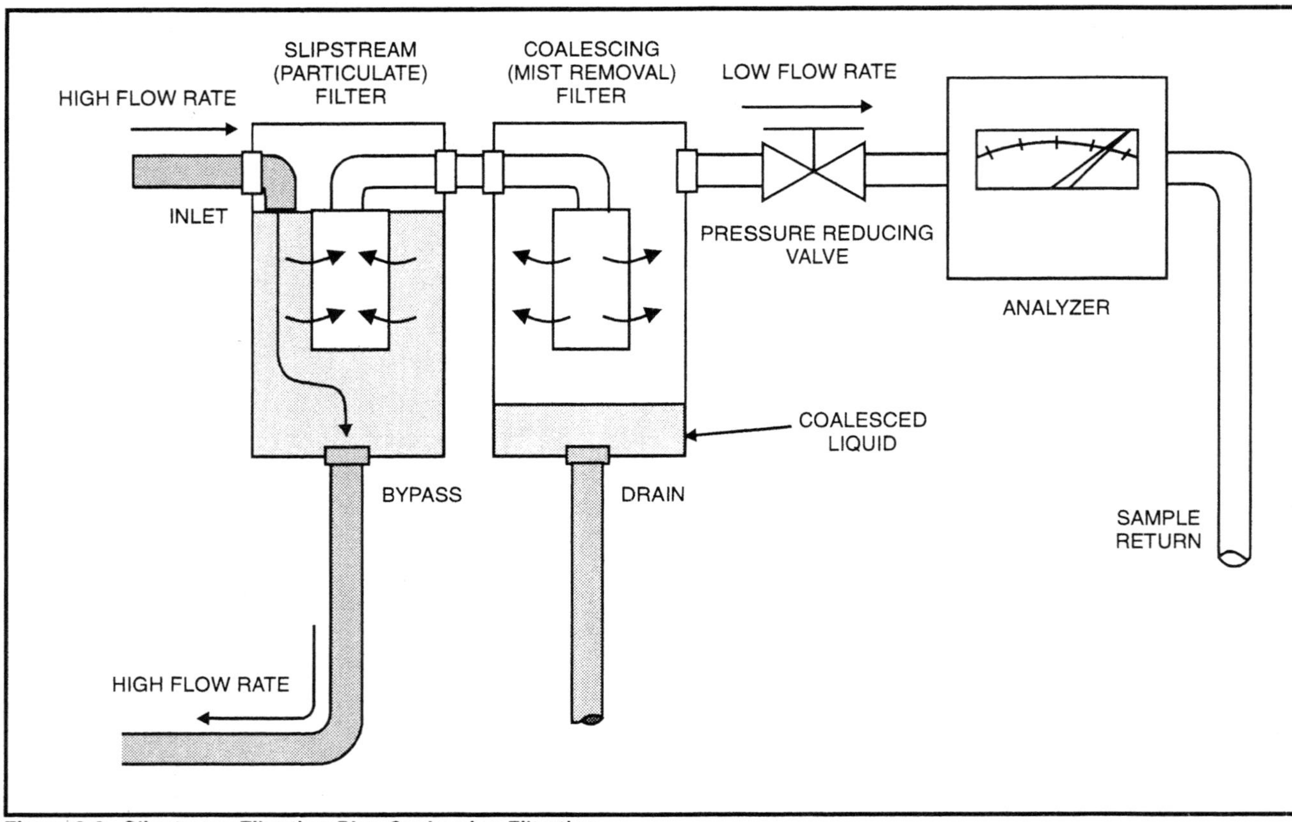

Figure 3-9. Slipstream Filtration Plus Coalescing Filtration
(Courtesy of Balston/Whatman, Inc.)

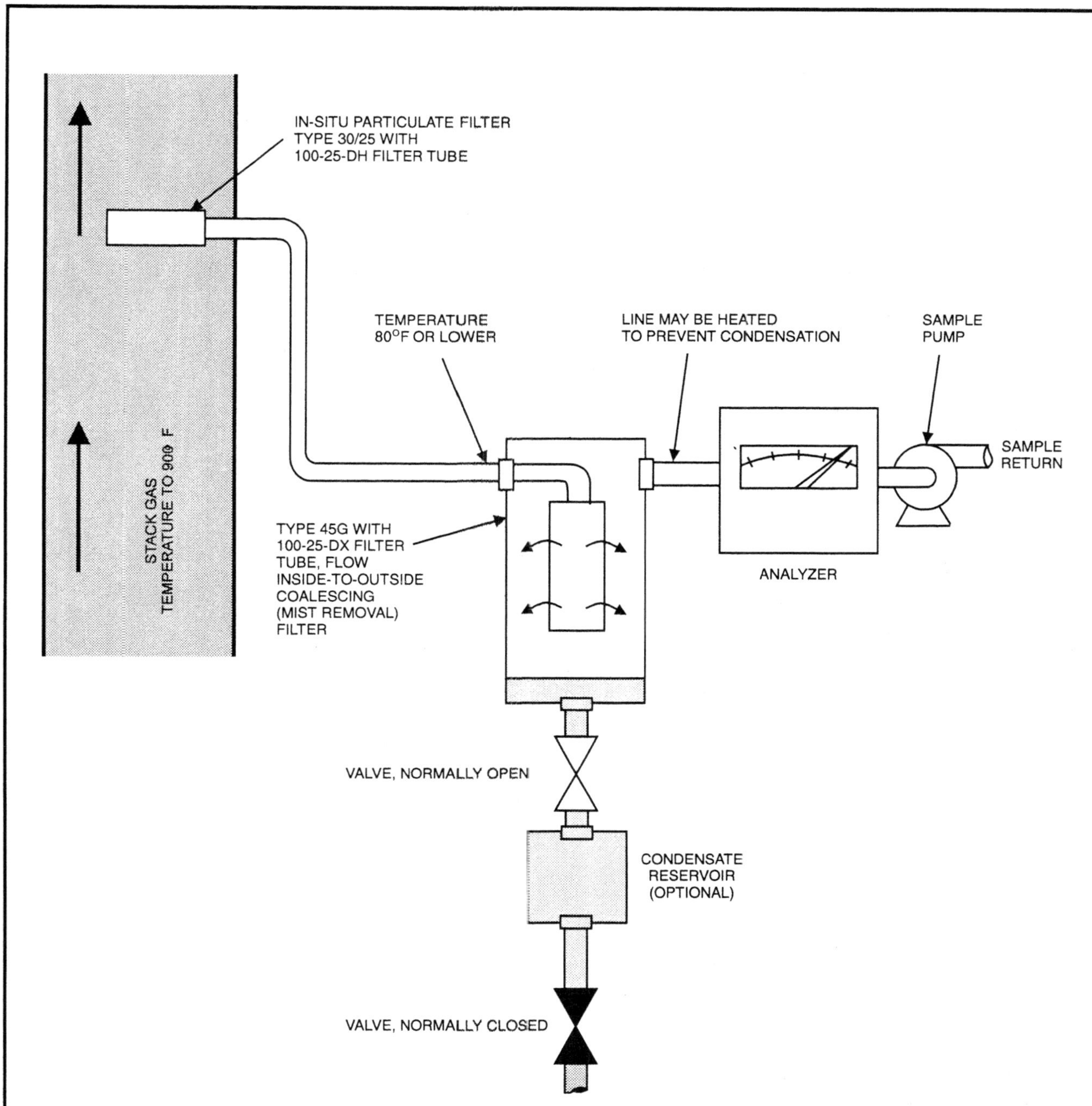

Figure 3-10. Extractive Coalescing Filtration with In-Situ Particulate Filter
(Courtesy of Balston/Whatman, Inc.)

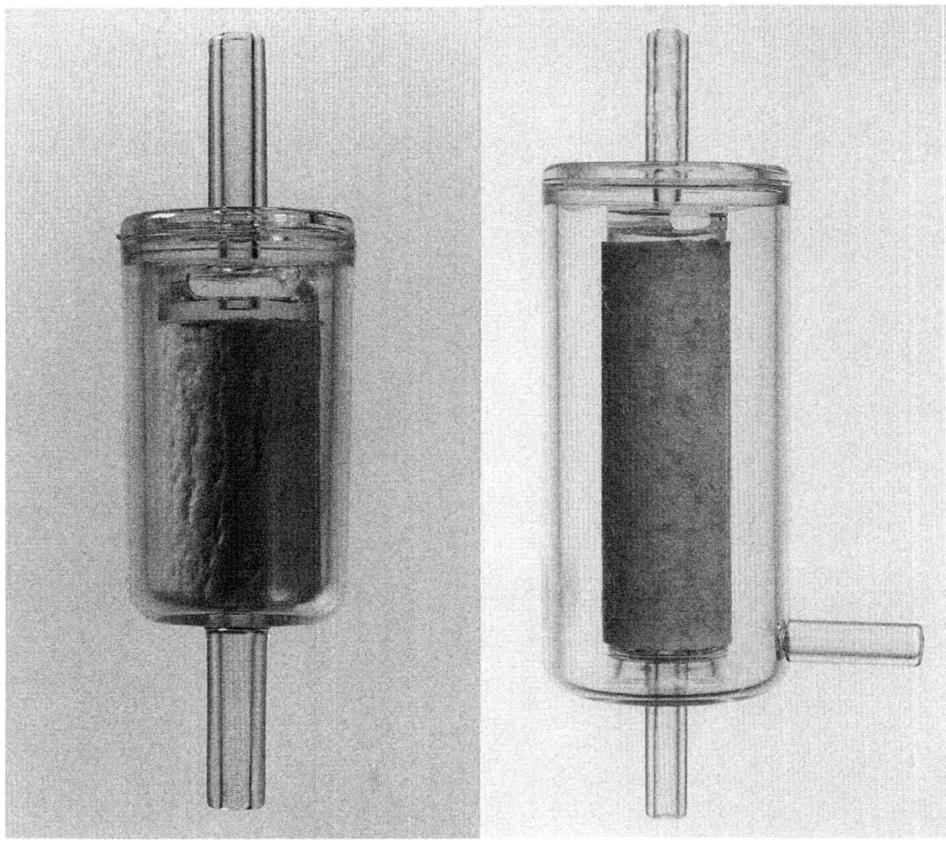

Figure 3-11. Depth Particulate Filters: DFU 9922-05, 9933-05 and Coalescing Bypass Filters: DFU 8822-11, 8833-11 *(Courtesy of Balston/Whatman, Inc.)*

OPERATION

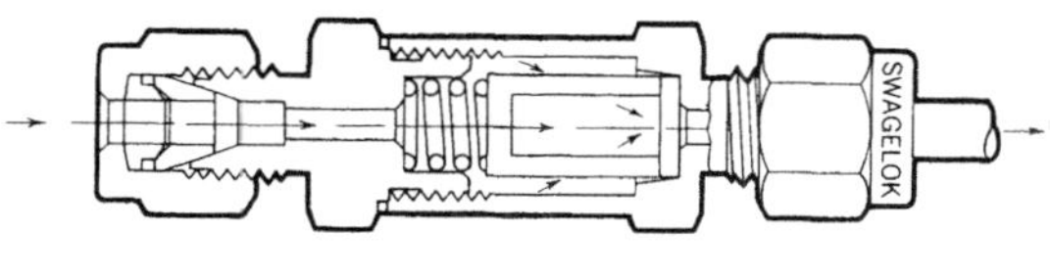

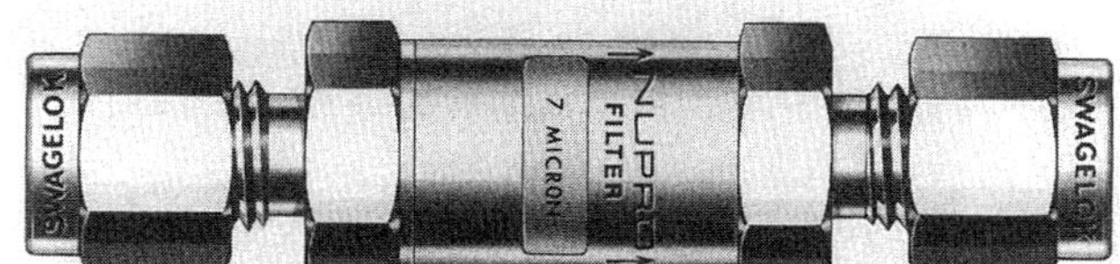

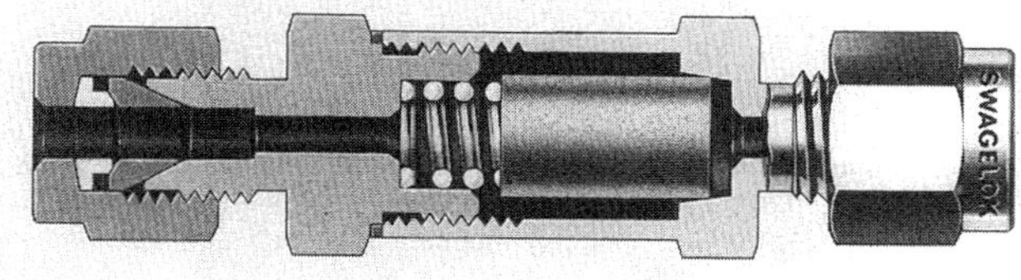

As the system fluid enters the NUPRO Inline Filter, the 316 stainless steel sintered element or strainer element stops the contamination and permits the cleaned fluid to pass through. Bypass flow is prevented by press fitting the element into a tapered bore in the female body piece. A precision metal gasket seals the body to prevent leakage to atmosphere.

MATERIALS*

Body — 316 stainless steel or brass

Metal Gasket — Silver-plated 316 stainless steel in stainless steel filters; aluminum in brass filters.

Spring — 302 stainless steel

Elements — 316 sintered stainless steel or 316 stainless steel wire mesh, silver-brazed.

All Other Parts — Same materials as body.

*NOTE: "4F4" filters in Alloy C-276 are also available on special order. All parts are made of Alloy C-276 including the sintered elements in 2, 7, 15, 60 and 90 micron sizes.

SPECIAL FEATURES

NUPRO "F" Series Inline Filters can be used in place of inline removable-type filters when elements do not have to be changed often or space is limited.

They are available with straight tube outlet connections which permit a direct connection to any valve or fitting with SWAGELOK® Tube Fitting and CAJON VCR® connections.

FILTRATION DEFINITIONS

Micron — One micron equals 1/25,000 inch (0.00004 inches/ 0.001 mm). Microns are used to indicate mean pore diameter of a sintered element or mean particle diameter of fluid contamination. **Filter Element** — That portion of a filter which actually traps fluid contamination. **Nominal Rating** — Initial removal of 95% to 98% of particles larger than a given size. **Effective Filter Area** — The actual area open to flow in a filter element. **Filter Cake** — The accumulation of solids deposited on a filter element from usage.

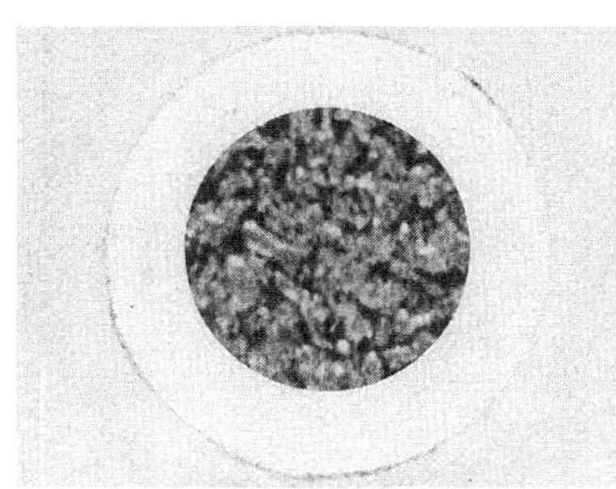

SINTERED CONSTRUCTION
Sintered 316 stainless steel filter elements in nominal micron sizes of 0.5 to 90, trap particulate contamination in the tortuous matrix. (Magnified 13X.)

ELEMENT MATERIALS

Sintered Elements — 316 stainless steel (0.5, 2, 7, 15, 60 and 90 micron available) — Alloy 400 and Alloy C-276 available on special order.

Figure 3-12. Compact Inline Filters: "F" Series
(Courtesy of Swagelok Co.)

Example 3-1
(Courtesy of Balston Inc., Haverhill, MA)

Coalescing Filtration: Separating Liquids from Gases:

Microfibre Filter Tubes efficiently separate suspended liquids from gases. The filter fibers capture the fine liquid droplets suspended in the gas and cause the droplets to run together to form large drops within the depth of the filter tube. The large drops are then forced by the gas flow to the downstream (outside) surface of the filter tube, from which the liquid drains by gravity. This process is called "coalescing." Since the coalesced liquid drains from the tubes at the same rate that liquid droplets enter the tubes, the tubes have an unlimited life when coalescing liquids from relatively clean gases. The filters operate at their initial retention efficiency even when wet with liquid. Note that the flow direction is inside-to-outside, to permit the liquid to drip from the outside of the filter element to the housing drain.

Since the coalesced liquid drips off the down-stream surface of the filter tube in the presence of filtered gas, it is important to avoid carryover, or re-entrainment, of liquid droplets by the gas leaving the filter housing. In all housings except the Type 95 models, the possibility of re-entraining coalesced liquid is minimized by using an X-Type filter tube. The X-Type filter tubes are constructed of two layers, an inner high-efficiency coalescing layer and an outer layer of coarse glass fibers. The coarse, rapidly-draining outer layer ensures that the liquid drips continuously from the bottom of the filter tube and minimizes the chance of liquid carryover. In the Type 95 models, the small internal volume does not permit use of the thick-wall X-Type tubes, and therefore a Q-Type tube must be used. Re-entrainment of coalesced liquid is avoided by ensuring that the gas flow rate through the housing is safely below the maximum.

Recommendations: When a large quantity of liquid in the gas is expected, two stage filtration is recommended for complete liquid removal. First Stage: Grade DX or DQ, Second Stage: Grade BX or BQ. If only a trace of liquid is expected (as would be the case with most cylinder gases), only one stage of liquid removal filtration is required, Grade BX or BQ.

To Remove Solid Particles from Gases:

If a depth filter is acceptable as a final filter, Microfibre Filter Tubes Grade AQ (99.9999+% efficient at 0.1μm) provide a combination of high efficiency, excellent solid-holding capacity, and low pressure drop. Any liquid in the gas should be removed (see recommendations above) before using Grade AQ as the final stage.

If absolutely no particle migration can be tolerated, a membrane final filter should be installed as close to the end use point as possible. If any liquid is present in the gas, Microfibre Grade BX or BQ prefilter is essential. Even if no liquid is expected, we recommend using a Grade BX or BQ prefilter to protect the membrane final filter and greatly extend its life.

Example 3-2
(Courtesy of Balston Inc., Haverhill, MA)

Coalescing Filtration: Separating Two Liquid Phases

In principle, Microfibre Filter Cartridges separate suspended droplets of a liquid which is immiscible in another liquid by the same process as they separate droplets of liquid from a gas. The liquid droplets suspended in the continuous liquid phase are trapped on the fibers and run together to form large drops, which are then forced through the filter to the downstream (outside) surface. The large drops separate from the continuous liquid phase by gravity difference, settling if heavier than the continuous phase and rising if lighter. The coalescing action of Balston filters is effective with aqueous droplets suspended in oil or other hydrocarbons, and also with oil-in-water suspensions.

In practice, liquid-liquid separations are much more difficult than liquid-gas separations. The specific gravity difference between two liquids is always less than between a liquid and a gas, and therefore a longer phase separation time is needed. Either the filter housing must be oversized or the flow rate greatly reduced to avoid carryover of the coalesced phase. As a rule of thumb, flow rate for liquid-liquid separation should be no more than one-fifth the flow rate for solid-liquid separation. Even at low flow rates, if the specific gravity difference between the two liquids is less than 0.1 units (for example, if an oil suspended in water has a specific gravity between 0.9 and 1.1), the separation time for the coalesced phase may be impractically long. In that case, if there is only a small quantity of suspended liquid, the filter tube can be used until saturated with the suspended liquid and then changed.

Another practical problem with liquid-liquid separations is that small quantities of impurities can act as surface-active agents and interfere with the coalescing action. For that reason it is not possible to predict accurately the performance of a liquid-liquid coalescing filter, and each system must be tested on site. The general guidelines for the system to start testing are to use Grade DX filter tubes (Grade DQ in Model 95 housings), and flow inside-out at very low flow rates. If the suspended liquid is lighter than the continuous phase, the housing should be oriented so that the drain port is up. In general, Microfibre Filter Cartridges should be used for liquid-liquid coalescing in slipstream sampling applications only.

Removing Gas Bubbles from Liquids

Microfibre Filter Cartridges readily remove suspended gas bubbles from liquid, eliminating the need for deaeration tanks, baffles, or other separation devices. Flow direction through the inverted filter body (housing) is outside-to-inside, and the separated gas bubbles rise to the top of the housing and are vented through the drain port. If slipstream sampling is used, the separated bubbles are swept out of the housing with the bypassed liquid. Grade DX or Grade DQ is a good choice for gas bubble separation.

Sample Coolers

Sample coolers can be air-powered or electric-powered.

The function of these sample coolers is to remove moisture from the sample which is passing through them. Typically, the sample cooler cannot operate at a surface temperature (of the heat exchanger) that is less than 32°F or the moisture being condensed (removed) from the sample will freeze and block passage of the sample gas. If sample flow is sufficient, this heat exchanger surface temperature can fall below freezing due to the heat transfer from the incoming sample gas but this is a risky situation.

This condensate (liquid) being removed from the sample stream must itself be removed from the sample cooler, preferably with minimum surface contact between flowing "dry" gas and condensed "condensate" liquid. This minimum contact consideration is to prevent dissolving gases from the sample gas stream into the condensate as well as to prevent rehydration of the sample gas by re-evaporating or entraining (physically carrying) condensed liquid out of the sample cooler.

Air-powered sample coolers can either be simple air-to-tube heat exchangers or vortex-tube (air pressure/expansion refrigeration) units. *Air-to-tube heat exchangers* leave little possibility for temperature control, typically only providing temperature reduction from some elevated value down towards an ambient value. *Vortex-tube exchangers* now have some semblance of temperature control (typically by some means of air flow/pressure modulation) but they still suffer from their temperature floating on excursions in ambient temperature (see Figure 3-13). In short, if your plant has 'wet' compressed instrument air (pressure dewpoint of instrument air below ambient temperature) and the ambient temperature falls to anywhere near freezing, your vortex-tube cooler is going to freeze and cease operation. You can counteract the effects of this ambient temperature by freeze-protect heat-trace-and-insulating the air supply line. Use only freeze-protect heat-trace since the efficiency of the vortex-tube relies on inlet (high-pressure) air temperature and the Joule-Thompson effect (pressure reduction/volume expansion is a cooling process).

Vortex Tube Cooling System Facts

Vortex Tubes provide a straight-forward and economical solution to industrial and laboratory spot-cooling problems. Utilizing only compressed air and employing no moving parts or electricity, Vortex Tubes can provide air temperatures ranging from –40 °F to +200°F in flow rates of 2 to 25 SCFM.

Vortex Tubes incorporate the proven, reliable principle of the Ranque-Hilsch tube. Compressed air is injected into a vortex generator so that it is uniformly flowing in a high speed circular motion. As the circular-moving-air progresses down a decreasing radius circular tube, it separates into two distinct swirling air streams which have very different temperature characteristics. One stream is slightly to moderately warmer, +20 F° to +70F°, than the supply air temperature. The other stream is significantly colder than the supply air temperature. The temperature of this cold air stream can be as much as–110 F° lower than the inlet supply air and can provide a continuous stream of cold air at –40 °F assuming an inlet temperature of +70°F and an inlet pressure of 100 psig.

FEATURES:

POSITIVE
- Fingertip adjustable temperature.
- Cooled air to –40 °F.
- 2-25 SCFM output ranges.
- Refrigeration up to 1500 BTU/hr. (440 watts).
- No moving parts, no maintenance.
- Compact, lightweight, portable.
- Compressed air operation — no electricity needed.
- No radio frequency interference.
- Low initial cost.

NEGATIVE
- Can cease operation with cold, wet air supply (freeze-up).
- Noisy (available muffler compromises efficiency).
- High operating cost (extremely poor thermal efficiency/usage of compressed air).

Figure 3-13. Transonix™ Cool Sure™ Enclosure Cooling Systems
(Courtesy of Transonix Corporation)

Electric-powered sample coolers can either be mechanical refrigeration or thermoelectric module (TEM or, more properly, Peltier-effect) units (Figures 3-14, 3-15, and Example 3-3). The *mechanical compressor* can either utilize cfc (chlorofluorocarbon) refrigerant or cfc-free (environmentally friendly) refrigerant. These units are classic refrigeration or air-conditioning units and work by mechanically compressing the refrigerant, cooling it to remove the heat of compression and expanding it into a large physical-area expansion coil which becomes cold due to the Joule-Thompson cooling effect of the refrigerant expansion from high pressure (compressed) to low pressure (expanded). The motors of these units are subject to the aggressive atmospheres of process plants as well as the electrical hazard rating of the area in which they are installed. Also, the mechanical vibration and physical abuse in process plants make these mechanical refrigeration units subject to leaks and subsequent loss of cooling capacity. The *TEM (Peltier-effect) units* suffer from a capacity limitation for BTU transfer and chemical corrosion considerations of TEM material but are otherwise near ideal for sample cooling purposes (Figure 3-16). The Peltier-effect is a phenomenon wherein a solid-state device heats or cools in proportion to the level of applied positive or negative direct current (DC) voltage.

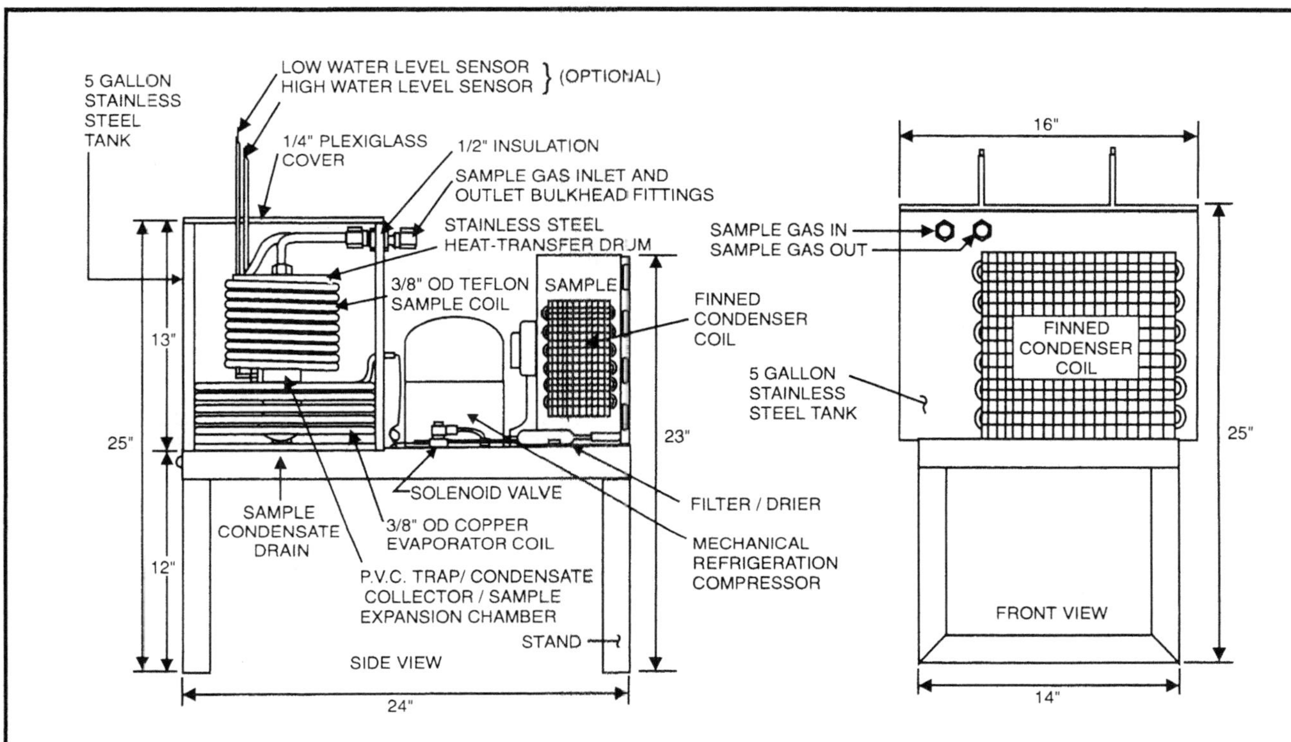

Figure 3-14. Typical Mechanical Compressor Refrigeration Sample Cooler (Pictorial)
(Courtesy of Lawler Manufacturing Corporation)

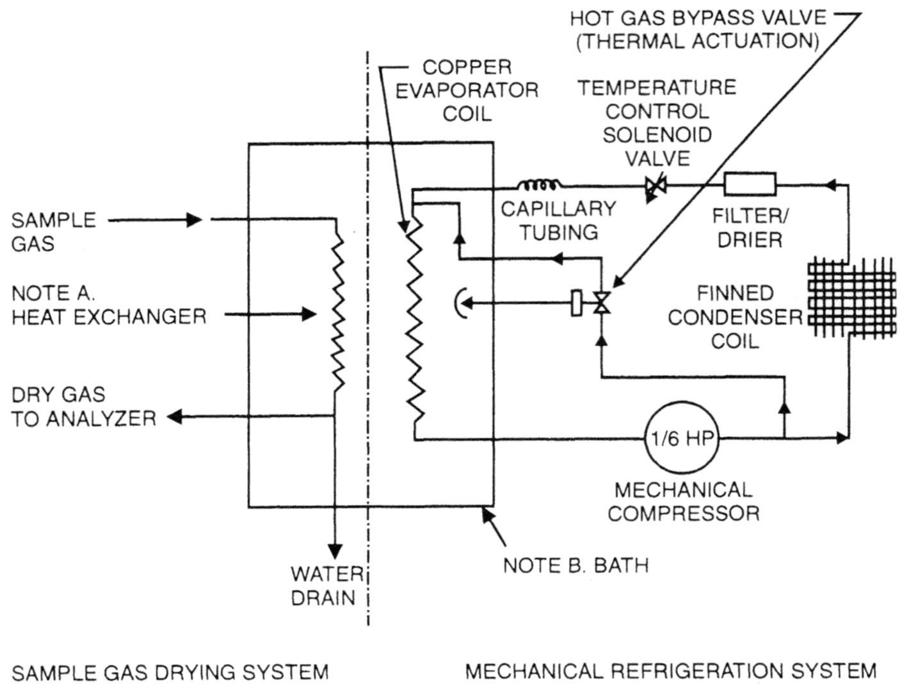

OPTIONS:
NOTE:
A. HEAT EXCHANGER
   1. SINGLE OR DOUBLE COIL HEAT EXCHANGER
   2. 3/8" OD SEAMLESS STAINLESS TUBING OR 3/8" OD TEFLON
   3. STAINLESS FITTINGS OR POLYPROPYLENE FITTINGS
   4. WATER SEPARATOR: STAINLESS STEEL OR PVC

B. BATH
   1. 5 (SINGLE COIL) OR 8 (DOUBLE COIL) GALLON STAINLESS STEEL LIQUID BATH
   2. DRY BATH

Figure 3-15. Typical Mechanical Compressor Refrigeration Sample Cooler (Schematic)
(Courtesy of Lawler Manufacturing Corporation)

Example 3-3. Basic Refrigeration Cycle
(Courtesy of McLean Midwest)

The compressor pulls the warm refrigerant from the evaporator and pumps it into the condenser, compressing it in the process. The compression causes the temperature of the refrigerant to rise sharply, along with the pressure. Then, because a great temperature difference exists, the condenser coil can easily pass this heat to the ambient air as the refrigerant condenses from a high pressure gas into a high pressure liquid.

Next, the capillary tube meters the flow rate of the refrigerant liquid into the evaporator. From the compressor to the capillary tube the refrigerant is under high pressure. As the refrigerant liquid is metered from the capillary tube into the evaporator, it enters the low pressure side of the cycle. The pressure drop on the evaporator side allows the liquid refrigerant to begin expanding, or evaporating, into a vapor.

During the process of vaporizing, the refrigerant absorbs heat. The heat source is the hot sample gas. The evaporator coil draws the heat from liquid inside the tank, where the sample gas heat is absorbed. Excess moisture in the sample gas is condensed during the cooling process (the moisture is discharged through the condensate drain tube).

After vaporizing, the low pressure gas is drawn back into the compressor, completing the refrigeration cycle. Before the gas enters the compressor, it passes through an accumulator, where any unvaporized liquid is separated from the gas.

One important control should be noted. It's the hot gas bypass valve. Its purpose is to limit the cooling of the sample cooler and eliminate evaporator coil freezing during periods of low heat load and low ambient temperature. If the evaporator gets too cold, the hot gas bypass valve opens and allows some of the high temperature gas to pass directly from the compressor into the evaporator, bypassing the heat removal that would normally take place by flowing through the condenser coil.

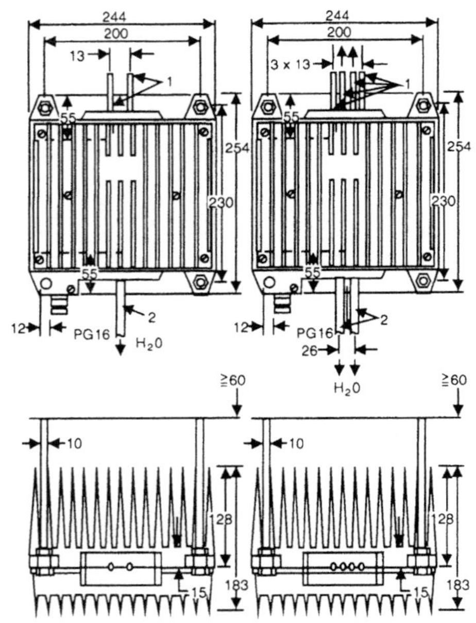

1 PVC hose of 10 mm outer diameter and wall thickness 1.5 mm
2 PVC hose of 15 mm outer diameter and wall thickness 2 mm

Fig. 8/35 Gas cooler with one gas duct (left) and with two gas ducts (right), dimensions in mm.

DESIGN
The gas cooler M52798 consists of a gas cooler and a power supply unit.

GAS COOLER
There are two electrically isolated (by a foil) cooling blocks between the cooling pocket and the rear wall of the housing in the cooler with one gas duct, and four cooling blocks in the cooler with two gas ducts. The cooling blocks are electrically in series. The cooling pocket and blocks are pressed against the cooling fins by two or four springs. The contact pressure is limited by the springs. The temperature sensor is attached to the front of the cooling pocket. The cooling pocket is protected against corrosion by a plastic coating.

POWER SUPPLY UNIT
The circuitry for the power supply to the cooling blocks and the temperature limiter is on one base plate. The sheet-metal case is dust-proof and splashproof.

MODE OF OPERATION
The Peltier effect (converse of the thermoelectric effect) is used to cool the sample gas. When an electric current flows across the junction between two different metals the temperature drops on one side of the junction, while on the other side the temperature increases. Several Peltier cells are combined to form one SIRIGOR cooling block. The gas cooler contains two or four of these cooling blocks. The cold faces of the cooling blocks lie against the cooling pocket through which the sample gas flows. The higher temperature of the hot faces is discharged to the atmosphere using large cooling fins. In order to prevent the cooling pocket from freezing at ambient temperatures below 30 °C, a temperature limiter (sensor and limit switch with transistors) is used to maintain the lowest temperature at 2 to 6 °C (±1 C°). The condensation which drops out must be collected in a condensation trap, which also forms the gas termination. Two condensation traps must be provided for the cooler with two ducts.

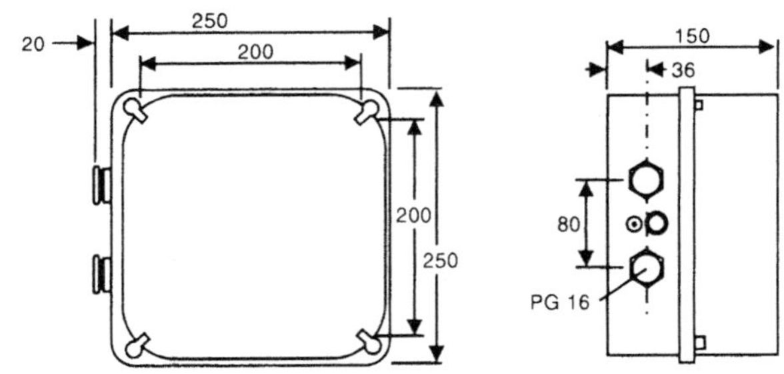

Fig. 8/36 Power supply unit, dimensions

TECHNICAL DATA

Gas cooler	With one gas duct	With two gas ducts
Temperature reduction (average value)	30°K (min. 28°K) at 2 l/min and	25°K (min. 23°K) at 2 × 2 l/min and
	... 100% relative humidity of the sample gas	
Absolute cooling limit (protection against freezing)	2 to 6°C (production tolerance)	
Dew point fluctuation	±1K°	
Permissible gas flow	Max. 2.5 l/min per duct	
Permissible operating pressure	Max. 5 bar	
Operating voltage and current	Approx. 6 VDC, 5 A	Approx. 12VDC 5 A
Permissible ambient temperature	5 to 55°C	
	to achieve constant sample humidity: max. 35°C	max. 30°C
Degree of protection	IP 54 to DIN 40 050	
Warm-up period	Approx. 10 min	
Power supply unit		
Mains consumption	220 V AC, 48 to 63 Hz	
Power consumption with gas cooler	500 VA	90 VA
Output	6 V DC, 5 A	12 V DC, 5 A
Permissible ambient temperature	10 to 55°C	
Permissible resistance of the lead to the Peltier cells	Max. 0.04 Ω	Max. 0.08 Ω
	(Total of outgoing and return leads)	
Degree of protection	IP 54 to DIN 40 050	

ORDERING DATA

	Ord. No.	Approx. weight in kg
Gas cooler, consisting of cooler and power supply unit		
With one gas duct		
With 1 condensation trap	M52798-A4	17.7
Without condensation trap	M52798-A5	15.2
With two gas ducts		
With 2 condensation traps	M52798-A6	20.8
Without condensation traps	M52798-A7	15.8
Spare parts		
For gas coolers M52798-A4 and -A5:		
Gas cooler	C71211-A6-A4	9.5
Power supply unit	C74280-A166-A1	5.7
For gas coolers M52798-A6 and -A7:		
Gas cooler	C71211-A6-A5	9.8
Power supply unit	C74280-A166-A2	6.0

Figure 3-16. Gas Coolers M52798 with Peltier cells
(Courtesy of SIEMENS)

Condensate Removal

Condensate removal is initially by gravity separation of the condensed liquid from the flowing sample gas stream at a temperature below the condensate's dewpoint and above the freezing point of the condensate. This physical removal is typically via continuous positive pressure or gravity flows or, in many cases where the sample pressure is at or near a negative value, condensate removal is accomplished by a reservoir and a peristaltic pump (see Figure 3-18).

If you have sufficient sample pressure and the sample is non-toxic and non-flammable, you can use sample pressure to force the condensate out of the system on a continuous basis. If you are lucky enough to have a drain below the level of the sample condensate exit, gravity will aid your sample pressure motive force. One can employ a float-operated trap (Figure 3-17) or a simple siphon to remove condensate from a low-pressure sealed system although most process sample condensates have components in them which will rapidly foul both of these devices. Where the sample pressure is at or near a negative value you must use a condensate reservoir and either sequenced solenoid valves or a peristaltic pump (continuously-operating or level-actuated) to remove collected condensate. The sequenced solenoid valves will only work where you have gravity to remove the condensate. The peristaltic pump will pump material against both a considerable vacuum on its inlet and a considerable liquid head on its outlet. The operating principle of a peristaltic pump is rollers sequentially compressing flexible tubing inside of a fixed cavity and moving "segmented packets" of liquid to the outlet of the pump ahead of its rollers.

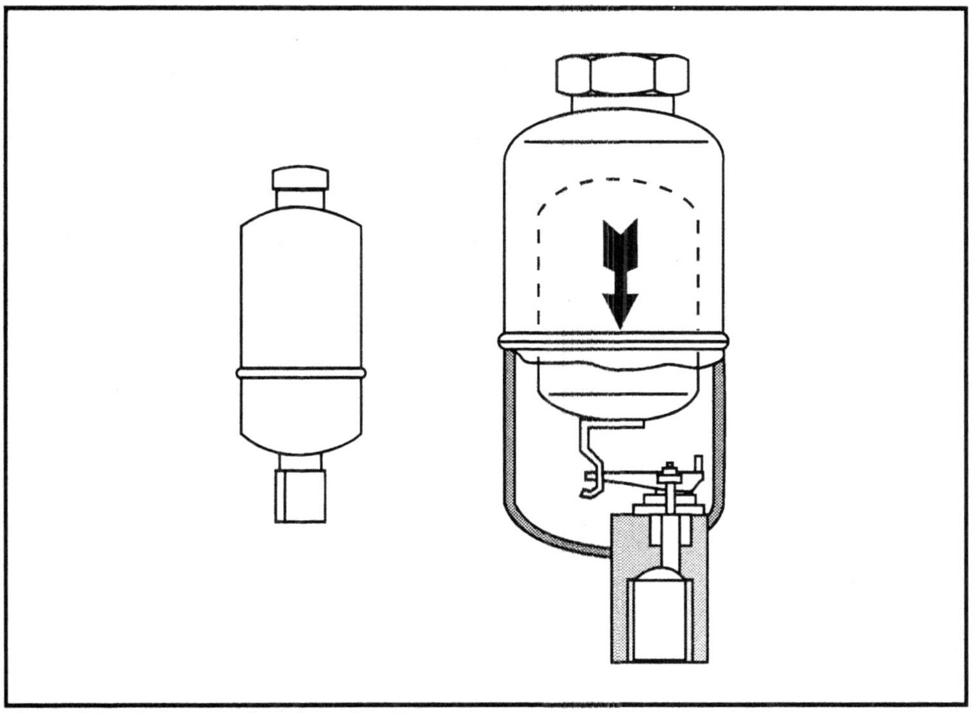

Figure 3-17. Automatic Condensation Trap
(Courtesy of SIEMENS)

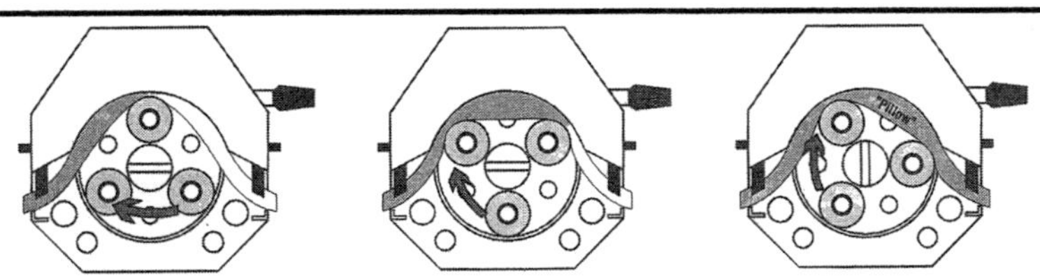

1. Pump head consists of only two parts: the rotor and the housing. The tubing is placed in the tubing bed—between the rotor and housing—where it is occluded (squeezed)

2. The rollers on the rotor move across the tubing pushing the fluid. The tubing behind the rollers recovers its shape, creates a vacuum, and draws fluid in behind it.

3. A "pillow" of fluid is formed between the rollers. This is specific to the ID of the tubing and the geometry of the rotor. Flow rate is determined by multiplying speed by the size of the pillow. This pillow stays fairly constant except with very viscous fluids.

ADVANTAGES OF PERISTALTIC PUMPS

1. Fluid does not contaminate pumps—the pump head never gets wet!
2. Pumps do not contaminate the fluid
3. Easy to clean and sanitize
   - —fluid contacts only the tubing
   - —simply change tubing to clean pumps; saves time and makes changing processes easy
   - —some tubing types are sterilizable
4. Positive displacement, no backflow
5. Good suction lift
   - —nonsiphoning and safely run dry
6. Low maintenance
   - —no seals to leak or valves to clog or wear
   - — easily serviced with only one moving part
7. Versatile
   - —gentle, low-shear pumping
   - —pump liquids, gases, or mixed phases
   - —handles high-viscosity liquids
   - —wide range of tubing formulations
   - —orientation-insensitive (mount the pump head in any position.)

Figure 3-18. Peristaltic Pumps
(Courtesy of Cole-Parmer Instrument Company)

Condensate Entrainment Protection

You must prevent condensate carryover beyond the gas outlet of the condensate removal device. This is typically accomplished by mechanical means (coalescing filter with drain or selective membrane device [see Example 3-4] with drain) or by electrical means (utilizing a capacitance/ionization-moisture sensing device with sample diversion valve). The coalescing filter is a typical bypass (slipstream) configuration filter which utilizes a special filter element where the sample gas flows from the inside of the filter element to the outside of the filter element. This combination of sample flow and special filter element make large liquid droplets out of very small liquid droplets. These large liquid droplets physically fall out of the gaseous sample stream and drain out of the bypass (slipstream) port of the filter housing, exiting the sample system by means similar to condensate removal devices above (often the coalescing filter bypass (slipstream)/drain flows to the condensate reservoir and condensate removal system described above). The selective membrane device is similar in action to the coalescing filter with the exception that the coalescing filter element is replaced by a gas-permeable hydrophobic membrane enclosed in a housing with a continuous liquid bypass (slipstream) port on the inlet side of the membrane. Hydrophobic membrane refers to a membrane which repels (phobic, phobia, fear of, repulsive) water (hydro). The *capacitance/ionization-moisture sensing device* is typically a low-volume flow-thru housing which contains a salt-impregnated piece of filter paper which is in contact with two DC current-carrying terminals. When moisture contacts this filter paper, the salt ionizes and carries a current between the terminals. This current typically trips a relay which lights a warning lamp, either local or local and remote, as well as actuates a sample diversion valve. This sample diversion valve causes sample flow to bypass the analyzer whose cell or detector may be damaged by moisture (in the sample) which has been detected by the moisture-sensing device.

Example 3-5 describes the benefits of Nafion® Gas Sample Dryers.

Example 3-4. Series 100 Genie Membrane Separators – Gas/Vapor Service
(Courtesy of A+ Corporation)

FUNCTION

Genie Membrane Separators remove entrained liquid and particulate in gas samples, thereby preventing contamination and/or damage to analyzers and sample system components.

DESCRIPTION

The separators consist of a housing with a porous membrane sandwiched between its two halves. The housings typically have ports labeled "inlet," "bypass" and "outlet." The "inlet" and "bypass" ports are located on the upstream side of the membrane, while the "outlet" port is on the downstream side.

Gas sample enters the housing through the "inlet" port, is able to flow easily through the membrane, and exits from the "outlet" port (see Fig. #1). Because entrained liquid cannot flow through the membrane, it is removed via the "bypass" port. If particulate is present, it is retained on the membrane. This membrane is supported by a sintered porous disk, located on the "outlet" half of the housing.

MEMBRANE

The membrane contains microscopic passages, which permit molecules of gas or vapor to flow through with ease. This is due to the fact that gas and vapor molecules are separated from each other, and therefore are able to flow through the passages individually. Liquids, on the other hand, consist of large numbers of molecules that cling tightly together. This property, called surface tension, is substantially greater for liquids than for gases or vapors. Bound tightly together as a "cohesive group," the liquid molecules cannot flow through the membrane's small passages under normal operating conditions. Because of the efficiency of this process, even the smallest aerosol droplets and particles are removed from a gas stream (see Fig. #2). And, since all of the gas or vapor molecules physically flow through the membrane, the composition of the sample gas is unchanged.

The membrane construction is proprietary. It is extremely inert, and recommended for applications with most process fluids, excluding hydrofluoric acid. Its extremely low absorption characteristics make it suitable for use in systems designed for PPB, PPM, and "percent level" component concentrations. Although the membrane is soft and pliable, it is very strong and durable.

FIGURE #1

FLOW DIAGRAM OF A TYPICAL SERIES 100 GENIE MEMBRANE SEPARATOR

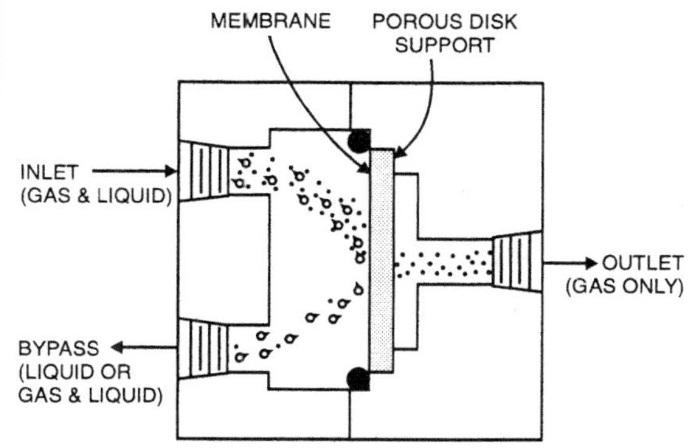

FIGURE #2

GENIE MEMBRANE PRINCIPLE OF OPERATION

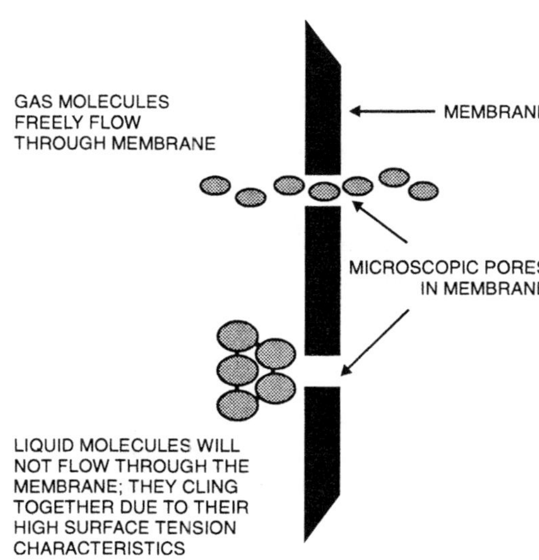

Example 3-4 continued.

GENERAL

The Series 100 Genie Membrane Separators are suitable for removing liquid and particulate in gas samples spanning a broad range of applications. Application notes with detailed information are available for many types of applications. Expert factory assistance is also readily available by phone or fax. The Genie Separators are ideally located immediately upstream of the analyzer or sample system component – such as a diaphragm gas pump – that is being protected (See Fig. #3). This location provides protection for the analyzer or gas pump, even if other sample system components were to fail. For example, failure of the following sample system components are frequently responsible for liquid damage to gas analyzers and pumps:

FIGURE #3

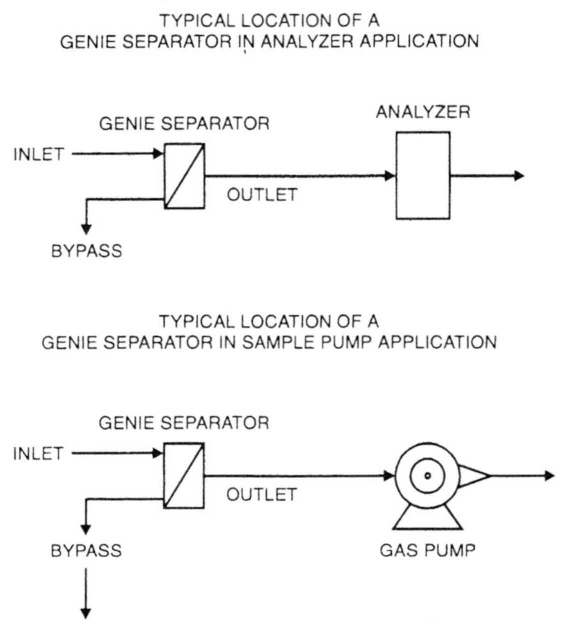

A SEAL LEG OR LIQUID PUMP CAN DRAIN LIQUIDS CONTINUOUSLY FROM THE BYPASS PORT WHEN THE SAMPLE PRESSURE IS SUB-ATMOSPHERIC

a) <u>Sample line heat tracing failure</u>
Sample components, which would otherwise remain in a vapor state when heated, may condense if the sample line is allowed to cool.

b) <u>Sample cooler failure</u>
Sample components, which are intended to be removed by condensation/draining, may pass through the failed cooler in vapor state, and then condense before entering the analyzer or pump.

c) <u>Liquid drain valve or liquid pump failure</u>
If liquid is not pumped or drained out of the system, carryover into the analyzer or pump will likely occur.

A second reason for locating the Genie Separator immediately upstream of the analyzer or pump is that the sample pressure is apt to be the lowest at this point. It is best to operate the Genie Separator in a manner that minimizes the differential pressure across the membrane. The factors which have the greatest influence on membrane differential pressure are gas flow rate and membrane area.

SELECTING THE APPROPRIATE TYPE OF MEMBRANE

There are two basic types of membranes for the Series 100 Genie Separators. The "Standard" membrane is suitable for separation of most liquids from gases. The "Hi-Flow" membrane is best suited for the separation of water and other high surface-tension liquids from gases.

To select a membrane type, first determine the type of liquid to be separated from the gas sample. If the liquid is water or a solution consisting primarily of water, sulfuric acid, caustic solution, glycols, oily liquids, or any other high-surface-tension liquids, then the "Hi-Flow" type membrane is recommended. The "Standard" type membrane will also separate the liquids referenced above, but will require greater membrane differential pressure than required for the "Hi-Flow" type membrane. For all other applications the "Standard" type membrane is recommended.

How to Select the Right Genie Model and Membrane Type

1) Determine the following application requirements:

A. Gas flow rate to the analyzer excluding the bypass flow.

B. Type of liquid to be separated and amount normally present in the sample.

C. Gas sample supply pressure at Genie inlet.

2) Select a Genie model and type membrane which meets the application requirements. Note that the membrane differential pressure for the model/membrane type selected must be lower than the available gas sample supply pressure.

Both types of membranes are excellent for filtration of submicron-sized particles. The "Hi-Flow" type membrane is normally selected for particulate filtration applications, due to its greater flow rate capacity.

Example 3-5. Nafion® Gas Sample Dryers
(Courtesy of Perma Pure® Inc.)

Gaseous sample analysis requires a clean, dry sample to produce accurate and reliable results. Filters will remove entrained particulates, but removing moisture from a sample without affecting the analytes is not so simple. Refrigeration dryers will remove water by condensation and collection, but water-soluble analyte gases will dissolve in the condensate and be removed as well. Desiccant dryers will absorb water, but also typically absorb analyte gases. Permeation dryers will preferentially remove water, but also remove some analytes.

Nafion gas dryers are a superior alternative. They continuously and highly selectively remove water from gaseous samples without affecting analytes, reaching final sample dew points as low as –45°C. Nafion gas dryers have no moving parts and require no maintenance.

Nafion® – the Perfect Drying Material

Nafion® is a copolymer of tetrafluoroethylene (Teflon®) and perfluoro-3,6-dioxa-4-methyl-7-octene-sulfonic acid.

Like Teflon®, Nafion® is highly resistant to chemical attack, but the presence of its exposed sulfonic acid groups confers unusual properties to it. The sulfonic acid has a very high water-of-hydration, absorbing up to 13 molecules of water for every sulfonic acid group in the polymer. Consequently, Nafion® will absorb 22% by weight of water.

Unlike membrane permeation, which is a relatively slow diffusion process, water absorption as water-of-hydration is very fast, proceeding as a first order kinetic reaction. Since it is a specific chemical reaction with water, it does not affect analyte gases.

Principle of Drying

When gas containing water vapor passes through Nafion® tubing, the water is absorbed by and moves through the walls of the tubing, evaporating into the surrounding air in a process called perevaporation. The remaining components in the gas are unaffected.

The reaction is driven by the humidity gradient until equilibrium is reached. If a dry purge gas flows over the exterior surface of the Nafion tubing, water vapor will be continuously extracted from the sample gas stream inside the tubing until the sample humidity matches that of the purge gas.

Since water is absorbed into the tubing from the vapor phase and evaporates from the exterior of the tubing into the vapor phase, there is no net change of free energy. No external energy source is needed to drive the reaction, other than a supply of dry purge gas. Perma Pure® gas dryers enclose one or more strands of Nafion® tubing in a shell with fittings to supply a countercurrent purge gas flow. Water vapor from the sample stream is carried away by the purge gas, and the sample is dried without being exposed to condensate or any material that might absorb analyte gases. A purge gas flow rate twice that of the sample flow rate is sufficient to achieve full drying performance.

Tips, Traps, Hints and Hazards

The following cursory treatment of Tips, Traps, Hints and Hazards provides guidelines and points-to-ponder when designing a sample conditioning system. The points addressed should be *flags*. If you are addressing the major subject (sample phase considerations) and the topic is gaseous samples then a major concern should be sample temperature (too hot or too cold) and this phenomena (physical characteristic) is related to dew point. A similar format is followed throughout this section.

The section "Sample Conditioning as a System" is an effort to bring all elements of this chapter together, emphasizing that *sample conditioning* must be a *system*! This is a summary of many concepts addressed in this chapter but here they are addressed as a *system*. Phenomena are addressed that occur as a result of several components' interaction as a *system*. Diagrams are provided for a *typical* Sample Conditioning System and the typical flows that would be found in this *system*.

Sample Phase Considerations

GASEOUS SAMPLES

Too hot and there is the possibility for polymerization of materials which are present as reactive gaseous elements in the sample stream. **Too cold** and there is certainty of condensation as well as the possibility of crystallization of gaseous elements in the sample stream. This phenomena is known as *dew point* (condensation) and *freeze point* (crystallization).

LIQUID SAMPLES

Too hot and there is the possibility for bubble formation within the liquid sample stream as well as the possibility for partial vaporization of the liquid sample stream (more vapor than bubbles). This phenomena is known as *bubble point*.

Too cold and there is the possibility of viscosity problems (hindrance of flow upon cooling) as well as solids/crystal formation. This phenomena is again known as *freeze point*.

GASEOUS SAMPLES

It is a totally erroneous conclusion that "If you condense liquids during the transport of a gaseous sample, they will vaporize while traversing the length of the heated sample transfer line."

The theory is it takes a large amount of heat to change phase from liquid to gas (latent heat of vaporization). Theory is the gas does not really spend a lot of time in the heated sample transfer line. The last theory is the molecules are relatively few and far between (that is why it is a gas and not a liquid). Thus, not much heat gets transferred to any liquid (condensate) flowing along with the gas (desired sample) in this heated sample transfer line.

A final point: if the condensate in this gaseous sample does vaporize during the transport of the sample, you will have spikes in composition as droplets of minor sample condensate (heavy, high-boiling material) vaporize and are carried within the major gaseous sample (light, low-boiling material).

ALL SAMPLES

All samples extracted for transport to a sample conditioning system should be representative, homogenous (uniform composition), and single-phase. No sample should ever be taken directly to an analyzer. At the very least, a particulate guard (depth) filter or a condensate guard filter/alarm device should be in the analyzer sample line to protect the analyzer sample cell and/or detector.

Sample Probe Considerations

GASEOUS SAMPLES

The most basic sample probe may be a simple tube (quill) or, if there are entrained (physically swept along) materials (particulate or condensate), you may use an inertial separation probe (see Figures 3-19 through 3-26).

The inertial separation probe has the sample entry end capped and several holes drilled in the downstream side of the sample extraction tube. This allows particulate and droplets to be swept over the tube and (theoretically!) only gas enters the downstream holes.

Sample probe entry should take place via side entry into a vertical run of pipe or via top entry into a horizontal run of pipe. Top entry should assure condensed liquids are on the bottom of the pipe and away from the probe.

Gaseous Samples

Sample probe entry should take place via side entry into a vertical run of pipe or via top entry into a horizontal run of pipe.

In high-particulate-loading sample probes you may wish to install a timed, heated sample probe blowback-to-process. The pressure and frequency of this blowback must be determined from experience but initial guidelines would be 75 psig pressure and 30 sec of blowback every 15 min (duration and frequency). Care must be taken to preclude condensation forming in the blowback air due to the Joule-Thompson effect (volume expansion/pressure drop cooling). This can usually be eliminated by (1) providing HT&I line for the high pressure (HP) air source; (2) wrapping the HP air source line around a local heated duct or vessel; or (3) coiling a calculated length of HP air line inside of the heated probe enclosure.

In extremely high-particulate-loading sample probes, you may wish to have a heated high-pressure air reservoir exhausting through a quick-acting, full-port, electric-pilot pneumatic-actuated ball valve for probe blowback. This equipment specification provides the following benefits:

- Precludes Joule-Thompson (volume-expansion pressure-drop) cooling and condensation of moisture from blowback air.
- Provides an initial high-volume high-pressure high-temperature air blast to the probe to dislodge particulate loading and blow it back into the process.
- Minimizes blowback air pressure and volume limitations caused by long air supply line runs.

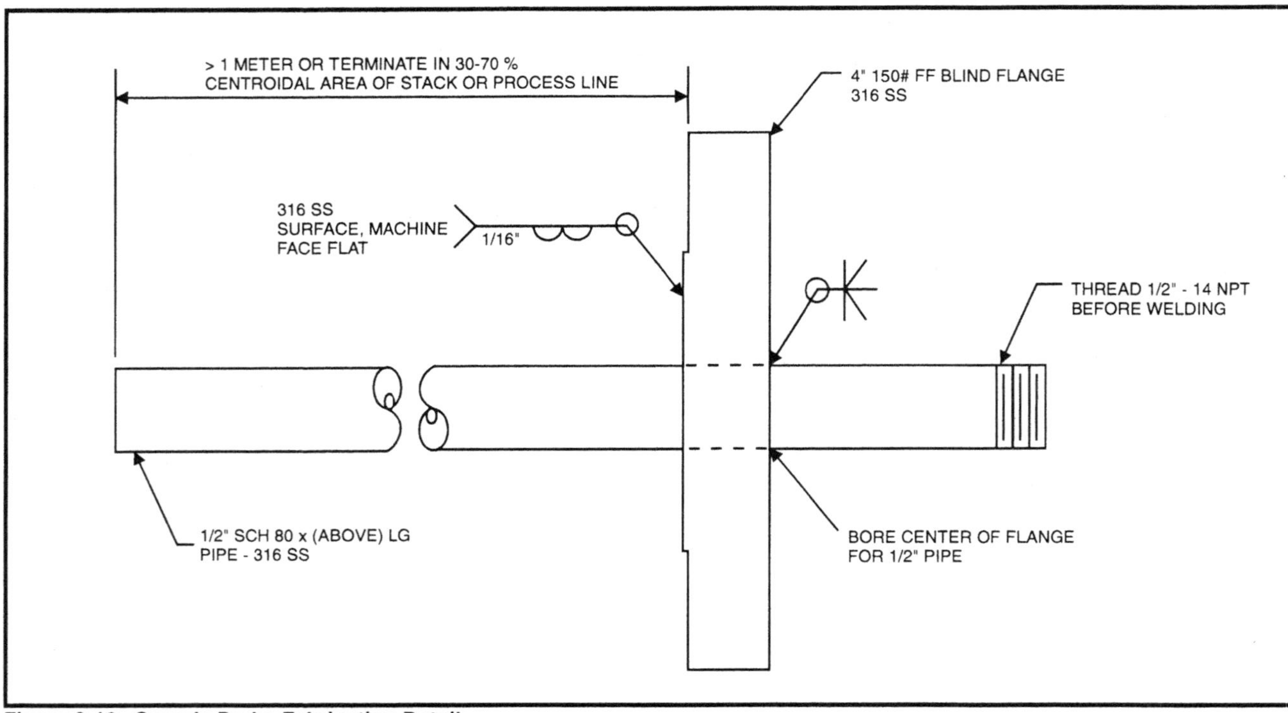

Figure 3-19. Sample Probe Fabrication Detail

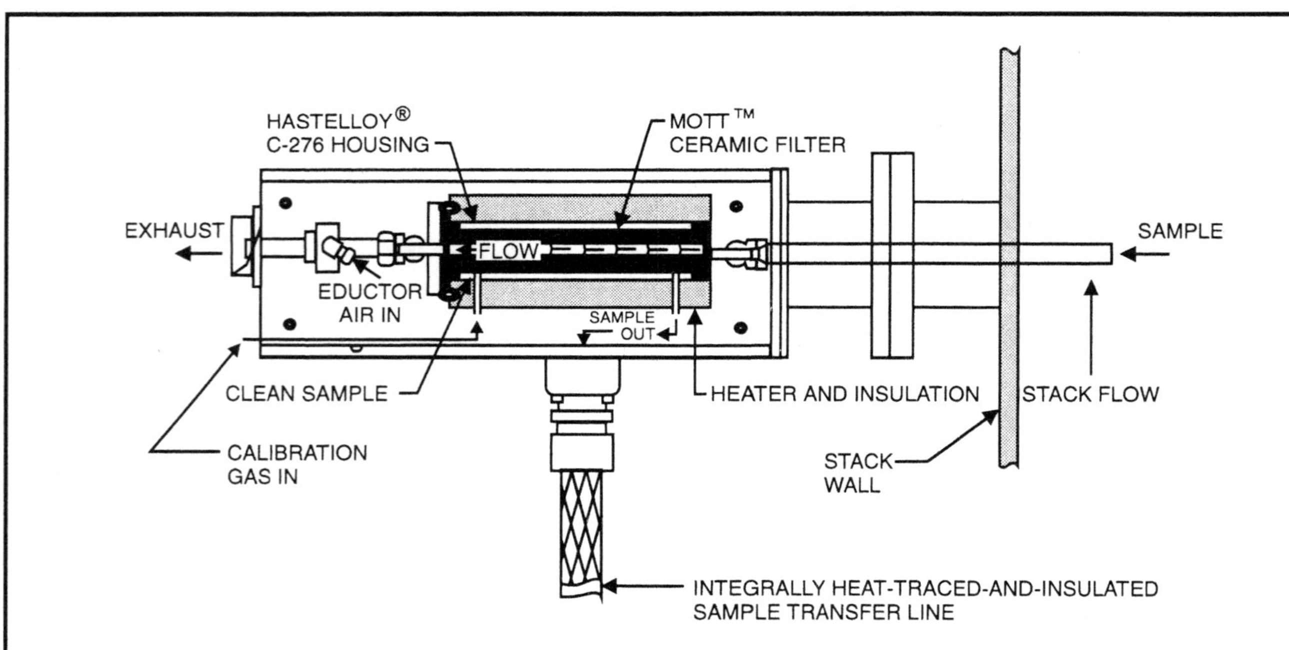

Figure 3-20. Sectional View of an Inertial Filter Probe

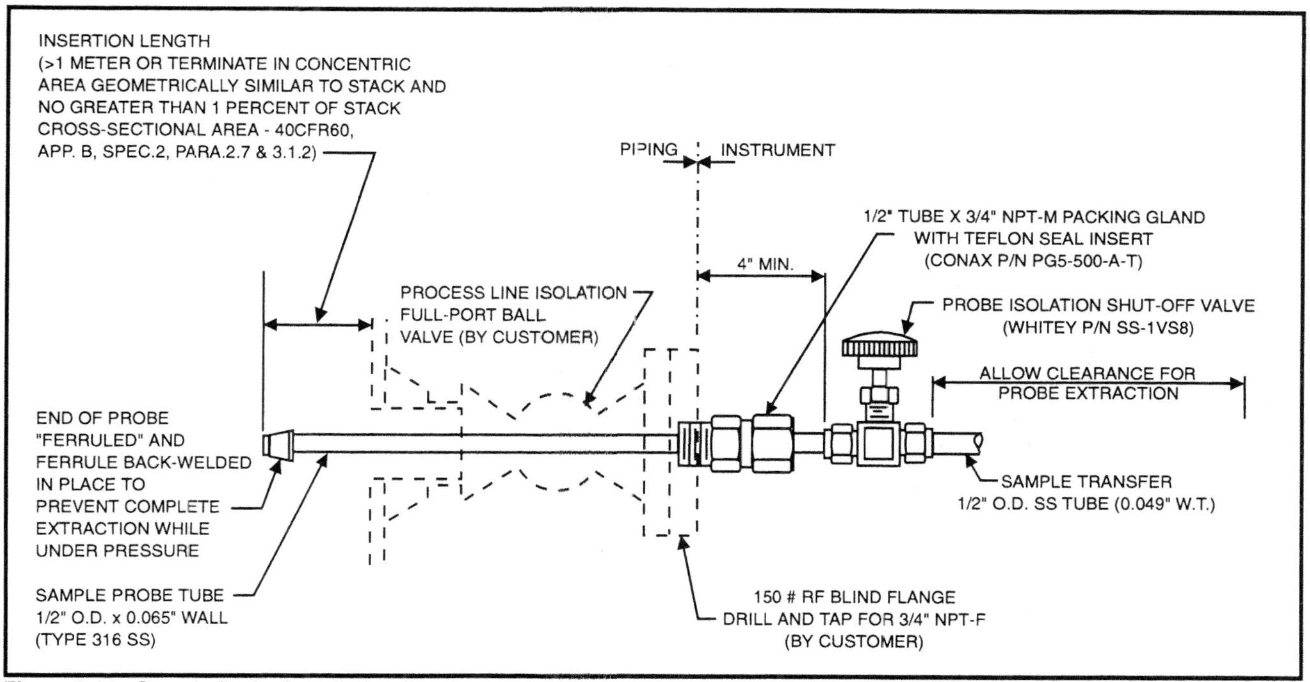

Figure 3-21. Sample Probe Installation Detail

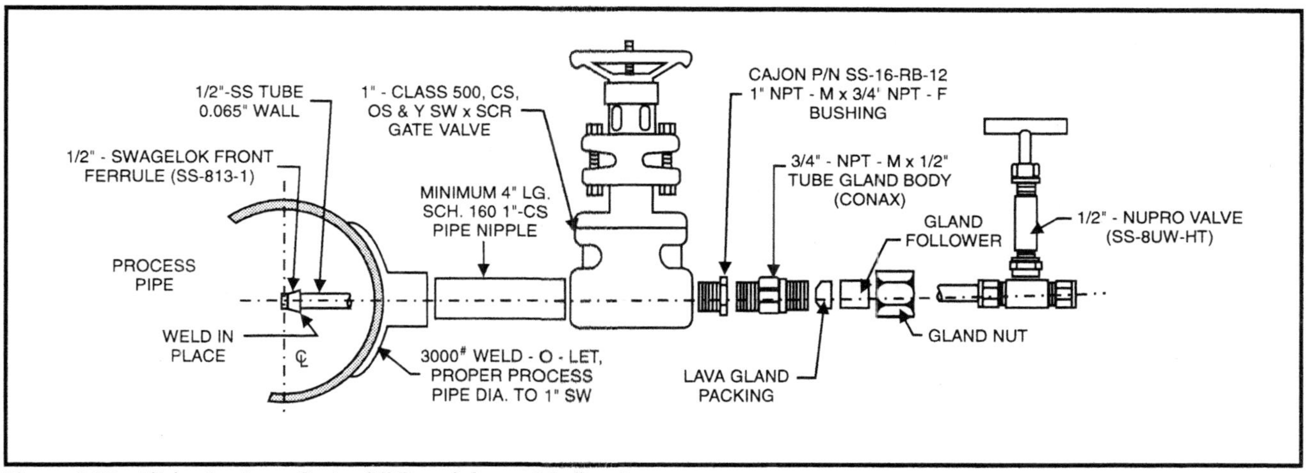

Figure 3-22. 1/2-in. Analyzer Probe (HI Temp) Assembly Detail

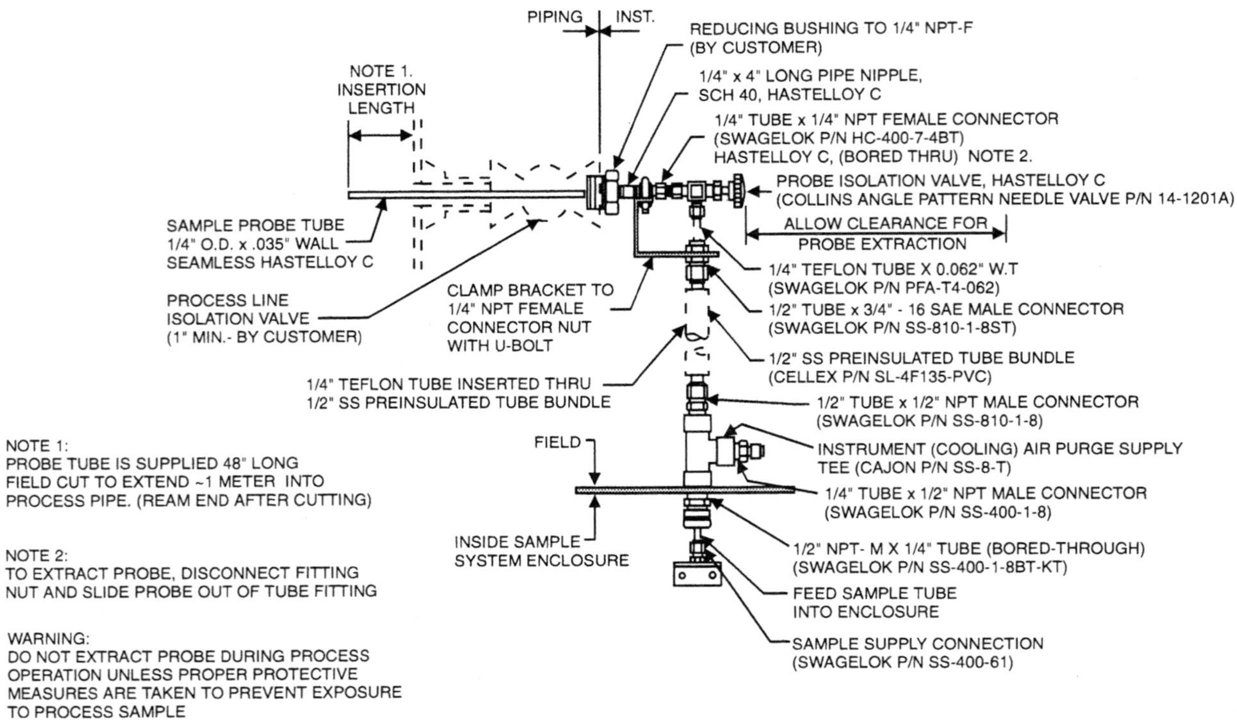

Figure 3-23. Process Sample Probe Design Detail

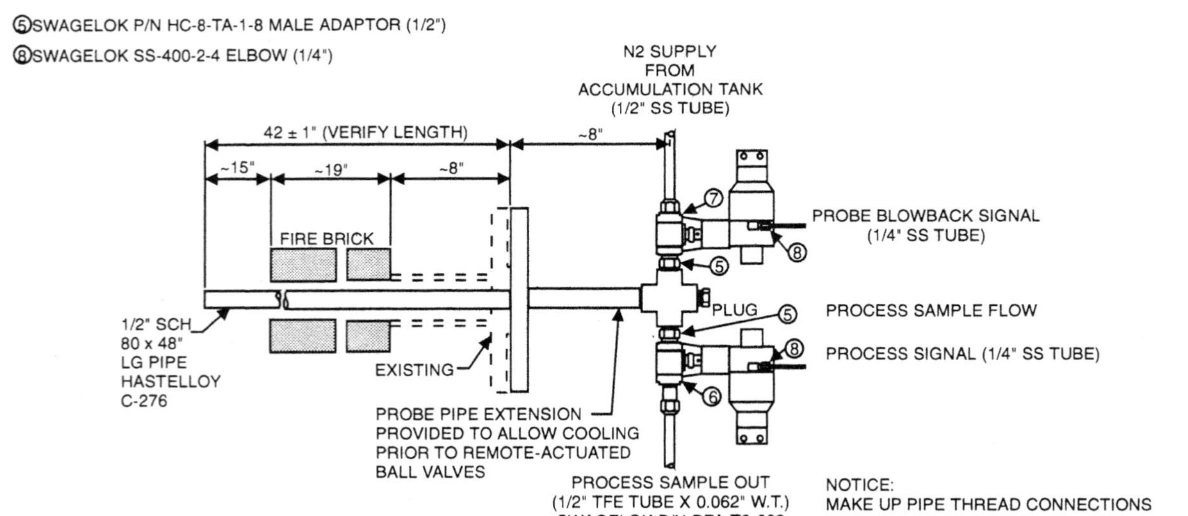

PROCESS SAMPLE DESIGN CONDITIONS
TAG NUMBER : (LATER) TEMPERATURE : 1832°F PRESSURE : 5" W.C. (VACUUM) PHASE : GAS (SATURATED)

SAMPLE PROBE DESIGN SPECIFICATIONS
⑥PROCESS VALVE : WHITEY HC-63TS8-133-NC (W/VITON O-RINGS ON ACTUATOR) ⑦BLOWBACK VALVE: WHITEY HC-63TS8-133-NO MAXIMUM TEMPERATURE : 450°F MAXIMUM PRESSURE : 100 PSIG @450°F

Figure 3-24. Process Sample Design Conditions and Sample Probe Design Detail

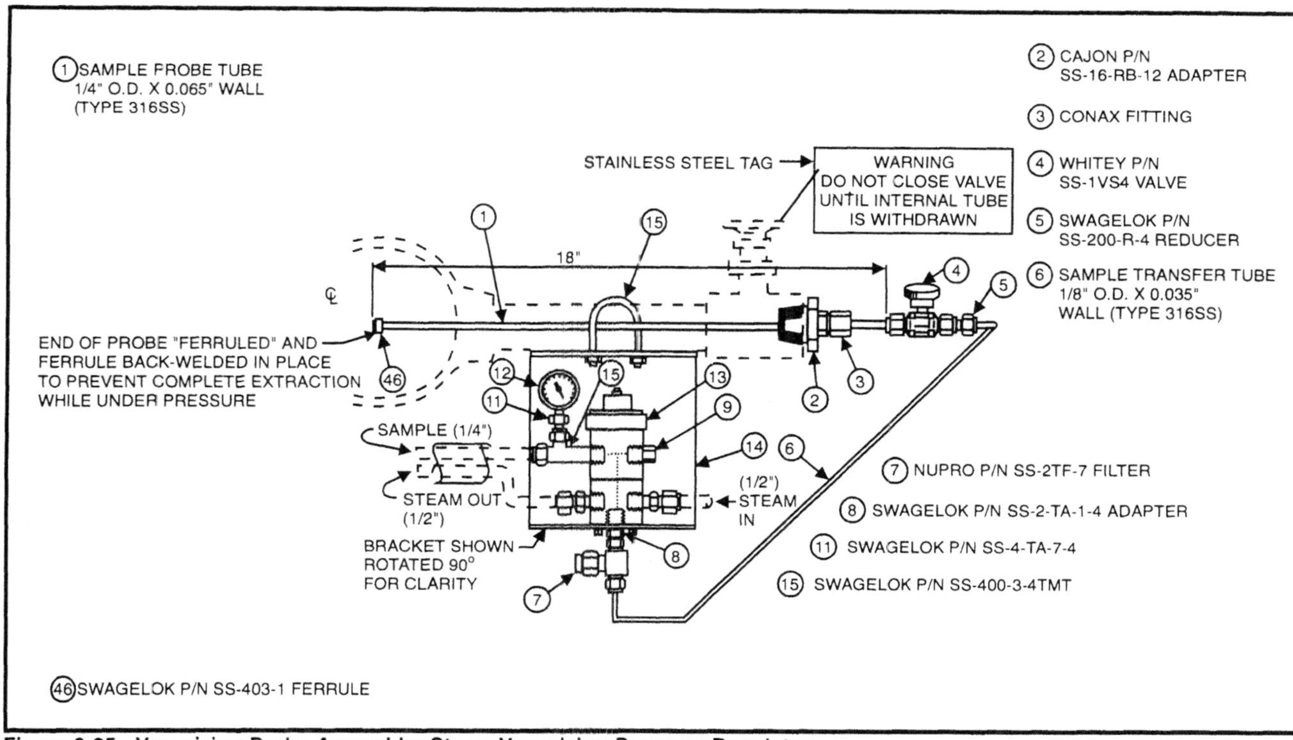

Figure 3-25. Vaporizing Probe Assembly: Steam Vaporizing Pressure Regulator

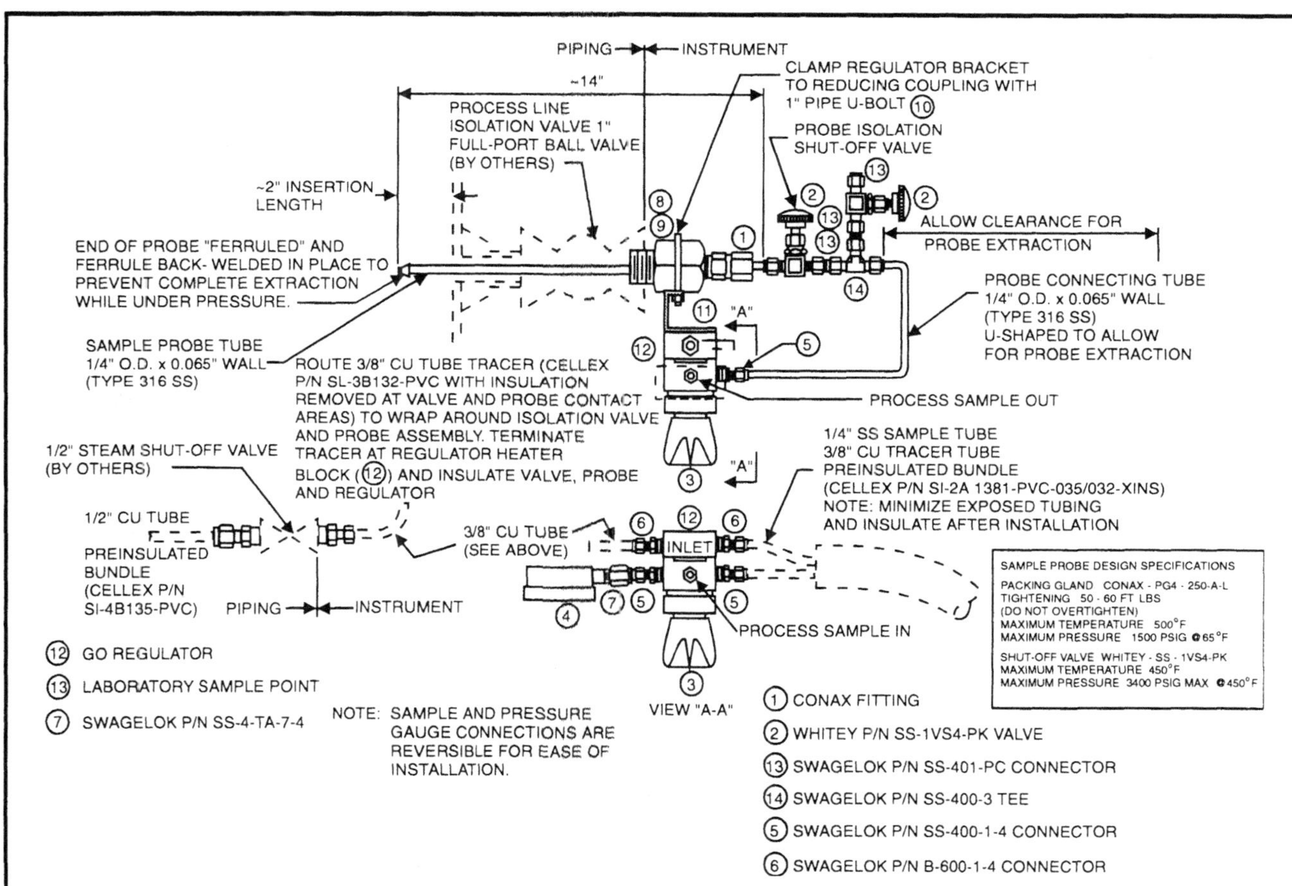

Figure 3-26. Sample Probe Design and Installation Detail

LIQUID SAMPLES

The most basic sample probe may be a simple tube (quill) or, if there are entrained (physically swept along) materials (particulate or condensate), you may use an inertial separation probe.

The inertial separation probe has the sample entry end capped and several holes drilled in the downstream side of the sample extraction tube. This allows particulate and immiscible droplets to be swept over the tube and only process liquid enters the downstream holes.

Liquid Samples

Sample probe entry should take place via side entry into either a vertical run of pipe or into a horizontal run of pipe.

Sample probe entry should take place via side entry into either a vertical run of pipe or into a horizontal run of pipe. Bottom entry should be avoided since particulate tends to collect around the probe and make probe removal difficult or impossible during operating conditions.

To preclude the entry of bubbles into a liquid sample stream you may insert the sample probe into a "stilling well" which is a reservoir created as a low point in the process system. Sample enters this "well" and the bubbles rise to the surface (and must be properly vented) and are relatively large in volume. Due to this stilling well the sample outflow is not representative of the process at that moment. This "sample lag" is the tradeoff for eliminating bubbles in the sample to the analyzer.

ALL SAMPLES

For particularly hazardous or toxic sample extraction, it is recommended to not use a retractable sample probe but to use a flanged-or-welded-installation thermowell, properly calculated and installed, with its tip drilled through or cut-off for sample transfer.

Sample Transfer Line Considerations

ALL SAMPLE LINES

Sample transfer line field run should have a continuous downward slope of at least 0.5 in/linear foot of line from sample probe to sample conditioning system cooler entry point. This will ensure no pockets, coils or low spots to collect condensed liquids or deposited solids.

ALL INTEGRALLY-HEAT-TRACED-AND-INSULATED (HT&I) SAMPLE LINES

Sample transfer lines which are integrally HT&I should never be coiled upon themselves if too long. This creates two problems: (1) condensed liquids/deposited solids entrapment; and (2) ***fire-hazard****.*

Since these HT&I lines are designed to be referenced to ambient temperature, when they are coiled upon themselves the heat buildup within the adjacent coiled surfaces is such that autoignition of the protective plastic sleeve will often result.

The entry of HT&I sample transfer line through a metal enclosure or cabinet wall should take place via a feedthrough sealed at a narrow ring only to the HT&I line's protective outer jacket. You should never terminate the HT&I line in a bulkhead fitting for transfer through a metal enclosure or cabinet wall. The use of a bulkhead fitting for gaseous sample transfer makes a radiator (heat sink) of the metal enclosure or cabinet wall. Since a heated gas has very little "sensible" heat (heat contained within its own molecules) the use of a bulkhead feed-through will almost always result in a liquid condensation problem or a combination liquid/solids accumulation which will almost always result in sample line blockage.

The routing of HT&I sample transfer line through a brick or concrete block wall should take place via a feedthrough sealed at a narrow ring only to the HT&I line's protective outer jacket. The reason for this arrangement is again that this type of line is designed to be referenced to ambient temperature. Lack of free air circulation to this type of line's outer protective jacket will usually result in HT&I

sample transfer line meltdown at the feedthrough point and occasionally will result in a **fire hazard** from **autoignition** of the HT&I line's outer protective jacket.

Sample Conditioning as a System

All gaseous analyzer samples must be:

- Single phase (no liquid or solids)
- Homogenous (sample extracted is uniform quality throughout)
- Representative (sample extracted is same as total stream composition)

Also, it is a foregone conclusion that the following will always occur:

Condensate + particulate = slurry

No single-phase homogenous sample can be extracted from a slurry. No filter can operate when presented with a slurry as a sample. Therefore:

- Gaseous samples must first have particulate removed.

  or

- Samples must be kept dry (water or potential liquid in gaseous form must be kept above dew point) until particulate is removed.

  and

- Particulate removal must be such that it does not clog the filter media.

  or

- Adequate filter media area must be provided to preclude blocking of flow within a reasonable maintenance interval.

If a gaseous sample must have its pressure dropped "by a large amount" and if this sample has a broad range of boiling point components, then a *vaporizing pressure regulator* must be used. This regulator will replace the sample's "latent heat of vaporization" from an added heat source (steam or electric) to create "flash evaporation" which will preclude condensate or liquid droplet formation due to the Joule-Thompson effect of a pressurized sample cooling upon volume expansion (pressure reduction).

After particulate is removed, a sample must have water or light liquids stay gaseous until "crash cooling" (cooling over a very short flow path). "Crash cooling" is the opposite of "flash evaporation." Both processes must occur over a very short flow path.

Condensate from sample cooling must have minimum surface contact with flowing sample gas (to prevent re-evaporation or "saturation" of the "dry" gas stream).

The condensate creator (cooler) must be the coldest spot in the sample conditioning system, in the analyzer and in the sample exhaust system. If this is **not** the case, additional condensation will occur (and havoc will be wrought with your gaseous-measurement analyzer). In defining the "coldest spot" you must take into account Joule-Thompson (gaseous volume expansion cooling). This can occur downstream of a cooler, especially at a flow measurement device (rotameter) or at a flow control valve. Once a gaseous sample has been cooled to its desired dewpoint (usually 34-36°F for water-based gaseous samples), heat may have to be added before or at flow-control or flow-measurement devices. Also consider where your sample vent line terminates! If it is outside and below 32°F you will build an ice block which will stop your sample flow. Cellex Manufacturing makes a nifty discrete heated sample vent line termination (PN: SE-3C1-18-0-PVC-062) self-limiting with 25 ft. plug-in cord.

Now we have a particulate-free (clean, no solids) dewpoint-controlled (dry, no liquids) gaseous sample. How can you be certain that it is homogenous? This consideration starts with the sample probe which (theoretically) extracts a uniform and representative (homogenous) gaseous sample from the process stream of interest.

Take a breather from theory now and consider the components that constitute a sample conditioning system (and I emphasize **system**) (see Figure 3-27). If all components do not work together, the general observation will be that "the analyzer does not work!" No one will make the observation that "The analyzer will properly zero and span; it is the sample conditioning system that is malfunctioning!"

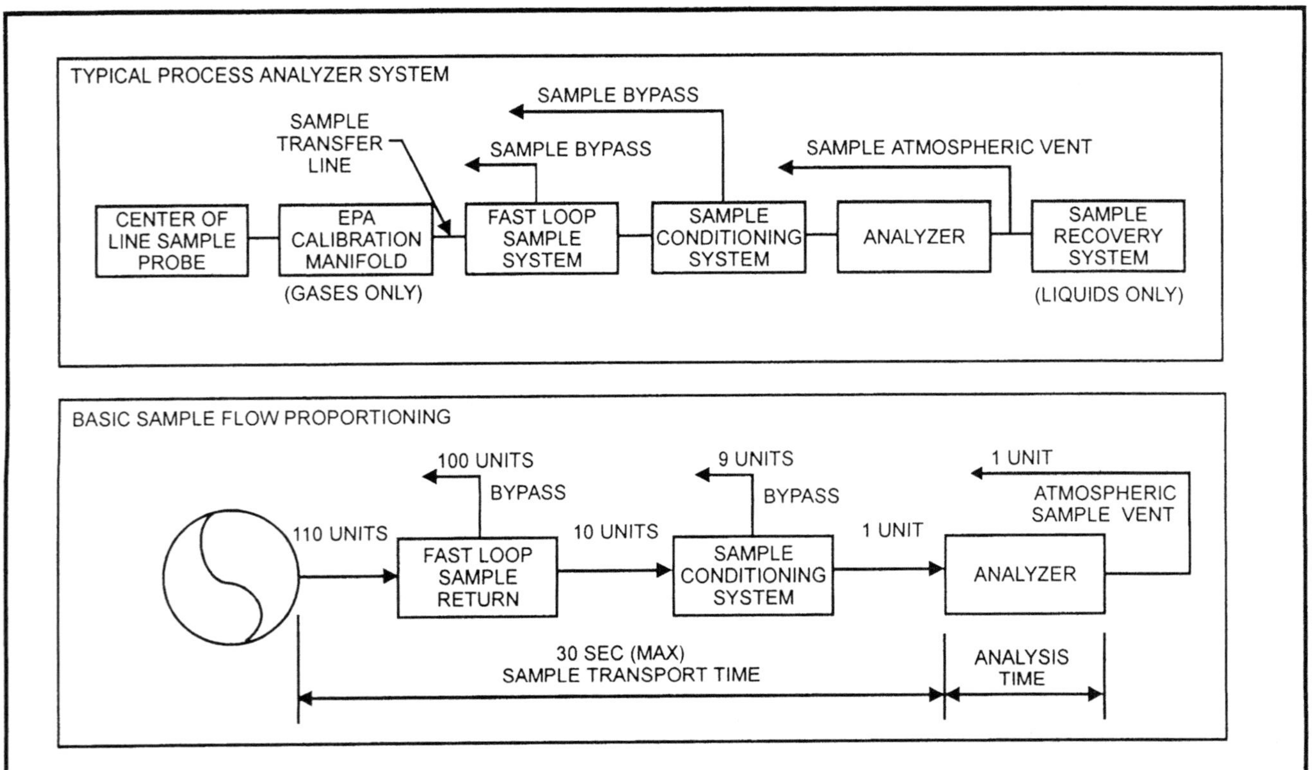

Figure 3-27. Sample Conditioning: Typical Process Analyzer System and Basic Sample Flow Proportioning
(Adapted from June 1989 Control)

Let's describe several sample system components which could conceivably be included in a *typical* sample conditioning system.

First, is the sample desired in gaseous or in liquid phase? We will discuss gaseous samples at this time.

What are the **process conditions**?

- Pressure (P)
- Temperature (T)
- Flow rate (Q

- Complete stream composition

  Percent (%)
  by volume (%v)*
  by weight (%w)
  Parts per million (PPM)
  by weight (PPMw)*
  by volume (PPMv)

  * most common

PRESSURE

- Positive or negative?
- Do we need a **pressure reduction device** (and a pump for long distance sample transmission)

  or

- Do we need a **pump** for how great of a negative pressure (vacuum)?

  or

- Can we use an **aspirator** (gas-powered jet pump)?

Is the pressure negative? Is the pressure fluctuating? If both of the aforementioned are true then we need to incorporate subatmospheric pressure regulators into the sample conditioning system and run the sample conditioning system at a stable pressure less than the lowest negative pressure noted for the system. The subatmospheric pressure regulators bleed in atmosphere or a bleed gas (nitrogen) at the aspirator or pump inlet to maintain the aspirator/pump inlet at a stable pressure, thus keeping the analyzer cell at a stable (albeit negative) pressure.

TEMPERATURE

- Positive or negative?
- System sealing elastomers become a problem for **too high** of a temperature [typically above 450°F (232°C) for VITON and TEFLON (TFE)]

  or

- System sealing elastomers become a problem for **too low** of a temperature [typically below −20°F (−29°C) for both VITON and TEFLON (TFE)].

FLOW RATE

- What is the **velocity**?
- Will it bend the probe (or erode it if hard particulate is present with **high** velocity)?
- Will flow be **laminar** (stratified), **turbulent** or **transitional** (in-between)?

STREAM COMPOSITION

- What are the **maximum, normal and minimum values** of all components present in the sample?
- *Trace* does not compute unless *close* is the level of **precision** you require!
- Notice the introduction of the word *precision*!

Precision is what most process control analyzer systems attempt to provide. Precision, as an example, is all shots grouped closely together without respect to how near the bullseye or center of the target these shots may be.

Accuracy is what all continuous emissions monitoring systems (CEMS) or regulatory analyzer systems must provide. Accuracy, as an example, is all shots near or at the bullseye or center of the target. Accuracy is a measure of how far you are from an absolute (known) value. In CEMS analyses, this absolute value is usually traceable to NTIS (formerly NBS) and based upon gravimetric composition (or mass of components).

References

1. Clevett, K.J., *Process Analyzer Technology*, Wiley-Interscience (1986) pp.774 - 780.
2. Houser, E.A., *Principles of Sample Handling and Sampling Systems Design for Process Analysis*, University Microfilms International-Ann Arbor, MI, USA (1984) pp. 4 - 15.
3. Nichols, G.D., *On-Line Process Analyzers*, Wiley-Interscience, 1988, 300 pp.
4. Callahan, F.J., *SWAGELOK® Tube Filter's Manual*, SWAGELOK Co., 1993, 282 pp.
5. Cayanus, L.E., and Taggart, R.C., *On-Line Analyzer Sample Conditioning Systems: A Guide for Sizing Sample Lines and Flow Indicators*, ABB Process Analytics, 1976, 16 pp.
6. Annino, R., and Villalobos, R., *Process Gas Chromatography: Fundamentals and Applications*, Research Triangle Park, NC: ISA, 1992.

About the Author

Robert E. Sherman has more than 25 years of experience in the design and commissioning of process analyzer and continuous emission monitor systems (CEMS). He has managed process analyzer projects including specification, engineering, procurement, fabrication, acceptance testing, field commissioning, documentation, operations and maintenance training.

Mr. Sherman has been on the Technical Advisory Boards of DuPont (photometric and moisture analyzers — now a division of Ametek), Applied Automation, Inc. (gas chromatographs — now a division of Hartmann & Braun who is owned by ABB), and Precision Scientific, Inc. (physical property analyzers — now a division of Varlen). He is an Invited Speaker to Pittcon, has presented many training sessions to industry technical personnel, and has published/presented more than 30 technical articles and papers relevant to process analysis, process control and process industry management issues. He is currently authoring a monograph, "Process Sampling Technology," to be published in June 2000.

His employment has included three years with Beckman Instruments, 12 years with Amoco Oil Company, five years with Fluor Daniel, three years with Applied Automation, and six years operating his own design/ fabrication/training and consulting firm, Process Analyzer System Technologists (CTI/PAST). Bob is currently employed with Dearborn Professional Services Inc. as Manager - Advanced Solutions Group. He can be reached via Internet at roberteshermann@compuserve.com or at res@dps-hq.com.

4

Principles of Process Analyzer Calibration

Measurement is a comparison process in most cases. Absolute measurements are rare and difficult; therefore the practice of ratioing an unknown signal to that of a standard is generally used. In the case of chemical concentration measurements, the standard is prepared from first principles such as weight, volume, and pressure. Procedures must be developed for preparation of calibration samples, introduction of the material into the analyzer system, treatment of the measurement signal, and adjustment of the testing device. The purpose of these activities is to remove bias from the resulting measurement so that an accurate representation of true component concentration is produced. Defining the scope of a calibration procedure helps to validate the reliability of the measurement.

Prerequisites

Validity of Measurement

It must first be established that the method being used truly responds to the target analyte in a manner proportional to its concentration. It is also necessary to establish how other components in the sample affect the measurement. Component isolation is the only way to ensure interference-free measurement. Environmental Protection Agency (EPA) protocols are good examples of procedures for calibration and documentation. United States Pharmacopeia (USP) and American Society for Testing and Materials (ASTM) validation procedures are excellent guidelines.

Stability of the System

The short- and long-term stability of the measurement system (Figure 4-1) will dictate the method of calibration. A procedure using control samples is necessary when long-term stability of a system cannot be achieved. The limit of short-term stability is the time required to make sequential measurements on an unknown and a standard. The criteria for an on-site automated chemical concentration analyzer should be to make the system meet the customer requirements. There is no point in calibrating a system that drifts beyond the accuracy limits as soon as the calibration is completed.

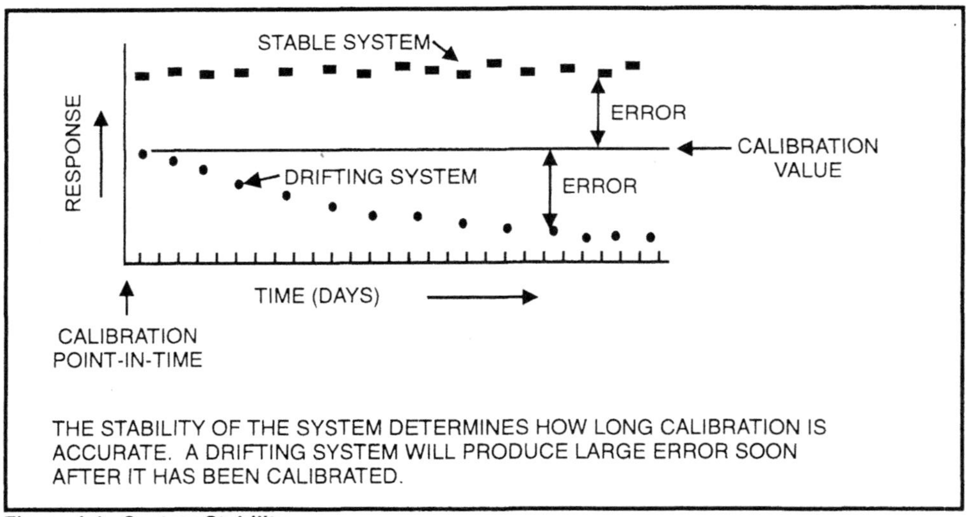

Figure 4-1. System Stability

Precision of Measurement

Precision (Figure 4-2) is defined by statistical treatment of repeated measurements on an unchanging reference material. It is obvious that precision cannot be determined on a measurement system that is "out-of-control." Control charts can be used to establish the precision of a system that has long-term stability. Care must be taken to use appropriate sampling intervals to avoid autocorrelation when applying control charts to process analyzers. The term "standard deviation (σ)" is a measure of the dispersion obtained on multiple measurements. The closer together a series of test results, the smaller the standard deviation. Control limits are usually set at three standard deviations from the mean.

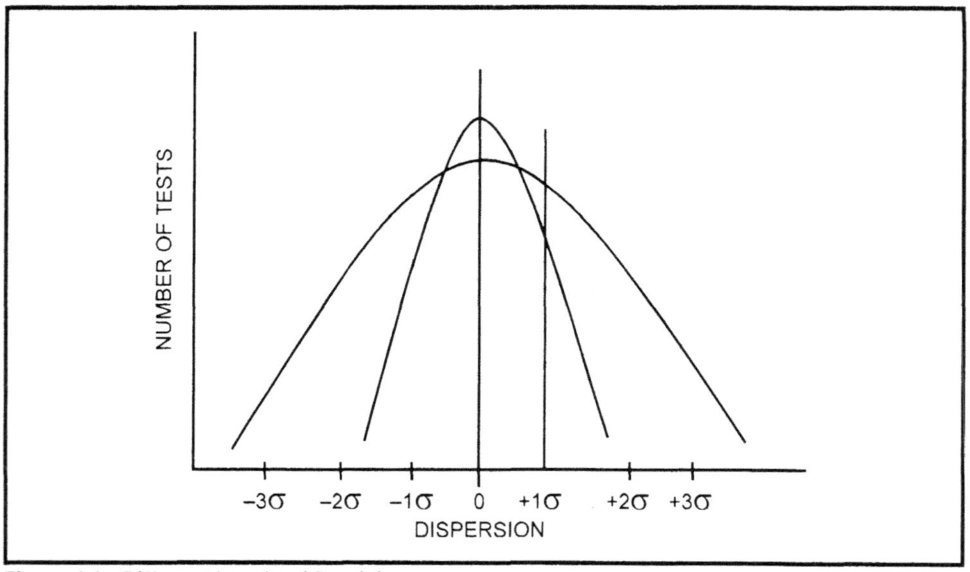

Figure 4-2. Different Levels of Precision

Capability of Measurement

Capability can be defined for a process when the mean, standard deviation, and specification limits are known. One generally defines capability for a measurement system by comparing the variance (standard deviation squared) of the measurement to the variance of the process being measured: $\sigma_m^2/\sigma_p^2 < 0.20$. This equation states that the ratio of variance between the measurement and the process must be less than 20% if the information produced can be useful for detecting changes in the process.

Another criterion is that the minimum detectable signal increment divided by $6\sigma_t$ must be less than 10%. These are only guidelines; the achievable capability must be acceptable to the customer per contract agreement.

Method of Calibration

External Standards

One way of making comparisons is to utilize measurement of the process relative to an external known material. The external standard must substituted for the process material within the analyzer. This places certain requirements on the system:

1. System response must not change during the period of running the standard.
2. Sample size must remain the same for each test.
3. Measurement of the process must be interrupted during calibration causing loss of information.
4. Cost of standard preparation and calibration adds to system maintenance cost.
5. The system is perturbed by changing from unknown to standard and back to unknown.
6. Several measurements of the standard must be made to establish stability and obtain an average value.

The quality of the results obtained using external standards is exceptionally good, considering all the possible sources of error. There are many arguments for and against using external standards. Quality and stability of standards will be discussed later.

Internal Standards

Comparisons can be made by introducing a known amount of reference component into the unknown sample. The concentration of the target analyte is calculated from the ratio of the responses of the reference and target components. These response factors must first be determined from measurements on a sample of known component concentration. In the laboratory, the internal standard is added by weight or volume and mixed with the sample material.

*An automated on-line analyzer **does not** sense whether the materials are premixed. **It is only essential that the relative amounts are constant.***

This allows the internal standard to be introduced at any appropriate time during an analysis. Several advantages are obvious:

1. System drift has less impact since the reference and target analyte are measured in close proximity.
2. Real time measurement of the process is not interrupted.
3. Neat material can be used (thus avoiding preparation of standards) and calibration can be automated.
4. The system is not perturbed by switching samples.

The relative size of the process and reference samples is critical and must not change significantly. This technique should be considered, and ways to overcome objections should be sought.

Control Samples

This technique is similar to internal standards except that a known standard containing the target analyte is introduced intermittently between unknown samples. Ratioing the unknown to the known eliminates determining a response factor. It is generally utilized when the measurement system cannot be stabilized easily. In chronography, it has been termed "deferred standard." The injection of the standard is delayed such that none of the separated peaks interferes with the peaks of the unknown. Two valves can be used, but relative sample size is critical. Alternate flow of unknown and known through a single valve eliminates sample size consideration, but risks cross-contamination. Manual preparation of standard is required, but introduction is automated. Real-time measurement is not interrupted, but time between unknown information update is lengthened. A single measurement on each of the materials alternately **may degrade the precision of both.** Care must be used when applying this technique.

A single measurement on each of the materials alternately may degrade the precision of both. Care must be used when applying this technique.

Secondary Standard

Sometimes a stable and representative primary standard is difficult or impossible to obtain. Normal practice has been to draw a sample of the process stream and analyze it in the laboratory. The process analyzer is then adjusted to match the laboratory results. This technique has been highly criticized and should only be used when there is no other choice or the absolute value of the analyzer is not critical. The lab measurement is not infallible, nor is it certain that the primary standard used by the lab is correct. One should carefully study ASTM D3764-92 ("Standard Practices for Validation of Process Stream Analyzers"). Statistical practices must be utilized to obtain the best possible calibration accuracy, no matter which method is selected.

Autocalibration

ALTERNATE INTRODUCTION OF STANDARD

A calibration standard sample is prepared, placed in a vessel, connected to the analyzer system, and introduced periodically by automatic control. The process sample is bypassed during the period of calibration. The procedure is exactly the same as the manual calibration except that a complex control scheme is required. Switching and identification add extra cost and complexity. Large amounts of standard must be prepared, and extreme care must be taken to prevent calibration shifts when the standard runs out or changes. This means verification of the calibration measurement prior to changing analyzer settings.

DEFERRED STANDARD INTRODUCTION

The deferred standard method developed by Guillemin et al. [Ref. 1] introduces the calibration sample during the unknown sample analysis cycle. It is delayed such that the components of the standard do not interfere with the components of the unknown sample. It was developed for chromatography, but can be applied to all analytical techniques. It is equivalent to running a control sample with each analysis where the unknown is determined by ratioing its response to the response of the standard. This technique is necessary when long-term stability is not achieved with the measurement system. It requires a significant amount of standard preparation and cost. Calculations are included in the operational program, thus eliminating the external controls.

INTERNAL REFERENCE INJECTION [REF. 2]

This technique, like the deferred standard, introduces a reference material during the unknown analysis cycle. The reference material is chosen to meet the detector requirements, and the quantity is set to produce a response of the same magnitude as the unknown component to be measured. A pure material or naturally occurring mixture (e.g., air) is used as the reference in order to avoid preparation of large quantities of standards. Once the system is in operation and stabilized, a calibration sample is utilized to determine the relative volumes of the reference and unknown samples as well as relative response factors of components measured. The system performance is then evaluated relative to the repeated response to a fixed quantity of reference. Control charts can be prepared from the reference sample data. Relative sample volumes must remain constant for calibration to remain accurate.

Calibration Sample Materials

Methods of Preparation

The term "calibration sample" implies absolute knowledge of component concentration based on the best practices, materials, and equipment used to prepare it from first principles such as weight, volume, pressure, or flow. Statistical treatment of multiple measurements will produce an average value with confidence limits. It is a good practice to prepare several calibration samples (standards) and test them for agreement. If four out of five agree and one does not, there is a good chance that the outlier is inaccurate. **Real confidence can be gained by having different persons prepare standards for comparison.** It is even better if several standards are prepared by different methods and agree. One must weigh the cost of preparation against the degree of confidence desired.

The term "reference sample" implies constant composition but does not claim knowledge of accurate component concentration. Such materials are utilized to determine precision or test instrument operation. They are not intended to set analyzer response for calibration purposes, but may be used to detect changes in performance. Such samples may be drawn from a process or mixed in a nonquantitative manner.

Procedures for preparation by weighing or volumetric combination of liquids are common knowledge. Weighing of gases requires special equipment, and mixing of gases by metered flow requires care and control. Preparation of gas standards by partial pressure requires inclusion of compressibility factors to obtain correct calculations.

Sources of Materials

Pure materials are best for preparing standards, but impure chemicals generally are all that can be obtained. It is important to procure materials that are issued with a certificate of analysis or to test impurity levels before using. Corrections must be applied to the concentration calculations. Unfortunately, this doubles the total error. Purchase of certified prepared standards is no guarantee that the actual concentration is the same as the stated value on the label. One can only compare several standards to see if they are consistent.

Stability of Samples

Changes in component concentration with time may appear as analyzer drift, indicating instrument instability. When such behavior is observed, a freshly prepared standard should be tested to differentiate between analyzer and reference sample stability. Shelf life of standards must be established and indicated on the container label.

Matrix Interferences

Other components in the sample may interfere with component measurement. The calibration sample should contain all components present in the unknown sample as near as possible. This is another argument for component isolation, since measurement of the target analyte in the unknown and in the calibration sample are both subject to matrix interference.

Methods of Certifying Calibration Samples

The sample must be prepared with materials and equipment which are certified correct. Multiple analyses statistically averaged will help to ensure correctness of calibration samples. Analysis by different and independent methods will also support certification.

Sample Introduction Techniques

Substitution for Process

The calibration sample can be introduced in place of the process material by manual or automatic means. The frequency of introduction should be dictated by the stability of the analyzer system as determined by control charts. A major concern is how much of the analyzer system should be calibrated. Experience has shown that most systems are designed to provide calibration sample introduction at the analyzer (Figure 4-3). This completely ignores calibration of the sample preparation system. EPA protocol for continuous emission measurement systems dictates that the calibration sample must be introduced at the point of sample extraction (Figure 4-4).

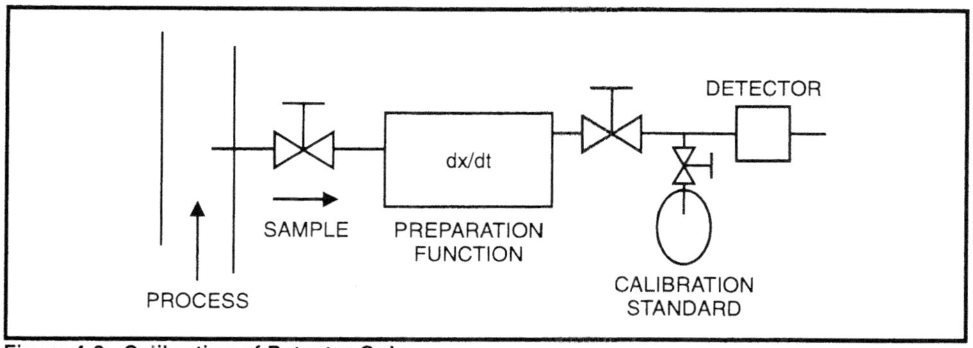

Figure 4-3. Calibration of Detector Only

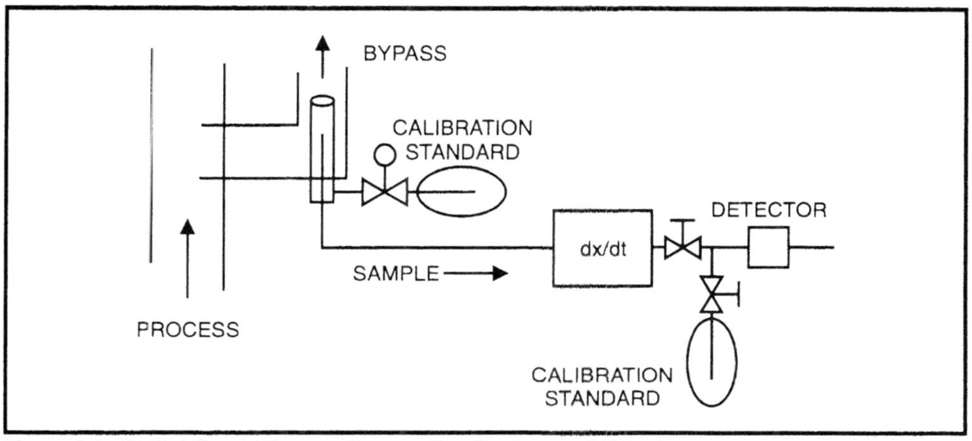

Figure 4-4. Calibration of Sample Preparation System and Detector

Provision is also made for regularly scheduled calibration and thorough recordkeeping. The government had to regulate procedures because industry chose not to. Maintenance practices and documentation for nonregulated process analyzers are now being altered because of the need to meet International Standards Organization (ISO 9000) requirements. The practical guidelines being developed by industry are several years late and will only mimic governmental and international dictates. Another problem is that the sample preparation and measurement functions have been somewhat confused. Every operation and manipulation that conditions the sample should be isolated from the measurement device. One can then calibrate the two systems and determine measurement error and sampling error. The capability of a process is usually determined from the "total" standard deviation, which is the sum of process, measurement, and sampling standard deviations:

$$\sigma_t = \sigma_p + \sigma_m + \sigma_s$$

If only the measurement device is calibrated, the variation due to the sample preparation system is buried in the process. We have no way to measure it and therefore no chance of improving the sample preparation performance. A good practice would be to introduce standard sample at each point between major components of the sample preparation system as well as just prior to the measurement device and the sample extraction point. This would provide diagnostic capability for maintenance personnel.

Real-Time Calibration

Introduction of an internal standard into the process sample provides real-time evaluation of analyzer performance. This practice can be applied to continuous or discrete sampling. Several different methods have been developed to implement this technique. The procedure shown in Figure 4-5 introduces a fixed quantity of sample and a fixed quantity of reference into the analyzer utilizing valves. Both materials are processed by the analyzer sample preparation function and measured by the detector. Ratioing the known and the unknown response allows the unknown to be determined. This procedure is called *real time* since it is performed each analyzer cycle. The second reference monitors only the detector.

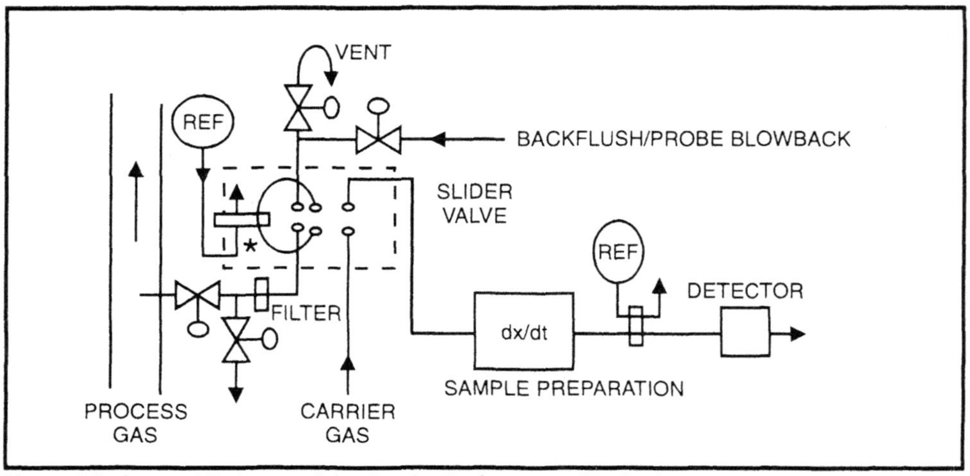

Figure 4-5. Internal Reference Injection for Performance Monitoring

Implementation of Gas Standard Introduction

Methods for introducing a known composition material into an analyzer include:

1. Direct flow of blended standard.
2. Discrete sample reference injection.
3. Permeation membrane generated standard switched into analyzer periodically.
4. High-boiling-point liquids or solids dissolved in solvent and totally vaporized POSSUM (**P**roducer **O**f **S**tandard **S**amples for **U**se by **M**anufacturing) — a device which forces the standard solution into a vaporizing chamber and is specifically designed for field application.
5. Controlled blending of carrier and analyte.
6. Controlled dilution of concentrated analyte.

Implementation of Liquid Standard Introduction

Methods for introducing known liquid composition material are primarily limited to:

1. Direct flow of mixed standard.
2. Discrete sample reference injection.

Calibration Records

Application requirements will dictate recordkeeping in most cases. Analyzers used in governmentally regulated applications such as EPA (40CFR) state or local permits and many others have well-defined recordkeeping protocol. ISO 9000 is having a significant impact on company management's attitude toward recordkeeping as well as regulatory agency standards. Adequate calibration practices, documented procedures, and thorough records are finally being recognized as integral to total quality management and as essential to customer satisfaction. The paradigm that service functions are "overhead" and that cost of funding should be minimized has not been eradicated, however. Company guidelines and individual preference still control nonregulated applications, so the analyzer specialist has a responsibility to implement good practices and play a leadership role.

Frequency of calibration can be set in several ways. Past practice often has been to calibrate frequently on startup and then to decrease frequency until the performance of the system indicates need. Organizations utilizing preventive maintenance programs generally calibrate on a time-based schedule. Control-chart-regulated calibration schedules may be more appropriate, but most analyzer systems are not routinely control-charted. Conventional practices for calibration require significant time and labor to collect chart data. The concept is supported in theory, but there is a lack of commitment in terms of personnel and money. This is changing very slowly! Customers are not happy with calibration programs that are based on demand by complaint. They may now accept time-scheduled calibration, but as they grow more knowledgeable, control charts will become a way of life.

Cross-testing between field and laboratory devices is one way to validate analytical results. Unfortunately, the perception in the past has been that the lab is always correct. Documentation has shown this to be untrue, and it has been recommended in ASTM D3764-92 that, upon disagreement, both systems need to be evaluated. This may be a victory for the people in the field, but most likely this practice will not be a reality for some time to come. When control charts are routinely maintained on all analytical systems and a sound procedure for finding the root cause is utilized, we may then take appropriate action without seeking blame.

Type of Records

CURRENT ANALYZER SETTINGS/READINGS

These need to be documented on a scheduled basis, and some form of security to prevent unauthorized changes must be provided.

CHANGES OF ANALYZER SETTINGS

These should not be made without good cause in order to prevent overadjustment, and they must be documented.

MAINTENANCE SERVICE PERFORMED

Records of diagnosis and apparent cause must be noted along with all action taken. A record of what was done to the system is needed in case it causes other problems.

CAUSES OF CHANGE AND REPAIR

When the system is returned to service, conclusions regarding the real cause and repairs made should be documented. The real "out-of-service" and "actual maintenance" times should recorded for performance evaluation.

Control Charts

CONTROL CHARTING

Manual or automated methods for preparing control charts may be utilized. Cost and personnel requirements are less for automated real-time charting. One must be very careful when using control charts to ensure appropriate practices and to prevent autocorrelation [Ref 3]. Internal reference injection is a technique for producing real-time chart data [Ref 4].

USE OF CHARTS

Plots of average response versus time obtained from a stable calibration or reference sample provide information on system stability and precision. Control limits are set at three times the standard deviation. As long as the response follows certain rules, the system is said to be "in control." The capability of the system is related to the magnitude of the standard deviation. Control charts are utilized to make decisions on changing system calibration and performance acceptability [Ref 3].

ACTION ON "OUT-OF-CONTROL" POINTS

When the system response exceeds the control limits or indicates trends or runs, cause must be found. Action must then be taken to correct the problem. It is essential to understand the sample preparation and measurement processes to be able to find the cause of the deviation from a controlled state. Skill and knowledge gained from proper training are needed in order to take appropriate action.

ALARMING CONTROL LIMITS

Manual preparation of control charts demands involvement by maintenance personnel and, therefore, awareness of when a point goes out of control. The major problem with automating control chart preparation is the loss of involvement and awareness. It becomes necessary to alarm an "out-of-control" point with an automated system so that someone has to respond. The alarm should be presented to both operating and maintenance personnel.

References

1. Guillemin, C. L., and Guiochon, G., "Quantitative Gas Chromatography for Laboratory Analysis and On-Line Process Control" in *Journal of Chromatography Library*, Vol. 42, New York: Elsevier Science, 1988.

2. Converse, J.G., "Internal Reference Introduction for Performance Verification, Diagnostics, and Calibration Validation" pp. 185-189 in *ISA/Proceedings of AID Symposium*, Vol. 21, 1985.

3. Pevoto, L.F., and Converse, J.G., "Decisions to Change Analyzer Calibration Based On Statistical Quality Control Charts" in *ISA Transactions*, 30: 1, 1991.

4. Converse, J.G., "Real Time Validation of On-Line Measurements Utilizing Internal Standard Introduction" pp. 1-9 in *ISA/Proceedings of AID Symposium*, Vol. 29, 1996.

About the Author

Jimmy G. Converse is an Analytical Services Technologist at Sterling Chemicals, Inc. Texas City, Texas (409/942-3740), has been involved in design, fabrication, installation, start-up, and maintenance of process analyzers for 28 years. He earned a Bachelor of Science degree (San Diego State) and a PhD (Iowa State) in Chemistry.

Mr. Converse developed analyzer systems for Monsanto (1968-1986) and received on-site instruction from maintenance personnel on how to design reliable and maintainable systems! He has authored many papers regarding on-site automated chemical analyzer systems, was ISA Analysis Division Director (1986-1987), and is currently involved with quality assurance, lab and field analytical services, and corporate technical support. He is an ISA Certified Specialist in Analytical Technology and received the ISA UOP Technology Award in 1994. He was elected an ISA Fellow October 1, 1995 and won the "Gilmer-Thomason-Fowler" Best AD Symposium Paper Award in 1996. Applications include processes such as maleic and phthalic anhydride, acetic and lactic acids, methanol and longer-chain alcohols, ethylbenzene and styrene, hydrogen cyanide, and acrylonitrile.

5

SPC/SQC for Process Analyzer Systems

A process is a combination of people, machines, and procedures, [Ref. 1]. Statistical process control (SPC) and statistical quality control (SQC) are the application of statistical methods to measurements on the process in order to draw objective conclusions that can be used to control the process to achieve the desired output. For convenience and clarity, we shall separate SPC from SQC, even though many of the manufacturing management methods that incorporate statistical methods do not have firmly established definitions. Statistical process control refers to measurements on the process used to produce the tangible output, and statistical quality control refers to measurements on the output itself.

General Principles

It is helpful to think of SPC as a set of data-driven decision-making tools used to monitor, control, and improve a process, and SQC as a set of tools to monitor and improve specific product quality characteristics. The acronym SPC/SQC is used because it is necessary to think of both the manufacturing process and its output as a "logical whole" from a statistical and mathematical standpoint.

It is important not to confuse SPC with conventional manufacturing process control that is characterized by a measured variable, a controlled variable, a manipulated variable, a set point, and disturbances, although the two must be dealt with at the same time when analyzers and other instrumentation are used to control the quality of measurements and product. It is also important to note that SPC/SQC can be used to improve manufacturing process control; that is the aim of this chapter, as it relates to process analyzers.

SPC/SQC must not be confused with conventional automatic process control of manufacturing equipment.

The linking of SPC and SQC implies a close relationship between controlling the process and controlling the quality of the product (output) of the process. The traditional practice of quality control was to concentrate measurement efforts on the raw material and the finished product. Though this is necessary, reliance on these measures, by themselves, is an expensive way to "inspect quality in." That is, upon inspection, "good" material is passed and "bad" material is rejected for rework or scrap. It is less costly and more productive to prevent the manufacture of "bad" material in the first place. That is the goal and motivation of SPC/SQC.

It is less costly and more productive to prevent the manufacture of "bad" material in the first place.

Good SPC/SQC eliminates or minimizes the need to "inspect quality in" after manufacture.

Good SPC/SQC eliminates or minimizes the need to "inspect quality in" after manufacture.

Measurements on the process include those made on the raw materials to the (physical) manufacturing process, on the products from the process, and on the people, machines, and procedures that make up the process. Control of the process includes letting the process run "as is" as well as changing the process when indicated by the measurements. Although changing the process is frequently required, this chapter has examples that illustrate the need to "let the process run."

In working with process analyzers, "the process" is not restricted to the manufacturing process that produces the commercial product or just to the periodic calibration of the analyzer. "Process" also refers to combinations of process analyzers, their service and maintenance procedures, the procedures for using the information that the process analyzer provides, the manufacturing processes, other process control devices, and the people who operate and service the "machines." Therefore, SPC/SQC, when applied to process analyzers, provides statistical tools to improve analyzer performance and to facilitate better maintenance and reliability of the analyzer. This chapter focuses on the application of SPC/SQC to process analyzer performance, maintenance, and reliability.

The mathematical tools of SPC and SQC were developed by Dr. Walter A. Shewhart of Western Electric Company during the 1920s [Ref. 2]. Though the tools and methods have been used ever since then, their use has become more widespread since about 1980 due to international manufacturing competition. This use will continue to increase. Managerial and business uses of SPC and SQC rightfully receive much attention. But it is important to understand that SPC and SQC are readily applicable in all technical situations, including work with process analyzers and other manufacturing process control situations.

The applicability of SPC and SQC to process analyzer work is especially appropriate because the quality and fitness-for-use of manufactured products frequently depends on their measured chemical characteristics.

The technical uses of SPC and SQC are directly and closely linked with their managerial and business uses because the objective in both cases is to make the best-quality manufactured products at the lowest cost to society.

The applicability of SPC and SQC to process analyzer work is especially appropriate because the quality and fitness-for-use of manufactured products frequently depends on their measured chemical characteristics.

Basic Statistics

A manufacturing process offers many opportunities for making measurements. In fact, our capacity to make measurements already far outstrips our capacity or willingness to make the best use of the available measurements. In many manufacturing plants, thousands of unused measurements are made daily. It may not be an exaggeration to state that these measurements accumulate in stacks of paper or are "swallowed" by computer memories, never to be seen by humans. The potential for many of the measurements could be realized by judicious application of descriptive and inferential statistics.

Descriptive statistics allow one to make generalizations or summaries about small groups of measurements with a few calculations. Once these generalizations are made, inferential statistics permits one to draw conclusions about large numbers of measurements, based on the descriptive statistics of the small groups.

Descriptive statistics allow one to make generalizations or summaries about small groups of measurements with a few calculations. Once these generalizations are made, inferential statistics permits one to draw conclusions about large numbers of measurements, based on the descriptive statistics of the small groups.

Although the distinction between descriptive and inferential statistics does not recur in this discussion, the concept of making decisions that affect a large process on the basis of a few measurements is critical to the correct application of SPC/SQC.

The most fundamental descriptive statistics in SPC/SQC are the average, range, and standard deviation. The term "mean" is often used instead of "average" and has identical meaning. The square of the standard deviation, called the variance, is another important descriptive statistic. Range is the difference between the greatest and least numbers in a series of measurements, irrespective of the other measurements in the series.

These ideas are illustrated with a numerical example. Let us say that a series of ten hourly measurements of the concentration of product in a solvent from a certain process are:

46% 49% 44% 46% 47% 45% 42% 46% 47% 45%

The average, or mean, of the measurements is their sum divided by the number of measurements:

$$\bar{x} = (46 + 49 + \ldots + 47 + 45)/10 = 45.70$$

where $\bar{x}$ is the mean or average.

The average, or mean, is noted symbolically as $\bar{x}$ since each measurement, x, represents the same measurement but at a different point in time or on a similar but different item, and the bar over the x indicates that the average represents more than one measurement. The three dots in the middle of the parentheses, "...", are called ellipses and are shorthand to indicate that some of the measurements have been left out of the written equation but are included in the calculations. The general equation for the mean of a series of measurements is:

$$\bar{x} = \Sigma_n/n \qquad (5\text{-}1)$$

where:

$\bar{x}$ = mean

Σx_n = sum of n measurements

The range of these ten measurements is $R = 49 - 42 = 7$ where R is the difference between the greatest number and the least number. The greatest number is 49 and the least number is 42. This number is useful "by inspection" for judging the performance of the process; using range "by inspection" can be misleading, but the range is important in other types of SPC/SQC calculations.

Assume 15 consecutive chemical concentration measurements are made on a manufacturing process. Consecutively, they are: 3.33%, 3.42%, 3.62%, 3.57%, 3.67%, 3.35%, 3.63%, 3.52%, 3.58%, 3.57%, 3.51%, 3.41%, 3.53%, 3.37%, and 3.44% (*n* = 15). Their average is 3.501%; their standard deviation and variance are 0.108% and 0.0117, respectively; their mode and median are 3.57% and 3.52%, respectively; and their range is 0.34%.

Note that the average does not tell anything about how the ten measurements, or numbers, are distributed about the average. That is, many other sets of ten measurements could have the same average, 45.70. For example, 5 is the average of 4, 5, and 6, but 5 is also the average of 1, 5, and 9. The standard deviations of these two sets of three numbers are 1 and 4, respectively. Therefore, the greater the standard deviation, the more "spread out" the numbers are in the set under consideration. The standard deviation of the ten concentration measurements is:

$$s = \sqrt{\frac{(46 - 45.70)^2 + (49 - 45.70)^2 + \ldots + (47 - 45.70)^2 + (45 - 45.70)^2}{9}}$$

$$= 1.89$$

where s is the standard deviation.

The denominator under the radical is one less than the number of measurements when "estimating" statistics from a sample; the denominator is called the "degrees of freedom." This is the number of "independent estimates" that can be made from the set of data. Since one "estimate," the mean, has already been made from the ten measurements, only nine can now be made. Further explanation of this is beyond the scope of this discussion. The general equation for standard deviation is:

$$s = \sqrt{\frac{\sum (x_i - \bar{x})^2}{n - 1}} \tag{5-2}$$

where:

s	=	standard deviation
x_i	=	ith measurement
$\bar{x}$	=	mean
n	=	total number of measurements
$n - 1$	=	degrees of freedom

Sigma indicates the sum of the squared differences between the mean and the individual measurements. The variance is the square of the standard deviation and is usually designated symbolically simply as s^2.

The average, or mean, is called a "measure of central tendency," or a number about which all similarly obtained measurements are grouped. Less reliable but often used measures of central tendency are the mode and the median. The mode is the most frequent measurement in the group; for the ten concentrations the mode is 46 since it appears three times and no other measurement appears more than twice. The median is the middle number of an odd number of measurements and the average of the two middle numbers of an even number of measurements, after all the measurements have been ordered from least to greatest. The median of the ten concentrations is 46. There are few sets of measurements in which the mean, mode, and median are all the same. The mean is regarded as the most reliable of the three since all measurements in the set are used to calculate it; mode uses only one measurement, and median uses at most two.

The range and the standard deviation/variance are called "measures of dispersion," or an indication of how all the numbers are distributed about the mean. There are other measures of dispersion, but these are the only ones of direct interest in SPC/SQC. Although the range is calculated from only two numbers and the standard deviation from all the numbers, the range is used in certain SPC/SQC calculations.

Normal Distribution

A critical underlying assumption in SPC/SQC is that no matter how precisely we try to control a process, there will always be random variation among the units produced by the process if we measure the process or its output with sufficiently accurate tools and methods. The common bell curve is the probability distribution upon which most SPC/SQC is based, because it tends to describe these random variations consistently and accurately. It is called the normal distribution because experience and experiment demonstrate that data from many kinds of measurements follow this distribution.

Mathematically, the normal distribution is described by:

$$y = \frac{1}{s\sqrt{2\pi}} e^{-1/2[(x-\bar{x})/s]^2} \tag{5-3}$$

where:

y	=	frequency of occurrence of the value x
s	=	standard deviation
x	=	measured value
$\bar{x}$	=	mean
π	=	3.1416
e	=	2.7183

The normal distribution is a theoretical concept usually expressed in terms of the measures of either an infinite population or a population so large that it is impractical to measure each individual in the population. When the form of Equation 5-3 is used to describe infinite or very large populations rather than small samples, the lowercase Greek symbols μ and σ replace $\bar{x}$ and s, respectively.

Since it is not possible to deal physically with such populations, Equation 5-3 is written in terms of a sample of a very large population. In this form it is of practical use in day-to-day manufacturing situations. The theoretical treatment is beyond the scope of this discussion; the References section at the end of the chapter includes detailed discussions of the theory of statistics and probability [Ref. 3-5].

As mentioned earlier, the normal distribution resembles a bell curve. Several such curves are illustrated in Figure 5-1. Note that all the curves in Figure 5-1 have the same mean but their standard deviations differ; the larger the standard deviation, the flatter the distribution. These curves not only have a common mean, but their conformance to Equation 5-3 assures that they each have the same proportion of their area under the same portion of the curve as measured in units of standard deviation.

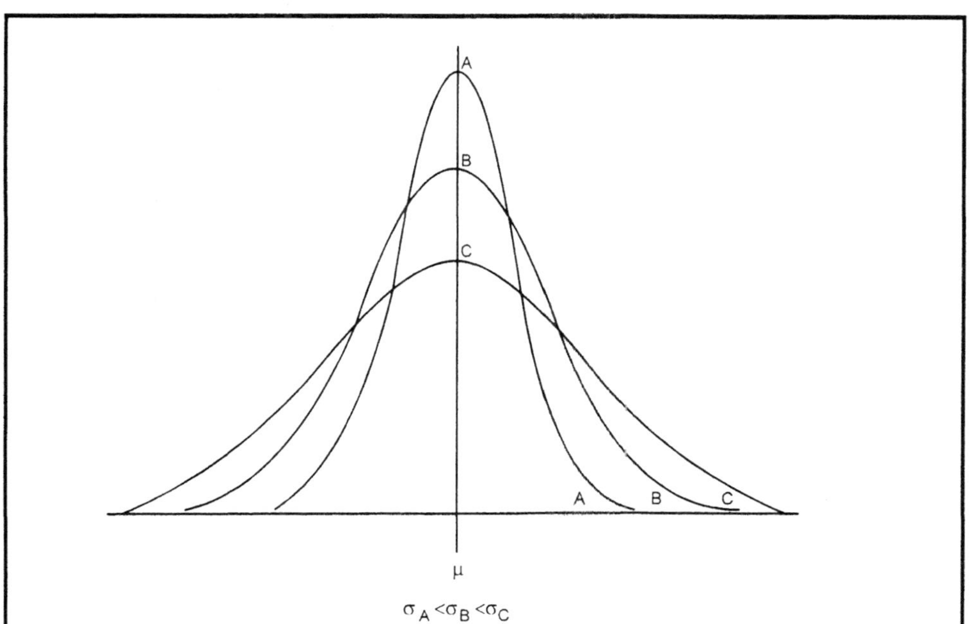

Figure 5-1. Three Normal Distributions With Identical Means but Different Standard Deviations

Describing the normal distribution in terms of the standard deviation is illustrated in Figure 5-2. In this figure, the symbol μ replaces $\bar{x}$ in reference to the mean and σ replaces s in reference to the standard deviation. The Greek symbols refer to entire populations and the normal alphabetic characters refer to samples of populations.

There is always random variation in any process. Any attempt to eliminate this random, "common cause" variation will result in making the process worse instead of better! This is the first of the two most serious statistical errors that are made.

Figure 5-2 has several important features. First, the vertical axis, *y*, corresponds to the frequency, calculated from Equation 5-3. The horizontal axis, x, corresponds to the values of the measured variable, such as concentration, length, or weight; the physical units of measure are converted to units of standard deviation. Second, the horizontal scale is in units of $+\sigma$ and $-\sigma$ with reference to the mean. Because this is a normal distribution, 68.3% of the measured values are between $\pm 1\sigma$. Similarly, 95.4% of the values are between $\pm 2\sigma$, and 99.7% of the measured values are between $\pm 3\sigma$. The 3σ limits are the most commonly used and are sufficient for most manufacturing work; the 3σ limits are used in the text and examples throughout this chapter. However, 3σ limits are not universal and there are instances where narrower or wider limits may be more appropriate. The $\pm 4\sigma$ limits include 99.994% of the measurements! These percentages are valid only if the population is normally distributed and the sample is truly representative of the entire population; they are used as an approximation in most quality control work.

The $\pm 1\sigma$, $\pm 2\sigma$, $\pm 3\sigma$, and $\pm 4\sigma$ limits include 68.26, 95.44, 99.74, and 99.99% of the data, respectively, for a normal distribution.

In absolute terms, this means that if 1,000 screws were made on the same set of machines, without adjustment to the machines, raw materials, and so forth, their weights would follow this distribution. That is, three of the 1,000 would weigh either less than three standard deviations or greater than three standard deviations from the mean. In common terminology, we would say that three screws were outside the 3σ limits. By analogy, we would also say that 46 of the screws weighed outside the 2σ limits, and so on. The screws outside the 2σ limits would also include those outside the 3σ limits. Therefore, there would be 43 screws between the 2σ and 3σ limits. A given measurement thus has a higher probability of lying closer to the mean than farther from the mean. It is important to note that this is a model of statistical behavior; a given set of measurements will likely show some difference from these specific numbers. But over a long period of time and a large number of measurements, the results would tend to fit the model.

This quick example used discrete parts, screws, because one can easily "try it out" with a box of screws and a laboratory balance. It is not necessary to do it with 1,000; two or three dozen should be adequate. But the same principle holds for making a series of concentration measurements on the product from a chemical process. Regardless of what is measured, it is important to note that the measuring instruments or tools also constitute part of the "system" or "process." Therefore, if the measuring instrument is not sufficiently accurate to detect the variation in only the manufacturing part of the process, then the variation, or lack of precision, of the measuring instrument will become an unwanted part of the measurements.

Don't measure with a micrometer, then cut-to-length with a broad axe.

This is analogous to finding the width of a tight tolerance machined part with a pocket tape measure; the tape measure may be good only to ±1/32 in., or about ±0.03 in. when one needs a measure to ±0.0002 in. The same principle is true in chemical analysis: The process analyzer must have the capability of measuring more precisely than the permissible variation in the manufacturing process.

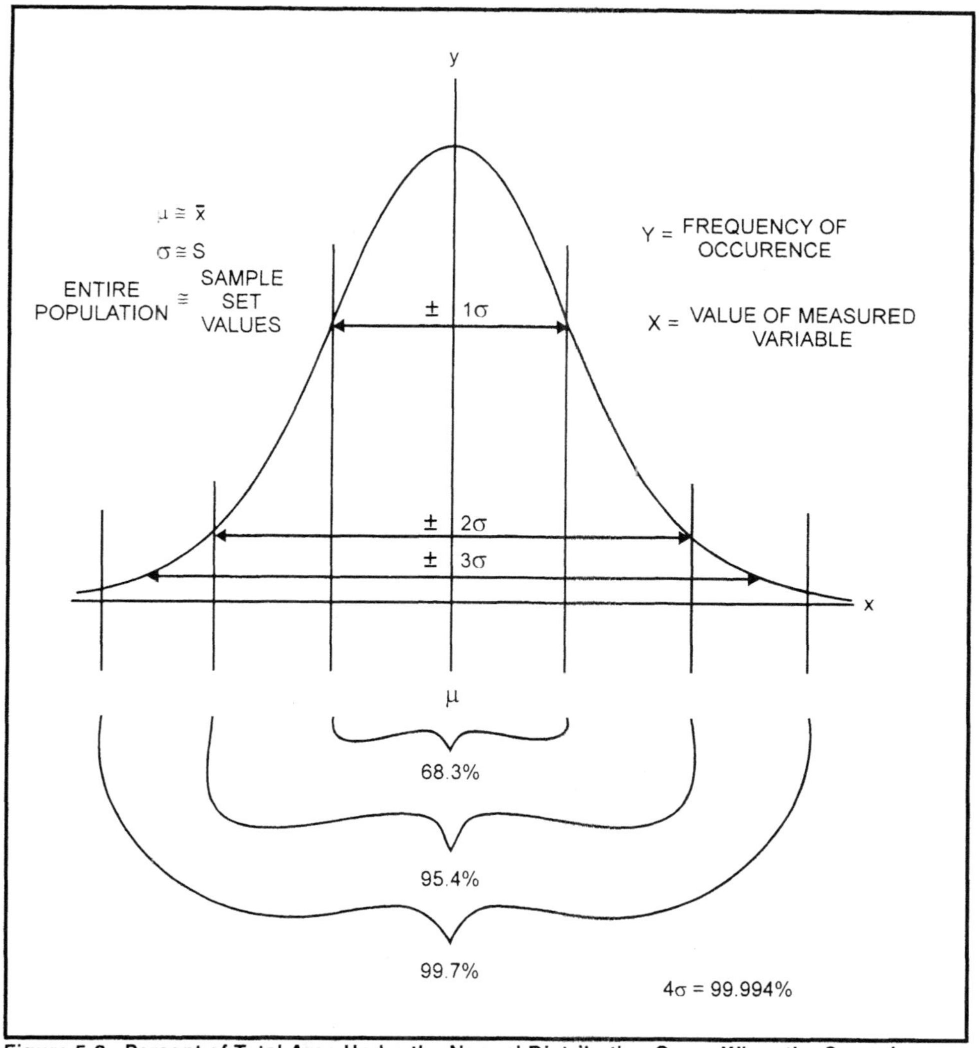

Figure 5-2. Percent of Total Area Under the Normal Distribution Curve When the Curve is Divided Into Units of Standard Deviation

It is critical that the variation in the measurement process be insignificantly small in comparison with the variation of the manufacturing process so that decisions based on the measurements do not increase the manufacturing variation.

Previously, it was noted that there are some cases for which the normal distribution does not describe the random variation in the process. If this is discovered to be the case, the Central Limit Theorem can be used to enable application of the normal distribution. The Central Limit Theorem states that if samples of size n are drawn from a population with a mean of m and a standard deviation of s, the distribution of sample means approaches a normal distribution as n increases. Many pocket scientific calculators have a so-called "random number generator" function. These random numbers are usually uniformly distributed; that is, there is an equal probability of getting a value close to the mean or of getting one far from the mean. These uniform random number generators can be used to obtain normally distributed random numbers if the means of subgroups of five or ten are calculated. The collection of means of the subgroups approaches a normal distribution. Larger and larger subgroups allow closer and closer approximation to the normal distribution.

The Central Limit Theorem can be used to deal with measurements that are not normally distributed.

The Central Limit Theorem was applied to create the simulated control charts used in this chapter. Although the process analyzer cases used here are real, the actual data are proprietary. Because the actual data followed the normal distribution, simulated data are valid for demonstrating the efficacy of SPC/SQC in dealing with process analyzer problems. Simulations are described in more detail in Appendix 2.

Control Charts

The control chart is the most widely used tool in SPC/SQC. A control chart is constructed from process data based on the assumptions that there is a certain amount of random variation in the data, that the data are normally distributed, and that data, when they are out of control, will exhibit some type of behavior on the control chart that can be detected by either statistical analysis or visual inspection.

Figure 5-3 is a typical control chart of in-control data. The control chart is characterized by the centerline (CL), upper control limit (UCL), lower control limit (LCL), the vertical axis, and the horizontal axis. This control chart might represent the process analyzer concentration measurements of a critical chemical (vertical axis) in the feed to a chemical process operation over time (horizontal axis). Each point on the chart would represent the concentration at some point in time, such as at the beginning of the day shift each day for 30 days. The numerical data are tabulated in Table 5-1. The centerline is the mean, or average, of all 30 data points. The upper control limit is the mean plus three standard deviations; the lower control limit is the mean minus three standard deviations. This process is said to be "in control" because all 30 data points lie between the UCL and LCL and appear to be distributed randomly about the CL.

Table 5-1. Hypothetical Concentration Data From Which the Control Chart in Figure 5-3 Was Prepared (Mean = 5.29, standard deviation = 0.86, n = 30)

Point	Value	Point	Value	Point	Value
1	4.8	11	6.0	21	7.2
2	5.9	12	6.0	22	5.4
3	6.4	13	6.0	23	4.0
4	6.4	14	4.6	24	5.1
5	5.5	15	4.7	25	5.0
6	6.7	16	4.9	26	5.9
7	4.0	17	5.8	27	4.2
8	5.0	18	5.0	28	4.7
9	5.2	19	4.9	29	3.4
10	5.5	20	5.7	30	4.8

Based on the assumption that the data are normally distributed, the probability that a data point from an in-control process is either above the UCL or below the LCL is 0.003, 0.3%, or three chances in 1,000. We must note here that there are more sophisticated ways to calculate control limits. The advanced ways give more "conservative" control limits, but require additional calculations and the use of numerical constants that must be looked up; statistical software usually incorporates them because of the ease and speed of computation using computers. The advanced methods are beyond the scope of this chapter. The method demonstrated here is adequate for most practical field work on process analyzers, but if the advanced ways are available or required within the organization, they will give satisfactory results.

Although it seems obvious from visual inspection of Figure 5-3 that the process depicted is in control, visual inspection is not always a reliable way to judge a control chart. One way to estimate whether the data are normally distributed is to construct a histogram. A histogram is a graphical representation of numerical data.

Figure 5-4 is a histogram of the data in Figure 5-3 and Table 5-1. Note in Figure 5-4 that the data have been grouped so that there are only a few "bars" to represent 30 pieces of data. The length of each bar represents the number of data points within a small part of the range of measurements represented. For most sets of data, the number of bars should be approximately the square root of the number of data points. The 30 data points in Table 5-1 are broken into six bars, or data classes, in Figure 5-4. Visual inspection of Figure 5-4 indicates that data are probably normally distributed, because the tops of the bars form a rough, bell-shaped curve. Since Figure 5-4 contains all 30 points and the UCL and LCL lie off the scale, we have further indication that the process is in control. It is important to note that this is only one criterion as to whether a process is in control; there are other "rules" for out-of-control processes. These other rules are discussed in the following section. The remainder of this section will be devoted to other important aspects of constructing and using control charts.

The data collected for making a control chart must be statistically independent from one another. This means that any given data point must not be dependent on any of the preceding data points. The existence of such a dependency is called autocorrelation. To extend the chemical concentration example used earlier, we shall assume that the feed stream for which the concentration is measured is either continuous or that a tank from which the feed is pumped is emptied and filled each day, then emptied and refilled before the next measurement is made. With either continuous feed (no tank) or a daily empty-and-fill cycle on the tank, we are assured that one day's data are independent of any previous day's data with respect to the measured concentration.

A histogram is a graphical representation of numerical data.

Visual inspection of data is okay as a first step, but it should not be used in place of more sophisticated statistical and mathematical tools.

A useful rule of thumb for constructing a histogram is to use a number of bars roughly equal to the square root of the number of individual data points. For example, five or six bars would be appropriate for showing 30 data points, about eight bars for 60 data points, and so on.

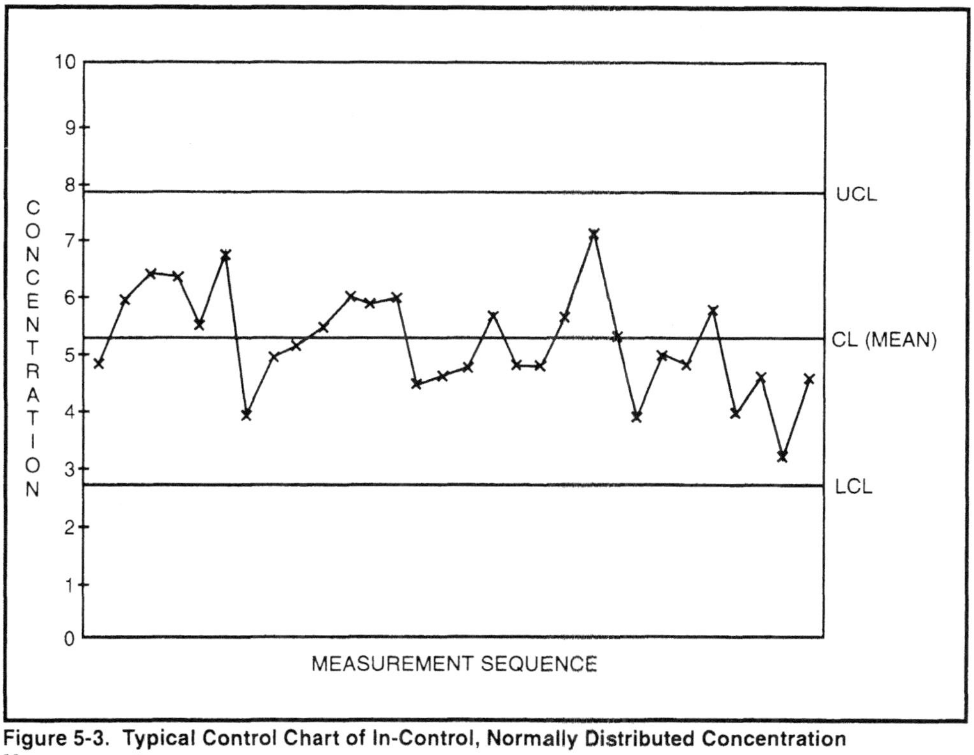

Figure 5-3. Typical Control Chart of In-Control, Normally Distributed Concentration Measurements

CL = Mean = 5.29
s = 0.86 n = 30
LCL = Mean − 3s = 2.71
UCL = Mean + 3s = 7.87

Mean = 5.29
$s = 0.86$ $n = 30$

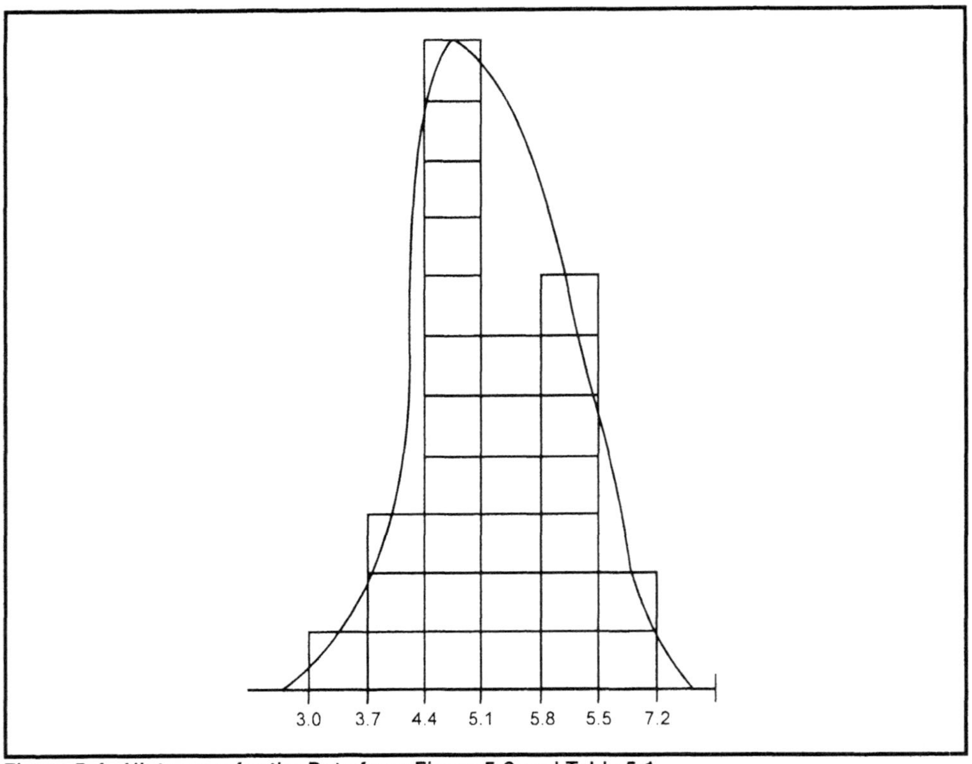

Figure 5-4. Histogram for the Data from Figure 5-3 and Table 5-1

Autocorrelated data can result in misleading statistics. In some cases, autocorrelation cannot be totally eliminated, but there are usually ways to minimize its effects.

To further illustrate, let us make a different, though reasonable, assumption. This time, let us assume that the feed tank is large enough to hold a 3-day supply of feed and that as feed is pumped out of the tank, new feed is pumped in from the upstream process. This means that, even with ideal mixing, a proportion of the feed that we are analyzing one day is the same as the feed analyzed on previous days. If daily measurements are made, successive measurements will be autocorrelated, rather than statistically independent. It further means that the proportion from previous days is "diluted" with feed that has been added to the tank since the last measurement. Because it is unlikely that either new feed is added or feed is pumped out at uniform rates, it is not possible to know and completely compensate for the dependence. If "on average," however, the feed tank is "turned over" every 3 days, it is acceptable to take groups of three consecutive daily analyses and average them to calculate one data point for the control chart. The data will still be autocorrelated, though not to as great an extent as with the daily measurement. Nothing can remove this disadvantage, since it is "built into the process" as a 3-day feed tank. One can argue that the 3-day averaging method is not theoretically correct since there is a constant filling, mixing, and emptying. But from a practical standpoint, this type of averaging will provide information upon which control decisions can be based.

Reducing the variability of in-control processes is called continuous improvement.

The control chart and histogram and the procedures for creating them do not, by themselves, provide improvement in reliability and performance of the process analyzer or the chemical process. This chapter focuses on the application of SPC/SQC to process analyzer maintenance and reliability, though the tools are the same. They are merely numerical, objective tools that are useful for understanding the process analyzer performance, the behavior of the chemical process being monitored, and the relationship between the two.

If the control chart indicates that the process is out of control, the process must be brought into control. If the process is in control as indicated by the control chart, the random variability must be reduced so that the control limits are narrowed. Reducing the variability of an in-control process is often more difficult than bringing the out-of-control process into control in the first place. Reducing the variability of in-control processes is called continuous improvement.

Use of statistical methods does not ensure a good manufacturing process. Statistics are only valuable technical and management tools to be used for continuous improvement.

After the process has been brought into control, we retain the same control limits while we are trying to improve the process. During this time of improvement, the data points should still be randomly distributed on both sides of the centerline, but they should generally be getting closer to the centerline. Once the in-control process has been improved, as evidenced by data points closer to the centerline, new control limits are calculated using only data points obtained after the process has been improved. Often, it is a matter of judgement when to calculate new control limits. However, it makes little sense to continue to use wide control limits that provide no incentive or information as to how to improve. Continuous improvement begins at this point. Continuous improvement means reducing variability until it is no longer beneficial or rewarding to the supplier or the customer to do so. Even then it is still wise to make (less frequent) measurements to assure the process is still in control.

Continuous improvement is usually more difficult than bringing a process into control to start with.

Here's a hypothetical example. Assume that the manufacturing process and the process analyzer used for monitoring the process are in control. The control chart for plant output as indicated by the measurements from the process analyzer might look like Figure 5-5. The data in Figure 5-5 have a mean of 65% and 3σ control limits of 62 and 68%, or ± 3%. The commercial customer for this product may compensate for this variation by varying their own processing rates.

Now, assume that process analyzer variability is linked to calibration and that it is routinely calibrated daily with a standard, which is also made up daily with a laboratory balance precise to ±0.001 g. It may be found that using an analytical balance precise to ±0.0001 g reduces measurement variability to ±2%. Therefore, the control chart for plant output after the new balance is used would look like Figure 5-6; note that the mean is still 65%, but the 3σ control limits are now 63 and 67%. Now the commercial customer does not have to vary processing rates and incur the associated expense as much to compensate for the lesser product variations.

Let us further suppose that investigation reveals that process analyzer measurement variability is partly due to temperature and humidity fluctuations throughout the day and night. If the process analyzer is enclosed in a climate-controlled shelter, with a heater/air conditioner, one might expect a further reduction in measurement variability. After this is done, the control chart for the process analyzer measurements might look like Figure 5-7, with a mean of 65% and 3σ control limits of 64 and 66%. The product variability that accompanies this change may not require the customer to change rates at all to compensate for the remaining ±1% variation.

Overall, the results might be less equipment wear and expense for the customer, and a higher price and market share for the manufacturer/supplier. Side-by-side placement of Figures 5-5, 5-6, and 5-7 graphically typifies continuous process improvement. Note that in this example the manufacturing process was not changed except for control changes indicted by the process analyzer measurement. The manufacturer may be able to realize reduction in manufacturing costs along with the customer.

Mean = 65.03
s = 1.06 n = 30
LCL = 61.85
UCL = 68.21

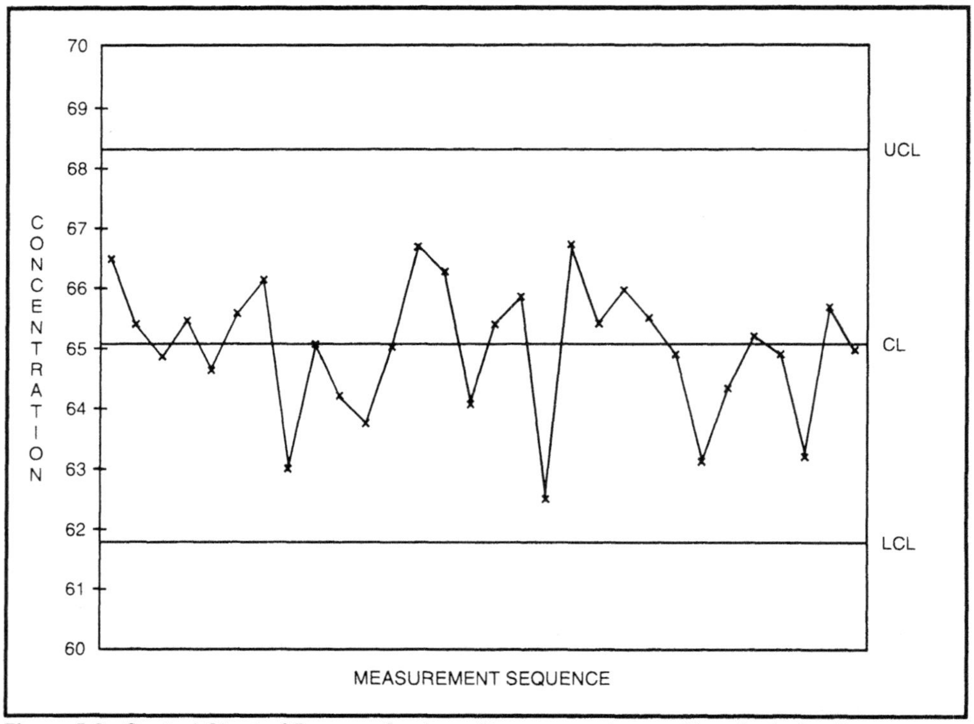

Figure 5-5. Control Chart of Process Analyzer Measurements on a Manufacturing Process (The process analyzer is calibrated daily using analytical standards made to a precision of ±0.001 g.)

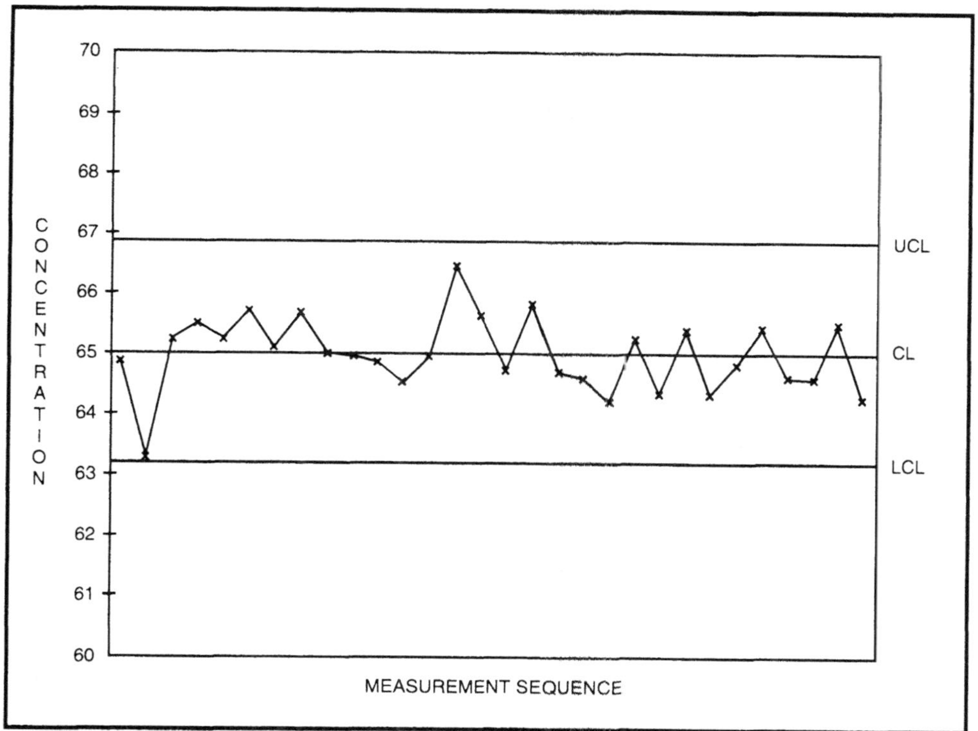

Mean = 64.96
s = 0.63 n = 30
LCL = 63.07
UCL = 66.85

Figure 5-6. Control Chart of Process Analyzer Measurements on a Manufacturing Process (The process analyzer is calibrated daily using analytical standards made to a precision of ±0.0001 g.)

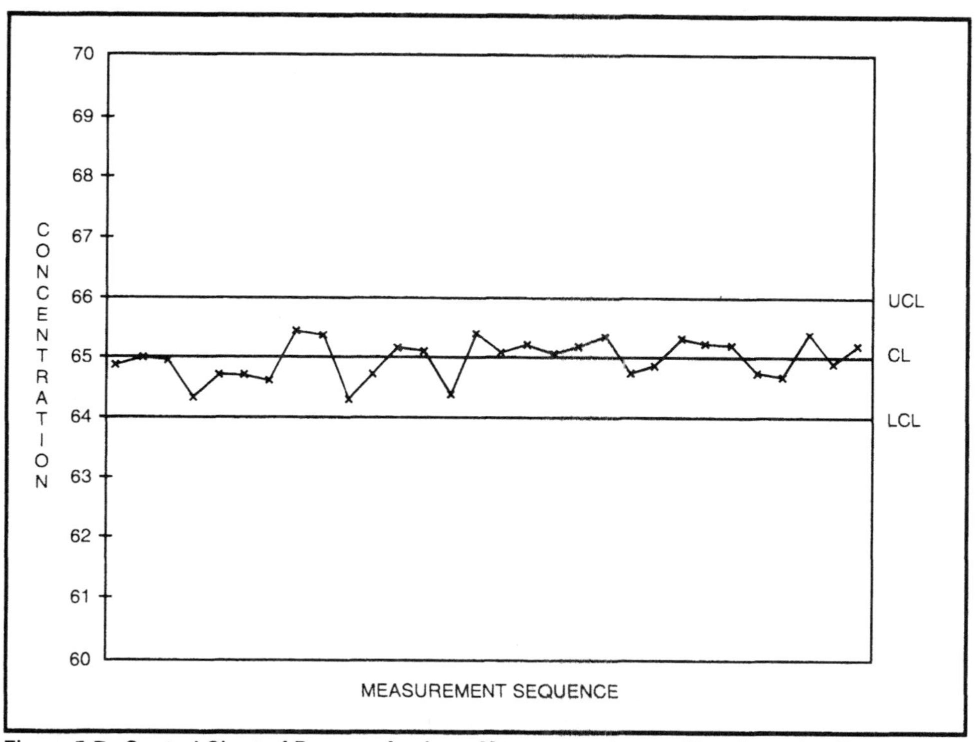

Mean = 65.00
s = 0.35 n = 30
LCL = 63.95
UCL = 66.05

Figure 5-7. Control Chart of Process Analyzer Measurements on a Manufacturing Process (The process analyzer is calibrated daily using analytical standards made to a precision of ±0.0001 g, and the analyzer is enclosed in a climate-controlled shelter.)

The important point of this example is that continuous improvement on only one part of a process, in this case, the process analyzer, can result in overall cost savings and improved profitability for both the manufacturer and the customer. At no time in this discussion were product specifications mentioned. Product specifications are very important considerations in continuous improvement and reducing variability. The relationship between control limits and product specifications is covered in the discussion on capability. It is important to understand that control limits and specifications are not the same set of numbers; treating them as interchangeable is one of the greatest causes of confusion and misuse of statistics in manufacturing.

Control limits and specifications are not the same. This is the second of the two most serious statistical errors that are made.

Out-of-Control Criteria

To this point, the terms "in control" and "out of control" have been used loosely and intuitively with no more specific definition than to refer to whether data points on a control chart are outside or inside the 3σ limits. Some workers argue that 3σ limits are too wide for analyzer calibration and that 2σ limits are more appropriate.

One way to set control limits is to examine the types of errors that are inherent in using statistical tools. A Type I error is made when average process performance has not changed and the statistics indicate that it has. A Type II error is made when a process performance change has, in fact, occurred, and the statistics fail to detect the change. So making a Type I error means that the control chart will more frequently indicate a problem when there is none, and a Type II error means that the control chart will more frequently miss a problem when one exists. Since the expense of investigating a problem when none exists can be as costly as missing a problem that does exist, a balance must be struck between making Type I and Type II errors. Unfortunately, if the likelihood of making a Type I error is made small, the likelihood of making a Type II error is automatically increased, and vice versa.

Therefore, some workers suggest using 2.5σ control limits to balance the occurrence of Type I and Type II errors. Control limits of 2σ, 2.5σ, and 3σ are associated with Type I error rates of 5, 1.2, and 0.3%, respectively. The theory behind error types and the derivation of these numbers is beyond the scope of this chapter; the reader is referred to the general statistics texts in the References and Bibliography.

Type I and Type II errors are inherent and unavoidable results of using statistics. Therefore, it is important to balance them properly, depending on the measuring and manufacturing processes that one wants to control.

Whether to use 2σ, 3σ, or more statistically sophisticated ways to set control limits is actually a management decision that should be based on customer needs, technological considerations, and the manufacturing process.

In general, process analyzers that are used in closed-loop control should have narrower control limits than those that are used for simple monitoring. Since this chapter is devoted to SPC and SQC, more appropriate terms would be "in statistical control" and "out of statistical control." Because the latter terms are wordy, it is understood that use of the shorter terms implies statistical or objective context rather than subjective criteria for judging process performance. Therefore, we define "in control" as the statistical measure of process performance which exhibits only "common cause," or random, variation in the process. By analogy, "out of control" refers to a process for which the variation is not random, but which exhibits one or more points outside the control limits or some type of discernible pattern upon statistical examination. This is called "special cause" variation. As pointed out earlier, acceptable random variation is usually normally distributed.

It is important to reemphasize that "the process" refers to more than the manufacturing equipment and the tangible product being produced; process refers to the combination of people, raw materials, equipment, and procedures. Each of these parts of the process has its own random variation that contributes to the overall random variation. In many cases, only the overall random variation can be measured. Therefore it is important, when the process is out of control, to determine systematically and objectively which individual part or parts of the process cause the out-of-control condition.

Although an "out-of-control" situation is cause for concern, it is helpful to regard such a situation as an opportunity to learn more about the process.

Deciding which parts of the process are out of control is the most difficult and time-consuming aspect of SPC/SQC. Many times, this type of problem cannot be resolved without the use of SPC/SQC. The calculations of SPC/SQC are simple; some workers find the arithmetic boring after a little practice and a few applications.

Manufacturing process control software, first produced in the 1980s, is capable of collecting manufacturing process data, preparing control charts, and making process changes with minimal operator intervention. It is still important for plant personnel to understand SPC/SQC, to know what the process control computer is doing, and to work with data and equipment that are not computerized and automated.

Preprinted SPC/SQC worksheets and pocket calculators suffice for most applications. Inexpensive, user-friendly personal computer programs are available for doing the calculations, storing the information, and recalling and comparing old data with new data.

To this point in the discussion, we have looked at only one control chart indication that a process is out of control: when one or more points are outside the 3σ limits. There are several other tests, or rules, one may use with a control chart to decide whether a process is out of control [Ref. 1]. These rules are illustrated in Figure 5-8. Figure 5-8 is characterized not only by the centerline and the 3σ control limits, but also by the 1σ and 2σ lines. The areas between adjacent standard deviation lines are called "zones." The zones are labeled A, B, and C, from the outside to the center. Using the three zones and the areas outside the 3s limits, the process is out of control if any one of the following conditions exists:

1. A single point lies outside the 3σ control limit (beyond Zone A).
2. Two of three consecutive points lie in Zone A or beyond.
3. Four of five consecutive points lie in Zone B or beyond.
4. Eight consecutive points lie on one side of the centerline.

Rule 1 is nearly universal. There are several variations of Rules 2, 3, and 4. But these four rules are convenient for getting started or until other statistically sound rules have been developed and implemented by the organization. Note that these rules apply to only one side of the centerline, or the other, for a given set of points; in-control points should alternate randomly across the centerline and be as close to the centerline as possible. For this reason, these four rules are sometimes given as seven or eight rules to avoid confusion. CUSUM (for "cumulative sum") charts are a more advanced form of control chart, but can be used without additional rules. CUSUM is not yet widely used, but it is becoming more popular. Reference 3 includes a good discussion of CUSUM.

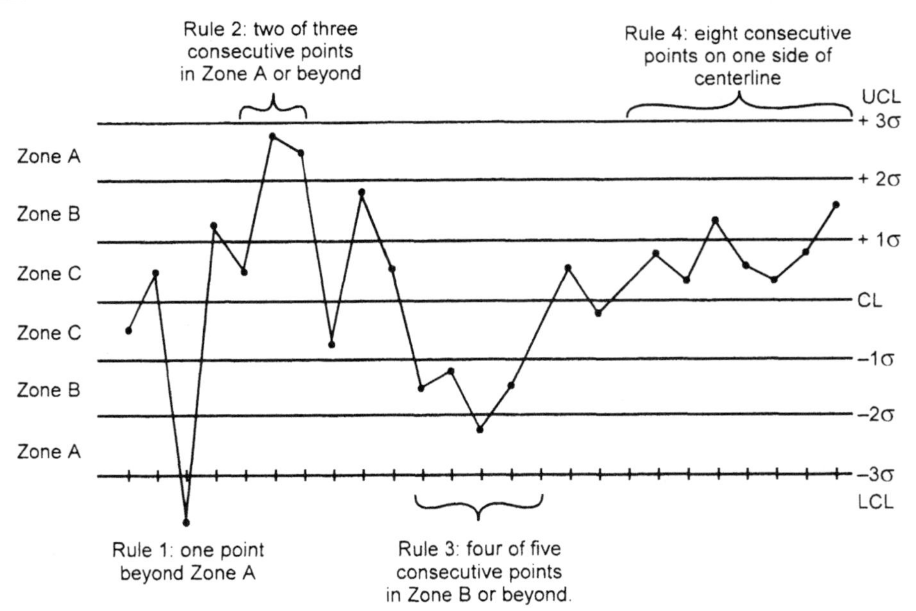

Figure 5-8. Control Chart Demonstrating the AT&T Out-of-Control Rules [see Ref. 1]

There are other patterns, or features, of control charts that are not described by the rules listed above but which nonetheless signal a potentially out-of-control process [Ref. 1]. These features or patterns are:

- Periodicity or cycles
- Upward or downward trends
- Bunching of measurements, consecutive or not, at one value
- Abrupt level shifts
- Gradual level changes
- Interactions among variables (tracked on separate control charts)
- Systematic variation
- Tendency of the control chart for one variable to follow the control chart for another variable.

For measurements on some variables—such as density and viscosity—interactions, systematic variations, and tracking are expected. Space does not permit us to describe each of these patterns; Reference 1 provides a good discussion of the details.

More important than knowing the definition of each out-of-control pattern is the ability to recognize an "unnatural" appearance on a control chart, list possible causes for the pattern, objectively investigate which possible cause(s) might be responsible for the pattern, and eliminate the cause from the process.

When working with a process in order to make it conform to some standard or "target" output, there is a natural tendency to constantly adjust the process based on the most recent data point so that the next data point will fall at the target value. For example, if the target value for a concentration is 35% and the process has most recently achieved 33%, the natural reaction is to change the process, based on experience, to produce at a 2% higher level. If the process then makes product at the 38% level, we adjust the process downward by 3%, again expecting to achieve the 35% target value. The series of adjustments goes on and on and, hopefully, over an extended period of time, the average concentration of the product will be 35%. Although this is the natural human reaction, it is wrong! This practice goes by the names "overcorrection," "overcompensation," and "tampering with the process." The negative connotations are intentional and appropriate.

Although this is the natural human reaction, it is wrong!

Overcorrection or "tampering" with a process is the same as trying to eliminate common cause variation. Common cause variation cannot be eliminated, and trying to do so invariably makes the process worse instead of better. This is one of the most serious errors one can make in applying SPC/SQC.

The fallacy in tampering with the process is that we ignore the random variation inherent in the process. The effect of tampering is to increase the variation from that which would normally be present and to actually defeat the goal of continuous improvement. Figure 5-9 shows the control chart of a process with only random variation. Figure 5-10 shows the same process and the same random variation, point for point, with tampering artificially added. Figures 5-9 and 5-10 are computer simulations, but the data used to generate Figure 5-10 have a correction added (or subtracted) to each point equal to the difference between the target value and the value of the immediately preceding point. Notice the greater variation in Figure 5-10 in comparison to that in Figure 5-9 as indicated by the standard deviation and the 3σ control limits. Both sets of data indicate that the process is in control; the difference is that part of the process charted in Figure 5-10 is the tampering that has been artificially introduced.

The increased variation that always results from tampering is avoided by "letting the process run" without interference. It is true that sometimes the process may not achieve the target value, but it is usually better to miss the target value by a relatively consistent difference than to achieve the target value, on average, with product that is inconsistent from one time to another. The reward for "letting the process run" is that a process that is consistently off the target by a constant amount is easier to bring on target if the variation is small rather than large. The gas chromatography example described later in this chapter illustrates the effects of "tampering," "letting the process run," and "continuous improvement" to a target value.

Mean = 5.16
s = 1.25 n = 30
LCL = 1.41
UCL = 8.91

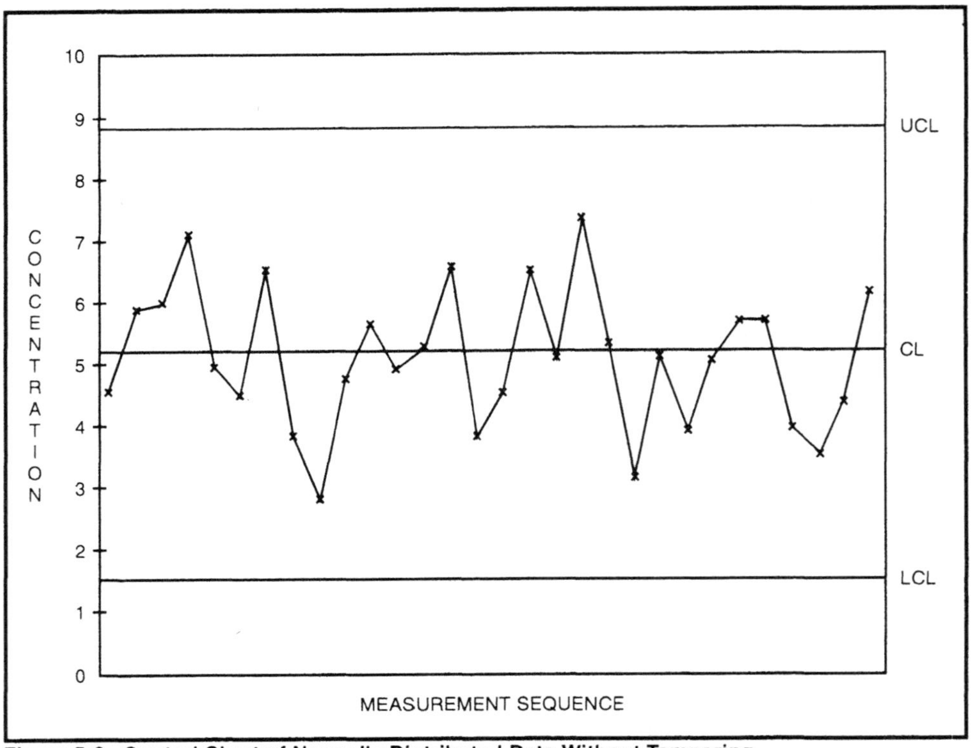

Figure 5-9. Control Chart of Normally Distributed Data Without Tampering

Mean = 5.12
s = 2.03 n = 30
LCL = −0.97
UCL = 11.21

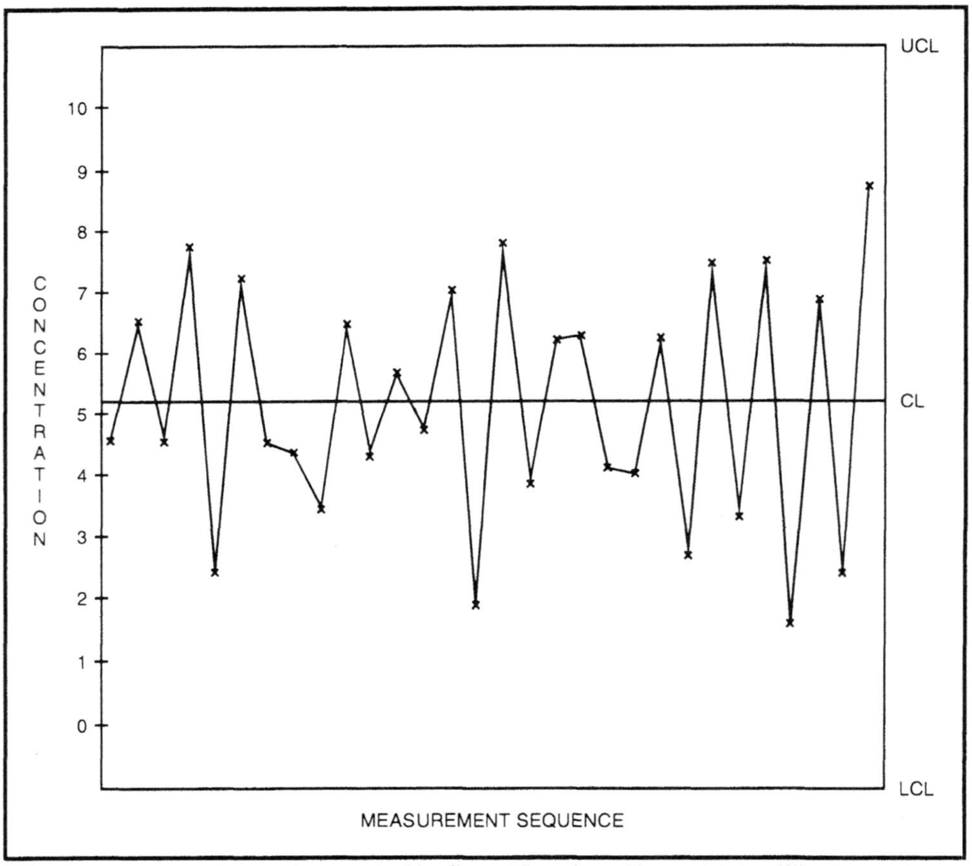

Figure 5-10. Control Chart of Normally Distributed Data from Figure 5-9, but With Tampering Added to Achieve a Target of 5.00

Other SPC/SQC Concepts

Statistics is an important aspect of process analyzer validation, addressed in depth in Chapter 6. Validation is mentioned here because a correct validation procedure is the first step in developing a meaningful statistical history of the performance of a process analyzer. Validation uses statistics to establish the validity of the results obtained by a process analyzer. Validation uses most of the statistical concepts and procedures discussed in this chapter, along with the t-test, F-test, and a few other concepts directly derivable from descriptive statistics and hypothesis testing. The reader is referred to Chapter 6 for the details.

Resisting the temptation to tamper with the process, to "let the process run," is one of the most difficult management and interpersonal tasks of implementing SPC/SQC.

Reference Standards and Line Samples

After validation, the performance of the process analyzer must be routinely checked against a reference with a known composition as a part of routine service and maintenance to ensure that the analyzer continues to provide quality data. The reference may be a certified reference standard, such as from the National Institute of Standards and Technology (NIST), called the reference sample method, or a process stream sample that has been analyzed in the laboratory by another analytical method, called the line sample method. Reference 6 is an excellent procedure for process analyzer validation.

Process analyzer validation upon commissioning and ongoing maintenance and service are important and closely related functions where the application of SPC/SQC results in superior analyzer performance.

Which method is used is dictated by the availability of reference standards, the chemical characteristics of the process stream, and the availability of a method for analyzing the reference. Unfortunately, there is often an assumption that the number reported for a reference is absolute; this is most assuredly not the case! It must be remembered that the reference sample or the line sample and the reference analytical method also constitute a "process," with their own common cause variations. Therefore, the variation from the reference sample must also be considered when setting control limits.

Reference standards and reference analytical methods inherently have their own common cause variations. These variations must be taken into consideration when they are used to check the performance of a process analyzer.

Fortunately, there is a way to incorporate the variation of the reference analysis with the variation of the process analyzer. This is done by using the additive property of variances. This property is stated as:

$$s_T^2 = s_L^2 + s_P^2 \tag{5-4}$$

where:

s_T^2 = total variance

s_L^2 = variance of the laboratory method

s_P^2 = variance of the process analyzer method

We are assuming that variations in the laboratory analysis and the process analysis on the same samples are independent. The combined, or pooled, standard deviation is then the square root of the variance obtained from Equation 5-4. Therefore, when making a control chart in which the process analyzer results and the laboratory results are compared, the control limits must be set using the pooled standard deviation to calculate the control limits.

Accuracy and Precision

Accuracy is a measure of the ability of the analytical system to agree with a target value.

Now the issue of accuracy versus precision becomes important. Accuracy is a measure of the ability of the analytical system to agree with a target value, such as the concentration reported for the reference using an independent analytical method. Precision is a measure of the ability of the analytical system to obtain the same value on the same reference or process sample regardless of any other measurement. Usually, when only a "trend" analysis is requested, precision is more important than accuracy. Naturally, one wishes both high accuracy and high precision. Although high accuracy and high precision are often theoretically attainable, the perceived need for both does not always justify the real expense of obtaining both.

Precision is a measure of the ability of the analytical system to obtain the same value on the same sample regardless of any other measurement.

Therefore, management must decide, based on product quality, process control, and legal liability considerations, whether accuracy, precision, or both are needed, and to what degree. This decision is important in setting the control limits for process analyzer performance and how they will be computed. If accuracy is critical, the control limits must incorporate the pooled standard deviation, calculated using the pooled variance from Equation 5-4. If only precision is important, then only the standard deviation calculated from repetitive measurements by the process analyzer on the same reference need be used; this is the process analyzer standard deviation, which is the square root of the second term in Equation 5-4.

Accuracy and precision are not the same. It is possible to have good accuracy and poor precision, or vice versa, although the two are not mutually exclusive. Which is more important is both a technical and a management decision.

Capability

To this point most of our discussion has revolved around the control chart and the 3σ control limits. There has been little mention of specifications for the process output. The distinction between specifications and control limits is very important, especially since newcomers to SPC/SQC tend to confuse the two. The two are not only different, but a calculated number called "capability" provides an objective way to compare the control limits and product, or output, specifications.

Capability is an objective statistical comparison of the control limits and the specifications and is a measure of how well the process meets the specifications.

Early in this discussion, the algebraic formula for the standard deviation and 3σ control limits was given. Recall that the 3σ control limits are calculated directly from data about the process output. This is the key to understanding why control limits and specifications are different concepts. The control limits are "natural" numbers that come from the process itself. In contrast, the specifications are numbers that people or circumstances set in order to achieve a desired result. Examples of specifications are chemical concentrations of manufacturing output which must be achieved in order to be able to sell product to a customer or to be able to use the product of the process safely and economically for some particular purpose. In the first case, a customer may not accept a product if the concentration of the key ingredient is not between 25.2 and 26.1%. In the second case, the product may perform best in the next production step if the key concentration is as close to 65% as possible, but may react explosively and destructively if it is above that value. The 25.2 and 26.1% values are upper and lower specification limits, and 65% is a single specification limit.

Capability is a measure of process potential if the process could be centered in its specification range. By definition, a process that is out of control as measured by the 3σ control limits is not capable. Therefore, capability calculations are done only when the process is in statistical control. There are several algebraically related equations for calculating capability. The measure of process potential is:

$$C_p = \frac{USL - LSL}{6\sigma} \qquad (5\text{-}5)$$

where:

C_p = process potential
USL= upper specification limit
LSL= lower specification limit
σ = standard deviation of the process data

The measurement of actual process performance is:

$$C_{pk} = \frac{min\,[\,(USL - \bar{x}),\ (\bar{x} - LSL)\,]}{3\sigma} \qquad (5\text{-}6)$$

where:

C_{pk} = process capability
$min\,[A, B]$ = minimum of A or B
$\bar{x}$ = mean of process measurements

By taking the lesser of the two values, one is "conservative" in stating the process capability. For a single specification, the process capability is simply:

$$C_{pk} = \frac{USL - \bar{x}}{3\sigma} \qquad (5\text{-}6a)$$

or

$$C_{pk} = \frac{\bar{x} - LSL}{3\sigma} \qquad (5\text{-}6b)$$

Both C_p and C_{pk} assume that the distribution is normal.

Because the objective of continuous improvement is to narrow the control limits (implying a smaller standard deviation) a higher capability is more desirable than a lower capability. Some customers require not only that the product be within their specifications, but also that the capability be some minimum value.

In process analyzer work, the specification may be to match the analyzer reading on the chemical process as closely as possible to the corresponding measurements made on a laboratory instrument. In this case, the specification would be zero—no difference between measurements made independently on two instruments. (This is a common "specification" in process analyzer work, although it is not always a wise one, as the ultraviolet analyzer case discussed later in this chapter will demonstrate. When this specification is used, one is using the laboratory instrument as a primary standard. For this practice to be valid, the laboratory analysis method must be in statistical control by itself!)

Another specification may be to minimize the drift on the process analyzer from one service period to another as measured against a primary standard. This situation could have a specification of "zero" or upper and lower operational limits of ±0.2% of the full scale of the analyzer range with respect to the concentration of the primary standard. In any event, the measurement process using the analyzer must be in statistical control before capability is calculated. If achieving statistical control is approached as an opportunity to learn more about the analyzer, and how it interacts with the manufacturing process, then setting specifications and using capability values can be done more logically and systematically.

Using a laboratory measurement as a primary or reference standard for a process analyzer is not wise if the shortcomings of the laboratory method and its common cause variations are not understood and accounted for.

If the analyzer is "not capable," that is, not accurate enough to satisfy manufacturing requirements, then statistical analysis on calibration data should be done. On the other hand, if the analyzer is "capable," the failure data and on-stream factor should be examined. In the latter case, it may be possible to reduce the maintenance load rather than to tighten the performance specifications on the analyzer. Unfortunately, there will always be disputes between production, who want tighter and tighter control limits, and engineering and maintenance, who want to stay with an agreed value and subsequently reduce maintenance costs.

R-Charts

Dispersion (Range) is the difference between the largest and smallest values in a sample set of data.

The R-chart is another useful statistical tool in SPC/SQC. Recall from earlier discussion that R is the symbol for range. We first used range as a measure of dispersion—the difference between the largest and smallest values in a sample set of data. As the $\bar{x}$-chart is used to chart shifts of the mean, the R-chart is used to chart shifts in dispersion. To understand how range is used in control chart applications, it is helpful to consider a control chart for which each point on the $\bar{x}$-chart is calculated from the mean of the ten values in a sample. Therefore, each point on the corresponding R-chart is calculated from the range of the same set of numbers. The control charts discussed in this chapter are constructed from single data points rather than from samples of two or greater. Therefore, this discussion will address only R-charts for a sample size of one. The reader is referred to the general texts listed in the References for details on R-charts for sample sizes of two or greater. Reference 1 is especially recommended for examples.

At this point, one might ask how a range can be calculated from a single value. For the purpose of constructing an R-chart, each adjacent pair of two points is treated as a sample. Therefore, the ranges for data of a sample size of one are the differences between point 1 and point 2, point 2 and point 3, point 3 and point 4, and so on. For an $\bar{x}$-chart of 30 data points, the corresponding R-chart has 29 points. Ranges are plotted on the control chart without regard to sign; ranges are never negative. Therefore, the individual ranges for a sample size of one are:

$$R_k = |x_k - x_{k-1}| \qquad (5\text{-}7)$$

where:

R = difference between adjacent measurements

x_k = any measurement

x_{k-1} = immediately preceding measurement

k = sequence number of the measurement

Although adjacent data points in the R-chart share one value in their calculation, this is not autocorrelation since the R-chart is based on differences rather than directly measured values. The centerline for the R-chart is simply the mean of all the ranges:

$$\overline{R} = \frac{R_1 + R_2 + \ldots R_n}{n} \tag{5-8}$$

where:

$\overline{R}$ = mean of all ranges

n = total number of ranges

The lower control limit (LCL_R) for R-charts for a sample size of one is always zero. The upper control limit is:

$$UCL_R = 3.27\overline{R} \tag{5-9}$$

where 3.27 is a constant derived from statistical theory.

Any point above the UCL_R indicates an out-of-control process. One chart may help identify the cause for the out-of-control situation when the other chart fails because the causes that affect the appearance of one chart do not always affect the other. Many SPC/SQC software packages automatically calculate and construct both charts at the same time. The author's experience has been mostly with $\bar{x}$-charts. Therefore, the case studies in this chapter use only $\bar{x}$-charts. In these case studies, the $\bar{x}$-charts provided sufficient information for the extent of process improvement that was necessary and justifiable when the studies were made. Note that the $\bar{x}$-chart is used to control the mean, and the R-chart is used to control variation.

... the $\bar{x}$-chart is used to control the mean...

... the R-chart is used to control the variation.

Cause-and-Effect Diagrams

Although the control chart is the foremost SPC/SQC tool, it tells only whether an out-of-control situation exists. It does not, by the numbers and patterns alone, tell why the process is out of control. Bringing the process back into control is, of course, the job of the people involved. Once the out-of-control situation is detected, the cause-and-effect diagram is a useful tool for finding out where the problem lies. Other names for cause-and-effect diagrams are Ishikawa diagrams (after the Japanese professor who popularized their use) and fishbone diagrams (after their physical appearance).

Once the out-of-control situation is detected, the cause-and-effect diagram is a useful tool for finding out where the problem lies.

Figure 5-11 is a "generic" cause-and-effect diagram that might be used for determining the cause of a process analyzer reliability problem. Note the fishbone appearance. Note also that the possible causes are broken down into four areas: materials, procedures, workers, and machines. A sample of possible causes for a process analyzer-related problem might be:

Materials

- Calibration standards
- Manufacturing process stream
- Chemical reagents
- Forms used for calculations and recordkeeping

Procedures

- Calibration procedure
- Calculations performed on the raw data
- Use of the raw data
- Maintenance procedure
- Repair procedure
- Maintenance recordkeeping
- Software routine
- Maintenance training
- Manufacturing operator training

Workers

- Level of experience
- Level of training
- Management support
- Fatigue
- Understanding of procedures
- Physical impairment
- Others interfering with work
- Inadequate staffing
- Type of supervision

Machines

- Appropriateness of the analyzer for the application
- Tools used for maintenance and repair
- Installation engineering and design
- Manufacturing process equipment
- Process control computer

The "4M's" of process improvement are Materials, Methods, Machines, and Manpower. These are the major branches of cause-and-effect diagrams.

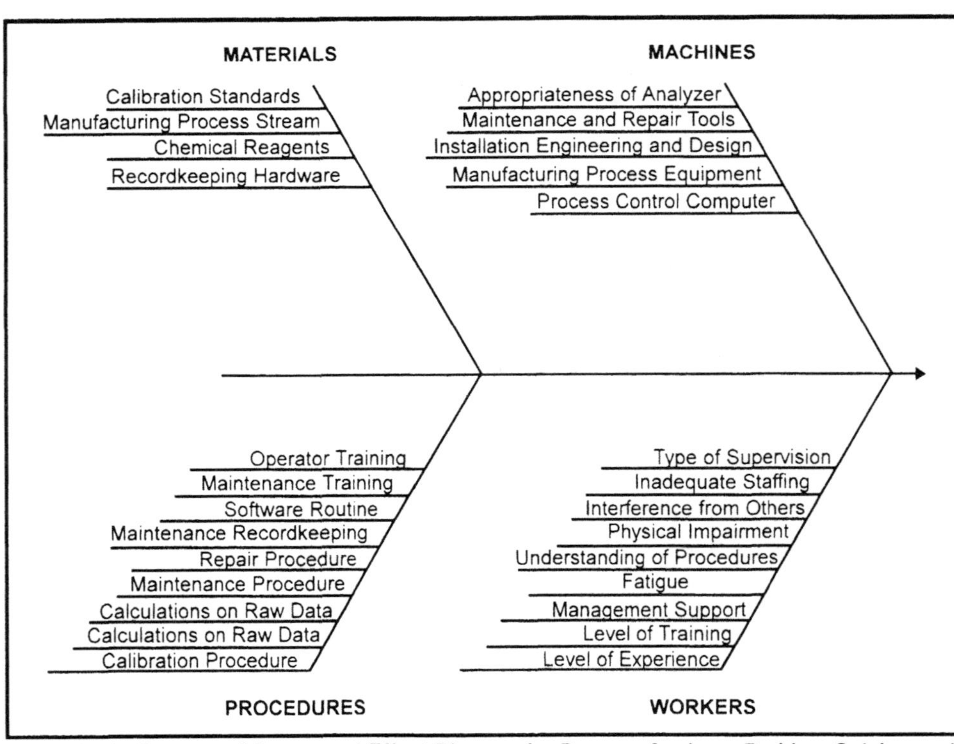

Figure 5-11. Suggested Cause-and-Effect Diagram for Process Analyzer Problem-Solving and Troubleshooting (This example can be used as a starting point and customized for particular process analyzer applications and situations.)

In practice, many of these possible causes can be used for addressing out-of-control situations. Some items can logically fit under more than one of these four categories. For example, a calibration standard (materials) may be the cause because it was not properly prepared (procedures or workers), because the reagents were bad (materials, again), or because the laboratory equipment used for its preparation was not working properly (machines). Some cause-and-effect diagrams include a fifth major branch called "environment" that includes uncontrollable variables. This is okay so long as the causes listed there are truly out of the control of the organization.

In order to cover all these possibilities, it is helpful to break down some of the "first-level" causes into one or two sublevels in order to develop highly specific possible causes that can be tested. "Testing" the possible causes to determine whether they actually caused the out-of-control condition is the next step. Testing may be as simple as asking key personnel a few questions or as involved as performing a laboratory or manufacturing process study. (Some process analyzer "problems" are actually problems with the manufacturing process that the analyzer is measuring; the process analyzer is only the "bearer of bad news!") There are no specific rules for this part of SPC/SQC. The members of the team must use their experience, training, common sense, and the tools at their disposal from this point forward in the problem-solving process. There are several good problem-solving methods that are adaptable to SPC/SQC. One of the better and more popular methods is Kepner-Tregoe [Ref. 7].

Teamwork and brainstorming are critical to good problem solving and decisionmaking.

The causes offered in Figure 5-11 are only suggestions or a starting point. The team involved in solving any specific problem should hold a meeting to brainstorm. Teamwork and brainstorming have two direct benefits for problem solving:

1. Everyone gets involved and therefore accepts some "ownership" in getting the problem resolved.
2. One person's comments often help others generate ideas that might not otherwise be suggested or recognized.

Using cause-and-effect diagrams and problem-solving procedures sounds formidable because, for many workers, this is a new approach. However, these diagrams and procedures are powerful tools that become easier with use.

Another valuable statistical tool is the linear regression, or least squares, plot. This is a method for determining how well a straight line describes the relationship between the independent and dependent variables. Linear regression is never a substitute for a control chart, but it is frequently an indispensable adjunct. Linear regression is described in Appendix 3.

Data Selection

In order to apply SPC/SQC methods, a lot of specific data must be gathered. This aspect is demonstrated in the next section of this chapter. In a manufacturing process, a lot of data is also generated in the normal course of operations. Therefore, it is important to be very selective about which data is subjected to statistical analysis. The examples in this chapter initially used only two sets of measurements for problem solving and reliability improvement; other measurements were introduced later to discover highly specific information or to verify what the initial control charts suggested.

... the goal of the SPC/SQC effort: improve the accuracy of the analyzer or reduce maintenance.

Too much data gathering can be overwhelming and confusing. It is thus important to control-chart as few variables as are necessary to complete the job. Using a limited number of variables keeps the team's attention focused on the immediate task and minimizes confusion from secondary or unrelated matters. At this point in the project, the team should decide the goal of the SPC/SQC effort: improve the accuracy of the analyzer or reduce maintenance? Though the two goals are not mutually exclusive, one is usually more important than the other at a given time.

Select only the data necessary to do the job. Unnecessary data collection and manipulation is frustrating for the personnel on the team and can result in their losing sight of the problem-solving objective.

Application to Process Analyzers

Gas Chromatography

In this case study, a process gas chromatograph was used to measure the concentration of a key component in the overhead stream from a distillation column in a plant that manufactures a common plastic monomer. The analysis was necessary to control the operation of the column. The sample was a solid at room temperature and a viscous liquid when heated to moderate temperatures. The sample was difficult to analyze by any process or laboratory methods. Since the sample could be vaporized at very high temperatures, process chromatography was the least difficult way to effect the analysis.

In addition to the process analysis, a laboratory analysis was performed on a process stream sample once every 12 hr. The laboratory chromatograph was calibrated to an external standard, but no calibration standard was suitable for the process chromatograph. Therefore, the results of the laboratory analysis were, by default, the standard to which the process chromatograph was calibrated.

The process chromatograph was frequently recalibrated to agree with the results of the most recent laboratory analysis. Plant management made the decision to recalibrate on the basis of the difference between the most recent laboratory results and the current process chromatograph reading; if the difference was greater than a certain amount, recalibration was performed. This is overcorrection. The project objective was to increase the reliability of the process analyzer. After some of the hardware was upgraded and replaced to assure analytical, mechanical, and electronic reliability, data collection and control-chart preparation began in order to assure effective use of the analytical data.

While the process analyzer was out of service, the distillation column was operated on the twice-daily results of the laboratory analyzer and the physical measurements (temperature, level, pressure, and flow) from the ordinary process instrumentation on the column. After the analyzer was restarted, the practice of recalibrating the process analyzer on the basis of the laboratory results (overcorrecting) was resumed. The only change made was that the time at which the laboratory sample was drawn was noted on the process analyzer recorder. If recalibration was necessary, the correction could be made on the basis of how the process was behaving at a single point in time rather than on analytical results from different times. (The assumption that the process would not change between the time the laboratory sample was taken and the results were returned to the plant had theoretical validity, but could not be experimentally demonstrated because of practical limitations.) This data-collection method provided "paired data"; that is, any differences between the process analyzer results and the laboratory analysis results were due only to differences between the two analytical methods and not to any changes in the manufacturing process.

The paired data were collected and a control chart was made of the difference between the two analytical results (process analyzer results minus laboratory analyzer results). This control chart is shown in Figure 5-12. Recalibrations, corresponding to the difference between the most recent data pair, were performed before the next data pair were collected. Ideally, the difference should be zero; that is, both analytical methods should represent the chemical process accurately and in the same manner. Although no analytical method is "ideal," the methods should track each other closely, with only a simple mathematical relationship between the two. Whether this relationship exists cannot be determined from Figure 5-12, but some variability between the methods is evident. It cannot be determined from the tampered data in Figure 5-12 whether this is normal variability or whether an additional source of variability is present.

As discussed earlier, changing a process or a piece of equipment frequently on the basis of one piece of information is called overcorrecting or tampering. The practice of recalibrating the process analyzer on the basis of only the most recent data pair is a specific example of tampering. When the tampering was discovered, plant management was persuaded to "let the process (analyzer) run" without the frequent recalibration; the results are shown on the control chart in Figure 5-13. Note that the difference between the two sets of analytical data, while still not zero, is closer than it was during tampering. Note also that the standard deviation is smaller and that the control limits are much narrower for this set of data than for the tampered data.

... changing a process or a piece of equipment frequently on the basis of one piece of information is called overcorrecting or tampering.

After letting the process run, the averages of the process analyzer data and the laboratory data were compared. Since there was no tampering, it was believed that the individual sets of data were representative of the behavior of the respective analytical methods. Therefore, the process analyzer was recalibrated by the amount of the difference between the averages of the untampered data. Figure 5-14 is the control chart for the next period of data collection. Note that the average differences for the paired data moved close to zero, but the standard deviation of the difference remained about the same as in Figure 5-13. Therefore, another calibration was performed on the basis of this set of data. Figure 5-15 shows the results of the final calibration. This time, the difference between the two sets of analytical data was less than the repeatability of either analytical method, indicating no further need to recalibrate.

It must be noted, however, that any change in either analytical method or in the manufacturing process that might affect the basis of either analytical method may result in the need to recalibrate. Examples of such changes are chromatograph column changes, new primary standards for the laboratory method, change of sample size in either method, new operating conditions in the distillation column, and lengthy (several days) shutdown of either the manufacturing plant, the process chromatograph, or the laboratory chromatograph. Furthermore, control charting should not be stopped when the two methods consistently "agree," as they did in Figure 5-15. Continuous control charting will detect such problems as instrument drift, sample line plugging (as evidenced by obtaining the same analytical result for several analytical cycles in succession), and some mechanical and electronic problems. Control charts of data from the individual analytical methods can also be helpful when related to other data from the manufacturing process, but these data are generally of greater use in operating the plant than in assuring process analyzer reliability. Table 5-2 is a summary of the mean differences, standard deviations, and control limits for the four sets of data. Note that the mean differences approach zero as the project progresses and that the standard deviations and control limits decrease after tampering is stopped.

Mean = 2.50
s = 3.24 n = 30
LCL = −7.22
UCL = 12.22

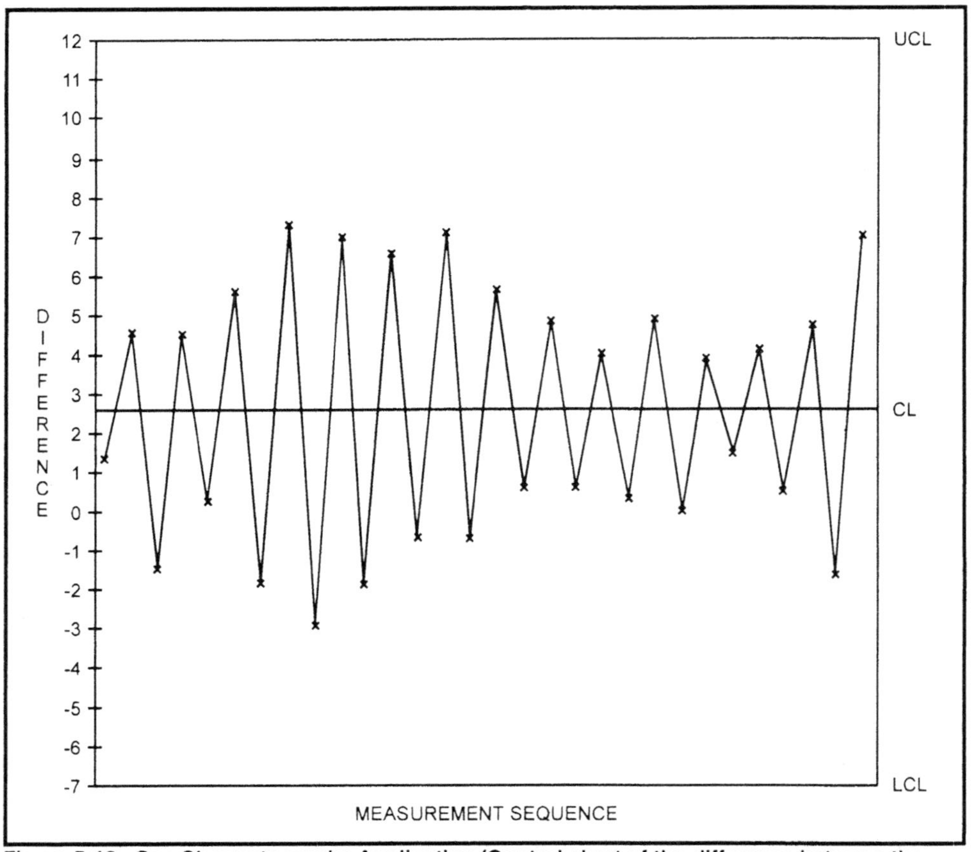

Figure 5-12. Gas Chromatography Application (Control chart of the difference between the process chromatograph measurement and the laboratory chromatograph measurement with tampering.)

Mean = 2.39
s = 0.93 n = 30
LCL = −0.40
UCL = 5.18

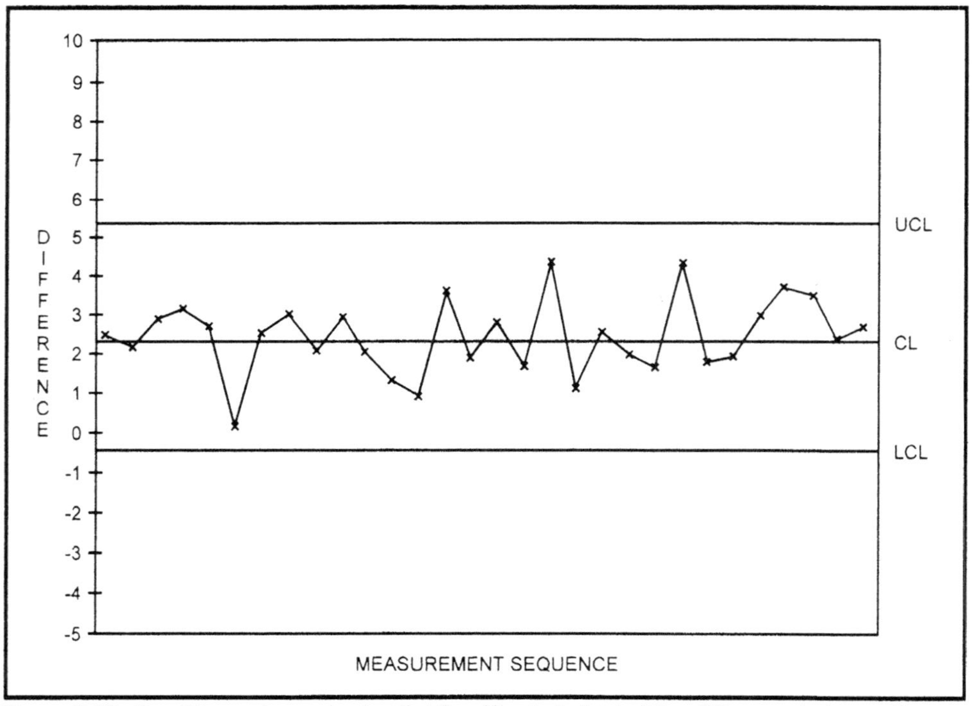

Figure 5-13. Gas Chromatography Application (Control chart of the difference between the process chromatograph measurement and the laboratory chromatograph measurement without tampering.)

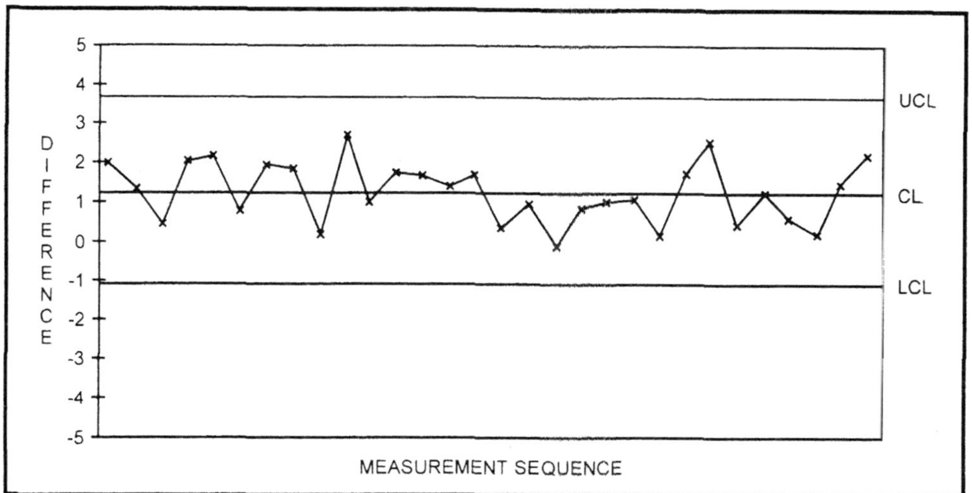

Mean = 1.29
s = 0.77 n = 30
LCL = −1.02
UCL = 3.60

Figure 5-14. Gas Chromatography Application (Control chart of the difference between the process chromatograph measurement and the laboratory chromatograph measurement without tampering and after one recalibration.)

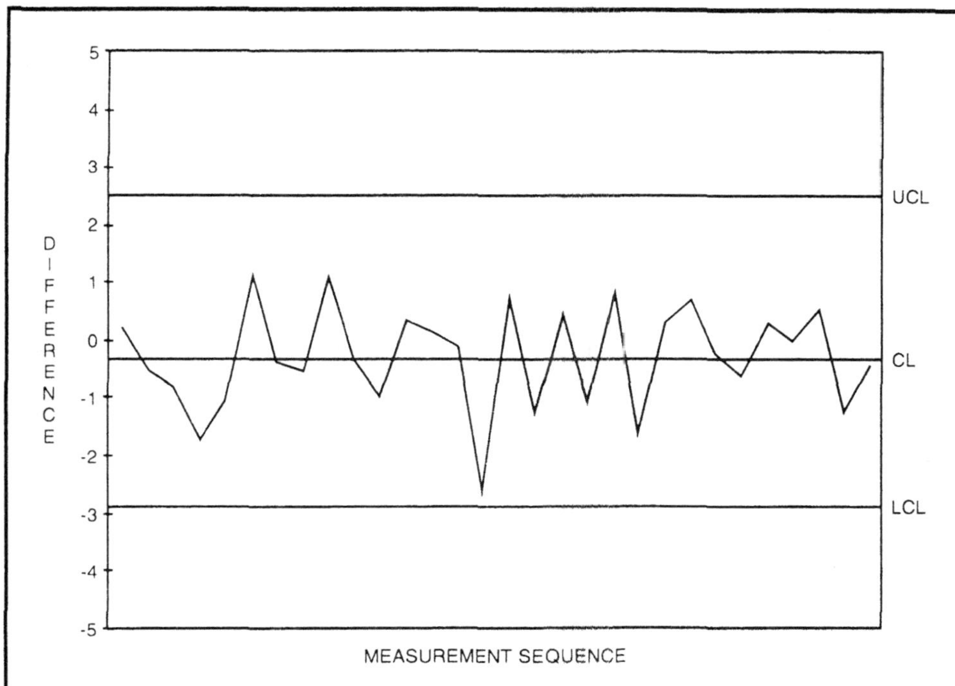

Mean = −0.29
s = 0.90 n = 30
LCL = −2.99
UCL = 2.41

Figure 5-15. Gas Chromatography Application (Control chart of the difference between the process chromatograph measurement and the laboratory chromatograph measurement without tampering and after two recalibrations.)

Table 5-2. Summary of Descriptive Statistics for Continuous Improvement of Process Gas Chromatograph

Data Set	Mean	Standard Deviation	LCL	UCL
Initial "tampered" data (Figure 5-12)	2.51	3.24	–7.22	–12.22
First "let the process run" data (Figure 5-13)	2.39	0.93	–0.40	–5.18
After first calibration (Figure 5-14)	1.29	0.77	–1.02	–3.60
After final calibration (Figure 5-15)	–0.29	0.90	–2.99	–2.41

Continuous control charting, even after the process is brought in control, is necessary to assure that the process stays in control and to facilitate continuous improvement.

In this project, no attempt was made to narrow the control limits other than to eliminate the tampering. Therefore, stating measures which could have been taken to narrow the control limits would be speculative. In general, though, reducing the variability in any part of the "process" carries the potential to reduce the variability in the results. In this case, it might be controlling more carefully the temperature in the process analyzer shelter, controlling more closely the air pressure in the pneumatic parts of the process analyzer, using an automatic sampler on the laboratory instrument, reducing the variation in the time between when the laboratory sample is taken and when it is run, and making sure that the laboratory sample is taken the same way at the same time on each shift of each day. Furthermore, these simulated control charts show no out-of-control points. The actual control charts for this project showed several such points. However, the reasons for the out-of-control points were not determined at the time under the assumption that it was more important to first eliminate the tampering and obtain better overall agreement between the process analyzer and the laboratory method.

Ideally, there should be zero difference between the means for the two analytical methods. The zero difference is a single specification. We can take the zero ideal difference and set specifications based on the repeatability of the analytical methods. If the repeatability of each method is ±1%, then the specifications become ±1.414%, based on the formula for calculating the variance of two independent variables using the additive property of variances:

$$s^2 = s_1^2 + s_2^2 \qquad (5\text{-}10)$$

where:

s^2 = total variance

s_1^2 = variance of the first set of independent data

s_2^2 = variance of the second set of independent data

This is an application of the pooled standard deviation discussed earlier and presented in Equation 5-4. The difference between the paired process and laboratory values implicitly uses the pooled standard deviation, since both sets of values were used to calculate the difference. However, note that this application of SPC/SQC to process analyzers and the following applications in this chapter, are for solving major reliability problems rather than for validation or ongoing maintenance monitoring.

Figure 5-16 is the scatter plot of the tampered data from Figure 5-12 and the data from Figure 5-15 after the two analytical methods were brought into agreement. These data can be helpful, but note that they are of limited value in determining how well the two methods agree and deciding whether to take any action. Note that the results of this application of SPC/SQC to this analyzer were, first, improved precision as a result of narrowing the control limits and data spread and, second, improved accuracy as a result of only infrequent analyzer calibration after precision was improved.

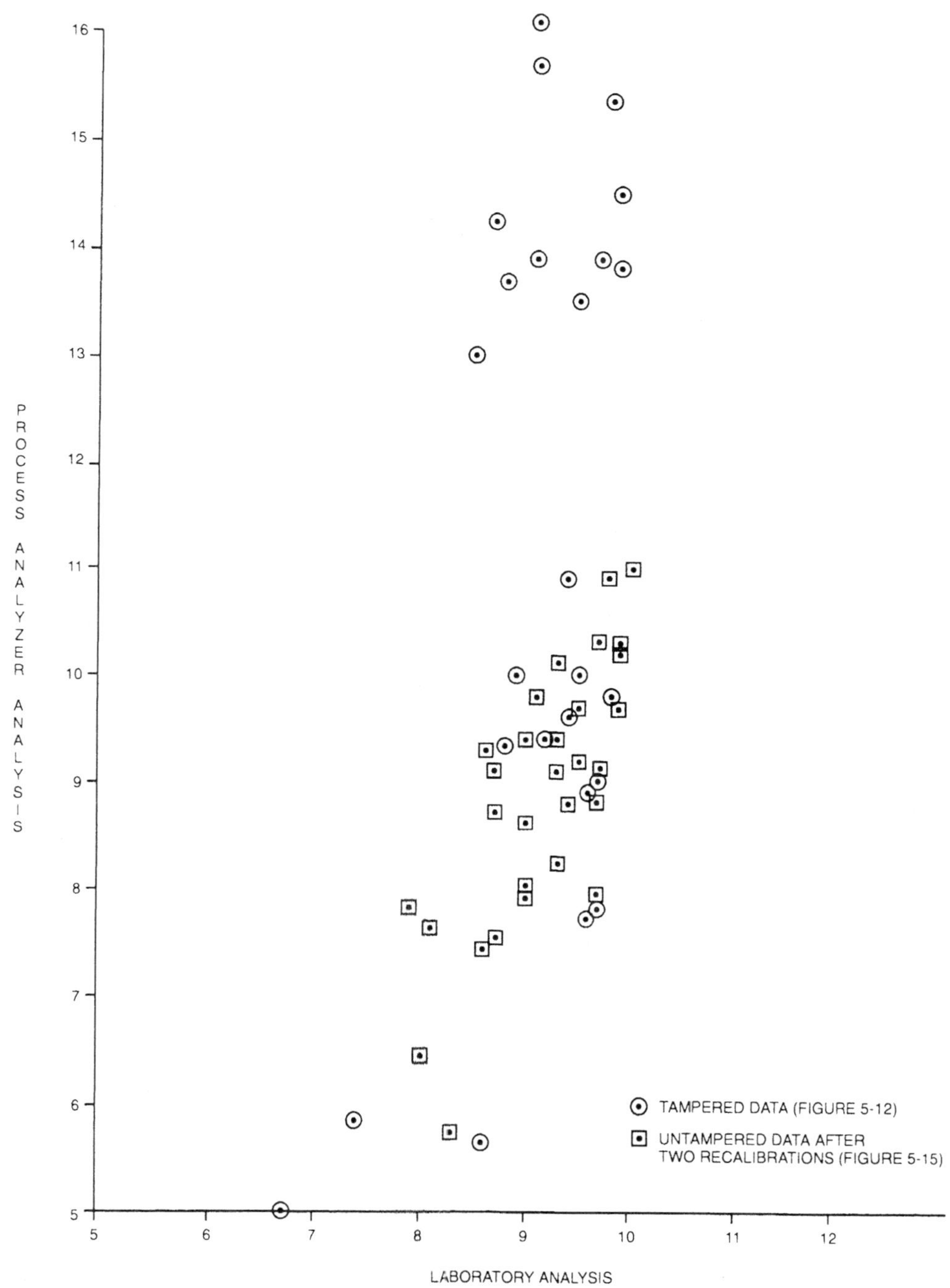

Figure 5-16. Gas Chromatography Application [(Scatter plot of process analyzer measurements versus laboratory analyzer measurements for tampered data (Figure 5-12) and untampered data after two recalibrations (Figure 5-15)]

Ultraviolet Photometer

This project involved an ultraviolet (UV) photometric analyzer that was intended to measure the concentration of component A in a two-component stream with component B. The two-component stream was fed to a polymerization reactor. The ratio of the two components determined the average molecular weight of the polymer. The reactor was operated in "campaigns" from several days to several weeks to produce a variety of products, differentiated primarily by average molecular weight. The time constant for the process, that is, for feed to the reactor to become salable product, was several days. Due to the long time constant, it was economically important to control the ratio correctly at the reactor. Although bad product could be reworked, the procedure was quite expensive because of the long time constant.

In addition to the process UV photometric analyzer, the reactor feed was monitored by two independent sets of mass flowmeters on components A and B before mixing, and by laboratory analysis of a daily composite sample taken at the process analyzer sample point. The two sets of mass flowmeters were adjacent to each other in the process piping but were based on different operating principles. The goal of the project was to improve the performance of the process analyzer, correlate the process analyzer results with those from the mass flowmeters, and decrease dependence on the daily laboratory analysis.

No hardware or software changes were made initially. The results of all four analyses were collected daily and individually control-charted. Care was taken to assure that readings from the process analyzer and the mass flowmeters were taken at the same time. The A:B ratio was calculated from the individual mass flowmeter readings. Since the laboratory analysis was run on a composite sample, this value was not time dependent. Ideally, the process should have run at steady state throughout a single campaign, but the long time constant, various adjustments, and process upsets prevented this from a practical standpoint. Therefore, a single campaign may have had associated with it differing A:B ratios from day to day.

The control charts for each of the four analytical methods are shown in Figures 5-17 to 5-20. Note that the process analyzer and the mass flowmeters demonstrate about the same response to the process changes; they show similar shifts, trends, and out-of-control patterns. However, the laboratory analysis differs considerably from the other three. It is important to note that these control charts demonstrate not only the behavior of the analytical methods, but also the behavior of the manufacturing process. The means, standard deviations, and control limits were not calculated for these control charts since the step changes in Figures 5-17 to 5-19 indicate out-of-control situations. Serious doubt is also cast on Figure 5-20.

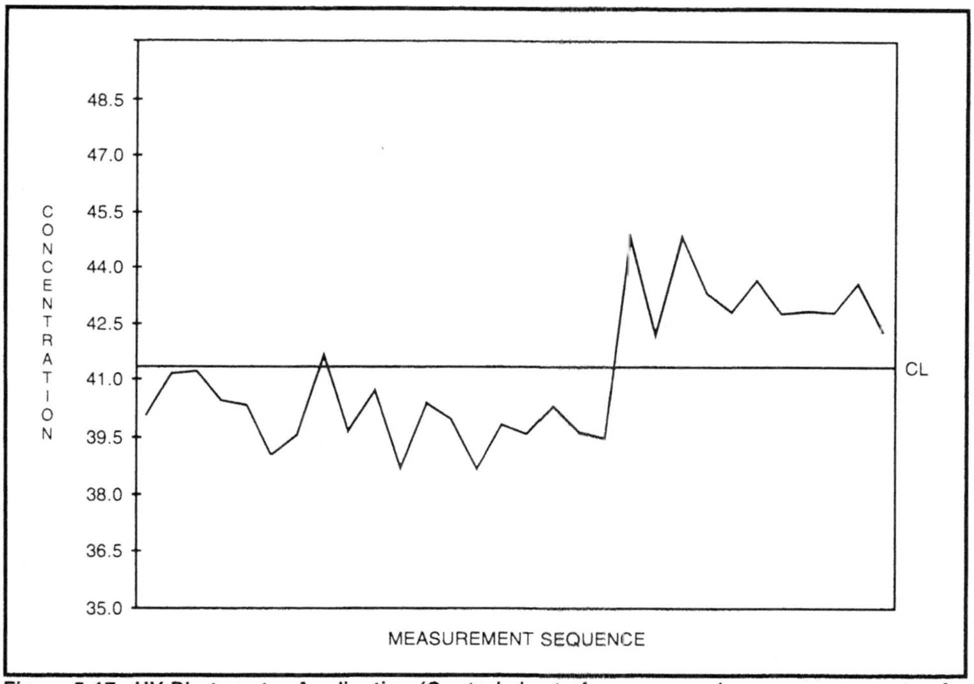

Mean = 41.2
n = 30

Figure 5-17. UV Photometer Application (Control chart of process analyzer measurements of the concentration of component A in component B.)

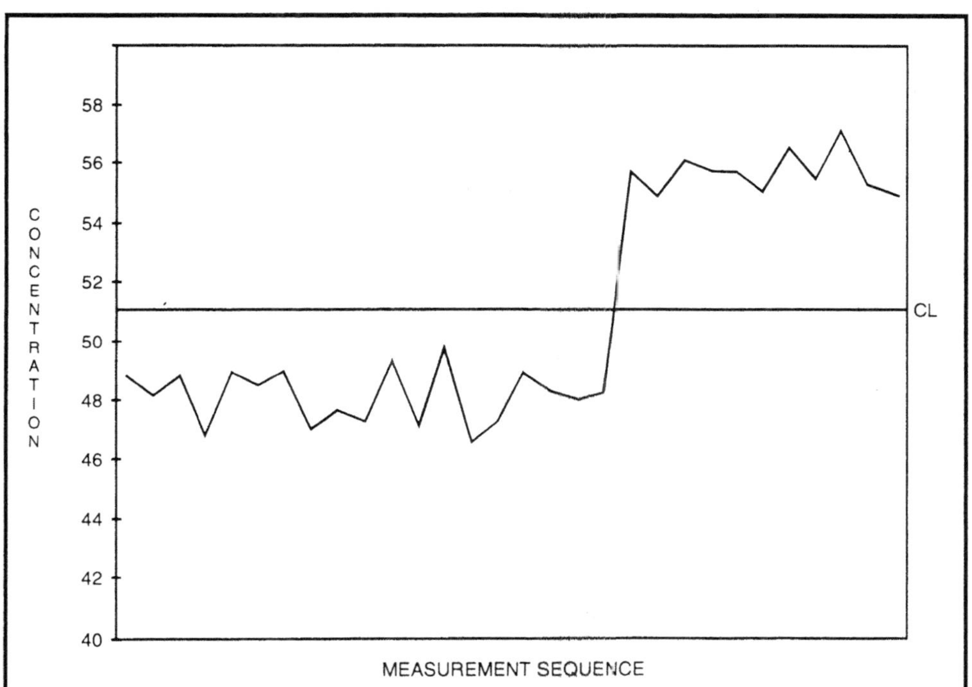

Mean = 51.0
n = 30

Figure 5-18. UV Photometer Application (Control chart of the calculated concentration of component A in component B based on measurements from the more reliable mass flowmeter pair.)

Mean = 46.9
$n = 30$

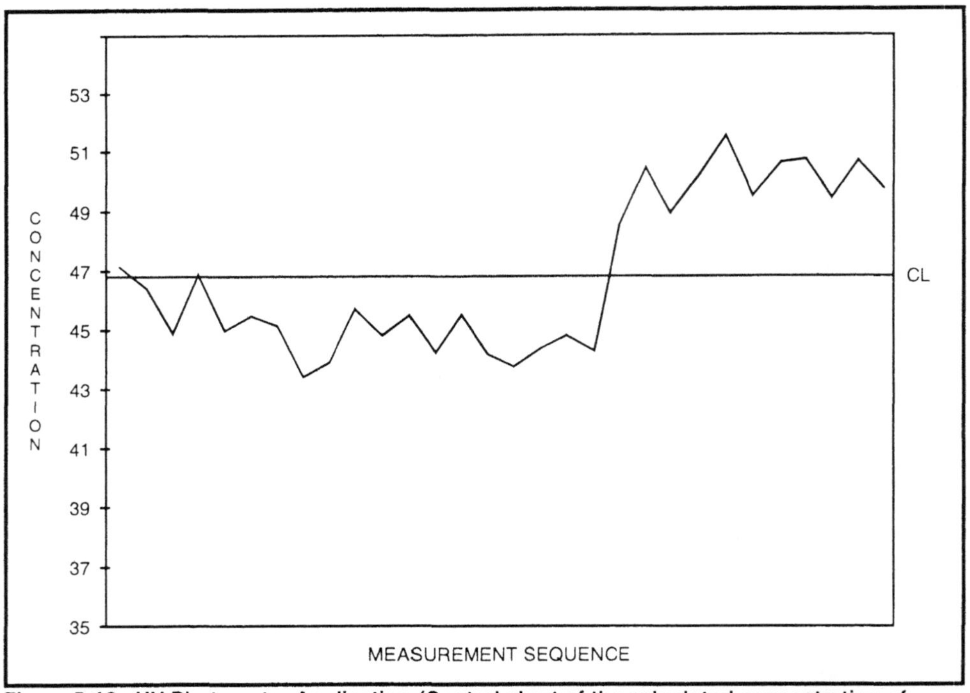

Figure 5-19. UV Photometer Application (Control chart of the calculated concentration of component A in component B based on measurements from the less reliable mass flowmeter pair.)

Mean = 43.2
$n = 30$

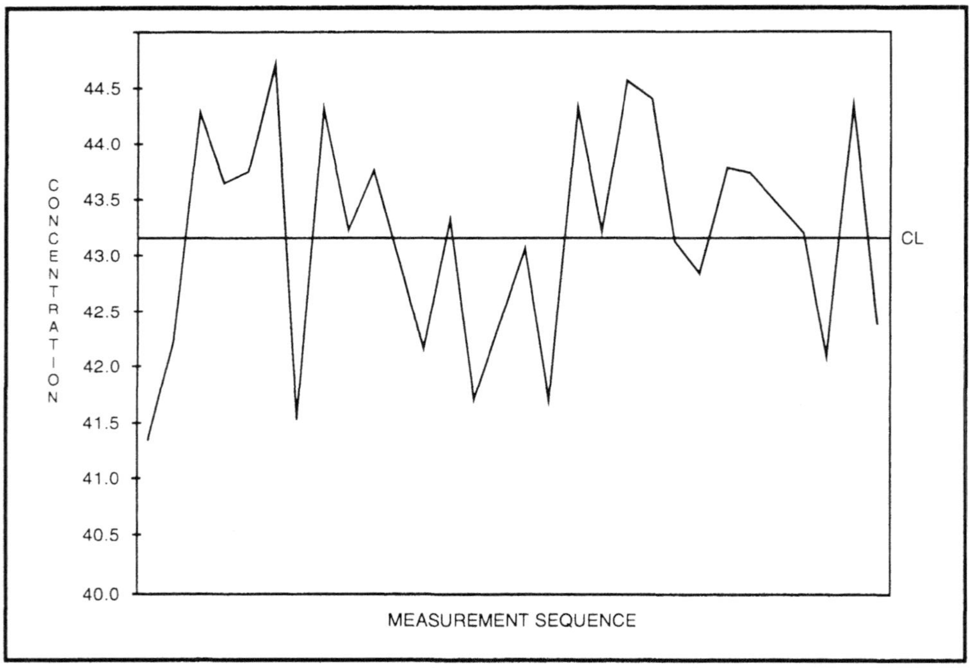

Figure 5-20. UV Photometer Application (Control chart of laboratory concentration measurements of component A in component B.)

In order to eliminate the effect of the process, it was necessary to compare the process analyzer results with those of the other methods. This was done by control-charting the difference between the process analyzer and the lab method and the difference between the process analyzer and the expected "more reliable" of the mass flowmeter methods. These comparisons are shown in Figures 5-21 and 5-22 using the same data as in Figures 5-17, 5-18, and 5-20. The two levels evident in Figures 5-17 to 5-19 correspond to campaigns for making different products. This necessitated changing the A:B ratio in the manufacturing process. If all analytical methods are valid, there should be no change in their performance regardless of the process change. But note the shifts in Figures 5-21 and 5-22 when a new campaign was begun. The shift appears more dramatic for the comparison of the process analyzer and the mass flowmeter method. Even given the assumption that the laboratory method is the least reliable of the methods since it did not follow the other three well, the shift is nonetheless apparent. Figures 5-23 and 5-24 are histograms of the data from Figures 5-21 and 5-22, respectively. Both histograms exhibit a bimodal distribution, an indication that the shift represents a change from one condition of the system to a different condition. Since the difference between two analytical methods should eliminate the effect of the manufacturing process, the only conclusion is that the analytical methods do not respond equally to the process change.

Based on knowledge of the operating principles of the mass flowmeters, there was no reason to believe that both sets were not responding linearly to the A:B ratio change. The tentative conclusion, therefore, was that the process analyzer was not responding linearly to concentration changes of the A:B stream. Additionally, comparison of the laboratory method to the other three methods indicated a performance problem with it as well. These two ideas were investigated simultaneously using primary analytical standards of A:B mixtures, prepared by weight in the laboratory. The mixtures were analyzed by both the laboratory method (titration) and in the process analyzer after it was taken off-line.

The process analyzer demonstrated nonlinear (non-Beer's law) behavior with respect to both the weights measured during preparation of the standards and the laboratory analysis results. Further investigation revealed that the applications work on the process analyzer installation had not taken into consideration the non-Beer's law behavior of the analytical system in the A:B concentration range in which the process was operated. The nonlinearity problem was quickly solved by installing an electronic linearizer card in the signal-processing part of the analyzer.

Because the laboratory analyses of the primary standards by titration, normally a very reliable method of analysis, were very close to the concentrations indicated by the weights, it was concluded that the analytical method, by itself, was not a problem. Investigation of this indicated that the composite sample-collection method was not reliable because one of the components in the A:B process stream is highly volatile. Since the sample compositor was neither cooled nor sealed to minimize evaporation, the daily laboratory analysis reflected the ambient temperature at the sample as much as it did the composition of the process stream for the collection period.

Since the sample compositor was neither cooled nor sealed to minimize evaporation, the daily laboratory analysis reflected the ambient temperature at the sample as much as it did the composition of the process stream for the collection period.

Lesser problems with the process analyzer installation were discovered and remedied during work on the major problems of nonlinearity and composite sample problems. This demonstrates that the use of SPC/SQC and other statistical methods is an opportunity to learn more about the process. The result of this project was that operations relied more on the process analyzer after the nonlinearity problem was demonstrated and solved. The laboratory analysis was dropped, at some cost savings, because the sampling problem could not be easily solved, and the process analyzer was, by then, reliable.

Frequently, installing a process analyzer and assuring its reliable performance removes the necessity and cost of laboratory analyses.

Mean = −1.96
s = 1.78 n = 30
LCL = −7.30
UCL = 3.38

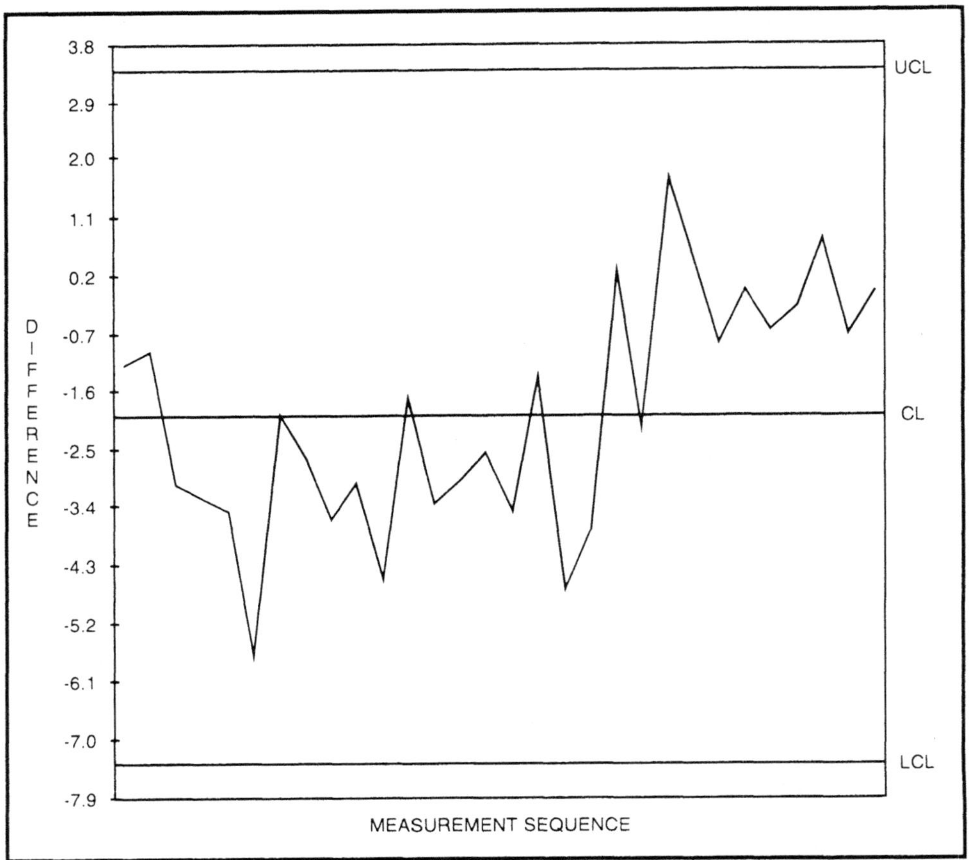

Figure 5-21. UV Photometer Application (Control chart of the difference between the process analyzer concentration measurement and the laboratory concentration measurement of component A in component B.)

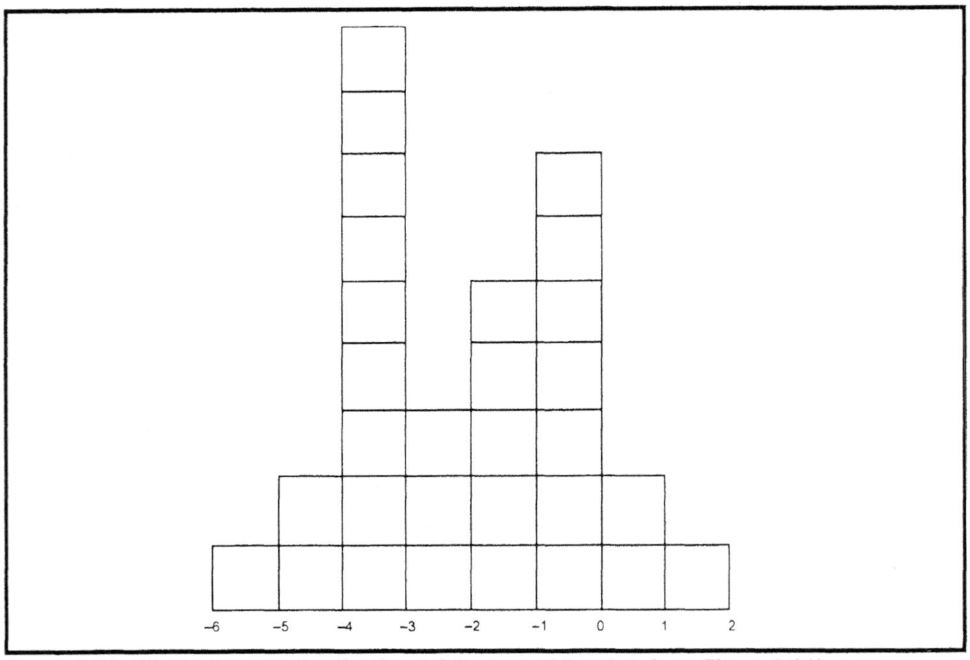

Figure 5-23. UV Photometer Application (Histogram of the data from Figure 5-21)

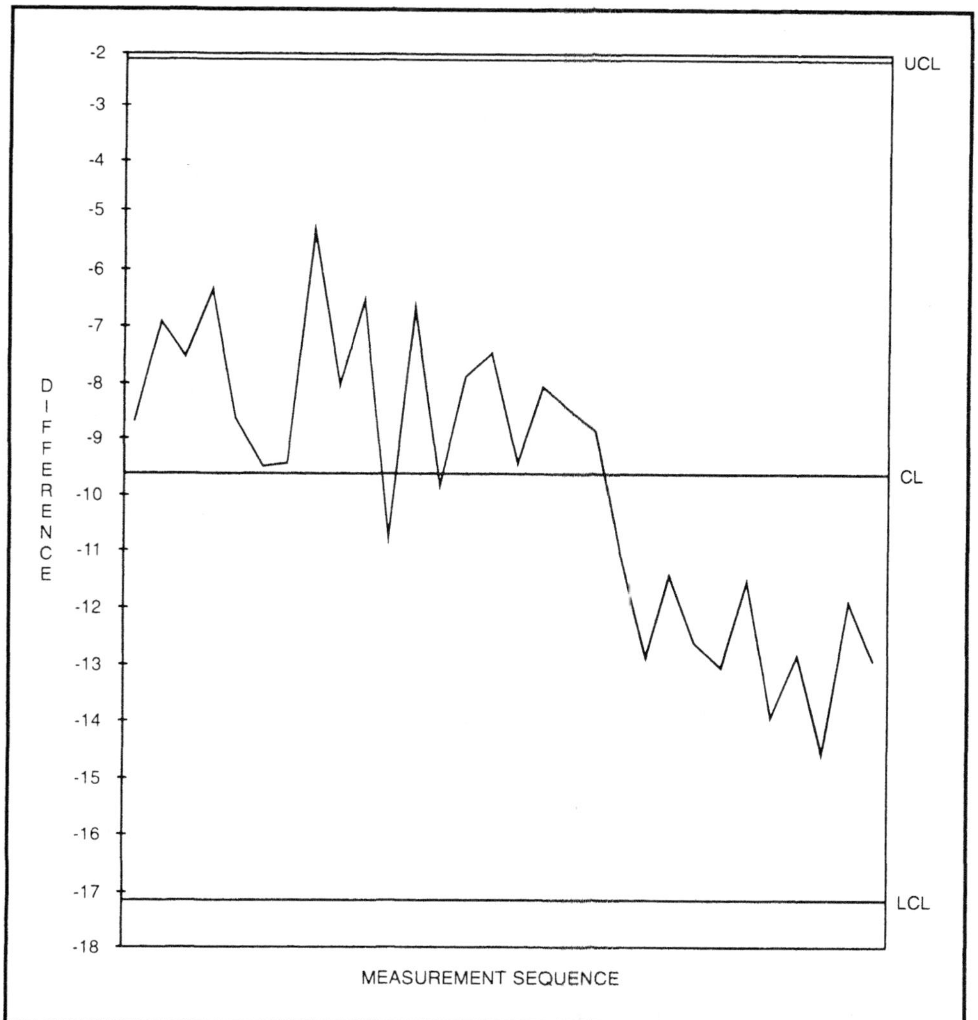

Mean = −9.71
s = 2.47 n = 30
LCL = −17.12
UCL = −2.30

Figure 5-22. UV Photometer Application (Control chart of the difference between process analyzer concentration measurements and the calculated concentration measurements from the more reliable mass flowmeter measurements of the concentration of component A in component B.)

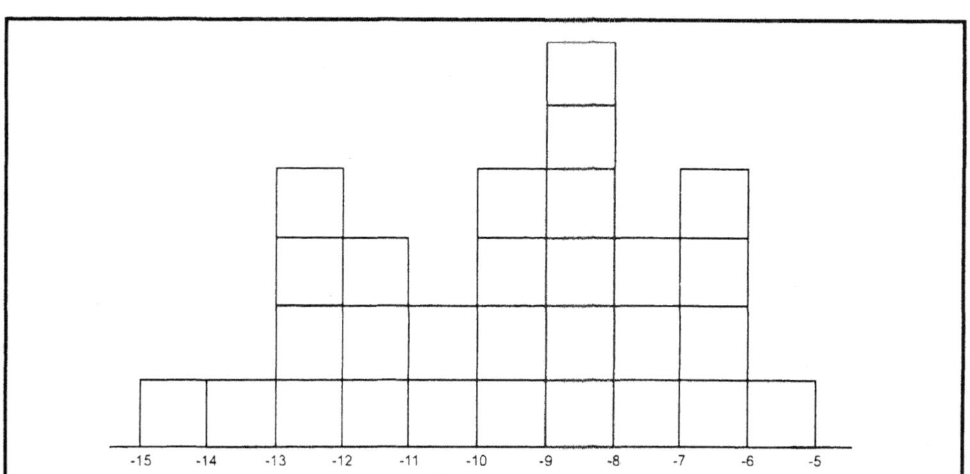

Figure 5-24. UV Photometer Application (Histogram of the data from Figure 5-22)

Infrared Photometer

A nondispersive infrared (NDIR) process analyzer was placed on a gaseous distillation offgas. A daily sample was taken to the laboratory for analysis by gas chromatograph. The laboratory analysis provided not only the concentration of the component that was measured by the process analyzer, but also the concentrations of all the other major stream components. The component of interest for the process analyzer comprised about 50 mol% of the stream. The laboratory GC analysis measured the concentration of five other stream components in addition to the one of interest. The GC analysis accounted for about 98 mol% of the stream, with the remaining 2 mol% being attributed to trace components and experimental error.

The process analyzer was considered acceptable for controlling the manufacturing process by general trends. However, the process analyzer results were not believed to be accurate and did not correlate well with the laboratory analysis. As a consequence, the process analyzer was frequently recalibrated on the basis of the laboratory results. The goal of the project was to determine why the process analyzer provided only general trends rather than accurate results and to improve analyzer performance.

A description of the stream is as follows, with a qualitative indication of the sensitivity of IR analysis to each one:

Component A
Key Component 50 mol% of stream. Only minor in references (from B and D).

Component	Type	IR Sensitivity
A	Unsaturated hydrocarbon	Active
B	Atmospheric component	Active
C	Unsaturated hydrocarbon	Active
D	Unsaturated hydrocarbon	Active
E	Atmospheric component	Active
F	Atmospheric component	Inactive

Component A was the component of interest (key component), which was usually about 50 mol% of the stream. Components B and D had absorbances at the same wavelength at which component A was measured. However, the usual concentrations of components B and D were too low most of the time to seriously affect the measurement of component A. Components C and E both had IR absorbance bands but not close to the measurement wavelength of component A. Component F was inactive throughout the IR region of the spectrum. None of the components had absorbances at the reference wavelength.

Therefore, only components B and D appeared to be of interest in finding out whether spectral interference was responsible for the questionable process analyzer results. Interestingly, the process analyzer was originally specified for measuring lesser concentrations of component A in an equal concentration of component B and a high balance concentration of component F. The original process analyzer specification was based on process calculations because the process was new and the plant had not been constructed. Surrogate compensation for the presence of component B was built into the process analyzer.

In order to speed data collection, the laboratory sample was collected and analyzed twice daily. The process analyzer results were compared with the laboratory analyses. A control chart of the difference between corresponding process analyzer measurements and laboratory measurements for component A is shown in Figure 5-25. Note that the first part of the control chart appears in control when taken by itself, as does the last part. Both parts taken together show no in-control points, though, due to level shift. Since the control chart follows the *difference* between the two analytical methods, we cannot tell, on the basis of this chart by itself, whether one method or the other changed, an external influence caused an artificial appearance of change, or if the concentration of component A actually changed and the two methods responded differently to the change. More information is necessary to determine what might have caused the level shift.

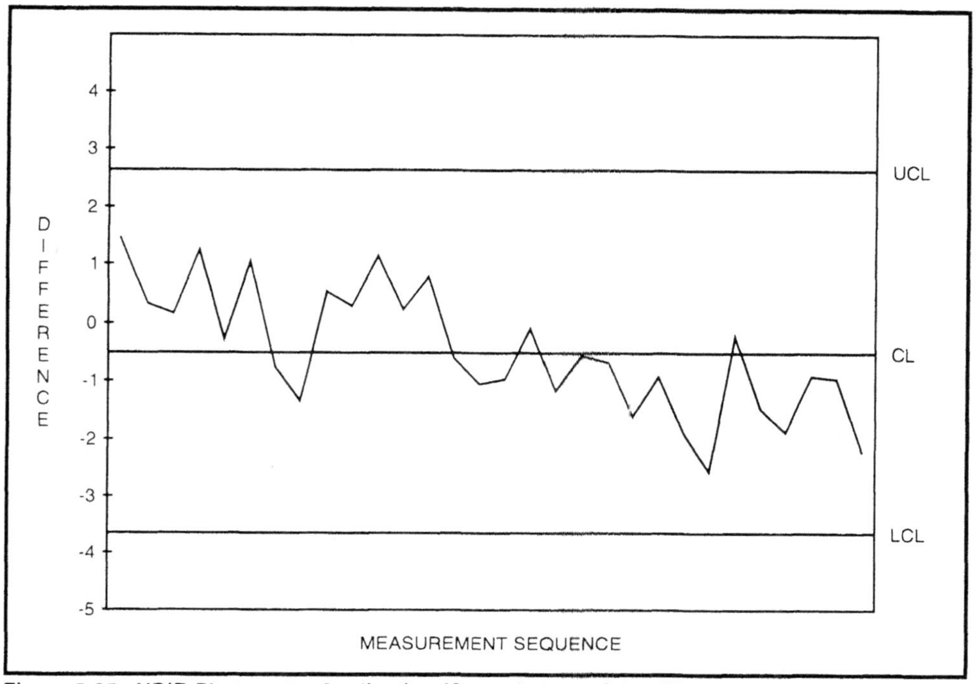

Mean = −0.50
s = 1.05 n = 30
LCL = −3.65
UCL = 2.65

Figure 5-25. NDIR Photometer Application (Control chart of the difference between process analyzer and laboratory measurements of the concentration of component A.)

First, it is helpful to look at the individual control charts for measurement of the concentration of component A by each of the analytical methods. These control charts are shown in Figures 5-26 and 5-27. Note that the chart for the laboratory method is in control but that the chart for the process analyzer shows the same level shift that the chart for the difference between the methods showed. Therefore, we can make the preliminary assumption that the manufacturing process changed and that something caused the process analyzer response to the manufacturing process to change as a level shift. This can be only a preliminary assumption because the control charts for the separate analytical methods do not cancel out other effects of the manufacturing process changes, as does the difference control chart. At this point, it is best to look for effects which could influence the process analyzer response and cause a level shift.

Since components B and D had absorbances at the same wavelength as component A, a look at the control charts for these components as measured by the laboratory method could be helpful. These control charts are shown in Figures 5-28 and 5-29. Note that each of these charts shows a level shift at the same point as the level shift in the difference control chart and the control chart for component A as measured by the laboratory method. The control charts for components C and E (Figures 5-30 and 5-31, respectively) which had absorbances near, but not at, the absorbance for component A, show similar level shifts. Figure 5-32 is a control chart of the laboratory concentration measurement on component F, the infrared-inactive stream component. The laboratory analyses for all stream components except component A showed a change in the manufacturing process to which the process analyzer was responding, although the response was undesirable.

Mean = 49.63
s = 0.76 n = 30
LCL = 47.35
UCL = 51.91

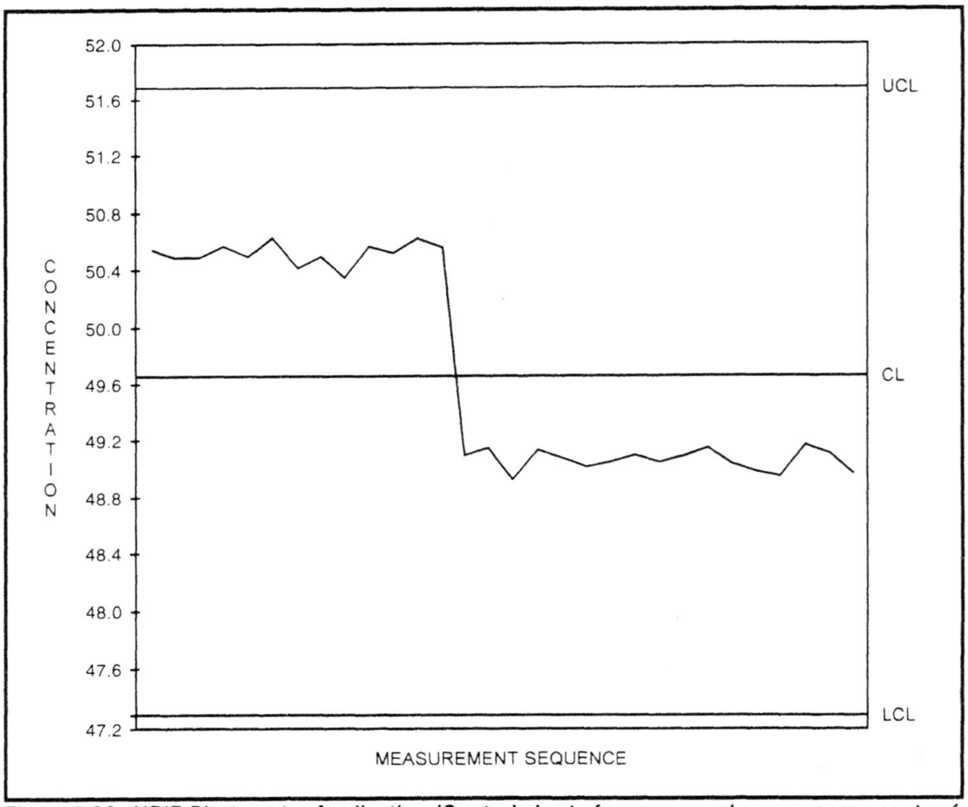

Figure 5-26. NDIR Photometer Application (Control chart of process analyzer measurements of the concentration of component A.)

Mean = 50.13
s = 0.71 n = 30
LCL = 48.00
UCL = 52.26

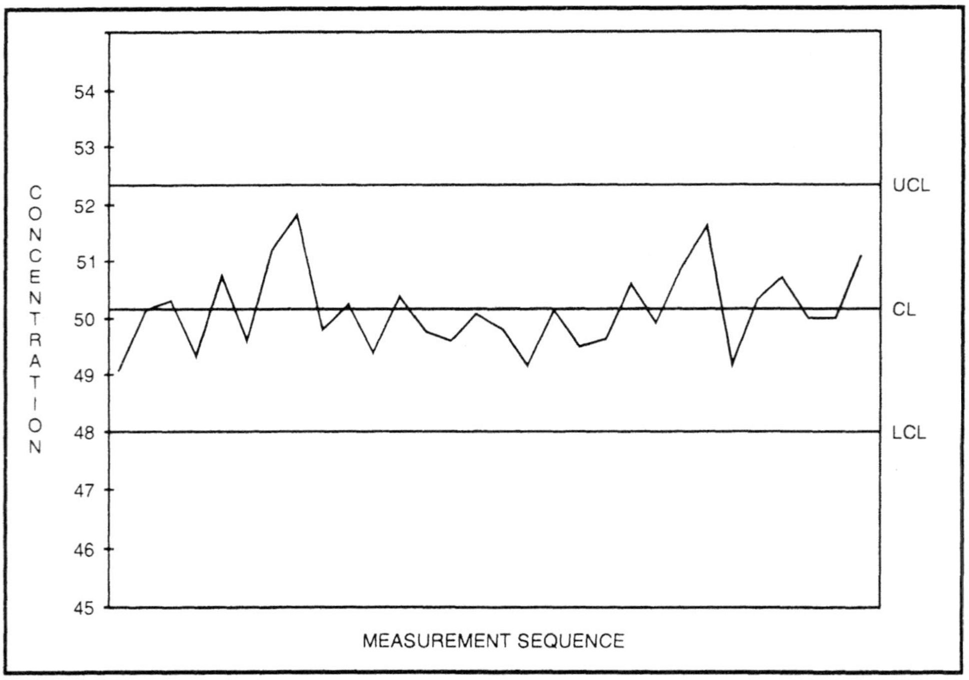

Figure 5-27. NDIR Photometer Application (Control chart of the laboratory measurements of the concentration of component A.)

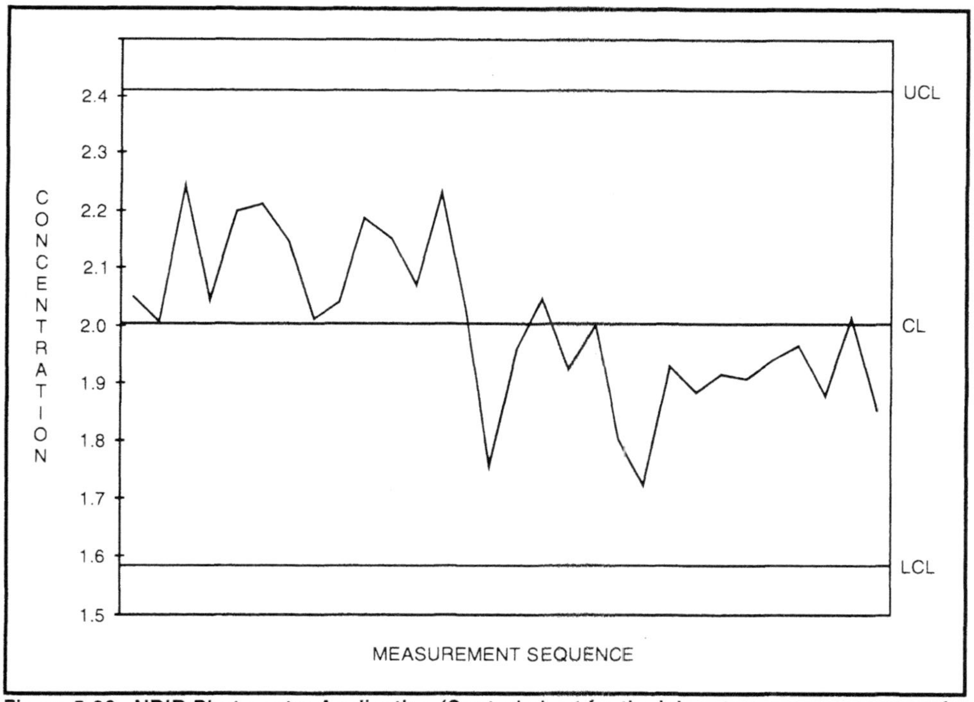

Mean = 2.00
s = 0.14 n = 30
LCL = 1.58
UCL = 2.42

Figure 5-28. NDIR Photometer Application (Control chart for the laboratory measurements of the concentration of component B.)

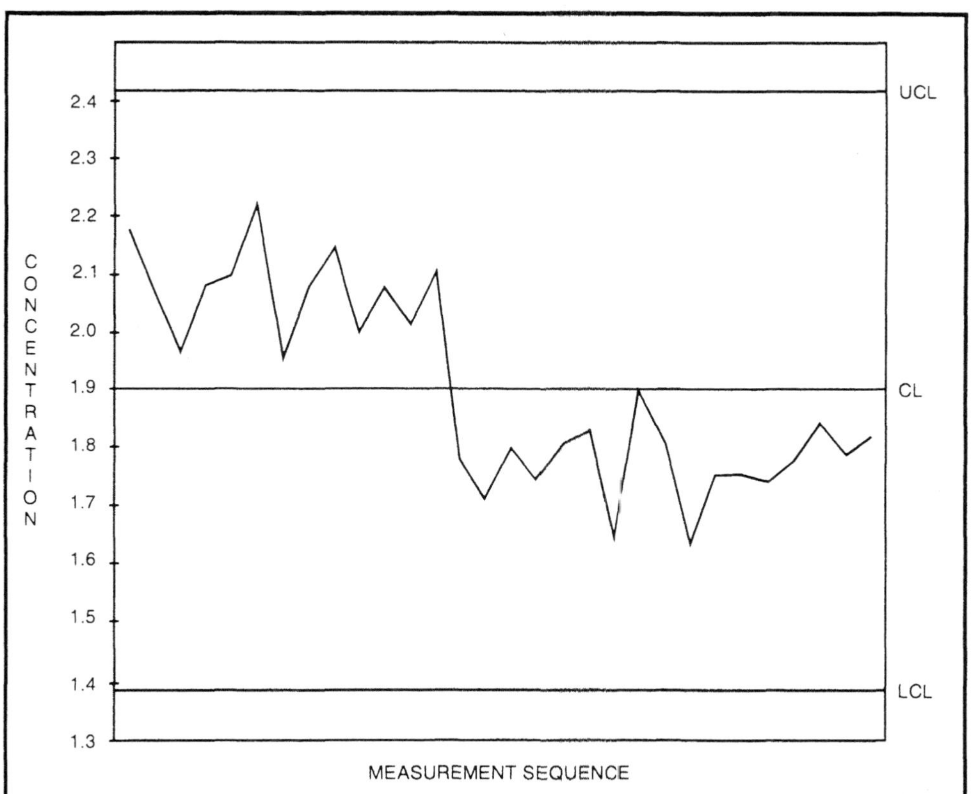

Mean = 1.90
s = 0.17 n = 30
LCL = 1.39
UCL = 2.41

Figure 5-29. NDIR Photometer Application (Control chart of the laboratory measurements of the concentration of component D.)

Mean = 31.23
s = 2.02 n = 30
LCL = 25.17
UCL = 37.29

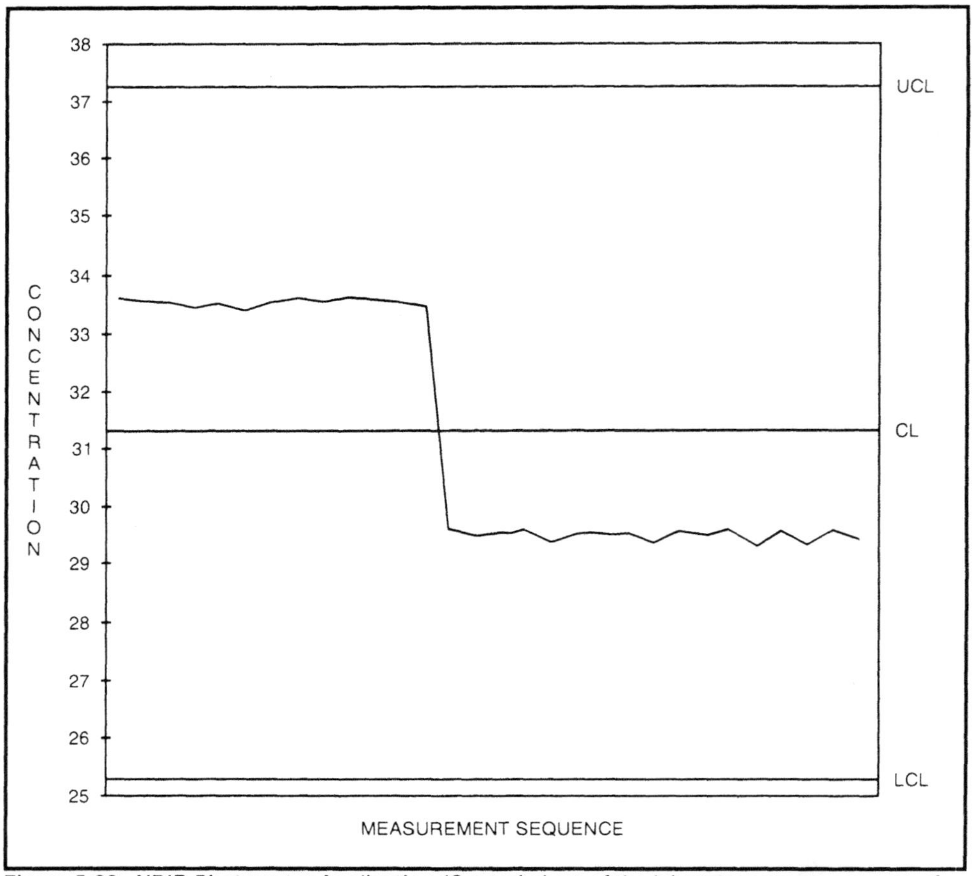

Figure 5-30. NDIR Photometer Application (Control chart of the laboratory measurements of the concentration of component C.)

Mean = 1.22
s = 0.25 n = 30
LCL = 0.47
UCL = 1.97

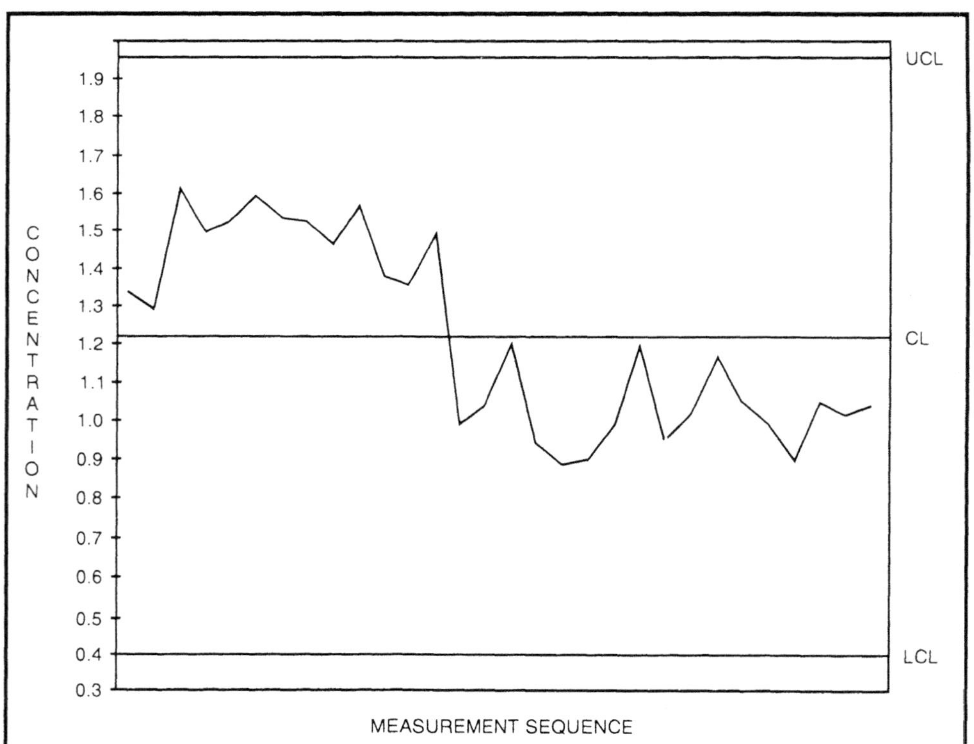

Figure 5-31. IR Photometer Application (Control chart of the laboratory measurements of the concentration of component E.)

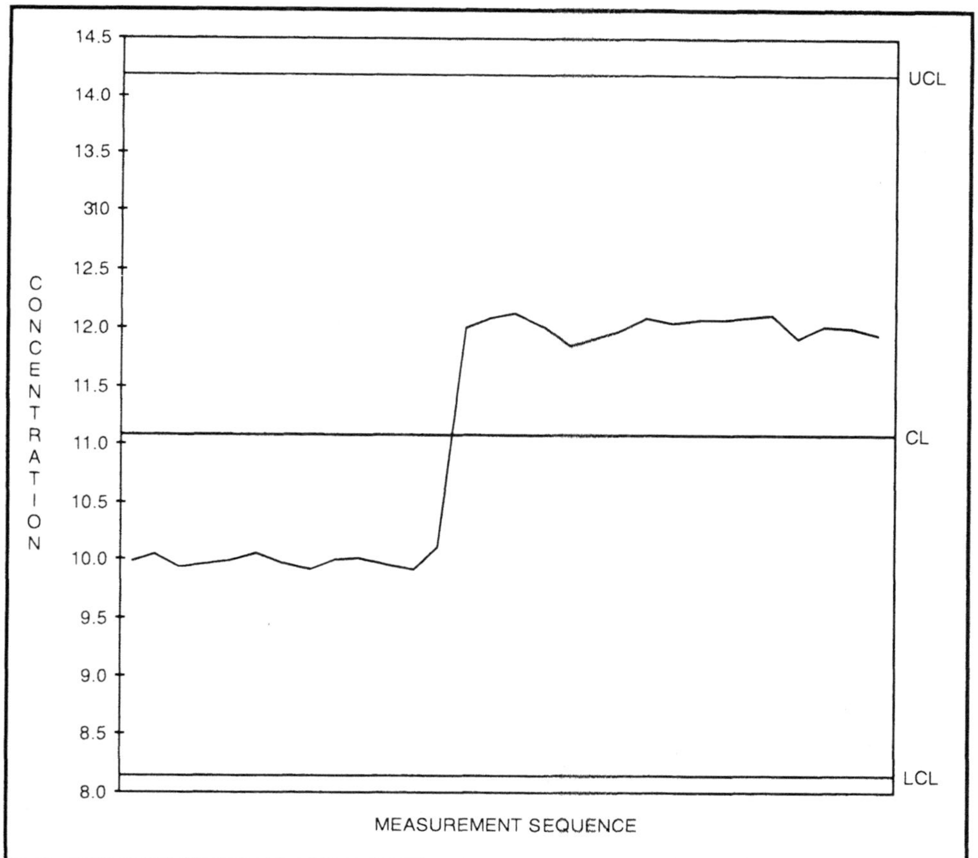

Figure 5-32. IR Photometer Application (Control chart of the laboratory measurements of the concentration of component F.)

Mean 11.13
s = 1.03 n = 30
LCL = 8.04
UCL = 14.22

Even though the concentration of component A, the component of primary interest in controlling the manufacturing process, did not shift, the concentrations of the other gases changed. The other gases included an unidentified 3% which was not measured or accounted for in either the laboratory analysis or compensated in the process analyzer design. Therefore, the problems with the process analyzer are:

The application of SPC/SQC to process and analyzer problems can reveal problems with application, design, engineering, installation, and maintenance of the process analyzer and aid in improvement after these problems are identified.

1. The process analyzer was originally designed for a stream other than the one which it was actually measuring.
2. There are two known interference gases (B and D) for which the assumption of insignificant influence may not be valid.
3. There are two other gases for which the assumption of noninterference may not be valid.
4. There is an unaccounted 3% of the stream for which the detector response (interference), if any, is not known.

Note that throughout this discussion, the laboratory chromatographic method has not been questioned. Investigating the results from the laboratory method would be the subject of another study, but nothing in this study indicates that the laboratory method should be questioned. A complete reapplication of the process analyzer, including consideration of different optical bench designs, was recommended. A process chromatograph was not acceptable for this application because the cycle/response time would be too slow.

References

1. *Statistical Quality Control Handbook*, Indianapolis: AT&T Technologies, 1984.
2. Shewhart, W. A., *Economic Control of Quality of Manufactured Products*, New York: D. Van Nostrand, 1931.
3. Ryan, Thomas P., *Statistical Methods for Quality Improvement*, New York: John Wiley & Sons, 1989.
4. Davies, O. L., Ed., *Statistical Methods in Research and Production*, 3rd ed., New York: Hafner, 1961.
5. Oakland, John S., *Statistical Process Control*, New York: John Wiley & Sons, 1986.
6. American Society for Testing and Materials (ASTM), *Standard Practices for Validation of Process Stream Analyzers*, D-3764-92, Philadelphia, PA, 1993.
7. Kepner, Charles H., and Benjamin B. Tregoe, *The New Rational Manager*, Princeton, NJ: Kepner-Tregoe, Inc., 1981.

Bibliography

1. Dixon, W. J., and Massey, F. J., Jr., *Introduction to Statistical Analysis*, 3rd ed., New York: McGraw-Hill, 1969.
2. Juran, J. M., and Gryna, F. M., Eds., *Juran's Quality Control Handbook*, 4th ed., New York: McGraw-Hill, 1988.
3. Nichols, G. D., "Use of Control Charts for Increasing the Reliability of On-Line Process Analyzers," *ISA/90 Proceedings*, Research Triangle Park, NC: Instrument Society of America, 1990, p. 349.
4. O'Connor, P. D. T., *Practical Reliability Engineering*, New York: John Wiley & Sons, 1991.
5. *PI Quality*, Carol Stream, IL: Hitchcock Publishing Company.
6. *Process Control and Quality*, Amsterdam: Elsevier Science Publishers.
7. *Quality Progress*, Milwaukee: American Society for Quality Control.
8. Mamzic, C.L., Ed., *Statistical Process Control*, Research Triangle Park, NC: ISA, 1995.

About the Author

Gary D. Nichols received his B. S. in chemistry from California Polytechnic State University at San Luis Obispo and his M. S. in analytical chemistry from the University of Georgia. He has worked in process analyzer development, applications, engineering, and maintenance with PPG Industries, BASF Corporation, Exxon Chemicals, and Sunoco. He is currently analyzer technologist with Sterling Chemicals in Texas City, TX. He is a Senior Member of ISA. In 1988, he published *On-Line Process Analyzers*. He has been active in ISA's Certified Specialist in Analytical Technology (CSAT) program and its SP76 Standards and Practices Committee for Composition Analyzers. He is also an ASQC Certified Quality Engineer (CQE). He is the News Editor for *Process Control and Quality* and is a contributing author to the *Encyclopedia of Analytical Chemistry*.

6

Validation of Process Analyzers

A new on-line process stream analyzer must be validated upon commissioning to assure that it produces the same average analytical results as a reference method, such as a lab method or a process analyzer that is being replaced. This process must be repeated from time to time during routine use. The validation process assures the accuracy of new process analyzers and any other process analyzers in use. It also assures that the precision of a new process analyzer meets the specification requirements determined by the process engineers. The process capability index, or C_{pk}, is used as a quantitative measure of the ability of an analyzer to meet control specifications.

Validation Process

The statistical tool known as Student's *t*-test is used for the validation process. A step-by-step example is given in this chapter to explain how an analyzer validation is typically performed. The major steps in the process are summarized here.

First, the newly installed on-line process stream analyzer is checked for performance against the manufacturer's specifications. A calibration sample is used to obtain the response factor (or factors) from an average of at least six repeated analyses. The calibration sample can be an artificial blend or a process sample of known composition. The linearity of the analyzer is checked over the range of interest.

Second, a reference sample is selected from an artificial blend or a process stream that has components with concentrations similar to typical operating values. The latter is recommended due to its availability and similarity to what is being measured. A portion (split or aliquot) of the reference sample is analyzed in quality assurance (QA) laboratories by a certified method at least six times, preferably in several laboratories using different operators and different test apparatus. The rest of the reference sample is analyzed by the on-line analyzer at least six times over a period of time, with routine testing in the interim. During each analyzer analysis of the reference sample, about three runs are conducted and the analytical results from the last one are taken for this validation process. This is to assure that the analyzer has enough on-line and purge time and run cycles to purge the sample loop and thus avoid sample cross-contamination. The on-line and purge time and cycles may vary from analyzer to analyzer.

Third, the data should be visually examined for outliers, and **Dixon's outlier test** should be applied if some values are suspect. After the outlier check, an ***F*-test** is conducted to ensure that the lab measurements are as precise as the historical precision of the lab method.

Finally, a **Student's *t*-test** is used to determine if the on-line analyzer and laboratory method are in agreement at a given allowable false signal rate (usually at 5%). If the average of process analyzer measurements is not significantly different from the average of lab measurements, the process analyzer is validated. Many statistical reference books (see, for example, [Ref. 1]) contain detailed information on the *t*-test; refer also to ASTM D3764 [Ref. 2].

The repeatability of the process analyzer is expected to be as good as the reproducibility (or, in most cases, repeatability if only one laboratory instrument is used) of a certified laboratory reference method. The repeatability of the process analyzer must meet the requirements of the process engineers and operations personnel who are the end-users of the analyzer data output.

The **process capability index** is used as a quantitative measure of the ability of an analyzer to meet specifications. C_{pk} is estimated based on the following formula:

$$C_{pk} = \frac{\text{Required repeatability}}{\text{Repeatability of process analyzer}}$$

The required repeatability is often equal to or less than the repeatability of a laboratory reference method. If C_{pk} is less than 1, then the requirements are not met; either an improvement of this analyzer or an alternate analyzer is required. The estimate C_{pk} varies with time, since it is usually based on a set of measurements.

The amount of variation in the estimated C_{pk} depends on the amount of data used to calculate it. The estimate of C_{pk} will vary less with large data sets.

It is recommended that this validation process be conducted for all analyzers where accuracy is important. Examples of such analyzers include those used for custody transfers, environmental monitoring, safety monitoring, and any other process analyses that are compared to laboratory methods.

Example of Analyzer Validation

The following is a step-by-step example of how this validation process is used to compare a flow injection analyzer (FIA) with a lab titration method for chlorine measurements.

Analyzer Calibration and Linearity

The analysis of chlorine in water is performed in the laboratory by the iodometry titration method. In order to transfer this method to process analysis, a flow injection analyzer was selected. The FIA was first calibrated using a gravimetrically-blended standard with an equivalent chlorine content of 15 ppm in water. The calibration was obtained by averaging six repeated measurements at this one point, that is, 15 ppm. The 15 ppm was chosen as approximately half the full range or span value of the process analyzer. The linearity of the analyzer was checked with the blended standards in the range of 7 to 25 ppm, which covered the range of the process stream. Figure 6-1 shows the linearity test. Each symbol contains five repeated measurements. The linearity requirements need to be established with the end data users and are not covered in this chapter.

Sample Analyses

If possible, the lab measurements should be performed in different laboratories and by different operators.

A water sample containing about 15 ppm of chlorine was chosen as the reference. It was split into two containers: one submitted for the lab titration analysis and the other for the FIA. Eight measurements (two more than the required six) were performed in both places over a period of a few days. The lab measurements were done in one lab by one analyst due to limited availability. If possible, the lab measurements should be performed in different laboratories and by different operators. The chlorine results of both lab titration and FIA data sets are given in Table 6-1.

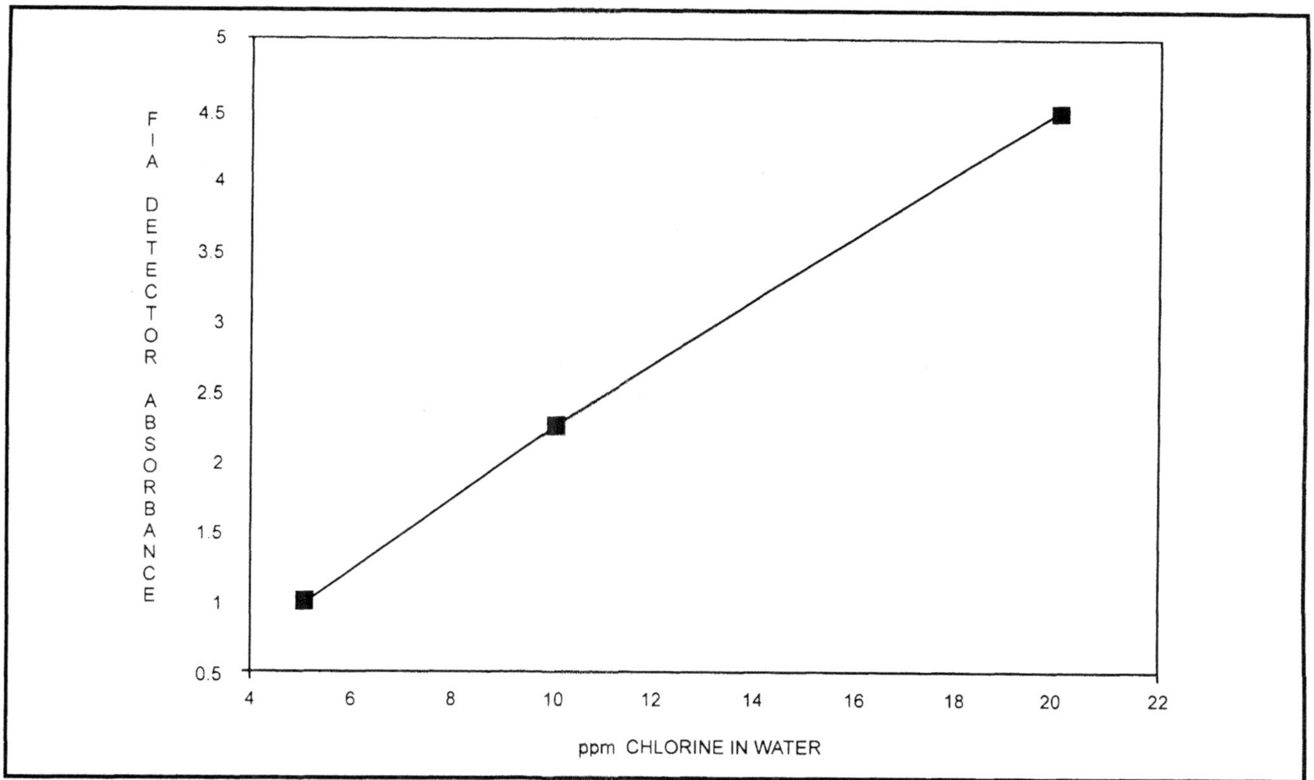

Figure 6-1. FIA Detector Linear Response Test for Chlorine Measurements

Table 6-1. Example of *t*-Test for Analyzer Validation

Parameter	Iodometry Titration Results, ppm (Lab Method)	FIA Results, ppm (Process Analyzer)
Measurement No.		
1	15.84	15.41
2	15.73	15.44
3	16.08	14.41
4	15.94	15.41
5	13.38	13.32
6	13.45	13.27
7	13.43	13.26
8	21.92*	13.27
Average, ppm	14.836	14.349
No. of measurements	7	8
Sample standard deviation	1.329	1.143
Sample variance	1.765	1.306
Degree of freedom**	13	
Critical***	2.145	
Calculated**	0.764	

* Not included in the calculations since this was determined to be an outlier according to Dixon's test.

** See text [Ref. 1] for the calculation equations.

*** Critical t is from Table A1.3 of ASTM D3764 [Ref. 2].

Note: Since the calculated t-value is less than or equal to the critical t-value, the FIA analyzer is validated. Therefore, the FIA analyzer can be expected to give essentially the same average results as the iodometry titration lab method at the quality level tested.

Outlier Test for Lab Measurements

Both the lab titration and FIA data sets were visually examined for outliers. The FIA data did not show any suspect outlier. The lab data, however had one unusually large value. The eight lab measurements were arranged in increasing order of magnitude as follows:

X_1	X_2	X_3	X_4	X_5	X_6	X_7	X_8
13.38	13.43	13.45	15.73	15.84	15.94	16.08	21.92

The largest value, 21.92, was suspected to be an outlier. This value was tested using Dixon's outlier test. According to Annex A-1 of ASTM D3764 [Ref. 2], different equations are used depending on the number of measurements taken. For eight measurements, the second equation in the document is used:

$$r_8 = \frac{(X_n - X_{n-1})}{(X_n - X_2)} = \frac{(X_8 - X_7)}{(X_8 - X_2)} = \frac{(21.92 - 16.08)}{(21.92 - 13.43)} = 0.689 \qquad (6\text{-}1)$$

The calculated r value was compared with the "r rejection limits" in Table A1.1 of ASTM D3764. For eight measurements (sample size N), the ***r* rejection limit** is 0.683. Since r_8 is greater than 0.683, the measurement X_8 was determined to be an outlier based on the Dixon test and thus was rejected from further calculations.

When X_{n-1} and X_n, or X_1 and X_2, are close to each other but are both outliers, the Dixon outlier test may fail to detect them as such. In this case, the use of a test for an outlier pair is more appropriate. Reference 3 describes a test for two outliers on either the high or low side.

Calculation of Standard Deviation

Equation 6-2 calculates the standard deviation of the lab reference test results, excluding the outlier, X8:

$$S_r = \sqrt{\frac{\sum_{r=1}^{N_r} (X_r - \bar{X}_r)^2}{(N_r - 1)}} \qquad (6\text{-}2)$$

where:

S_r = standard deviation

X_r = individual lab test results on the reference sample

N_r = number of test results (excluding the outlier)

$\bar{X}_r$ = arithmetic average of the test results

In this example, the number of test results, excluding the largest measurement (which was determined to be an outlier), is $N_r = 7$. The average of these seven results is 14.836. The standard deviation, S_r, is calculated to be 1.329.

Calculation of the standard deviation can be simplified by using a spreadsheet program with a function that assumes its arguments are a sample of the population. For example, function @STDEV should be used instead of @STDEVP in Microsoft® Excel™ spreadsheets.

F-Test

The precision of this set of laboratory measurements needs to be compared to the historical precision of this laboratory test method. If the variation of this set of data is significantly greater than the historical precision for the method, the set may be suspect. If the data are suspect, then corrective action must be taken. After corrective action, a minimum of five more measurements should be taken on the reference sample until acceptable lab test precision is obtained. The comparison for acceptable precision is done using the *F*-test as follows:

$$F = \frac{S_r^2}{\sigma_t^2} \tag{6-3}$$

S_r^2 is the **variance** of this set of measurements. It is $1.329^2 = 1.765$ in the example (refer to Equation 6-2). The term σ_t is the **historical standard deviation** of this type of lab measurement. If σ_t is not available and a reproducibility for this method is given in an ASTM document, the value of σ_t can be calculated. If the set of measurements in question comes from multiple labs, as recommended by ASTM, then σ_t is the reproducibility divided by 2.77. If the set of measurements in question comes from one operator running one piece of test equipment in one lab, then σ_t is the repeatability divided by 2.77.

In this example, the historical standard deviation of this iodometry titration is 0.95 based on the lab QA records. The *F*-value is calculated as:

$$F = \frac{S_r^2}{\sigma_t^2} = \frac{1.329^2}{0.95^2} = 1.956 \tag{6-4}$$

Compare this calculated *F*-value with the critical *F*-value in the *F*-distribution table (Table A1.2 of ASTM D3764 [Ref. 2]). To find the *F*-value from the table, one needs to know the **degrees of freedom** for this set of measurements. The numerator degrees of freedom is the number of measurements in S_r^2 minus 1. We have seven measurements in this example (excluding the outlier); hence, there are six numerator degrees of freedom. It is best to use the true denominator degrees of freedom, which is the number of data values used to calculate σ_t^2 minus 1. However, ASTM D3764 suggests that 20 denominator degrees of freedom be used. For our example, the *F*-value from the *F*-distribution table is 2.60 for six degrees of freedom in the numerator and 20 degrees of freedom for the denominator. The calculated *F*-value, 1.956, is less than the table limiting *F*-value, 2.60. Therefore, we can conclude that the precision of this set of measurements is not worse than the historical precision of this lab method, and we can proceed with the validation process.

Outlier Test for Analyzer Measurements

The eight process analyzer readings (see Table 6-1) are also checked for outliers according to Dixon's test. The calculation is the same as that given in the section "Outlier Test for Lab Measurements." There are no outliers in this example. The standard deviation, Sa, of this set of process analyzer measurements is calculated as follows:

$$S_a = \sqrt{\frac{\sum_{a=1}^{N_a} (X_a - \bar{X}_a)^2}{\sqrt{N_a - 1}}} \tag{6-5}$$

Note that the STATGRAPHICS® [Ref. 4] Two-Sample Analysis procedure will perform all the remaining statistical calculations in this section.

where:

X_a = individual process analyzer results on the reference sample

N_a = number of process analyzer test results

$\overline{X}_a$ = arithmetic average of the process analyzer test results

In this example, X_a is the process analyzer results in Table 6-1. The number of process analyzer test results, N_a, equals eight. The arithmetic average of these eight process analyzer results is 14.349. The **standard deviation**, S_a, calculated by using Equation 6-5 or a spreadsheet, is 1.143.

t-Test

The Student's *t*-test is applied here to compare the lab method measurements with the process analyzer measurements to determine whether they produce essentially the same average result. There are two ways to calculate the *t*-values for two different situations. The two situations are determined by applying the *F*-test as follows:

$$F = L/M \tag{6-6}$$

where:

L = larger variance of the two sets of measurements

M = smaller variance of the two sets of measurements

In our example, the variance of the lab measurements, S_r^2, is 1.765, and the variance of the process analyzer measurements, S_a^2, is 1.306. Therefore, L is 1.765, M is 1.306, and

$$F = L/M = 1.765/1.306 = 1.35 \tag{6-7}$$

This calculated *F*-value is compared with the *F*-value in the *F*-distribution table (Table A1.2 of ASTM D3764 [Ref.2]). The degrees of freedom for the numerator and degrees of freedom for the denominator in the *F*-distribution table correspond to the degrees of freedom of the variances used in the numerator and denominator of Equation 6-7, respectively. Accordingly, to find the *F*-value from the table, the degrees of freedom for both the lab and process analyzer measurements must be determined. The degrees of freedom for the lab measurements is six (number of lab measurements minus 1). The degrees of freedom for the process analyzer measurements is seven (number of analyzer measurements minus 1). Thus, there are six degrees of freedom for the numerator and seven degrees of freedom for the denominator in this example. The *F*-value from the *F*-distribution table is 3.87.

The calculated *F*-value is compared with the table critical *F*-value to determine whether or not it is reasonable to assume that the population variances of the process analyzer results and the lab method results are the same. In this example, the calculated *F*-value of 1.35 is smaller than the table *F*-value of 3.87. Therefore, we can conclude that the variances of the process analyzer and lab methods are not significantly different, and Equations 6-8 and 6-9 for Student's *t*-test and degrees of freedom are used. The procedure to be followed if the calculated *F*-value is greater than the table *F*-value is discussed later (beginning with Equation 6-13).

$$t = \frac{\left|\overline{X}_a - \overline{X}_r\right|}{S_d} \tag{6-8}$$

$$S_d = \sqrt{\frac{(N_a - 1) * S_a^2 + (N_r - 1) * S_r^2}{N_a + N_r - 2} * \left(\frac{1}{N_a} + \frac{1}{N_r}\right)} \qquad (6\text{-}9)$$

where:

t = value used in the t-test

$|\overline{X}_a - \overline{X}_r|$ = absolute value of the difference between the average process analyzer result and the average reference lab result

S_d = standard deviation of the difference between X_a and X_r

N_a = number of process analyzer results

N_r = number of lab results

$\overline{X}_a$ = arithmetic average of the process analyzer results

$\overline{X}_r$ = arithmetic average of the lab results

S_a^2 = variance of the process analyzer results (Equation 6-5)

S_r^2 = variance of the lab results (Equation 6-2)

In our example, $N_a = 8$, $N_r = 7$, $X_a = 14.349$, $X_r = 14.836$, $S_a^2 = 1.306$, and $S_r^2 = 1.765$. The **standard deviation, S_d,** is given by Equation 6-9:

$$S_d = \sqrt{\frac{(8 - 1) * 1.306 + (7 - 1) * 1.765}{8 + 7 - 2} * \left(\frac{1}{8} + \frac{1}{7}\right)} = 0.638 \qquad (6\text{-}10)$$

From Equation 6-8:

$$t = \frac{|14.349 - 14.836|}{0.638} = 0.763 \qquad (6\text{-}11)$$

This calculated t-value is compared with the critical t-value in Table A1.3 of ASTM D3764 [Ref. 2]. To use this table, the degrees of freedom must be determined:

$$\text{Degrees of freedom} = N_a + N_r - 2 = 8 + 7 - 2 = 13 \qquad (6\text{-}12)$$

The t-value from the table is 2.160 for 13 degrees of freedom. Since the calculated t-value, 0.768, is smaller than the table t-value, 2.160, we can conclude that the average FIA process analyzer output is not significantly different from the average lab iodometry titration output at the 5% level of statistical significance. Therefore, the process analyzer is validated. The 5% significance level is the same as the 5% probability level used in Table A1.3 for the t-value. The 5% significance level means that in repeated applications of this procedure when the analyzer is truly valid, we will falsely conclude that the analyzer is invalid 5% of the time. Most statistics' texts provide critical t-values for different probability levels (see, for example, [Ref. 5]). Table 6-1 summarizes the t-test for the FIA process analyzer validation for chlorine analysis in drinking water by comparison with a lab iodometry titration method.

Using a smaller probability level, such as 1%, means that we will make fewer false conclusions that a process analyzer is bad.

However, using a smaller probability level also means that it will be more difficult to detect a bad analyzer, and, in fact, more bad analyzers will pass the test.

To balance the trade-off between concluding that good process analyzers are bad, and concluding that bad process analyzers are good, statistical practitioners commonly recommend using the 5% probability level. This is based on an accumulation of experience in decisionmaking using statistical techniques when confronted with variable data and uncertainties about natural phenomena.

If the calculated F-value from Equation 6-6 is greater than the table F-value, we can conclude that the variances of the process analyzer results and the lab method results came from two populations with different (but unknown) variances. In this case, the following equations for the calculation of t and degrees of freedom are used:

$$t = \frac{|\bar{X}_a - \bar{X}_r|}{S_d} \qquad (6\text{-}13)$$

$$S_d = \sqrt{\frac{S_a^2}{N_a} + \frac{S_r^2}{N_r}} \qquad (6\text{-}14)$$

$$d_f = \frac{[(S_a^2/N_a) + (S_r^2/N_r)]^2}{[(S_a^2/N_a)^2/(N_a - 1)] + [(S_r^2/N_r)^2/(N_r - 1)]} \qquad (6\text{-}15)$$

where d_f is the **degrees of freedom** rounded to the nearest integer. Note that Equation 6-15 is slightly different from the corresponding degrees of freedom calculation given in ASTM D3764 [Ref. 2]. Both are approximations; however, we prefer Equation 6-15 because it is used in statistical packages such as SAS® [Ref. 6] and STATGRAPHICS® [Ref. 4].

For purposes of illustration, let's change the variance of the lab measurements of our example from $S_r^2 = 1.765$ to $S_r^2 = 5.10$. The other parameters are kept the same. The larger variance means there is a bigger spread of the data measurements centered around the same mean value of 14.836 ppm. Referring to Equation 6-6 for the calculation of F, we have $L = 5.10$ and $M = 1.306$.

Equation 6-7 gives:

$$F = \frac{L}{M} = \frac{5.10}{1.306} = 3.91 \qquad (6\text{-}16)$$

This calculated F-value of 3.91 is greater than the limiting F-value of 3.87 from Table A1.2 of ASTM D3764 [Ref. 2]. We can conclude that the lab method produces more variable results than does the process analyzer. The t-test for this situation, where the variances of the process analyzer results and the lab method results came from two populations with different (but unknown) variances, is sometimes referred to as the **Aspen-Weltch test**, or **modified t-test** [Ref. 1]. In this case, Equations 6-13 to 6-15 in this section are used for the calculation of t and the degrees of freedom. Equations 6-13 and 6-14 give:

$$t = \frac{|14.349 - 14.836|}{\sqrt{(1.306/8) + (5.10/7)}} = 0.516$$

Equation 6-15 gives the degrees of freedom:

$$d_f = \frac{[(1.306/8) + (5.10/7)]^2}{[(1.306/8)^2/(8 - 1)] + [(5.10/7)^2/(7 - 1)]} = 9$$

The critical t from Table A1.3 [Ref. 2] is 2.262 for nine degrees of freedom. The calculated t of 0.516 is smaller than the critical t of 2.262. From the t-value comparison, we can conclude that the average FIA process analyzer output is not significantly different from the average lab iodometry titration output at the 5% significance level. Therefore, the analyzer is validated.

Acknowledgment

The author would like to acknowledge the contributions made by G. A. Stenback and J. D. Jobe, both formerly of Shell Development Company.

References

1. Sung, C. Choi, *Introductory Applied Statistics in Science*, Englewood Cliffs, NJ: Prentice-Hall, 1978.
2. ASTM D3764-92, "Standard Practice for Validation of Process Stream Analyzers," *Annual Book of ASTM Standards*, Philadelphia: ASTM.
3. Beyer, W. H., Ed., *Handbook of Tables for Probability and Statistics*, Cleveland: Chemical Rubber Co., 1968.
4. STATGRAPHICS®, IBM Version 5.0, Rockville, MD, STSC, Inc.
5. Iman, R. L., and Conover, W. J., *A Modern Approach to Statistics*, New York: John Wiley & Sons, 1983.
6. *SAS/STAT® User's Guide, Version 6*, 4th edition, vol. 2, Cary, NC: SAS Institute Inc., 1989.

About the Author

Fusheng Wang is a 1968 Chemical Engineering graduate of Tsinghua University, China, and has received his Ph.D. in analytical chemistry from the University of Houston, Texas. He works in the control engineering area for Shell Oil Products Company. His main work interests include statistical quality control of analyzers, evaluation and development of new analyzer technologies, and technical support to Shell refineries and chemical plants.

7

Specification and Purchase of Process Analyzer Systems

Specifications form a critical part of the purchase document. They must be not only well written but also succinct and complete. The only way to assure that you will obtain the analyzer system that you require is to develop a document that describes what you need — not what you think you might require someday or what will cover you in all eventualities, laying all blame on the system supplier. To be effective, your relationship with the supplier must be based on open communications from the initial request for proposal (RFP) through commissioning.

To be truly effective, a specification must be easy to understand and contain only the information required for the purchase of the analyzer system under consideration. Toward this end, it is a good idea to break your specifications into sections covering such topics as general purchase requirements, documentation, engineering and design, shelters, and so on. Each section should also have a corresponding guide for internal use only, clearly stating why the specified option was selected over other options considered at the time of the latest revision. This then gives the design engineer an opportunity to understand why things are being done as they are and also to investigate more recent cost-effective alternatives for inclusion in the next revision.

All specifications should be reviewed at a minimum of once every 5 years and, considering today's rapid changes in technology, probably more frequently than that.

Don't cross-reference too many other specifications, because it won't take long before you will find that one section contradicts another.

Don't cross-reference too many other specifications, because it won't take long before you will find that one section contradicts another.

Use of Specification and Data Sheets

Specification sheets represent a summary of all the text included in the specification document and thus must be clear and concise. The ISA SP 20 data sheets are the result of the combined efforts of many individuals over an extended period of time and should, as a minimum, be used as a guideline for your data sheets. ISA has not yet issued a series of data sheets for process analyzers, although they are under development.

Since specification sheets are used in this way, they should be well organized and contain no extraneous information. This must be balanced against the potential requirement of different data sheets or series of data sheets for each analytical technology being used.

An example of a process analyzer data sheet* that contains critical information for a variety of applications but yet remains somewhat generic is shown in Figures 7-1 to 7-4, which correspond to pages 1 to 4 of a complete data sheet (printing this on double-sided 11 x 17-inch paper keeps all data conveniently together). Sheet 1 (Figure 7-1) identifies the main analyzer environment conditions as well as the expectations of the analyzer detector unit, including such parameters as response time. Sheet 2 (Figure 7-2) concentrates more on the electronics aspect of the equipment and the process stream(s) themselves. Sheet 3 (Figure 7-3) covers sample system parameters as well as stream composition data for multistream analyzers and unique applications such as radioactive devices. Sheet 3 also contains a section for listing reference documentation should someone be interested in additional data on the analyzer system. Sheet 4 (Figure 7-4) is a "blank" sheet to be used for sketches, column/valve switching configurations and other generic notes not contained elsewhere. Obviously, multiple copies of this final sheet can be prepared as required.

An example of a process analyzer shelter data sheet* is shown in Figures 7-5 to 7-8. This data sheet contains critical item categories and descriptions to elicit necessary information. Shelter sheet 1 (Figure 7-5) identifies the project data necessary for communications, as well as the availability and quality of utilities and piping/tubing specifications. Shelter sheet 2 (Figure 7-6) identifies shelter size and analyzer contents. The multiple-choice listing provides either generic or industry-standard categories of equipment required in a functional process analyzer shelter. Sheet 3 (Figure 7-7) continues the generic/industry-standard listing of required process analyzer shelter equipment. Space is provided for comments regarding the client's selections from the equipment listing. Shelter sheet 4 (Figure 7-8) provides a format for analyzer shelter exterior and interior physical arrangement sketches. This will (hopefully) preclude having a door on a delivered process analyzer shelter open directly at a superstructure support column in the process plant! The suggested directional orientation arrows provide a gentle reminder to the client to provide this important information. Of course, nothing beats an actual site visit or video footage of the actual location after award of the contract and before construction to assure that both parties fully understand the intended use of the system. Space is also provided for a device legend so that the sketches are intelligible to the everyone.

Since it is commonly held that 85% of all analyzer system problems are the result of the sample-handling system, it is imperative that critical process conditions be well stipulated on the data sheet.

One thing all data sheets do require is a good overview of the original process sample and area conditions as well as the conditioned sample conditions as presented to the analyzer. Since it is commonly held that 85% of all analyzer system problems are the result of the sample-handling system, it is imperative that critical process conditions be well stipulated on the data sheet.

Data sheets should also contain a notation if the sample conditions are estimated from a process simulation or model rather than actual samples. This will provide the design engineers some indication of how much variation to incorporate into their design.

Data sheets should also contain a notation if the sample conditions are estimated from a process simulation or model rather than actual samples. This will provide the design engineers some indication of how much variation to incorporate into their design.

In summary, a data sheet is meant to be concise yet complete, to serve as the starting point for a complete understanding of the analyzer system. It should give the person using it an idea of not only how the system works but also where to go to find additional information should it be required.

---

* (Adapted from copyrighted material of Consulting Technologists, Inc./Process Analyzer System Technologists [CTI/PAST], P.O. Box 0978, Homewood, IL 60430-0978.)

GENERAL

- INSTRUMENT TAG NO
- SERVICE DESCRIPTION
- CRITICAL INSTRUMENT
- P & ID NO
- VESSEL NO
- VESSEL TRIM SPEC

ANALYZER ENVIRONMENT

OUTDOOR/INDOOR		
AMBIENT TEMP (°C)	MINIMUM	
	NORMAL	
	MAXIMUM	
RELATIVE HUMIDITY (%)	MINIMUM	
	NORMAL	
	MAXIMUM	
BAROMETRIC PRESSURE (psia)	MINIMUM	
	NORMAL	
	MAXIMUM	
AMBIENT ATM CONDITIONS		
INTERNAL ISA SEVERITY		
CORROSIVE GAS	LEVEL (ppm)	

UTILITIES AVAILABLE

POWER SUPPLY	VOLTS	VARIATION
	FREQ. (Hz)	VARIATION
PURGE	FLUID	
	PRESS (psig)	
	TEMP (°C)	
INSTRUMENT AIR	DEWPOINT	
	PRESS (psig)	
	TEMP (°C)	
STEAM	QUALITY	
	PRESS (psig)	
	TEMP (°C)	
COOLING WATER	PRESS (psig)	
	TEMP (°C)	

ELECT.

- EXTERNAL AREA ELECTRICAL CLASS
- ANALYZER AREA ELECTRICAL CLASS
- ELECTRICAL APPROVAL — REQUIRED
- INTRINSICALLY SAFE

REFERENCES

- APPLICABLE SPECS
- LOCATION DRAWING NO
- LOOP DRAWING NO
- V.P. DRAWING NO
- REFERENCE DRAWINGS

PURCHASE

- MANUFACTURER
- MODEL NO
- P.O. NO
- M/R NO
- SERIAL NO

ANALYZER FIELD UNIT

MEASUREMENT PRINCIPLE			
ENCLOSURE TYPE			
MOUNTING			
INDICATING/BLIND			
DETECTOR	REMOVABLE		
	UNIT DIST (ft.)		
SCALE	RANGE	-	
	UNITS		
ANALYSIS TIME (sec.)	DURATION		
	CYCLE		
RESPONSE TIME (sec.)	50 % FS		
	90 % FS		
TEMP COMPEN	REQ'D		
	BASE (°C)		
	SEP/INT		

TROUBLE ALARMS

RATING	VOLTS	VAC	VDC
	AMPS		
LOAD TYPE			
CONTACT CONFIG			
ACTION	SWITCH #1		
	SWITCH #2		
	SWITCH #3		
DEADBAND RANGE	SWITCH #1	-	
	SWITCH #2	-	
	SWITCH #3	-	

BATTERY BACKUP (hr.)		
CHART	SIZE (in.)	
	TYPE	
CHART DRIVE	TYPE	
	SPEED (in./min)	

CONNECTION

PROCESS	SIZE (in.)	TYPE
	RATING	FACING
CONDUIT SIZE (in. NPT)		
PURGE SIZE (in. NPT)		
CARRIER GAS SIZE (in. NPT)		

OUTPUTS

NO 1	RANGE	-
	UNITS	
NO 2	RANGE	-
	UNITS	
PROGRAMMER REQ'D		

CARRIER GAS	CYLINDER
	REGULATOR TYPE
	SINGLE/DUAL
SPAN GAS/LIQ.	CYLINDER
	REGULATOR TYPE
ZERO GAS/LIQ.	CYLINDER
	REGULATOR TYPE

NO.	DATE	REVISIONS	BY	INST. APPR	PROC APPR

NOTES

PROCESS ANALYZER

JOB NO.

DATA SHEET NUMBER — REV.

DS

Sheet 1 of 4

Figure 7-1. Analyzer Data Sheet 1 (Adapted from copyrighted material of Consulting Technologists, Inc./Process Analyzer System Technologists [CTI/PAST], P.O. Box 0978, Homewood, IL 60430-0978.)

REMOTE RECEIVERS				
	INTEGRATOR	TYPE		
		NO OF DIGITS		
	SCALE	TYPE		
		RANGE	UNITS	
		SIZE (in.)		
	CONTROLLER	TYPE		
		RANGE		
	OUTPUT	TYPE		
		DESTINATION		
	SPECIAL CABLE			
	OUTPUT FOR CONTROL/INDICATION			
	APPLICATION NO			
	INTERNAL ELECTRICAL	AREA ELECTRICAL CLASS		
		ELECTRICAL APPROVAL	REQUIRED	
		INTRINSICALLY SAFE		

LOCAL SIGNAL DEVICES			
	MEMORY AMPLIFIER		
	PEAK PICKER		
	RECORD; Cont, Bar, Trend		
	ANNUNCIATOR PANEL	NO OF INPUTS	
		NO OF OUTPUTS	

CHROMATOGRAPHS				
	COLUMN	TYPE (eg. Cap, Packed)		
		DIAMETER (in.)		
		MATERIAL		
		LENGTH (ft.)		
		PACKING	MATERIAL	
			LENGTH (ft.)	

STREAM OPERATING CONDITIONS			
	TEMPERATURE (℃)	MINIMUM	
		NORMAL	
		MAXIMUM	
		RATE/MIN	
	SAMPLE PRESSURE (psig)	MINIMUM	
		NORMAL	
		MAXIMUM	
	VENT PRESSURE (psig)	MINIMUM	
		NORMAL	
		MAXIMUM	
	S.G. @ OPER TEMP	MINIMUM	
		NORMAL	
		MAXIMUM	
	MOLECULAR WEIGHT	MINIMUM	
		NORMAL	
		MAXIMUM	
	VISCOSITY (cp)	MINIMUM	
		NORMAL	
		MAXIMUM	
	pH	MINIMUM	
		NORMAL	
		MAXIMUM	
	PHASE	VAPOUR	
		LIQUID	IBP (℃)
			FBP (℃)
	SOLIDS	VOLUME PERCENT	
		STRATIFICATION	
		COATING	
		PARTICLE SIZE RANGE (micron)	
	STABILITY	POLYMERIZE	
		HYDRATES	
		RADIOACTIVE	
	CORROSIVE		

STREAM COMPONENTS & ANALYSIS

COMPONENT	ANALYSIS REQUIRED				STREAM 1			STREAM 2		
	YES	ACCURACY	RANGE	UNITS	MAX	NORMAL	MIN	MAX	NORMAL	MIN
WATER			-							
			-							
			-							
			-							
			-							
			-							
			-							
			-							
			-							
			-							
			-							
STREAM SOURCE - LINE NUMBER										

NO.	DATE	REVISIONS	BY	INST. APPR	PROC APPR

NOTES

	PROCESS ANALYZERS	JOB NO.	
		DATA SHEET NUMBER	REV.
	Sheet 2 of 4	DS	

Engineering Standard Form

Figure 7-2. Analyzer Data Sheet 2 (Adapted from copyrighted material of Consulting Technologists, Inc./Process Analyzer System Technologists [CTI/PAST], P.O. Box 0978, Homewood, IL 60430-0978.)

SAMPLE SYSTEM & CONDITIONING			
SAMPLE STREAM(S)	NUMBER OF STREAMS		
	MAX	DIST TO ANALYZER (ft.)	
		TAP SIZE (in.)	
		ALLOW SAMPLE TIME (sec.)	
	SLIPSTREAM RETURN	LOCATION LINE NO	
		DISTANCE (ft.)	
		TAP SIZE (in.)	
	TEMP CONDITION	COOL / HEAT	
		MEDIUM	
	CONNECTION TUBING	MATERIAL	
		SIZE (in.)	
SAMPLE	FLOW	RATE / UNITS	
		CONTROL	
	PROBES	REQUIRED	
		TYPE	
	FILTRATION / COALESCER		
	PRIME MOVER (Pump, Eductor)		
	DRY/HUMIDIFY		
	ENCLOSURE TYPE		
	MODEL NO.		

RADIATION DEVICE		
SOURCE	MATERIAL	
	STRENGTH (mCi)	
HOLDER MATERIAL		
RADIATION AT 30 cm (mCi)		
SYSTEM STABILITY		
SOURCE DECAY COMP.		
INTEGRATION PERIOD (sec.)		
UPDATE PERIOD (sec.)		

NOTES

REFERENCES		
APPLICABLE SPECS		
LOCATION DRAWING NO		
LOOP DRAWING NO		
V.P. DRAWING NO		
FIELD TERMINAL BOX	ELECTRONIC	
	ALARM	
REFERENCE DRAWINGS		

STREAM COMPONENTS & ANALYSIS

COMPONENT	STREAM 3			STREAM 4			ZERO	SPAN	CARRIER
	MAX	NORMAL	MIN	MAX	NORMAL	MIN	FLUID	FLUID	FLUID
WATER									
LINE NUMBER									

BYPASS FLOW	
ST	
1	
2	
3	
4	
5	
6	
7	
ZERO	
SPAN	
CARRIER	
UNITS	

NO.	DATE	REVISIONS	BY	INST. APPR	PROC APPR

NOTES

PROCESS ANALYZERS	JOB NO.	
	DATA SHEET NUMBER	REV.
Sheet 3 of 4	DS	

Engineering Standard Form

Figure 7-3. Analyzer Data Sheet 3 (Adapted from copyrighted material of Consulting Technologists, Inc./Process Analyzer System Technologists [CTI/PAST], P.O. Box 0978, Homewood, IL 60430-0978.)

NOTES AND REMARKS

NO.	DATE	REVISIONS	BY	INST. APPR	PROC APPR

	PROCESS ANALYZERS	JOB NO.	
		DATA SHEET NUMBER	REV.
	Sheet 4 of 4	DS	

Engineering Standard Form

Figure 7-4. Analyzer Data Sheet 4 (Adapted from copyrighted material of Consulting Technologists, Inc./Process Analyzer System Technologists [CTI/PAST], P.O. Box 0978, Homewood, IL 60430-0978.)

SYSTEM IDENTIFICATION DATA

CUSTOMER		UNIT	
ADDRESS		STATE, ZIP	

PROJECT IDENTIFICATION

TITLE			
PROJECT NO.			
E & C CONTRACTOR			
MAILING ADDRESS			
CITY		STATE, ZIP	
DATES PREPARED	APPROVAL	QUOTE	PURCHASE
PURCHASED BY	☐ SYSTEM HOUSE	☐ CUSTOMER	☐ E & C CONTRACTOR
PRE-SHIP INSPECTION DATE		DELIVERY ON-SITE DATE	
START-UP DATE			

UTILITIES AVAILABLE

ELECTRICAL	STEAM	AIR ☐ INSTR. ☐ PLANT (MAX NORM MIN)	WATER (MAX NORM MIN)
VOLTS (AC)	PRESSURE (psig)	PRESSURE (psig)	PRESSURE (psig)
FREQUENCY (Hz)	TEMPERATURE	DEW POINT	DEW POINT
PHASE	SUPERHEAT (YES/NO)		
NOTES:	NOTES:	NOTES:	NOTES:

PIPING AND TUBING

PROCESS TUBING	COOLING WATER TUBING	AIR TUBING	STEAM PIPING
316SS (MINIMUM) (SEAMLESS, BRIGHT ANNEALED)	☐ PVC-COATED COPPER ☐ 316 SS	☐ PVC-COATED COPPER ☐ 316 SS	SUPPLY AND CONDENSATE LINES ARE TO BE CARBON STEEL SCHEDULE 40 PIPE.
☐ OTHER ______ ANALYZER TAG NO. ______	FITTINGS ☐ SWAGELOK ☐ OTHER ______	FITTINGS ☐ SWAGELOK ☐ OTHER ______	MAIN HEATING SYSTEM(S) FOR SHELTER ARE TO HAVE TEMPERATURE FEEDBACK CONTROL VALVES.
☐ OTHER ______ ANALYZER TAG NO. ______	☐ BRASS ☐ 316 SS	☐ BRASS ☐ 316 SS	EACH SAMPLE SYSTEM AND HEATER ARE TO HAVE INDIVIDUAL TRAPS.

NOTES

NO.	DATE	REVISIONS	BY	INST. APPR	PROC APPR	NOTES

	ANALYZER SHELTER	Sheet 1 of 4	JOB NO.	
	Unit/Plant ______		DATA SHEET NUMBER	REV.
	Prepared By: ______	Checked By: ______	DS	

Engineering Standard Form

Figure 7-5. Shelter Data Sheet 1 (Adapted from copyrighted material of Consulting Technologists, Inc./Process Analyzer System Technologists [CTI/PAST], P.O. Box 0978, Homewood, IL 60430-0978.)

GENERAL SPECIFICATION DATA

SHELTER IDENTIFICATION	WIDTH	WEIGHT LENGTH	HEIGHT
ANALYZER TYPE			
ANALYZER MODEL			
MANUFACTURER			
TAG NUMBER			
TAG NUMBER			
TAG NUMBER			
TAG NUMBER			
TAG NUMBER			
TAG NUMBER			
TAG NUMBER			

ALL FOLLOWING DETAILS ARE TO APPLY TO PROCESS ANALYZER SHELTER AS NOTED BY SELECTION (DARKENING SQUARE) OF THAT ITEM OR REQUIREMENT. SEE EXAMPLE BELOW.

■ SELECTED ITEM ☐ REJECTED ITEM

BUILDING

☐ CLEAR-SPAN, SELF-FRAMING METAL ☐ STEEL (18 GAUGE) ☐ ALUMINUM
☐ CLEAR-SPAN, SELF-FRAMING FIBERGLASS ☐ OPEN BASE BUILDING
☐ BASE FRAME OF: ☐ I-BEAMS ☐ C-CHANNEL ☐ FIBERGLASS MEMBERS ☐ STEEL MEMBERS
☐ NON-SLIP PAINTED CHECKER PLATE FLOOR (FLAT) ☐ NON-SLIP TEXTURED FIBERGLASS FLOOR (FLAT)
☐ CENTER DRAIN (IN FLOOR/SLAB) TO EXIT THRU LONG SIDE BEAM/BASE OPPOSITE SIDE OF GAS CYLINDERS
☐ 3" INSULATION IN WALLS AND CEILINGS ☐ 3" INSULATION UNDER FLOORS (OR EQUIVALENT)
☐ 1" INSULATION IN WALLS AND CEILINGS ☐ OTHER INSULATION ______
☐ 18 GA. INTERIOR PANELING ☐ INTERIOR GELCOAT
☐ UNISTRUT VERTICAL ON 16" CENTERS, 7FT. LONG, CENTERED FLOOR TO CEILING
☐ UNISTRUT HORIZONTAL AT ______ FT. AND ______ FT. ELEVATIONS AROUND ALL SHELTER WALLS
☐ GA TER) ☐ GAS CYLINDER HOOD ______ FT. ______ IN. LONG
☐ 3'-0" X 7'-0" OUTWARD OPENING DOOR(S) EACH WITH 12" SQ. SAFETY GLASS WINDOW AND RETENTION HARDWARE
DOOR(S) ☐ 1 ☐ 2 DOOR TYPE ☐ STANDARD METAL ☐ STAINLESS STEEL ☐ FIBERGLASS
☐ BUILDING PAINTED INSIDE AND OUTSIDE (AFTER PRESCRIBED PREPARATION) WITH ______
WALLS, CEILINGS, ROOF TO BE ______ BASE BEAMS TO BE ______
☐ OTHER PA ______

☐ ELECTRICAL HAZARD RATING - EXTERIOR CL 1 DIV 2 GRP D ☐ OTHER CL. ______ DIV. ______ GRP. ______
☐ ELECTRICAL HAZARD RATING - INTERIOR CL 1 DIV 1 GRP D ☐ OTHER CL. ______ DIV. ______ GRP. ______
☐ MAIN POWER BREAKER: RAW POWER ☐ MAIN POWER BREAKER: UPS POWER
☐ SEPARATE CIRCUIT BREAKER IN POWER DISTRIBUTION PANEL FOR EACH ANALYZER, OUTLET AND HEAT TRACE
☐ OTHER CIRCUIT BREAKER ARRANGEMENT (DESCRIBE IN NOTES & DIAGRAM)
☐ PREFERRED BRAND OF ELECTRICAL DEVICES: ______

NOTES

NO.	DATE	REVISIONS	BY	INST. APPR	PROC APPR	NOTES

ANALYZER SHELTER	Sheet 2 of 4	JOB NO.
Unit/Plant ______		DATA SHEET NUMBER / REV.
Prepared By: ______ Checked By: ______		DS

Engineering Standard Form

Figure 7-6. Shelter Data Sheet 2 (Adapted from copyrighted material of Consulting Technologists, Inc./Process Analyzer System Technologists [CTI/PAST], P.O. Box 0978, Homewood, IL 60430-0978.)

GENERAL SPECIFICATION DATA (continued)

☐ BUILDING ELECTRICAL HAZARD RATING: OUTSIDE: CLASS ______ DIVISION ______ GROUP(S) ______
INSIDE: CLASS ______ DIVISION ______ GROUP(S) ______

☐ LIGHT FIXTURES: ______ ☐ LAMPS: ______
☐ AUXILIARY OUTLETS: ______ ☐ SIGNAL JUNCTION BOX: ______
☐ RAW POWER JUNCTION BOX: ______ ☐ UPS POWER JUNCTION BOX: ______
☐ TERMINALS, MANUFACTURER: ______ ☐ TYPE(S): ______
☐ 10% SPARE TERMINALS ☐ 20% SPARE TERMINALS

☐ AIR CONDITIONING BY: ______ ☐ HEAT PUMP BY: ______
☐ ELECTRIC HEAT ☐ STEAM HEAT
☐ BUILDING PRESSURIZING SYSTEM ☐ INTAKE PRESSURIZATION FAN(S) ☐ 1 ☐ 2
☐ LOUVER(S) IN WALLS ☐ LOUVER(S) IN DOOR(S)
☐ PROCESS BULKHEAD ☐ UTILITY BULKHEAD
☐ TEMPERATURE ALARM AT ______ INTERIOR TEMPERATURE ☐ HYDROCARBON ALARM FOR SHELTER INTERIOR
☐ ALARM LIGHT FOR: ______ ☐ ALARM HORN FOR: ______
☐ ALL LINES LABELED WITH STAINLESS STEEL STAMPED TAGS ON EXTERIOR OF EACH SHELTER WALL PENETRATION
☐ ALL LINES LABELED WITH 2-PLY PLASTIC BLACK LETTERS/WHITE TAG ON INTERIOR OF EACH SHELTER WALL PENETRATION AND ON EXTERIOR OF ALL SAMPLE CONDITIONING SYSTEM ENCLOSURES ☐ OTHER TAGS
☐ BLOCK VALVES ON EACH LINE AT EXTERIOR OF EACH PENETRATION ☐ OTHER VALVES
☐ ENGINEERING DRAWING PACKAGE (DESCRIBE IN NOTES BELOW) EXTERIOR / INTERIOR RATING
ELECTRICAL HAZARD RATING MUST BE MOST HAZARDOUS OF TWO DIVISIONS CONTROLLING SHELTER (EXTERIOR / INTERIOR RATING)

NOTES

NO.	DATE	REVISIONS	BY	INST. APPR	PROC APPR	NOTES

	ANALYZER SHELTER	Sheet 3 of 4	JOB NO.	
			DATA SHEET NUMBER	REV.
	Unit/Plant ______		DS	
	Prepared By: ______	Checked By: ______		

Engineering Standard Form

Figure 7-7. Shelter Data Sheet 3 (Adapted from copyrighted material of Consulting Technologists, Inc./Process Analyzer System Technologists [CTI/PAST], P.O. Box 0978, Homewood, IL 60430-0978.)

ANALYZER SHELTER EXTERIOR

ILLUSTRATED DEVICES TO INCLUDE (BUT NOT BE LIMITED TO) MAIN CIRCUIT BREAKER PANEL, POWER JUNCTION BOX, SIGNAL JUNCTION BOX, SAMPLE AND UTILITY BULKHEADS, AC UNIT, CALIBRATION AND CARRIER CYLINDERS, DOOR LOCATIONS, PRESSURIZING FANS, FAN AND LIGHT SWITCHES, ALARMS AND HORNS, ETC.

N

☐

CHECK CORRECT ORIENTATION

☐N

DIMENSIONS AS NOTED

AC UNIT MUST BE LOCATED AT LEAST 5 FT. AWAY FROM ANY HYDROCARBON VENT OR BUILDING OPENING AS PER API RP500A

ANALYZER SHELTER INTERIOR

DEVICES TO INCLUDE (BUT NOT BE LIMITED TO) ANALYZERS, CIRCUIT BREAKER PANELS, DISTRIBUTION PANELS, HEATERS, SAMPLE CONDITIONING SYSTEMS, FAN AND LIGHT SWITCHES, LIGHTS, AUXILIARY OUTLETS, ETC.

N

☐

CHECK CORRECT ORIENTATION

☐N

DIMENSIONS AS NOTED

NO.	DATE	REVISIONS	BY	INST. APPR	PROC APPR	NOTES

	ANALYZER SHELTER	Sheet 4 of 4	JOB NO.	
			DATA SHEET NUMBER	REV.
	Unit/Plant ______		DS	
	Prepared By: ______ Checked By: ______			

Engineering Standard Form

Figure 7-8. Shelter Data Sheet 4 (Adapted from copyrighted material of Consulting Technologists, Inc./Process Analyzer System Technologists [CTI/PAST], P.O. Box 0978, Homewood, IL 60430-0978.)

Single-Sourcing Versus Open-Bidding Techniques

In today's environment of alliances and preferred vendor lists, it is becoming more common to "sole-source" analyzer equipment from one supplier rather than use an open-bidding process for each installation. Table 7-1 lists a number of advantages and disadvantages regarding the use of preferred vendors and alliances.

Table 7-1. Advantages and Disadvantages of Single-Sourcing

Advantages	Disadvantages
Familiarity of equipment by facility staff, including maintenance and design	Restricts technology to what source vendor has to offer and reduces innovation
Reduced training	Less competition for your business
Input on new designs and improvements to equipment, including manufacture of "custom" units as standards	You may not have the latest technology since vendors tend to leapfrog one another in innovations
Removes the "we" versus "them" relationship	Loss of familiarity with other vendors' equipment
Reduced time spent on purchasing decisions	Vendor can become complacent
Reduced spare parts inventory	
Pricing discounts	
Increased familiarity of your process by your chosen vendor	

Quality Relationships with Vendors

To develop a quality analyzer system requires a quality relationship between all parties working on the project. This means both sides must trust each other and work together in an open atmosphere to complete the task at hand (rather than each side claiming its turf up front to assure that if and when something goes wrong it has enough paper to cover itself for the legal fight to follow). The cooperative approach results not only in a better project but in a happier one as well. The role of specifications and standards is not to cover every contingency should something go wrong but rather to obtain the best system possible for the best price. Remember, someone bidding on your project may not have read every last nuance of your specification-by-weight-or-volume, but when they see it, warning bells go off. They buy "insurance" to cover the eventual disagreements by adding incrementally to the price. You are the one who is paying for this vendor insurance.

Remember, someone bidding on your project may not have read every last nuance of your specification-by-weight-or-volume, but when they see it, warning bells go off. They buy "insurance" to cover the eventual disagreements by adding incrementally to the price. You are the one who is paying for this vendor insurance.

System Concept Versus Individual Analyzer Procurement

You may have noticed that the term "analyzer system" has been used throughout this chapter. This has been deliberate since no analyzer operates by itself, but instead normally must interact, as a minimum, with a sampling system and a data-acquisition system. Often, of course, the data-acquisition system then uses these data to act on something else in turn. Everything in the loop — from the initial sample to the final control element — represents the "system." Of course, we are primarily concerned with the smaller components of the analyzer system: the sample conditioning, the measurement, and the transmitted output signal.

Consider the concept of purchasing analyzers as systems rather than as individual components that are simply placed together with the hope that they will happily coexist and work. In light of the move by many organizations in the '90s towards downsizing and contracting out specialty skills, many companies no longer have the expertise and experience to properly put together the pieces to make the congruent whole. As a result, several system houses have evolved which specialize in just this sort of work. Many analyzer vendors have expanded their roles in this area as well. As a caution, please note that no one knows your process better than you. You should critically review any system house or analyzer vendor design for proper interface with your process.

Unless you have the experience to design the sample system (which causes 85% of the problems), it may be well worth the additional cost to have someone with demonstrated expertise to design it.

Unless you have the experience to design the sample system (which causes 85% of the problems), it may be worth the additional cost to have someone with demonstrated expertise to design it.

Typical Standards Specifications

Occupational Safety and Health Administration (OSHA) exposure limits are very important standards that define the amount of exposure a person can have to a substance. These limits are important and should be considered during any project. A related item is the material safety data sheet, which specifies all the criteria on a particular substance, including exposure limits, toxicity, and preventive measures.

The International Organization for Standardization (ISO; the world's largest standards organization) has developed an easy way to access a multitude of standards. On January 16, 1995, ISO launched an Internet-based service to provide a catalog of standards, background information on ISO and its activities, listings of technical committees and member bodies, and a calendar of upcoming events. The Universal Resource Locator (URL) is http://www.iso.ch/.

About the Author

Ian Verhappen is a professional engineer in the province of Alberta where he works as an engineering associate for Syncrude Canada Ltd. As part of his duties at Syncrude, he is responsible for developing and maintaining the company's analyzer specifications.

Mr. Verhappen graduated from the University of Alberta in 1982 with a B.Sc. in Chemical Engineering. He had previously received a B.Sc. in Environmental Science in 1979. In addition to being a professional engineer, Mr. Verhappen is one of the original Certified Specialists in Analytical Technology (CSAT), a member of ISA where he presently serves as both the Analysis Division Director and newsletter editor, a member of the American Institute of Chemical Engineers, and received the 1994 ISA Analysis Division Member of the Year award. Mr. Verhappen serves on numerous technical standards committees including ISA's SP76 - Process Analyzers, SP5.1 - Symbols, SP51 - Terminology, and SP20 - Forms. He is also an active member of Canada's IEC SC65D Technical Committee.

8

Process Analyzer System Training Considerations

It is estimated that in the decade of the 1990s, more analyzers will be installed than since the industry began. The decade started with a shortage of personnel, and, if the training activity that has transpired to date is an indicator, the decade will end with an even greater shortage of qualified analyzer technicians.

Currently, a large investment is being made in the purchase and installation of modern and highly-complex analyzer systems. The return on this investment will never be realized unless the analyzers are properly installed and commissioned. The solution, of course, is to have qualified analyzer personnel. The problem, of course, is where to get them! This chapter will discuss an approach to determining the number of personnel required and methods of recruiting and training process analyzer maintenance personnel.

Quantifying Analyzer Personnel Requirements

The number of analyzer personnel required is a function of the number and type of analyzers installed — or, more precisely, the number and type of analyzers being productively maintained. Therefore, the first task is to quantify the maintenance function on an analyzer-category basis and determine the exact manpower requirements.

To establish norms and assure a high degree of accuracy, the database search is restricted to a typical refinery in North America. Further, analyzers are categorized as "simple" (oxygen, pH, conductivity, density, etc.) and "complex" (gas chromatographs, mass spectrometers, FTIR analyzers, etc.). Other categories include environmental and physical property analyzers. The norms are as follows:

Analyzer Type	Total Maintenance Function Man-hours/Week/Analyzer
Simple	1.7
Complex	3.8
Environmental (H_2S, combustibles, etc.)	2.0
Physical property	2.7

In support of these recent numbers, there has been a general agreement for the past several years that one analyzer technician can maintain 11 gas chromatographs. It should be noted that this time includes the maintenance reporting system or "paperwork," periodic or preventive maintenance (calibrations, filter changes, inspections, etc.), and unscheduled or crisis maintenance (breakdowns).

As stated, these are the current norms for a refinery in North America. By contrast, a typical chemical plant in North America will show an almost 50% decrease in maintenance man-hours while having a greater population of complex analyzers. A principal reason for this disparity is better trained (and more efficient) personnel, resulting from ongoing and more aggressive training programs necessitated by the installation of more complex and newer (high-technology) analyzers. Other contributing factors noted in recent years are streamlined (computerized) maintenance and data reporting systems, apparent high morale in the analyzer group, and more delegated responsibility (e.g., an individual is delegated total responsibility for a particular group of analyzers). Another factor is the growing trend toward establishing certain analyzer personnel as "experts" on a particular class or type of analyzer.

In an example of a typical refinery, the types of analyzers are as follows:

Analyzer Type	Quantity
Simple (pH, conductivity, oxygen, temperature, flow)	53
Complex (GC, IR, moisture, color)	29
Environmental toxicity, combustibles opacity	83
Physical property (and misc.)	14
TOTAL	179

To determine the maintenance personnel requirement, we need to factor in the weekly man hour requirements:

Analyzer Type	Quantity	Man-hours/Week Required	Total Maintenance Hours/Week
Simple	53	1.7	90.1
Complex	29	3.8	110.2
Environmental	83	2.0	166
Physical property	14	2.7	37.8
TOTAL			404

The total maintenance hours per week requirement is 404 hours, which divided by 40 gives the number of personnel required to "do the job." Therefore, 404 ÷ 40 = 10.1 (10) analyzer technicians are required to maintain the workload.

This formula can be used to quantify analyzer personnel requirements at almost any facility.

It is strongly suggested that new analyzers be factored into the manpower requirements as soon as possible — at least as early as the engineering phase. All too often, new analyzers command an inordinate amount of maintenance personnel time during the installation and commissioning phase, as well as during the training phase. This is often to the detriment of installed analyzers.

Some organizations are never able to fully recover from "falling behind." Most attempts to do so result in less intensive maintenance programs, with the installed analyzers simply ignored or shut down in-place. In any case, productivity is minimized.

Even when an analyzer is shut down or removed from service, costs continue in the form of utilities, shelter maintenance, and the stocking of spare parts. Further, once an analyzer is shut down for a year, seldom can it be recovered economically.

Even when an analyzer is shut down or removed from service, costs continue in the form of utilities, shelter maintenance, and the stocking of spare parts. Further, once an analyzer is shut down for a year, seldom can it be recovered economically.

Qualitative Definition of Analyzer Maintenance Personnel

Once the number of maintenance personnel needed has been determined, consideration should be given to the quality of the technician who will be hired and trained to perform the work.

The factors that make a good analyzer technician are varied and, in most cases, arguable. Education (specifically, is the person willing to be educated?), experience and/or ability, self-confidence, and the ability to adapt to an ever-changing organization while maintaining efficiency are the real keys to a good analyzer technician. Before any qualitative evaluation can be made, a norm must be established. Many studies in this direction have been performed, and the consensus is as follows.

First, it must be recognized that morale is a nonqualitative factor; the concept of "high" or "low" morale does not affect individuals to the same degree. Morale, as a mental state, appears to be unique to an individual, especially in a small group and where the individual involved believes he or she can change policy or the situation. More often than not, good or reasonable management can control the morale factor. Also, it has been noted that in a small group, morale issues are mostly transitory in nature, and the reasons stated for low morale are seldom the real cause. (Low pay is the most frequently mentioned factor.)

Most, if not all, of the elements that make up a good analyzer technician are intrinsic. However, the organization can provide assistance, motivation, and inspiration.

Most, if not all, of the elements that make up a good analyzer technician are intrinsic. However, the organization can provide assistance, motivation, and inspiration.

An intangible, but essential, characteristic is the desire to be educated. This is especially necessary when a general upgrade of the workforce is being considered. Unfortunately, more than a few people believe that high school, trade school, or college marks the end of the educational experience. Today's workplace demands constant attention to changing technology. Candidates for upgrade training should have demonstrated not only a thirst for education, but also the ability to embrace new concepts.

Another consideration is experience versus ability. In the common workplace, all jobs can be mastered to an acceptable level in a given amount of time. Individuals who have acquired an inordinate amount of experience without gaining proficiency at a particular job are usually poorly placed. Such workers were able to "hang on" in the past. Today's workplace, however, demands greater knowledge as well as greater efficiency.

If it is true that various organizations use identical or similar technology and hardware, it follows that the greatest profit will be earned by the organization with the best people.

A good candidate will exhibit confidence and self-esteem about himself and his job. Individual pride in a job well done and the confidence to assume new duties or take on a new type of job are all characteristics of the successful and competent analyzer technician.

Individuals who cannot cope with the changing scene will become distinct liabilities and provide very little return on investments of any type.

In today's workplace, changes — in policy, hardware, and even location — are constant. Individuals who cannot cope with the changing scene will become distinct liabilities and provide very little return on investments of any type.

Historical surveys indicate that pre-1985 maintenance personnel tended to be settled into their jobs and expected few changes. Records show that a typical instrument technician attended fewer than three formal training programs in his entire career. In cases where new technology was introduced, qualified personnel were hired and the overall workforce was increased. Because this approach was followed throughout process industry, product prices were increased and profits were unaffected.

Since 1985, technology and the "system engineering concept" have swung to extremes. Some organizations embraced the newer ideas and took steps to merge with the new technology, while others took a "wait and see" approach. Those who prepared for the future now have a monopoly (at least!) on trained and better qualified analyzer personnel. Although there is currently a shortage of qualified personnel, organizations with a vision are functioning efficiently and profitably.

The type of analyzer technician that will be needed in the future depends on the complexity of the analyzers that will be introduced into a particular organization.

The type of analyzer technician that will be needed in the future depends on the complexity of the analyzers that will be introduced into a particular organization. Based on several independent surveys, Table 8-1 projects the future (1995 - 2000) analyzer requirements of a typical refinery (chemical plants tend to be more individualistic).

Table 8-1. Projected Analyzer Use in a Typical Refinery (1995 - 2000)

Analyzer Type	Anticipated Increase Based on Current Total Analyzers,%
Gas chromatograph	26.2
Oxygen	5.7
Boiling point	7.4
Distillation (physical property)	6.6
Distillation (by GC plus vapor/liquid ratio and total amines)	14.2
Octane (by NIR)	2.6
Combustion and Toxic	6.1
Supercritical fluid GC (replacing vacuum distillation analyzers)	4.6
Infrared analyzers	6.1

Note: All other typical refinery analyzers are expected to increase by a nominal 1 to 5%.

There are some who believe that the shortage of personnel and resultant higher salaries will attract college graduates who prefer to "work with their hands."

Based on these data and the current and expected state of the art, the new analyzer technician will be a specialist in electronics and possess a good understanding of basic chemistry and physics. Typically, he obtained a good background in mathematics, chemistry, and physics in high school, attended a junior (community) college, and obtained an associate degree in electronics or computer technology. Alternatively, he completed a comprehensive trade or commercial school or an extensive apprenticeship program. There are some who believe that the shortage of personnel and resultant higher salaries will attract college graduates who prefer to "work with their hands."

Existing data for North American refineries show that the vast majority (84%) of analyzer technicians started as instrument mechanics/technicians. After some time (1 to 5 years) and based on above-average performance, the technician was moved into the analyzer group as a trainee or apprentice. Progression from trainee to senior analyzer technician appears to be a variable from company to company, but there is considerable evidence to show that an influx of newer technology has resulted in rapid advancement.

The average experience profile indicates that a senior analyzer technician has more than 10 years of experience, of which 6.4 years have been totally in analyzers. Typically, this person also has almost 4 years as an instrument technician. The apprentice or trainee period is almost always 1 or 2 years, depending on the location and the organization.

Potential Sources for Analyzer Technicians

There are basically two ways to fill the need for analyzer technicians: (1) hire skilled personnel, (2) develop or implement a comprehensive training program.

Hire Skilled Analyzer Personnel

In today's marketplace, this is almost impossible — at any salary. There are not enough qualified analyzer personnel to satisfy current demands, and if this trend continues, there will be fewer in the future. Contrary to some beliefs, the shortage is not caused by distribution, but rather by a simple case of demand exceeding supply. As previously discussed, the number of refinery analyzers will increase by approximately 20% in the period from 1995 to 2000, which corresponds to an increased need for qualified personnel. The lowest projection from the authorities who project future requirements is a 15% increase in the analyzer population.

The lowest projection from the authorities who project future requirements is a 15% increase in the analyzer population.

Among available analyzer personnel, the competition is very high. Most available personnel are currently working for contractors, and the salary required to lure them away is beyond the pay scales of most, if not all, refineries.

Hiring analyzer personnel from the available pool will be almost futile. Most of the personnel hired will be expensive and will leave when a better offer appears. This will only increase tensions and lower the overall productivity of the personnel who remain.

One option, however, may be to hire technicians from the United Kingdom or India. Both countries have educational systems and apprenticeship programs that always produce more personnel than required. This is a very limited option because of the federal bureaucracy controlling immigration and general opposition of the unions. Companies with operations in the U.K. are able to transfer (import) personnel with some ease.

Our experiences with technicians from both countries have been favorable. Extremely well-qualified personnel at competitive rates are readily available for specified contract lengths. It should be noted that education claims and relevant experience should be independently verified.

Hiring foreign personnel is a viable alternative, albeit a lengthy and cumbersome effort fraught with endless paperwork. Most companies with overseas subsidiaries prefer to hire technicians locally and transfer them to a U.S.-based operation <u>for a specified contract period</u>. In some cases, reciprocal agreements between countries (e.g., the United States and Great Britain) make this course of action reasonable.

Develop or Implement a Comprehensive Training Program

The possibilities here are many and varied — ranging from in-house training to having specific courses designed and taught by a local junior or community college or other educational institution. Also available are "canned" courses from specialty training organizations such as ISA and Measurementation. The next section will discuss possible approaches.

Training Programs

Personnel Selection

The ideal approach is to make the training investment in personnel who are committed to the organization or who can be retained.

For the moment, assume that a training program exists. Who will be trained? From where will these candidates come? The ideal approach is to make the training investment in personnel who are committed to the organization or who can be retained. In some cases, organizational constraints and labor agreements will preclude using personnel who are already on the payroll.

Some companies have successfully avoided labor conflicts by establishing an exempt group of analyzer technicians. It has been observed that a company going through a major upgrade that increases the quantity of analyzers and the overall technology typically will engage a contractor to do the installation, commissioning, and preventive maintenance for a period of time. When the contract expires, the company will hire the contractor personnel, who will continue the maintenance program (which is, in most cases, clearly established outside of union jurisdiction).

Some 10 years ago, one major company established a policy of having the entire analyzer staff composed of one-third union personnel, one-third non-union contract personnel, and one-third exempt employees. Although this scheme was originally designed to minimize the effects of prolonged labor disputes, the current shortage of analyzer personnel has a minimal effect. Another company has expanded its definition of the exempt group to include "para-engineers/ professionals." Now senior or exceptionally well-qualified individuals in this group "bridge the gap" between the engineering group and the field — in all matters from project conception through analyzer system commissioning.

Unfortunately, wholesale hiring and training of people "off the street" is a numbers game. Without too much fear of contradiction, it can be estimated that of ten people completing a comprehensive training program, three will be above average, three will be average, three will be below average, and one will fail outright. Further, at least two of the three above-average people and one of the three average people will find other jobs making more money. This represents an immediate 30% loss on investment, in addition to being an inestimable setback in attaining the training goals. For those who think that success can be achieved with the "residue," consider the following. As the need for analyzer personnel becomes more critical, most of the "residue" will appreciate and become more valuable to some other organization, because normal payroll policies will have prevented these personnel from being internally advanced to meet the competition. The result is simply that ten people have been trained, and in 3 years, two or three of them remain.

There appears to be no shortage or projected shortage of apprentice-type instrument or electrical personnel.

The rule is: If you train them, you've got to keep them. One scheme would be to select training candidates from the instrument and electrician crafts. Selection of employees with 10 years or more of service who exhibit an aptitude and desire for upgrading would make room for younger replacement personnel at lower pay scales. There appears to be no shortage or projected shortage of apprentice-type instrument or electrical personnel.

Training Program Content

The primary problem in training analyzer personnel is simply that there is no such thing as a standardized training program. Few, if any, educational institutions offer a comprehensive training program titled "Analyzer Maintenance Technicians." Therefore, the first task is to define the curriculum that meets the organization's exact requirements.

A typical curriculum would include the applicable chemistry, physics, basic electronics, and digital techniques needed to understand analyzers in general. Once this basic training is complete, the detailed operation and maintenance of the various installed analyzers can be taught. At this point, the company should begin to realize a return-on-investment, and as the employees gain experience, the rate of return should increase accordingly.

Based on typical course outlines, 14 to 16 weeks appear to be the average time needed to cover basic chemistry, physics, electronics, and digital techniques. The duration of operation and maintenance courses on the various analyzers will vary as a function of complexity.

About the Author

Buzz Goins currently acts as an independent process analyzer system consultant after a successful career in the petrochemical industry. He has significant experience in analyzer operations and maintenance training.

9

Process Analyzer System Maintenance: Personnel and Organizational Issues

Before any company commits extensive resources in the pursuit of new process analyzer technology, maximum effectiveness from existing technology should have been obtained. More often than not, it is the application of fundamentally-proven process analyzer technology that may have the greatest effect on a company's competitive position. The ability of a company to implement and manage proven technology can be significantly affected by nontechnical issues. The effective management of these personnel and organizational issues can to a large extent determine the success or failure of a process analyzer program. In a technical field that is capital intensive, the benefits obtained by effectively managing these issues require little or no capital.

Little effort has been made to address personnel and organizational issues because of their inherently subjective nature. However, a great deal of experienced-based information on these issues has been accumulated over the years by industries actively involved in the implementation of process analyzer technology. Experiences related to personnel and organizational issues are fairly universal in nature. A critical examination of these experiences may help place the issues of personnel organization in a more objective light.

Personnel Issues

Investing in process analyzer technology is not without risk. Based on past experience, most companies that have used process analyzers can point to analyzer installations that did not meet requirements. The field of process analyzers is replete with stories of missed opportunities and wasted resources. This is not to minimize the tremendous benefits that process analyzers have had on the process industries. However, it can be argued that there is not a single company that cannot improve productivity in their management of the business of process analyzers. With increased competitive pressures and environmental, health, and safety concerns, this is not a time for using analyzer technology resources in a wasteful fashion. The first line of defense in the productivity effort is the quality, experience, and knowledge of the personnel who work in the process analyzer area.

The fundamentals of personnel issues in process analyzers is really little different from in any other technical field — with the exception that it is normally a specialized field resident within a larger organization. Analyzer personnel must be technical experts in their own right, but they must focus their efforts on business goals of the company using the analyzers. There is a fine line between what is in the best interests of the company and what is in the best interests of increasing the effectiveness of process analyzers.

Team Issues

Strong-minded individuals who cannot see past the current crisis will eventually cause damage to an analyzer program. There is not a single individual in the process analyzer field who has complete knowledge of the total body of process analyzer technology. The field is simply too large. Therefore, the most effective programs rely not on the abilities of a single individual but rather on a team effort.

An analyzer team can weather many of the personnel problems associated with making difficult decisions. A team decision that has been carefully thought out with long-term goals and the company's best interest in mind can still be the subject of intense scrutiny. It is good business practice to question decisions before resources are committed. Normally, if the decision is a sound one, it will stand up well under such scrutiny. However, if a dissenting "analyzer expert" can be found, then an alternate decision favoring the "analyzer expert" can prove tempting. This is an example of normal business practice that works well for most applications, but can be disastrous in the analyzer business. Persons not familiar with the analyzer business may not understand the finer points of a disagreement among "analyzer experts." Indeed, in the past this has been a serious problem. This personnel issue must be corrected or serious consequences could result.

Experience Issues

The analyzer field is experience-based; both analyzer personnel and management must accept this. The "analyzer expert" and his management must test opinions. Valid concerns must be addressed; creating an effective program based on reality is the goal. When in doubt, do not ignore the opinion that says it can't be done. Even after all the dissenter's objections have been addressed, continue to be skeptical. When trying to effectively utilize resources in the analyzer business, 95 times out of 100 it is better to abandon a bad idea than to install it. A good practice to follow is to question the easy decisions at least as hard as the difficult decisions. If the dissenter's motive is one of not wanting to repeat past mistakes, his or her concerns should be addressed. Experience lessons are hard learned, but not easily forgotten.

Based on experience, especially in the project area, following the wrong "expert" advice could be very costly. Many idle analyzers in plants are examples of the end result when this scenario is played out. These monuments to poor decisions are vivid reminders of the downside of the analyzer business. Experiences like this will often lead to resentment and distrust of the decision-making process. It is difficult to build any type of effective analyzer team when this scenario is repeated.

Is there a foolproof way to be sure this mistake is not duplicated? No. However, certain tests can be applied to ensure that a more objective decision is reached.

- The first test is **ownership**. Who will get the unglamorous job of guaranteeing the analyzer's performance?
- The second test is **knowledge**. Who has the greatest experience and knowledge of the analyzers and process streams in question?
- The third test is **credibility**. Who has a proven track record of producing results in your plant?

It is far better to try and prevent this issue from ever occurring than to attempt to rectify the situation in midstream. This issue will also be addressed in the section "Organizational Issues."

Interfacing Issues

Interfacing with other segments of the organization is absolutely essential [Ref. 1] in the pursuit of an effective analyzer program. Having the judgement and maturity to work in a team environment is as important as technical qualifications. An individual with judgement and maturity can be taught the technical aspects of the business. An effective analyzer program requires an organization to make difficult decisions. A little diplomacy can go a long way when working with those in other parts of the organization to attain sound technical decisions. Thorough technical knowledge is essential for these decisions. However, technical knowledge alone is not enough.

Thorough technical knowledge is essential for these decisions. However, technical knowledge alone is not enough.

Maximum benefits occur when the proper decisions are made. If these decisions are hampered by individuals showing a lack of judgement or maturity, at whatever level in the organization, then one of three things will happen. The individual will be controlled or bypassed, he or she will be removed from the decision-making process, or the whole program will suffer. These comments are not made casually and are offered in light of experience based on significant problems that have occurred in various companies. These are tough questions, but if an effective analyzer program is to be achieved, a hard look at personnel must be taken. A good analyzer team effort will carry an organization far beyond what individual effort alone will achieve. Team-building efforts should be encouraged.

To accomplish the end result, the different teams and individuals that find themselves involved with analyzers must learn to work together. Individual compromises can be agreed to, but overall goals cannot be sacrificed for the good of team unity. This is an example of where good judgement goes hand-in-hand with thorough technical knowledge. All team members must be willing to work together to achieve the best solution. Under no circumstances should a team decision be made that repeats a previous mistake!

When working in new territory, some brainstorming may be acceptable. However, it must be done within the framework of being sure not to repeat common errors. Quite often, decisions are inadvertently made to repeat past mistakes. In fact, care must be taken that brainstorming does not endorse an idea that resulted in a previous failure. Proper judgement as to what compromises can produce acceptable risk is essential. It must be remembered that suggesting the organization make the same mistake twice happens more frequently than one might imagine!

Nothing is perhaps more devastating to a program than to repeat a past mistake. In fact, the single most important thing, as simple as it sounds, is to avoid repeating past mistakes.

This may sound trivial. However, because what may constitute good practice in other areas could constitute a serious mistake in the analyzer area, the statement bears repeating.

Rotational Issues

A highly efficient analyzer program requires seasoned, experienced personnel. These individuals have usually spent more than 8 hrs per day working in the analyzer field, either from overtime work or continued involvement with analyzer technology outside the work environment. It is not reasonable to expect personnel with no prior experience to excel in the process analyzer field. Generally, the more experience a person working in the analyzer area has, the more productive that person will be. There is a general consensus among analyzer professionals that several years of experience are necessary before an engineer is fully capable. Short tours or rotational assignments in the process analyzer area in either professional, paraprofessional, or mechanical areas are generally unprofitable. The analyzer business suffers while the individual is in training. Additionally, a brief stint in the analyzer area rarely produces the desired flexibility.

One company made a practice of rotating some young instrument/process engineers into the analyzer area for 6-month tours. Several years later when trying to fill internally an analyzer engineer opening, none of the originally rotated engineers expressed an interest in the position. Rotation in the professional, paraprofessional, or craftsman ranks can work with individuals already having several years' experience in a particular plant. However, this arrangement works only because the experienced individuals in question are familiar with nearly all technical aspects of a particular plant's operation. These people are usually some of the most respected individuals in the plant and are the key to an effective competitive edge.

Getting the most out of each of them may depend to a large extent on their individual talents and personal preferences.

If you want to prove a point, certainly you can find individuals in your organization whom you can rotate into a highly technical field. These individuals may not enjoy the rotation and may not provide the depth necessary to maintain a competitive edge, but business constraints may dictate this course of action. It is obvious that in most cases a rotational policy does not allow the depth of knowledge to expand to a level that would create a competitive edge. After an individual has established himself in the analyzer field, there is a tendency for that person to remain in the field. The field is of definite interest to someone who likes a continuing challenge. This is especially true in a plant or facility that is highly dependent on analyzer results.

From a personnel point of view, the needs of the individual are not really far removed from the needs of the organization. Analyzer personnel at all levels should be left in the same position long enough to become knowledgeable and productive. When productivity gains begin to run out, a rotational assignment should be considered for the benefit of the individual and the organization.

Therefore, the ability of an individual to function at an acceptable level should not be determined at the very beginning or at the tail-end of his work cycle. Often an individual is prejudged initially or after the job has ceased to hold any challenge for him. It is the responsibility of the organization to provide an initial training/transition period prior to making a definitive judgement. Also, as a bare minimum, an individual who is capable of performing in an exceptional fashion should be given additional opportunities to increase his skill and contribution to the company after thoroughly learning the job. The right move generally produces remarkable benefits for both the person involved in the move and the business area that experiences the changes. It is curious that frequently quite the opposite policy is followed. As an individual matures in a certain job, too often the possibility of movement to another area diminishes.

Of course, a smaller organization has greater limitations in terms of such personnel moves. An alternative may be to share parts of a job/work area with another individual. If the situation deteriorates to unacceptable levels, then regardless of the organizational limits, action must be taken. In this case, resources outside the organization should be considered (e.g., a contractor) as a temporary solution to the problem. It is preferable to anticipate the problem and make these types of outside contacts before the situation affects the process analyzer program.

Qualification Issues

Personnel selected for analyzer positions or possible rotation should be chosen from related fields. The following are possible choices that have worked in the past. However, individual talents and preferences must still be considered.

Professional (analyzer engineer)

- Instrument engineer, project or technical service
- Laboratory chemist (analytical discipline preferred)
- Process engineer (chemical engineering background)

Paraprofessional

- Instrument/electrical foreman
- Instrument/electrical inspector
- Digital technician

Mechanic

- Instrument mechanic
- Process operator/controller
- Laboratory technician

These are not the only possibilities, especially if a potential candidate expresses a desire to work in the analyzer area. The desire and the interest to learn the analyzer business can in many cases make up for a lack of in-depth background. This can be especially true in the analyzer field, where no one comes equipped with the perfect background to initially perform all functions expertly. Even though an outstanding individual from a completely different field may do well in the analyzer field, cases are rare. Generally, the following characteristics must be present:

- Compatible background
- Desire to work in the analyzer area
- Demonstrated potential to work for extended periods towards long-term goals
- Interest in using process analyzers to achieve business goals
- Ability to work in an unsupervised mode
- Ability to communicate with a wide variety of people in the organization
- Ability to attack problems with an open mind
- Innovative in electrical, mechanical and analytical methodology areas.

It may not be possible to obtain an individual with these characteristics from the current organization. A suitable alternative would be to hire someone from another plant. Although there is a lot to be said for homegrown talent, isolation for an analyzer professional can lead to stagnation and a feeling that all good technology comes from one's own plant. By the same token, an individual with an outstanding track record at another company may experience difficulty in achieving results in a different organization. Both situations are to be avoided; either could affect long-term growth.

Generally, most paraprofessionals/mechanics are selected at the plant site location. Selection of mechanics follows the same vein as selection of potential analyzer engineers. A more experienced individual in a related discipline is preferred. A typical mechanic's job description is shown in Table 9-1. In smaller plants or areas where a mechanic is required to maintain analyzers and other equipment as well, it is preferable that the mechanic be proficient in a related field, such as instrumentation, before learning the analyzer business. It may not be possible to choose the best qualified person to assume the duties of an analyzer mechanic. In this case, closer supervision of individual activities is required. The partition of duties or the gradual addition of more responsibility is also a viable alternative. More routine work initially will help the mechanic trainee adjust to analyzer maintenance.

Table 9-1. Analyzer Mechanic Qualifications

Technical Skills
Received factory training on all analyzers in his/her area of responsibility
Has been trained in plant processes, chemistry, physics, and electronics
Has been cross-trained on analyzers in other parts of the plant
Is familiar with computing systems in use at the plant
Work Habits
Uses his/her time well
Orders his/her own parts and keeps area documentation up to date
Understands the impact of the analyzers he/she maintains on plant operations
Does performance tracking of his/her analyzers
Understands that solutions can be applied to similar situations
Suggests solutions on his/her own to solve recurring problems and accepts input/decisions from others
Is familiar with and uses resources available
Willing to self-train
Exhibits flexibility in his/her work habits
Takes an active role to keep abreast of other areas
Team Orientation
Can work in a team relationship with individuals from operations, engineering, other maintenance disciplines, analyzer manufacturers, laboratory, drafting, etc.
Shares information with other mechanics
Possesses good interpersonal skills
Is able to train other mechanics
Is well thought of by peers and supervisors

The analyzer engineer/coordinator performs an absolutely critical role when alternatives and changes to a system are considered. This person is in the best position (assuming that he or she possesses the experience and data) to determine the overall effects of a given action. Selection of the proper individual to fill this position is critical. Regardless of whether the prospective candidate is acquired internally or externally, an adjustment time will normally be required for the individual to get up to full speed.

Recently, ISA has announced a certification program of "Analytical Technologist." The basis of the certification is knowledge and experience in the process analyzer field. In preparation, ISA has assembled a list of skills the analytical technologist should possess [Ref. 2]. To date, a better listing of the skills involved in process analyzer work has not been prepared. As the field of process analyzers matures, formal training may herald the coming of a time when analytical technologists graduate trained in the field. Until then, years of experience, a proven track record of performance, and the respect of one's peers are the prime qualifications.

While we are busy defining the necessary qualifications of process analyzer personnel, those qualifications are changing. Process analyzer personnel will need to respond to different concerns as we move through the 1990s [Ref. 3]. Figure 9-1 illustrates some of the potential changes that analyzer specialists will face in basic job skills over the present decade. Different skills will need to be emphasized as times and business priorities change. One advantage that process analyzers give a business is some degree of flexibility to respond to different business conditions.

One quality that is desirable in analyzer personnel is the ability to design plant analyzer systems so that future changes are possible with minimal disruption.

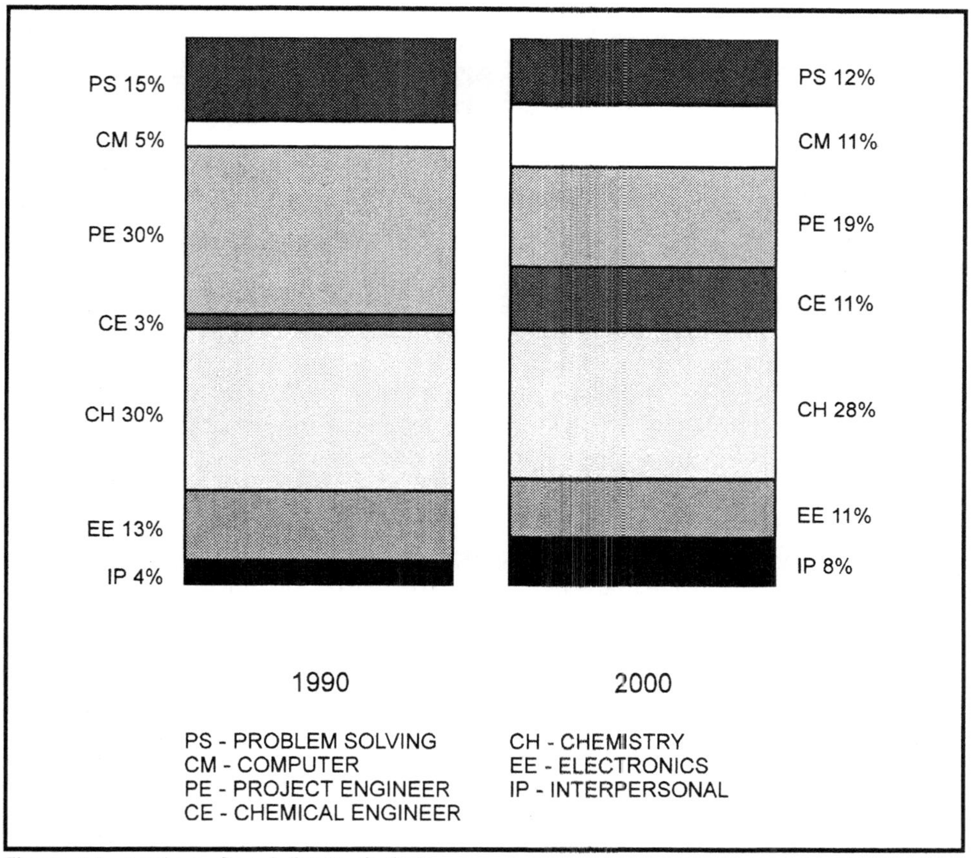

Figure 9-1. Analyzer Specialist Basic Skills

Development Issues

Although there is really no substitute for experience in the process analyzer area, appropriate technical training will accelerate the learning process. Outside training is also beneficial in order to gain another perspective and thus avoid the "not invented here" complex. Business training conducted by one's company may help an experienced new hire get feet on the ground faster. Those needing work with interpersonal skills and so forth can benefit from these types of courses.

Mentoring, either formally or informally, is an excellent way to gain needed experience. Analyzer personnel cannot be the type of individual who seek to enhance their position or make themselves irreplaceable by not sharing information. Allowing an individual with this type of attitude to gain control of an analyzer area or program is to be avoided. Sharing learned information benefits the team effort and lays the groundwork for the next level of performance.

The field of process analyzers has been hurt because of the inadequate supply of thoroughly knowledgeable professionals available. Some individuals are genuinely too busy to share information and concern themselves only with getting as much accomplished as possible. It is tempting to let these workaholics alone to accomplish what they will. Indeed, the best analyzer people will be somewhat consumed by their work. However, implementing analyzer technology is still a team effort. Therefore, the efforts of these lone workers need to be assimilated and aligned by the team.

"Networking," or meetings/communication with others in the process analyzer business, can be very worthwhile from a development point of view. Younger engineers can benefit from the experience and contacts of more experienced engineers. Smaller sites can benefit from the experience of meeting periodically with others at larger sites. A central engineering/research organization can review corporate goals and directions. If no such organization exists, the networking committee, with management guidance, can decide directions and approaches. An effective networking program can greatly enhance the effectiveness of all the individual plant programs involved.

Networking is sometimes viewed in a negative light. Indeed, network meetings conducted at long intervals (e.g., every 5 years) may actually be disruptive. Such meetings offer limited time to express concerns and address problem areas. So much activity and so many personnel changes have occurred that usually there is little meaningful exchange of technology or information. Network meetings conducted more frequently than twice a year typically offer no additional advantages and may even contribute to a decline in the efficiency of the individual plant sites because of the efforts necessary to prepare for the meetings. If properly monitored and controlled, periodic network meetings can also help address many of the personnel issues that have already been discussed.

Another way to gain a wider perspective is association with professional societies that focus on process analyzers. The obvious organization that comes to mind is ISA. ISA and other smaller local organizations offer opportunities to associate with experienced analyzer engineers at reasonable cost to the company. Contacts generated in professional societies can be useful for a variety of business purposes.

The analyzer engineer/coordinator will have his or her hands full maintaining technical competence. It is difficult to supervise and remain technically competent. If the analyzer engineer spends more than 40% of his or her time on issues other than technical matters, technical competence may suffer. Developing the technical competence to guide a plant's analyzer program requires a great deal of time and effort. However, mixing supervision with the job is preferable to rotation from technical to supervision and back to technical. Considering the extensive list of skills the analyzer professional should have, the organization and the individual needs to take advantage of any reasonable opportunities to develop these skills.

Communication Issues

Communication and negotiating skills are very important in the business of process analyzers. There is no way to minimize the importance of these skills in the team approach or in interfaces with other parts of the organization. Everyone is normally anxious to avoid failure. It is very important to build on failures in the business of process analyzers. If the failure is due to a breakdown in the team because of a communication problem, then an effort must be made to correct the situation. As mentioned before, repeated failures demand significant action. If the first failure was explained away as a fluke, then even more attention needs to be directed. Often, due to the long lead time of implementing process analyzer technology, the same type of error is repeated by a different set of personnel. For this reason, consistency in personnel, to the maximum extent possible, is essential.

Analyzer personnel need to communicate clearly what existing analyzer technology can and cannot do. With the analyzer technology available today, it is possible to achieve high service factors in a great many applications. However, this requires correct initial selection of equipment, good initial system design, effective preventive and on-call maintenance procedures. If the required service factor is not being achieved in a given application, then one of the above factors is at fault and a study should be made to upgrade or improve the situation as required. Communication of process analyzer concerns should be proactive and not simply a matter of waiting for a crisis to develop.

Career Issues

After a core base of analyzer personnel has been developed or acquired, the concern becomes keeping this resource intact. Any program may be damaged by a loss of personnel. The smaller the group and the more productive the program, the greater the risk. One way to minimize this risk is a slow, but purposeful rotation. Ideally, this rotation should involve career advancement for the most proficient and experienced people. As mentioned before, rotation for the sake of rotation is to be avoided. If an individual is doing well in an area and a great deal is being accomplished, rotation often should be avoided. If a great deal of networking and sharing of information is occurring, rotation accomplishes little.

If a program were to be established to single out analyzer personnel for special career treatment, then additional responsibility would of course be appropriate. Special career enhancement might also justify a higher penalty if a perceived failure occurs. Special career treatment may be desirable during the initial building stages of a program. Normally, this type of special career treatment is difficult to administer on a long-term basis. It may also be resisted for other fundamental reasons, especially when a collective bargaining organization is involved. Such a program may not be necessary in order to attract and keep high-quality personnel.

Every individual, regardless of the type of job, will over time ask two questions:

1. Does the current job offer the same degree of advancement as another job for which I am also qualified?
2. If the chances for advancement are the same, does the other job offer a better responsibility/risk ratio?

These two questions should be objectively addressed from time to time by those who control and administer the analyzer program. The pulse of the program can, to a certain extent, be taken by determining how those in and those influenced by the analyzer program perceive career opportunities.

Lack of recognition and career advancement possibilities can be a problem in the analyzer field. The key is to make sure that career opportunities are identical to those offered in other parts of the organization. Obviously, making programs identical is very difficult; however, administration of career opportunities is minutely scrutinized by all interested parties. If there is even a hint that analyzer personnel are not given equal treatment, then the fallout could be twofold. First, good quality analyzer personnel will begin thinking about making a move. Second, attracting quality personnel into the analyzer area will become more difficult.

A second possible problem is the difference in the responsibility/risk ratio, either perceived or real. If an individual feels that by being in the analyzer area he takes on greater responsibility than another comparable job in a related area, he may have second thoughts. However, increased responsibility alone might not deter an individual from working the analyzer area, and can even be a positive motivator. However, if it appears that the performance of analyzer personnel is more closely scrutinized and that mistakes/failures are viewed in a more unfavorable light than similar mistakes in another discipline, then a high risk factor may be perceived.

When comparing jobs, a mental process occurs wherein the advantages and disadvantages are weighed against the likelihood of success. Unless an individual believes that he can overcome any potential problems, no matter how difficult and numerous, he will appraise the responsibility/risk ratio. For this reason, expectations in the process analyzer area must not be perceived as unrealistic. This does not mean that expectations cannot be high. If everyone who made an incorrect decision during his career was removed from the process analyzer field, then only the newest personnel would be working in the field.

Recognition Issues

Recognition of the efforts of all analyzer personnel by their peers and fellow workers should also be equitably distributed. Once again, the object is to put together an effective team, one where self-respect for a job well done is possible and where others respect and recognize superior work. Recognition problems are not unique to the process analyzer business; however, a couple of special cases need to be covered.

At one end of the spectrum are individuals who are recognized by both their fellow analyzer personnel and those outside the program as deserving of recognition. At the other end of the spectrum are individuals recognized by the same two groups as needing improvement. In the middle are two unique groups:

1. Individuals who receive little outside recognition but consistently demonstrate superior performance and have a high degree of respect from their peers. These individuals, for whatever reason, have an aversion to "selling" their accomplishments.

2. Individuals who are recognized by those outside their area as excellent performers but whose performance is viewed by their peers as average, subpar, or even damaging. Thankfully, such individuals are few in number, and most operate from the misconceptions of inexperience which will be corrected over time.

An experienced I/E Supervisor once commented about a young and enthusiastic analyzer engineer, "He has about as many ideas as there are stars in the skies, and some of them are just about as bright!"

Every individual in the workplace probably fits into each of these categories at one time or another. In the highly specialized field of process analyzers, it is difficult for the unfamiliar observer to distinguish between superior performance and the appearance of superior performance.

There is no magic formula for avoiding the problems related to recognition issues. However, it is helpful to keep in mind that one of the primary purposes in putting together an effective analyzer program is to reduce risk and minimize failures. If too much attention is focused on failures, the end result could be to reduce the effectiveness of the analyzer team and actually increase your risk rather than reduce it. Of course, gross failures and significant problem areas cannot be ignored, but the focus should be to learn from these mistakes.

Also, customer satisfaction can sometimes be an extremely poor indicator of performance in the process analyzer business. There are cases where below-average programs are well thought of within the companies where they reside. There are also cases where excellent programs are always questioned and all accomplishments are viewed as not good enough. Neither situation is desirable. Placing ultimate confidence in a program that is subpar and wasteful of the resources devoted to it will not enhance your company's competitive position.

And failing to recognize excellent performance can negatively affect the morale of personnel who have done all that can be done with resources they have been given.

Organizational Issues

Managing organizational issues at the individual plant level is the key to developing a competitive position in the business of process analyzers. Properly managed, process analyzers can meet and in many cases exceed original business expectations. Improperly managed, process analyzers can consume company resources without producing the desired results.

A successful analyzer program is difficult to develop and requires careful management in order to keep it successful. One way of viewing a successful analyzer program is to think of it as "the goose that laid the golden egg." This simple idea helps to illustrate some of the issues that must be faced when trying to organize an effective process analyzer program. Like the goose that laid the golden egg, an analyzer program must be nurtured over the long haul to continue producing results.

Structure Issues

When the analyzer program in a plant is well organized, then a certain amount of independence should be granted to it. This independence is necessary to assure that the overall program is not skewed toward a limited agenda. Conflicting priorities for analyzer resources can frequently arise from different parts of the organization; for example:

- Operations
- Projects
- Computer Control
- Laboratory
- Environmental
- Maintenance

There are advantages and disadvantages to administratively placing analyzers under different parts of the organization. All possibilities are susceptible, to a large extent, to the same general organizational issues. None has a very good chance of achieving success unless the analyzer personnel who are most familiar with the plant have a vested interest in making the analyzers work. The following are some of the advantages and disadvantages that may be encountered when the analyzer program is administratively placed in different areas:

- <u>Laboratory</u>: Process analyzer maintenance resources may be used to focus on lab analyzers, which require different technical skills and thought processes. The laboratory organization may be held responsible for all measurement values reported, which may minimize conflicts. This arrangement is a possible consideration for smaller plants with small technical staffs and plants where lab versus process analyzer conflicts are not resolved with statistical approaches [Ref. 4].

- <u>Computer Control</u>: Exotic applications, representing significant control application enhancements, may be emphasized to the detriment of less risky applications. Control personnel tend to realize the importance of process analyzers, and the organization as a whole may benefit from this arrangement.

- Project Engineering: This arrangement may lead to neglect of maintenance concerns. However, more emphasis may be placed on detailed up-front review of the analyzer portion of a project by the process analyzer staff.
- Operations: Under this structure, stagnation of the process analyzer technical staff may occur, but could be offset by networking. The analyzer staff may have a greater feeling of belonging to the front-line part of the organization. It is advantageous when most plant priorities are set by the local operations organization.
- Separate analyzer group: This is the most efficient use of analyzer resources. However, it may produce alienation from the rest of the organization and may also result in more interface conflicts.
- Maintenance: This arrangement may not produce a stable program since maintenance is the area most subject to reorganization. It may result in alienation from the rest of the organization.

Due to size constraints, some organizations will be limited as to where the analyzer program can be placed for administrative purposes. However, placing the analyzer program under one of the above groups could cause the priorities of work to be weighted to that group. The best solution is to place the analyzer program in a part of the organization where an independent manager sets the group's priorities.

What about splitting the analyzer program functions between groups? The same general principles apply, but now the focus is even narrower. For instance, if analyzer personnel report to only one part of an operational organization, then the priorities of work will favor that specific operation's analyzers. In recent years, the trend has been to subdivide organizations into smaller, self-sufficient work groups. This arrangement offers certain organizational advantages. However, for specific technical fields, like process analyzers, it can be a very inefficient way to operate.

For example, one large plant reorganized into three separate groups, splitting along product lines. The decision was made to have no organizational or functional ties at all between the three groups. This arrangement resulted in significant increases in analyzer personnel, primarily in the professional and paraprofessional ranks. The isolation fostered by this concept reduced the exchange and sharing of resources and ideas. The new groups may have felt that they were getting more control of their piece of the analyzer business. Indeed, initially more attention from the professional ranks may have resulted in a greater feeling of control of the analyzer business. However, committing more people to solve complex technical issues does not always produce the desired results. The story goes that when a certain manager who was behind on the development of a new product told his lead engineer to add twice as many people to improve the schedule, the engineer responded by stating that development was more like hatching an egg than digging ditches. Two chickens on a nest will not hatch the egg twice as fast!

Product Development . . .
. . . more like hatching an egg than digging a ditch!
. . . two chickens won't hatch it any faster!

Organizational changes are bound to occur, and retaining an effective analyzer program may not be a top priority when these changes are implemented. However, one of the prime reasons for organizational changes is to be more competitive. In this respect, change may benefit the analyzer program. A change in corporate culture that had previously hindered the development of an effective analyzer program may now make the development of such a program possible.

Technical Service Issues

An analyzer program can be divided into three fairly distinct roles:

1. Technical service
2. Maintenance
3. Projects

Although the same individuals may have responsibility for one or more of these roles, we will discuss them separately. Generally, control of the analyzer program should rest with the technical service function. Technical service should also be the single point contact for all the upstream and downstream interfaces necessary to a successful analyzer program. As Peter Reeves has stated [Ref. 1], "Practical achievement demands an understanding of the roles of the different technical disciplines on a site and the need for solving interfacing problems between site departmental technical functions." Technical Service might be viewed as the glue that binds the whole program together. Some of the key Technical Service responsibilities are:

- Coordinate the analyzer quality program
- Audit plant analyzer equipment and functions
- Assure conformance to industry standards in environmental, safety, and ethical practices as related to analyzers
- Specify documentation/standards requirements
- Troubleshoot analyzer failures
- Improve analyzer reliability
- Establish goals and strategies
- Interface with analyzer vendors
- Function as analyzer technology "gatekeeper"
- Recommend and coordinate replacement of obsolete analyzers
- Coordinate turnaround activity
- Review analyzer projects
- Coordinate network activities
- Recommend/coordinate training for maintenance organizations
- Monitor analyzer performance
- Forecast future resource needs (personnel, hardware, and training)
- Provide estimates for analyzer-related facilities
- Evaluate the cost/benefit ratio of all analyzer-related activities
- Steward analyzer spare parts, calibration gases, and informational databases
- Interface with all other parts of the organization regarding analyzer issues

Analyzer personnel must have enough independence to move from one activity to another to ensure that a program balance is maintained. It is unusual for an organization to focus on all the technical service activities at the appropriate times or at the appropriate priority levels. The goal is not to be the best in all categories of technical service all of the time. However, the technical service activities have to be balanced so that they are addressed in the most effective and efficient manner to support the base business. The later section on "Cyclic Issues" will help to illustrate that cycling in and out of these activities is the best approach.

The following should not be part of a typical plant's analyzer technical service activities:

- Building analyzers
- Developing mobile analyzers
- Running an applications lab

In the earlier days of process analyzer development, some companies chose to build their own analyzers. Indeed, we owe the development of several existing types of analyzers to these companies. However, given the state of the process analyzer business today, such activities cannot compete with analyzer vendors, especially at the plant level. Companies that still engage in development activities should carefully examine the benefits that are obtained for the resources that are invested. They should only be undertaken after all commercially available options have been explored.

Maintenance Issues

Although maintenance, projects, and technical service issues are covered separately in this chapter, as a practical matter the three functions are completely interrelated. There are some estimates that, after an extensive period of project activity, as much as 50% of a plant's maintenance resources are spent on modification of project equipment. Many maintenance organizations are under constant pressure to reduce manpower. However, in the area of process analyzer technology, the biggest influences on manpower — projects and technical service activities — typically are not directly controlled by maintenance.

Any maintenance organization can have well-trained, qualified personnel and still be faced with challenges such as:

- One of every kind of process analyzer ever manufactured
- A plant full of obsolete and maintenance-intensive analyzers
- A large number of recently installed project analyzers with serious sample system and application problems
- General craft first-line supervisors with no process analyzer experience

A maintenance organization will never be able to operate at peak efficiency unless the technical service and project functions are monitored. Every decision in the area of process analyzer technology must be evaluated for its potential impact on plant maintenance.

Three developments over the last decade have caused many plants to view their maintenance organization as an "opportunity center" rather than as a "cost center":

1. Increased process computer control
2. Increased emphasis on product quality
3. Increased environmental regulations (analysis and reporting)

In many cases, these developments have necessitated radical changes in maintenance personnel and organizational structure. If a plant is heavily involved in one or all of these activities, then the "maintenance activity" must be viewed in a different light. Maintenance support must be tailored to a plant's analyzers and those analyzers' requirements. The above additional requirements may be viewed as limiting traditional flexibility in the maintenance organization. However, with the proper technical support program, many specific requirements may be used to meet more than one goal. For example, analyzers installed to meet environmental requirements might in some cases be used for plant control [Ref. 5]. Instead of focusing completely on how the "maintenance activity" can be reduced, efforts should also focus on how the plant can benefit more from existing maintenance activities. This "single dimension" view of the maintenance activity could affect corporate competitive position as external forces continue to be a predominate factor during the 1990s.

In many plants, process analyzers represent the top of the pyramid in technology. Maintenance personnel need an overview of what is expected and a vision that goes beyond daily problems. Rather than tolerating unreliable equipment conditions that are a drain on valuable maintenance resources, they should "attack" reliability issues. The maintenance organization often is in the best position to judge field conditions that require upgrades or changes to improve reliability.

Projects Issues

It is difficult to minimize the complexity of a plant project that involves a large number of analyzers. Many plants have relics of failed analyzer projects. Therefore, an effective method of implementing analyzer projects must be developed.

The driving force to install analyzers can come from any number of sources. Some examples of these installation incentives:

- A young engineer may have an overwhelming desire to try unproven technology
- A plant manager may feel that a certain device <u>must</u> be installed
- Laboratory personnel may want to place one-of-a-kind lab instruments in the field and have field technicians take care of them
- Users of a particular complex analysis device may become dissatisfied with its performance and be willing to try an alternative
- A computer control project may contain dozens of new analyzers
- Obsolete analyzers may need to be replaced
- A customer may insist on an innovative on-line measurement
- Environmental regulations may require the installation of new on-line measurements
- New analyzers may need to be installed to solve problems with existing analyzers

People often think that an analyzer can be found for anything. Analyzer personnel need to be involved in project activity from the very start. The effects of an influx of new project analyzers that have not been carefully thought out can cause serious damage to an existing program. The best way to handle this problem is a rigid set of standards and expectations.

Perhaps the best way to illustrate the potential consequences of analyzer projects is to discuss some actual industry cases. Cases 1 to 4 illustrate the concern that even analyzer technical service personnel have about analyzer project activity.

CASE 1

In a large subdivision of a chemical plant was a working core group of analyzers produced by a single manufacturer. A computer control project installed a large number of another manufacturers' analyzers without regard to the years of effort that had been spent to make the existing analyzer applications successful. The project remained open for 5 years while the analyzers were being debugged. Forewarning of the potential disaster by the plant's experienced analyzer personnel (at all levels) was ignored, and struggles to assign blame went on for years.

CASE 2

A small chemical plant site actually consisted of several even smaller units. Each unit contained at least one furnace. Each furnace had an oxygen analyzer. Each oxygen analyzer was from a different manufacturer. This was the consequence of using different project engineers on each furnace job, with no input from plant analyzer personnel. Not surprising, the maintenance staff had a hard time keeping all the different oxygen analyzers running and calibrated.

CASE 3

A chemical plant unit wanted to measure a metallic catalyst at high pressure and temperature. Lab testing revealed that sampling was possible. More than $250,000 was spent on the installation. More than a man-year of analyzer personnel effort went into attempting to make the x-ray fluorescence analyzer work. The sample had to be cooled, degassed, and introduced into the analyzer at no greater than 2 psig. After a very significant effort, the analyzer would run for 6 hr before plugging up while sampling a process stream of over 3,000 psig pressure and high temperature. Finally, the entire sample line plugged completely from one end to the other. Later, an obscure literature reference was found that indicated that the conditions at the analyzer prompted the formation of a solid metallic chemical species. The analyzer was abandoned in place, much to the chagrin of the analyzer manufacturer and the plant personnel.

CASE 4

The plant manager at a large chemical plant wanted a lab robot. Plant laboratory personnel could not find a suitable application, so extensive modifications were required to adapt a lab test. The robot had been successfully adapted to analysis of pharmaceutical tablets, but no comparable chemical plant analyses could be found. After considerable effort, the robot was abandoned.

When a good analyzer project plan is in place and properly supported with adequate resources, the results are different. Two cases of successful approaches to analyzer project development are now illustrated.

CASE 5

A new plastics plant was brought on-line and was staffed from a well-established base of experience. Efforts were made to train the technicians working on the analyzers prior to start-up. An analyzer specialist (paraprofessional) was assigned and the original project analyzers were well specified. This plastics plant became a quality and technology leader, largely due to the efforts of the analyzer personnel. (However, over the years the experienced technicians were moved and the specialist was assigned other duties. The specialist became overwhelmed and the analyzers begin to deteriorate as no resources were made available to resolve problems. The specialist eventually found employment elsewhere.)

CASE 6

An entire chemical plant production unit was relocated from one part of the country to another. Although money was an issue, an improvement in the analyzer technology was sought to update the plant. Using an outside vendor and local maintenance resources that were to be devoted to the relocated plant, the amount of analysis information was roughly tripled. The relocated plant came on-spec 2 months earlier than originally planned due to the availability of the analyzers during the start-up. The project manager gave the local analyzer personnel 'carte blanche' in putting the analyzer plan and system together.

Survey Issues

One of the first activities an analyzer program should be concerned with is the overall condition of the total plant analyzer system. This can be accomplished by critically examining its individual elements. A useful tool is a thorough survey of the analyzer functions, including:

- Safety issues
- Environmental issues
- Housekeeping/corrosion damage
- Service factors/analyzer statistics
- Idle analyzers
- Obsolete analyzers
- Problem analyzers
- Maintenance procedures/practices
- Documentation/records
- Manpower factors (i.e., experience level, training)
- Recently completed projects
- New projects

Each of these survey items must be carefully examined in depth to determine the field effectiveness of the analyzer program and where to focus additional efforts. Such surveys are best done on individual process units. The intervals between surveys can vary, but for areas experiencing a great deal of activity, every 2 years is recommended. The intervals may be extended if relatively little activity or few changes have occurred.

These types of surveys should be conducted by someone knowledgeable in process analyzer technology and familiar with company policy and goals. Each analyzer should be inspected in the field. The more detail the survey contains, the better. Plans for improvements in various areas can be formulated directly from a detailed survey.

Effective surveys are one of the best tools for managing a process analyzer program.

Manpower Issues

Manpower is a primary management concern and one that is usually very closely monitored. How much manpower is sufficient (professionals, paraprofessionals, and mechanics)? What is needed to provide a proper base to take care of routine items and provide resources to advance profitable business issues? Mechanics are typically pool-controlled by the maintenance department and often are subject to broad policies that limit flexibility. Ideally, local mechanic policy should not interfere with the ability to create and maintain an effective process analyzer program. Having too few qualified mechanics will probably affect the bottom line more than having too many mechanics.

Mechanic manpower levels require constant attention. Once a stable situation exists, as can be determined by audits, manpower levels should be allowed to stabilize. In practice, it is not unusual to see manpower levels in a constant state of flux, especially in a large plant. As priorities change, manpower resource levels must change to keep goals on track.

The number of analyzers to be maintained by one technician is a much-discussed subject. The general consensus seems to be that an experienced technician should handle from 10 to 15 analyzers. However, the relative complexity of the analyzers must be considered. Some companies "weigh" analyzers in terms of their relative complexity in order to assess maintenance man-hours required. A chromatograph, wet chemical analyzer, or moisture analyzer might be considered complex; an infrared, oxygen, or density analyzer, of medium complexity [Ref. 6].

A rough idea of appropriate mechanic manpower levels can be determined from audit results. Once the necessary equipment density is known, such things as Soloman Associates Inc. survey results and typical maintainability factors for different types of analyzers can be used to estimate appropriate levels. In practice, however, actual manpower levels are best determined from end-results and by focusing on meeting performance requirements rather than numbers. In a well-run and monitored program, the numbers will take care of themselves. We must ask ourselves if our business goals are being met. If they are, then our manpower levels are at least adequate. If our goals are not being met, then changes are necessary. Tracking analyzer performance will illustrate how this determination can be made. Many individual factors actually contribute to "effective" manpower.

Avoid creating a manpower "cycle." New administrations frequently want to make changes that affect the bottom line. Simply cutting manpower in most well-run organizations is not the answer. Indiscriminate cuts create potential personnel problems by sending the signal that this segment of the business is not in control. Organizational changes in a well-running program should be made only after thorough examination. The lead time associated with making improvements in the analyzer area also increases when resources are reduced. An extended period of time may pass before end-users feel the full impact of manpower cuts. By then, the damage has been done, requiring even more resources to return the program to its former level. This is not to say that manpower levels cannot be carefully examined and changed when business needs dictate.

The number and type of analyzers in use and your performance goals are the determining factors for analyzer mechanic manpower requirements. Direct productivity gains are made by:

- Increasing the experience level of mechanics
- Implementing analyzer/system modifications which reduce maintenance
- Altering maintenance practices (e.g., calibration frequency, preventive/ predictive maintenance)

If an analyzer program is sound, a sudden reduction in manpower eventually will result in reduced performance. Therefore, if a certain level of performance is required, then a reduction in mechanic manpower levels means a reduction in the number of analyzers that can be maintained. All too often, manpower is reduced without a corresponding reduction in the number of analyzers, and the whole program gets progressively worse. The correct response is to reduce the number of analyzers maintained if a reduction in manpower cannot be avoided.

The most efficient analyzer mechanic "work cell" consists of four or more skilled personnel. This allows adequate backfill for vacation, overtime, sickness, and so on. If the work area is not large enough to support at least four mechanics, then consideration should be given to reorganizing the area. After the mechanics have been trained, this size group will show a significant improvement in productivity over smaller groupings. Direct supervision should be by a first-line supervisor who is familiar with process analyzers.

Determining the correct number of paraprofessionals and professionals is somewhat more difficult. It can be argued that both paraprofessionals and professionals may not be needed. This may be true, especially in a small plant. The number of paraprofessionals and professionals depends to a large extent on the amount of project work and field changes occurring. An analyzer population that is stable would require less of these services than one undergoing radical changes. It is almost impossible, and highly undesirable, to completely separate technical service from project work when considering paraprofessional/ professional resource levels.

Cyclic Issues

The analyzer business is typically cyclic in nature. Attention is focused on certain specific goals — sometimes at the neglect of others. Certain cycles are unavoidable and necessary, such as efforts to correct problems uncovered in an audit or a large project workload. Other cycles, such as across-the-board manpower reductions, have a detrimental effect. Lead times in the analyzer business are typically longer than in other business areas. It may take many years to recover from indiscriminate cutbacks or changes.

A balanced approach is the best way to avoid negative cycle effects. A manager must know how to monitor the health of his analyzer business. Even the best programs can be improved, and some of the worst programs have positive sides. Focusing on only one aspect of the business may result in great improvement in that area for a short period of time. But this improvement will probably be at the expense of other areas, and there may be no lasting results (even in the area that was focused on). The 'Quality Approach' to resolving problems and normal cycles will be covered later.

Surveys fit very well with the cyclic nature of the analyzer business and the review of process analyzer resources, especially in large plants. Surveys should be timed to coincide with some deterioration in performance. After the survey is conducted, recommendations are made to improve performance and address the list of audit areas of concern. Resources are then committed to carry out the recommendations, and the area should be able to run for some period of time at a high performance level, taking advantage of the improvements made. The survey process is not begun again until a drop-off in performance is detected. Ideally, different areas of the plant are in different stages of the audit cycle. Resources are adjusted to meet area survey requirements, and no area is allowed to deteriorate to unacceptable levels.

You cannot afford to tie up a whole organization over one problem. Technical problems of one sort or another are often cyclic in the analyzer business. An important point is that these problems may occasionally "fix" themselves if left alone. One process unit periodically experienced a quality problem with product odor. One of the engineers who had worked in the area for several years was asked what solutions worked best for him. He responded that he thought about the problem real hard until it went away! Wasting valuable resources on transient, cyclic problems is not the key to an efficient operation. It is sometimes difficult to determine which problems need immediate correction and which will fix themselves. Sometimes the hardest and best approach is to do nothing at all and wait until available vendor technology catches up to the problem.

Some problems can be "... thought about real hard until they go away!"

It is often tempting to take on difficult applications that pose a high risk. Instead, file the application and wait for the process analyzer technology to come to you! Using this approach, in 1 to 3 years rapidly developing technology may make the application much less of a risk. Overall, the best approach is a structured, balanced approach. The more you know about your analyzer business, the better you can identify and respond correctly to cyclic trends. Ask yourself, where is your organization in the cycle? Is real and lasting progress being made?

The time interval required to achieve even small permanent improvements in the analyzer business typically can be weeks or months, not days. Trying to force a quicker solution, no matter how great the business driving force, can sometimes be counterproductive. Few organizations have sufficient resources to drop everything to work on one problem without the rest of the business suffering. Sometimes this situation is unavoidable, especially when coming from an external source (such as environmental or legal issues, or a serious customer concern). However, long-term resolution to serious technical problems can seldom be resolved by sheer will. Repeated disruptions of this magnitude can wreak havoc on an analyzer program!

One cycle that analyzer personnel cannot control, and yet which has a significant impact on the development of new technology, is the interest cycle. During the initial stages of the development of a new analyzer project, there is intense interest on the part of those in the organization who would benefit most. As field problems occur and corrections are made over a period of time, interest wanes. This cycle of interest and support is outlined in Table 9-2. This is one reason why analyzer personnel are hesitant to work on applications that require continual modification in the field before a useful analysis is obtained. It is often difficult to predict the amount of time and resources needed to solve a difficult problem.

Table 9-2. Cycles of Interest and Support in Applying New Technology

Psychological State	Reality	Funding Support*
Acceptance		
Mild interest	Introduction	0
Increased interest	First successes	+
Intense interest	Successes published	++
Excitement	First field failures	+++
Rejection		
Disbelief	More field failures	++
Caution	Business setbacks	0
Betrayal	Random success/failure	–
Disappointment	Addressing of field failures	– –
Rebirth		
Ignore	Systematic approach	–
Toleration	High field success rate	0
Mild interest	Continued successful operation, but limited by funding	+

* +, positive; 0, neutral; –, negative.

Another cycle that produces its own issues is the management cycle. In all plants, changes in management are fairly typical and, in certain business climates, frequent. Oftentimes, managers are called upon to supervise personnel working in the process analyzer area when they have little if any knowledge of the technology. This situation can be a bad experience for all involved. However, changes in management and organization do not have to be detrimental. If all parties are willing to make the most of a new opportunity, change can actually be positive. The new manager needs to realize that:

- If he/she is inheriting an effective program, observe it carefully before making radical changes. Past failures must not be repeated.
- Management of communication for business needs may require modification to fit the nature of process analyzer technology.
- Process analyzer technology is very important to competitive positioning.

On the other hand, analyzer personnel need to realize that the new manager is the primary link to upper-management strategies for how to position the business. Such input is needed, and often a fresh look may produce new ways to make the program more effective.

Dealing in an efficient way with these cycles may be very difficult within certain corporate cultures. This will not change the timing or cyclic nature of the process analyzer business, but it is probably unreasonable to expect the climate of a company to change just to satisfy analyzer requirements. In cases of unresolvable conflict, it might be best to have additional resources (such as contractors) brought in to assist in a difficult situation.

Quality/Performance Issues

The advent of quality programs in the process industries has been a boon. In fact, the quality process is one of the most significant developments in the process analyzer field in the last 10 years. One of the greatest benefits has been the improvement in communication, allowing better understanding of the requirements for improving analyzer effectiveness. It is unfortunate that it has taken us so long to take full advantage of these powerful tools. The real benefit of applying a successful quality program to analyzers is attained in combination with a total plant quality assurance program. The increased profitability and the increased competitive edge can be impressive. However, these benefits do not come without a great deal of commitment and training.

In implementing a quality program, the following concepts need to be accepted by the organization [Ref. 7]:

- All processes, good and bad, result in variable product.
- Controlling variability increases quality and productivity.
- Emphasis is on consistent quality — on target, not just within specifications.
- The common language of quality assurance is statistics.
- The key is prevention, not after-the-fact inspections. "Do it right the first time."
- Everyone in the organization is responsible for quality.
- Production of quality products is essential for remaining competitive in the current world economy.
- Arbitrary adjustment (i.e., calibration) of a process in control could be responsible for causing an out-of-control condition.
- The use of statistical process control (SPC) results in reduced overall quality costs.

"Do it right the first time!"

The competitive incentives for implementing a plantwide quality assurance program with a focus on analyzers are substantial and include:

- Reduction in reprocessing costs and wasted capacity due to off-specification product.
- Reduction in tankage inventory used for intermediate storage of off-specification product.
- Reduction in routine laboratory workload.
- Reduction in product variability as a result of improved process control.
- Minimization of the difference between marketing and manufacturing specifications as a direct result of variability reduction.
- Improved maintenance through predictive and preventive maintenance.
- Reduction in unnecessary calibrations since calibration frequency is determined statistically.

One plant goal should be to produce all products on the basis of analyzer results against those specifications that can be measured on-line. Ultimately, the long-range goal should be to certify/release products on the basis of on-line analyzer results whenever possible.

Statistical process control is the most powerful tool in the implementation of an effective quality program. SPC can be used to [Ref. 5]:

- Determine if analyzer hardware is operating in a controlled state
- Examine analyzer capability
- Improve calibration procedures
- Validate on-line analyzer and laboratory data

- Determine if a continuous process is in control
- Determine the limits of a process
- Keep a process in statistical control

Statistical methods are also important in the area of predictive maintenance. These methods are discussed in-depth in Chapter 5.

Networking Issues

Networking can best be described as a sharing of technical information. One of the primary benefits to a strong networking program is tapping into outside sources of reliability improvement ideas. Although networking activities must be monitored, allowing analyzer personnel to discuss their mutual successes and failures is usually of significant benefit to the individual programs. Generally, the smaller the analyzer group, the greater the benefit.

Networking fosters healthy professional competition and excellence.

Networking in the best sense can cross all sorts of organizational lines, and information/equipment can be moved to parts of the organization that are in the most need. A well-run network in a large organization can produce remarkable results. It can foster a professional atmosphere that encourages striving for excellence in all facets of an analyzer program. Good networking programs can multiply the resources a company can use to improve its analyzer program. However, as has been noted before, this level of teambuilding takes time. The first network meeting is likely to be disruptive as approaches and ideas are debated.

Dramatic differences can sometimes exist, even between different units within the same plant, as to how analyzer technology is managed. These differences are often greater between plant sites. There is a temptation to examine the different programs and adopt the most effective as the corporate standard. This approach could result in a discrete level of improvement in the company's analyzer program. Another level of improvement would be to adopt the most effective parts of each of the existing individual programs. One of the primary purposes of networking is go one level further and champion the best available technology approaches that can be worked into the existing programs.

If an analyzer group exists in name only, or is scattered organizationally throughout a single site, then the most basic form of networking must be allowed. It is a serious mistake to discourage the sharing of information and resources at a single plant site. For example, one large plant reorganized into three separate business groups. There was literally no organizational contact between individuals down the hall from each other. The result was a dramatic increase in analyzer staff at the professional and paraprofessional levels. This separation precluded the normal sharing of information and resources essential to a well-run program. Getting singular control of a piece of the analyzer workforce may result in initial success as the controlling part of the organization gets more attention paid to its immediate problems. However, this warm feeling can disappear when no greater volume of work gets done.

Political Issues

Even well-run programs will face circumstances that are completely beyond the programs' control. In such cases, the only alternative is to cut your losses and regroup. An example of such a situation is illustrated by the following incident.

A refinery in the southern California air-quality district paid a tremendous amount of attention to its air-quality monitors. Every analyzer outage of more than 1 hr had to be reported to the air-quality board. Analyzer personnel lived close to the plant and could be on-site in 20 min. During a 1-year period, because of the fast response and an extensive preventive maintenance program, all the air-quality analyzers ran with no recorded downtimes. Unfortunately, the refinery had to be sold, in part because of anticipated increased costs due to related environmental expenditures. Part of the refinery was not sold, and process analyzers from this part of the plant were moved to another refinery site.

Planning for this type of incident is obviously difficult and impractical. However, given the current business and regulatory climate, such radical changes may occur. Some of the causes could be:

- Extended strike
- Plant sold to another company
- Indiscriminate layoffs

One plant had three different owners over an 8-year period. The plant had a senior analyzer engineer and an efficient analyzer maintenance staff. Throughout the 9-year period, the analyzer program continued to function and, although manpower was very tight, the plant analyzers continued to meet requirements. The lesson here is to stick to the fundamentals. Always strive to provide basic value service first.

Vendor Issues

A serious analyzer failure can result in serious misgivings about a vendor and can lead to bad decisions. For instance, an engineer once had a bad experience with a particular analyzer vendor. Even 15 years after the incident, the engineer would still do anything within his power to avoid purchases from this vendor. Vendors are essential to a successful program; excluding one of them, except on purely technical grounds, because of past experiences can lead to serious consequences. Vendors can provide valuable input to an analyzer program. Several companies have used the "quality partnership" idea with vendors to improve relations and assure successful field installations.

Developing long-term relationships with vendors is a technical service function. However, traditional project development does not acknowledge the special need in process analyzer projects to consider the cost impact on the analyzer program that a project decision may have. The importance of a good working vendor relationship is illustrated by the following hypothetical project attempted by "Plant A" and "Plant B."

Both Plant A and Plant B were roughly the same size and had identical analyzer personnel resources. Both had budgets of $5 million in fixed costs. This included salaries, installation of a new environmental analyzer, and replacement of two obsolete analyzers. Both decided to spend $5 million on a new computer control project. There were 30 possible streams for analyzer applications. Of these 30 streams, 10 had been done before at other plants within the company. Fifteen applications had not been done before, but upon detailed review could be accomplished with significant engineering effort. Five of the applications were questionable, three because of the difficulty of obtaining a representative sample and two because they required modification of complex laboratory equipment.

During the scoping stage, according to organizational procedure, Plant A's analyzer engineer worked with the project and technical personnel to develop the project specifications. With management support, vendors with good track records were selected and, with requirements defined, reasonable costs were negotiated. After much discussion, it was agreed not to attempt the five questionable applications as part of the project, and they were replaced by two other instrument/analyzer alternatives. During the start-up phase, additional manpower was added. Three major problems occurred. One was an equipment problem, resolved by the manufacturer. The second was a major sample system modification to one of the analyzers. Both caused operations a lot of problems, but did not affect the start-up schedule. The third problem occurred after 2 weeks of run time, and required continual maintenance to keep on-line. After 4 months the problem was isolated, and after 9 months the problem was corrected. Another subtle software problem caused a plant upset 6 months later. However, all the original corporate goals were met or exceeded.

Plant B's analyzer engineer had the same concerns as Plant A's engineer. However, no organizational support mechanisms were in place. All 30 of the analyzers were purchased from the lowest bidder, who had never installed analyzers in this plant before. The analyzer engineer protested to anyone who would listen, but to no avail. Due to analyzer problems (and instrument problems; they had been procured in the same fashion), the plant could not be started up on schedule. Fifteen analyzers were eventually started up by the plant analyzer engineer, analyzer maintenance personnel, and vendor personnel.

This example illustrates the single largest waste of resources in the field of process analyzers. By failing to resolve vendor/project issues, the following can occur:

Project costs exceed estimates.

Project schedules are not met.

Significant company and vendor resources are devoted to project bid specifications and documentation of project change orders.

Vendors lose potential future business.

Company funds are spent on projected analyzer benefits that are never realized.

Good vendor relationships are not easily developed. We have seen the reasons for fostering such relationships from the plant's point of view. Developing a stable relationship is definitely in the interest of the analyzer vendor from a competitive point of view. Vendors also recognize that they share some of the liability for selling analyzers into applications that are unworkable. Many misunderstandings result from failing to appreciate the other party's business requirements. Vendor analyzer designs that are incompatible with other plant analyzer systems limit plant flexibility and result in increased costs and missed opportunities.

References

1. Reeves, Peter, "On-Site Acceptance of Process Analytical Instrumentation Interfacing between Technical Disciplines," *Trends in Analytical Chemistry* 6(1): v, 1987.
2. "Certified Specialists in Analytical Technology: Role Delineation," Research Triangle Park, NC: Instrument Society of America, 1992.
3. Hardy, W. M., "Process Analyzers — Predictions for the 1990's and Beyond," *ISA 90, International Conference and Exhibit Proceedings*, October 1990, 391-400.
4. Hardy, W. M., "Prevention of On-Line Analyzer Validation Problems," *Proceedings of the 34th Annual ISA Analysis Division Symposium*, 23: 111-121, April 1988.
5. Makansi, Jason, "Move Toward Process Control for CEMS Natural but Slow," *Power*, p. 10, August 1989.
6. Kilgore, J. L., and Holland, E. M., "Benefits of Centralized Maintenance," *Instrumentation Technology*, p. 47-49, January 1973.
7. Hardy, W. M., "On-Line Analyzer Applications of Statistical Process Control," *Proceedings of the Sixth Annual Control Engineering Conference*, p. 442-446, May 1987.

About the Author

Bill Hardy has a B.S. degree in Chemistry from the University of North Carolina and an M.S. degree in Chemistry from the University of Georgia. He has over 20 years of experience in the petrochemical industry.

Bill worked over 14 years with Exxon Chemical, the last nine years dealing exclusively with process analyzers. He was Chairman of Exxon's (North America) Process Analyzer Technical Committee and the Chemical Control Specialist Network (Process Analyzer Subcommittee). He has been with Shell Oil and now Shell Chemical, dealing exclusively with process analyzers, for almost seven years. He is currently the Analyzer Specialist for several chemical and refining process units at three separate sites. He is the site representative for the company-wide Analyzer Technical Direction Team.

Bill is a Certified Specialist in Analytical Technology (CSAT #10026) and a Senior ISA member. He was the first elected President and is a current active member of "The Process Analyzer Association" (a local organization of over 100 individuals involved in process analyzer technology). Bill has 11 technical publications and has made several presentations at technical conferences and professional meetings.

Part II

Process Analyzer Application Fundamentals

Typical Measurement Principles

Analyzers in common use may be broadly grouped into four general categories. *Physical property analyzers* usually monitor simple physical properties of a process stream. *Electrochemical analyzers* measure ionic species in a process stream or reactions of ions in the stream when a voltage or current is applied. *Spectrophotometric analyzers* measure inherent optical properties of the molecules in the process stream (either an absorbance by, or emission from, the molecules in the stream). *Compositional analyzers* can measure one, several, or all of the components individually in a gas or liquid stream.

Physical Property Analyzers

There are numerous types of physical property analyzers, but by far the most common involve a measurement technique which results in data that must be correlated to an off-line ASTM laboratory measurement. Often the physical property is temperature or pressure dependent and a correlation can be developed relating the change in the physical property with temperature or pressure. These types of analyzers are covered in Chapters 10 through 23.

Electrochemical Analyzers

Numerous types of electrodes have been developed to measure ionic species in aqueous solutions or solids. Most of these are specific to ions such as pH (for H^+) ammonia ($NH4^+$), and residual chlorine (Cl^- or OCl^-). These types of analyzers are often applied to the monitoring of waste effluent and are covered in Chapters 24 through 27.

Spectrophotometric Analyzers

Spectrophotometric analyzers have been employed in numerous applications to measure either a specific component in a process stream (for example, CO in furnace stack gas analysis), or a class of similar compounds (such as alcohols in MTBE). The target molecules in a process stream absorb light at specific frequencies. The energy gained may be released in the form of heat or by emission of light (at a wavelength of light characteristic of the molecule). Measurement techniques may record the absorption of light at a specific wavelength, or the subsequent emission of light (fluorescence). Analyzer spectrophotometers have been developed over the entire electromagnetic spectrum, from high energy x-rays to very low energy infrared radiation. Chapters 28 through 32 will cover the more common types of spectrophotometric analyzers.

Composition Analyzers

Many techniques have been developed to separate and measure components of process streams. The most common types employed are gas chromatography (GC) and mass spectrometry (MS). A limited number of applications for liquid chromatographs (LC) also have been developed. Literally thousands of GC applications have been developed, enabling this technology to be widely employed in the manufacturing facilities of almost every industry.

Gas chromatographic applications have been and continue to be developed to replace the historically complex and difficult to maintain physical property analyzers. Simulated distillation GC applications have been available for years as an alternative to ASTM boiling point analyzers, and newer applications include the measurement of total olefin or aromatic content of complex hydrocarbon streams. Manufacturers are also adapting their basic GC hardware to enable measurements of the total sulfur and total nitrogen content of streams to meet increasingly tight hydrocarbon product specifications.

Applications of mass spectrometers have been limited in the past by the need to employ high vacuum systems as a part of the sample preparation/introduction system for mass spectrometers. Recent technological advancements have greatly reduced this front-end requirement, and techniques are being developed to allow measurements to be made at atmospheric pressure. Applications of mass spectrometers are expected to greatly increase in the future. Compositional analyzers are discussed in Chapters 33 through 36.

Physical Property Process Analyzer Systems

There are two basic categories of physical property analyzers, non-ASTM and ASTM. The non-ASTM-type directly measure a physical property of the process stream. *Densitometers* and *viscometers* are examples of analyzers that directly measure known physical properties. Typically no correlation to lab techniques are required, therefore, there is no need to develop a model from a large number of samples of known value for the measurements to be accepted as valid. The ASTM-type requires correlation to the accepted laboratory analytical technique.

Generally speaking, few ASTM lab methods can be easily automated as a process analyzer. Often, a completely different measurement technique is employed in an on-line analyzer which is expected to yield ASTM data. A correlation to an off-line ASTM laboratory test is either developed by the user, or in a few cases, supplied by the manufacturer.

For example, a typical on-line boiling point analyzer continuously heats a flowing sample, fractionating it into a vapor stream and liquid stream. The vapor is then cooled back to a liquid and the heating adjusted until the fraction (% off) flowing overhead is at the desired fractionation point (or cutpoint). The analyzer then reports the temperature of the nonvaporized liquid portion as the temperature of the sample at the cutpoint. Similarly, the fraction (% off) boiling overhead at a fixed cutpoint temperature may be obtained. The data from this process analyzer must then be correlated to the data obtained from the laboratory. The corresponding laboratory method involves the use of a glass assembly where the sample is heated and the overhead vapor is then condensed. The temperature is recorded at specific vaporized fractions as viewed on a thermometer in the distillation assembly by a laboratory technician. A correlation between these two fundamentally different techniques must be developed.

Manufacturers have developed several measurement techniques for each type of physical property analyzer (both ASTM and non-ASTM). For example, techniques utilized for measuring density include measuring the change in resonant frequency of a vibrating tube as a function of density of a process stream flowing through the tube. Other techniques employed include the absorption of x-rays from a radioactive source as well as measurement of the momentum of gas through a curved path.

Other physical property analyzers employ combinations of techniques to measure the desired property or component. For example, a total sulfur measurement may be obtained by combusting a sample at a very high temperature to convert the sulfur species to SO_2, and then measuring the SO_2 by one of several methods. One detection type involves reaction with ozone and subsequent measurement of the resulting fluorescence or light emitted at a specific wavelength as the molecules relax.

Typical analyzers in this physical property group include the following:

non-ASTM	ASTM
Boiling Point	Distillation
Gas and Liquid Density	Vapor Pressure (RVP and KVP)
Gas and Liquid Specific Gravity	V/L (Vapor/Liquid Ratio)
Gas and Liquid Viscosity	Flash Point
Thermal Conductivity	Cloud Point
Dew Point	Pour Point
Hydrogen Sulfide	Freeze Point
Total Sulfur	Viscosity
Refractive Index	Octane and Cetane Number
Combustible Gas	Color

There are many types of physical property analyzers. The following chapters describe the more common analyzers of this type most widely used throughout various industries:

Chapter 10	Boiling Point Analyzers
Chapter 11	Cloud Point and Pour Point Analyzers
Chapter 12	Flash Point Analyzers
Chapter 13	Freeze Point Analyzers
Chapter 14	Refractive Index Analyzers
Chapter 15	Thermal Conductivity Analyzers
Chapter 16	Combustible Gas Analyzers
Chapter 17	Viscosity Analyzers (non-ASTM)
Chapter 18	Moisture/Dew Point Analyzers
Chapter 19	Hydrogen Sulfide Analyzers
Chapter 20	Total Sulfur Analyzers
Chapter 21	Octane and Cetane Analyzers
Chapter 22	Vapor Pressure Analyzers
Chapter 23	Vapor-to-Liquid Ratio (V/L) Analyzers

10

Boiling Point Analyzers

Distillation has been an important operation in industrial processes for many years. The discovery of crude oil in 1859, however, led to the rapid development of distillation methods to separate and purify petroleum products. More recently, the development of continuous refining processes, rather than batch-type operations, necessitated more refined laboratory techniques and, subsequently, on-line analyzers.

ASTM Standard Test Method D-86. This laboratory test was developed many years ago to cover the atmospheric distillation of many petroleum products. A 100 mL sample in a glass distillation flask is heated with a Bunsen burner to vaporize petroleum liquids, and the temperature of the vapor is measured with a mercury thermometer. The vapor passes through a condenser immersed in an ice-cooled bath, and the condensate is collected in a 100 mL graduated cylinder. The vapor temperature at which the first drop of condensate is observed to fall into the cylinder is recorded as the initial boiling point (IBP). As distillation continues, the temperatures are recorded at which specific volumes of distillate collect, such as 5, 10, 50, 90, and 95 mL. Since the original volume of the sample was 100 ml, these volumes will correspond to the percents distilled (% OFF). The temperature at which the last drop of liquid falls into the graduated cylinder is recorded as the final boiling point (FBP).

With volatile samples, such as gasoline or light naphthas, the final volume of liquid collected will not exactly equal the 100 mL charge. Some of the light ends will not condense in the ice-cooled condenser and will be lost. To correct for this loss, the difference between the volume of sample charged to the flask and the total volume of recovered distillate is added to the individual percents recovered to give percents distilled.

This is a tedious, dangerous, and time-consuming procedure that requires careful attention to detail to obtain reliable results. Consequently, automatic equipment has been developed which uses controlled electric heaters to heat the sample, thermocouples to measure and record the vapor temperatures, and special devices to detect the level of the liquid in the receiver. This method can be used for samples with FBPs as high as 600°F (315°C).

ASTM Standard Test Method D-1160. For higher-boiling-point materials, a vacuum distillation procedure is available. The basic method is similar to the ASTMD-86 test, but is conducted under vacuum at pressures as low as l mm Hg. It may be applied to samples with boiling points as high at 750°F (400°C). at l mm pressure, a temperature equivalent to 1000°F (540°C) at atmospheric pressure.

Four other ASTM methods are available for boiling point analyses. Method D-216 is used for natural gasoline, D-285 for crude petroleums, D-850 for aromatic hydrocarbons, and D-1078 for volatile organic liquids.

On-line Process Analyzers

TOTCO 1269 Initial Boiling Point Analyzer

This instrument could monitor any preselected point on the ASTM distillation curve from the initial boiling point of a hydrocarbon stream to approximately 30% of its boiling point range.

The process sample was fed continuously into a boiler and fractionating column and was distilled into a preset ratio of distillate to bottoms. The column operated above atmospheric pressure, so the corresponding boiling points were higher than those from the ASTM D-86 distillation. However, correlations could be obtained with reproducibility equal to or better than the laboratory tests.

The analyzer system included an integrated sample-conditioning unit in the left-hand cabinet. In the sample-conditioning system (Figure 10-1), the sample was filtered to remove any solid particles and passed through a coalescer to remove any entrained moisture. The in-line strainer (#1 in the diagram) received sample from the process and was part of a fast loop to minimize dead time in the analyzer response to process changes. The eductor (#10) in the fast loop developed a partial vacuum to suck the waste sample from the analyzer back into the loop.

Filtered sample from the in-line strainer passes to one of two coalescers (#2 and #3), which removes any entrained moisture from the sample. Two coalescers are provided so one can be cleaned without having to shut down the system. Bypass streams from the bottom of the coalescers flow through the flowmeter (#4), and this flow is controlled by the manual drain valve (#5) at a rate sufficient to carry away any moisture from the sample. Clean, dry sample passes to the pressure regulator (#8), which maintains a constant sample pressure into the analyzer. The pressure gage (#9) indicates the controlled sample pressure.

Effluent sample from the analyzer passes through the check valve (#12) joins the coalescer bypass stream, and returns to the fast loop through the eductor (#10). The vacuum supplied by this eductor is sufficient to dispose of the waste sample, and thus no separate sample recovery system is required.

Details of the analyzer are shown in Figure 10-2. The pretreated sample enters the system at the lower left corner, passes through a preheater coil in the bottom of the boiler, and then into the fractionating tower through a valve in the top of the tower. The sample passes through the float guide tube that surrounds the float rod and enters the fractionating tower through an opening about halfway down the guide tube. A float in the boiler controls this valve so that the boiler fills to a fixed level. A fixed-wattage heater provides heat to the boiler. Required wattage, which can run from 400 to 3,200 W, is determined by the boiling range of the sample and the fractionation point to be monitored.

Vapor from the boiler passes through the fractionating tower, and condensate refluxes to the boiler. Distillate from the top of the tower flows through the overhead orifice. The bore of this orifice determines the percent distilled. A stream-of-bottoms flows through a check valve, which prevents distillate from flowing backwards down the bottoms line into the boiler. The bottoms stream and the distillate flow to the sample valve.

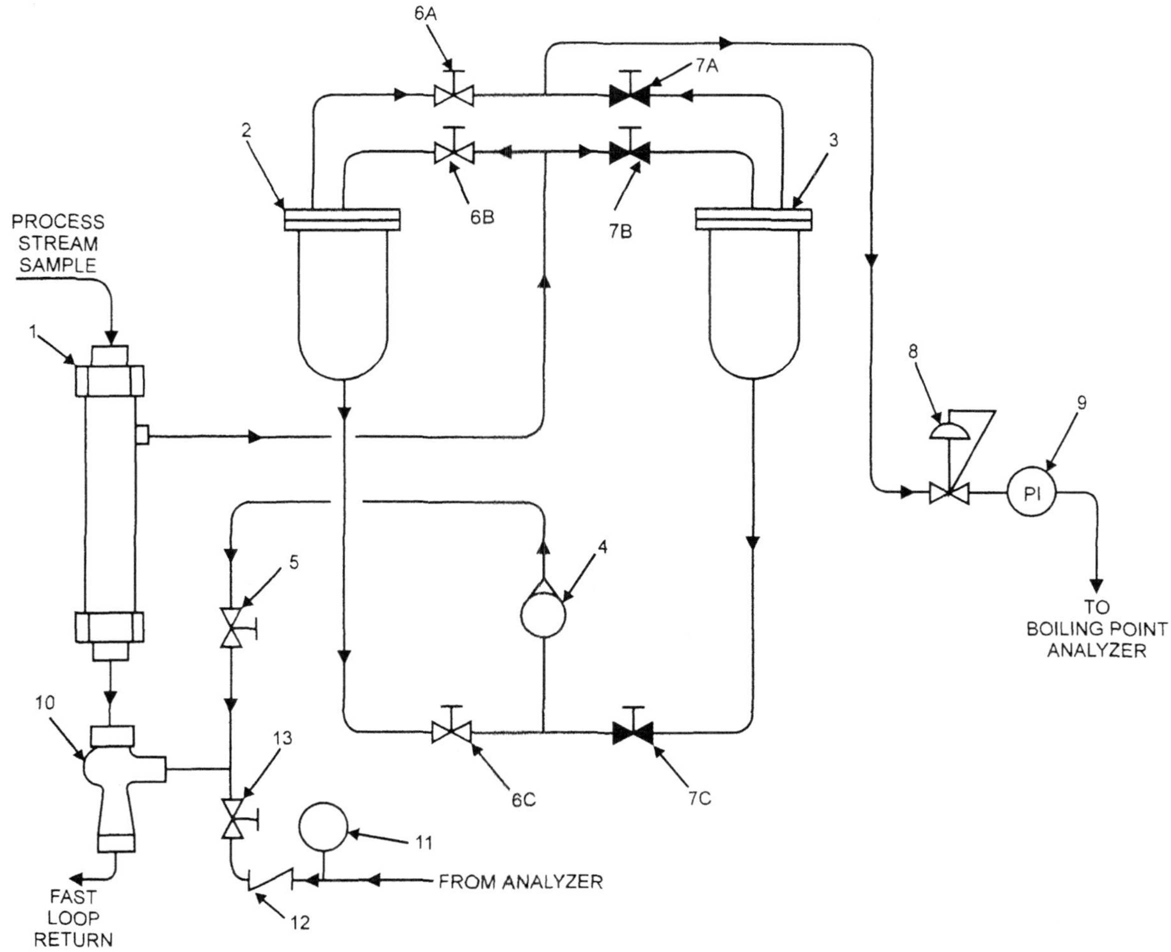

1	INLINE STRAINER	7A	OUTLET BLOCK VALVE
2	LH FILTER-COALESCER	7B	INLET BLOCK VALVE
3	RH FILTER-COALESCER	7C	DRAIN VALVE
4	FLOWMETER	8	PRESSURE REGULATOR
5	COALESCER DRAIN FLOW VALVE	9	PRESSURE GAUGE
6A	OUTLET BLOCK VALVE	10	EDUCTOR
6B	INLET BLOCK VALVE	11	COMPOUND GAUGE
6C	DRAIN VALVE	12	BOTTOMS CHECK VALVE
		13	RETURN BLOCK VALVE

Figure 10-1. TOTCO Model 1269: Sample Conditioning Flow Diagram
(Courtesy of Sun Refining and Marketing Co.)

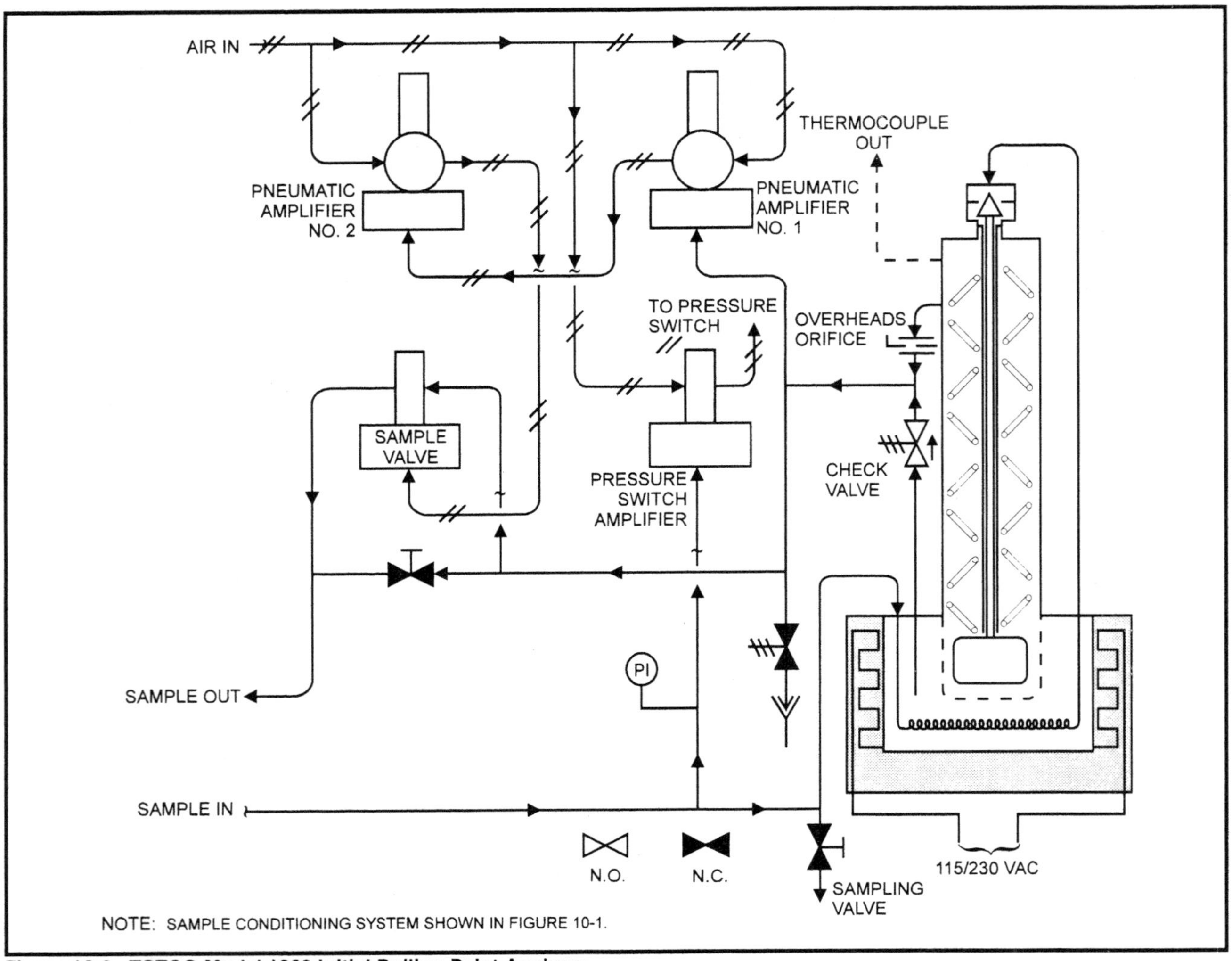

Figure 10-2. TOTCO Model 1269 Initial Boiling Point Analyzer
(Courtesy of Sun Refining and Marketing Co.)

The system is controlled by a series of pneumatic valves. Amplifier 1 senses the pressure of the distillate/bottoms streams flowing from the tower. When the pressure exceeds the set point of the amplifier, instrument air flows through amplifier 1 to amplifier 2. This system generates enough air pressure to open the sample valve and allow a fixed ratio of product to exit. Immediately, the pressure in the outlet line decreases, the sample valve closes, and the pressure rises again. Thus, the opening and closing of the valve every few seconds allows an essentially constant ratio of overhead and bottoms liquids to exit the system, and maintains a constant pressure in the tower. The vapor temperature is measured by a thermocouple in the top of the tower and can be correlated with the ASTM D-86 temperature.

TOTCO Model 6500 Final Boiling Point Analyzer

This instrument was similar in many ways to the Model 1269 Initial Boiling Point analyzer. Details of the system are shown in Figure 10-3. It could monitor any point from about the 30% recovered point to the FBP of the sample.

Sample entered the fractionating tower through the inlet valve on top and the guide tube as in the IBP analyzer. Since the boiler temperature was usually much higher than that of the IBP analyzer, the sample did not pass through a preheater coil in the bottom of the boiler.

The pneumatic control system was identical to that in the IBP analyzer. The pressure switch amplifier sensed incoming sample pressure and shut off the heat if the sample pressure dropped below 30 psig.

Distillate from the top of the tower flowed through an air-cooled condenser to the sample valve. Bottoms sample from the boiler passes through the orifice and then joined the distillate before passing through the sample valve. The sample valve cycles open and closed every few seconds as in the IBP analyzer. The temperature of the bottoms in the boiler was measured with a thermocouple, and it correlated with the corresponding ASTM D-86 point.

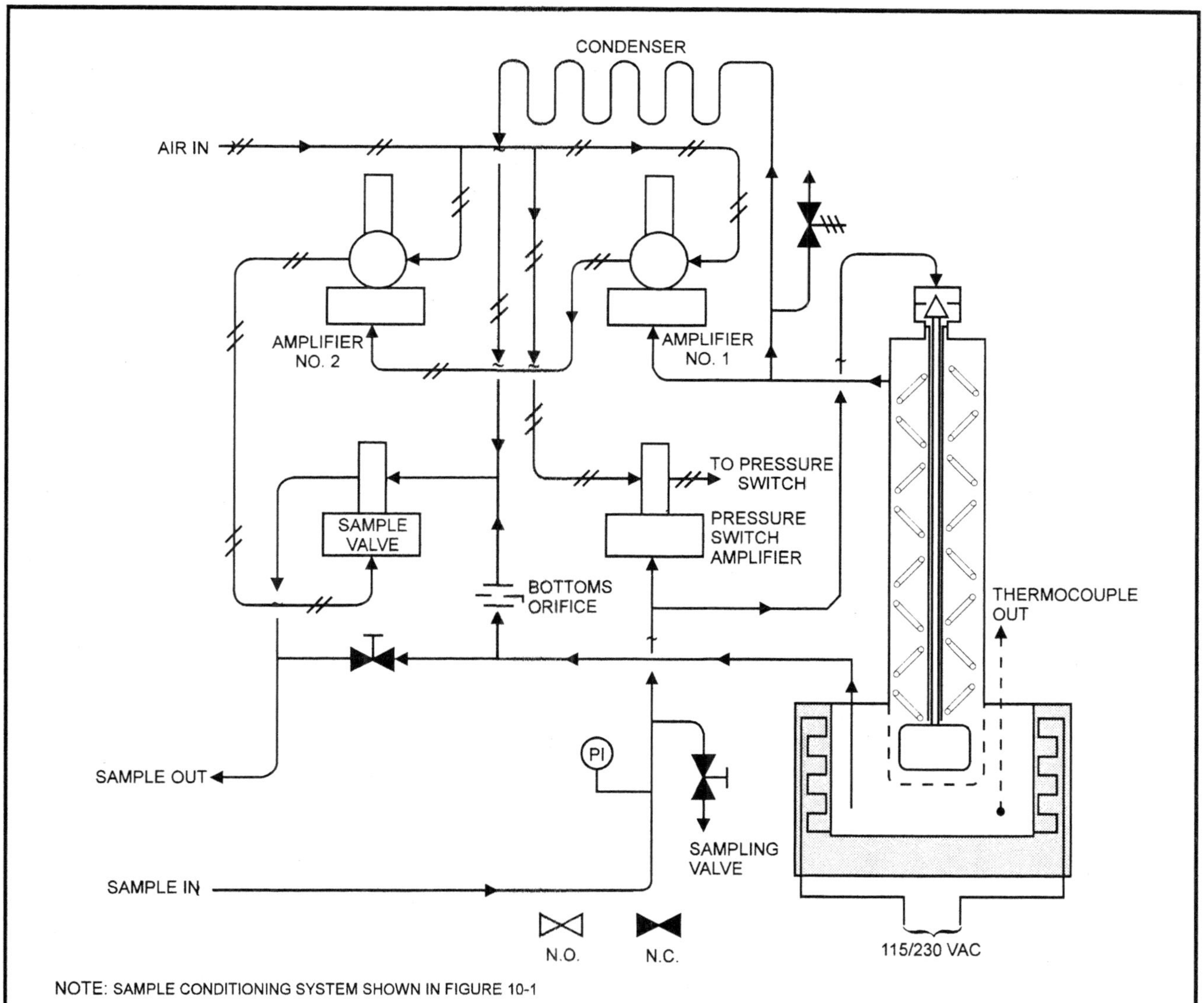

Figure 10-3. TOTCO Model 6500 Final Boiling Point Analyzer
(Courtesy of Sun Refining and Marketing Co.)

Precision Model 41468 Vacuum Distillation Analyzer

This instrument is essentially the original TOTCO Model 1468 vacuum distillation analyzer. The reboiler and fractionating tower are similar to those of the other TOTCO analyzers, but since it operates under a vacuum as low as 10 mm Hg absolute, the tower is larger.

The process sample is filtered and coalesced in a system similar to that used for the IBP and FBP analyzers. The detailed system is illustrated in Figure 10-4. Treated sample flows through the float-actuated inlet valve at the top of the column, through a preheater in the boiler, and into the side of the packed fractionating tower. The boiler fills until the float rises and closes the inlet valve. Overhead distillate flows to the water-cooled condenser. Part of the distillate is returned to the fractionating tower as reflux, and the remainder is removed by the overhead pump, a constant volume gear pump. Bottoms from the reboiler enters a second gear pump. The relative pumping rates of the two gear pumps can be adjusted to provide the desired percent distilled fraction. The bottoms pump removes liquid from the reboiler, allowing the float to drop and admit fresh sample.

The vacuum in the system is produced by the vacuum pump and is controlled by the pressure transmitter and pressure controller. The latter actuates the valve in the line to the vacuum pump. A rupture disk on the condenser protects the system from excessive pressure if a component should fail or if cooling water should leak into the condenser.

The reboiler temperature is recorded by the thermocouple in the reboiler. The ratio of overheads to bottoms can be set by adjusting the two gear pumps. Any point between 5 and 95% distilled can be monitored by this adjustment. The bottoms temperature can be correlated with the ASTM D-1160 distillation data. Samples with the equivalent of atmospheric boiling points between about 650 and 1000° F (345 and 540°C) can be monitored. Repeatability is within 1% of the sample boiling range.

Model 41464 is designed for operation on 60-cycle AC. Another unit, Model 41466, is available for operation on 50-cycle AC.

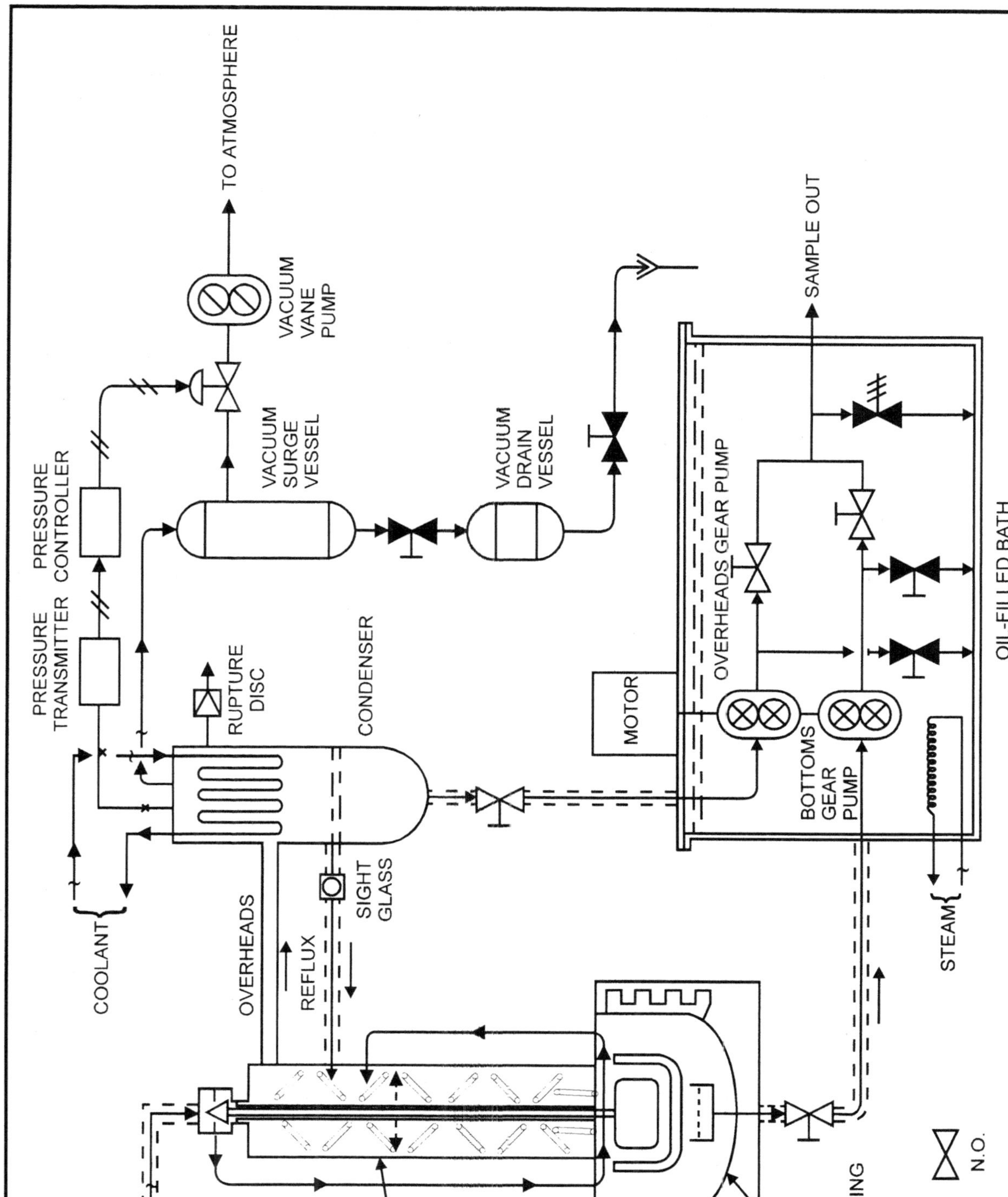

Figure 10-4. Precision Model 41468 Vacuum Distillation Analyzer
(Courtesy of Precision Scientific Petroleum Instruments Co.)

Hallikainen Model 1463 Distillation Analyzer

This analyzer was the first on-line Hallikainen analyzer. Details are shown schematically in Figure 10-5, and a simplified flow diagram is presented in Figure 10-6. This instrument measured the boiling point temperatures of petroleum products from the 5% point to the 97% point, in the temperature range of 150 to 700°F (65 to 370°C). The results correlated very well with the ASTM D-86 data, and the analyzer responded rapidly to changes in the incoming sample.

A preconditioned sample from a sampling system was metered to the analyzer by one head of a duplex metering pump. The sample flowed through preheating coils and then to a flash cup in the top of the evaporator. Sample flowed down over the surface of a vertical heated element, which evaporated a portion of the charge. The residue collected in the bottoms cup, where a portion was removed by the second diaphragm pump. The liquid level in the bottoms cup was detected by a level sensor, which in turn controlled the electrical power to the vertical heating element. Enough heat was supplied to vaporize the portion of the product that was not removed by the sample-out pump.

The temperature of the residue in the bottoms cup was measured with a thermocouple. By adjusting the pumping ratio of the sample-out pump, any point in the distillation curve between the 5% and the 97% boiling point could be monitored.

The float housing was connected to the bottoms cup so that the liquid levels in each were identical. The float moved a flag up and down between a lamp and a photocell. The photocell output, in turn, controlled the circuit that supplied the power to the heater. If the liquid level rose, more power was supplied to the heater and more sample vaporized, thus reducing the liquid level.

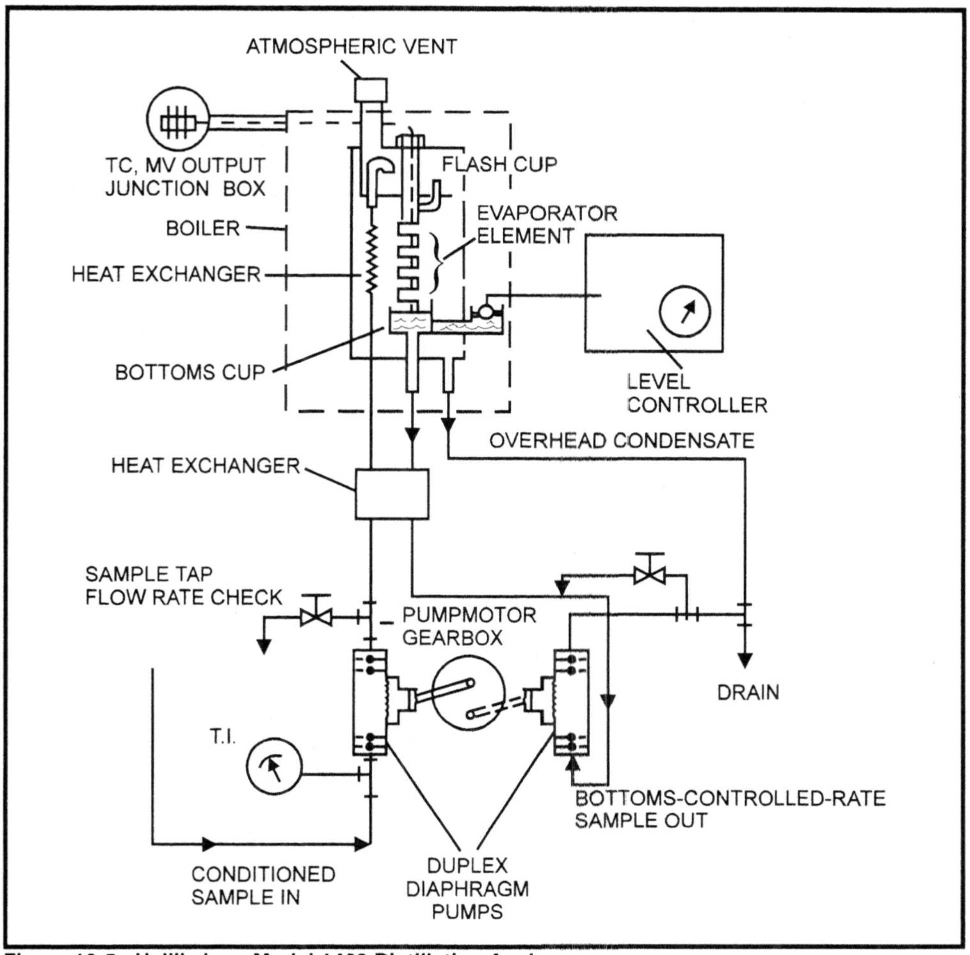

Figure 10-5. Hallikainen Model 1463 Distillation Analyzer
(Courtesy of Fluid Data, Inc.)

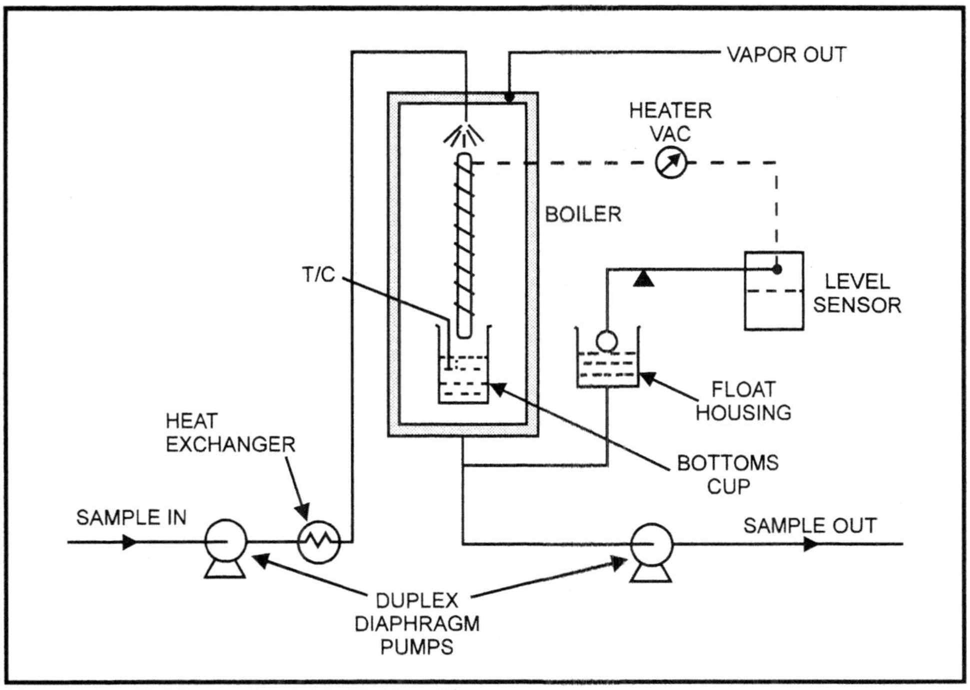

Figure 10-6. Hallikainen Model 1463: Flow Diagram
(Courtesy of Fluid Data, Inc.)

Rexon Model 52000 Boiling Point Analyzer

The Rexon analyzer (Figure 10-7) was essentially a miniaturized version of IBP and FBP analyzers. A sampling system was contained in the left-hand enclosure, and the distillation column and the control valves were in the right-hand enclosure.

The details of the system are shown in Figure 10-8. The general operation was virtually identical to that of the TOTCO instruments, except that by placing orifices in both the overhead and the bottoms lines and by choosing the proper orifice sizes, it could monitor any point on the distillation curve from the 5% to the 95% point, over a temperature range of 100 to 650°F (40 to 345°C). A digital meter indicated the actual boiling point temperature.

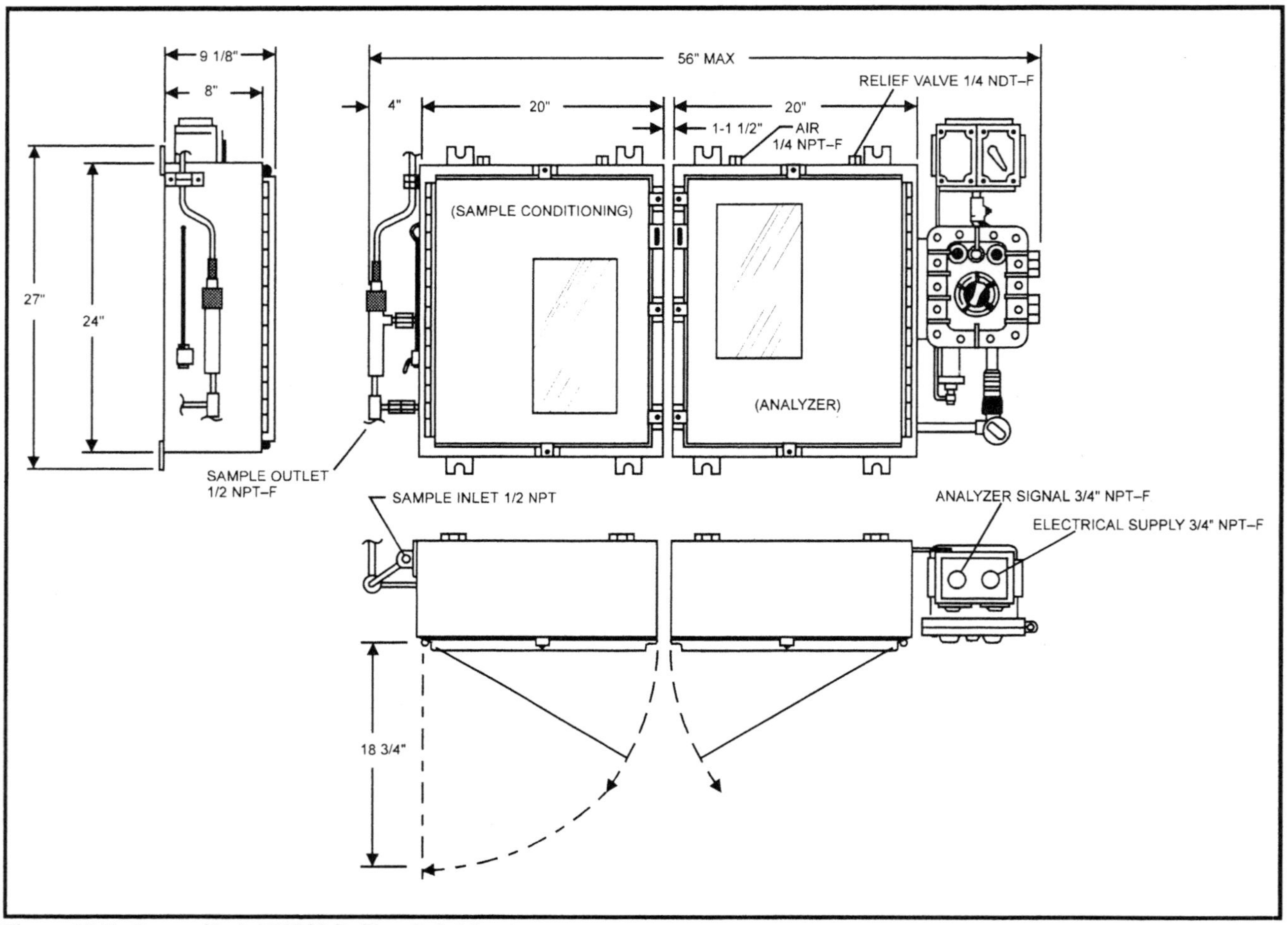

Figure 10-7. Rexon Model 52000 Boiling Point Analyzer
(Courtesy of Sun Refining and Marketing Co.)

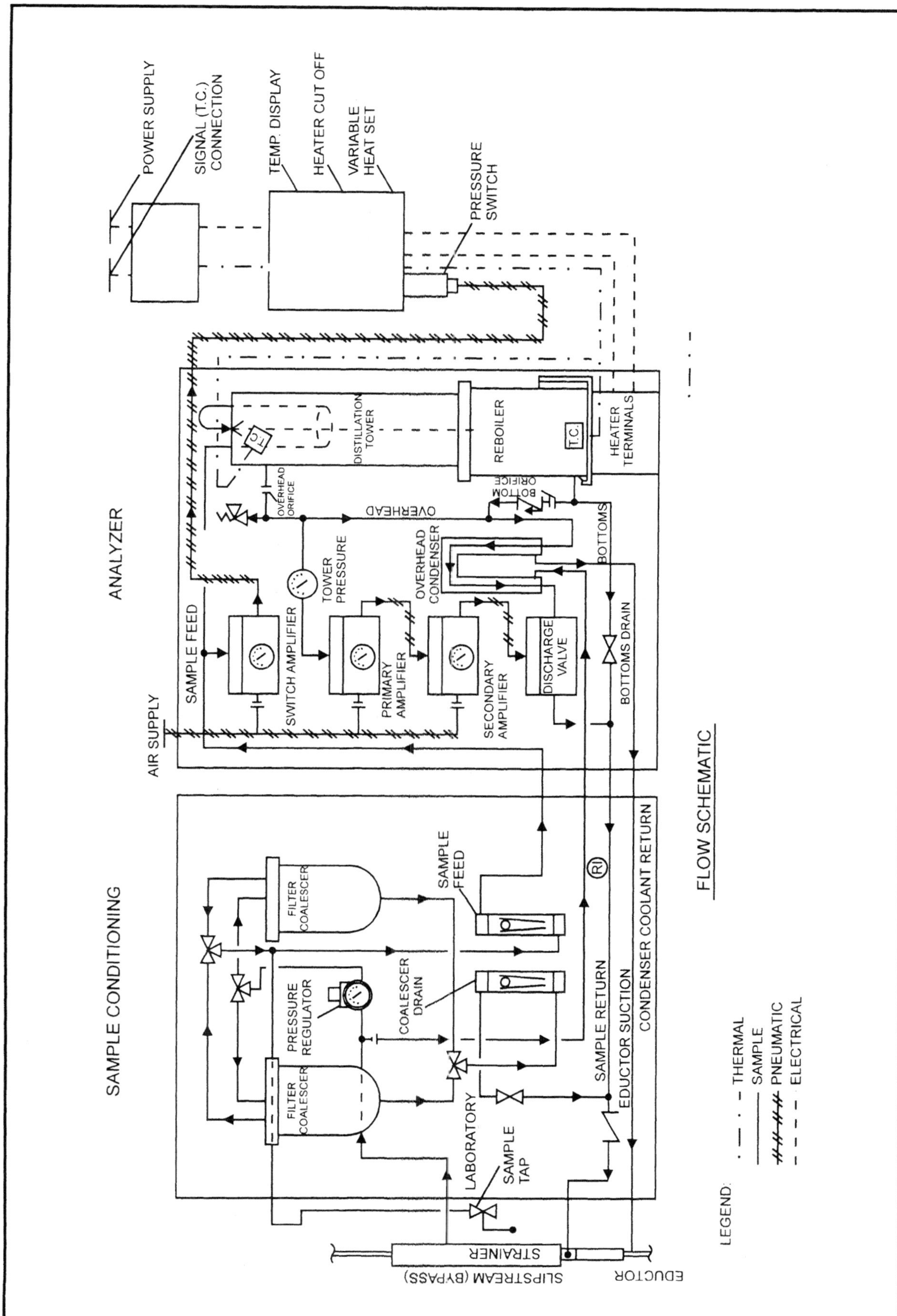

Figure 10-8. Rexon Model 52000: Flow Diagram
(Courtesy of Sun Refining and Marketing Co.)

Fluid Data Model 1463M Distillation Analyzer

This unit is a later version of the 1463D boiling point analyzer (which was an updated version of the Hallikainen Model). The boiler unit and metering pumps are virtually identical, but a microprocessor is used to control the heater. Details are shown in Figure 10-9. Temperature in the bottoms cup is measured with a resistance temperature detector (RTD) instead of a thermocouple. The microprocessor can be programmed to monitor any point from 5 to 95% distilled in the range of 100 to 650°F (40 to 345°C). A pressure transducer measures the sample level in the bottoms cup and controls the boiler heater through the microprocessor.

Alternatively, the system can be programmed to monitor a sequence of percent distilled points. It can also maintain a constant temperature in the bottoms cup by varying the pumping rate of the bottoms stream to determine the percentage distilled at one or more fixed, chosen temperatures.

Fluid Data Model 1463H Distillation Analyzer

In the 1980s, it was decided that the microprocessor was an unnecessarily expensive feature, and Model 1463H was introduced. This model can monitor boiling points from the 5% point to the 97% point at temperatures from 150 to 680°F (65 to 360°C). Alarms are provided for malfunctions, and a digital readout indicates the boiling point temperature.

Model 1463H conforms to the National Electrical Code (NEC) Class I, Group D, Division 1 (ID1) specifications. Another model, 1463G, conforms to the European CENELEC standards.

Precision Model 47500 BPA-II Boiling Point Analyzer

The Precision BPA-II is a modernized version of the Rexon boiling point analyzer. In this unit, the sampling system is contained in the housing under the analyzer assembly, so the system requires less floor space. This analyzer can monitor any point on a distillation curve from 5 to 95% and can do the work of both the TOTCO IBP and FBP models and the Rexon analyzer.

Its operation (Figure 10-10) is virtually identical to that of the TOTCO and Rexon analyzers. Process sample from a fast loop is filtered and coalesced to remove solid particles and water, then enters the small fractionating tower/reboiler through a float-operated inlet valve. The pneumatic control system is also basically the same as that for the Rexon system. The analyzer can handle samples with boiling ranges from 100 to 650°F (40 to 345°C), with a repeatability of 1% of the sample boiling range.

Two models are available. Model 47500 is designed for operation on 115 vAC and Model 47501 for 230 vAC.

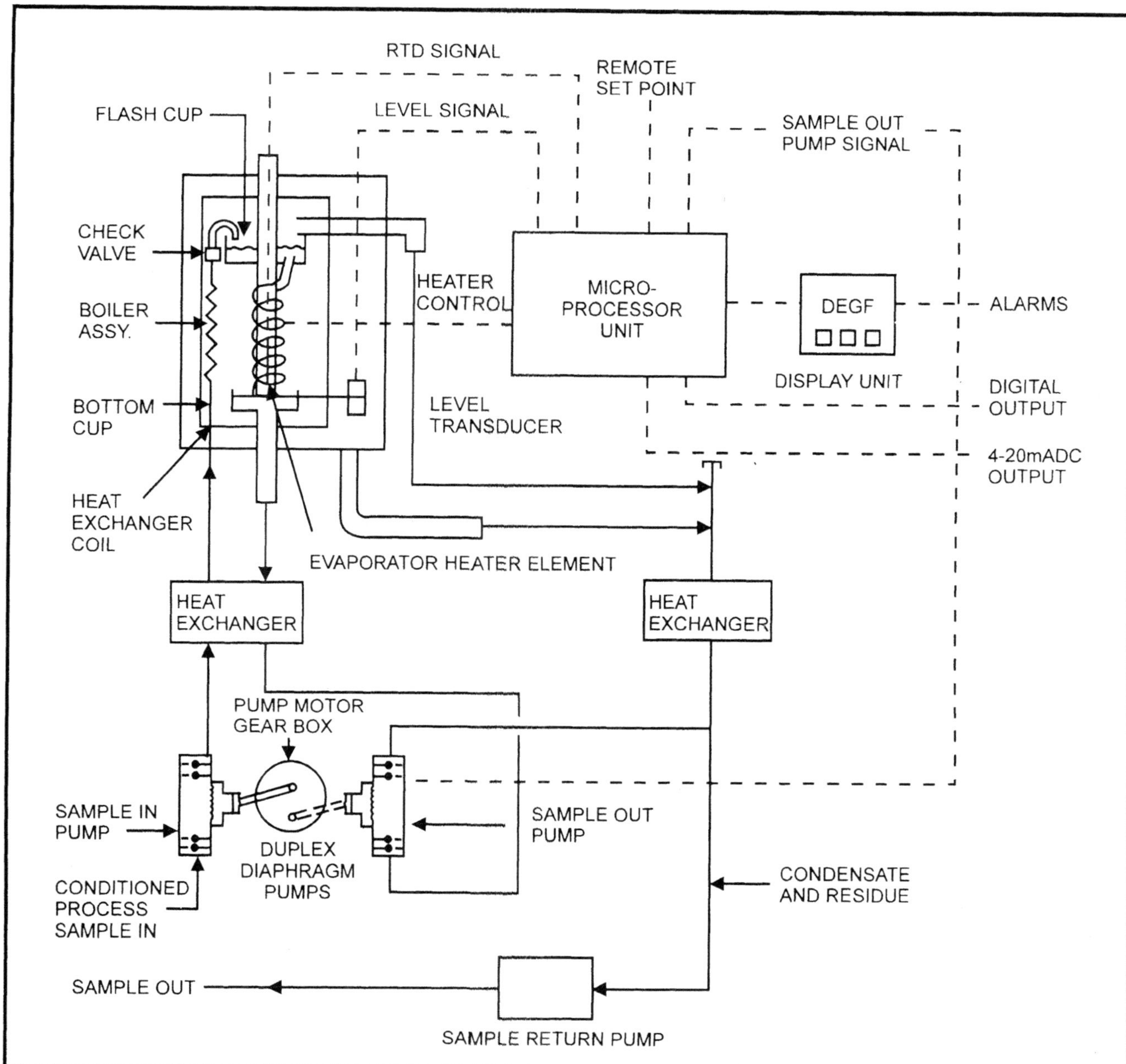

Figure 10-9. Fluid Data Model 1463M Distillation Analyzer
(Courtesy of Fluid Data, Inc.)

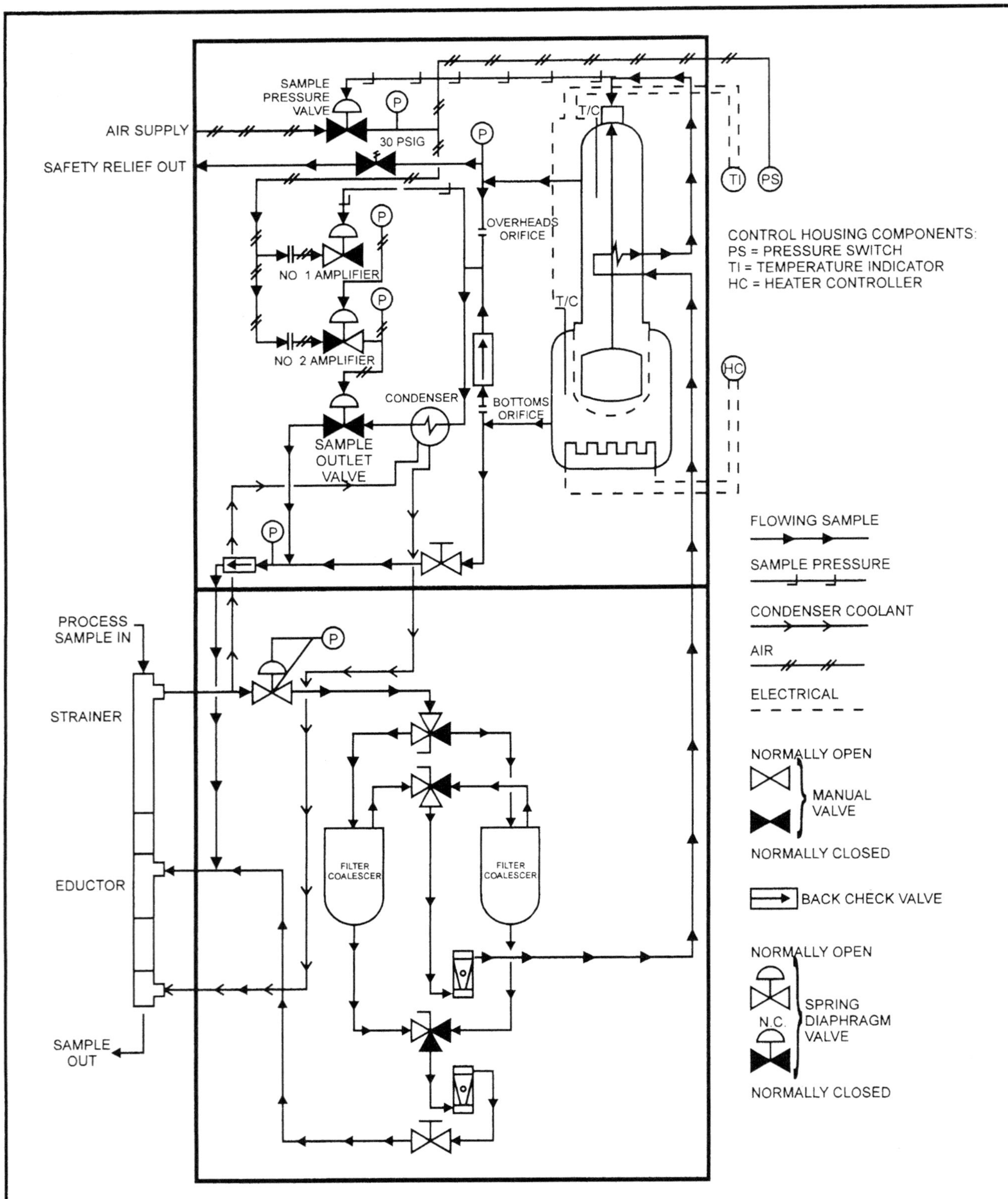

Figure 10-10. Precision Model 47500 BPA-II: Flow Diagram
(Courtesy of Precision Scientific Petroleum Instruments Co.)

Precision Model 44520/44521 Distillation Analyzer

This analyzer is a microprocessor-controlled instrument. Sample is pumped by a feed pump, at a constant rate (Figure 10-11), onto a compact column. It flows down over an electric heater, which provides a large contact area for the sample and produces an efficient separation of sample fractions.

A bottoms fraction (see Figure 10-11) is removed, at a constant rate, by the bottoms metering gear pump, which is adjusted to remove from 5 to 95% of the feed. The remaining feed vaporizes, condenses in a heat exchanger, and then joins the bottoms fraction. Both are removed by the exhaust gear pump. Any point on the distillation curve from the 5% to the 95% boiling point can be monitored in the range of 100 to 650°F (40 to 345°C). The analyzer can also be programmed to monitor multiple distillation points. Microprocessor diagnostics monitor all analyzer functions and utilities. Alarms can be programmed to detect excessive distillation temperatures or system malfunctions. Programming can be done from the keyboard or from a remote terminal.

Models 44520/44521 are designed to meet NEC Class I, Group D, Division 1 (ID1) hazardous-area classifications. Models 44522/44532 meet the European CENELEC standards.

Cyclic Analyzers

In addition to the boiling point analyzers described thus far, two cyclic models have been developed, which essentially duplicate the ASTM D-86 laboratory method. One is the Carlo Erba Model PD155 process distillation analyzer and the other is the Hone-Addison "Distillar." However, since these are not continuous instruments, they will not be detailed here.

Common Applications

Continuous on-stream boiling point analyzers have been used mainly in the petroleum refining and petrochemical industries since the 1950s. Most of the process streams on which they are used are complex mixtures of many organic compounds that could not be analyzed by methods such as spectrometry or gas chromatography.

Before the advent of continuous process analyzers, all feedstocks and intermediate and final products of a process had to be analyzed by laboratory methods. Several hours might elapse before the results were available to the process operators, during which time the streams could be off-specification. It would be necessary to recycle this material through the process to meet specifications, or to blend it into other products that might have a better value. Continuous on-stream analyzers have allowed plant operators to maintain product specifications closer to their desired limits and, consequently, to increase yields.

Initial boiling point (IBP) analyzers are used to monitor the production of naphthas, which will be used as charge stocks to catalytic reformers to convert these stocks to aromatics (high-octane components for gasolines) and charge stocks for petrochemicals. The pentanes, which do not convert to aromatics, can be removed from the naphtha stream and used as blend stock for high-octane gasoline.

Final boiling point (FBP) analyzers are used to monitor the end points (EP) of blended gasolines in order to maximize the yields in the summer (when gasoline demand is greatest) without including higher-boiling-point components, which could form coke deposits in the engine combustion chambers.

Vacuum distillation analyzers are used to monitor products from vacuum distillation columns in order to meet specifications for fuel oils and lubricating oils. Some typical applications follow.

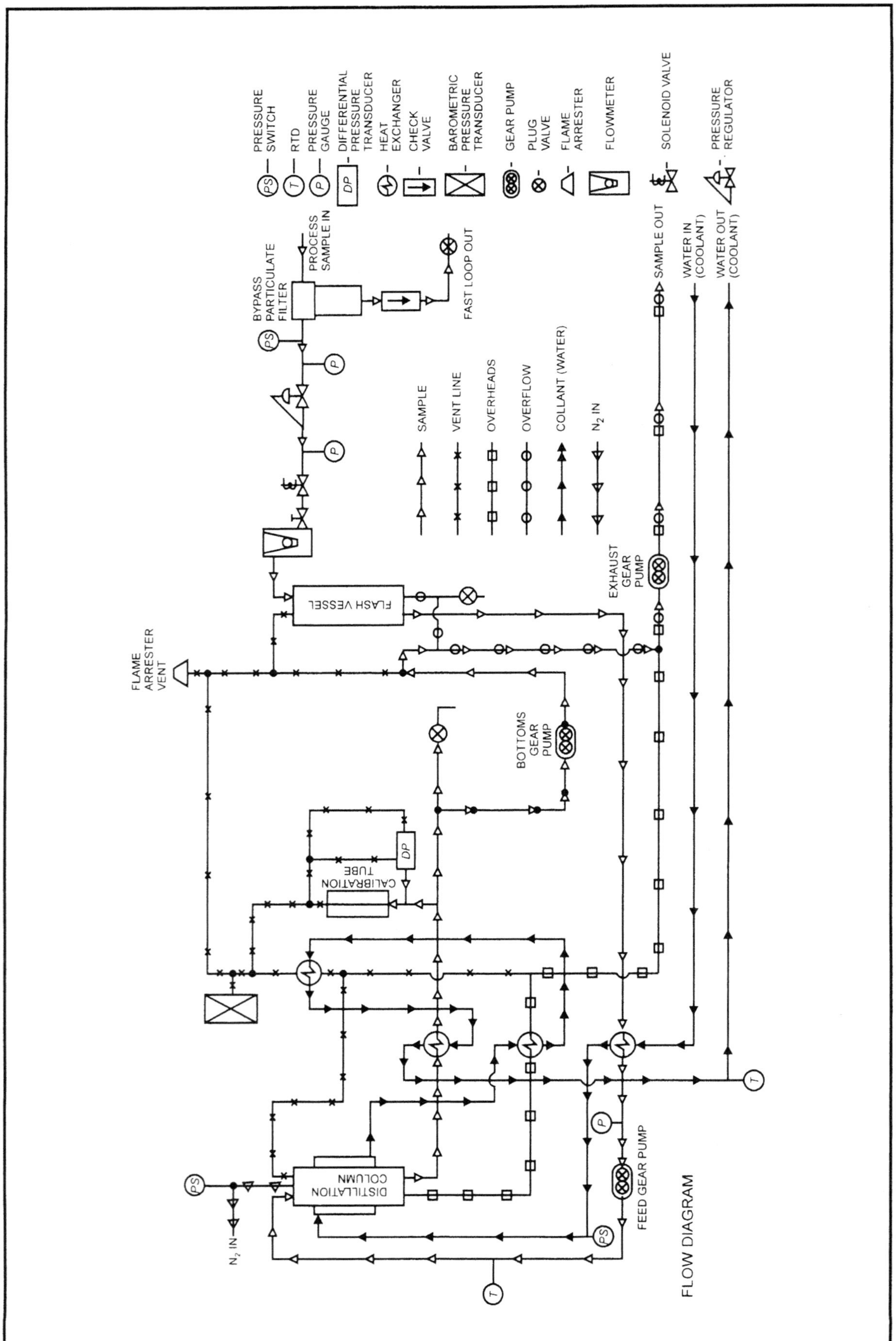

Figure 10-11. Precision Model 44520: Flow Diagram
(Courtesy of Precision Scientific Petroleum Instruments Co.)

Crude Distillation Units

A crude distillation unit in a refinery separates crude oil into gases, light and heavy naphthas, vacuum distillates, and residues. The components are processed further to produce fuel gas, gasoline, jet fuels, diesel fuels, kerosene, lubricating oils, and so on. The boiling range of each of these components and products is important to maintain maximum yields of the most valuable products. The desired boiling ranges can vary, depending on market conditions and the time of the year.

Light Naphtha

An IBP analyzer can maximize the yield of fuel gas (methane, ethane, propane, and butane) and avoid the loss of valuable pentanes from the light naphtha. An end-point (EP) analyzer can maximize the yield of the light naphtha and avoid contamination with heavier components.

Heavy Naphtha

Much of the heavy naphtha is treated in a catalytic reformer to increase its octane number, and this reformate is blended back into gasoline. Monitoring the 5% boiling point of the feed to a reformer with an IBP analyzer and the 95% boiling point with an end-point (EP) analyzer can improve the performance of the reformer. The IBP analyzer can eliminate low-boiling-point components that would not be converted to higher-octane products and would consume reformer capacity. The end-point (EP) analyzer can eliminate the components boiling higher than 400°F (205°C). These components can contaminate the reformer catalyst, shorten its life, and require more frequent, time-consuming, and expensive on-line regeneration.

Kerosene

Much of the kerosene is consumed as jet fuel. Monitoring the initial boiling point (IBP) can maintain better control on the desired flash point (FP). Monitoring the end point (EP) can eliminate undesirable heavier components, which can cause coking and contaminate jet engines.

Atmospheric Gas Oil

Much of this fraction will be used as a charge to a catalytic cracking unit, such as a hydrocracker, to produce the more valuable light (low-boiling) products. Use of an end-point (EP) analyzer can minimize contamination of the cracking catalyst and reduce on-line regeneration times.

Vacuum Gas Oil

This cut from the crude tower will usually have an end point of about 900°F (480°C). Much of this material will also be used for charge to a hydrocracker, so justification for the vacuum distillation analyzer on this stream is similar to that for the atmospheric gas oil. The vacuum gas oil can also be used as feedstock for a vacuum fractionating tower in the production of lubricating oils. Sidestreams with different boiling ranges and viscosities are produced in the tower and can be processed further into lubricating oils. Vacuum distillation analyzers can monitor the initial boiling point and end point of each sidestream to assure it has the correct boiling range for further processing.

Ranges of Variables

Most available boiling point analyzers, which operate at atmospheric pressure, can handle samples with boiling ranges from about 100°F (40°C) to 650 to 700°F (345 to 370°C). Samples with higher end points will cause excessive coke deposits in the boiler and require frequent analyzer cleaning.

Vacuum distillation analyzers can operate at pressures as low as 10 mmHg absolute. Boiler temperatures should not exceed about 650°F (345°C) to minimize coking. Under these conditions, the corresponding atmospheric pressure distillation temperature would be about 1000°F (540°C).

Options and Accessories

All boiling point analyzers require sample-conditioning systems to remove a representative sample from the process stream and render it suitable for the analyzer. In most cases this is included as part of the analyzer system, but some additions may be necessary because of specific conditions in the process. Most analyzers will also require a sample-recovery system to dispose of the waste sample. A number of suitable systems are available.

Limitations

Distillation analyzers are available that can monitor points on distillation curves from 5 to 97% distilled. Those operating at atmospheric pressure, or slight positive pressure, can handle samples with end points up 650 to 700°F (345 to 370°C). Vacuum distillation analyzers can handle samples with equivalent atmospheric pressure boiling points up to about 1000°F (540°C).

Installation Hints

Most boiling point analyzers are designed to meet NEC requirements for installation in Class I, Group D, Division 1 (ID1) hazardous areas. However, most require protection from the weather for reliable operation and convenient maintenance. Therefore, analyzers should be installed indoors or in a suitable climate-controlled analyzer shelter. This may prevent installation close to the most desirable sampling point, where any change in the product properties could be detected quickly, but this lag will be balanced by the improved precision and enhanced maintenance afforded by the climate-controlled environment.

Specific Safety Issues

Although most analyzers are constructed to meet NEC requirements for hazardous areas, the boiler for many of them operates at high temperatures and could be dangerous if volatile samples are involved. The analyzers should be installed in climate-controlled buildings away from the most hazardous areas, ventilated with a positive-pressure fan to remove explosive vapors, equipped with a vent stack extending out of the top of the building to discharge vapors outside, and have hydrocarbon detectors/ flame detectors monitoring the analyzer area.

Unique Sampling System Requirements

Since most analyzers will be installed in shelters away from the most hazardous areas, they will require sampling systems with fast loops to minimize the dead time in getting the process sample to the analyzer. Sometimes a process pump can be used to circulate the sample through the fast loop. Sample is taken from the discharge side of the pump to the analyzer and is returned to the suction side of the pump. Frequently, however, in order to bring the temperature to a level that is safe for the pump, process streams are passed through heat exchangers or coolers. These large units could increase the dead time required to get the sample from the process to the analyzer and increase the response time to sample changes. It may be desirable to remove the sample from a point ahead of the heat exchanger. In this case, a separate sample cooler and pump should be added to deliver the process sample to the analyzer and to return it downstream of the original sample point at the process.

Some samples, such as lubricating oils or streams from vacuum towers, may be so viscous at ambient temperatures that they cannot be pumped through the sampling system. These sample lines should be heated with steam or electrical tracing to maintain the sample in the liquid phase.

Typical Calibration Requirements

Analyzers can be calibrated by removing a sample periodically from the fast loop and sending it to the laboratory for analysis. A more convenient method is to install a calibration system at the analyzer location, in which a "standard sample" can be stored and introduced periodically into the analyzer in place of the process sample.

The calibration sample could be a sample of the process stream that has been analyzed in the laboratory or, in some instances, a pure chemical whose physical properties are known. Sufficient calibration sample should be provided to calibrate the analyzer for a period of at least several weeks. The calibration sample must be stable and maintained in suitable liquid phase for use.

Troubleshooting and Maintenance

All analyzers should be examined periodically by an instrument technician to assure proper operation. The analyzer manufacturer's operating and maintenance manuals should provide detailed inspection procedures.

For example, the daily routine checklist for the TOTCO analyzers is as follows:

1. Check the output temperature record for evidence of malfunction.
    a. Too high a temperature may indicate plugging of the inlet line or the inlet valve, causing the boiler to empty of the process sample.
    b. A high temperature could also be caused by plugging of the bottoms line or the bottoms orifice. Heavy ends of the sample would accumulate in the boiler and gradually increase the boiler temperature. The analyzer must be shut down and allowed to cool until the tower pressure drops to zero before the system can be opened.
    c. If the temperature cycles up and down over a range of about 10 to 50F° (−10 to 10C°), the float may be binding and cannot accurately follow the liquid level in the boiler. Binding could be caused by a burr

on the top end of the float rod that interferes with its smooth operation in the inlet valve, or a small bend in the float rod could interfere with its free movement in the tube in the center of the fractionating tower. Check the analyzer manual for instructions on how to clean or repair these components.

2. Check the system for sample leaks. Liquid sample leakage at the vent port of a pneumatic valve indicates a ruptured diaphragm.
3. Check the sample pressure delivered to the analyzer, usually 40 psig.
4. Check the outlet valve pulsation rate, about once per second.
5. Check the operating pressure, exactly 10 psig.
6. Check the bottoms line, hot to the touch.
7. Check the jacket of the effluent sample cooler (if used) cool to the touch.
8. Check for no leakage through the bulls-eye below the relief valve.
9. Open the bottoms drain valve for 10 s to blow back the bottoms orifice.

Similar procedures should be outlined in the particular instruction manual for each analyzer.

About the Author

Conard Donnell received a Bachelor's and a Masters degree in Chemistry and Physics from Temple University. His career with automatic process analyzers began and ended at Sunoco Refinery where he developed on-line instrumentation to measure the physical properties of process streams automatically and much faster than laboratory analysis. Since retirement, Mr. Donnell has worked as a consultant for Sunoco, Fluid Data, Precision Scientific Co., Lockwood and McLorie, and DuPont working on projects that included design and installation of process analyzer systems in refineries worldwide.

11

Cloud Point and Pour Point Analyzers

Pour points and cloud points are important characteristics of many petroleum products. The pour point is the temperature below which a liquid sample will not flow. The cloud point is the temperature at which wax crystals form a hazy cloud when the sample is cooled under controlled conditions. Jet fuels, diesel fuels, and household fuel oils usually contain small amounts of dissolved waxes which, if they precipitate, could plug fuel filters and fuel pumps. Laboratory methods are available to measure the lowest points at which these products can be handled without pumping problems or filter plugging.

The laboratory pour point test, now ASTM D-97, has been used since 1914. Originally, the cloud point test was included in the ASTM D-97 procedure. In 1967 it was established as a separate test, ASTM D-2500.

The ASTM D-97 pour point test and the ASTM D-2500 cloud point test can be used on any petroleum oil. However, the flow characteristics of some waxy oils, such as fuel oils, can depend on their thermal history in handling and storage. The repeatability and reproducibility of the tests can be quite poor for thermally-cycled products.

ASTM Standard Test Method D-97. In this test an oil sample is placed in a test jar with a thermometer in the center of the sample. The sample is heated to 115°F (45°C) and then placed in a cooling bath maintained at 30 to 35°F (–1 to 1.5°C). Each time it cools 5F° (3C°), it is removed from the bath and tilted slightly to determine whether there is movement. If so, it is replaced and allowed to cool again. If the sample is still liquid when it reaches 50°F (10°C), it is transferred to a second bath maintained at 0 to 5°F (–20 to –15°C). For samples with very low pour points, additional baths could be necessary. Each bath is maintained 30F° (15C°) below the previous one.

If the sample does not flow when the jar is tilted, the jar is moved to a horizontal position and held for 5 s. If the sample still does not flow, the sample temperature is noted as the solid point, and a temperature of 5F° (3C°) higher is recorded as the pour point. The repeatability of the test for most samples is 5F° (3C°) and reproducibility is 10F° (6C°).

ASTM Standard Method D-2500. This method applies to products that are transparent and have cloud points below 120°F (50°C). The sample is heated to a temperature at least 25F° (15C°) above the expected cloud point and poured into a clear, flat-bottomed glass test jar with a thermometer centered in it and nearly touching the bottom of the jar.

The jar is immersed in a cooling bath maintained between 30 and 35°F (–1 and 1.5°C). The sample cools, and at each 2F° (1C°) interval the jar is removed and inspected for cloudiness. If none is observed, the jar is replaced and cooling continues. When a distinct haze is observed, the sample temperature is recorded as the cloud point. If no haze appears to 50°F (10°C), the jar is placed in a second bath maintained at 0 to 5°F (–20 to –15°C), and the process is repeated.

These procedures are time-consuming, tedious, and subject to human error. Process analyzers perform this test automatically and with much greater accuracy.

On-line Process Analyzers

Precision Model 44670 "Pour Pointer"

The Pour Pointer is a cyclic instrument that can measure pour points as low as –58°F (–50°C) on samples with viscosities as high as 45 cSt (centiStokes, a viscosity measurement unit) at 100°F (40°C). A flow diagram of the analyzer is shown in Figure 11-1.

Sample from the fast loop, on the lower left of Figure 11-1, enters a sample cell through valve 6. It fills the sample cell to the overflow line and is discharged through valve 7. Valves 4 and 8, which are open, allow instrument air to enter the test cell and the sensor probe, to prevent sample rising into these areas. High- or low-pressure air can be used, depending on the viscosity of the sample. Sample flow continues until the sample cell is flushed sufficiently and all valves close.

Valves 1 and 2 open to vent any excess air from the sample cell and to produce atmospheric pressure on both sides of the differential-pressure sensor above the sample cell. Valve 10 opens and momentarily traps low-pressure air in the tubing between valves 3 and 10. The Peltier unit (T/E [Thermo Electric] Module - TEM) is energized to cool the sample.

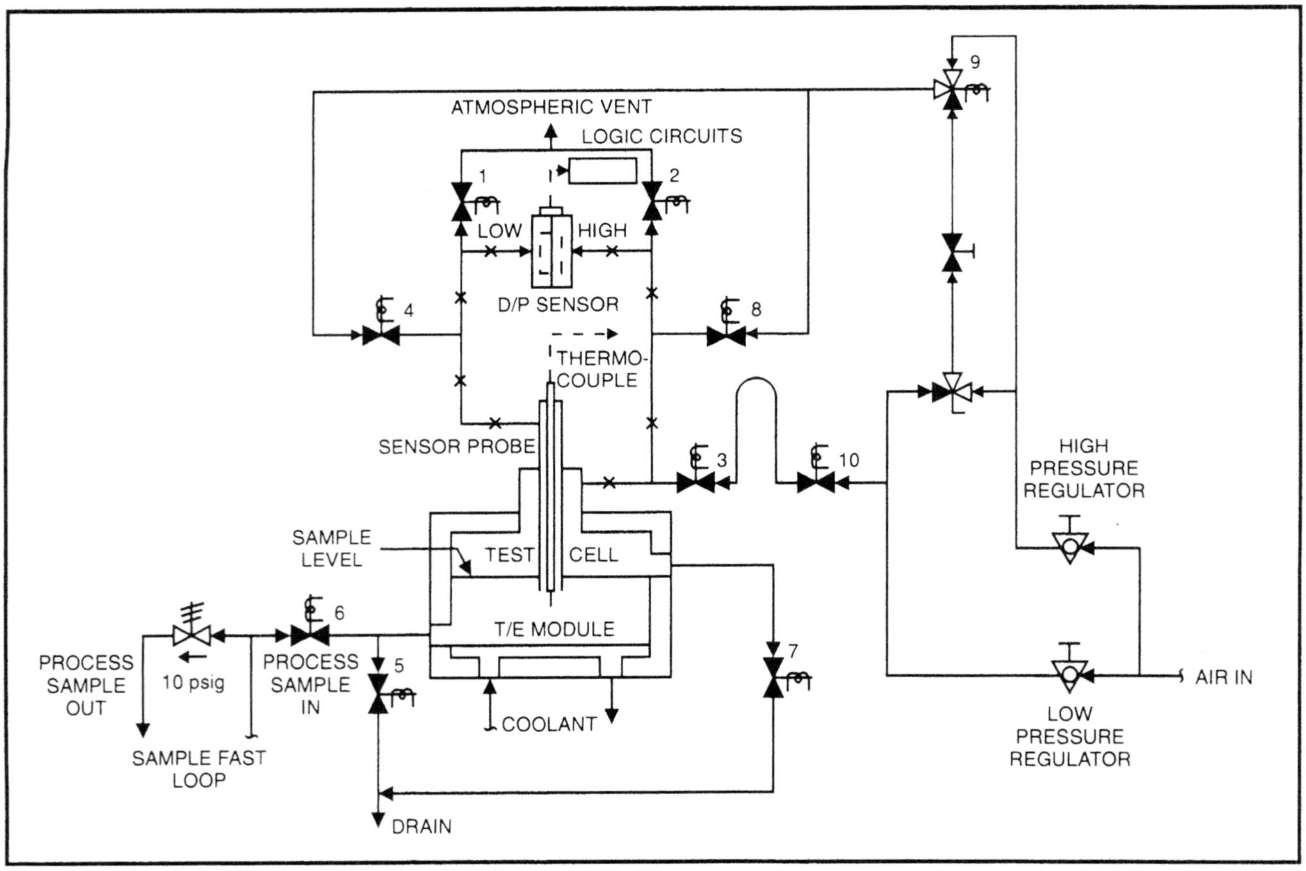

Figure 11-1. Precision Model 44670: Flow Diagram
(Courtesy of Precision Scientific Petroleum Instruments Co.)

Valve 3 opens to allow the trapped low-pressure air to expand into the test cell. A temporary pressure is produced on the liquid surface outside the sensor probe and against the high-pressure diaphragm of the differential-pressure sensor. The sensor immediately generates an output voltage because of this pressure pulse. The pressure on the surface of the liquid also pushes liquid upward into the probe, and air in the probe is compressed. This pressure reduces the output signal from the sensor. After a short delay, the sensor output is measured. If it is below a fixed value, because the liquid has risen into it, the sample cell is allowed to continue to cool. Valve 10 opens again momentarily, and the pulse sequence is repeated. As the sample cools, its viscosity increases until it can no longer compress the air inside the sensor probe. The output from the sensor remains high, and the cell temperature is recorded as the pour point.

Current to the thermoelectric module (TEM) is reversed to warm the cell. Valve 5 is opened to drain the sample cell. Valves 4 and 8 then open to allow air to purge the cell before another cycle begins.

The upper enclosure contains the electronics and the lower one the analyzer unit. This system meets National Electrical Code (NEC) Class I, Group D, Division 1 (ID1) requirements for hazardous locations.

A sample flow of 50 mL/min at a maximum pressure of 20 psig is required. Since spent sample must be discharged to atmospheric pressure, a sample-recovery system may be required. Sample should be at least 20F° (10C°), but not more than 120F° (65C°) above the expected pour point. The analyzer measuring range is –50 to 50°F (–45 to10°C), and repeatability is 5F° (3C°). The cycle time depends on the pour point, but it can be selected manually at 10, 30, or 60 min. If a refrigerated cooling system is used, pour points as low as –58°F (–50°C) can be detected.

Fluid Data Model 444B Hallikainen Pour Point Analyzer

This analyzer is an on-line instrument designed to measure automatically the ASTM pour point of process samples. It is a cyclic unit, which can produce two to three results per hour. A schematic diagram of the system is shown in Figure 11-2.

The analyzer has a cup that rotates slowly and which contains a trapped sample in which a tungsten carbide ball is suspended. Sample from a fast loop enters the system at the bottom and flows continuously through two (2) three-way valves (SV1A and SV1B in Figure 11-2) to fill the 23 mL reservoir.

To start a cycle, the valves are energized to open from port A to port B. The trapped sample drains into the rotating cup and the valves are deenergized. The tungsten carbide ball is suspended by a thin steel rod, like a pendulum, and will remain vertical as long as the sample is liquid. A thermocouple welded to the ball measures the sample temperature. The sample is cooled by a Peltier (TEM) unit. As the sample cools, its viscosity increases and the drag on the ball increases. When the "solid point" of the sample is reached, the ball moves far enough to trip an optical detector mounted on the top of the pendulum. This optical system consists of a lamp, a shutter on the pendulum, and a photocell. The shutter interrupts the light beam to the photocell and actuates a detection circuit, which records the temperature. This temperature is the solid point. By definition, the ASTM D-97 pour point is 5F° (3C°) above the solid point, so the equivalent of 5F° (3C°) is added to the output signal to indicate the pour point. Repeatability of the analyzer is 2F° (1C°).

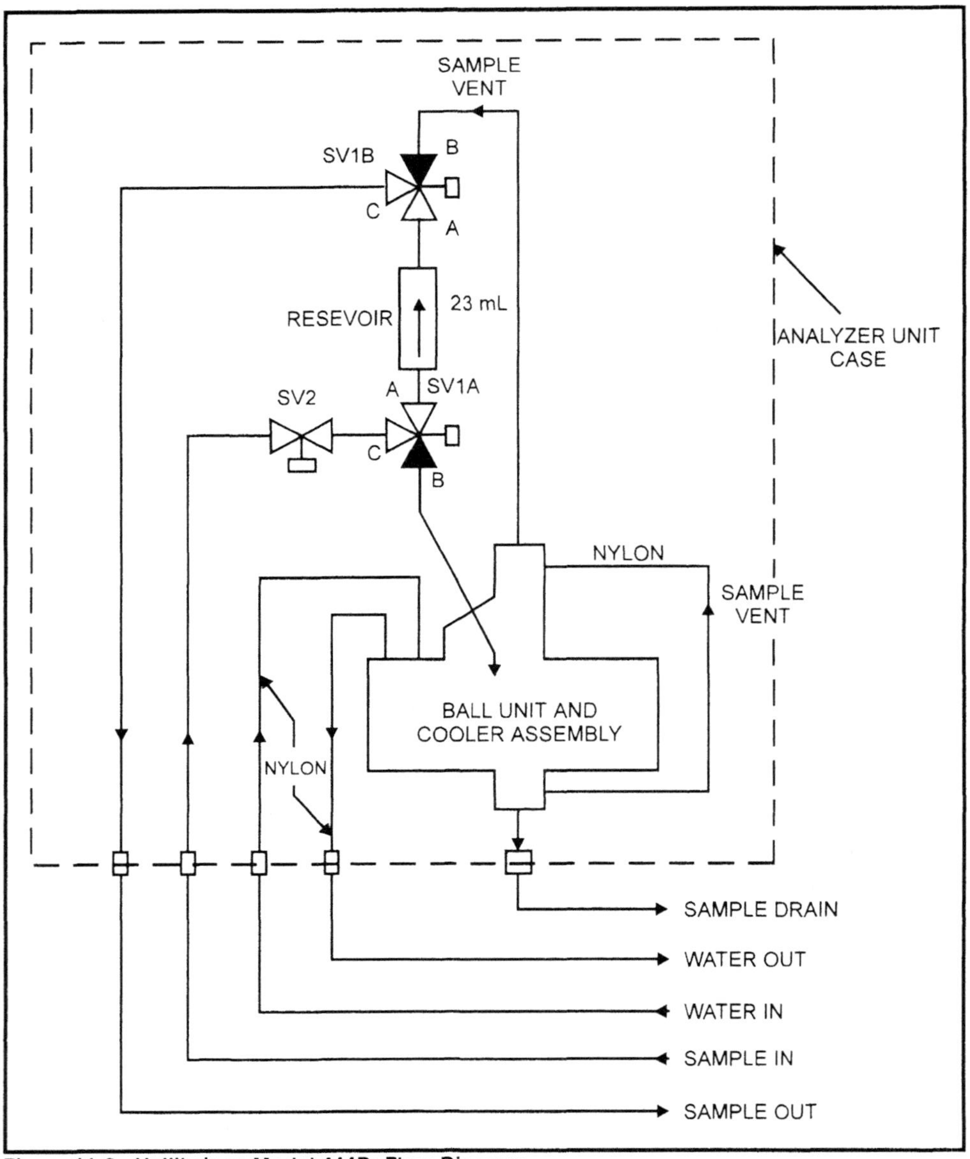

Figure 11-2. Hallikainen Model 444B: Flow Diagram
(Courtesy of Fluid Data, Inc.)

After the pour point has been recorded, the rotation of the cup stops and the Peltier (TEM) unit reverses to become a heater. The sample melts and the heating continues until the sample temperature reaches 115°F (45°C). At this point, a drain valve opens to remove the sample and to allow a fresh sample to flow for 1 min to flush the cup. The system can now begin a new cycle.

The cycle time is 30 min for samples with pour points above 32°F (0°C). A period of 20 min of cooling is followed by 10 min of heating. A longer cycle with 30 min of cooling and 12 min of heating can be programmed to extend the possible range to –10°F (–25°C).

The analyzer requires filtered sample, flowing at 10 to 100 mL/min with a temperature less than 160°F (70°C). The required sample pressure depends on the viscosity of the sample, but usually 15 psig is sufficient. This instrument can handle samples with viscosities as high as 250 cSt (centiStokes) at 120°F (50°C).

The cooling water temperature determines the minimum attainable pour point, but it should be less than 80°F (25°C). For very low pour points, a refrigerated sample cooler can be used.

Precision Model 42668 Cloud Point Analyzer

This cloud point analyzer is a newer version of a TOTCO instrument (see Chapter 10). This instrument can measure cloud points ranging from –20 to 77°F (–30 to 25°C). The principles are illustrated in Figure 11-3.

Sample in a glass cell is illuminated with light that has been polarized by passage through an optical polarizing filter. A second polarizing filter is mounted perpendicular to the first one. As long as the sample is clear, the second filter blocks passage of virtually the entire beam of light to a photocell. As the sample cools, wax crystals eventually begin to form in the liquid sample. These crystals reflect and refract some of the beam, which then passes through the second polarizing filter to the photocell. When the photocell detects the beam, a thermocouple in the liquid senses the sample temperature, and this temperature is recorded as the cloud point.

The analyzer cycle is controlled by a microprocessor, which first opens the sample solenoid valve to flush the sample cell and then introduces fresh, warm liquid. The thermoelectric cooling modules (TEMs) on the sides of the sample cell are activated, and they cool the sample until wax crystals are detected by the photocell. When the sample temperature at this point is recorded, the cycle is repeated.

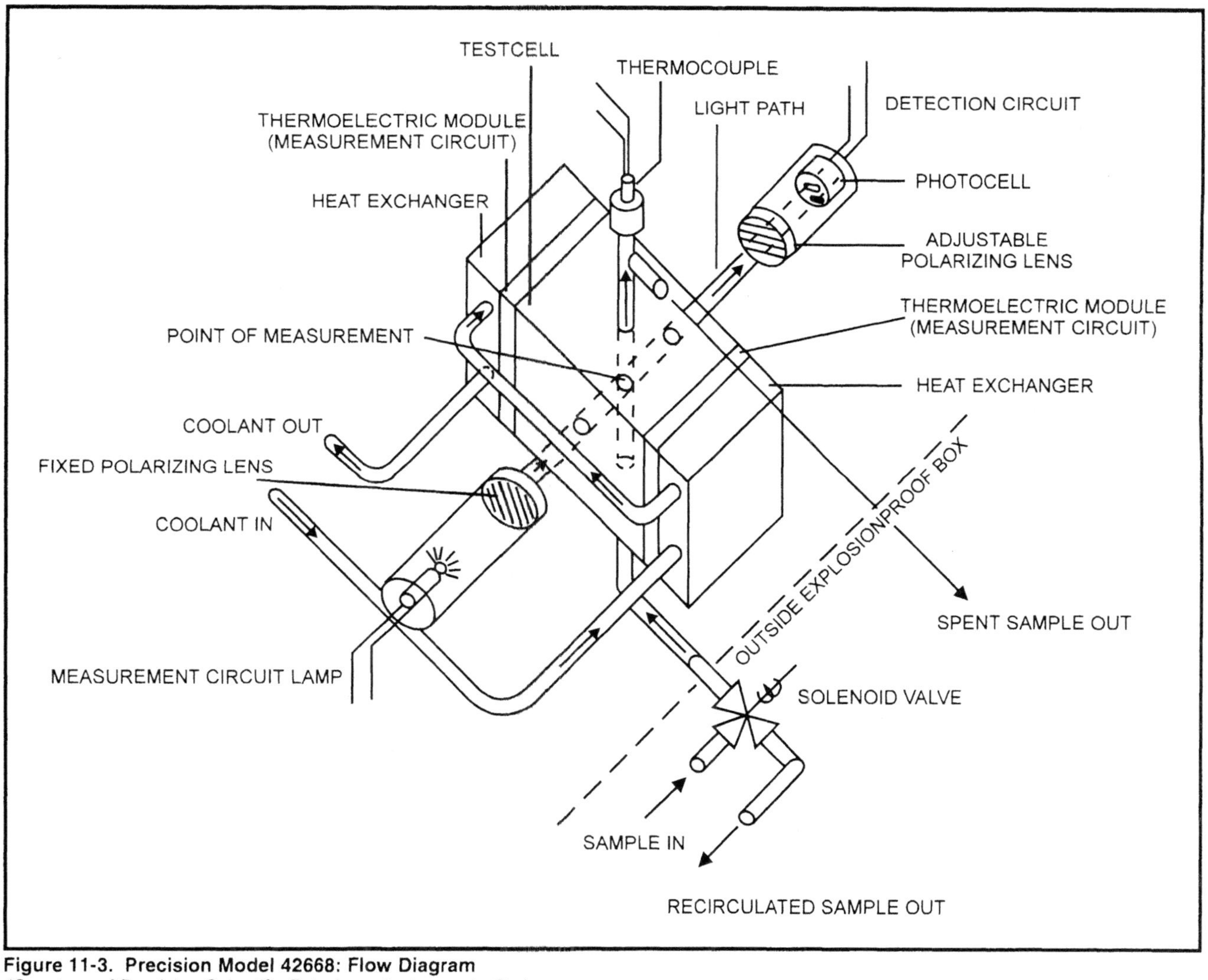

Figure 11-3. Precision Model 42668: Flow Diagram
(Courtesy of Precision Scientific Petroleum Instruments Co.)

The analyzer can be operated in three modes. In Mode 1, the flushing cycle is 30 s, and the thermoelectric modules (TEMs) are powered to provide cooling at a rate which detects the cloud point in about 90 s after the flush cycle is completed. A new cloud point is detected about every 2 min. If a cloud point is not detected in this time interval, the flushing cycle repeats.

In Mode 2, the analyzer simulates the laboratory procedure. The sample cell is flushed for 60 s and then cooled slowly as in the lab test. If a cloud point is not detected in 20 min, at maximum cooling power, the sample is discharged and the cycle repeated.

In Mode 3, the analyzer performs 120 cycles in Mode 1, followed by a single cycle in Mode 2. The power required to reach the cloud point in Mode 2 is sensed by the microprocessor, and this value is used for cooling in the next Mode 1.

The sample cell unit and the microprocessor are housed in two explosionproof enclosures rated for use in Class I, Group D, Division 1 (ID1) locations. The required sample flow rate is 200 to 300 mL/min at a pressure of 5 to 25 psig. Sample temperatures must be at least 20F° (10C°) above the expected cloud point temperatures. Clean, dry air is required to purge the enclosures to cool the electronics and to prevent moisture condensation on the cold measuring components.

Sample color up to ASTM 8 and moisture content up to about 5000 ppm will not affect analyzer performance. The signal output is the cyclic millivolt signal from the thermocouple, but a peak-holding device is available to provide a continuous 4 to 20 mADC output.

Hone "Cloudar" Cloud Point Analyzer

This analyzer is an updated version of the first Hone cloud point analyzer produced in the late 1960s. In the first model, light passed through a glass cell and, as a wax cloud formed, the detector responded to a decrease in the light level. In the present version of the Cloudar, the detector is located at right angles to the light beam (Figure 11-4). As long as the sample is clear, no light is detected. When the wax crystals form, they reflect light in all directions, and some of it is received by the detector. A thermocouple in the test cell then records the cloud point temperature. The cycle is controlled by a microprocessor, which initiates a fill cycle. The cell is flushed with fresh, warm sample for 20 to 40 s, then cooled by thermoelectric (TEM) units until the next cloud point is detected. A 1-s "read" cycle senses and stores the cloud point temperature and updates the output signal. If no cloud is detected in 25 min, a timer automatically starts a new cycle.

Cloud points 140F° (80C°) below the cooling water temperature can be detected. Cycling time is 1 to 4 min, and repeatability is 1F° (0.5C°). Sample is required at 5 gallons per hr (GPH) at 15 psig and at a temperature at least 45F° (25C°) above the expected cloud point. Cooling water temperature must be below 75°F (25 °C). The analyzer can handle samples containing up to 1000 ppm moisture.

The components are mounted in three explosionproof housings on a freestanding frame. The test cell is in the upper right enclosure, the control unit in the upper left one, and the main power supply in the lower left. LED status indicators in the housing on the lower right indicate sample control valve operation, cell cooling, cloud point detection, light source failure, overtemperature, and standby operation. The enclosures are suitable for Class I, Group D, Division 1, (ID1) hazardous areas.

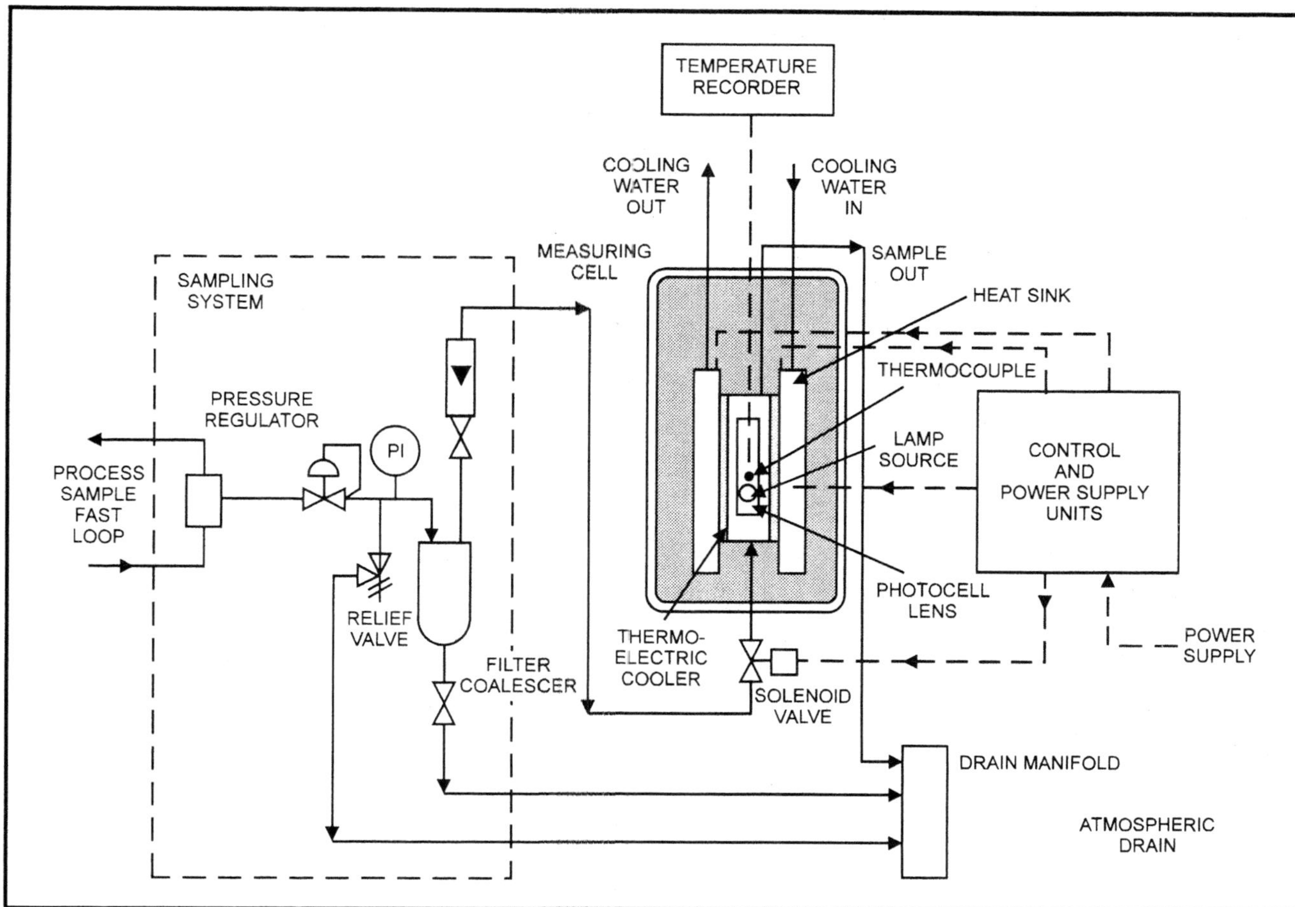

Figure 11-4. Hone "Cloudar": Flow Diagram
(From Process Analyzer Technology, Kenneth J. Clevett, New York: John Wiley & Sons)

Guided Wave Near-Infrared Octane Analyzer

This instrument can monitor cloud points of diesel fuels.

Common Applications

Cloud and pour point analyzers are useful in the petroleum industry mainly for monitoring feedstocks, or in-line blending operations, for jet fuels, kerosene, diesel fuels, and lubricating oils and their finished products. By operating closer to the pour point and cloud point specifications, on-spec production can improve significantly, with considerable savings on valuable fuel oils and blend stocks.

Domestic fuel oils or diesel fuels can be prepared by blending heavy naphtha with catalytic gas oil. An analyzer permits fine tuning of this blending system to maintain the cloud point of the fuel to within 1 to 2F° (0.5 to 1C°) of the specification and save appreciable amounts of heavy naphtha. The heavy naphtha that is saved can be used in the production of more valuable materials, such as petrochemicals.

Ranges of Variables

The Precision Model 44670 Pour Pointer can measure pour points as low as –60°F (–50°C) on hydrocarbon streams with viscosities as high as 450 cSt (centiStokes) at 100°F (40°C).

The Fluid Data Model 444B pour point analyzer can handle samples with pour points as low as –20°F (–30°C) with cooling water at 60°F (15°C), and as low as –70°F (–55°C) with 40°F (5°C) cooling water. Viscosities can be as high as 250 cSt at 120°F (50°C).

The Precision Model 42668 cloud point analyzer can detect cloud points in the range of –22 to 77°F (–30 to 25°C) for jet fuels, diesel fuels, and heating oils.

The Hone Cloudar can detect cloud points 140F° (80C°) below the temperature of the cooling water supplied to the instrument. The maximum water temperature recommended is 77°F (25°C), which would allow the detection of a pour point of –60°F (–50°C). Lower points could, of course, be detected with colder water.

Options and Accessories

A version of the Hone Cloudar cloud point analyzer is available that meets the European CENELEC safety requirements.

Limitations

Pour point and cloud point analyzers have limits on the viscosities of the samples they can handle. Samples must be clean and dry. Maximum sample temperatures are also limited, and refrigerated sample coolers may be required. Consult the individual instruction manuals for specific details.

Installation Hints

Pour point and cloud point analyzers should be installed indoors or in a climate-controlled analyzer shelter to protect them from the weather and to facilitate maintenance. The enclosures should be purged with clean, dry air to cool the electronic components and to keep the moisture in humid air from condensing in the enclosure.

Specific Safety Issues

All analyzers are rated for installation in NEC Class I, Group D, Division 1 (ID1) areas. The samples processed are not particularly hazardous for the temperatures at which the analyzers operate.

Unique Sampling System Requirements

The pour point and cloud point analyzers require 3 to 5 GPH of sample at 5 to 25 psig. The temperature of the sample should be at least 20F° (10C°) above the expected pour or cloud point. Since the analyzer will usually be installed indoors or in a climate-controlled analyzer shelter, sample systems with fast loops should be used.

Typical Calibration Requirements

Cloud points of samples are measured in the laboratory by ASTM Method D-2500. This test is time consuming, tedious, and requires meticulous attention to detail. Consequently, it is subject to appreciable human error. In addition, the reproducibility of the laboratory test can be as high as 8 to 10F° (4.5 to 5.5C°), whereas the reproducibility of process cloud point analyzers is about 2F° (1C°). A precise correlation of laboratory tests and analyzer results can be very difficult; therefore, an on-line calibration sample system is desirable.

Troubleshooting and Maintenance

All pour point and cloud point analyzers require regular examination for correct operation and reliable results. Consult the individual manufacturer's operation and maintenance manuals for specific details.

Procedures for the Fluid Data pour point analyzer include checking the position of the ball in the test cup. The vane should pass freely between the lamp and the photocell. The conical pivots should be adjusted and locked in position so that there is no free play and negligible friction. Sample volume should be approximately 23 mL. The ball should travel along a line 8 mm from the center of the test cup. When the sample is fluid, the shaft should be in a vertical position, with the ball nearly touching the wall of the cup. The solid point is detected when the ball is just past the centerline of the cup, and the photocell housing should be adjusted so that the detector just triggers when the ball is in this position.

About the Author

Conard Donnell received a Bachelor's and a Masters degree in Chemistry and Physics from Temple University. His career with automatic process analyzers began and ended at Sunoco Refinery where he developed on-line instrumentation to measure the physical properties of process streams automatically and much faster than laboratory analysis. Since retirement, Mr. Donnell has worked as a consultant for Sunoco, Fluid Data, Precision Scientific Co., Lockwood and McLorie, and DuPont working on projects that included design and installation of process analyzer systems in refineries worldwide.

12

Flash Point Analyzers

Flash point is the temperature at which the vapor of a flammable liquid will ignite when exposed to a flame or spark. Applications are mainly in the petroleum industry. Too high a flash point may indicate that products, such as kerosene, gas oils, and diesel fuels, will not ignite properly. Too low a flash point may indicate that the product is contaminated with lower-boiling-point material and could be a hazardous fuel (ignite prematurely). Three laboratory test methods are used to measure flash points: ASTM D-56, D-92, and D-93.

ASTM D-56 Tag Closed-Cup (TCC) Test Method. This test is the oldest method. It can be used on any mobile liquid with a flash point below 200°F (95°C) and a viscosity less than about 5 cSt (centiStokes) at 100°F (40°C). The system is shown in Figure 12-1.

With the test cup at least 20F° (10C°) below the expected flash point, the sample is poured into the cup, the lid is closed, and heat is applied. If the sample flash point is below 140°F (60°C), the heating rate is 2F°/min (1C°/min), as indicated by the cup thermometer. For samples with higher flash points, the rate is increased to 5F°/min (3C°/min). When the sample temperature reaches about 10F° (5C°) below the expected flash point, the lid of the cup is lifted and a small test flame is inserted momentarily into the cup. The flame is reinserted after each 1F° (0.5C°) rise in sample temperature (2F° [1C°] for flash points above 140°F [60°C]) until a distinct flame is observed in the cup. This temperature is recorded as the flash point.

ASTM D-93 Pensky-Martens (P-M) Closed-Cup Test Method. This method can be used for fuel oils, lubricating oils, and other viscous materials with flash points up to 700°F (370°C). The flash cup is similar to the one used in the TCC test. The sample is stirred continuously to keep the more viscous materials uniformly mixed. A Bunsen burner heats the flash cup in the Precision Scientific Petroleum Instrument Company P-M Model 74545, and an electric heater is used in Model 74537.

The sample, whose temperature is at least 30F° (15C°) below its flash point, is poured into the test cup and heated at a rate of about 10F°/min (5C°/min). If the flash point of the sample is below 230°F (110°C), a thermometer with a range of 0 to 230°F (–20 to 110°C) is used, and the test flame is applied every 2F° (1C°) until the flash is observed in the cup. For samples with higher flash points, a thermometer with a range of 200 to 700°F (95 to 370°C) is used and the flame is applied every 5F° (3C°).

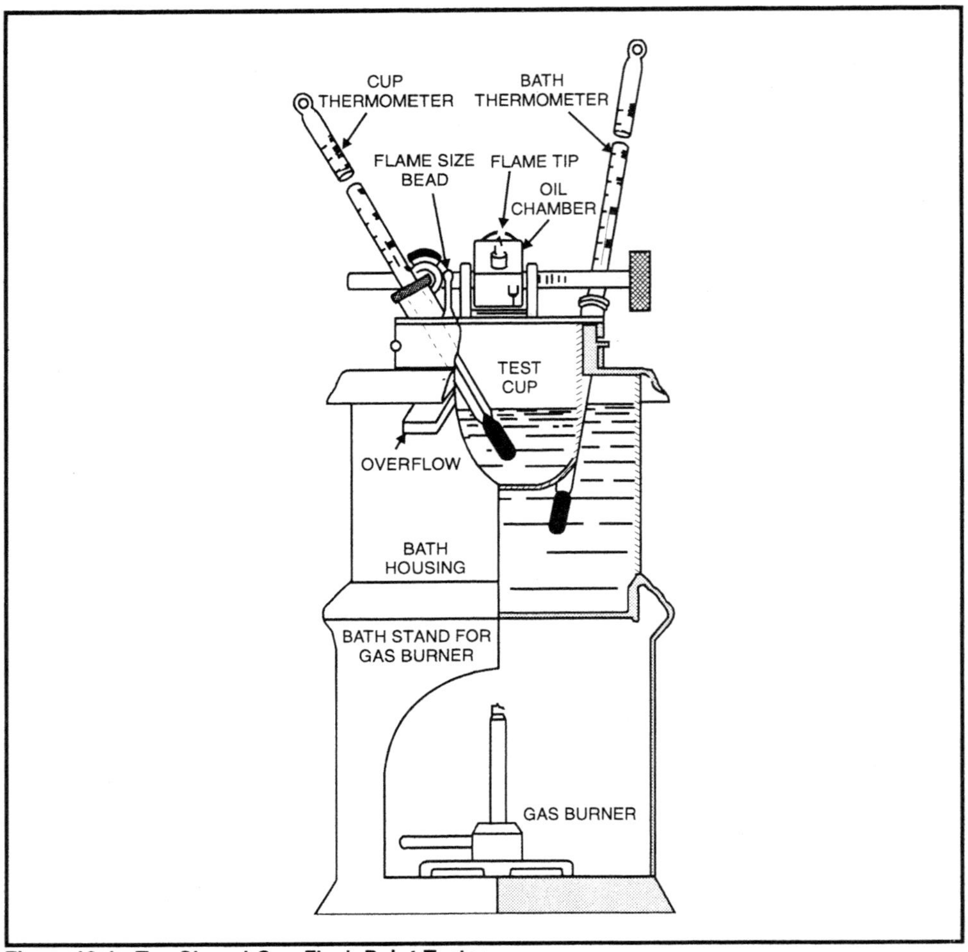

Figure 12-1. Tag Closed-Cup Flash Point Tester
(From Process Analyzer Technology, Kenneth J. Clevett, New York: John Wiley & Sons)

ASTM D-92 Cleveland Open-Cup (COC) Test Method. This test method can be used for all petroleum products with flash points below 175°F (80°C), except fuel oils. A sample with a temperature at least 100F° (55C°) below the expected flash point is placed in the open cup. Heat is applied to raise the sample temperature at about 10F°/min (5C°/min). Every 30 s (i.e., every time the temperature increases by 5F° [3C°]), the test flame is passed across the cup. The temperature at which the sample first ignites is the open-cup flash point.

The heating cycle can continue, with the flame passing over the sample periodically, until the sample ignites and continues to burn for at least 5 s. This temperature is recorded as the ASTM D-92 fire point. Flash point testers that perform and record these same functions automatically are available for laboratory use.

On-line Process Analyzers

The first on-stream flash point analyzer was developed by Precision Scientific Development Company in 1958 in cooperation with Shell Development Company. This instrument, the Model 44614 high-range flash point monitor, became very popular. Since then, four improved models have been developed. Hone-Addison, in England, manufactures the "Flashar." Carlo Erba, in Italy, markets the "Flash-O-Matic," which will not be included in this discussion, because it essentially reproduces the laboratory test (a cyclic or batch basis).

Precision Model 44650 General Flash Point Monitor

This analyzer is the first updated version of the original Model 44614 monitor. It can handle streams having viscosities of up to 2000 cSt (centiStokes) at 100°F (40°C) and flash points from 140 to 600°F (60 to 315°C). Its output correlates well with ASTM D-92 and ASTM D-93 laboratory tests. A schematic flow diagram is shown in Figure 12-2.

The components are mounted in three enclosures, one in the rear, on a freestanding steel frame. This arrangement simplifies installation. The sampling system is enclosed in the housing on the back of the frame and, if necessary, can be heated to maintain viscous samples in a fluid state. The gear pump delivers sample into the measuring assembly at a constant flow rate of 25 mL/min. Waste sample is returned to the process stream by the second gear pump. The throttling valve and the check valve comprise the sample fast loop, which minimizes the time required for sample to get from the process to the analyzer. The throttling valve provides positive pressure at the suction side of the feed pump and controls the sample velocity through the fast loop at a rate of about 5 ft/s. The check valve prevents recirculation of waste sample through the analyzer. The filter is a motorized, self-cleaning unit, which protects the sample return pump from any coke that forms from testing the sample. The motors actuating the pumps and filter are installed in an air-purged, explosionproof housing above the sampling system enclosure.

Sample from the feed pump continuously passes through the horizontal heater in the bottom of the measuring chassis, the lower explosionproof housing on the front of the frame. Instrument air passes through a coil and flows around the sample heater, carrying sample into the bottom of the flash cup. Sample fills the cup to the level of an overflow orifice. The excess sample flows from the cup through a filter and is removed continuously by the return gear pump. The air and any hot vapor rise past the electrodes and exit from the top of the cup into the separator. Air and any noncondensable vapor exit through the explosionproof (flame arrester) atmospheric vent.

Power to the sample heater is controlled by a setpoint potentiometer in the control chassis, which is located in the upper explosionproof housing. The potentiometer is rotated by a reversible motor. This controller applies a continually increasing voltage to the heater until the heated vapor is ignited by a spark between the electrodes. The spark, which lasts a half second, is generated every second until the sample ignites. A thermocouple in the liquid in the flash cup records that temperature as the flash point. The pressure switch detects the pressure generated by the ignition and deenergizes the spark circuit. The setpoint potentiometer motor reverses to decrease the setpoint temperature to the heat controller, and a solenoid valve opens to admit a high flow of unheated purge air into the flash cup. This purge air cools the sample cell, flushes vapor from the flash cup, and minimizes coke buildup on the electrodes.

The cooling cycle is quite brief. The sample temperature in the flash cup drops only a few degrees below the flash point and a new heating cycle begins. Therefore, the system maintains the sample temperature very close to the flash point at all times.

The analyzer design permits easy maintenance. The control chassis can be removed intact for required bench work. Quick-disconnect clamps seal all sample components, and adjustable limit switches prevent excessive coking in the event of an analyzer controller failure. Since the output temperature continuously reflects the measured variable, it is compatible with control, alarm, and data-acquisition systems without additional signal conditioning. The system is designed for National Electrical Code (NEC) Class I, Group D, Division 1 (ID1) hazardous areas.

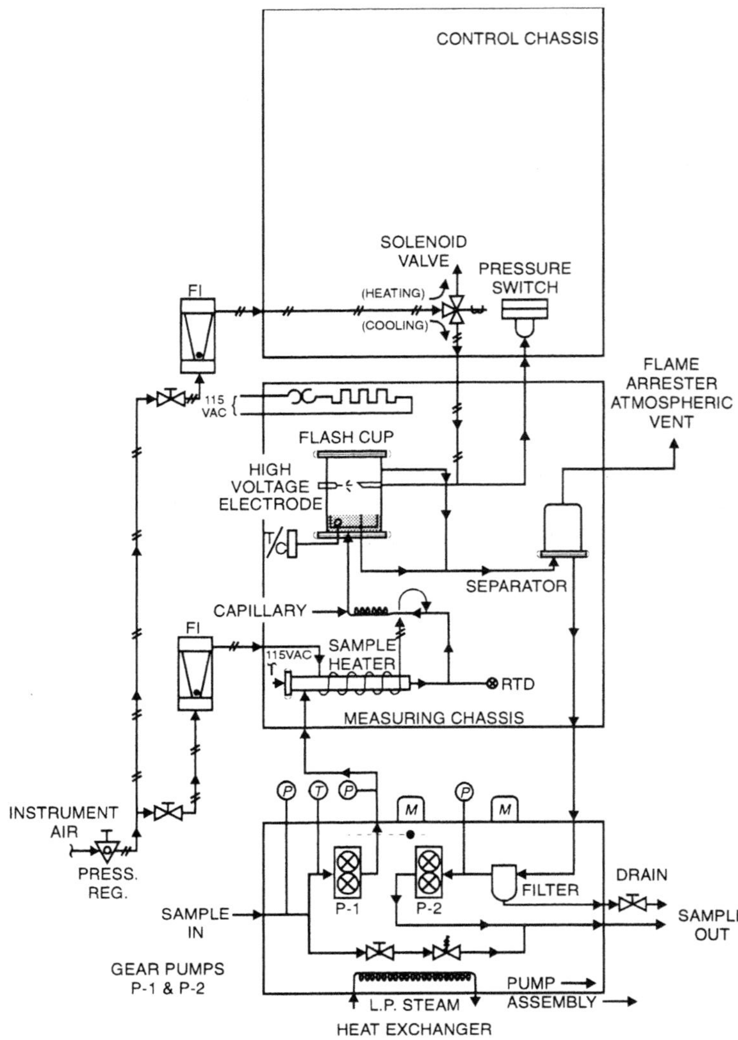

Figure 12-2. Precision Model 44650: Flow Diagram
(Courtesy of Precision Scientific Petroleum Instruments Co.)

Precision Model 44607 "Flash Pointer"

This analyzer is designed to monitor lighter hydrocarbon streams (lower flash point) than the general flash point monitor. It can handle samples with flash points from 50 to 250°F (10 to 120°C) and viscosities up to 215 cSt (centiStokes) at 100°F (40°C). The results correlate very well with the ASTM D-56 TCC procedure and can be correlated with the ASTM D-93 P-M closed-cup method within the measuring range of the analyzer.

Sample is fed into the flash cup (Figure 12-3) at a rate of 30 to 50 mL/min and at a temperature at least 30F° (15C°) below the expected flash point. Air, at the rate of 600 to 800 cm^3/min, mixes with the sample and goes to the flash cup. The liquid passes around the liquid measurement thermocouple and forms a small pool in the bottom of the cup. Excess liquid overflows through the drain. The overhead vapor and air mixture passes around the splash shield, the vapor thermometer, and the high-voltage electrodes. The flash cup is heated by a vertical heater in the bottom of the cup. A heater-control board applies continuously increasing power to the heater during the heating cycle, and the fuel-air mixture is heated at a uniform rate. A 5000 vDC spark is generated across the electrodes once per second. The liquid and vapor thermocouples register essentially the same temperatures until a spark ignites the vapor. The heat of explosion increases the vapor temperature dramatically. The temperature sensed by the liquid thermocouple is recorded as the flash point and vented vapor condenses and passes to the drain. A detection-circuit board is energized to operate a relay, which initiates a cooling cycle. In the cooling cycle, the voltage to the heater is removed, the spark circuits shut down, and the heater-control board lowers the setting. The incoming fuel-air mixture cools the flash cup. The cooling cycle time is adjusted manually to allow the incoming liquid to cool about 30F° (17C°). The maximum temperature reached by the liquid thermocouple represents the flash point. Since the registered temperature oscillates during the cycle, a signal converter with current output and peak-holding capability is required for alarm or closed-loop control applications.

The Flash Pointer is explosionproof and can be mounted on a wall or a rack. It should be protected from the weather.

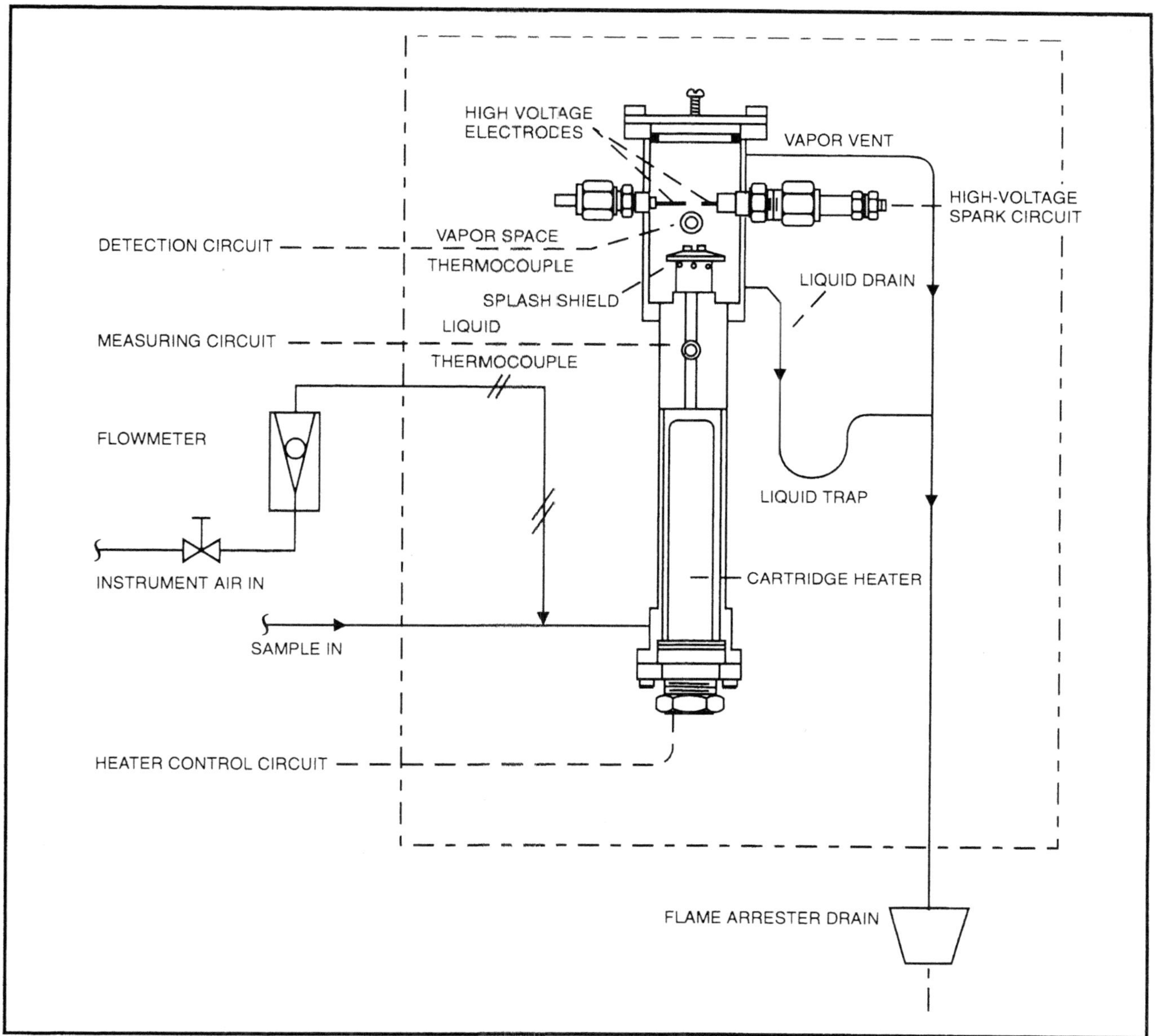

Figure 12-3. Precision Model 44607: Flow Diagram
(Courtesy of Precision Scientific Petroleum Instruments Co.)

Precision Model 44624 Integrated Flash Point Monitor

This system includes the Model 44607 Flash Pointer, a sample-conditioning system, and a refrigeration unit to cool samples with low flash points. It also includes a fast loop (Figure 12-4) to minimize deadtime in supplying the process sample to the analyzer, a solenoid valve to allow the remote shutoff of the sample, and a filter-coalescer to remove solid particles and entrained water from the process hydrocarbon stream flowing to the analyzer.

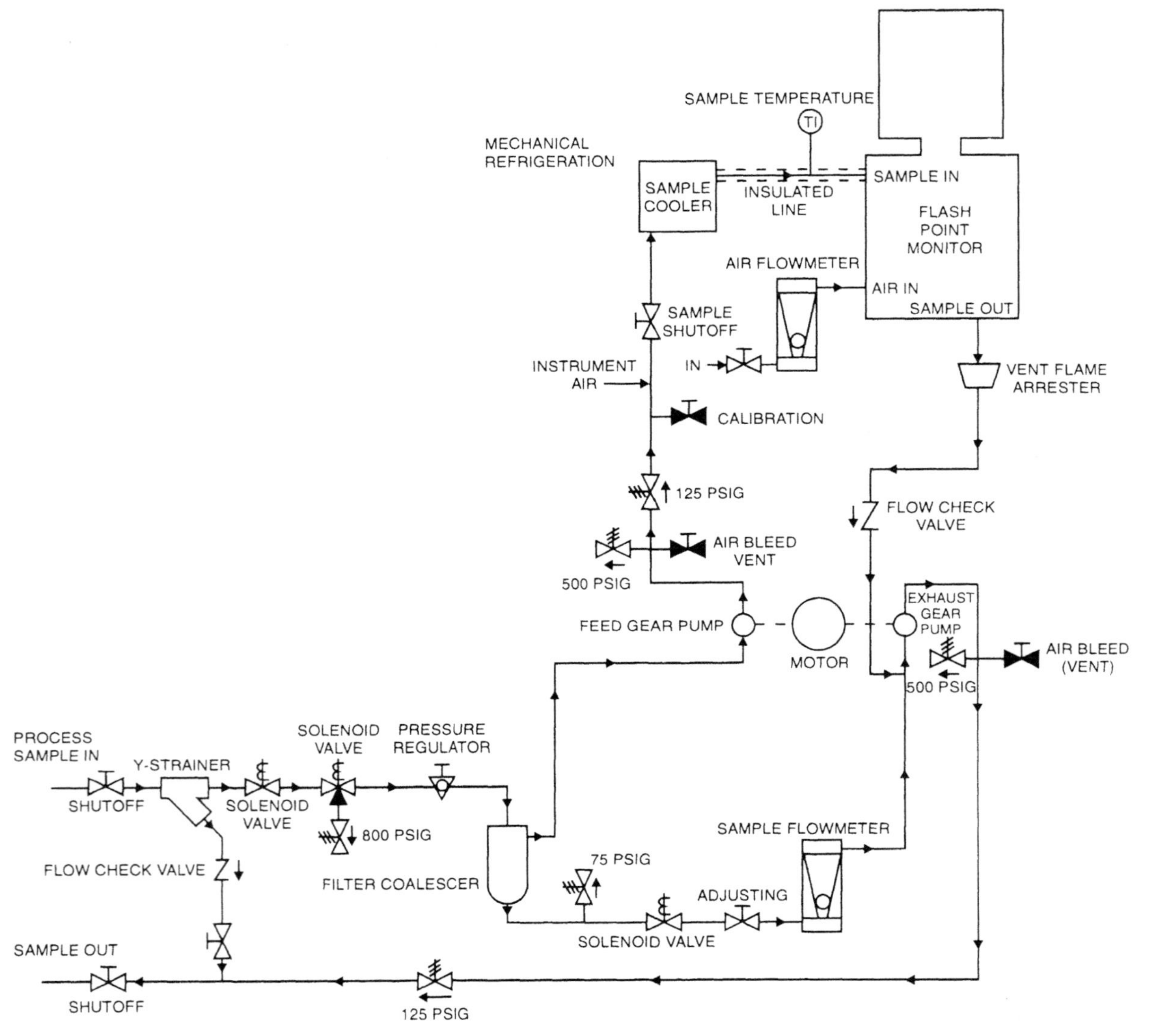

Figure 12-4. Integrated Flash Point Monitor: Sampling System
(Courtesy of Precision Scientific Petroleum Instruments Co.)

Precision Model 44620 Pipeline Flash Point Monitor

This system is essentially the Model 44624 integrated flash point monitor, which has been modified to permit operation on intermittent service. It also provides for local or remote start-up and for initial warm-up on a standby fuel sample. The system includes an air compressor, a fuel reservoir, and the standby sampling system (Figure 12-5). Precision developed this analyzer system specifically for pipeline operations.

Actuation of the "standby" or recycle switch energizes three solenoid valves. Two of these establish circulation of the standby sample in the reservoir, and the third closes off the coalescer bypass flow to prevent recirculation of this stream into the reservoir. The reservoir has a capacity of 4.5 gal, which at 50 mL/min allows for operation as long as 5 hr without a process sample.

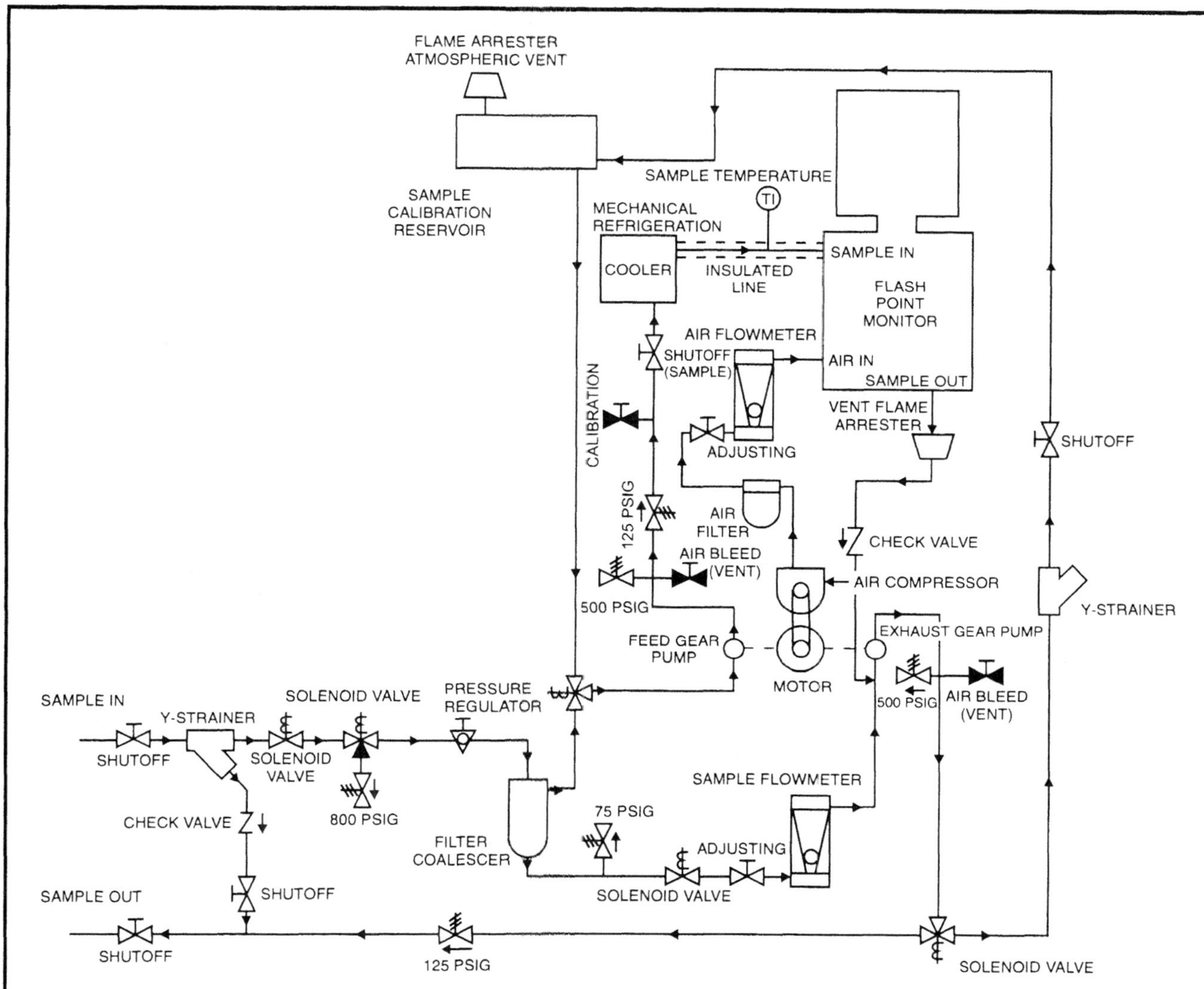

Figure 12-5. Pipeline Flash Point Monitor: Flow Diagram
(Courtesy of Precision Scientific Petroleum Instruments Co.)

Hone-Addison "Flashar" Flash Point Analyzer

The Flashar is designed to monitor flash points of petroleum products at temperatures as high as 480°F (250°C), with an accuracy and repeatability equal to or better than the ASTM D-93 method. Because of its unique operating principle, it is a truly continuous analyzer. Its operation is based on the ability of platinum and palladium to catalyze the oxidation of hydrocarbons in the presence of air and, when the temperature is high enough, produce ignition without using a spark.

A mixture of hydrocarbon vapor and air exists in the atmosphere over a liquid hydrocarbon. As the liquid is heated, the concentration of hydrocarbon vapor increases until the temperature of the mixture reaches its flash point. Since the higher-boiling-point hydrocarbons have higher flash point temperatures, the hydrocarbon/air ratio in the vapor is essentially constant for all hydrocarbons. A temperature-control system in the Flashar maintains the liquid sample at a temperature at which its overhead vapor can be ignited by a platinum/palladium catalyst. The temperature of the liquid is recorded as the flash point and can be correlated with flash points determined by ASTM D-93.

Figure 12-6 is a flow diagram of the Flashar system. The process sample, which has been filtered and coalesced to removed solid particles and moisture, is pumped by a metering pump to a heat exchanger at a rate of about 35 mL/min. Here the sample is preheated with heat from the effluent sample. Preheated sample flows through the flash unit box, which contains a sample heater, maintained at a suitable constant temperature, to minimize the effects of variations in ambient temperatures. The sample goes to the mixing chamber, where it is mixed with instrument air that has been preheated by passing through a coil around the sample cup. Excess sample passes to the overflow chamber, through the heat exchanger to preheat incoming sample, and then through a flame arrester before emptying into a drain. This vapor/air mixture in the mixing chamber goes through the detector housing and then to an atmospheric vent.

The catalytic detector in the detector housing produces an output voltage that varies with the concentration of hydrocarbon in the mixture and thus with the temperature of the mixture. The output voltage is used to control the detector temperature at the value at which the vapor/air mixture ignites. A platinum resistance thermometer (Pt100RTD), immersed in the liquid in the mixing chamber, measures the sample temperature, which is recorded and correlated with the laboratory flash point.

The system is mounted in explosionproof enclosures on a freestanding frame. The upper right enclosure contains the flash unit and the upper left one the control unit. The control unit contains a transmitter that converts the signal from the thermocouple into a 4 to 20 mADC output signal, which can be ranged to suit the sample being analyzed. The power supply is in the lower left enclosure. An on-off switch, fuses, and the drive motor for the metering pump are at the lower right.

In the event of sample loss or detector failure, a high-temperature limit switch cuts off the power to the sample heater. In addition, an interlock prevents power being supplied to the main heater until the detector housing has reached its correct operating temperature. A standby switch is provided that shuts off the power to the main heater, but maintains power to the detector housing enclosure. This setup reduces start-up time considerably when the analyzer is used in a batch-type operation. Hone-Addison can also supply custom-engineered sampling systems for their Flashar analyzer.

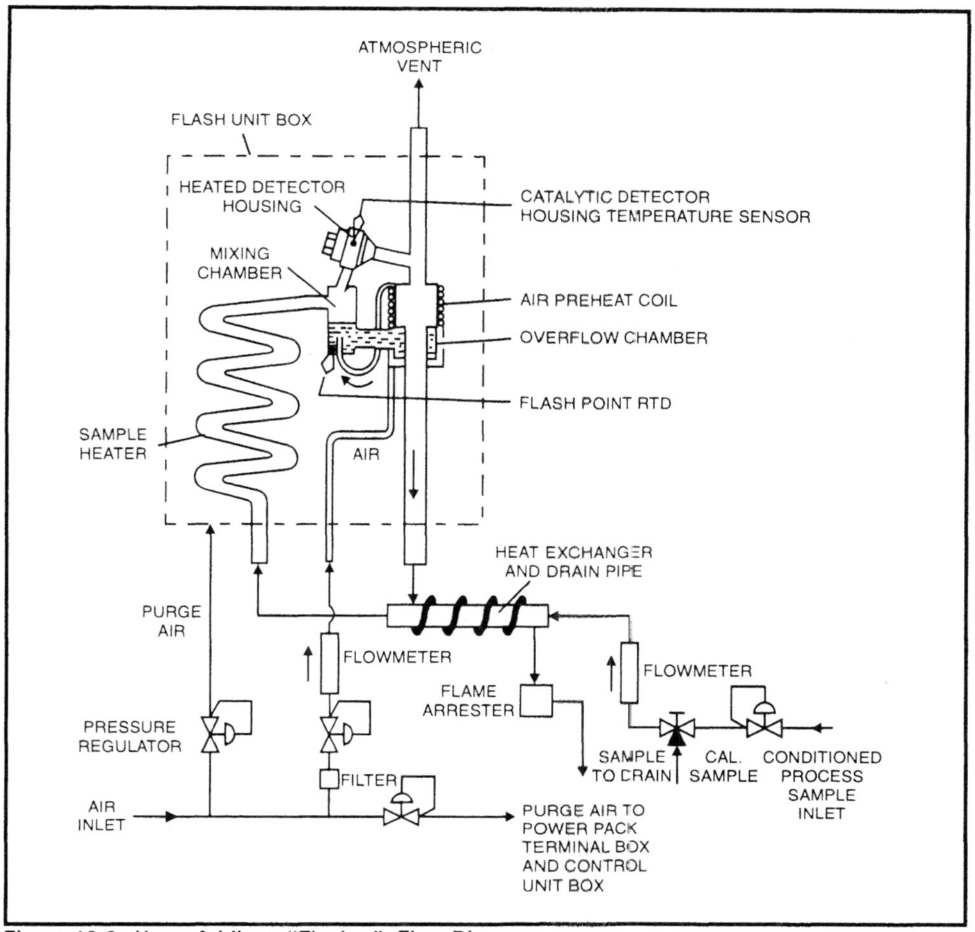

Figure 12-6. Hone-Addison "Flashar": Flow Diagram
(From Process Analyzer Technology, Kenneth J. Clevett, New York: John Wiley & Sons)

Guided Wave Near-Infrared Octane Analyzer

This instrument can be used to monitor flash points of aviation gasolines.

Common Applications

Flash point analyzers are used mainly in the petroleum industry to detect contamination of products, such as gasoline in kerosene, which could produce a hazardous material with an unexpectedly low flash point. Many common applications are on crude distillation units. Light naphtha streams are monitored to maximize the yield of fuel gas while preventing the loss of valuable pentanes from the naphtha. Other sidestreams, which will ultimately be used to produce kerosene, jet fuel, diesel fuel, and heating oils, are monitored to assure the safety of these products and to maximize the yields of the most valuable ones. Products from heat exchangers can be monitored to detect contamination with lighter products due to leakage between sections of the exchanger. The products from solvent refining operations, such as furfural extraction, are monitored to detect contamination with volatile solvents, which may be hazardous as well as expensive. In-line blending systems that produce finished products are monitored to maintain product quality and to maximize the yields of the most valuable products.

Ranges of Variables

Precision Scientific flash point analyzers can handle samples with flash points from 50 to 250°F (10 to 120°C) and with viscosities as high as 215 cSt (centiStokes) at 100°F (40°C). The Hone-Addison Flashar can handle samples with flash points as high as 450°F (230°C).

Options and Accessories

The Precision Scientific pipeline flash point analyzer and the Hone-Addison Flashar are equipped with standby facilities for batch operation. The Precision analyzer is equipped with a reservoir from which sample can be fed to the analyzer for as long as 5 hr while the process is shut down. On the Flashar, the main heater can be turned off while the detector housing is still maintained at its operating temperature, thereby reducing future start-up time.

Limitations

The analyzers require clean, dry samples. The Precision Scientific analyzer is limited to samples with viscosities below 215 cSt (centiStokes) at 100°F (40°C).

Installation Hints

Flash point analyzers should be installed indoors or in a climate-controlled analyzer shelter to protect them from the weather and to facilitate their maintenance.

Specific Safety Issues

Although the analyzers are rated for installation in NEC Class I, Group D, Division 1 (ID1) areas, their volatile samples and the high temperatures involved can produce hazards. The analyzer should be installed in a climate-controlled building, vented with a positive pressure fan for exhausting out of the upper wall area of the building, and have hydrocarbon detectors/flame detectors monitoring the analyzer area.

Unique Sampling System Requirements

Flash point analyzers require clean, dry samples, which the sample conditioning system must supply. The Precision Scientific instruments include a pump to return the waste samples to the process. However, the waste sample from the Flashar must be discharged at atmospheric pressure, so a sample-recovery system is usually required.

Typical Calibration Requirements

The laboratory tests on highly volatile samples may not be accurate because of loss of volatile light ends. An on-stream calibration system designed to minimize these losses is desirable.

Troubleshooting and Maintenance

All flash point analyzers require routine maintenance. For example, the recommended procedure for the Precision Scientific general flash point analyzer includes:

1. Check the incoming sample pressure and temperature. If they are not at specification, check the sample-conditioning system.
2. Check the pumping rates of the feed and discharge pumps; service them if the rates are not correct.
3. Check the combustion air rate, which should be 1000 mL/min. Adjust if necessary.
4. Check that the spark plug is firing once per second.
5. After the ignition point has been detected, check that the air purge rate is 8 L/min.
6. Clean the motor-driven filter in the inlet line to the return pump. Refer to the instruction manual for details.

About the Author

Conard Donnell received a Bachelor's and a Masters degree in Chemistry and Physics from Temple University. His career with automatic process analyzers began and ended at Sunoco Refinery where he developed on-line instrumentation to measure the physical properties of process streams automatically and much faster than laboratory analysis. Since retirement, Mr. Donnell has worked as a consultant for Sunoco, Fluid Data, Precision Scientific Co., Lockwood and McLorie, and DuPont working on projects that included design and installation of process analyzer systems in refineries worldwide.

13

Freeze Point Analyzers

The freezing point is the lowest temperature at which a fuel remains free of solid hydrocarbon (wax) crystals. Solids in a system can restrict the flow of fuel through filters. The freeze point of aviation jet fuels is a critical property, because at high altitudes the temperature of the fuel in an aircraft tank decreases markedly due to the very low ambient temperatures.

ASTM Standard Test Method D-2386. This test measures the freeze point and defines it as the temperature at which the last wax crystals disappear when a cold sample, containing the solid particles, is warmed. The test is conducted in a double-wall glass vessel with the space between the double walls filled with dry nitrogen or air at atmospheric pressure (Figure 13-1). A cork stopper in the top of the vessel supports a thermometer to measure the sample temperature, and a stirrer keeps the sample uniformly mixed.

The sample vessel is suspended in a double-wall vacuum vessel that contains acetone or alcohol. Solid carbon dioxide is added to the vacuum vessel to cool the sample, and the stirring rod is pushed up and down to mix the sample. A cloud may occur at about 15°F (−10°C). If it does not increase in intensity as the temperature decreases further, the cloud is caused by a small amount of water in the sample and should be disregarded. Cooling and stirring continue until crystals of hydrocarbon (wax) appear. The temperature at this point is recorded. The sample is removed from the coolant and allowed to warm up, while stirring continues. The temperature at which the crystals completely disappear is also recorded. The difference between the two recorded temperatures should be less than 5F° (3C°). If the difference is greater, the cooling and warming should be repeated. When the difference is less than 5F° (3C°), the temperature at which the crystals disappear is recorded as the freeze point of the sample.

On-line Process Analyzers

Precision Model 42680 Freeze Point Analyzer

This analyzer uses a sample cell and optical system identical to that of the Model 44668 cloud point analyzer (see Chapter 11). Sample is trapped in a test cell, which is cooled by a Peltier cooler (thermoelectric module - TEM) (Figure 13-2). A light beam from an incandescent lamp passes through a polarizing filter, which limits its wave action to one plane. The beam passes through the sample and then through a second polarizing filter mounted optically perpendicular to the first one. The second filter limits the passage of any nonrefracted light onto a detector (photocell). When the sample is cooled to its freezing point, solid crystals appear. These crystals refract the light beam in such a way that they pass through the second filter and are detected by the detector (photocell).

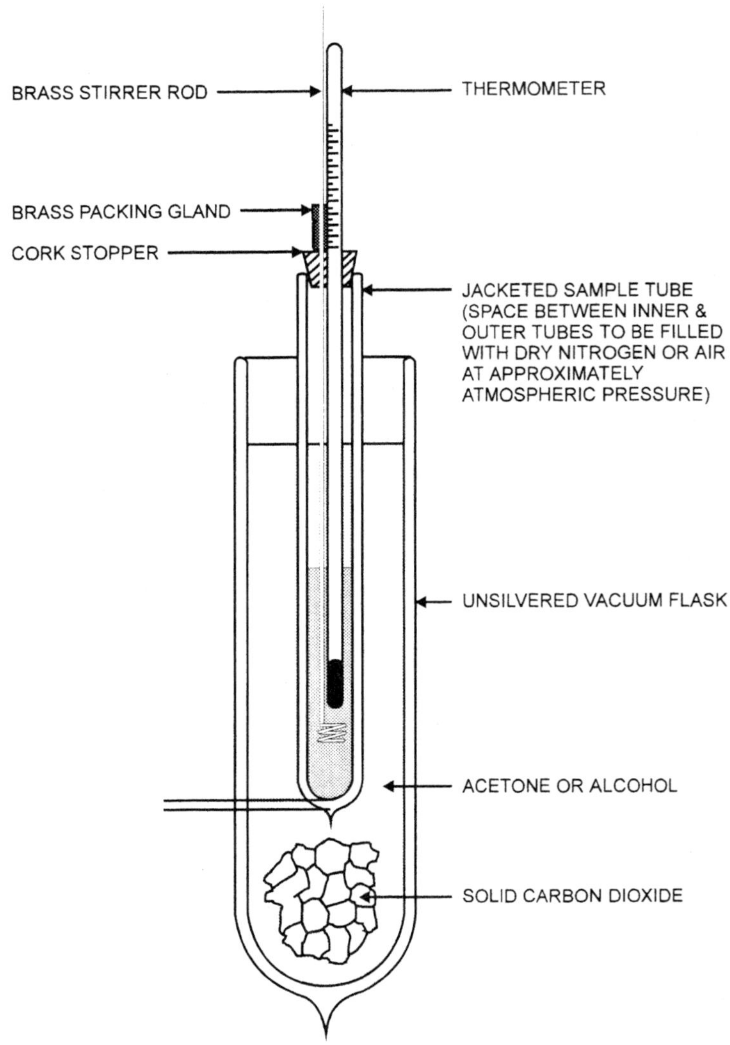

Figure 13-1. ASTM D-2386 Test Apparatus
(Courtesy of ASTM)

In operation, the trapped sample is cooled by the Peltier (TEM) unit until crystals are detected. Then the sample is warmed until the crystals just disappear and reduce the amount of light that reaches the photocell. This temperature is recorded as the freezing point. The sample cell is flushed with fresh sample, and a new cycle starts.

The equipment is mounted in two explosionproof housings on a freestanding frame. The sample cell is in the lower housing, and the electronics are in the upper one. This process analyzer system meets the National Electrical Code (NEC) Class I, Group D, Division 1 (ID1) standards for hazardous equipment.

Common Applications

A major application for the freeze point analyzer is in monitoring aircraft jet fuels. Because of the very low temperatures encountered at the high altitudes where jet planes fly, the freezing point of the jet fuel is critical. By reducing the production of off-spec fuel, the analyzer installation can pay for itself in a matter of months.

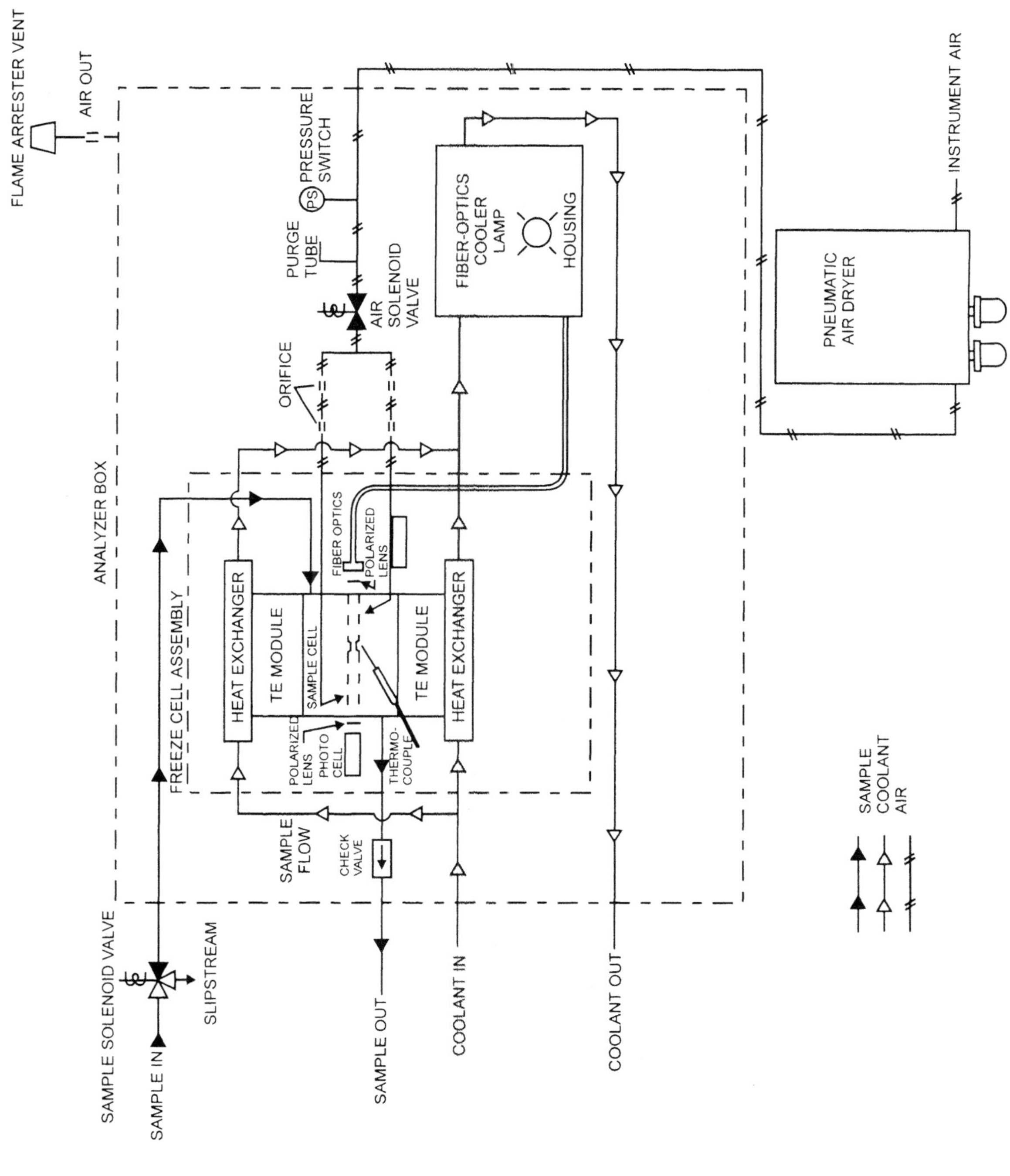

Figure 13-2. Precision Model 42680: Flow Diagram
(Courtesy of Precision Scientific Petroleum Instruments Co.)

Ranges of Variables

The Precision freeze point analyzer can measure freezing points from –67 to 10°F (–55 to –10°C) with coolant at 50°F (10°C). The analyzer completes each measuring cycle in 3 min.

Options and Accessories

A sample-recovery system to return waste sample to the process line is recommended. An eductor system should not be used because the spent sample must be discharged from the sample cell at atmospheric pressure. There should be provisions for introducing a standard sample into the analyzer for calibration. Precision Scientific can supply suitable calibration and sample recovery systems for its analyzers.

Limitations

Sample temperature at the analyzer inlet should be between 50 and 100°F (10 and 40°C). A cooler may be required. The cooling water to the Peltier (TEM) units should be 40°F (5°C) or lower. Precision Scientific can supply an explosionproof water chiller for this purpose.

Installation Hints

Freeze point analyzers should be installed indoors or in a climate-controlled analyzer building to shelter them from the elements and to facilitate maintenance. Since ambient air temperature should be maintained between 40 and 90°F (5 and 30°C), heaters or air conditioners could be needed in the buildings.

Specific Safety Issues

Freeze point analyzers are constructed to meet NEC requirements for hazardous areas. However, they should be housed in buildings away from such areas, if possible, and ventilated with a fan to remove any combustible vapors. The analyzer should be purged with filtered, dry –40°F (–40°C) instrument air or nitrogen to prevent the accumulation of moisture on supercooled components.

Unique Sampling System Requirements

A fast-loop sampling system is required to minimize the dead (lag) time in getting the sample to the analyzer.

Typical Calibration Requirements

Provision should be made to introduce a calibration sample into the analyzer periodically. Since the reproducibility of the process analyzer is much better than that of the ASTM D-2386 laboratory test, the calibration sample should be analyzed four to five times in the laboratory and the results averaged for comparison with the value determined by the analyzer.

Troubleshooting and Maintenance

The Precision freeze point analyzer has a microprocessor that provides versatile and highly reliable performance. It should require a minimum of maintenance, but a periodic examination is necessary to ensure proper cycling. Consult the manufacturer's operation and maintenance manuals for detailed maintenance procedures.

About the Author

Conard Donnell received a Bachelor's and a Masters degree in Chemistry and Physics from Temple University. His career with automatic process analyzers began and ended at Sunoco Refinery where he developed on-line instrumentation to measure the physical properties of process streams automatically and much faster than laboratory analysis. Since retirement, Mr. Donnell has worked as a consultant for Sunoco, Fluid Data, Precision Scientific Co., Lockwood and McLorie, and DuPont working on projects that included design and installation of process analyzer systems in refineries worldwide.

14

Refractive Index Analyzers

Light incident on a solid object, a liquid, or a gas is either reflected, totally absorbed, or partially reflected and/or transmitted. The response of each is a characteristic property of the media involved. In addition to varying with the medium, the response changes with the concentration of dissolved solids. Study of this property of light has developed into the science known as refractometry.

A refractometer determines the concentration of a solution by measuring the optical refractive index. Ernst Abbé is credited with having invented the refractometer in 1874 when he published a description of an apparatus for determining the refractive index of solids and liquids.

Note:
Suspended solids are different from dissolved solids.

This property of light is illustrated in many simple effects. One of the most common is the apparent bending of a rod in a glass of water. This bending illusion is due to distortion of the light rays passing through the interface of water, glass, and air. The amount of the apparent bending is characteristic of the air/water interface and the amount of dissolved solids (Figure 14-1). It should be noted that such "bending" is not affected by the presence of bubbles and suspended solids in the water. Other examples of light bending include mirages in hot deserts, distortion of objects behind heat sources, and color changes in large bodies of water.

This property of light transmission has been studied for many centuries. The laws governing the reflection and refraction of light were studied by Willebrod Snell, a professor of mathematics at Leyden University. He was able to define the "blending" of light in the form of an equation known as Snell's law:

$$\frac{\sin i}{\sin r} = N_v = 1.00027$$

where i and r are the angle of incidence and angle of refraction of a beam passing from a vacuum into the plane surface of the medium. For ease, it is an accepted practice to refer to air as a standard for refractive index rather than to vacuum. The refractive index relative to air is denoted by n and is related to N_v by (Figure 14-2):

$$\frac{N_v}{n} = N_{air} = 1.00027$$

Light travels at different speeds in different media. The more dense a medium, the slower the speed of light in that medium. When light passes from one medium to another at any angle other than 90° it not only changes speed, it also changes direction at the boundary between the two media.

The refractive index (n or n_D) of a medium is defined as the speed of light in air divided by the speed of light in that medium. This method is referred to as the transmission method of determining refractive index. It is useful in the laboratory, but not for process measurements.

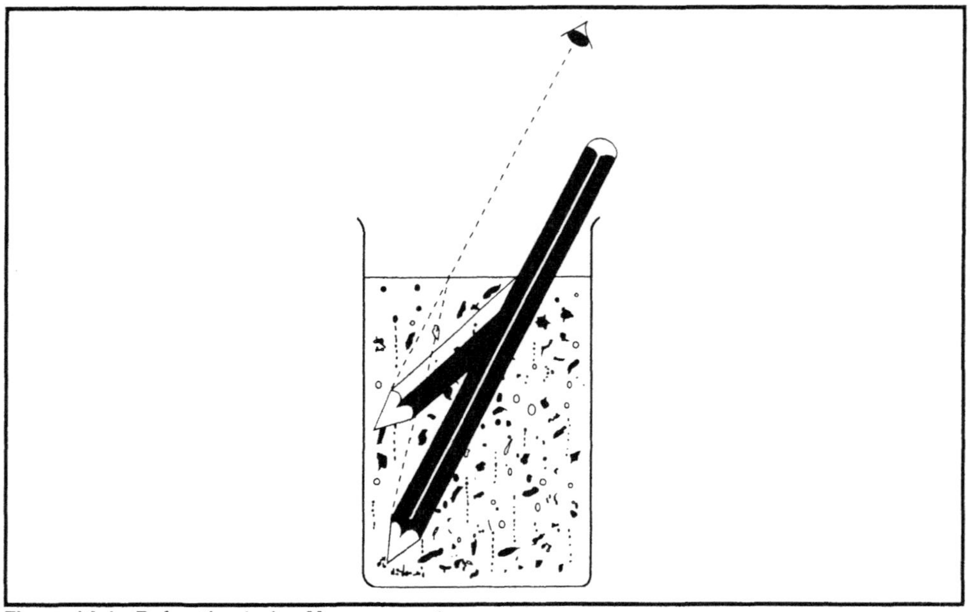

Figure 14-1. Refractive Index Measurement
(The amount of apparent bending is characteristic of the air/water interface and the amount of dissolved solids. "Bending" is not affected by bubbles or suspended solids.)

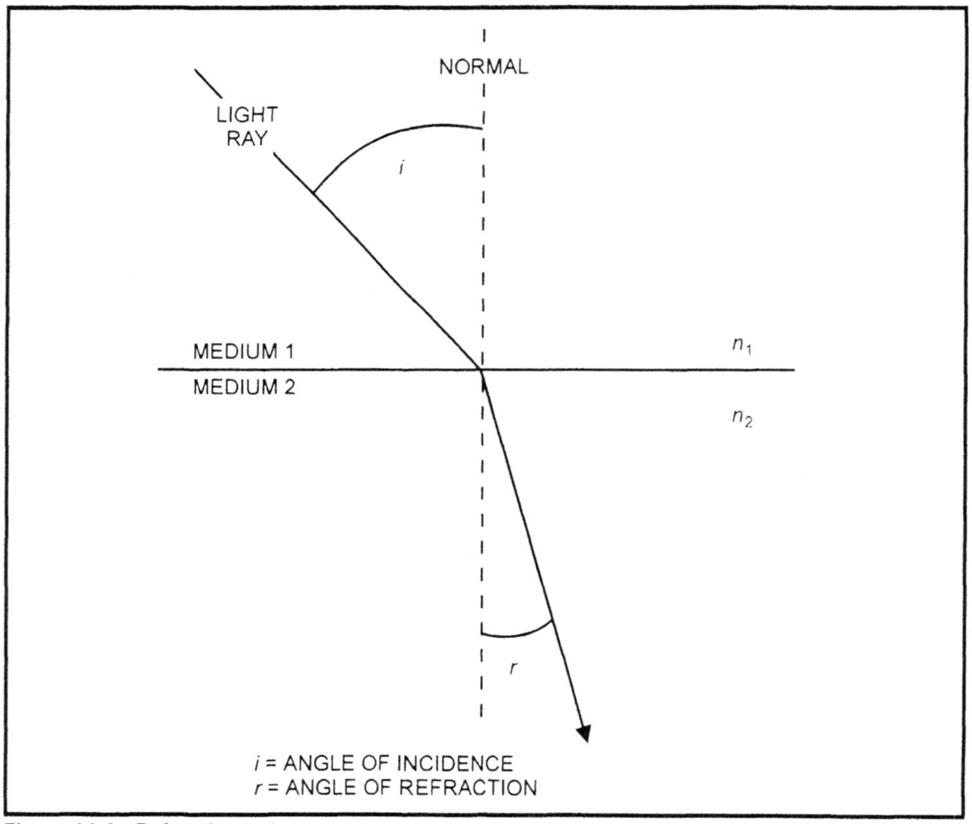

Figure 14-2. Refractive Index

Instrument Operation: Internal Reflection and Critical Angle

When light passes from one medium to a more optically dense medium, both reflection and refraction will occur for all angles of incidence. Starting with a small angle of incidence, there will be a weak internally refracted ray and a strong externally refracted ray. As the angle increases, the angle of refraction also increases; at the same time, the intensity of the reflected ray becomes stronger than that of that of the refracted ray. Finally, at the "critical angle" of incidence (c), the angle of refraction becomes 90°. Since it is impossible to have an angle of refraction greater than 90°, it follows that for all angles of incidence greater than the critical angle, c, the light will experience total internal reflection. This is expressed mathematically as:

$$\frac{\sin c}{\sin 90^\circ} = \frac{1}{n} \text{ (air to glass)}$$

$$\sin 90^\circ = 1$$

Therefore,

$$n = \frac{1}{\sin c}$$

The critical angle, c, is the transition point from total reflection to where refraction begins.

In order to utilize this principle to make a useful process measurement, the reflected/refracted image must be created, focused, and projected onto a measuring element. This measuring element is a prism. As shown in Figure 14-3, the light from source L is passed into the prism in contact with the process fluid. The light contacts the prism/process interface at various angles. The rays A are totally reflected, and the rays B are refracted. The point C defines the transition point between total reflection and refraction; this is the critical angle and defines the refractive index of the process fluid.

As the concentration of dissolved solids in the process fluid changes, the refractive index changes. As a result, the position of the critical angle will change. An increasing concentration of dissolved solids increases the amount of refraction and reduces the amount of reflection. As the amount of dissolved solids decreases, the amount of refraction decreases and the amount of reflection increases. From these changes in the optical image, the measurement is made by conversion to an electrical signal. The form of this signal conversion is what differentiates the various types of process instrumentation.

An increasing concentration of dissolved solids increases the amount of refraction and reduces the amount of reflection.

Installation Considerations

This section outlines the necessary application considerations for using a process refractometer. It is presented as a general guideline for anyone wanting to use refractive index for process measurement.

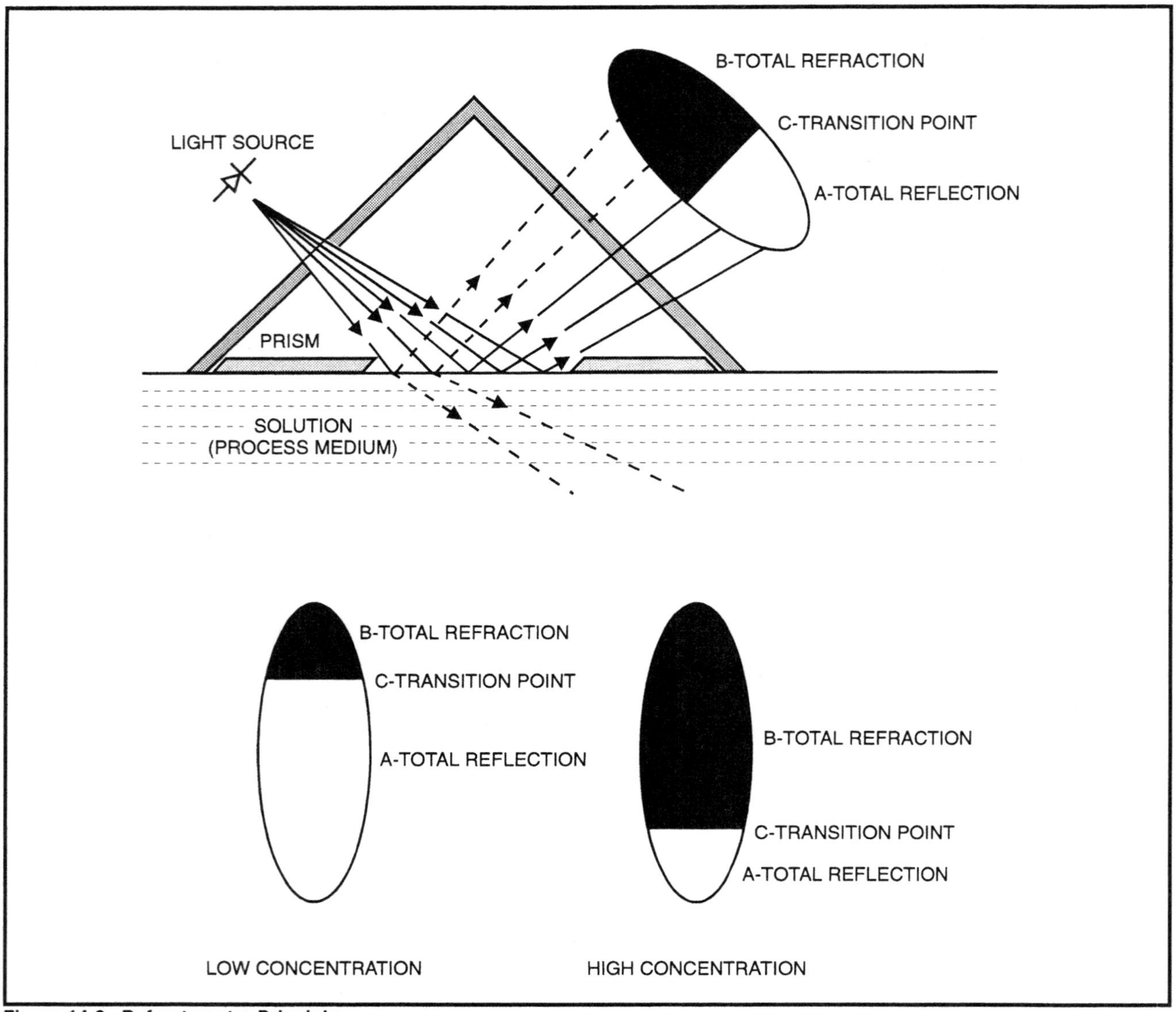

Figure 14-3. Refractometer Principle

Temperature Compensation

Temperature is a very important parameter in order to properly relate concentration to refractive index using a process refractometer (Figure 14-4). In general, refractive index is inversely related to temperature. Typically, an increase in temperature of 1C° will decrease the refractive index by 0.0002 RI units. This may appear to be a very small change, but in terms of concentration the change is about 0.1 wt%. Over a measurement span from 0 to 10 wt%, this can yield an error of 1% of span. If the temperature is incorrect, the final concentration will be incorrect. The final amount of error will be directly related to the inaccuracy of the temperature input under steady-state conditions.

Because of the significant relationship between temperature and refractive index, the location of the temperature-sensing element is critical for proper process measurements.

Because of the significant relationship between temperature and refractive index, the location of the temperature-sensing element is critical for proper process measurements. The optimum location of the sensor is in the process. Temperature profiles across the pipe make it undesirable to locate the medium's temperature-sensing element close to the pipe's outer wall.

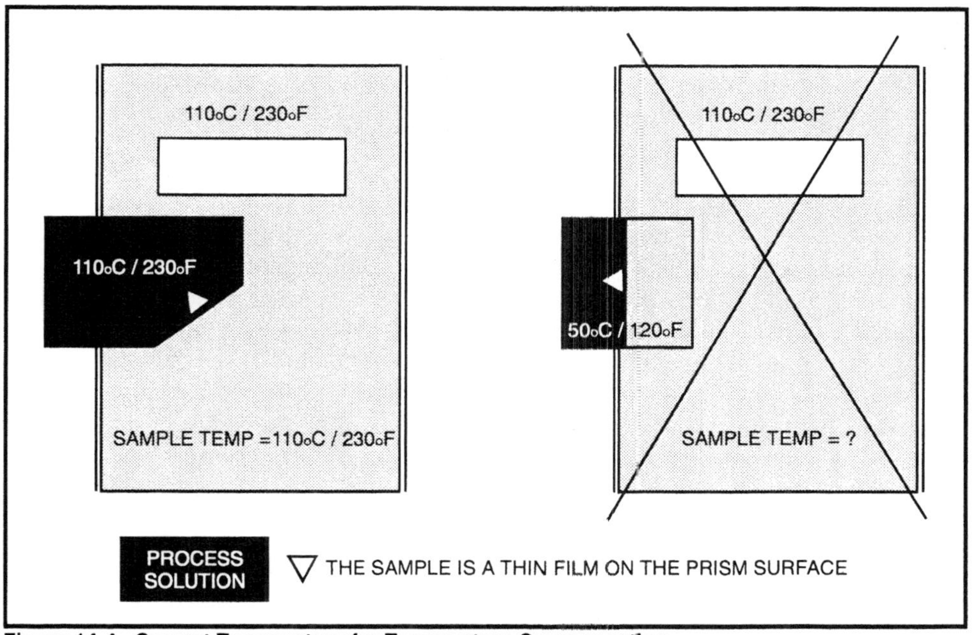

Figure 14-4. Correct Temperature for Temperature Compensation

An external temperature input can be used if the instrumentation design does not allow the temperature sensor to be located in the process. However, the temperature at the prism, where the actual measurement is being made, will still be different due to the temperature gradient of the pipe. This may still result in errors due to incorrect temperature compensation.

The error related to incorrect temperature measurement can be significant. A difference in temperature of 50C° can translate into an error of 0.001 RI units, or about 5 wt%. The table below relates temperature to the potential errors in refractive index and, in direct proportion, concentration:

Temperature Differential, C°	Error, %
5	0.5
10	1
20	2
50	5
100	10

Instrumentation Construction

Process refractive index instrumentation consists of several components: transmitter, sensor, and readout device. In some designs, several of the components may be packaged together. All known designs, however, include these three elements.

There are two major design philosophies: integral and discrete component (separate) design. Maintenance and repair are easier on separate components. If required, the transmitter can be moved to a safe location and the sensor need only have a hazardous rating. (This will be covered in more detail later). Not all process piping is at ground level or accessible. A component-type design allows the transmitter to be located conveniently for recalibration, while the sensor is located in the process piping. The typical distance of maximum separation is about 300 ft

(90 m). Sometimes it is desirable to remove the sensor under process flow. A separate design allows this with the use of an isolation valve. The separate design offers many advantages for industrial installations. Proper interconnect-cable choice is important in process plants due to EFI/RFI considerations.

SENSORS

The measurement of concentration using refractive index is a laboratory technique that has made the transition into process measurement. This transition has required that laboratory sensors be industrialized. Sensors need to be constructed from material that can survive the process fluid, such as stainless steel, exotic alloys, isotropic mineral crystals or single-element crystals.

The sensor and its material of construction must be able to operate under the process conditions. The sensor's limitations are important considerations. If the sensor is unable to operate at the process temperature and pressure, a special installation will be necessary. If the temperature limits of the sensor are exceeded, cooling will be required. This makes the process measurement dependent on the supply of coolant, which raises questions as to instrumentation reliability. Coolants are best avoided, if possible.

Coolants are best avoided, if possible.

Since the refractive measurement is based on light, the light source is another important component of the refractometer. The light should have the following properties for superior performance: long lifetime, narrow spectrum, and homogeneous distribution.

A long lifetime can be accomplished using a light-emitting diode (LED) source. An LED is a solid-state device that, unlike an incandescent light source, will not degrade.

An LED also has a narrow spectrum. The typical wavelength spread of an LED light source is about 100 nm. The refractive index line image quality suffers with broad spectrum light sources, such as filament bulbs, due to dispersion effects. A narrow spectrum will yield a sharper image than a broad spectrum light source (Figure 14-5).

An LED light source has a more homogeneous light distribution than does a filament lamp. This has the benefits of both a consistent regular image and an increased signal-to-noise ratio of the refractometer.

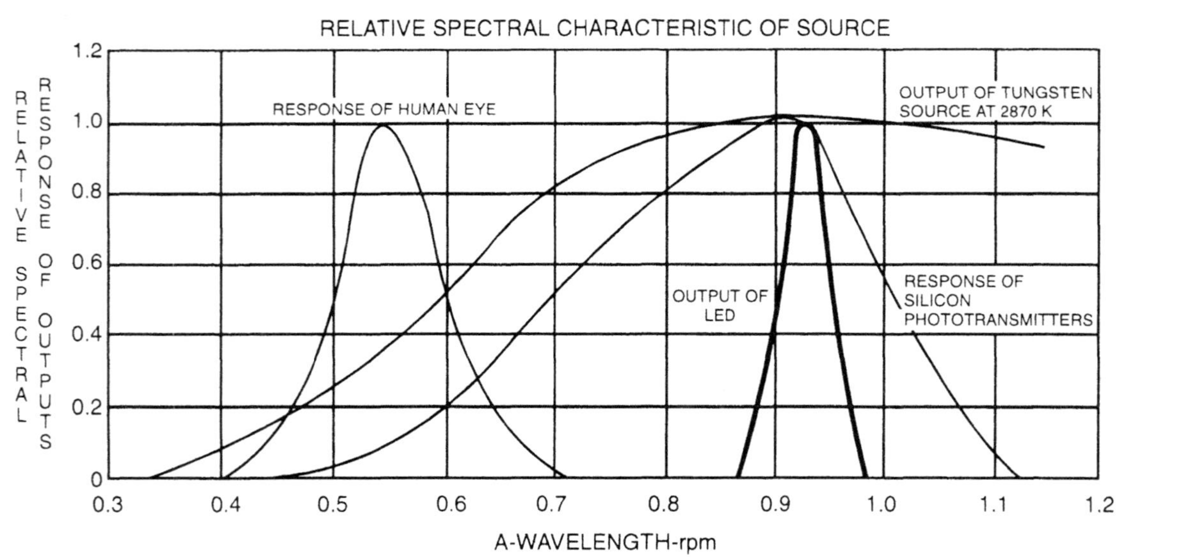

Figure 14-5. Typical Characteristics of Various Light Sources

Optical Prism

The material and design of the prism can influence the operation of the refractometer. The material must be optically-clear to allow the transmission of light, chemically-resistant to prevent damage, mechanically-strong to resist abrasion while in the process, and commercially available. Glass, for example, is optically clear and chemically resistant, but is not very mechanically hard (with a Rockwell hardness of 5.5). This makes it unsuitable because of its tendency to scratch and abrade.

Two materials that have all the desired properties are sapphire and spinel. Both materials are optically clear and have excellent chemical-resistance properties. Sapphire has a hardness factor of 9, slightly higher than spinel's hardness factor of 8. Spinel is a lower-cost, man-made material that has an isotropic nature; all of its properties are equal in all directions. Spinel's isotropic nature is attributed to its cubic crystal structure. Sapphire can also be man-made, but it is not truly isotropic because its crystal structure is hexagonal. Its hardness may vary with orientation of the crystal lattice. Thus, spinel offers advantages over sapphire.

Prism design also dictates that the number of optical surfaces be minimized. Using the least amount of optical surfaces reduces light losses and eliminates backscatter interferences. The ideal number of optical reflective surfaces is one.

Outputs

Instrumentation, depending on its sophistication, will provide several types of outputs. The most common outputs are direct readouts along with an analog milliamp current (4-20 mADC) output. The development of microprocessors has increased the number of output options. The microprocessor has given the instrumentation the capability to display engineering units as well as transmit data via the RS-232/422/485 communication ports. These features increase the utility of the measurement.

Sensor Coating

Coating of the sensor's prism by process fluid is a variable that can impair the measurement made by a refractometer. Prism coating can be controlled by prism washing systems, either water or steam. Steam is used if the monitored liquid is particularly sticky or tenacious. In an application where the liquid has the potential to "bake" onto the prism, such as starch, high-pressure water is preferred over steam.

It is very important that the prism wash be consistently applied.

During washing periods, it is preferred that the instrument reading remain in a "sample-and-hold" mode in order to prevent a false reading. This is also true of either of the outputs: analog or digital. The benefit of any washing system will be negated if this is not a feature of the analyzer system.

Self-Diagnostics

Any instrument that checks itself or provides the operator with an easy method of doing so is highly desired. In many cases, special test instruments may be required to check the analyzer's operations. Operating conditions do not always allow for the use of special equipment. If all self-diagnostics are contained within the analyzer system, downtime can be minimized. This can translate to increased production, saving both time and money.

Intrinsic Safety Requirements

The location or nature of the process fluid can require that the instrument be intrinsically safe. Intrinsic safety is accomplished through the use of a zener optical-isolated barrier. Such barriers prevent the possible surge of high voltage to the sensor from the transmitter. Systems can then be used in hazardous areas.

Accuracy, Repeatability, Stability, Measurement Method, and Circuitry

The accuracy required by an application is a function of the relationship between refractive index and concentration of the process fluid. The measurement accuracy of refractive index is a function of the how the instrument makes its measurement. The instrument's overall accuracy will depend on the interaction between these relationships and how the technique employed is influenced by suspended solids and bubbles.

The production of the image created by the refraction and reflection of light is what must be converted to an electrical signal. In the early 1950s this image was first analyzed by the relative intensity of light incident upon a photoelectric cell, that is, a light meter. This technology continues to be used in some present-day instrumentation.

This light-intensity technique can be influenced by suspended solids and bubbles. Light incident upon suspended solids and bubbles can be reflected and backscattered into the measuring area. The presence of these interferences may occur randomly and cannot be "zeroed" out by a calibration offset. Therefore, the best technique to reduce this error is to use a technique unaffected by these interferences.

Refractive index itself is not influenced by suspended solids or bubbles. An instrument that measures by digitizing the optical image truly determines refractive index and concentration. This can be done by measuring the image digitally and determining the transition point to total internal reflection or the critical angle (Figure 14-7). Interferences caused by backscattering from suspended solids and bubbles are ignored because they occur outside the critical angle region. High instrument-scan rates can be used to validate this technique by analyzing a sufficient amount of data to give statistical significance to the measurement. This advance to digital signal processing yields a high accuracy and stable dissolved-solids concentration measurement unaffected by typical interferences.

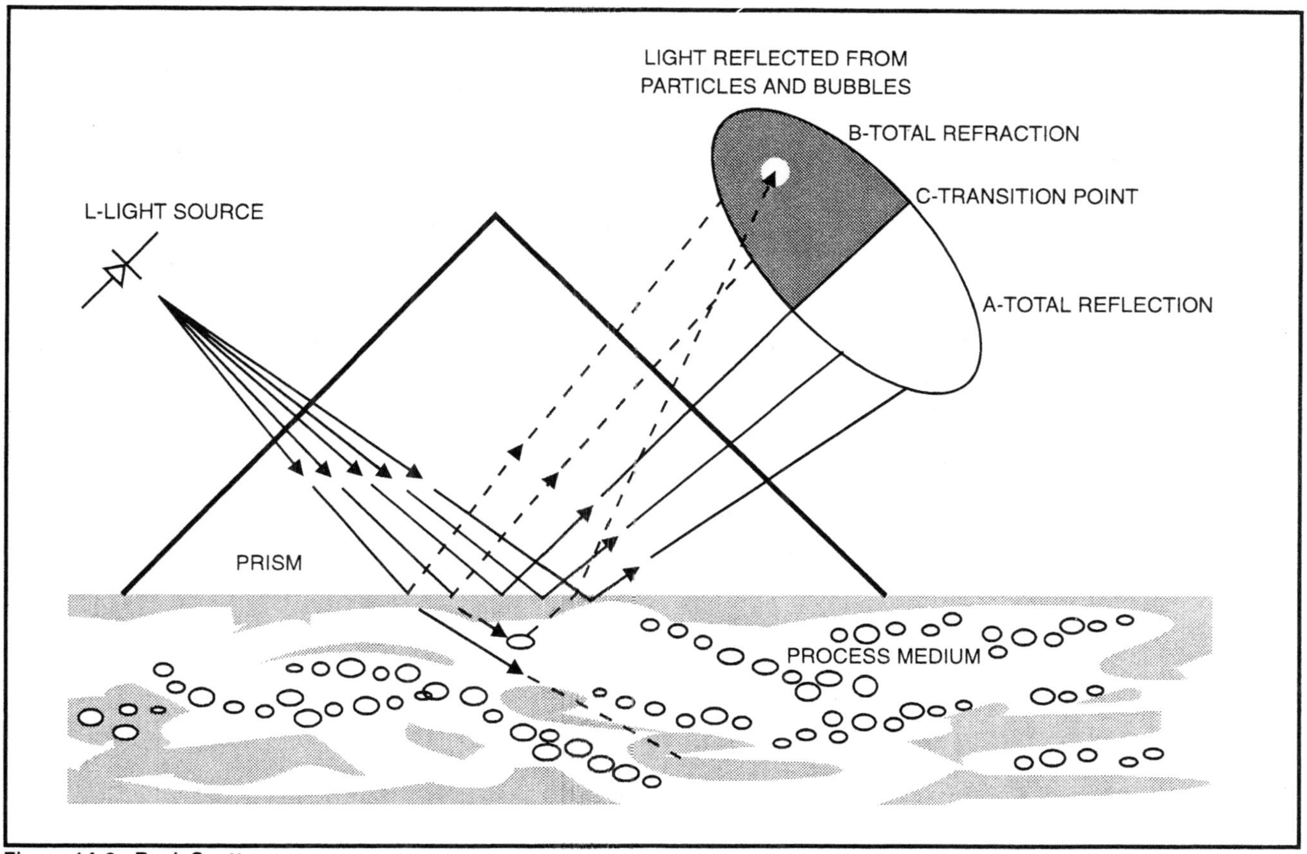

Figure 14-6. Back Scatter

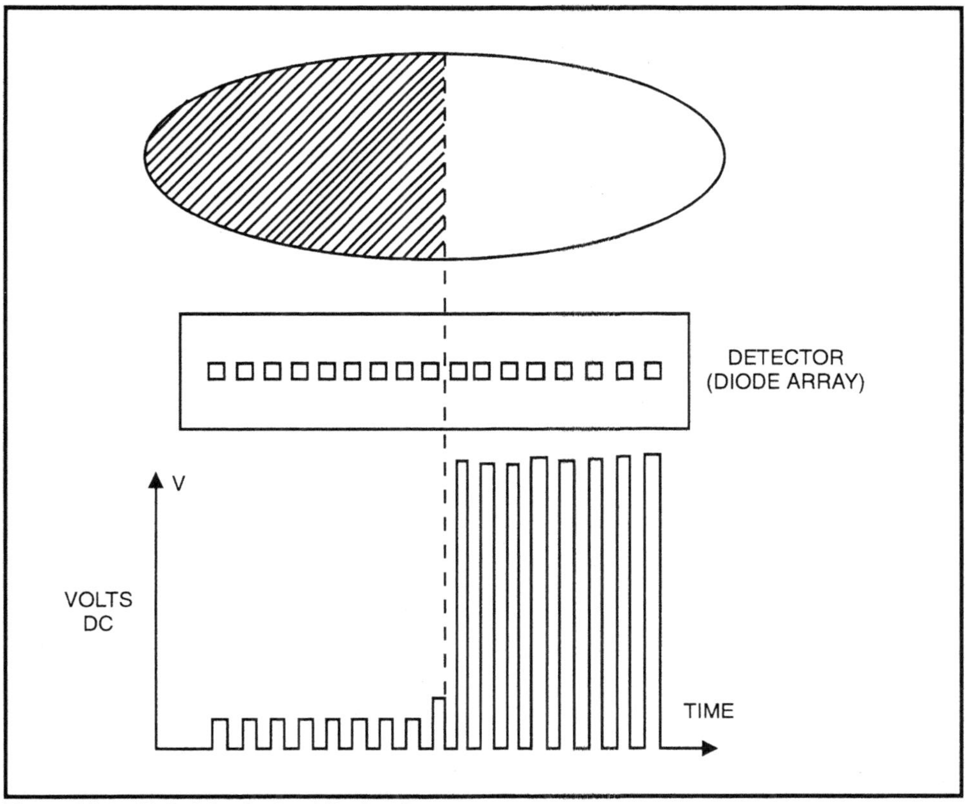

Figure 14-7. Digital Analysis of Optical Image

A well-designed and properly applied process refractometer can measure refractive indices over a wide range with high accuracy. A preferred range of operation would be a range of refractive index between 1.3 to 1.6, the range of typical process fluids. In this range, the resolution of 0.0002 RI units would yield an accuracy of 0.1 wt%. The table below relates instrument resolution to typical accuracy:

Resolution, RI	Typical Accuracy, wt%
0.0001	0.05
0.0002	0.10
0.0010	0.50
0.0020	1.00

Table 14-1 provides some insight into the design features found in different manufacturers' products.

Table 14-1. Comparison of Competitive Design Features

Features	Manufacturer				
	A	B	C	D	E
Sensor					
Prism Mounting	Flat	Flat	Angled	Flat	Flat
Prism Material	Spinel	Sapphire	Spinel	Sapphire	Sapphire
CCD Image Detector		X	X		X
Image Detector via Fiber Optics			X		
LED Light Source		X	X	X	X
Standard Process Connections		X	X	X	
Humidity Indicator in Sensor			X		
Prism-Cleaning System	X	X	X	X	X
Temperature Sensor at Prism		X	X		
Hazardous Area Approval			X		
Indicating Transmitter					
Separate Transmitter and Sensor	X	X	X	X	
Complete Diagnostics			X		
Display Optical Image			X		
Nonlinear Temperature Compensation			X		
Menu-driven Software Available		X	X		X
RS-232 and 4-20 mADC outputs	X	X	X	X	X
Outputs Isolated		X	X	X	X

Industrial Applications

Industrial applications using refractometers usually involve using refractive index to determine some component's concentration while in-line or to determine a process variable that has some critical importance to the process. The applications are only limited by the fact that the measured component must be a dissolved solid or an aqueous solution, acids and/or bases.

The determination of refractive index itself is typically not the goal of the in-line process refractometer. This extracted concentration information can be used as the input to a distributed control system and feedback control implemented in the production process.

Accuracy and repeatability of this measurement are important in control applications. By accurately measuring and using process control, a reduction in the standard deviation (σ) from the mean or setpoint is typically achieved. The magnitude of the reduction will depend on the process variable and how accurately it can be measured.

How can improved accuracy improve production? When producing a product, the manufacturer has a setpoint or specification concentration that they are trying to maintain. A normal bell-shaped distribution will exist, centered around this setpoint. The manufacturer also has an absolute lower limit that they do not want to fall below in 98% of all possible cases. This 98% represents 3 standard deviations (σ- sigma) from the setpoint. If the product is expensive, or if the final customer cannot accept a lower concentration, this lower setpoint is very important. The operating setpoint must be increased so that it is higher by three times the σ value from the lowest acceptable value of the original average. This translates to the fact that an over- or under-specification product is being produced.

This is best explained with an example (Figure 14-8). If the concentration of the measured component is 10% and the accuracy of variation of the measurement is 1%, a normal bell-shaped distribution can be drawn around 10%. The 3 σ points in this case are 9% and 11%. This means that some of the time a concentration that is too low is being produced and other times a concentration that is too high is being produced. Since production cannot tolerate any production below 10%, the actual operation distribution must be a bell-shaped curve that has 98% end points (3σ) at 10% and 12%, with an average of 11%. If an improvement in accuracy will generate a steeper measurement distribution, a bell curve that has 98% end points (3σ) at 10% and 11% is produced. The new average will be 10.5%. This is a net reduction of 0.5% without sacrificing the product's specification! Product savings of this type can be used to justify the cost of improved equipment.

Refractometers are used in many industries. The applications are limited only by the fact that the measured component must be a dissolved solid, aqueous solution, acids, and/or bases.

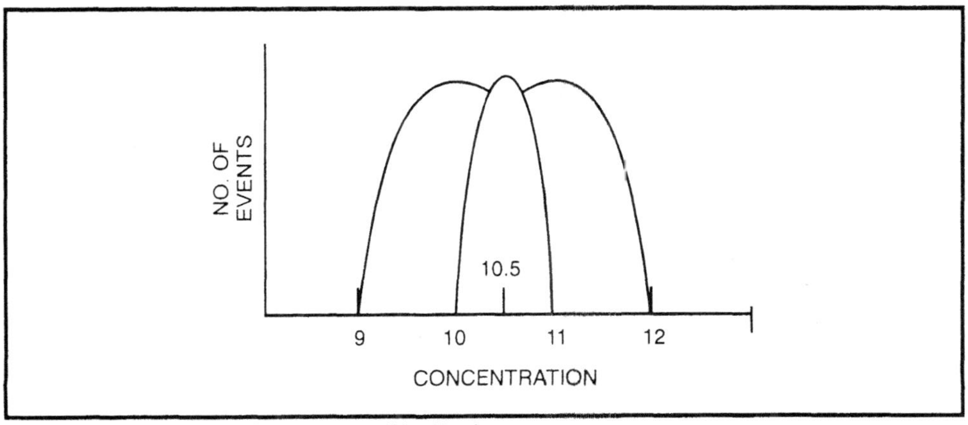

Figure 14-8. Statistical Measurement Distribution

Sugar/Sweeteners

Sugar is typically produced from either beets or cane. Refractometers have been used widely in both categories. The sugar industry has its own measuring scale, called Brix. The Brix scale is based on the refractive index of sugar (glucose). Applications are encountered all along the sugar production process.

An important application for refractometers is beet sugar crystallization. The refractometer is used to determine the seeding point and to control the drop in the mother liquor concentration after seeding. Fine control is important here to control the shape of the sugar crystals. If saturation is too low, crystals will not form; if saturation is too high, crystals will grow too fast and have irregular shapes.

Other refractometer applications in the sugar industry include:

Beet Sugar

- Diffuser control
- Press water
- Evaporation control
- Green syrup and molasses
- Desugarization of molasses

Cane Sugar

- Affination
- Decolorization
- Evaporation
- Vacuum pan
- Recovery pan

The production of sweeteners such as glucose syrup, dextrose syrup, fructose syrup, and sorbitol also uses refractometer control.

Refractometer applications in the sweetener industry include:

- Starch-sweeteners process
- Ion-exchange control
- Evaporation control
- Crystallizer for dextrose and fructose syrup
- Separation of fructose and glucose

Pulp and Paper

In the pulp and paper industries, refractive index is used to measure key dissolved solids in the black liquor going to the recovery boiler. This is an important measurement. It is a recommended practice that the solids content of the black liquor not fall below 58%. If the solids content does fall below this setpoint, it is recommended that the black liquor stream be diverted. This is typically done with an automatic divert valve. In order to prevent coating, a steam wash is used. The use of refractometers is well established in this application.

Another major application is paper sizing. Various methods are used to improve the properties of paper. The treatments can be divided into internal sizing, surface sizing, and coating. The refractometer is used in the preparation and addition of the sizing and coating liquids. The most common material used in sizing solutions is starch. Typically, a relatively low-concentration starch solution is used (5 to 15 wt%). High temperatures will cook the starch to the lens; high-pressure water is used as the wash media.

Other refractometer applications in the pulp and paper industry include:

- Green liquor
- Washing control
- Evaporation
- Size-press control

Chemicals/Petrochemicals

Control of a chemical's final concentration is important and can be very difficult. Chemical producers and refineries do not typically run consistent feedstocks or crude blends. Therefore, the individual process units constantly see variations in their feeds, and those variations can often be measured by refractive index. For example, the final viscosity of lube oil is related to the refractive index. If the lube oil production is not within its viscosity specification, an in-line refractometer can alert the process operator. Process conditions such as lower temperature and the feed ratio can be changed to compensate for these changes. The feedback and response time is reduced from hours via laboratory measurements to minutes via on-line process analyzers.

Other refractometer applications in the chemical/petrochemical industry include:

- Polyamide
- Formaldehyde
- Hydrogen peroxide
- Sulfuric, Nitric, and Hydrochloric acids
- Urea
- Sodium hydroxide
- Ethylene glycol
- Ethanol
- Ammonium nitrate
- Monosodium glutamate
- Polycarbonate
- Caprolactam
- Glycerin

Food

Most applications in the food industry involve measuring dilution or concentration. Refractometers also measure the sugar content of beverages and soft drinks. It is important to note that if the measuring technique is influenced by bubbles, this technique may be undesirable, especially if the measurement is to be made after carbonation.

Refractometer applications in the food industry include:

- Beer blending
- Beverage and soft drink production
- Jam and marmalade processes
- Tomato paste
- Fruit juice
- Rum distillation

Bibliography

1. Clevett, Kenneth J., "Measurement of Refractive Index," Chapter 17 of *Process Analyzer Technology*, J. Wiley & Sons, 1986.
2. "Safe Firing of Black Liquor in Black Liquor Recovery Boiler," Black Liquor Recovery Boiler Advisory Committee, Recommended Good Practice, August 1962, revised April 1991.

About the Author

John G. Groetsch, Jr. is currently the General Manager for K-Patents, Inc. and directs the sales, marketing, service and technical support of the K-Patents process refractometer for the US and related markets. The device is used to measure concentration of fluids in the chemical, food, pulp and paper, and sugar industries. He has a B.S. degree in Chemical Engineering from the University of Maryland and is an active member of both ISA and AIChE. Related experience includes marketing of process instrumentation, process engineering in the refining and chemical process industry, and research and development of metallurgical processes. He has written multiple technical publications/presentations along with being responsible for patents in these areas.

15

Thermal Conductivity Analyzers

The ability to conduct heat at a unique rate is one of the oldest methods used for gas analysis. Although this method has been overshadowed by more recent techniques, it is still used in several applications and is a cost-effective method of testing for gas purity. Thermal conductivity analyzers are relatively easy to install and maintain, and typically require little attention. This type of analyzer is simple, with excellent stability, rapid response, and high sensitivity.

This type of analyzer is simple:
- **excellent stability**
- **rapid response**
- **high sensitivity**

The British developed the first instrument for measuring the thermal conductivity of a gas about 1915. In the 1930s and 1940s, thermal conductivity filament detectors became the principal detector for gas chromatographs. Several technical refinements have been made to the electronics, the detectors, and the sampling systems, but the basic principles and method of operation have remained the same.

Measurement Principle

One of the most basic properties of a gas is its ability to conduct heat. The thermal conductivity analyzer uses the thermal conductivity of a gas (in the form of heat loss) to measure composition.

There are three types of heat loss by gases: thermal conductivity, convection, and radiation. The loss of heat by convection and radiation can be reduced to negligible proportions by proper analyzer design, and in most cases will contribute less than 1% to the analysis. Therefore, thermal conductivity of the gas alone can be measured. Each gas has a different thermal conductivity; by comparing this with a reference gas, a quantitative analysis can be made. Studies have shown that the thermal conductivity of a gas varies inversely with the square root of its atomic weight. Also, the value varies with density and viscosity of the type of gas present.

A typical analyzer consists of two or four detectors arranged in a Wheatstone bridge configuration (Figure 15-1). The detectors are located in individual, sealed, metallic cavities and are exposed to the reference and active gases. An instrument amplifier takes the signal generated by the Wheatstone bridge and passes it to a meter or other indicating device.

To perform an analysis of a specific gas, an analyzer uses two detectors. The reference detector is used to measure the conductivity of a reference gas, while the active detector is used to measure the conductivity of a sample gas (the gas to be analyzed).

Typically, the detectors are electrically-heated wire filaments or thermistors, and the analyzer measures the change in resistance of these wires. When a gas passes through a detector, heat is transferred from a hot element to the cavity wall. Depending on the gas, more or less heat will be transferred, which in turn will cause the detector to heat up or cool down relative to its initial state. The reference detector will "see" only the reference gas, which is nonchanging, and its output should be constant. The active detector will "see" the varying sample gas, and its output should vary according to the thermal conductivity of the sample gas. If the reference and sample gases are the same, then the reference and active detectors will be at the same temperature and their outputs will be the same; if the sample gas is different, then a difference will occur between the active and reference outputs. The measured differences in the detector outputs are proportional to the changes in the thermal conductivity of the sample gas. These can be scaled electronically to give a quantitative measure of the gas being analyzed.

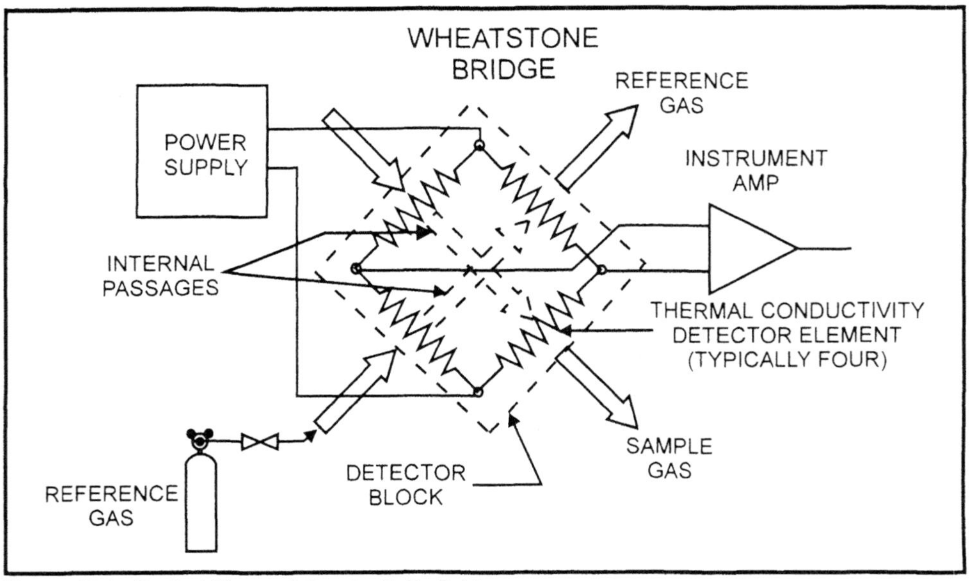

Figure 15-1. Typical Thermal Conductivity Bridge

Detectors

Detector filaments are the heart of thermal conductivity analyzers and the major detection system of most chromatographs. Two major types of thermal conductivity detectors are in use today: hot-wire sensors and thermistor-bead sensors.

Hot-wire sensors are coaxial-cylinder coils of wire. Typical wire used may be tungsten, tungsten with 3% rhenium alloy, nickel-iron alloy, platinum, or nickel. Depending on the application, the wire can also be plated with a variety of materials, including nickel, gold, and glass. The wire can be operated with currents between 150 and 600 mADC. Coil temperatures may vary between 500 and 1692°F (900°C). (High temperatures may help prevent coating of the wire by heavy-molecular-weight samples; see the section "Maintenance.") Hot-wire sensors have a positive temperature-coefficient-of-resistance. Such sensors are very stable and are often used as four detector elements in a Wheatstone bridge.

Thermistor-bead sensors are small glass devices that enclose oxide materials, such as manganese oxide, cobalt oxide, and nickel oxide. The beads are supported by small-diameter wires. These sensors are "cold devices" that are driven with low electrical currents: a few milliamperes to up to and over 15 mADC. The beads have less mass than hot-wire filaments and thus can change temperature faster; this means faster response to changes in thermal conductivity. Because of the faster response, however, the beads are subject to more noise from changes in carrier gas flow, temperature, and pressure.

Thermistors vary dramatically in resistance for different applications and analyzer manufacturers. The most popular is 30,000 ohms, but 8,000 and 100,000 ohms are often used (as measured at 77°F [25°C]). Normally, thermistors are installed in two-arm Wheatstone bridges. (Hot-wire filaments can generally be adjusted in production to fall within classes for matching of four units within desired tolerances. This is difficult with thermistors due to variations of the beta characteristic's dependence upon the mix in each lot of thermistor production.)

Thermistor bead's have a negative temperature coefficient of resistance (increase in temperature, decrease in resistance); this is the opposite of hot-wire filaments. Thermistors experience a large shift in resistance with a moderate increase in temperature, while hot-wire filaments do not. To help prevent hydrogen "poisoning," thermistors are typically double-coated or "double-dipped" to reduce migration of hydrogen along the leads into the metal oxides. The hydrogen would react with the oxides and cause a zero drift and loss of sensitivity. The upper temperature limit for thermistors is 615°F (325°C).

Detector Blocks

The detectors of the thermal conductivity analyzer are typically located in a metallic block that is elevated in temperature. The blocks are made from nonreactive metals such as stainless steel or nickel, although some manufacturers use brass or aluminum, depending on the application. The flow of the gas through the block is normally maintained such that the desired response time of the instrument is as short as possible yet turbulent flow does not occur at the detectors. Some manufacturers employ stainless steel sintered filters to ensure laminar flow and cut down on turbulence. Both diffusion and flow-through cavities are used, and each type has minor advantages and disadvantages. The blocks are normally massive compared to the volume of the detector cavity to assure that the temperature of the block remains constant for small variations of the sample gas temperature. The temperature of the block is normally maintained to within ±0.5F° (0.3C°).

Analyzer Electronics

The electronics package used by the basic thermal conductivity analyzer consists of a power supply, an instrumentation amplifier, some type of output display, and some type of output signal conditioner (Figure 15-2). More elaborate types of analyzers are available.

The power supply is typically a constant-voltage-output type that both generates power for the electronics package and drives the detector block. However, a constant-voltage-driver for the detectors can cause damage if the reference/sample flow is interrupted. Some manufacturers have opted for constant-current-bridges, which provide additional protection for the detectors but lose output at high drive levels. Finally, some have added constant-temperature-bridges, which give the most protection to the detectors but which inject additional noise for very low-level signals.

The instrumentation amplifier can be in the form of an integrated package or discrete components. In fact, at least one analyzer consists of only a packaged integrated instrument amp and one discrete transistor (not counting the power supply). The span and zero (calibration) circuitry are normally associated with the instrument amplifier. Some of the high-end analyzers use a dedicated microprocessor along with the instrument amplifier to provide control and data acquisition.

The normal local-output-device for a thermal conductivity analyzer is a mechanical movement meter. However, more manufacturers are adopting digital displays because of their inherently better resolution and long-term reliability.

Most thermal conductivity analyzers provide a 1 to 5 VDC and a 4 to 20 mADC output along with at least two alarm switches (form C contacts). The analog outputs are typically not isolated (floating), a fact that must be considered when installing the analyzer.

Limitations

Two principal limitations of the thermal conductivity analyzer are:

- Only binary gases or samples that can be treated as binary mixtures of gases can be analyzed with present technology.
- Only selected concentration ranges of gases can be analyzed. Consult published information for concentration ranges.

These limitations are very evident in the determination of hydrogen in helium. The theoretical response of hydrogen should be positive (cooling) since its thermal conductivity is greater than that of helium. Actually, an anomalous reading is obtained.

Small quantities of hydrogen in helium appear to have a thermal conductivity less than helium. In chromatography this gives rise to peaks responding in both directions as the concentration of hydrogen is increased — so-called "W" peaks. No complete explanation exists for this effect. Depending on the gases and the application, it is evident that thermal conductivity is not the dominant mechanism of heat loss in this case and that some other mechanism, such as forced convection, may enter into the picture.

No complete explanation exists...

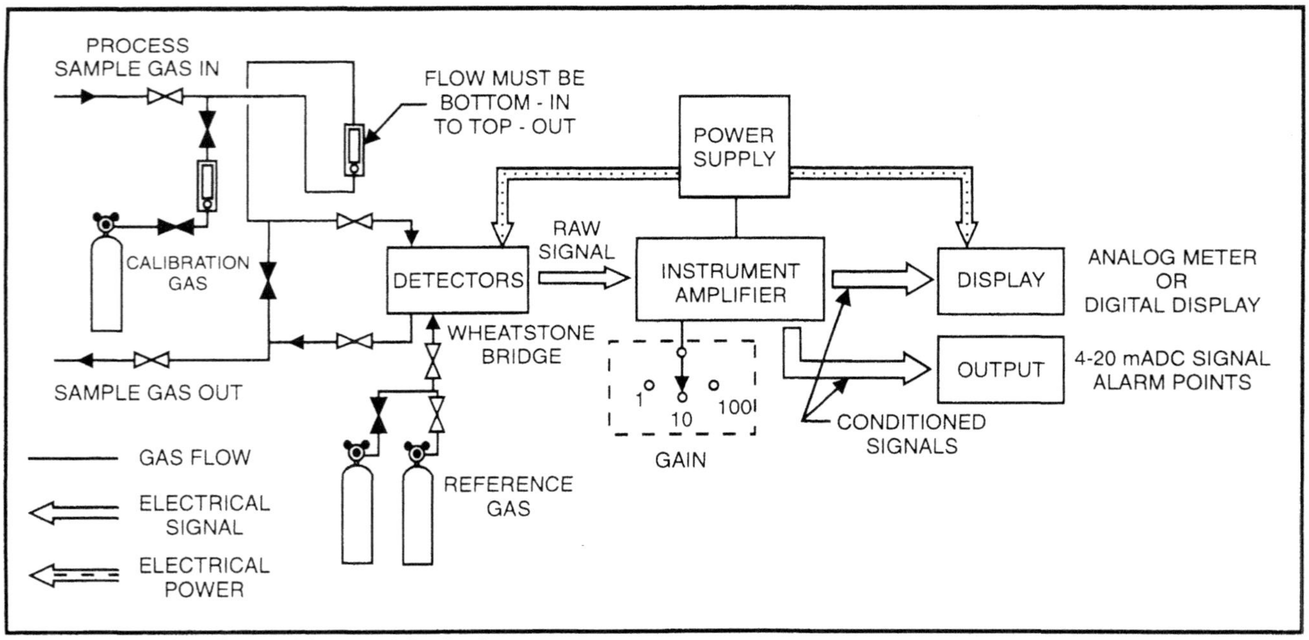

Figure 15-2. Typical Thermal Conductivity Analyzer

Typical Applications

- Purity measurements of binary mixtures of argon, hydrogen, helium, neon, methane, nitrogen, oxygen, carbon dioxide, ethylene, and other gases
- Ammonia plants: hydrogen in nitrogen; ammonia in hydrogen; ammonia in air
- Steel plants: top gas analysis of hydrogen in blast furnace gases
- Chemical plants: contamination control; hydrogen purity; monitoring of purging operations
- Power-generating plants: hydrogen cooling gas in turbine generator housings
- Petroleum refineries: hydrogen in hydrocarbon gases
- Refrigeration: ammonia, chlorofluorocarbon, or carbon dioxide in air

Typical ranges include:

- 50 to 100% hydrogen in hydrocarbon gases
- 0 to 5% ethane in methane
- 0 to 500 ppm hydrogen in argon
- 0 to 3% hydrogen in argon
- 99 to 100% oxygen in argon

These percent measurements can be either volume/volume (v/v) or mole (mol) depending upon the calibration method.

Maintenance

Thermal conductivity analyzers are normally simple to maintain and should require only periodic replacement of the reference gas. Preventive maintenance should be conducted according to the recommendations of the equipment manufacturer. In addition, the following steps will be helpful in maintaining the system.

Reference Gas Replacement. Never allow the reference gas flow (or the active gas flow) to stop while the analyzer is on. If the reference gas distribution system has multiple cylinders, always tie-in the new cylinder and assure that it is supplying gas before blocking off the old cylinder. Purge as much of the tie-in tubing with reference gas as possible so that no air is allowed to enter the system. If only one cylinder is available for reference gas or if system maintenance is to be performed, turn off the analyzer. Reestablish flow of both the reference and the active gases (and completely purge the flow lines) before reenergizing the analyzer.

Detector Replacement. Of course, follow the recommendations of the manufacturer. Check for defective filaments (or thermistors) before installation into the block. Do not touch the detectors or allow them to "bump" the detector block wall, as this may deform them. Once installed, confirm that the detector block mounting nuts are firmly tightened and leak test the outside of the nuts. Do not use "Snoop®" or other types of soapy water to check for leaks in the internal diameter of the detector nuts; this may cause noise on the detector output.

Troubleshooting

If a thermal conductivity analyzer fails, it is assumed that the normal procedures used by today's technicians will find and correct those problems not associated with the detectors and the sampling system. Problems associated with the detectors may not be so obvious or easily corrected.

Grossly Imbalanced Bridge (full-scale deflection of signal). (1) Failure of a detector (open detector): With the power off, check each filament for resistance. (2) Loss of bridge power: Confirm that power is being supplied to the bridge.

Long-Term Drift (hours, days). Confirm that the current being supplied to the bridge is within the manufacturer's suggested tolerances. Current that is too high will cause the bridge to drift in one direction and will lead to premature failure of the detectors. Confirm that no leaks have developed in the analyzer or the sample-conditioning system. A leak may allow the introduction of oxygen, which can attack the filaments.

Short-Term Drift (minutes, seconds). Confirm that the flow rates through the analyzer are constant and that the process conditions are not drifting. Confirm that the reference gas is flowing properly. Pressure swings and changes at the sample point can appear as short-term drift to the analyzer. Confirm that the analyzer electronics are not being subjected to wide changes in temperature. Opening and closing enclosure doors several times a day can cause the electronics to alternately heat and cool. Check the temperature of the detector block and confirm that it is constant (usually within ±0.2F° [0.1C°]). Check the wiring to the temperature sensor in the block to confirm that it is solid and making a good connection.

Poor Quality Signal Output/No Response. Confirm that the detector block is at its proper temperature. A very hot block (1290°F [700°C] for hot-wire filaments and 570°F [300°C] for thermistors) will cause detector sensitivity to drop to zero. Confirm that the flow through the detectors is within the manufacturer's suggested range. Too high a flow rate will not give the detectors sufficient time to "see" any changes and will cause the level of heat loss due to conduction to drop. Too low a flow rate will not give the block a chance to purge itself to "see" new sample changes. Confirm that the block, detector cavity, sample lines, and exhaust lines are not plugged or clogged. Confirm that the bridge is getting proper excitation power. Low or complete loss of bridge power will make the detector circuit go dead. Confirm that the electronics package is set to the correct amplification. Low or no amplification will appear as poor response. Confirm, check, and reset any amplifier jumpers associated with setting of the amplifier gain; over a long period of time these connections may become corroded. As a last resort, open the detector block and check the condition of the filaments. These can become coated or corroded, especially in low-temperature applications or with highly reactive samples. Coating or uniform corrosion can cause loss of response.

Too high a flow rate will not give the detectors sufficient time to "see" any changes and will cause the level of heat loss due to conduction to drop.

Too low a flow rate will not give the block a chance to purge itself to "see" new sample changes.

Signal Output Noise. Confirm that the signal from the bridge is relatively quiet (stable). Noise associated with the amplifier can normally be traced to malfunctioning adjustable resistors, poor connections, or overheated components. Confirm that all connections to the detector bridge are secure and tight. Loose or corroded wires in crimp-on connectors and in terminal blocks can cause significant noise. Confirm that the Wheatstone bridge current is within the manufacturer's suggested range. High current will cause a detector to become noisy before it begins to deteriorate or drift. Check that the sample system has no leaks. Small leaks (very small leaks) can cause small fluctuations in flow, which can cause noise. As a last resort, open the detector block and check the condition of the filaments. These can become coated or corroded, especially in low-temperature applications or with highly reactive samples. Uneven coatings and normal corrosion can cause hot spots on detectors, resulting in noise.

About the Author

Milton L. Jackson has over 40 years of experience in analytical instrumentation. He was active in the initial introduction of chromatographs in the U.S. as an engineer for CEC in the early 1950s. He later became the head of engineering of Process Instruments and Analysis (PAI) which manufactured several different lines of chromatographs and thermal conductivity analyzers. He has worked with several E&C companies as a consultant on process analyzers. He has published several articles on gas analysis and process control, mostly covering applications for chromatographs. He holds a patent on a Multi-Component Thermal Conductivity Analyzer. Mr. Jackson received a B.S. in Electrical Engineering from the University of Texas and currently is the Vice-President and CEO of Analytical Detectors Corporation.

John M. Konopacki has 18 years of experience in industrial instrumentation, mostly in gas processing, chemical, and refining. He has been the president of Analytical Detectors Corporation for 15 years. (ADC supplies thermal conductivity detectors to major chromatograph manufacturers and other physical property analyzer manufacturers. It also provides a variety of other miniature sensors for process analyzers.) He has conducted extensive research in low volume, high sensitivity T/C detectors for packed, megabore, and capillary columns. He has also developed several very low volume, very low noise detectors. Mr. Konopacki received a B.S. in Aerospace Engineering and an M.E. and D.E. in Interdisciplinary Engineering from Texas A&M University.

16

Combustible Gas Analyzers

One of the earliest known areas with a need for combustible gas monitoring was in the tunnels of coal mines. Early on, the mines used a caged canary as a gas detector. If the canary got sick or passed out, it was an indication of bad air. It served as a multigas detector. Although primarily used as a warning of the asphyxiation hazard of reduced oxygen, the canary would respond to any toxic gas. The most common asphyxiant in the mines was an odorless, tasteless, and highly-combustible gas called methane. The canary may not have always provided a very early warning of the presence of a potential (or even actual) combustible mixture, but it did serve to warn of a high concentration of methane or other asphyxiating gas.

Early on, the mines used a caged canary as a gas detector. If the canary got sick or passed out, it was an indication of bad air. It served as a multigas detector.

Another detector used in the mines at this time was the lantern flame. By monitoring the flame's color or temperature (either absolute or changes), an observant miner would have an indication of the air quality and the possible presence of combustible gases. The mines could also conduct routine chemical tests on the atmosphere, but these would not have warned of rapidly changing conditions.

Another detector used in the mines at this time was the lantern flame. By monitoring the flame's color or temperature (either absolute or changes), an observant miner would have an indication of the air quality and the possible presence of combustible gases.

In the 1950s, a catalytic-bead type of combustibles-sensing instrument was developed at the Safety in Mines Research Establishment in Sheffield, England. This has been in use since 1960. This technology had the ability to continuously monitor for and sense predanger levels of most combustible gases and to warn of their presence. The rapid acceptance of this technology and the development of additional sensing technologies indicate that there was (and still is) a great need for reliable combustible gas monitors. See Table 16-1 for a partial list of available combustible gas monitoring equipment.

Table 16-1. Selected Combustible Gas Analyzer Vendors

Company	Bead	IR	Flame	Semi-C
Astro		X		
Capteur Sensors & Analysers Ltd.				X
Control Instruments Corp.	X		X	
Detector Electronics Corp.	X	X		
National Draeger Inc.	X	X		
Enmet Analytical Instruments	X			X
Exidyne Institute Technologies (EIT)	X			
Figaro USA Inc.				X
Foxboro		X		
Gas Tech Inc.	X			
GP-Elliott Electronic Systems Ltd.		X		
Mine Safety Appliances (MSA)	X	X		
Sensidyne	X			
Sieger	X	X		X
Sierra Monitor Corp. (SMC)	X			
Taguchi				X
Teledyne	X			

Measurement Principles

Various types of combustible gas monitoring instruments currently are available that utilize several different methods to measure the presence of the desired combustible gas. No one method is superior in all applications. The total process environment must be considered when selecting the best instrument for a particular application. Besides sensitivity to the desired combustible gas or gases, all potentially interfering or influencing conditions must be considered. These can include other flammable gases, detector-inhibiting or poisoning gases, gas temperatures, gas densities, process area classification, and general environmental factors (humidity, temperature, and wind).

The following is an overview of some of the more widely utilized combustible gas monitoring techniques. Many analytical techniques that are not covered here can also be used for combustible gas monitoring. One example would be a gas chromatograph used to monitor a process atmosphere and programmed to alarm and report combustible materials' concentrations. Although such an installation may monitor and report a combustible mixture, it may not meet specific standards or legal requirements for combustible gas detection instruments. If the instrument is being used to fulfill a legal safety-monitoring requirement, make sure the instrument and installation conform to the legal instrument/installation requirements for your location. A few of the more widely used standards regarding combustibles monitoring will be referred to in the section "Specific Safety Issues."

If the instrument is being used to fulfill a legal safety-monitoring requirement, make sure the instrument and installation conform to the legal instrument/installation requirements for your location.

Methods discussed in this chapter are intended to detect combustible gases and vapors. They are not intended to measure or indicate the presence of either combustible dusts or fibers.

Bead-Type Sensors

The bead-type sensor typically utilizes two beads in a sensing head (Figure 16-1). One serves as an active sensing element and the other as a reference. The active bead is often composed of a small coil of platinum wire embedded in a ceramic bead (typically alumina) that is coated with a catalyst (typically palladium or rhodium). The catalyst coating encourages the oxidation of combustible gases. The reference bead is usually the same construction and the same thermal mass as the active bead, except that it is given an inert coating (such as a thin glass coating) to make it insensitive, instead of a catalyst coating. The coils are then heated, by passing a current through them, to achieve a temperature at which the active bead is considered to be highly effective at oxidizing combustible gases. If any combustible gases are oxidized at the active bead, the heat of reaction should raise the active bead's temperature relative to the reference bead.

The beads are typically monitored by either a thermocouple, an RTD, or a Wheatstone bridge resistance-detector circuit. In the case of the thermocouple and RTD detectors, the temperature of the beads is directly monitored. In the case of the Wheatstone bridge detector, the resistance of the coil is monitored and a change in resistance is observed as the bead/coil temperature changes.

These types of detector circuits are used in many other types of instruments (thermal conductivity detectors are one example) and will not be reviewed in more detail here. Your maintenance manual should cover your specific detector circuit in detail, and a more in-depth review of thermal conductivity detector circuits can be found in Chapter 33.

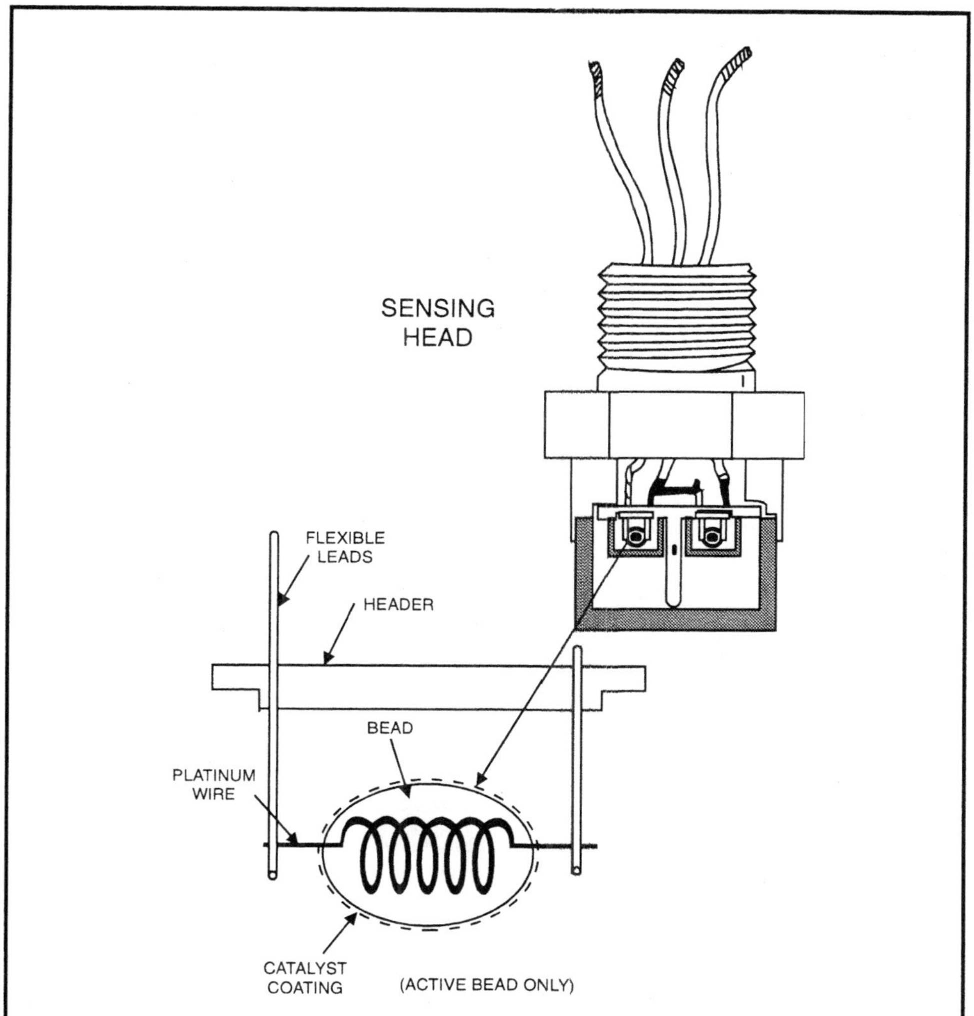

Figure 16-1. Typical Bead and Sensing Head Design
(Courtesy of Sensidyne Inc.)

Hot-wire detectors oxidize the gas directly on a hot wire and are otherwise very similar to bead detectors, except that they are normally operated at higher temperatures and tend to burn out sensing elements faster. For this reason, bead detectors have replaced many hot-wire systems. This is the only reference to hot-wire systems that will appear in this chapter. If you haven't yet acquired an instrument, ask your vendor to sell you a manual or to loan you one to help you make your instrument selection.

Infrared Sensors

Infrared (IR) sensors operate very much like IR spectrophotometers. Detailed discussions of these technologies (sources, detectors, filters, etc.) can be found in the section "Spectroscopic Analyzers" in this volume. Here we will only touch on the general concept of the IR sensor and how it is applied to combustible gas monitoring.

Hydrocarbons, water, carbon dioxide, and many other gases absorb infrared radiation at specific wavelengths. An IR combustible gas detector takes advantage of this principle by exposing the gas sample to IR radiation and monitoring the intensity of that radiation at wavelengths that will provide both good reference and good measure radiation intensity readings for the gas components that are anticipated to appear in the beam and be measured. These instruments are packaged in two versions: a traditional point sensor and an open-path version. The point sensor (like the other combustible gas monitors discussed in this chapter) measures the gas at the sample head. The open-path version monitors the average level in the optical path of the instrument (up to 650 ft [200 m]). Figure 16-2 presents examples of IR combustible gas instrument packages.

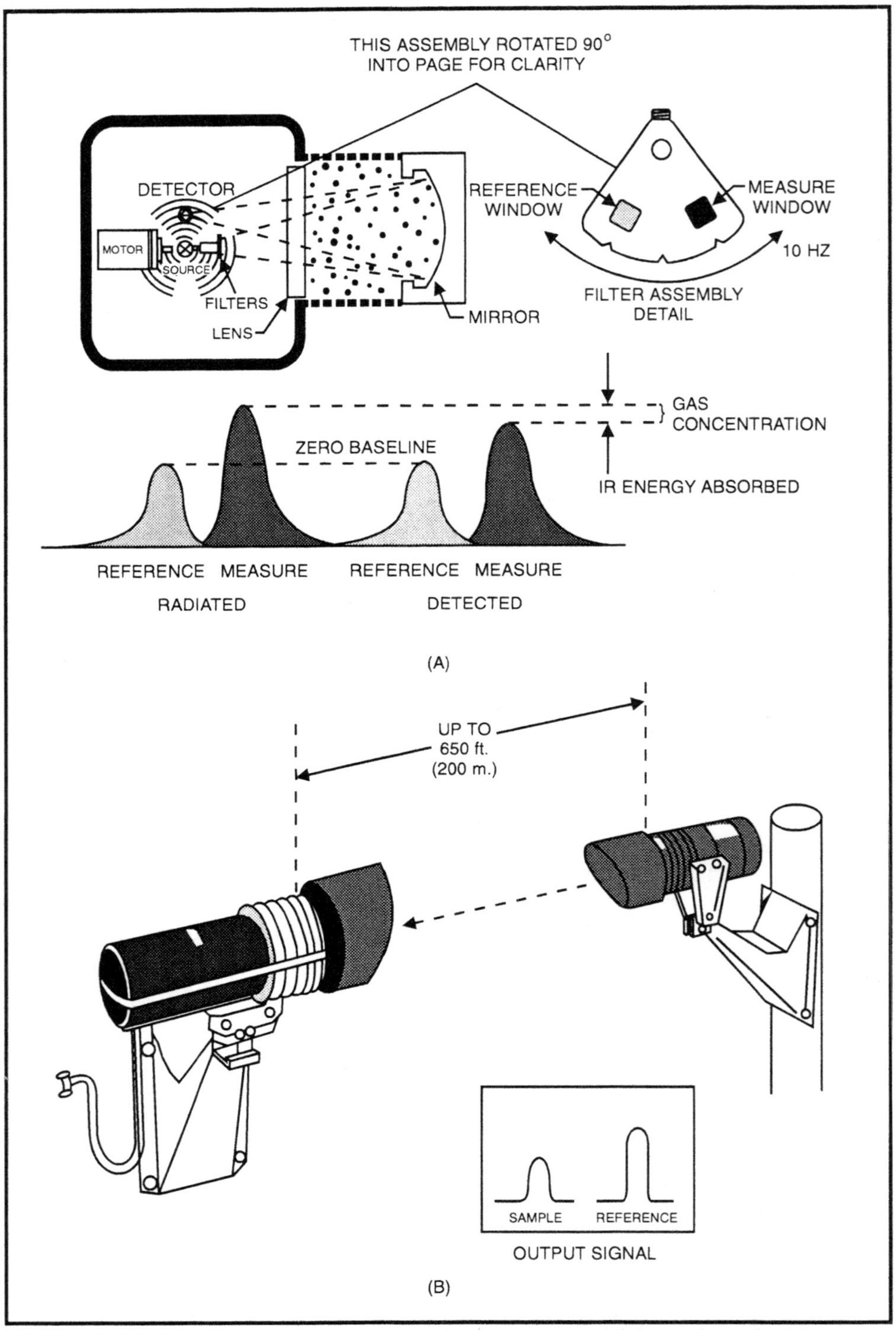

Figure 16-2. Infrared Combustible Gas Monitoring Approaches. (a) Astro Model 5600DG IR point sensor (Courtesy of Astro International Corp.); (b) Sieger Searchline Open-Path IR Sensor (Courtesy of Sieger North America).

Semiconductor Sensors

In principle, semiconductor sensors are very similar to bead sensors. They monitor the change in resistance of metallic-oxide semiconductor (MOS) when exposed to a target gas. The resistance changes are assumed to be caused by the chemisorption of a target gas into the oxide layer. The sensors are made by depositing a thin metal oxide film (usually tin oxide or another transition or heavy-metal oxide) on a silicon wafer (a piece of semiconductor-grade silicon). The sensor is heated and the electrical resistance of the oxide layer is monitored. If a combustible gas is absorbed onto the oxide layer, it results in a change in the resistance of the oxide layer. This change can be correlated to the combustible gas' concentration. These sensors are very sensitive and are primarily used for very low concentrations. Figure 16-3 shows a typical semiconductor sensor design.

The sensors are made by depositing a thin metal oxide film (usually tin oxide or another transition or heavy-metal oxide) on a silicon wafer (a piece of semiconductor-grade silicon).

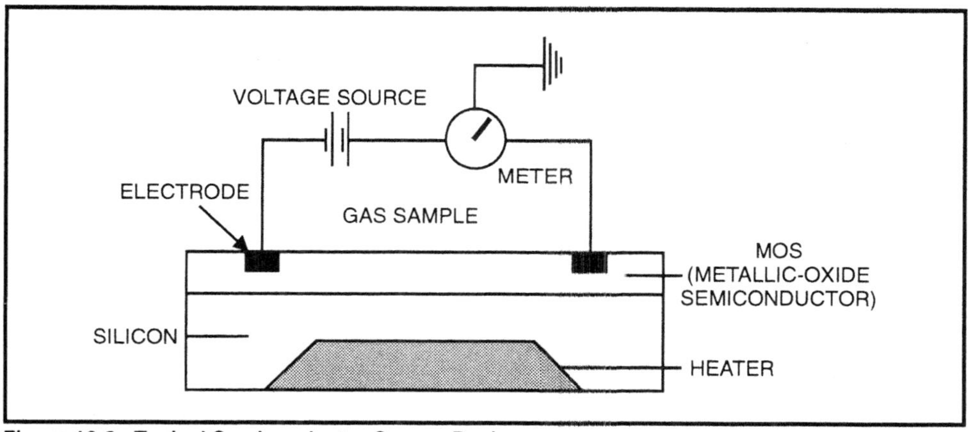

Figure 16-3. Typical Semiconductor Sensor Design
(Courtesy of Sieger North America)

Flame Sensors

Flame sensors date back to the days when miners would look at the flames of their lamps to assure an adequate supply of oxygen. Such sensors also were possible indicators of methane gas. Today, flame detectors are automated, requiring no manual monitoring. They also do not just expose the flame to the sample gas, but introduce the sample gas into the flame.

Generally, combustible-gas flame detectors can be divided into two categories: those that monitor the flame's exhaust gases and those that monitor the flame itself. Sensors that monitor the flame's exhaust gases utilize a hydrogen flame and a flame ionization detector (FID) similar to those used in a gas chromatograph. Most hydrocarbons and other combustible gases form ions when burned in a flame. The ability of the FID to sense a change in the ion concentration between the burner tip and a collector can be used to monitor the sample gas concentration. (Hydrogen and other gases that do not form ions in the flame will not be detected. A more detailed description of an FID detector can be found in Chapter 33.)

Generally, combustible gas flame detectors can be divided into two categories: those that monitor the flame's exhaust gases and those that monitor the flame itself.

The combustible gas analyzers that monitor the flame itself typically utilize either an optical sensor or a temperature sensor. As the concentration of combustibles in the sample gas changes, it results in a change in the total fuel content of the flame; the temperature, size, shape, and color of the flame may also change. By monitoring the appropriate flame characteristics, you can measure the concentration of the gas that has caused this change Many changes are quite specific to the gas causing them and, as such, appropriate calibrations with specific target gases are required.

Inert gases can have a negative effect on the flame and its ability to measure combustibles, unless adequately compensated for by introducing excess air/O_2 into the flame.

Inert gases can have a negative effect on the flame and its ability to measure combustibles, unless adequately compensated for by introducing excess air/O_2 into the flame.

Common Applications

Combustible gas monitoring applications can be divided into two major categories: fixed installation and portable.

Combustible gas monitoring with permanently-installed instruments is employed whenever a combustible gas safety hazard could routinely exist and/or whenever routine monitoring for a flammable gas is desired. (Legal requirements for monitoring do exist, but since they vary from location to location and from industry to industry they will not be covered in this chapter.) Most coal mines monitor for methane and other combustible gases to assure a safe work atmosphere. Most petroleum refineries and chemical plants that work with flammable materials utilize combustible gas detectors to warn of the presence of a potentially dangerous leak. Many process analyzer shelters have combustible gas detectors monitoring the atmosphere inside the shelter and also in the forced-air intake to detect sample leaks in the shelter and/or to prevent flammable gas-containing atmospheres from entering the shelter. Utilities such as sewer systems and natural gas distribution systems utilize combustible gas monitors. The alarms from these instruments can be used to warn of a dangerous condition; to shut down equipment that could be dangerous if operated in a flammable atmosphere; to shut down air-handling equipment to prevent dangerous fumes from entering a control room, analyzer house, or other protected area; to turn on fire-suppression systems; or for any other useful purpose.

Continuous-duty portable combustible gas instruments should have sufficient battery capacity to operate continuously for at least 8 h.

Portable combustible gas monitors are used for nonroutine/nonrepeat applications, such as emergency response, confined-space entry checks, maintenance hot-work permit approval, special work/maintenance activities, leak point identification, and temporary area monitoring. Portable monitors are generally classified into one of two categories: continuous-duty or intermittent-duty. Both should be transportable. Continuous-duty portable combustible gas instruments should have sufficient battery capacity to operate continuously for at least 8 h. This type of instrument might be carried by an individual working in a potentially hazardous area or temporarily set up at a location where special activities are taking place or risks exist. An intermittent-duty instrument is designed to operate only for periods of a few minutes at a time at irregular time intervals. This type of unit may be used by safety personnel checking an area.

An intermittent-duty instrument is designed to operate only for periods of a few minutes at a time at irregular time intervals.

Many portable combustible gas monitors are combined with oxygen analyzers and toxic gas detectors in a single portable instrument package. This simplifies the issuance of work permits for confined spaces (where at least oxygen and combustible gas monitoring would be required), since a single instrument at the work site can perform all of the required gas-sampling tests.

Portable combustible gas monitors should be tested with a known gas sample daily as an absolute minimum. Some applications may require testing prior to each use or even more frequently.

Ranges of Variables

Although some combustible gas monitors are used as gas concentration monitors and are calibrated and programmed to report an actual gas concentration, a combustible gas monitor that is reporting as a combustible gas monitor usually reports percent LFL. What does this mean? LFL is an abbreviation for Lower Flammable Limit, which is generally defined as the lowest concentration of a combustible gas in air that, if ignited, can sustain a flame. Flammable range is the limited band of gas/air concentrations (usually reported in volume percent units) of a specific combustible gas that can produce a combustible mixture. Percent LFL is a percentage of a percentage-gas-concentration that can be a different concentration for each flammable gas. Each combustible gas has a flammable range. Figure 16-4 is a graphic depiction of the concept of flammable range and LFL.

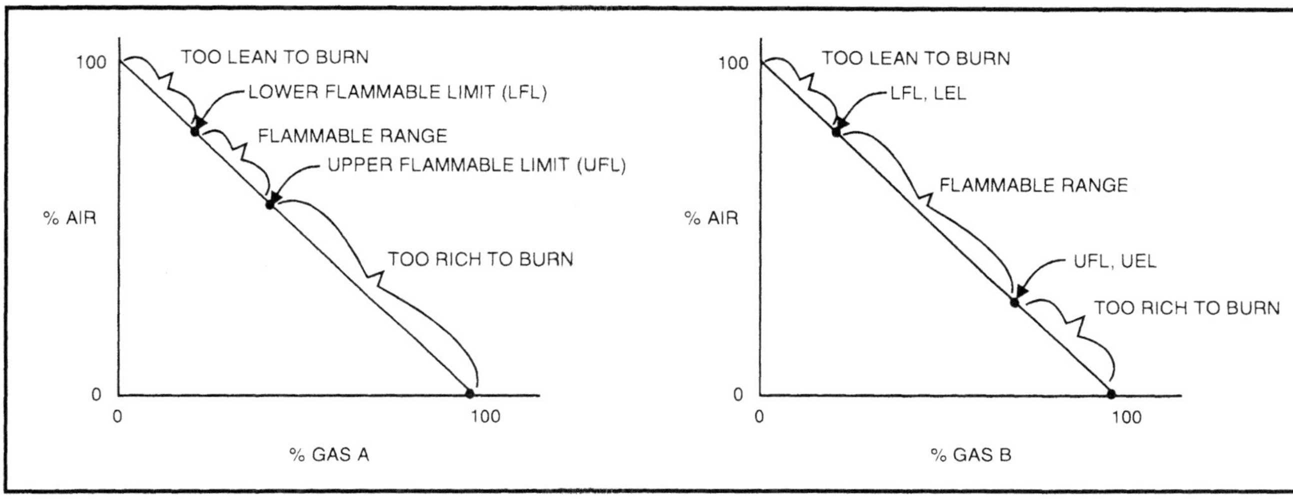

Figure 16-4. A Graphic Depiction of the Concept of Flammable Range

The principle behind the concept of flammable concentration range can be described from the perspective of the basic fire triangle. Three components must be present for combustion to take place: adequate fuel, adequate oxygen, and a source of ignition. If any one of the three is missing, combustion will not take place. In our situation, the combustible gas is the fuel, air supplies the oxygen, and we assume that there is an ignition source. In our hypothetical combustible gas and air mixture, the flammable range is that range of gas concentrations that are adequate to support combustion. Lower concentrations are too lean to burn (insufficient fuel), and higher concentrations are too rich to burn (insufficient oxygen).

Three components must be present for combustion to take place: adequate fuel, adequate oxygen, and a source of ignition.

The lower limit of the range is commonly referred to as the lower flammable limit (LFL) and the upper limit as the upper flammable limit (UFL). For combustible gas monitoring purposes, the terms LFL and UFL are generally deemed to be synonymous with lower explosive limit (LEL) and upper explosive limit (UEL). Some documents may differentiate, so it is always best to verify the definition that is to be used in each standard, regulation, or manual.

Enriched or inert atmospheres have volume-percent flammable ranges that are different from those in air. Also, combustible gas detectors may not give proper percent LEL readings, unless specifically calibrated in the correct and appropriate atmosphere for both the sample and the sensor.

Let's look at the Gas A example in Figure 16-4. The flammable range is 20 to 40 vol%, Gas A in air. That means the LFL of Gas A is 20 vol% and the UFL is 40 vol% in air. The value of 10 vol% of Gas A in air, then equals 50% of the LFL. This (50% LFL) is the number that would normally be reported by a combustible gas analyzer for a 10 vol% Gas A in air mixture.

Selection of the appropriate calibration gas and concentration alarm set points is critical to a successful application.

Since the sensors can and do respond to a variety of combustible gases, a unit like percent LFL allows the presence of any combustible gases to be reported and alarmed. Unfortunately, because different gases can have different detector response factors and different flammable ranges, you may only be able to accurately report the concentration of the one that was used to calibrate the instrument. For this reason, selection of the appropriate calibration gas and concentration alarm set points is critical to a successful application. It is highly recommended that these instruments only be used to detect those combustible gases for which they are calibrated.

To use the instrument to detect mixtures, a variety of gases, or gases for which you have no relative response data or standards, consult the instrument manufacturer.

Formulas like the following are available to help you approximate the limit for known gas mixtures [Ref. 1]:

$$L = \frac{100}{\frac{P_1}{N_1} + \frac{P_2}{N_2} + \frac{P_3}{N_3} + \ldots}$$

where

L = mixture flammability limit (vol% or pct_v gas in air)
P = proportion of each combustible gas present (where P1 + P2 + P3 + ... = 100)
N = appropriate flammability limit in air for each appropriate individual combustible gas

An example of how to use this formula would be a natural gas mixture comprised of the following:

Gas	Proportion, %	LFL in Air, %
Methane	80	5.32
Ethane	15	3.22
Propane	4	2.37
Butane	1	1.86

$$LFL = \frac{100}{\frac{80}{5.32} + \frac{15}{3.22} + \frac{4}{2.37} + \frac{1}{1.86}} = 4.55\% \text{ for the mixture}$$

Approximation formulas assume a known-and-constant ratio of concentrations of flammable constituents in the flammable mixture being monitored. They are not applicable to monitoring samples where the concentration ratios between the flammable gases vary.

The concentrations that define the flammable range are usually different for each combustible gas and should be verified in the latest edition of a legally-recognized source document for your location. In the U.S., you can usually reference National Fire Protection Association (NFPA) 325-M, "Flammability Characteristics of Combustible Gases or Vapors." The use of approximation formulas may not be legal for your application, and the appropriate standards for your location should be referenced before attempting to apply them.

Options and Accessories

Options and accessories are somewhat unique to each type and brand of instrument. One option that is fairly common is a poison-resistant bead sensor. The traditional bead sensor is subject to poisoning by many common industrial materials, such as silicones, fluorocarbons, sulfur compounds, and others. Most vendors of bead-type combustible gas sensors have poison-resistant sensors that offer an extended life when exposed to a poison. These typically incorporate modified bead structures and/or catalyst modifications. Other areas where vendors often offer options include calibration kits, sampling aids, materials of construction, and sensor/transmitter-communications devices. Special communications devices can allow you to adjust and calibrate the unit without opening the housing or without requiring a hot-work permit. Some of them incorporate magnetic switches; some utilize remote transmitters, similar to your television remote control; still others offer removable, intrinsically-safe sensors that can be calibrated in your shop.

The traditional bead sensor is subject to poisoning by many common industrial materials, such as silicones, fluorocarbons, sulfur compounds, and others.

Limitations

A basic limitation of the principle of combustible gas monitoring is the assumption of a binary gas/air mixture when determining the LEL gas concentration. It is quite conceivable to expect to encounter oxygen-rich and oxygen-deficient atmospheres in process equipment and vents. It is also rare to encounter a simple binary gas mixture. Therefore, calibration of the instrument and determination of the LEL can be quite complex, unless you are willing to risk making assumptions. Possible assumptions could include things like using the lowest gas concentration LEL for the gas mixture and the least responsive gas for calibration. Bead-sensor-based instruments that have been exposed to sample mixtures above their LEL have been known to give erroneous readings and can create a false sense of security. If the potential for this is anticipated, a dilution-based sample system or alternate method of detection should be chosen.

Bead-sensor-based instruments that have been exposed to sample mixtures above their LEL have been known to give erroneous readings and can create a false sense of security.

Bead-type sensors rely on oxidation and the presence of an adequate supply of oxygen to make their measurement. Consequently, they should not be used in oxygen-deficient atmospheres or for gas/air concentrations above the LEL.

Infrared and flame detectors do not always assure the presence of oxygen or the possibility of combustion. They may simply measure the presence of gases that absorb radiation in a certain spectral region or that burn in a flame. This makes them very useful for measuring the presence of combustible gases in inert samples or atmospheres or for combustible gas measurements that may exceed the LEL or even UFL, but not as reliable indicators of the actual combustible nature of the sample as presented to the analyzer. They are also quite useful in atmospheres that would poison a bead sensor. The bead, on the other hand, actually combusts the sample and, as such, actually indicates a combustible situation. It also can respond to most combustible gases, not just those for which it was set up to measure (it is often non-specific).

Open-path IR sensors measure/report a path-average concentration.

Open-path IR sensors measure/report a path-average-concentration. This offers the advantage of monitoring a larger area, not just a point, and the disadvantage of not knowing the concentration at any one point along the path. In theory, you can have a path with an average below the LEL and still have several points on that path above the LEL. Sighting and path layout for this type of technology are critical.

Fog, steam, snow, and rain can interfere with the operation of open-path IR monitors.

Fog, steam, snow, and rain can interfere with the operation of open-path IR monitors.

Instrument response data under varying conditions (oxygen-rich, oxygen-deficient, above LEL, condensing sample, poisons, inhibitors, etc.) should be obtained from your instrument manufacturer and/or located in the instrument's operation and maintenance manual before you commission the instrument.

Installation Hints

Read your vendor documents and follow their recommendations when applicable.

Since this category of analyzer comprises several measurement technologies and brands of instruments, the best installation guidance that can be given is to read your vendor documents and follow their recommendations when applicable. Also follow any regulations, standards, or practices that may apply to your installation. A few of the more general standards will be mentioned in the section "Specific Safety Issues."

Remember to keep in mind your total application when selecting an instrument. The reason there are so many combustible gas monitoring technologies is because no single method is the best (or even good) for the wide range of potential applications. Most vendors can propose a way to make your measurement with their instrument. It is up to you to select the one that is best-suited to your application.

When selecting an installation site for the measurement, consider factors such as accessibility, compatibility of materials of construction, vibration (which can loosen wire connections and damage sensitive instruments), ambient temperature and weather conditions, air currents, sample vapor density and pressure, sample temperature, potential leak and ignition source locations, and gas source dispersion (too close and you may not have enough oxygen to allow a bead sensor to make the measurement; too far and a low-volatility sample may not be detectable).

Specific Safety Issues

The whole topic of combustible gas monitoring is a safety issue. Sample point selection, instrument selection, and use of the measurement data can be safety issues. Sometimes a detection will simply trigger an alarm, sometimes the process will be shut down, and other times a whole range of safety responses (sprinkler systems, inerting systems, etc.) may be initiated. The appropriate response cannot be recommended in a handbook, but rather requires an appropriate safety audit and risk assessment of the total-site situation.

When exposed to gas concentrations above the LEL, many combustible gas monitors will yield an initial high-positive reading, which then goes erroneously low, or just a uniform false low reading.

Most combustible gas alarm systems utilize fail-safe latching relays (required by some codes, standards, and practices). Varying conditions (winds, drafts, leak rates, etc.) can cause varying gas concentrations at the sensor. Changing on/off alarm status can confuse personnel who should respond to an alarm condition. A latched alarm can help reduce this confusion. Local alarms for many fixed, confined-space entry situations (such as analyzer shelters, compressor-system shelters, below-ground containments, etc.) utilize a go/no-go type of display. An example of such a system would be a red and green light operated in a fashion that allows the green light to be on under a nonalarm condition and the red light under an alarm condition. If both lights or neither light is on, a fault exists and an unsafe condition is assumed. Additionally, many systems utilize both an audible and a visual indicator for an unsafe condition. Alarm systems are somewhat unique to each installation (responder requirements, danger level, noise levels, lighting, equipment layout, etc.) and require a good safety-audit review by personnel who are knowledgeable about the local situation and about appropriate regulations, codes, standards, and practices.

Table 16-2 provides a partial list of reference documents that may be helpful during the selection, design, installation, and maintenance phases of a combustible gas analyzer project. Obviously, make sure also to consult your vendor manuals and site-safety policies.

Alarm systems are somewhat unique to each installation (responder requirements, danger level, noise levels, lighting, equipment layout, etc.) and require a good safety-audit review by personnel who are knowledgeable about the local situation and about appropriate regulations, codes, standards, and practices.

Table 16-2. Selected Reference Documents for Combustible Gas Analyzer Installations

I.D. Number	Title
ANSI/ISA-S12.13, Part I (Standard)	Performance Requirements, Combustible Gas Detectors
ISA-RP12.13, Part II (Recommended Practice)	Installation, Operation, and Maintenance of Combustible Gas Detection Instruments
CSA-C22.2 No. 152	Combustible Gas Detection Instruments
BS 6020, Parts 1-3	Instruments for the Detection of Combustible Gases
BS 6959	Selection, Installation, Use and Maintenance of Apparatus for the Detection and Measurement of Combustible Gases (Other than Mining Applications or Explosives-Processing and Manufacture)
BS 5345, Parts 1-9	Code of Practice in the Selection, Installation, and Maintenance of Electrical Apparatus for Use in Potentially Explosive Atmospheres (Other than Mining Applications or Explosives-Processing and Manufacture)
US Code of Federal Regulations, 49 CFR, Part 154	USCG Standard for Gas Detection, Sections 1345 and 1350
ANSI/NFPA 70	National Electrical Code, Articles 500, 501, and 504
ANSI/NFPA 325M	Fire Hazard Properties of Flammable Liquids, Gases, and Volatile Solids
IEC 79, Part 16 and others	Electrical Apparatus for Explosive Gas Atmospheres
IEC 1285	Industrial-Process Control — Safety of Analyzer Houses

Note: Always consult the most recent version of the document.

Unique Sample System Requirements

Combustible gas monitoring usually involves monitoring open areas. At the detector, combustible gas samples should be clean and dry (noncondensing). Most process analyzers monitor a sample that is contained in a process. This difference in sample type often influences the sample system approach that is chosen. Some of the most popular sampling approaches used in combustible gas monitoring are gas diffusion, pumped/aspirated sampling, and dilution systems. See Figure 16-5 for examples of the multipoint-diffusion, pumped-multistream, and dilution approaches.

Single-point and multipoint (multisensor) diffusion systems are probably the most prevalent combustible gas systems in the field. They utilize gas-diffusion-based sampling at a sensor head. The multipoint systems simply monitor and report on multiple sensors through one device. The multipoint unit may look at the individual sensors sequentially, multiplexed, or simultaneously. Some people feel that such a unit offers added redundancy in that you have multiple field sensors; if one fails, you can still detect a leak on another. Some feel the opposite, because a common alarm or other output failure may impact several sampling points. Take a look at each proposed system and select the best options for your needs. Be it single-point or multipoint sampling, diffusion sampling and the positioning of the sensor heads require careful consideration of the process layout, possible leak-and-ignition source locations, prevailing weather and ventilation conditions, and the particulars (including diffusion characteristics) regarding the gas or gases being monitored.

How do you determine where to position your sensor heads/sample points? For economic reasons, you can't monitor everywhere, so points should be positioned based on the highest probability that a flammable gas leak will reach a sensor and provide adequate warning to prevent a combustible mixture reaching an ignition source. Issues regarding plant layout, potential leak sources, prevailing winds, ventilation, and the like are so special to each application that they cannot be covered in adequate detail here. Gas temperature, vapor pressure, and density, however, are general enough that they will be addressed.

Hot gases will tend to rise and cold gases will tend to drop. If the gases mix rapidly with the surrounding atmosphere, they will quickly reach atmospheric temperatures and neither rise nor drop because of temperature. In many cases, gas temperature is a factor only for a short distance from the potential leak source.

The vapor pressure and boiling point of the material being monitored are important because a leak may result in an extensive vapor cloud, a condensing liquid, or a puddle with a very limited vapor cloud above it. Low-volatility samples require monitoring near the source.

Gases with densities close to that of air show little tendency to mix rapidly and are often transported significant distances as a cloud by air currents.

The density of the flammable gas greatly influences how it diffuses and where and how you monitor for it. Generally speaking, the diffusion rate of a gas into the air is inversely proportional to its relative gas density. Gases with densities less than air tend to rise and become diluted quickly. Gases with densities heavier than air tend to settle and collect across the floor or in low spots such as trenches, pits, basements, and drains. Gases with densities close to that of air show little tendency to mix rapidly and are often transported significant distances as a cloud by air currents.

Once you have selected your sample points, take a look at your vendor's optional sampling equipment for additional help. Most vendors offer options in terms of materials of construction, collecting cones (to help concentrate lighter-than-air gases), wind shields, weather-protection housings, special filters and mounting assemblies, flame arrestors, calibration attachments, and more.

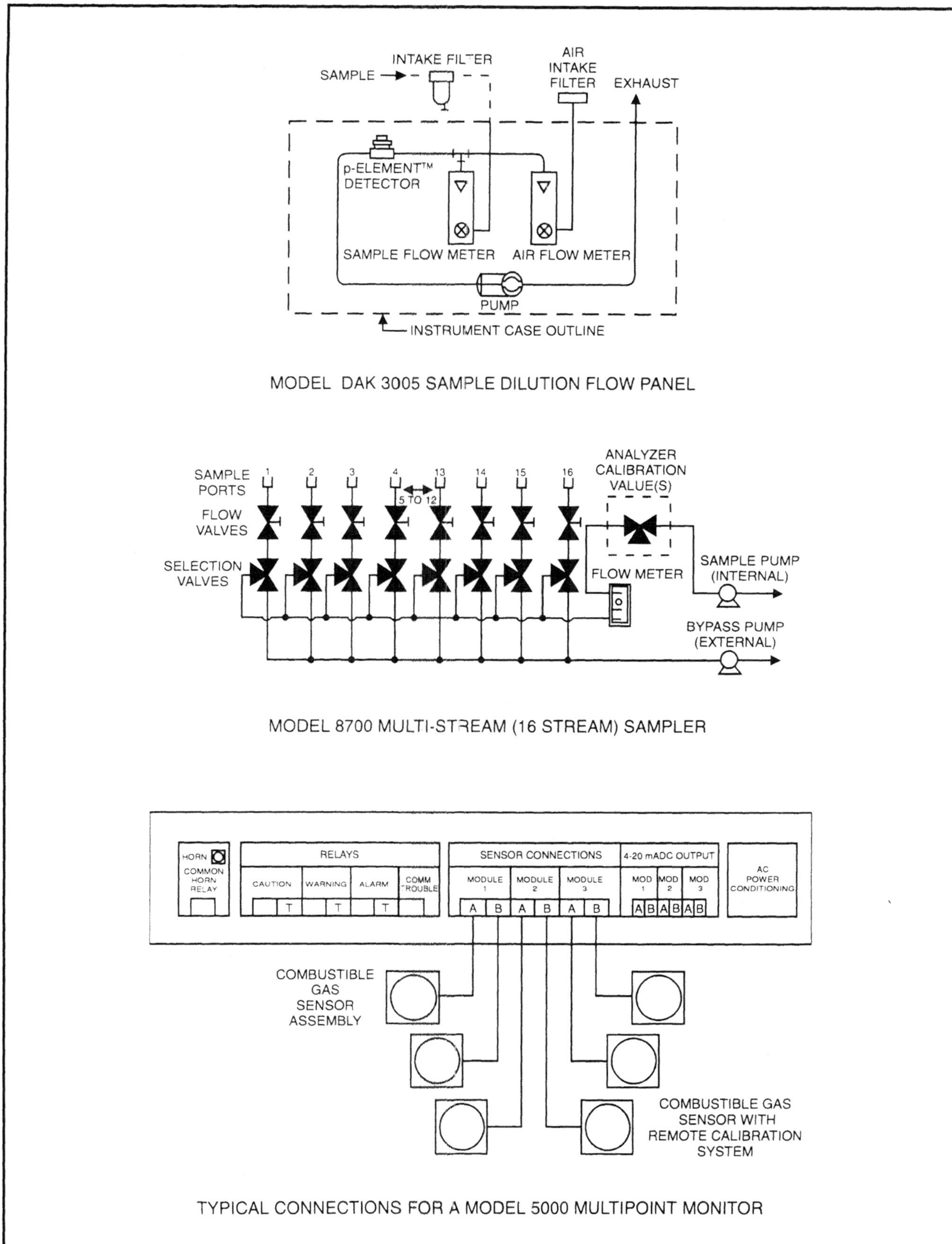

Figure 16-5. Examples of Sampling Approaches Used for Combustible Gas Monitoring
(Courtesy of MSA Instruments)

The multiple-stream (or remote-stream) approach to sampling involves transporting the sample(s) to the sensor. The field sample-point still needs to be located using the same considerations that were used to position the field sensors as just described. The sensor is positioned in the sample line shortly before or after the sample pump (sometimes a block-and-bleed valve assembly is used with the sensor). The analyzer/sample selection system must be capable of identifying and reporting the correct analytical result in accord with the correct stream (sample port). These systems are generally used where sample-point access problems exist, where ambient conditions prohibit direct-sensor installation, or where multiple sensor cost can justify the expense of the sample system. The cost of running tubing to the sample points, pumps, valves, and controller is not insignificant when compared to the installed cost of many field-sensor systems.

The single-point remote-stream sampling approach is also used extensively in monitoring for confined-space entry purposes. It offers the added safety benefit that the sample port/point can be easily inserted (using an extension probe) into the confined space without exposing the tester to the space's atmosphere. When testing confined-space areas where gases/vapors may be stratified in layers rather than mixed, it is best to make spot checks at different layers (levels) with an extension probe.

Dilution-based systems are used primarily when the sample is in an inert atmosphere, its combustibles content is expected to exceed the LEL, and the user wants to use a sensor that does not perform in these inert atmospheres. The sample is diluted in a precisely-controlled ratio with air to produce an LEL range that is within the measurement range of the chosen sensor. It should be remembered that a special calibration or a multiplication of the sensor result must be completed to reflect the gas concentration in the original sample. This result may reflect the combustible gas content of the original sample, but not necessarily its combustibility (because the original sample may have been in an oxygen-deficient or enriched atmosphere).

Typical Calibration Requirements

Because of the wide range of applications and sensor types available, it is impossible to provide detailed calibration requirements here. Combustible gas detectors can be calibrated for general-purpose use or for a specific combustible gas or vapor of interest. If possible, the instrument should be calibrated with the particular gas or gas mixture of interest. All instruments should clearly display how and with what they were calibrated. The safety audit of the instrument project should specify how often and with what each instrument will be calibrated (which will help the audit team determine whether an instrument is appropriate for the desired measurement).

The frequency of calibration is also somewhat specific to each application. In the case of intermittent use and/or portable combustible gas analyzers, it is generally recommended that at least a calibration check with a known test gas be done on each power-up of the instrument. (This is in addition to your vendor's calibration recommendations.) Instruments used in potentially toxic atmospheres will require more frequent checks and/or calibrations. Those used in potentially inhibiting atmospheres will require extra attention and sometimes additional monitoring (for the inhibiting agent).

Although it is well known that calibrating with the actual gas to be measured is the only way to assure an accurate calibration for a combustible gas measurement, this isn't always practical or possible. Some instruments are used for general-purpose monitoring and/or to monitor for multiple gases. (Instruments should never be used for gases other than the one for which calibrated unless custom response curves or factors are provided for those other gases.) One commonly asked question with regard to calibrating bead-type sensors is: How can I calibrate an LEL sensor with a readily available gas when I am really trying to detect another gas? The answer that was provided to the author [Ref. 2] takes the form of this equation:

$$Y = C \times (K_{cg}/K_d)$$

where:

Y = corrected reading in %LEL of gas to be detected
C = %LEL value of calibration gas being used
K_{cg} = K factor of the calibration gas
K_d = K factor of the gas to be detected

Note that K factors are theoretically unique for every gas sensor manufacturer. Their accuracy should be given by the manufacturer supplying the factor, but are generally believed to be ±15%LEL at best. Since factor definitions may vary from vendor to vendor, you should use an equation from the vendor supplying the factors.

Many vendors further simplify the calibration transfer procedure by reducing this formula to a look-up table. Figure 16-6 shows a portion of one of these tables and several typical example calculations. Recall that this is not a recommended alternate calibration procedure and that the tables are specific to each manufacturer's sensor. The actual numbers in this example are not to be used with your sensor, unless they are also specifically supplied to you by your vendor.

Infrared sensors look only at specific absorption bands. They are calibrated to detect a particular gas or gases that absorb in those bands. They will not detect any gases that do not adequately absorb at these bands and may experience interference problems with other gases that absorb at either the measure or reference bands.

Troubleshooting and Maintenance

Maintenance concerns such as calibrations, catalyst poisons, accessibility, and several others have been covered in previous sections of this chapter. It is sufficient to say that combustible gas analyzers need to be subjected to routine tests and recalibrations.

Preventive maintenance (PM) is required for such instruments. Don't wait until you know there is a problem. By then, you will have been operating without protection. Each type of instrument is somewhat unique. It is suggested that you start with the vendor-recommended PM program and adjust it, based on your location's experience. Some installations may fall under legal or plant safety policy requirements and may have specified minimum PM programs above and beyond those recommended by your vendor.

Infrared sensors will require window cleaning and source replacements. Flame sensors will need pilot gas supplies (H_2, propane, air, etc.). Routine process area maintenance activities, corrosives, improper materials of construction, and atmospheric conditions can result in sample filters and sintered metal flame arrestors becoming plugged or rendered useless by steam, water, dust, dirt, mud, sandblast materials, paint overspray, oil, sea spray, and so on.

NOTE

The following caibration methods and tables are not FMRC approved.

WARNING

It is recommended that calibration always be performed using a known %LEL concentration of the particular substance to be detected in order to attain the most accurate detection of that particular substance.

CHARACTERISTICS AND SELECTIVITY FACTORS OF SELECTED COMPOUNDS

A -- %LEL CONCENTRATION OF SUBSTANCE EQUIVALENT TO 1 %LEL METHANE
B -- %LEL CONCENTRATION OF SUBSTANCE EQUIVALENT TO 30 %LEL METHANE
C -- %LEL CONCENTRATION OF METHANE EQUIVALENT TO 30 %LEL SUBSTANCE
D -- %v CONCENTRATION OF METHANE EQUIVALENT TO 30% LEL SUBSTANCE

SUBSTANCE	MOLECULAR WEIGHT	100 %LEL (%v)	A	B	C	D
METHANE	16.0	5.0	1.0	30	30	1.50
ISOBUTYL BENZENE	143.2	.8	3.1	94	10	.48
ISOPENTANE	72.2	1.4	2.2	65	14	.69
ISOPROPYL ETHER	102.2	1.4	2.3	68	13	.66
METHYL ACETATE	74.1	3.1	2.0	60	15	.75
METHYL ALCOHOL	32.0	6.0	1.2	35	26	1.29
METHYLAMINE	31.1	4.9	1.3	39	23	1.16
METHYLCYCLOHEXANE	98.2	1.2	2.3	68	13	.66
METHYL ETHER	46.1	3.4	1.6	48	19	.95
METHYL ETHYL ETHER	60.1	2.0	2.3	68	13	.66
METHYL ETHYL KETONE	72.1	1.4	2.4	73	12	.61
METHYL FORMATE	60.0	4.5	1.5	45	20	1.01
METHYL PROPIONATE	88.1	2.5	2.0	59	15	.77
METHYL PROPYL KETONE	86.1	1.5	2.4	73	12	.61
NAPTHALENE	128.2	.8	2.9	88	10	.51
NITROMETHANE	61.0	7.3	1.7	52	17	.87
n-NONANE	128.3	.8	3.2	97	9	.46
n-OCTANE	114.2	1.0	2.7	81	11	.55
n-PENTANE	72.2	1.5	2.2	65	14	.69
PROPANE	44.1	2.1	1.8	55	17	.82

Calibration can be performed using methane by alternative methods to achieve approximate direct reading of the %LEL concentration of another combustible substance. Typical error can be 20-30%. The methods are not warranted since there is no way to assure accuracy. There are two possible methods.

METHOD A: Perform calibration with 30%LEL methane, and adjust the transmitter to read the theoretical equivalent %LEL concentration of another combustible substance. **EXAMPLE:** To read directly the approximate %LEL concentration of n-octane, calibrate with 30%LEL methane and adjust the transmitter to indicate 81%LEL n-octane; refer to column B for the %LEL concentration of other common combustible substances theoretically equivalent to 30%LEL methane.

METHOD B: For 30%LEL concentration of another combustible substance, determine the theoretical equivalent %LEL concentration of methane. Perform calibration with the theoretical equivalent %LEL concentration of methane, and adjust the transmitter to read 30%LEL concentration of the other substance. **EXAMPLE:** For n-octane, 11.1%LEL methane is the theoretical %LEL concentration of methane equivalent to 30%LEL n-octane. To read directly the approximate %LEL concentration of n-octane, calibrate with 11.1%LEL methane and adjust the transmitter to indicate 30%LEL n-octane; refer to column C for the %LEL concentration of methane equivalent to 30%LEL concentration of other common combustible substances. Column D gives the %v concentration of methane equivalent to the %LEL concentration cited in Column C.

Figure 16-6. Calibration Transfers Between Gases
(Courtesy of Sensidyne Inc.)

In the case of a diffusion sensor, a layer of water or oil can act as a diffusion membrane and a barrier to normal gas diffusion. Even worse, it may act as a medium that absorbs and holds the combustible gas of interest. Flame arrestors are safety devices that cannot be modified or aggressively cleaned without endangering their safety certification. Diffusion-sensor heads cannot be supplied with low-flow alarms, so PM checks are the only way to assure that they are operating correctly. A poisoned sensor must be replaced, but an inhibited sensor can be rejuvenated by the appropriate procedure. Use only those procedures approved by your instrument vendor.

The sintered metal covering over most bead and semiconductor types of detectors serves multiple purposes (filter, wind shield, flame arrestor, etc.). If your installation is in an electrical-hazard-rated area, do not remove the cover (flame arrestor) without the appropriate hot-work permits.

These instruments are electrical and subject to all of the problems (emf, power quality, wiring quality, grounding problems, noise, drift, etc.) experienced by any other electrical instrument.

References

1. The Nature of Combustible (Flammable) Gases, Section 1, Sieger Ltd., Sutton-in-Ashfield, Notts, U.K., p. 26.
2. Roarty, Daniel R., Exidyne Instrumentation Technologies (EIT), personal communication, 6 June 1995.

Bibliography

1. ANSI/ISA-S12.13, Part I, "Performance Requirements, Combustible Gas Detectors," Research Triangle Park, NC: Instrument Society of America, 1986.
2. BS 6959, "Selection, Installation, Use, and Maintenance of Apparatus for the Detection and Measurement of Combustible Gases," London, U.K.: BSI Publishing, 1989.
3. Clevett, Kenneth J., *Process Analyzer Technology*, New York: Wiley Interscience, 1986.
4. ISA-RP12.13, Part II, "Installation, Operation, and Maintenance of Combustible Gas Detection Instruments," 1987.

About the Author

James F. (Jim) Tatera is a Senior Analytical Systems Specialist with the Dow Corning Corporation. His experience has included both US and international assignments in analytical research, process engineering, project engineering, production management, and maintenance management. More than 15 of the last 23 years of his career have been involved working with process analyzers. He has published several papers and participated in several professional conferences as a presenter, session developer, panelist, and workshop leader.

Mr. Tatera received a B.S. in Chemistry and Math from the University of Wisconsin in 1968, attended a graduate program in Chemical Instrumentation and Analysis at the University of Minnesota, and graduated from the US Army Officers Candidate School. He also received an MBA from Central Michigan University in 1982. He is one of the original Certified Specialists in Analytical Technology (CSAT) and a member of the ISA, ACS, and AWMA. He is active in the US and international standards activities, and he is the ANSI USNC Technical Advisor to IEC SC65D (Analyzing Equipment in Industrial Process Measurement and Control). He has also served as the ISA SP76 (Composition Analyzers) committee chairman.

17

Viscosity Analyzers (non-ASTM)

Viscosity is a property that determines a fluid's rate of flow. As a fluid flows through a tubular pipe, the fluid next to the wall of the pipe will tend to remain stationary or move at a slow rate. The fluid moves faster as its distance from the wall increases, and reaches a maximum rate at the center of the pipe. The more viscous the fluid, the greater the variations in flow across the pipe.

Viscosity in fluid flow is analogous to the friction generated between solids that are in contact and in motion with one another. It is a measurement of the mechanical energy that is related to the friction generated when the molecular layers of a fluid move relative to one another.

Different liquids can behave quite differently when they are flowing. If the flow velocity increases in direct proportion to the pressure exerted, it is classified as Newtonian flow (Figure 17-1a). Water, gasoline, kerosene, mineral oils, and lubricating oils display this behavior. Other fluids (e.g., catsup) exhibit varying viscosity under differing shear/flow conditions; their flow is classified as non-Newtonian.

If the flow velocity increases in direct proportion to the pressure exerted, it is classified as Newtonian flow.

A true plastic-like material (Figure 17-1b), when subjected to a constantly increasing stress, will flow only after a definite yield-point temperature is reached. Tar from crude oil is an example. Mixtures such as printer's ink or paper pulp experience pseudo-plastic behavior (Figure 17-1c), in which some flow occurs at lower pressures but increases quite rapidly as the pressure increases. Other materials, such as asphalt, show dilatant behavior (Figure 17-1d), in which the initial flow rate under pressure is quite high but then decreases as the pressure continues to increase. Other types of flow behavior occur in materials which, for the most part, are not petroleum products.

Measurement Principles

Process viscometers typically measure the fluid viscosity by one (or a combination) of two methods: exposing a constant-area moving surface to a constant thickness of fluid and measuring the forces that resist the movement of the surface; or exposing a constantly moving thickness of fluid to a fixed surface and measuring the forces that resist the fluid flow.

Several viscometer methods determine the time required for a specific volume of sample to flow through a capillary tube, a nozzle, or an orifice. Other viscometer procedures use a metal ball falling through a vertical tube filled with sample, or a piston falling in a cylinder filled with sample. The torque produced by sample resistance in a rotating cylindrical cup, or by a rotating cylinder or disk in a stationary cup, is the basis for other methods.

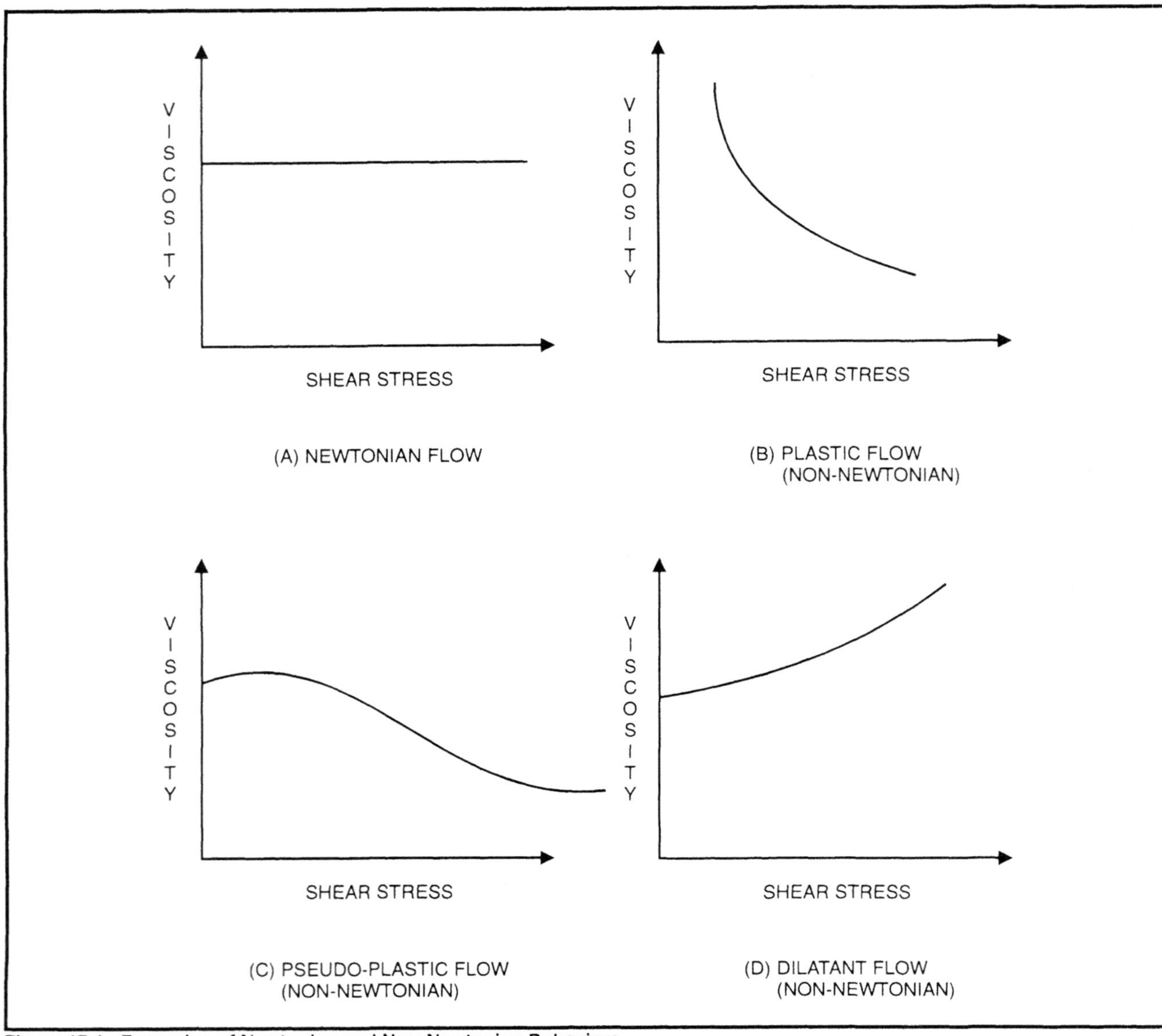

Figure 17-1. Examples of Newtonian and Non-Newtonian Behaviors

Capillary Viscometers

The capillary viscometer is based on the Hagen-Poiseuille law, which describes the absolute viscosity of a transported fluid at the wall of a capillary tube as:

$$\text{Absolute viscosity} = \frac{\text{Shear Stress}}{\text{Shear Rate}} = \frac{PR/2L}{4Q/\pi R^3} = \frac{\pi P R^4}{8QL}$$

where:

P = pressure differential across the fluid in the tube
R = inside radius of the tube
L = tube length
Q = volume rate of fluid flow

The instrument itself usually consists of a sample-conditioning system (sample bypass, sample filter, temperature-controlled oil bath, and heat exchangers) and a measurement system (metering pump, capillary, differential-pressure transmitter, temperature transmitter, and viscometer controller/transmitter). Essentially, a precision metering pump passes a constant-temperature process stream continuously through a capillary restriction at a constant flow rate. The pressure drop across the capillary is measured by a differential-pressure transmitter, and the measured differential pressure is directly proportional to the fluid's absolute viscosity at the unit's operating conditions.

Figure 17-2 depicts a basic capillary viscometer configuration. This type of process viscometer was designed and developed by British Petroleum and Hallikainen Instruments Ltd. It has been manufactured by several instrument companies including Fisher Controls Ltd., Fluid Data Inc., Servomex, and GCA/Precision Scientific.

A number of capillary viscometers are available that require only a small volume of sample for measurement. A typical U-tube viscometer is shown in Figure 17-3. In it, the viscosity of a sample is compared with that of a standard sample, such as water. The viscometer is placed in a constant-temperature bath, and sample is poured through tube A into chamber E and allowed to stand until it reaches bath temperature. Then, with tube C closed, the sample is sucked up through the capillary tube F to the upper edge of bulb D. Tubes A, B, and C are then opened, and the time required for the liquid level to drop from level m_1 to m_2 is measured. This time is compared with that for a standard sample of known viscosity to derive the kinematic viscosity.

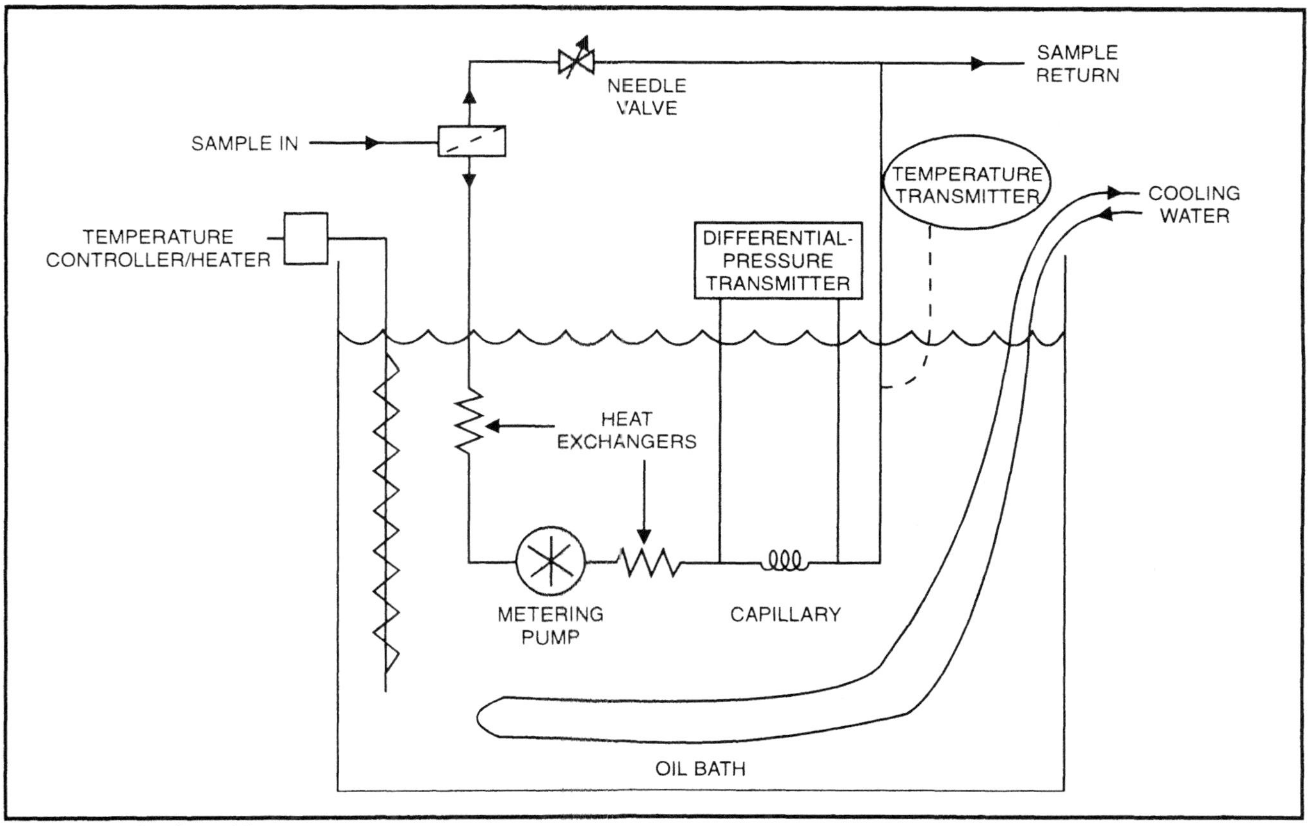

Figure 17-2. Capillary Viscometer

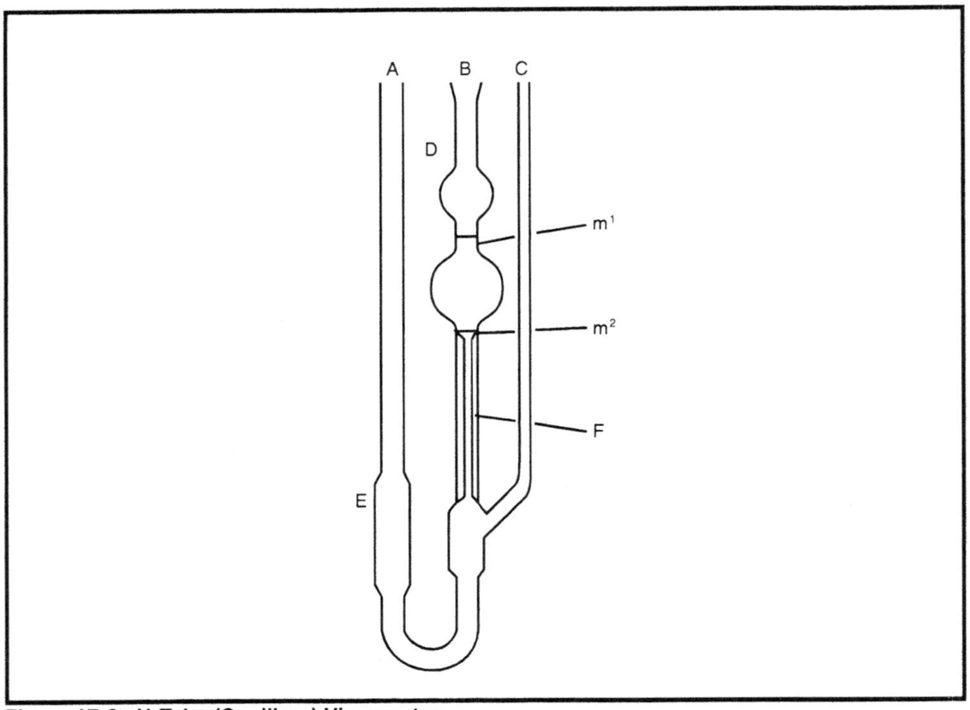

Figure 17-3. U-Tube (Capillary) Viscometer
(From Process Analyzer Technology, Kenneth J. Clevett, New York: John Wiley & Sons)

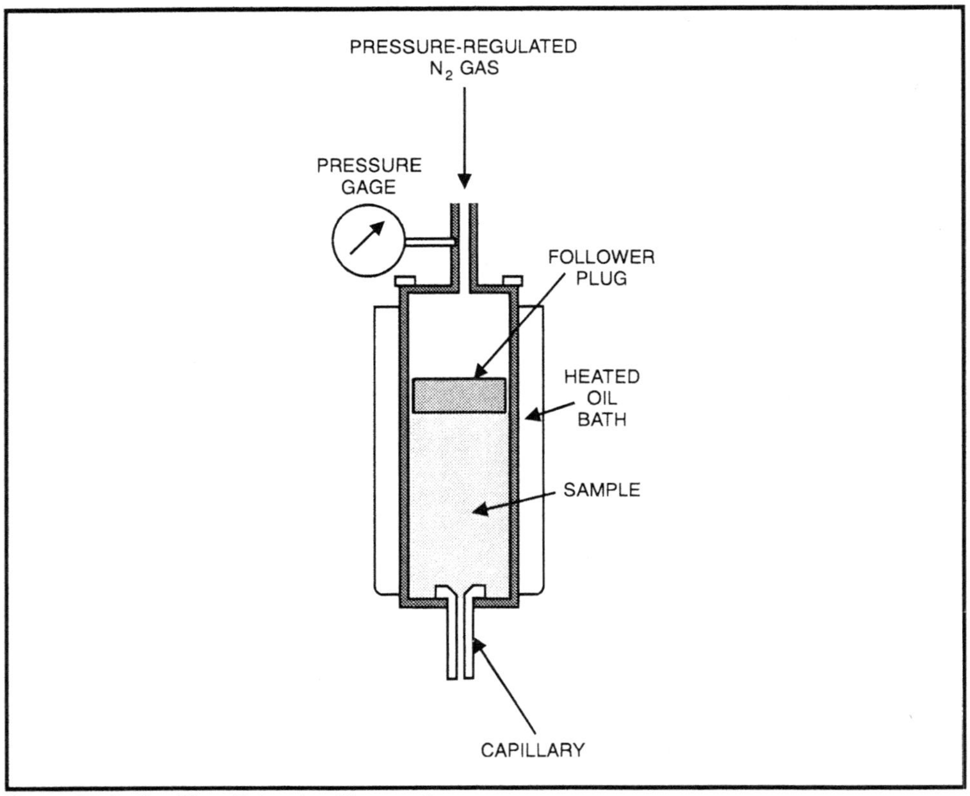

Figure 17-4. Capillary Extrusion Viscometer
(From Process Analyzer Technology, Kenneth J. Clevett, New York: John Wiley & Sons)

Capillary Extrusion Viscometers

Most glass capillary viscometers are limited to the measurement of Newtonian fluids with low viscosities. However, several models are available that can handle non-Newtonian fluids with very high viscosities. One of these (Figure 17-4) uses compressed nitrogen to force the sample through the capillary in a temperature-controlled cylinder. A free-floating follower-plug on top of the sample distributes the nitrogen pressure evenly over the surface of the liquid. The viscosity is calculated from the weight of sample that is extruded over a measured period of time.

The viscosity is calculated from the weight of sample that is extruded over a measured period of time.

Falling Sphere Viscometers

An object falling through a liquid encounters a resistance to its motion that is related to the viscosity of the liquid and the velocity of the body. A laboratory viscometer for this type of measurement is shown in Figure 17-5.

The sample is contained in the inclined tube, F, which is immersed in a constant-temperature bath, R. The tube has two calibration lines, A and B. The ball is raised above the upper line, A, and allowed to sink through the liquid. The viscosity of the sample is calculated from the time required for the ball to pass between the two calibration lines. The method is suitable only for transparent samples.

The viscosity of the sample is calculated from the time required for the ball to pass between the two calibration lines.

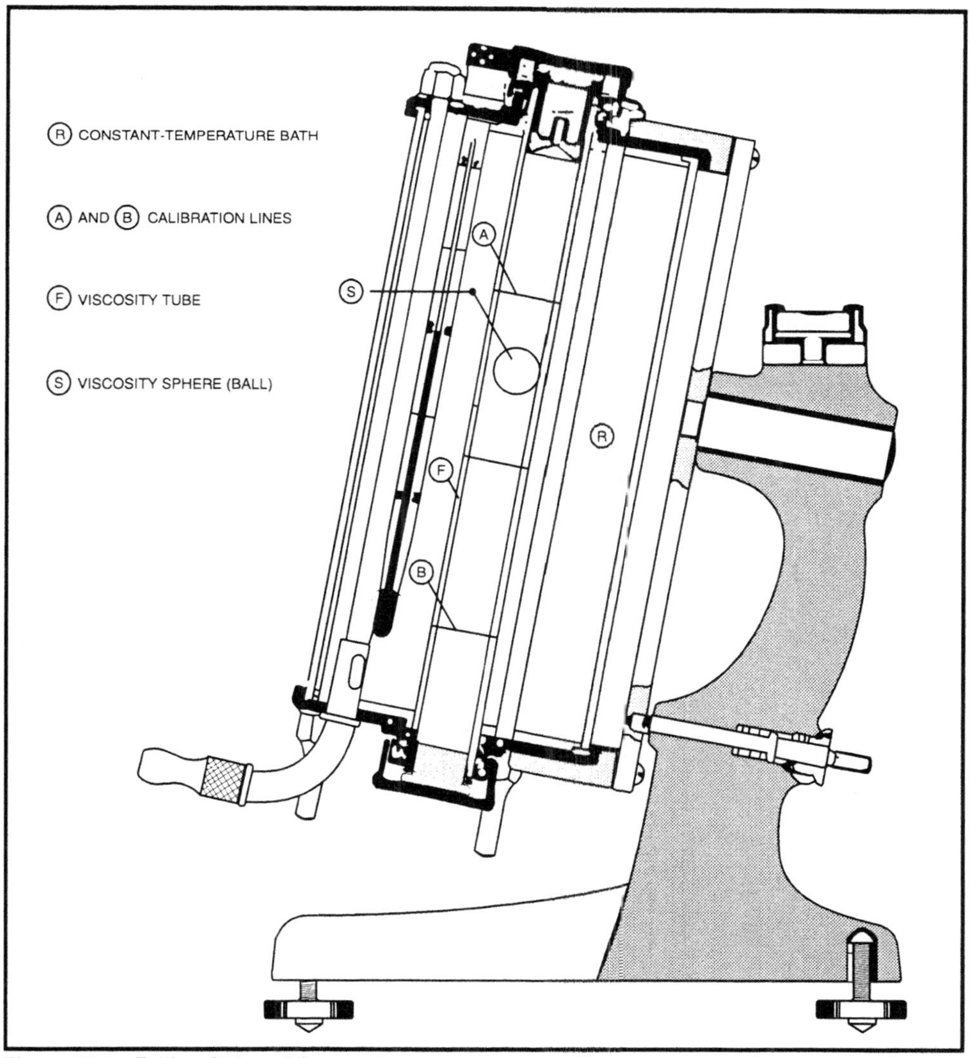

Figure 17-5. Falling-Sphere Viscometer
(From Process Analyzer Technology, Kenneth J. Clevett, New York: John Wiley & Sons)

Rotational Viscometers

These instruments usually use the measurement of a spring's deflection or a small angular deflection to indicate the viscosity of the sample.

Rotational viscometers are based on the principle of sensing the viscous drag imposed on a cylinder or disk rotated at a constant speed or, in a variation, sensing the drag torque applied to a stationary element when a cylinder or disk is rotated in its presence. These instruments usually use the measurement of a spring's deflection or a small angular deflection to indicate the viscosity of the sample.

Figure 17-6 shows one of the more popular rotational viscometer design configurations. A variety of optional shapes and sizes are available for both cylinders and disks. These options, combined with rotational speed variation, provide this style of viscometer with a large and dynamic measurement capability. Rotational analyzers are available in a variety of installation configurations, making them applicable for in-tank, in-line, and bypass installations. Viscometers utilizing this measurement concept are manufactured by several instrument firms, including Brookfield Engineering Laboratories Inc., Haake Mess-Technik GmbH Co., Contraves AG, Eurcontrol Kalle AB, and others.

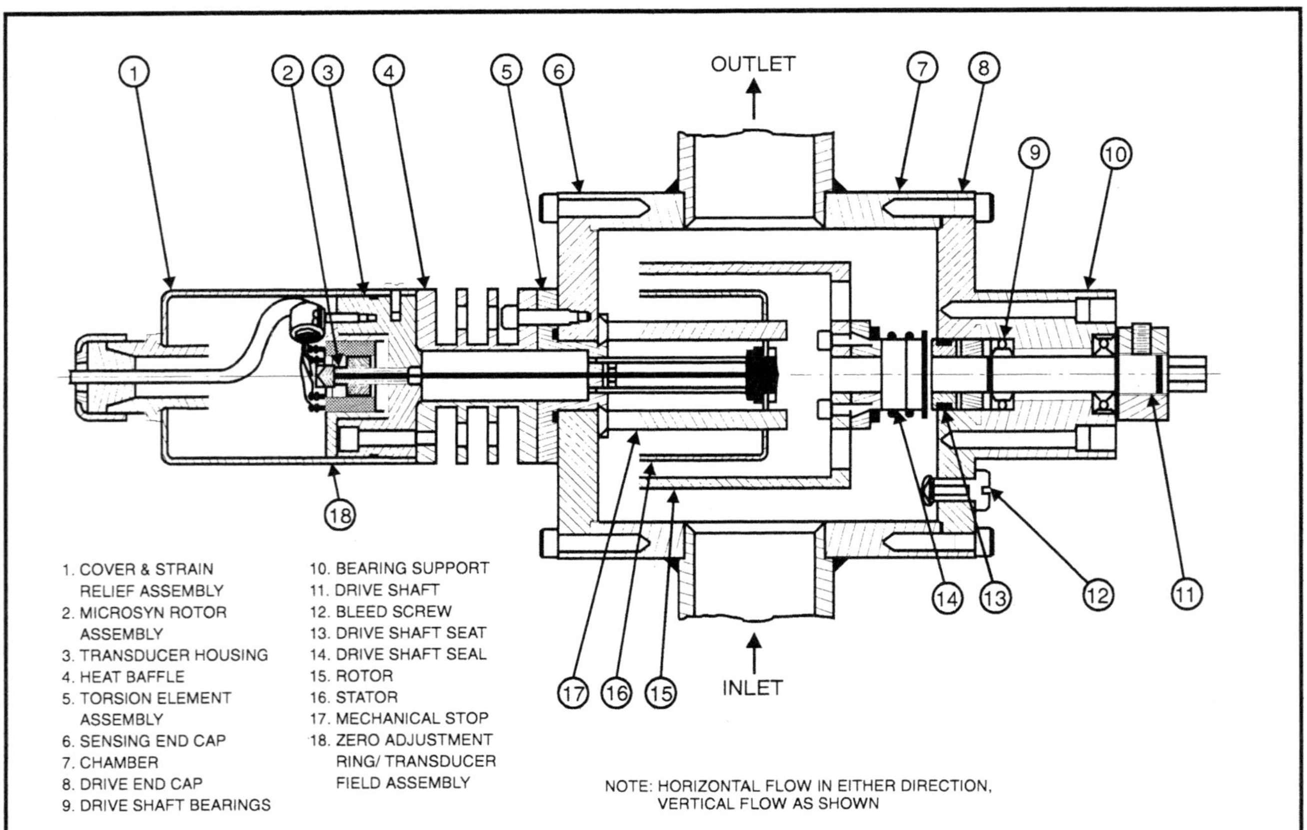

Figure 17-6. Brookfield TT100 Rotational Viscometer
(Courtesy of Brookfield Engineering Laboratories)

Vibrating and Oscillating Viscometers

In vibrating and oscillating viscometers, a probe is vibrated at a specified frequency and the viscous drag of the fluid dampens the amplitude of this vibration. A measurement of this damping is related to the viscosity of the fluid in which the probe is immersed. In some cases, the damping measurement is a direct measurement of the amplitude of the vibration; in others, it is the measure of the required frequency of reenergization of the oscillation to keep the vibrational amplitude above a predetermined level; and in still others, it is related to the power required to maintain a constant amplitude of oscillation.

The basic design of these sensors is that of an insertion probe concept, which allows flexibility of installation positions (in-vessel, in-line, bypass, and many physical orientations). Some of these instruments come with an optional internal temperature sensor built into the probe, and others require an independent temperature sensor input for temperature compensation.

Figure 17-7 depicts a style of vibrating viscometer. Viscometers utilizing these measurement concepts are manufactured by several instrument firms, including Nametre Company, ABB Process Analytics, Automation Products Inc., Haake Division of Fisons Instruments Inc., and others.

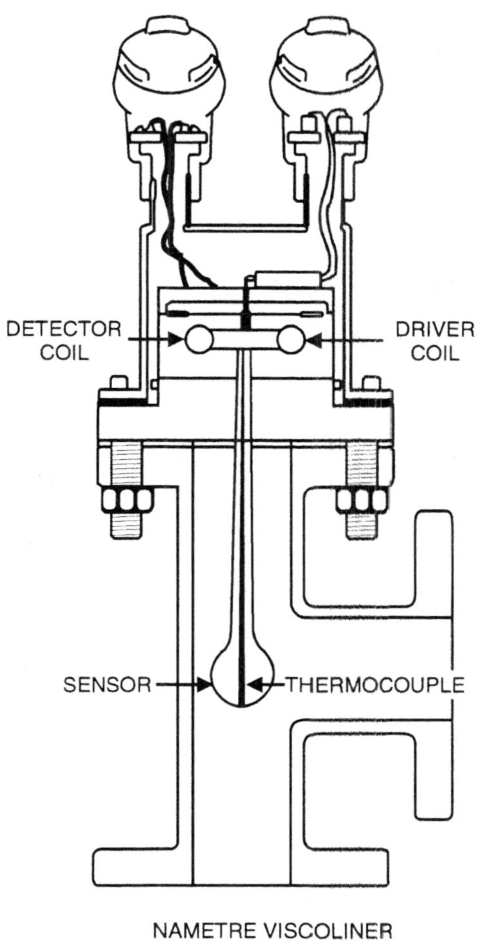

Figure 17-7. A Typical Vibrating Viscometer Design
(Courtesy of Nametre Company)

Moving Piston/Ball Viscometers

In moving piston/ball viscometers, the fluid viscosity is correlated to a time of movement. In the case of a Norcross Model 8B (Figure 17-8a), a piston rod and piston (inside a cylinder) are periodically raised, drawing fluid into the space under the piston. The piston rod and piston are then allowed to fall by gravity (expelling the fluid out the same path through which it was drawn into the space). The instrument measures the time-of-fall to a predetermined level and converts this time into the fluid's viscosity (with the radial clearance between the piston and the cylinder wall forming the measuring orifice).

In the Cambridge moving piston (Figure 17-8b), two coils built into the wall of the sensor move a piston (suspended in the fluid) back and forth by magnetic force and also detect the two-way motion. Movement of the piston is resisted by the viscosity of the fluid, so a measurement of the piston travel time is an indication of fluid viscosity. Inward and outward travel times are averaged to minimize flow-force effects on the measurement.

The basic design of the falling ball and falling piston types of instruments (where gravity is a factor) requires a more restrictive installation philosophy and generates data in a cyclic (batch) mode, rather than continuously. Designs like the Cambridge magnetic unit offer much more flexibility with regard to installation restrictions and are nearer to a continuous measurement. In these measurement approaches, the density difference between the piston/ball and the fluid must also be factored into the viscosity calculation.

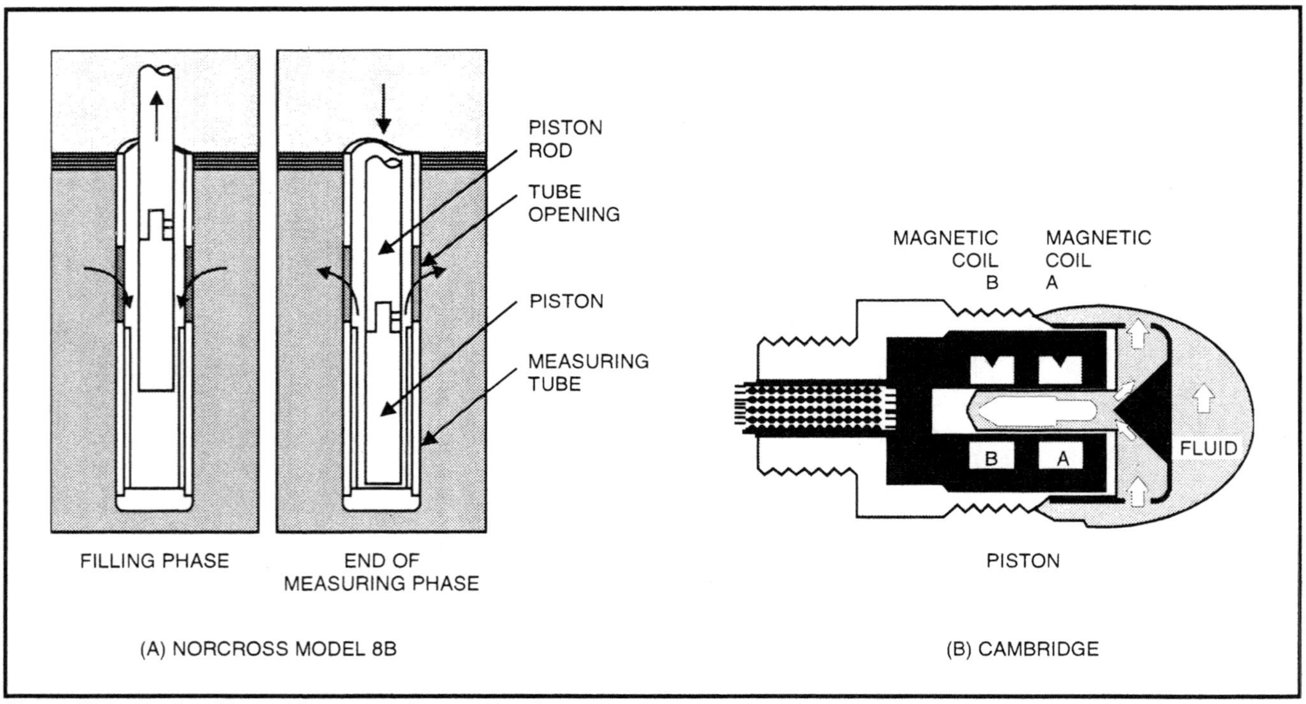

Figure 17-8. Moving Piston Viscometer Sensor Heads
(Courtesy of Norcross Corporation and Cambridge Applied Systems)

Rheometers

The graphical representations previously shown in Figure 17-1 are commonly called rheograms. From these depictions it is easy to see that a rheometer does more than measure a single viscosity. It has the capability to vary the shear rate that is applied to the fluid and report the corresponding viscosities. The sensor can interact with the fluid/material in an interactive way, providing much more information than a simple viscosity measurement (including information about stretch, elasticity, reaction to shear, and more). If the fluid is Newtonian, a simple viscosity measurement is usually all that is needed; the monitoring and study of more complex materials may benefit from the detailed characterizations that can be done on a rheometer.

If the fluid is Newtonian, a simple viscosity measurement is usually all that is needed; the monitoring and study of more complex materials may benefit from the detailed characterizations that can be done on a rheometer.

Most rheometers are similar in concept to the viscometers already discussed. Some are like the capillary viscometer, with a variable-speed pump that varies the shear rate and which has an expanded measurement range. Some are like rotational viscometers, with variable-speed drive motors and more sophisticated data-processing capabilities. Regardless of design, they basically measure a relationship between the deformation of a material and the force associated with that deformation. Rheometers are manufactured by Gottfert Werkstoff-Prufmaschinen GmbH, Rheometrics Inc., Seiscor Technologies Inc., and other companies.

Other Techniques

Besides an actual viscosity measurement, there are many ways to get an indication of viscosity from other process information (given appropriate fluid and process parameters). These can be especially useful if all you need is a trend or ballpark value. A change in pressure drop across an orifice plate under steady temperature and flow conditions, a change in the load on a pump under constant flow and temperature, or the change in the load on an agitator under constant conditions could all indicate a change in viscosity.

Rosemount is currently proposing the use of their Coriolos-principle mass flowmeter along with a differential-pressure transmitter to provide mass flow, volumetric flow, temperature, density and viscosity. Such a method is limited only by your knowledge of the process, the materials, and the available sources of information. Remember, you may not need a viscometer to obtain adequate viscosity information about your fluid.

Remember, you may not need a viscometer to obtain adequate viscosity information about your fluid.

Common Applications

Capillary viscometers are used mainly in the petroleum industry to control the blending of lubricating oils, fuel oils, and the dilution of viscous residual fuels. The vibrating, rotational, and falling-piston types are better suited to non-Newtonian materials and are usually used in the petrochemical, plastics, drug, and food industries. Table 17-1 lists several applications that are currently using on-line (process) viscometers.

Table 17-1. On-Line Process Viscometer Applications

Industry	Application
Petroleum	Fuel oil viscosity control Lubricating oil viscosity control Asphalt and wax viscosity control
Chemical	Polymerization process control Dilution level control Material blending control Reaction termination Pastes and gels manufacturing Resin bodying
Plastics	Injection molding feeds (melt) control Polyester fiber production (extrusion) Film thickness control
Food	Monitoring and controlling the dilution of syrups, concentrates, batters, etc. Thickening of materials such as catsup, pudding, ice cream, salad dressing, cottage cheese, yogurt, frostings, etc.
Pharmaceuticals	Concentration/dilution control of slurries, suspensions, and liquids
Printing	Ink viscosity
Paint	Solvent control for spraying and dipping Ability to coat
Paper	Starch cooking process control Paper-coating fluids control Paper pulp processing
Power	Control of coal slurries and fuel oil fed to the boiler

Note: Other popular applications include automotive (paint and sound dampening coatings), personal care products (toiletries, lotions, gels, etc.), household products (detergents, paints, adhesives, etc.), textile (dyes), cement (slurries), and pottery and china (clay slips, coatings, suspensions, etc.).

When designing a viscosity measurement system, the rheological behavior of the fluid must be given careful attention in all areas of the design (process conditions, upset conditions, sample conditioning system, analyzer design, etc.).

Most viscometers, except for capillary viscometers and rheometers, are relatively inexpensive. The wide variety of sample types accounts for the wide variety of viscometer instrument designs. This is an area where an instrument cannot be judged by its price. The price is more often related to the viscometer's design and cost of production than to its accuracy or capability. Some work better on slurries, non-Newtonian fluids, pastes, and gels. Others are more adapted to hazardous area installations, high-temperature or high-vibration process areas, or in-line (in-pipe) installation.

In the case of non-Newtonian fluids (where viscosity is a function of shear rate), every viscometer design is likely to give a different viscosity value, since they generally operate at different shear rates. Even different models of the same basic design can generate different shear rates. Simple changes in sample flow rates can result in different viscosity values for the same material.

In the case of a new application where data on the process and material are limited, it is best to provide a sample to vendors for evaluation in their laboratories on their equipment. If the application is of a proprietary nature or may offer your company a competitive advantage, obtain appropriate confidentiality agreements with vendors before divulging the application and/or sending out samples. The fact that a vendor can make the measurement in their lab is not a guarantee that the instrument will work well on your sample or in your plant. If it doesn't work in the lab, you can be sure it won't work in your plant. Several factors could negate a successful vendor lab trial: sample degassed, solids settled out, single phase at lab/sampling conditions and multiphase at process conditions, insufficient run time, samples not representative of all upset conditions, process vibration, full range of process and ambient conditions, etc.

If the application is of a proprietary nature or may offer your company a competitive advantage, obtain appropriate confidentiality agreements with vendors before divulging the application and/or sending out samples.

Temperature variations in a process sample can be automatically compensated in many on-line process viscometers. Because temperature has an impact on viscosity, this factor must be seriously considered in most applications. Temperature compensation/control capabilities vary dramatically from instrument to instrument. Some instruments (e.g., capillary viscometers) measure sample temperature in the sample cell, while others take an external temperature from another point in the process. Some can only use simple linear correction relationships, while others can handle complex nonlinear relationships.

Are you interested in viscosity at process conditions in terms of its impact on processing or to meet a product specification? The answer to that question could help determine whether you need temperature compensation. If so, you must first determine the effect that temperature has on your sample at the other processing conditions. A simple linear relationship affords many options. A few viscometers have some nonlinear temperature compensation capability; this should improve as more analyzers are upgraded to microprocessor controllers.

Some measurements may go beyond the scope of simple in-analyzer temperature compensation. In one case where a family of products was made in the same process, the temperature compensations were quite complex and differed for each product. Compensation was done off-line, in the process control computer. The uncompensated viscosity and temperature were fed into the control system, along with information about the viscosity/temperature relationship of the product being run; the system calculated the correct viscosity, which was used for control. If you can't acquire all the capability you need in the viscometer itself, consider enhancing it with data processing power as needed.

Viscosity measurements are often reliable on pure liquids, solutions, and some slurries. Caution must be exercised when making measurements on emulsions and multiphase systems.

Ranges of Variables

Capillary viscometers can cover viscosities up to 4,000 cP and operating temperatures up to 302°F (150°C). Other types can also be used for materials (e.g., plastics and slurries) that have viscosities in the millions of centipoise. Some units can operate at temperatures as high as 700°F (370°C).

Limitations

Each type of viscometer and each model has limits on the acceptable types of samples, ranges of viscosities, and operating temperatures and pressures. Consult the manufacturer's instruction manual for operating specifications.

Installation

Design considerations include sample flow, sample temperature and pressure, instrument mounting, full wetting of the sample cell, bubbles in the sample, sample filtration, area classification, process vibration, and ambient conditions.

Installation is a critical part of any successful project. In the case of a viscometer project, the installation design is also part of the analyzer selection process. Systems such as the capillary viscometer or the falling piston design are not compatible with an in-line (in-pipe) installation. Because they don't hang on a pipe or vessel, they require a bypass loop, an appropriate installation area, and proper orientation. Design considerations include sample flow, sample temperature and pressure, instrument mounting, full wetting of the sample cell, bubbles in the sample, sample filtration, area classification, process vibration, and ambient conditions. If these conditions are a problem, they can often be changed. The installation can include a sample conditioning system design that can modify many of them.

Your vendor must understand your process installation limitations and you must understand the limitations of the instrument's design. If possible, always follow the vendor's installation recommendations. In the example installation guidelines shown in Figure 17-9, the vendor advises how to orient the probe with respect to sample flow. This recommendation is based on the sensor design, and it is easy to see that the analyzer's performance would deteriorate if the flow was pushing against the sensor (especially against the direction of vibration). The analyzer performance specification assumes that the only damping impact on the vibrational amplitude is the viscosity of the fluid; flow impacts would be reflected as errors in the measured viscosity. Don't get carried away with saving capital money by installing viscosity probes in an in-line pipe tee or at the back of a vessel in an inaccessible area. Sometimes it is more economical in the long run to put in a bypass loop or to reroute some piping. Your maintenance personnel will need convenient access to these analyzers.

In most cases, the model 1800 Viscometer may be installed with little or no rework to existing facilities. However, the instrument measures sample conditions at the point of exposure to the sensor, so attention should be given to the method and point of sampling to assure that desired results are obtained.

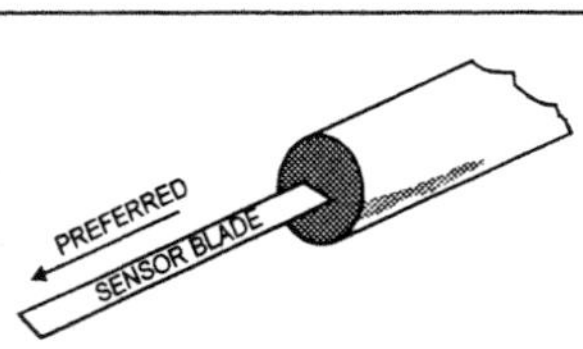

Sensor Installation in a Pipe

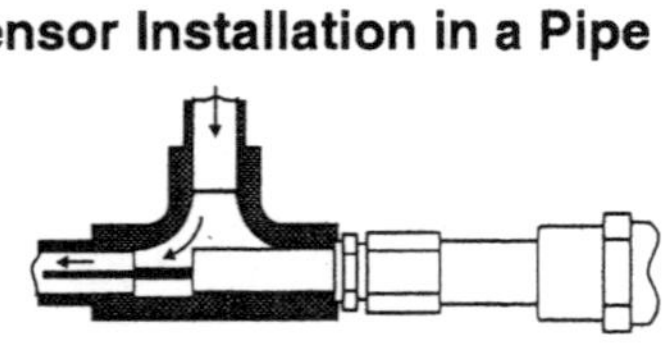

With an adequate connection, the sensor may be installed in any tube or pipe with a diameter of 9/16" or more. A standard 3/8" female pipe tee may be used to mount the 6-1/2" long sensor. The ideal mounting is to weld a 1" coupling on the centerline of a pipe elbow through which the sensor is installed. The sensor should penetrate a minimum of 2-1/2 × pipe diameters. If the sensor must be mounted at right angles to the pipeline, a sensor guard should be used and the blade positioned with respect to the flow.

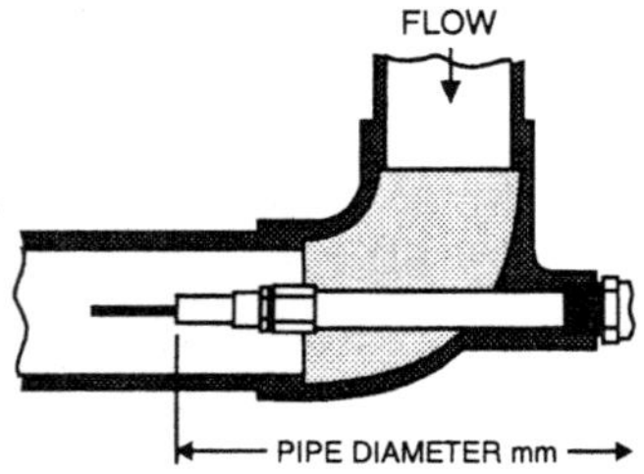

Automatic Reference Temperature Compensator Sensor Mounting

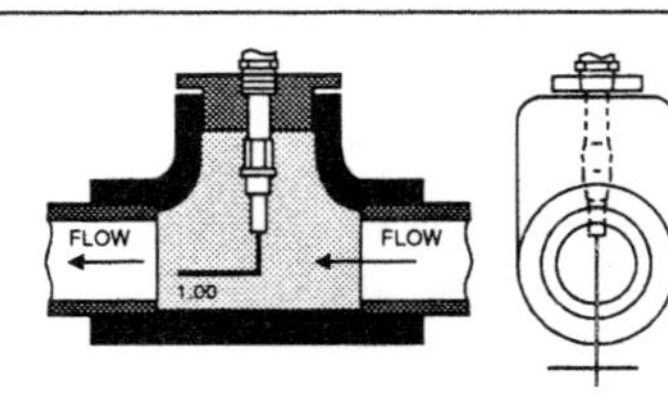

The RTC sensor is not "position sensitive" with respect to flow, so the direction of mounting is not important. However, it must be exposed to the same fluid temperature as the viscosity sensor.

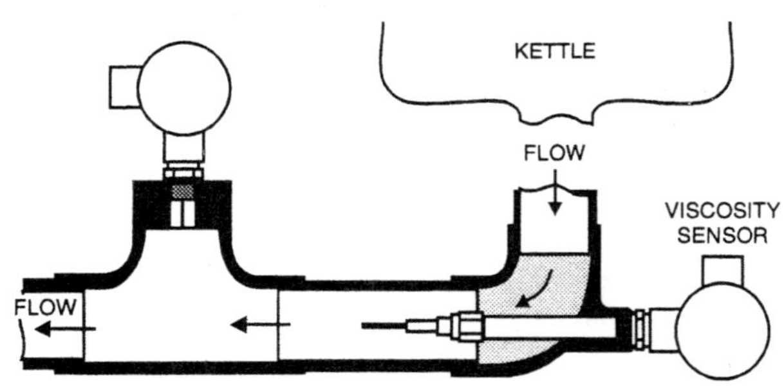

Viscosity Sensor Installation in Vessels

In kettle installations using pumps for agitation, mount the sensor in the exit line. If agitation is accomplished with impellers, the sensor should be exposed to minimum turbulence.

Figure 17-9. Vibrational Process Viscometer Installation Guidelines
(Courtesy of ABB Process Analytics)

Specific Safety Issues

Most field viscosity units are rated for installation in National Electrical Code (NEC) Class I, Group D, Division 1 areas. The control units, however, may require installation in electrically non-hazardous areas. The oil baths of capillary units that operate at elevated temperatures should be allowed to cool before they are lowered.

Unique Sample Conditioning System Requirements

If samples contain particulate matter, a filter should be considered for installation in the fast loop of the sampling system. Highly viscous samples may require heat-traced lines to bring the sample to the analyzer. For some operations, an in-line viscometer would be more appropriate.

Typical Calibration Requirements

Calibration systems that use prototype samples with known viscosities may be practical for some viscometers, such as capillary units monitoring relatively low-viscosity materials. On the other hand, calibration samples that have very high viscosities and which require high temperature in order to remain in a liquid state can decompose between periods of use. In such cases, the analyzer should be calibrated with samples that have been analyzed recently in the laboratory.

Calibration samples that have very high viscosities and which require high temperature in order to remain in a liquid state can decompose between periods of use.

Troubleshooting and Maintenance

Each type of analyzer requires some degree of routine maintenance to assure accurate results. The recommended procedure for one capillary viscometer model is as follows:

1. At least once a week, observe the level of the liquid in the oil bath through the sight glass. Add additional oil, if needed.
2. Once a month, stop the pump motor and check the level of oil in the gear case by removing a level plug on the side of the case. If the level is low, remove the vent plug and add oil.
3. At regular intervals, depending on sample condition, stop the pump motor. Remove the inlet filter screen in the Y-strainer and clean it or replace it.
4. Once a month, stop the pump motor and remove the capillary assembly. Remove the capillary inlet screen, if used, and clean or replace it. Inspect the bore of the capillary and clean it, if necessary. When reassembling, use new O-rings on the capillary tube and the capillary assembly.
5. At 6-month intervals, check the pumping rate of the metering pump. It should be 67 mL/min for 60 Hz power or 50 mL/min for 50 Hz. It should not vary more than 1 mL/min from these values.

The sample itself will influence analyzer performance and maintenance requirements. A sample that is nonlubricating to the metal gear pump that is usually found in a capillary viscometer will result in more frequent pump maintenance than a sample that is lubricating. A sample that contains abrasive particles will increase not only pump maintenance, but also maintenance to all parts of the instrument's measuring orifice that are exposed to the sample. Some samples may be corrosive. Others may polymerize and coat the viscometer, necessitating frequent equipment cleanouts. Some viscometer designs contain delicate parts in areas where they can be damaged easily during vigorous cleanouts.

Process conditions can also influence maintenance. Process viscometers often are selected with only "normal" process conditions in mind. What happens during process startups and shutdowns? What happens during process upsets? Will a maintenance person be required each time the unit is started up and/or shut down? Is the process area subject to high levels of vibration? Vibration not only tends to shake things loose, but in close-tolerance moving equipment it also often leads to mechanical damage and increased maintenance.

Bibliography

1. Clevett, Kenneth J., *Process Analyzer Technology*, New York: Wiley Interscience, 1986.
2. Considine, Douglas M., *Process Instruments and Controls Handbook*, New York: McGraw Hill, 1957.
3. Kalotay, P., and Schaffer, D., "On-Line Viscosity Measurement with Coriolis Mass Flowmeters," in *ISA/91 Conference Proceedings*, Research Triangle Park, NC: Instrument Society of America, 1991, p. 1029.
4. Moore, Ralph L., *Process Analyzers and Recorders – Basic Instrumentation Lecture Notes and Study Guide*, Research Triangle Park, NC: Instrument Society of America, 1982.
5. Nichols, Gary D., *On-Line Process Analyzers*, New York: Wiley Interscience, 1988.

About the Author

James F. (Jim) Tatera is a Senior Analytical Systems Specialist with the Dow Corning Corporation. His experience has included both US and international assignments in analytical research, process engineering, project engineering, production management, and maintenance management. More than 15 of the last 23 years of his career have been involved working with process analyzers. He has published several papers and participated in several professional conferences as a presenter, session developer, panelist, and workshop leader.

Mr. Tatera received a B.S. in Chemistry and Math from the University of Wisconsin in 1968, attended a graduate program in Chemical Instrumentation and Analysis at the University of Minnesota, and graduated from the US Army Officers Candidate School. He also received an MBA from Central Michigan University in 1982. He is one of the original Certified Specialists in Analytical Technology (CSAT) and a member of the ISA, ACS, and AWMA. He is active in the US and international standards activities, and he is the ANSI USNC Technical Advisor to IEC SC65D (Analyzing Equipment in Industrial Process Measurement and Control). He has also served as the ISA SP76 (Composition Analyzers) committee chairman.

Conard Donnell received a Bachelor's and a Masters degree in Chemistry and Physics from Temple University. His career with automatic process analyzers began and ended at Sunoco Refinery where he developed on-line instrumentation to measure the physical properties of process streams automatically and much faster than laboratory analysis. Since retirement, Mr. Donnell has worked as a consultant for Sunoco, Fluid Data, Precision Scientific Co., Lockwood and McLorie, and DuPont working on projects that included design and installation of process analyzer systems in refineries worldwide.

18

Moisture/Dew Point Analyzers

This chapter describes the methods and technologies for selective moisture detection at trace levels. The parameter that is measured is the partial pressure of water within the sample matrix. Moisture values are expressed as units of pressure, dew point (DP), frost point (FP), or calculated values incorporating independent parameters. The requirements of sample conditioning system design, including materials compatibility, are discussed in relation to the physical and chemical behavior of water. Important mathematical relationships are presented to aid in understanding certain applications and the behavior of hygrometers for process analysis.

Hygrometers are the focus of this chapter. Moisture can also be measured in gases and liquids by adapting Karl Fischer titration, gravimetric techniques, gas chromatography, mass spectrometry, and dispersive and nondispersive infrared spectroscopy. Sample handling, required skill level, and expense tend to go against the general application of these analytical methods. The goal in process applications is to reduce the moisture measurement to a continuous display or signal output with a minimum of operator involvement or maintenance.

Measurement Principles

Dew Cups

Moisture will condense when cooled sufficiently. The formation of dew or frost on a surface is related to the partial pressure of water and the temperature at which an equilibrium is achieved. The operation of a dew cup involves running a sample gas through a chamber with a viewing window, with the facility to chill the instrument. Cooling progresses by means of a dry ice/acetone mixture, or liquid carbon dioxide. Observation of condensate is noted against the temperature to identify the dew point or frost point (for values below 32°F [0°C]).

A dew cup, or the several commercial adaptations of this basic principle, is used for point testing of process conditions.

Chilled Mirrors

The electronic versions of the dew cup automate the performance of the condensation experiment. A Peltier device (in single or multiple stages) replaces the chilling fluid. For greater temperature depression, the additional cooling capacity of a water heat exchanger or chlorofluorocarbon refrigeration is utilized to complement the thermoelectric cooler. The sensing mirror temperature is depressed at a programmed rate. In the static or equilibrium-type instruments, the condensate is detected by sensitive optical elements and maintained by feedback control of the mirror temperature (equilibrium dew mirror).

An alternative method is to log the temperature for the event of condensation, and initiate an immediate dew removal cycle. Process moisture levels are monitored by a regular sequence of programmed chilling, dew detection, and reheat (dynamic dew mirror). The dynamic optical hygrometer forms supercooled water rather than frost below 32°F (0°C), as is the case with the equilibrium dew mirror instruments. This is important to recognize, as dew point and frost point temperature tables deviate in relationship to water partial pressure below 32°F (0°C).

Electrolytic Hygrometers

Absorbed moisture on a layer of phosphorus pentoxide (P_2O_5) may be dissociated to hydrogen and oxygen. The adsorption rate is driven by the partial pressure of water vapor in a controlled sample flow and is proportional to the magnitude of the direct electrical current. Moisture is typically displayed as parts per million by volume for gas phase, and total pressure is considered in the calibration.

Phosphorus pentoxide sensors are damaged by water and other aqueous solvents. The sampling system design must incorporate features to physically protect the sensor from liquids to maintain accurate calibration.

Piezoelectric Hygrometers

In these instruments, the piezoelectric sensor is calibrated against the resonance frequency of the sensing element (a quartz crystal) which varies with the amount (level) of adsorbed moisture.

Silicon Oxide Moisture Sensors

The sensing element in these analyzers is a nonporous silicon oxide layer which varies impedance with adsorbed moisture. Advantages are rapid offgassing for dry/moisture cycle measurements in point-sampling applications. Disadvantages involve limited sensitivity to partial pressures of moisture in the dew point range below –76°F (–60°C) due to limited surface area in the absence of significant porosity. Calibration involves the assumption that zero moisture is a value below the response of the sensor and that full scale is ambient humidity. There is a temperature coefficient to the calibration. The temperature of the sensor must be controlled to obtain a calibration response specific to absolute humidity.

Aluminum Oxide Hygrometers

Sensing elements of hygroscopic aluminum oxide change in dielectric value with partial pressure of water vapor. Aluminum oxide is a durable coating that shows a greater electrical impedance for drier conditions. The porosity and thickness of the layer influence the range and manner of the sensor response to the partial pressure of water vapor. The thin-film version of the aluminum oxide sensor is suitable for applications drier than –166°F (–110°C) DP/FP for absolute humidity. The temperature influence and hysteresis of calibration is minimized to within the accuracy specifications (±5F° below –85°F, ±4F° to 68°F ±3C° below –65°C, ±2C° to 20°C for the standard probe).

Aluminum oxide is insoluble in water and is compatible with a variety of noncorrosive gases and liquids. The thin-film aluminum oxide sensor may be viewed as a selective pressure transducer for moisture, and DP/FP measurements are measured independently of the total pressure, temperature, and matrix of the process sample stream.

Dew Point / Frost Point Conversion

Table 18-1 gives moisture conversions relating dew point to vapor pressure, ppm by volume, relative humidity, and ppm by weight in air. The relationship of dew point/frost point is modeled in the very complicated function known as the Goff-Gratch equation. Note that moisture over the range –166 to 68°F (–110 to +20°C) DP/FP shifts in partial pressure (and parts per million values) by more than 10^7. The accuracy of moisture calculations is no better than two significant figures, or ±20% of parts per million value for a band of ±2C° DP/FP tolerance.

$$\frac{P_{H_2O}}{P_{total}} \times 10^6 = \text{parts per million by volume}$$

This calculation is a consequence of the ideal gas law, where equal mole fractions of gases occupy the same volume. Parts per million by volume is a ratio without units.

Conversion of ppm by volume to ppm by weight in gas-phase applications involves the ratio of the molecular weight of water to the molecular weight of the balance gas. The average molecular weight of air is 28.96 g/mol; the value for water is 18.016 g/mol.

$$\left([\text{ppm by volume}] \times \frac{MW_{H_2O}}{MW_{\text{Balance Gas}}} \right) = \text{ppm by weight}$$

$$\left(\text{ppm}_v \times \frac{18.016}{28.96} \right) = (\text{ppm}_v \times 0.622) = \text{ppm}_w \text{ for air}$$

Most conversion tables express the partial pressure of water as frost point for condensation temperatures below 32°F (0°C). In dew cup experiments and the operation of chilled mirror devices, the first visible condensate may be supercooled water rather than frost. Refer to Table 18-2 to note the differences. For example, at –40°C FP, the corresponding dew point is –43.7°C. This 3.7C° shift is a change in moisture content from 127 ppm_v to slightly more than 80 ppm_v, about a 37% difference.

Accuracy for each type of hygrometer is based on the analysis of errors related to the principle of operation. Further, the definition of the readout of the instrument must be understood in relation to conversions to other moisture parameters and calculated values.

Table 18-1. Moisture Conversions

Dew Point °F	Dew Point °C	Vapor Pressure (Water/Ice In Equilibrium) mm Hg	ppm on volume Basis at 760 mm Hg	Relative Humidity at 70°F (20°C), %	ppm on Weight Basis in Air
−238	−150	7×10^{-15}	9.2×10^{-12}	...	5.7×10^{-12}
−220	−140	3×10^{-10}	4.0×10^{-7}	...	2.5×10^{-7}
−202	−130	7×10^{-8}	9.2×10^{-5}	...	5.7×10^{-5}
−184	−120	10×10^{-8}	1.3×10^{-4}	5.4×10^{-7}	8.1×10^{-5}
−180	−118	0.00000016	0.00021	0.0000009	0.00013
−177	−116	0.00000026	0.00034	0.0000014	0.00021
−173	−114	0.00000043	0.00057	0.0000023	0.00035
−170	−112	0.00000069	0.00091	0.0000037	0.00057
−166	−110	0.0000010	0.00132	0.0000053	0.00082
−162	−108	0.0000018	0.00237	0.0000096	0.0015
−159	−106	0.0000028	0.00368	0.000015	0.0023
−155	−104	0.0000043	0.00566	0.000023	0.0035
−152	−102	0.0000065	0.00855	0.000035	0.0053
−148	−100	0.0000099	0.0130	0.000053	0.0081
−144	−98	0.000015	0.0197	0.000080	0.012
−141	−96	0.000022	0.0289	0.00012	0.018
−137	−94	0.000033	0.0434	0.00018	0.027
−134	−92	0.000048	0.0632	0.00026	0.039
−130	−90	0.000070	0.0921	0.00037	0.057
−126	−88	0.00010	0.132	0.00054	0.082
−123	−86	0.00014	0.184	0.00075	0.11
−119	−84	0.00020	0.263	0.00107	0.16
−116	−82	0.00029	0.382	0.00155	0.24
−112	−80	0.00040	0.526	0.00214	0.33
−108	−78	0.00056	0.737	0.00300	0.46
−105	−76	0.00077	1.01	0.00410	0.63
−101	−74	0.00105	1.38	0.00559	0.86
−98	−72	0.00143	1.88	0.00762	1.17
−94	−70	0.00194	2.55	0.0104	1.58
−90	−68	0.00261	3.43	0.0140	2.13
−87	−66	0.00349	4.59	0.0187	2.84
−83	−64	0.00464	6.11	0.248	3.79
−80	−62	0.00614	8.08	0.328	5.01
−76	−60	0.00808	10.6	0.430	6.59
−72	−58	0.0106	13.9	0.0565	8.63
−69	−56	0.0138	18.2	0.735	11.3
−65	−54	0.0178	23.4	0.948	14.5
−62	−52	0.0230	30.3	0.123	18.8
−58	−50	0.0295	38.8	0.157	24.1
−54	−48	0.0378	49.7	0.202	30.9
−51	−46	0.0481	63.3	0.257	39.3
−47	−44	0.0609	80.0	0.325	49.7
−44	−42	0.0768	101	0.410	62.7
−40	−40	0.0966	127	0.516	78.9
−36	−38	0.1209	159	0.644	98.6
−33	−36	0.1507	198	0.804	122.9
−29	−34	0.1873	246	1.00	152
−26	−32	0.2318	305	1.24	189
−22	−30	0.2859	376	1.52	234
−18	−28	0.351	462	1.88	287
−15	−26	0.430	566	2.30	3.51

Table 18-1 (cont'd). Moisture Conversions

Dew Point °F	Dew Point °C	Vapor Pressure (Water/Ice In Equilibrium) mm Hg	ppm on volume Basis at 760 mm Hg	Relative Humidity at 70°F (20°C), %	ppm on Weight Basis in Air
−11	−24	0.526	692	2.81	430
−8	−22	0.640	842	3.41	523
−4	−20	0.776	1,020	4.13	633
0	−18	0.939	1,240	5.00	770
+3	−16	1.132	1,490	6.03	925
+7	−14	1.361	1,790	7.25	1,110
+10	−12	1.632	2,150	8.69	1,335
+14	−10	1.950	2,570	10.4	1,596
+18	−8	2.326	3,060	12.4	1,900
+21	−6	2.765	3,640	14.7	2,260
+25	−4	3.280	4,320	17.5	2,680
+28	−2	3.880	5,100	20.7	3,170
+32	0	4.579	6,020	24.4	3,640
+36	+2	5.294	6,970	28.2	4,330
+39	+4	6.101	8,030	32.5	4,990
+43	+6	7.013	9,230	37.4	5,730
+46	+8	8.045	10,590	42.9	6,580
+50	+10	9.209	12,120	49.1	7,530
+54	+12	10.52	13,840	56.1	8,600
+57	+14	11.99	15,780	63.9	9,800
+61	+16	13.63	17,930	72.6	11,140
+64	+18	15.48	20,370	82.5	12,650
+68	+20	17.54	23,080	93.5	14,330
+72	+22	19.83	26,092	Over 100	16,200
+75	+24	22.38	29,447	Over 100	18,284
+79	+26	25.21	33,171	Over 100	20,596
+82	+28	28.35	37,303	Over 100	23,162
+86	+30	31.82	41,868	Over 100	25,996
+90	+32	35.66	46,921	Over 100	29,133
+93	+34	39.90	52,500	Over 100	32,597
+97	+36	44.56	58,632	Over 100	36,405
+100	+38	49.69	65,382	Over 100	50,596
+104	+40	55.32	72,789	Over 100	45,195
+108	+42	61.50	80,921	Over 100	50,244
+111	+44	68.26	89,816	Over 100	55,767
+115	+46	75.65	99,539	Over 100	61,804
+118	+48	83.71	110,145	Over 100	68,389
+122	+50	92.51	121,724	Over 100	75,579
+126	+52	102.09	134,329	Over 100	83,405
+129	+54	112.51	148,039	Over 100	91,918
+133	+56	123.80	162,895	Over 100	101,142
+136	+58	136.08	179,053	Over 100	111,175
+140	+60	149.38	196,553	Over 100	122,040

Table 18-2. Dew Point/Frost Point Relationships [Ref. 1]
Below 32°F (0°C), dew point hygrometers measure the frost point temperature rather than the dew point. This table permits conversion from dew point to frost point.

English Units (°F)

FP	DP	FP	DP	FP	DP	FP	DP
+32	+32	+10	+7.4	−12	−16.7	−34	−40.3
+31	+30.8	+9	+6.3	−13	−17.8	−36	−41.4
+30	+29.7	+8	+5.2	−14	−18.9	−36	−42.4
+29	+28.6	+7	+4.1	−15	−20.0	−37	−43.5
+28	+27.5	+6	+2.9	−16	−21.1	−38	−44.5
+27	+26.4	+5	+1.8	−17	−22.2	−39	−45.6
+26	+25.2	+4	+0.7	−18	−23.3	−40	−46.6
+25	+24.1	+3	−0.4	−19	−24.3	−41	−47.7
+24	+22.9	+2	−1.5	−20	−25.4	−42	−48.7
+23	+21.8	+1	−2.5	−21	−25.4	−43	−49.8
+22	+20.7	0	−3.7	−22	−27.5	−44	−50.8
+21	+19.6	−1	−4.8	−23	−28.6	−45	−51.9
+20	+18.5	−2	−5.8	−24	−29.6	−46	−52.9
+19	+17.4	−3	−6.9	−25	−30.5	−47	−54.0
+18	+16.2	−4	−8.0	−26	−31.7	−48	−55.0
+17	+15.1	−5	−9.1	−27	−32.8	−49	−56.1
+15	+14.0	−6	−10.2	−28	−33.9	−50	−57.1
+15	+12.9	−7	−11.3	−29	−35.0	−51	−58.2
+14	+11.8	−8	−12.4	−30	−36.1	−52	−59.2
+13	+10.7	−9	−13.5	−31	−37.2	−53	−60.3
+12	+9.6	−10	−14.6	−32	−38.2		
+11	+8.5	−11	−15.6	−33	−39.3		

Metric Units (°C)

FP	DP	FP	DP	FP	DP	FP	DP
0	0	−12	−13.4	−24	−26.6	−36	−39.4
−1	−1.2	−13	−14.5	−25	−27.7	−37	−40.5
−2	−2.3	−14	−15.6	−26	−28.8	−38	−41.6
−3	−3.4	−15	−16.7	−27	−29.9	−39	−42.6
−4	−4.5	−16	−17.8	−28	−30.9	−40	−43.7
−5	−5.6	−17	−18.9	−29	−32.0	−41	−44.7
−6	−6.8	−18	−20.0	−30	−33.0	−42	−45.8
−7	−7.9	−19	−21.1	−31	−34.1	−43	−46.8
−8	−9.0	−20	−22.2	−32	−35.2	−44	−47.9
−9	−10.1	−21	−23.3	−33	−36.2	−45	−49.0
−10	−11.2	−22	−24.4	−34	−37.3	−46	−50.0
−11	−12.3	−23	−25.5	−35	−38.4		

Calibration Traceability

Manufacturers of hygrometers maintain calibration laboratories that can provide certification of traceability to the National Institute of Standards and Technology (NIST) for their instruments and probes. These data are available with the hygrometers or provided with the recalibration services. As time goes on, more manufacturers are gaining registry to ISO-9000 certification for design, manufacturing, and calibration.

Calibration systems are available in saturation/dilution, permeation tube, and other types. Reference 2 provides details on the design, accuracy, and operation of commercial calibrators. In the range drier than −58°F (−50°C) DP/FP, offgassing times are extended to remove absorbed moisture from the plumbing and sampling systems, as well as from the hygrometer sensor and components. The routine is to pre-dry the system on pure purge gas for 24 h before commencing a calibration from dry to wetter values. Repeats of the procedure and records of data are compiled with application of statistics to create a more defensible calibration table.

Validation

Field calibration systems find greater application to validation (as opposed to calibration) of hygrometer performance. The detailed calibration of the sensors and instrument is the province of a metrology laboratory, either at the manufacturer or within the end user's organization. At some point in the normal operation of a hygrometer, the question of accuracy will arise. Dew point/frost point mixtures wetter than –58°F (–50°C) can be generated by portable calibrators for comparison to the hygrometer readout. While less than a full calibration routine, the demonstration of measurement within the process moisture conditions can be extrapolated to acceptance of the hygrometer performance.

There are limits to the usefulness of cylinder gases as moisture references. A calibration check may be made using a certified analysis cylinder calibration gas. A mixture of 50 ppm_v water in a balance of nitrogen is a reasonably stable and convenient blend for testing a hygrometer and sampling system. Drier mixtures will suffer a greater significance of the adsorption and desorption of water vapor from the surface of the cylinder with temperature and pressure changes.

The upper limit to the moisture fraction is a consequence of pressurization. Condensation of water at 2,000 psig will occur at room temperature for an atmosphere not much wetter than 150 ppm_v, assuming ideal gas law relationships. That is equivalent to a mixture at –36°F (–38°C) when expanded to atmospheric pressure.

The calibration table for aluminum oxide sensors expresses moisture as dew point, but is calibrated to frost point for values below freezing. The calibration check gas of 50 ppm_v is approximately –54°F (–48°C) FP, even though the tables and hygrometer displays will indicate a dew point reading.

With a dew cup or a chilled mirror device, the first evidence of condensation may be supercooled water. In corresponding to the dew point, the 50 ppm_v check standard will be determined to be –67°F (–55°C) DP at atmospheric pressure.

Be aware of the principle of operation for the hygrometer under test and the factors that affect the conversion of moisture parameters to the calibration scale. This ambiguity is inherent to the different principles of hygrometer operation when comparing the readout of one instrument to validate the performance of another type.

Ten Considerations For Making Moisture Measurements In Gases [Adapted From Ref. 3]

1. The sampling system (and process) should be allowed to equilibrate for 24 h or longer.
2. A flowmeter aids in determining the equilibration point and will indicate the integrity of system seals.
3. If a hygrometer reading changes directly with flow rate:
   - Equilibrium has not been achieved for offgassing
   - The system is leaking
   - Both of the above

4. Measurements made at system pressure or the highest possible pressure minimize outgassing of metal parts, wall effects, and so on.
5. For sample systems that vent to atmosphere:
   - Back diffusion of atmospheric moisture is always present. This changes with ambient conditions.
   - Precise measurement of low (–76°F [–60°C] DP) moisture levels may be unachievable in an open-ended system. Back diffusion can be minimized (but not eliminated) with a "pigtail" coil of small-diameter tubing or capillary and maintaining the flow rate at a high practical value.
   - Return sample gas to the process, creating a closed-loop system to eliminate access of atmospheric moisture.
6. Any cross-checking between two or more similar instruments should be done at identical times and conditions. This consideration is particularly important in comparing dissimilar instruments.
7. Sampling techniques and setup procedures should be standardized. Manufacturer's specifications on probe cleaning, sampling, measuring techniques, and interpretation of data should be rigorously followed.
8. Recalibration should be performed every 6 months to 1 year. Spare probes should be maintained in "hot spare" inventory to check against one another, and to keep the hygrometer in service.
9. Become as knowledgeable as possible on the difficulties of moisture measurement, application of the equipment, and interpretation of the data.
10. Define problems carefully. The right approach to equipment selection and measurement techniques will instill confidence in the end results.

Sampling System Design

Sampling system materials for hygrometry are selected with regard to their adsorption of and permeability to moisture. An excellent study of materials compatibility is discussed in a series of articles by Ohmi et al. [Ref. 4-6]. Applications involving ultrahigh-purity gases require chemically-polished stainless steel tubing, orbitally-welded fittings, and bellows valves. The object is to mimic the conditions that might be obtained within a perfectly smooth, inert, and totally-sealed vessel.

Process conditions require some compromise of that ideal. Use stainless steel components whenever feasible in terms of cost and availability. Copper and other metals are impermeable to water vapor, but have more of an active oxide layer that absorbs moisture. Plastics such as nylon should be avoided as they are both permeable to water and slow to offgas surface-absorbed moisture.

Polyethylene, polypropylene, Teflon, and elastomeric tubing may be useful if lengths are short and moisture levels are wetter than –22°F (–30°C). Plastics allow diffusion of ambient moisture into the bore of the sampling lines that is offset by higher flow velocities than are available with metal tubing. In low dew point applications, the infiltration of moisture will cause a significant offset of readings to falsely wet values.

In low dew point applications, the infiltration of moisture will cause a significant offset of readings to falsely wet values.

Fabrication techniques range in quality from orbital-welding with helium leak detection to sleeved fittings. In between, metal gasket/glanded connectors (such as the Cajon VCR™ fitting) are closest to welding. Swagelok® double ferrule fittings are the standard for most applications in terms of integrity of the seal, ease of use, and cost. Other types of compression fittings — inverted flare, threaded, and O-ring systems — may be effective if properly specified for one-single-seal application.

Sampling system components may be necessarily constructed of glass, Teflon, or other materials. Whenever possible, it is advantageous to position flowmeters, gages, sampling ports, and other items downstream of the hygrometer sensing element to minimize the effects of moisture infiltration and offgassing.

Desiccant Air Dryer Sampling System

The design of a sampling system for a desiccant dryer application can be as simple as an adapter bushing to install the hygrometer sensor in a process "tee." This approach eliminates the issues of leaks, back-diffusion of ambient air, and offgassing of sampling tubing. The risks involve contamination and destruction of the sensor during upset conditions of high humidity.

A sampling system incorporating a filter/coalescer (Figure 18-1) is needed to protect the hygrometer sensor in most industrial dryer applications. A good design includes an upstream needle valve for isolation and flow control to downstream components. Next in the path is a filter/coalescer to capture particulate contamination, with a drain valve for continuous removal of liquid. The line pressure can be maintained at the filter by throttling flow to the moisture sensor chamber with an intermediate needle valve. The downstream needle valve may be operated fully-opened to normalize the pressure at the sensor to 0 psig. Venting coils of 1/8 or 1/4 in. (3.2 or 6.4 mm) tubing which are 5 to 10 ft (1.5 to 3 m) in length minimize back-diffusion of ambient moisture.

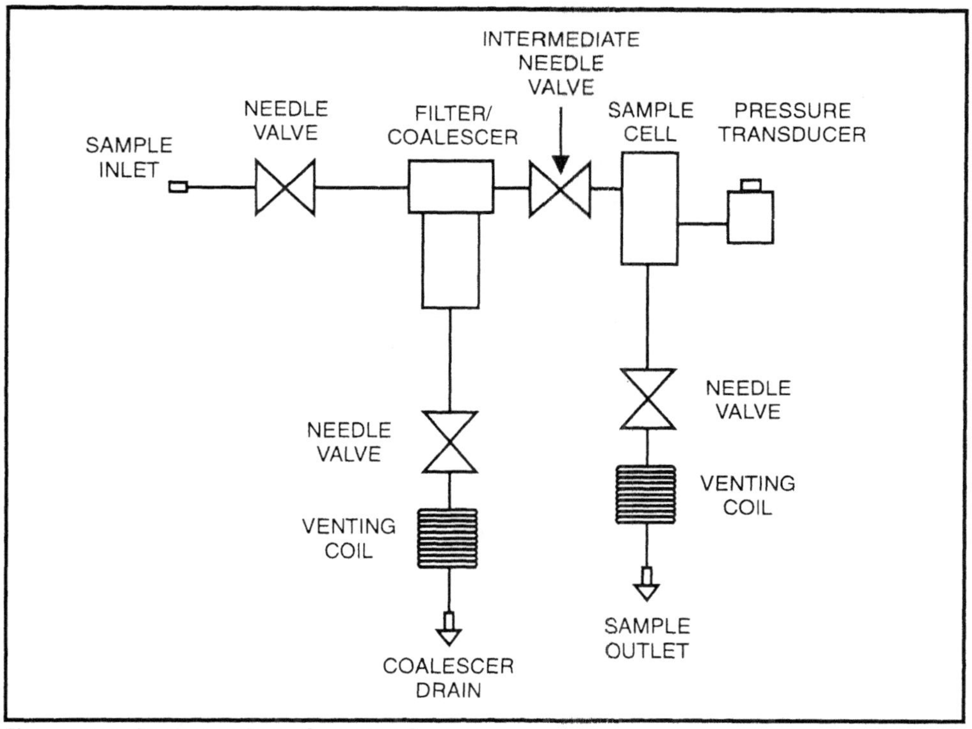

Figure 18-1. Desiccant Dryer Sampling System

By maintaining the filter at line pressure, and operating the sensor chamber at atmospheric pressure, there can never be a condition of 100% relative humidity (RH) at the sensor. Dew point depression (to a drier value) occurs whenever the sample gas pressure is relieved. By coalescing and removing the mist in the filter, the dew point at the sensor will be lower than the temperature of the sampling system filter. This eliminates exposure of the sensor to saturation, and the rust or dust that is carried with condensate. The hygrometer elements will tend to dry out faster, without operator attention, than the desiccant system does after an overload or other process failure.

A couple of principles apply to this technique. The first is to recognize that condensation (100% RH at saturation) occurs when the dew point matches the temperature of the system. Another is to treat moisture as an independent component (Dalton's law) in the total pressure and reduce the ideal gas law of state to a ratio of pressures.

For the sake of simplifying the calculations, an example of a system running at 90 psig, with atmospheric pressure being rounded to 15 psia represents a 7-to-1 ratio. It is calculated as follows:

Correcting for common units:

90 psig = (90 psig + 15 psia) = 105 psia

The pressure ratio is:

105 psia/15 psia = 7 to 1

In the worst case of a dryer failure, continuous mist would drain as liquid water from the filter coalescer at line pressure. At an ambient temperature of 75°F (24°C) the vapor pressure of water, taken from Table 18-1, is 22.38 mm Hg. The dew point depression occurs from throttling the flow with the needle valve between the filter and the sensor by the pressure ratio:

22.38 mm Hg/7 = 3.20 mm Hg

which is a value of approximately 25°F (–4°C) FP (Table 18-1). The relative humidity from this effect is the inverse of the pressure ratio (1/7 = 14% RH) for the condition with the sensor at atmospheric pressure.

The conversion of any dew point measured with a sensor at atmospheric pressure can be calculated back to line pressure values. For example, a hygrometer reading of –40°F (–40°C) FP is a vapor pressure of 0.0966 mm Hg. Multiplied by the pressure ratio:

0.0966 mm Hg X 7 = 0.676 mm Hg

which is a value of approximately –8°F (–22°C) FP at a system pressure of 90 psig pressure.

In desiccant dryer systems that are properly-sized and well-maintained, DP/FP values can be much dryer than –40°F (–40°C) at system pressure. The dew point depression to atmospheric pressure may be undesirable with the greater significance of offgassing and moisture infiltration to the sampling system. The pressurization of the sensor chamber is done by controlling flow with the downstream needle valve and fully opening the upstream and intermediate valves.

In this manner, the hygrometer will directly read moisture as DP/FP at line pressure.

In either case, the moisture specification for the process must be identified and defined in terms of DP/FP and pressure conditions. The choice depends on the moisture levels that are reasonable for the dryer system.

Nonideal Gas Applications

Dalton's law assumes that the moisture and balance gases do not interact chemically. At higher pressures, water and gases can co-condense [Ref. 7]. Conversions of DP/FP, pressure, and parts per million by volume may not follow the ideal gas law calculations above a certain pressure for the particular mixture in the process. Hygrometry of natural gas is nonideal at low pressures and includes special problems of contamination.

The natural gas sampling system (Figures 18-2 and 18-3) includes a coalescer that continuously removes glycol. Flow enters the bottom of the coalescer assembly. The coalescer is like a distillation column with a packing of glass beads and a length with sufficient theoretical plates for separation of the condensable mist from the noncondensable natural gas. The system may include a transducer as well (at the gage) so that pressure values can be read directly by the hygrometer. Since most installations are in the field, the components are assembled and mounted in an insulated-and-heated weatherproof enclosure.

The coalescing of the glycol removes the primary interference and contaminant. Dust is controlled by a fine-element sintered stainless steel filter. The sampling system is operated at process pressures to allow return of the natural gas (for economy) and glycol (for environmental reasons) back into the pipeline. The return point is selected downstream of a valve or regulator (or across a pump) to gain a pressure differential to drive flow through the sampling system.

The DP/FP conversion to industry units requires compensation for pressure and the nonideal behavior of natural gas with water. The Natural Gas Table [Ref. 8] in parts per million by volume or pounds per million cubic feet can be used for the conversion. Microprocessor-based hygrometers are now available with software to display the calculated parameters directly from moisture and pressure inputs.

Figure 18-2. Natural Gas Sampling System

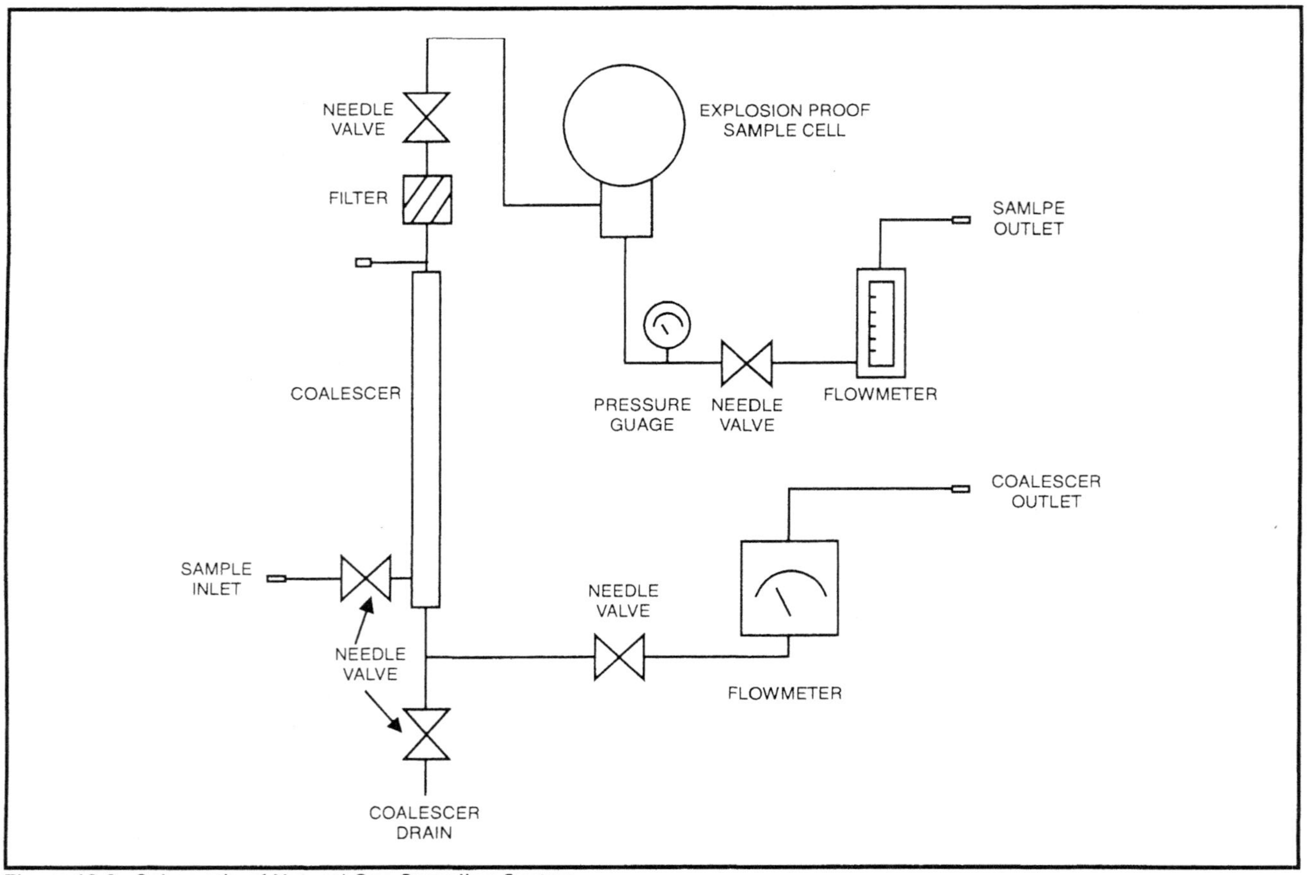

Figure 18-3. Schematic of Natural Gas Sampling System

Moisture in Organic Liquids [Ref. 9]

Aluminum oxide sensors can be applied to the measurement of moisture by direct immersion in organic liquids. Henry's law describes the mass of gas (such as water vapor) that is dissolved in a given volume of solvent as directly proportional to the pressure of that gas at equilibrium. That pressure is expressed by the partial pressure of water determined as dew point. The equation is akin to a relative humidity calculation applied to the saturation value for moisture in the organic liquid. This is done by the measurement of dew point and the temperature within the organic liquid by a dual-sensor probe assembly.

In Henry's law applications of the hygrometer, there are advantages gained by on-line measurement. A properly-designed sampling system eliminates contamination of the fluid that is otherwise endured in handling and transport of a specimen for lab analysis. Direct readout of parts per million by weight values are available, calculated within the microprocessor-controlled hygrometers from tables of saturation versus temperature for the fluid and the partial pressure versus dew point temperature. The equation is:

$$C_w = C_s \times \frac{P_w}{P_s}$$

where:

C_w = ppm_w of moisture that is being calculated from the measured dew point and temperature of the liquid

C_s = saturation value (in ppm_w) of water in the liquid at the measured temperature

P_w = partial pressure of water converted from the measured dew point in the liquid

P_s = partial pressure of water for saturation at the temperature of the liquid

Applications for Henry's law moisture measurement include organic liquids that are not fully miscible in water (e.g., oils, fuels, hydrochlorofluorocarbons, chlorofluorocarbons, and solvents such as tetrahydrofuran). Criteria include sensor stability in the liquid and a saturation table for the moisture to be determined. These factors can be tested by the chemistry lab at the hygrometer manufacturer or at the end-user's facilities. A sample system for hydrocarbon liquid is shown in Figures 18-4 and 18-5. The applicability of Henry's law analysis to various liquids is outlined in Table 18-3. Tables 18-4 and 18-5 list saturation values for hydrocarbons and chlorofluorocarbons, respectively.

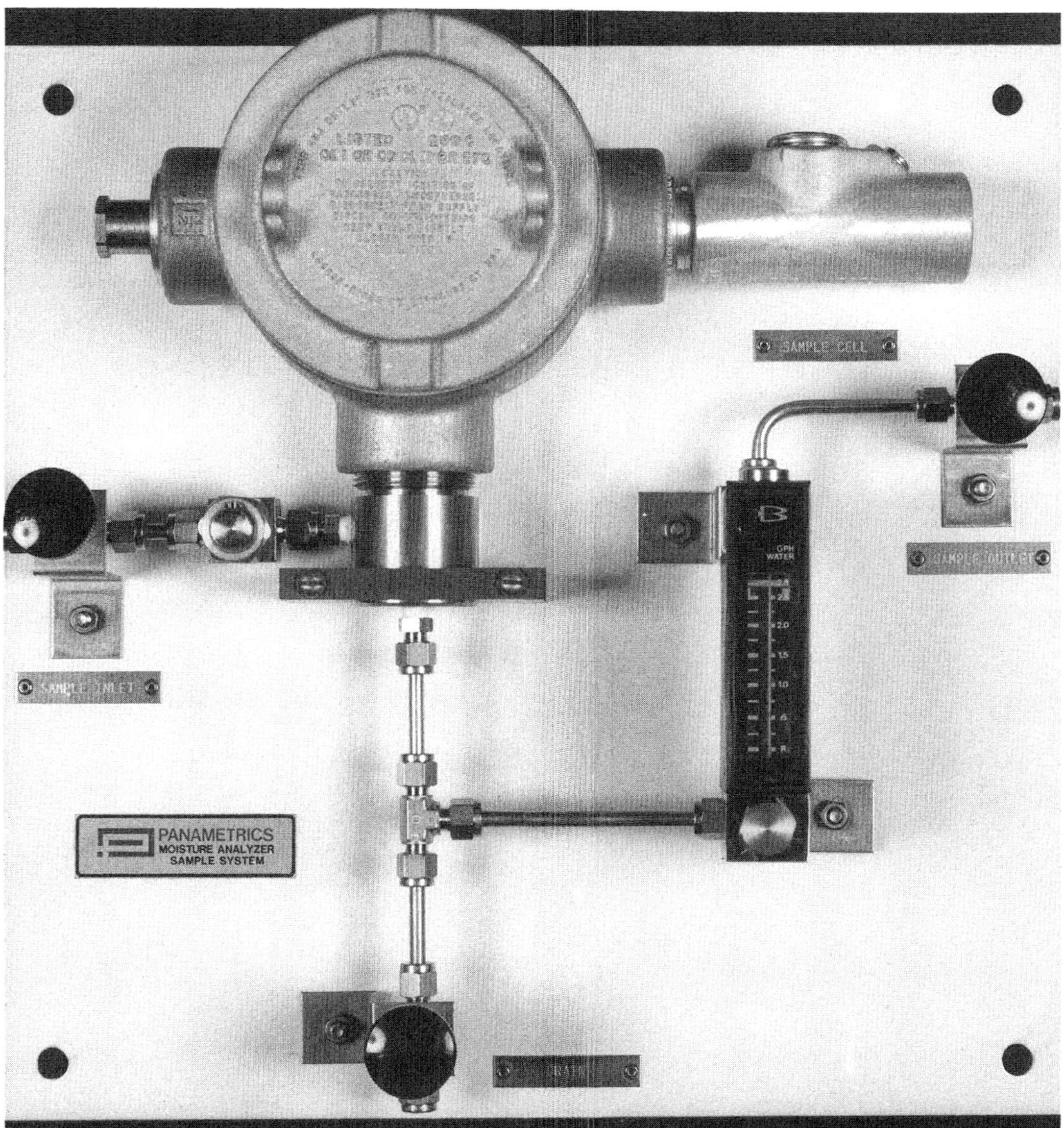

Figure 18-4. Hydrocarbon Liquid Sample System

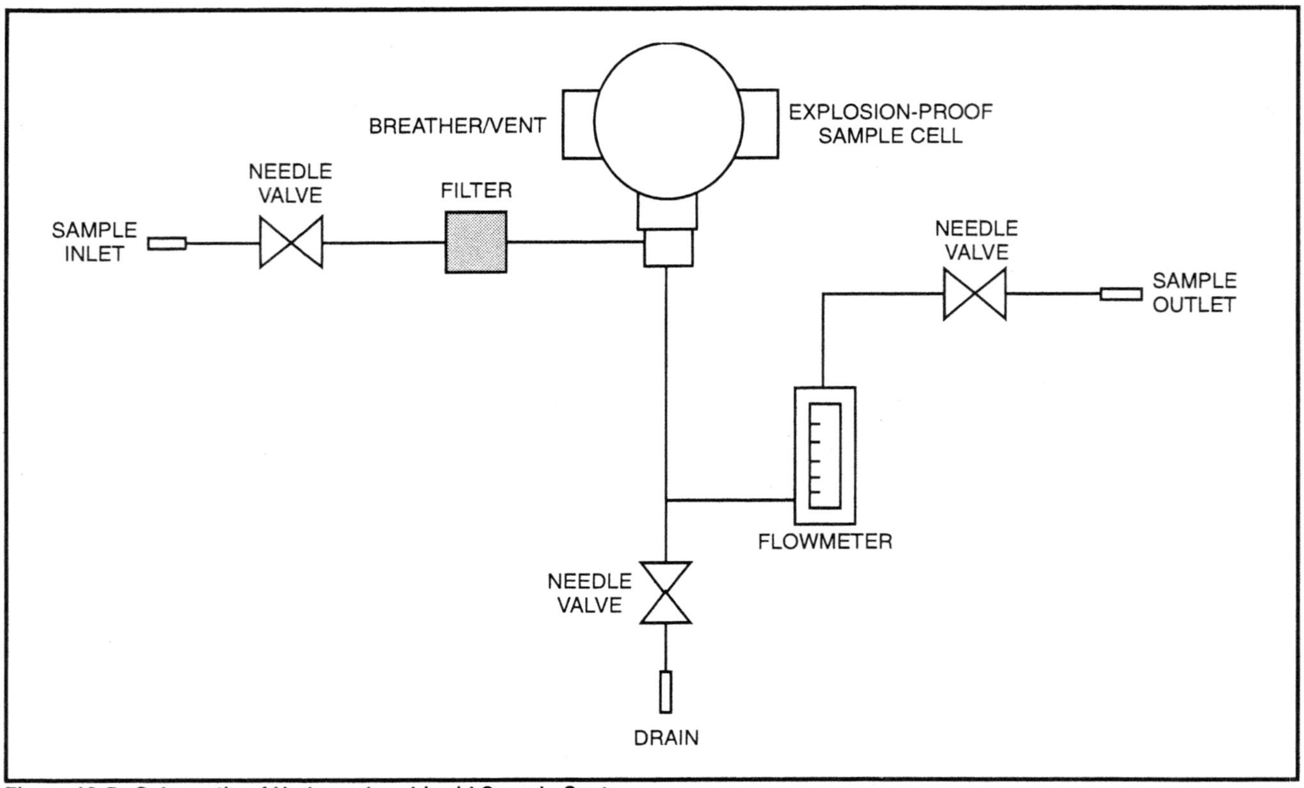

Figure 18-5. Schematic of Hydrocarbon Liquid Sample System

Table 18-3. Applicability of Henry's Law to Analysis of Various Liquids

Liquid	Henry's Law Type Analysis Applicable	Empirical Calibration Required
Acetaldehyde		yes
Acetonitrile		yes
Butylacetate	yes	
Butyl chloride	yes	
Butyl methacrylate	yes	
4-Butyrolactone		yes
Carbon disulfide	yes	
Castor oil	yes	
Chlorobenzene	yes	
Chlorofluorocarbons	yes	
1-Chlorohexane	yes	
1,2-Dichloroethane	yes	
1,2-Dichloroheptane	yes	
Diesel oil	yes	
Diethyl ether		yes
Diethylene glycol		yes
Dioxane		yes
N,N-Dimethylacetamide		yes
N,N-Dimethylaniline		yes
Dodecanol	yes	
2-Ethylbutylacetate	yes	
Ethylene glycol monomethyl ether		yes
Epichlorohydrin		yes
Furfural		yes
Gasoline	yes	
Hexanol	yes	
Hydraulic fluid	yes	
Hydrocarbons	yes	
Isoamyl chloride	yes	
Isobutanol		yes
Jet fuel	yes	
Kerosine	yes	
N-Methylaniline		yes
Methyl cellosolve		yes
Methyl ethyl ketone		yes
Methyl formate		yes
Methyl methacrylate	yes	
Mineral oil	yes	
Monoethanolamine		yes
Morpholine		yes
Nitrobenzene	yes	
Perchloroethylene	yes	
Petroleum ether	yes	
Petroleum naptha	yes	
Propyl acetate	yes	
Propyl chloride	yes	
Propylene carbonate		yes
Silicone oil	yes	
Tretrahydrofuran		yes
Transformer oil	yes	
Triethylene glycol		yes
Triglyceride blend	yes	
Vinyl acetate		

How to Use this Table:

(1) Look up compound of interest under "Liquid" column. If exact compound is not listed, select compound of similar organic structure.

(2) A "yes" in the "Henry's Law Type Analysis Applicable" column indicates that a knowledge of the saturation value of the liquid and the measured dew point will enable one to directly calculate the moisture content of the liquid.

(3) A "yes" in the "Empirical Calibration Required" column indicates that a calibration of dew point versus known moisture contents in the organic must be conducted.

(4) Lists of saturation values for specific hydrocarbons and chlorofluorocarbons are given in Tables 18-4 and 18-5, respectively.

Table 18-4. Saturation Values of Hydrocarbons (Parts per million by weight)

Hydrocarbon Liquid	Temperature, °F (°C)					
	32 (0)	50 (10)	68 (20)	86 (30)	104 (40)	122 (50)
Benzene		454	639	870	1178	1570
Bicyclo (2.2.1) hepta-2,5-diene			295	383		670
Butylbenzene		234	331	448		
sec-Butylbenzene		226	317	426		
tert-Butylbenzene		205	292	389		
Butylcyclopentane		56	95	151		
Cumene	156	219	303	407	550	710
Cycloheptatriene				630	773	990
Cyclohexane		67	122	194	317	490
Cyclohexene		252	317	424	562	
Cyclopentane	46	86	142	249	398	
1,4-Bix (cyclopentyl) butane		88	139	241		
2-Cyclopentyloctane		48	75	122		
Decalin			63	105	164	
Diethylbenzene	159	226	319	431	574	756
Diisobutylene		145	191	274		
2,3-Dimethylbutane	29	58	110	192	323	516
2,3-Dimethyl-1-butene				459		
2,6-Dimethylheptane		53	91	160	301	465
2,4-Dimethylhexane		53	98	180		
2,7-Dimethyloctane		48	87	152		
7,8-Dimethyltetradecane			77	134	219	344
Ethylbenzene		275	373	502		
Ethylcyclopentane		71	119	186		
1-Ethyl-2-pentylcyclopentane		103	168	273		
Heptane	27	54	96	172	308	480
1-Heptene		186	249	375		
Hexadecane			69	123	209	332
Hexane			101	179	317	
1-Hexene				477		
Hexylcyclopentane		52	84	141		
Isopropylcyclopentane		59	102	159		
p-Isopropyltoluene		223	305	415		
2-Methylbutane	32	59	112			
2-Methyl-2-butene			435			
Methylcyclohexane		61	116	179		
2-Methylhexane		56	103	182		
Methylcyclopentane		73	131	205		
2-Methyloctane		52	90	156		
3-Methyloctane		50	87	155		
1-Methyl-2-phenylcyclopentane		105	173	289		
Octane		51	160	184	315	507
1-Phenyl-5-methyl-1-cyclopentane				306		
2-Phenyl-2,4,6-trimethylheptane		96	154	252		
Propane				180 (23°C)		
Propylene				416 (23°C)		
Propylidenecyclopentane				383		
1,3,5-Trimethyl-2-allylbenzene			246	331	438	
1,3,5-Trimethylbenzene			291	393	519	

Table 18-4. Saturation Values of Hydrocarbons (Parts per million by weight) Continued

Hydrocarbon Liquid	Temperature, °F (°C)					
	32 (0)	50 (10)	68 (20)	86 (30)	104 (40)	122 (50)
2,2,3-Trimethylbutane	27	57	160	184	315	507
1,3,5-Trimethyl-2-ethylbenzene			259	350	461	
1,3,5-Trimethyl-2-propylbenzene			255	343	455	
2,2,4-Trimethylpentane	31	59	115	201	332	538
Toluene		316	460	615	750	965
m-Xylene		289	402	536		

Table 18-5. Solubility of Water in the Liquid Phase of Chlorofluorocarbons (Parts per million by weight)

Temperature, °F (°C)	Freon-11	Freon-12	Freon-21	Freon-22	Freon-502	Freon-113	Freon-114
100 (38)	168	165	1980	1800	810	168	148
90 (32)	140	128	1700	1580	690	140	120
80 (27)	113	98	1410	1350	580	113	95
70 (21)	90	76	1150	1140	490	90	74
60 (16)	70	58	960	970	400	70	57
50 (10)	55	44	800	830	335	55	44
40 (4)	44	32	645	690	278	44	33
30 (–1)	34	23.3	512	573	225	34	25
20 (–7)	26	16.6	398	472	180	26	18
10 (–12)	20	11.8	314	384	146	20	13
0 (–18)	15	8.3	260	308	115	15	10
–10 (–23)	11	5.7	205	244	90	11	7
–20 (–29)	8	3.8	158	195	69	8	5
–30 (–34)	6	2.5	118	152	53	6	3
–40 (–40)	4	1.7	90	120	40	...	2
–50 (–46)	3	1.1	66	91	30	...	1.5
–60 (–51)	2	0.7	49	68	22	...	1
–70 (57)	1	0.4	35	50	16	...	0.6
–80 (–62)	0.8	0.3	25	37	11	...	0.4
–90 (–68)	0.5	0.1	17	27	7.5	...	0.2
–100 (–73)	0.3	0.1	12	19	5	...	0.1

Note: Freon is a registered tradename of E.I. DuPont de Nemours & Co., Inc.

References

1. List, Robert I., *Smithsonian Meteorological Tables*, 6th ed., Publication No. 4014, Washington, DC: Smithsonian Institution.
2. Scelzo, Michael, and Pierce, Ralph, "A High Precision Saturation/ Dilution Calibration System for Water Vapor in a Carrier Gas," in *Proceedings of International Symposium on Moisture and Humidity*, Research Triangle Park, NC: Instrument Society of America, 1985, p. 569-576.
3. Application Note 4/89, Waltham, MA: Panametrics Inc., 1989.

4. Sugiyama, K., and Ohmi, T., "Ultraclean Gas Delivery Systems — Part I," *Microcontamination*, November 1988, p. 49-54.

5. Kanno, Y., and Ohmi, T., "Ultraclean Gas Delivery Systems — Part II," *Microcontamination*, December 1988, p. 23-30.

6. Sugiyama, K., Ohmi, T., Okumura, T., and Nakahara, F., "Electropolished, Moisture-Free Piping Surface Essential for Ultrapure Gas System," *Microcontamination*, January 1989, p.37-40, 60-65.

7. Hammon, Robert H., "Aluminum Oxide Hygrometry and Water Enhancement Factors," in *Proceedings of International Symposium on Moisture and Humidity*, Research Triangle Park, NC: Instrument Society of America, 1985, p. 843-848.

8. Research Bulletin No. 8, Institute of Gas Technology, 1700 S. Mount Prospect Rd., Des Plaines, IL 60018-1804.

9. Ardis, M. M., "Moisture Sensors for Hydrochlorofluorocarbons and Hydrofluorocarbons," in *ISA/92 CANADA General Program*, Research Triangle Park, NC: Instrument Society of America, 1992, p. 73-79.

About the Author

Dennis Yavorsky has been with Scientific Pittsburgh for over 15 years and currently is a Manufacturer's Representative. He has 20 years of experience in technical and analytical instrument sales. Mr. Yavorsky's education includes a Bachelor of Science in Chemistry from the University of Maryland and a Master of Science in Human Physiology from Georgetown University. He has worked at the National Institutes of Health (NIH) in Bethesda, Maryland performing artificial lung research.

19

Hydrogen Sulfide Analyzers

Hydrogen sulfide analysis is important for corrosion protection of pipelines, catalyst protection in refinery operations, and environmental compliance. When hydrogen sulfide exists along with water in a pipeline, stress-corrosion cracking can occur due to hydrogen embrittlement. Producer shut-in levels are normally one-fourth to one-half grain of H_2S/100 SCF (standard cubic foot), which is equivalent to 4 to 8 ppm/v (volume). Catalysts are used in refineries to enhance the productivity of reformers. The catalyst materials are subject to poisoning and deactivation when exposed to sulfur. Initial attempts to control sulfur exposure to reformer catalysts utilized a hydrogen sulfide measurement in the recycle hydrogen line. Concentration ranges of interest typically are 0-1 to 0-5 ppm/v full scale. Environmental concerns regarding hydrogen sulfide center on its conversion to sulfur oxides during combustion and the subsequent formation of acid rain. The fuel gas H_2S range regulatory requirements are 0 to 300 ppm/v full scale, with an alarm at 160 ppm/v.

The discussion in this chapter centers around the lead acetate [Pb $(OAc)_2$] method for hydrogen sulfide analysis. In 1930, E. C. Truesdale observed faint, reproducible color changes at 100 ppb concentration levels when he exposed H_2S to a piece of filter paper wetted by an aqueous lead acetate solution. From this early beginning as an analytical method, Rubicon developed an on-line H_2S analyzer. Analysis times were about 30 min and utilized transmitted light to obtain the color change. In 1964, Houston Atlas introduced a lead acetate H_2S analyzer based on color change but using incident instead of transmitted light. In 1974, Houston Atlas refined the detection scheme by employing a rate-of-change of color technique. Both the change of color and the rate-of-change of color measurement approaches will be discussed in the section "Operating Principle." Microprocessor-based lead acetate H_2S analyzers will also be discussed.

Operating Principle

Lead acetate-based analyzers detect H_2S in a gas sample. Two approaches to obtaining a hydrogen sulfide measurement using lead acetate detection are (1) change-of-color and (2) rate-of-change of color. Each approach uses the same chemical reaction to produce the color change, which is indicated by:

$$H_2S + Pb(OAc)_2 \xrightarrow{H_2O} PbS + 2HOAc \qquad (19\text{-}1)$$

Lead acetate, $Pb(OAc)_2$, a white crystalline material, is impregnated into a paper tape. A roll of tape is 100 ft (30.5 m) long and is supplied in a sealed metal-foil envelope. Water for the reaction is supplied by passing the sample gas through an aqueous bubbler solution. The 5 vol% aqueous acetic acid bubbler solution is used to humidify the gas prior to its exposure to the sensing tape. Lead sulfide, PbS, is the product of the hydrogen sulfide/lead acetate reaction (Equation 19-1). A light- to medium-brown color is produced when the analyzer is working within its detectable range. Overexposing the sensing tape will cause it to turn black.

Open the lead acetate tape or sensing tape envelope when installing it into the analyzer. The metal-foil seal prevents the lead acetate from reacting with hydrogen sulfide that may be in the ambient environment.

Normal analyzer operation occurs when the quantity of H_2S exposed to a segment of lead acetate tape is small compared to the amount of lead acetate in the exposed segment. Under these conditions, the change of color and the rate-of-change of color on the tape are proportional to the concentration of H_2S in the sample. For the case of the change of color, the mathematical relationship describing the operation is

$$\text{Change of color} = \text{Constant} \times [H_2S] \qquad (19\text{-}2)$$

In change-of-color lead acetate detectors, the time of exposure is controlled. This can be done by means of a timer coupled with a tape segment isolation system or by moving the sensing tape at a constant speed. Leakage and inadvertent tape exposure are minimized when the sensing tape is in continuous motion. Figure 19-1 schematically indicates the principle of operation of the change-of-color lead acetate H_2S detector.

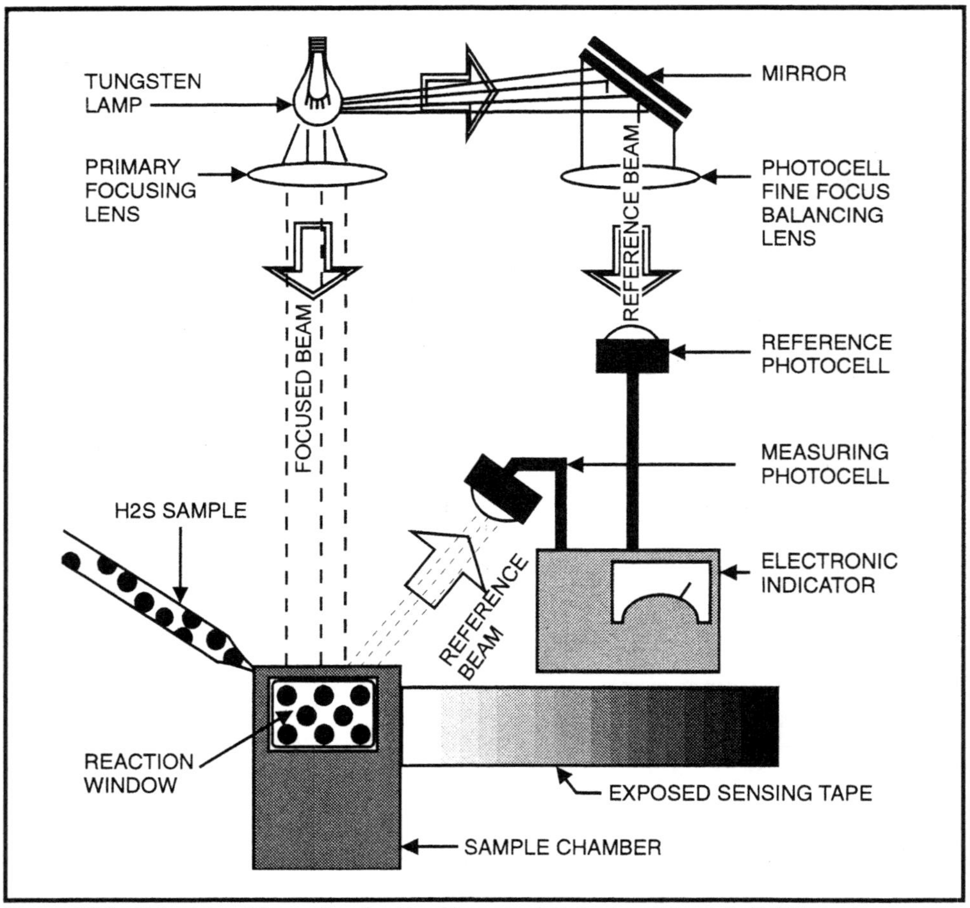

Figure 19-1. Principle of Operation for the Change-of-Color Lead Acetate H_2S Detector
(Courtesy of Houston Atlas)

When the need for greater accuracy and lower limits of detection grew, the rate-of-change of color lead acetate H_2S detector was introduced. The rate-of-change of color is based on the kinetics of the hydrogen sulfide/lead acetate reaction. Since PbS is the color-forming species in Equation (19-1), the mathematical expression for the rate-of-change of color is equivalent to the rate-of-formation of lead sulfide. As with the change-of-color approach, the amount of H_2S exposed to a segment of sensing tape during normal operation is small in comparison to the amount of lead acetate on the exposed segment. Operating the rate-of-color approach in this way simplifies the chemical kinetics from second order to pseudo-first order. Mathematically, the reaction kinetics is expressed by:

$$\frac{d\,(\text{change of color})}{dt} = \frac{dPbS}{dt} = k'\,[H_2S] \qquad (19\text{-}3)$$

where k' is the pseudo-first order reaction rate constant. Contrasting the change-of-color approach, the sensing tape remains stationary during the measurement cycle. At the end of the cycle, a fresh piece of sensing tape is moved into the reaction window. A schematic representation of the rate-of-change of color lead acetate H_2S detector is shown in Figure 19-2.

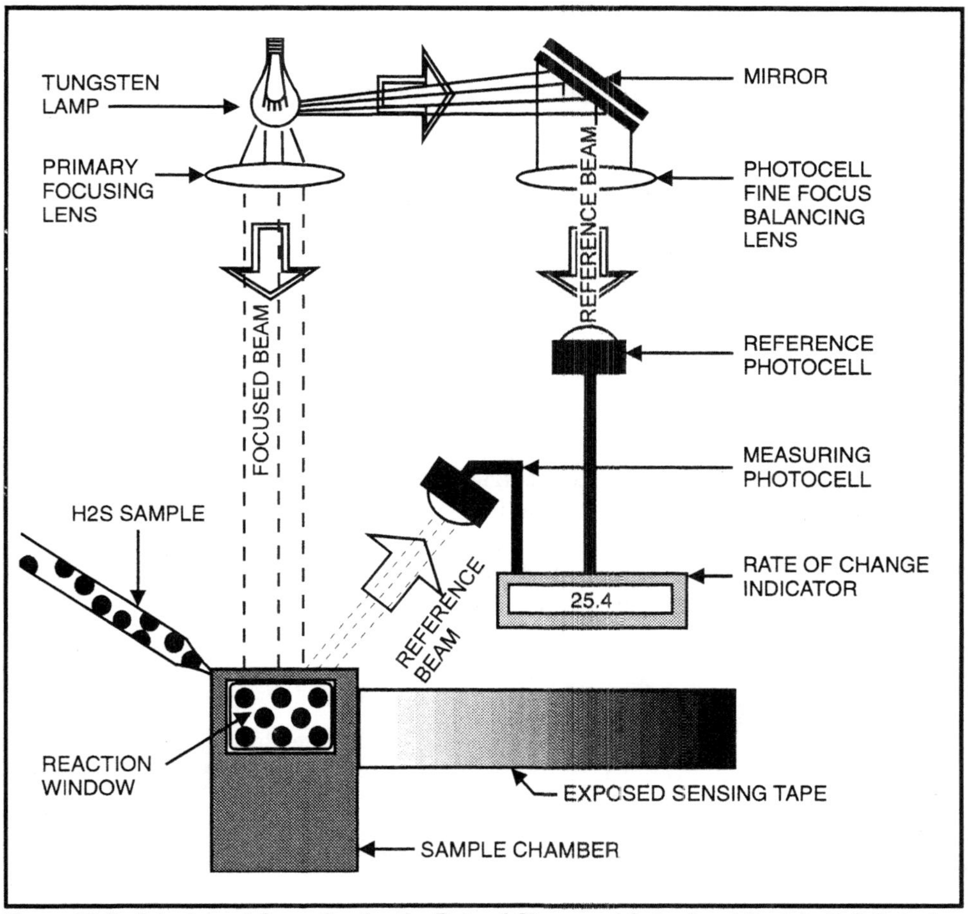

Figure 19-2. Principle of Operation for the Rate-of-Change of Color Lead Acetate H_2S Detector
(Courtesy of Houston Atlas)

Analyzer systems using lead acetate detection exposed to large variations in ambient temperatures must be supplied with a means of temperature control. Very low temperatures significantly reduce the degree of tape humidification and pose the risk of freezing the bubbler solution. A cabinet supplied with a heater, thermostat, and insulation prevents low-temperature exposure of the detector. Special insulation may be required for especially frigid conditions. Very high temperatures can cause excessive tape humidification, which can result in tape breakage. Mounting the detector in the shade or in an analyzer shelter or providing detector cooling by means of a vortex cooler are ways to prevent high-temperature exposure.

Microprocessor-based systems take advantage of the large capacity, speed, and relatively small size of computers to numerically determine the rate-of-change of color. Because of the numerical approach, only one photocell is required instead of the two needed on the non-microprocessor-based lead acetate H_2S detectors. By incorporating fiber-optic technology, the need for lenses is eliminated. The lamp and the photocell can be rigidly affixed into mounting blocks, as shown in Figure 19-3. A filter optimizes the response of the photocell to the incident light.

Using a microprocessor permits a greater degree of parameter flexibility coupled with increased sensitivity and accuracy, even though the reaction chemistry is unchanged. Data on the hydrogen sulfide/lead acetate reaction are generated by means of a photocell. As the tape darkens, the photocell responds by generating a voltage signal. The analog-to-digital converter partitions the available 10 VDC response band into 250,000 divisions with an absolute uncertainty of ±80 divisions. As the tape darkens, the voltage generated by the photocell proceeds from the bottom of the 10 VDC span to the top.

The software permits the user to define the concentration represented by the top of the span, the amount of time of the measurement cycle, the number of intermediate readings collected during the measurement cycle, and the length of time the tape is advanced. The analyzer output at the end of the measurement cycle becomes the average of the intermediate values collected during the cycle. Two consecutive intermediate readings above an alarm set point are required to trigger an alarm. An overrange condition is detected and outputted instantaneously. Depending on the value of the parameters selected, a 100-ft (30.5-m) roll of sensing tape can be consumed as quickly as a matter of hours or as long as 2 or more months. By packaging the sensing tape in cassette form, tape changing, with a little practice, can be done with one hand in about 15 s. Figure 19-4 shows the two methods for packaging the lead acetate sensing tape.

Like any other detector, the lead acetate H_2S detector, in its variety of forms, has an optimal range of operation. Staying within the inherent linear range of the detector maintains a simpler design. Broadening the measuring range is possible by means of sample conditioning. More information about how to achieve a greater detection range can be found in the sections "The System" and "Rangeability." Typical H_2S analyzer systems for natural gas service using analog and digital electronic packages are shown in Figures 19-5 and 19-6, respectively.

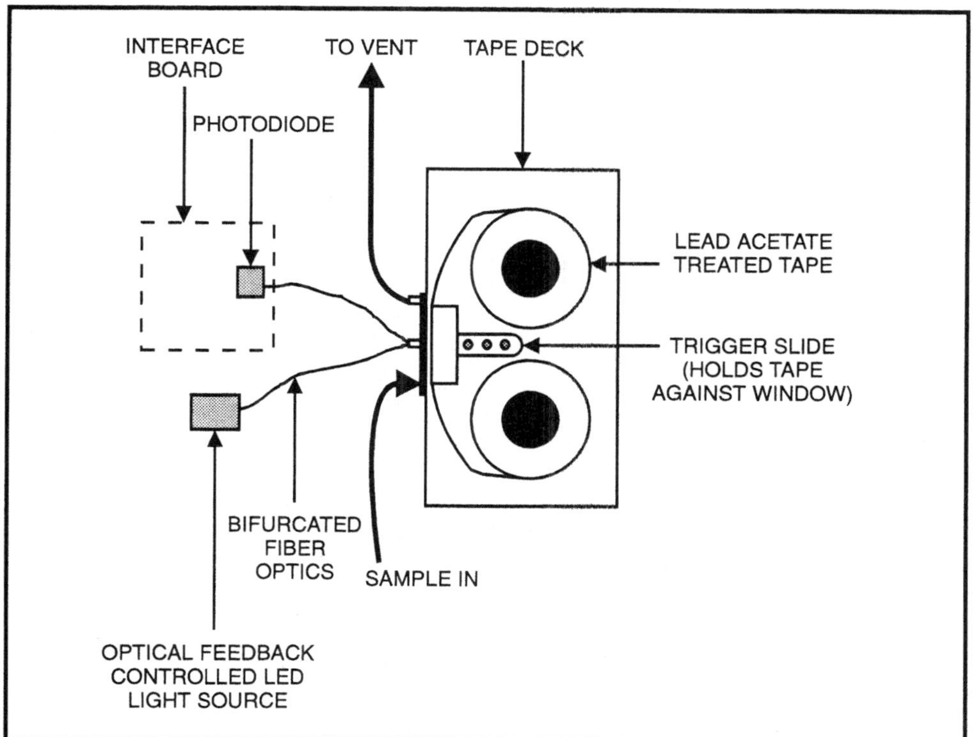

Figure 19-3. Schematic of Microprocessor-Based Lead Acetate H_2S Detector Utilizing Fiber-Optic Light Transmission
(Courtesy of Houston Atlas)

Figure 19-4. Installation of Lead Acetate Sensing Tape. (a) reel-to-reel roll. (b) Cassette packaging technique
(Courtesy of Houston Atlas)

Figure 19-5. Typical Natural Gas H_2S Analyzer System Using Either Change-of-Color or Rate-of-Change of Color Electronic Packages
(Courtesy of Houston Atlas)

Figure 19-6. Typical Natural Gas H_2S Analyzer System Using a Microprocessor-Based Electronic Package
(Courtesy of Houston Atlas)

The System

As with any on-line analyzer system, sample conditioning is the most critical portion. Without a properly designed sample-conditioning system, the information rendered by the detector has little or no value. Proper design of the sample-conditioning system requires knowledge of the stream composition, its temperature, and its pressure.

One of the first concerns to the sample system designer is the sample dew point. Knowledge of the dew point determines whether or not sample temperature-conditioning is required. At a given pressure, the temperature at which condensation first starts is called the dew point temperature. When water and hydrocarbons coexist, the dew point for each is normally calculated independently. Dew point control techniques are available to perform H_2S analysis on sample streams where water and/or hydrocarbon condensation can occur. These techniques center on maintaining the sample temperature above its dew point, both in the sampling train and in the analyzer. When the sample is mixed with a noncondensing gas, such as nitrogen, the dew point temperature is lowered. The dew point temperature can also be lowered by reducing the sample pressure. When the full stream composition is known, the amount of noncondensing gas needed to lower the dew point temperature below the normal operating temperature of the detector can be calculated.

At a given pressure, the temperature at which condensation first starts is called the dew point temperature.

In most H_2S analysis systems, the only issues requiring attention are filtration, pressure reduction, and assuring a minimum residence time in the sample tubing. When water and/or hydrocarbons heavier than C4s are present, condensation may occur. Depending on the quantity, the condensate can result in a loss of H_2S and/or a significant reduction in the volumetric flow of sample.

A rule of thumb in designing dew point control systems is to maintain a 10F° (5C°) cushion between the desired control temperature and the calculated dew point temperature.

In addition to dew point, the sample system design must also consider questions of sample composition stability. If the composition of a sample varies widely, then a sampling system must be used that is blind to the compositional changes. Fuel gas H_2S applications are an example of a widely variable stream composition. At times, the fuel gas contains almost pure hydrogen. At other times, the fuel gas contains almost pure butane. Combustible and noncombustible gas can be present in virtually any proportion.

One way to sample a stream without compositional variations affecting the quality of the detector output is to do so volumetrically. Volumetric sampling is accomplished using a fixed sample volume gas chromatographic injection valve. When the pressure and the temperature in the port of this fixed-volume gas sampler are controlled, a fixed number of moles is collected during each sampling cycle. The ideal gas law (equation of state) expresses the relationship between the pressure, temperature, and volume of a gas:

Volumetric sampling is accomplished using a fixed sample volume gas chromatographic injection valve.

$$PV = nRT \tag{19-4}$$

where:

P = pressure, atmospheres
V = volume, liters per mole
n = moles
R = gas constant, 0.082 dimensionless value
T = absolute temperature, degrees Kelvin (°C + 273)

Equation (19-4) in its more general form includes the compressibility factor designated by Z. At the conditions typical of the port conditions in the port of the injection valve, ideal gas conditions exist. Under ideal conditions the compressibility factor is equal to 1. The analyzer must be calibrated on a volume ($\%_v$) or mole ($\%_m$) basis in order for it to operate independently of composition.

The time between scheduled maintenance periods can be extended by decreasing the frequency of injection. A more elegant method to reduce maintenance is to eliminate the moving part.

Volumetric sampling has a major drawback in that it contains moving parts. Eventually maintenance is required. The time between scheduled maintenance periods can be extended by decreasing the frequency of injection. A more elegant method to reduce maintenance is to eliminate the moving part. This is accomplished by employing a gas transfer unit in the sampling train. A gas transfer unit uses a membrane permeable only to H_2S. The membrane separates the process sample stream containing the H_2S from the pure carrier gas stream. Nitrogen is typically used as the carrier gas. Air can be used in principle; in practice, the single jeopardy case of a ruptured membrane can cause explosive mixtures to form when analyzing combustible gases. Additionally, the presence of oxygen may cause membrane aging over time. During operation, H_2S permeates across the membrane at a fixed rate from the process sample to the carrier gas. The carrier gas then flows to the lead acetate detector, where the analysis is performed. Because the gas transfer unit uses no moving parts, mechanical wear and maintenance are reduced.

The presence of normal butane, and to a much lesser extent normal propane, in the sample stream can create some difficulties. Normal butane accelerates the transfer rate across the membrane. The greater the concentration of normal butane in the sample, the greater the rate of H_2S transfer across the membrane. If the concentration of normal butane remains relatively constant, it can be added to the calibration gas in the desired quantity to achieve a stable calibration. When the normal butane concentration varies radically, the possibility exists that the analyzer output variations are due to changes in the normal butane concentration and not to changes in the H_2S concentration. In terms of full-scale calibration, a 1% of full scale increase in the analyzer output is roughly equivalent to a 1 mol% change in the normal butane concentration. The change in signal caused by normal propane is approximately 10% of the change caused by normal butane.

Calibration

With any analyzer, calibration is required to establish a base response and to provide a means for interpreting the output signal. During normal operation, a calibration verification is sufficient. Whether performed manually or automatically, calibration procedures require using a standard. Standards can be purchased or made at the time of calibration. For gaseous samples, blends of H_2S in a variety of matrices are available from compressed gas suppliers.

An alternative to a standard gas blend is the permeation tube calibrator. A permeation tube contains a quantity of a pure compound. A diffusion barrier is installed to control the rate of release. When connected to a pressure-regulated known flow rate of clean and pure gas at a regulated temperature, a calibration blend is produced. Calibration procedures stipulate the flowing of a known H_2S gas concentration across the detector and then adjusting the signal output to an appropriate value. The calibration blend is called a span gas. Zero gas, which is H_2S-free carrier or sample gas, is passed through the detector before the span gas to establish the zero baseline response. Environmental applications require a zero and span (range) verification on a daily basis. Using a timer with solenoids as needed, standard gas cylinders and permeation tubes can provide the span gas needed for daily verification.

With microprocessor-based lead acetate detectors, an actual calibration can be accomplished instead of a calibration verification. The microprocessor's software can be programmed to accept a new output value representing the analyzer's response to a span gas. To prevent erroneous recalibration, such as might occur when the span gas bottle is depleted, the software should contain fault logic that prevents recalibration beyond a 10% change from the results obtained by calibrating manually.

To prevent erroneous recalibration, such as might occur when the span gas bottle is depleted, the software should contain fault logic that prevents recalibration beyond a 10% change from the results obtained by calibrating manually.

Gas calibration blends can also be made manually using a gas calibration kit, which volumetrically adds a known quantity of pure H_2S to a known quantity of clean carrier gas. A typical gas calibration kit is shown in Figure 19-7. Operation of the calibration kit for making a gas standard in air is depicted in Figure 19-8.

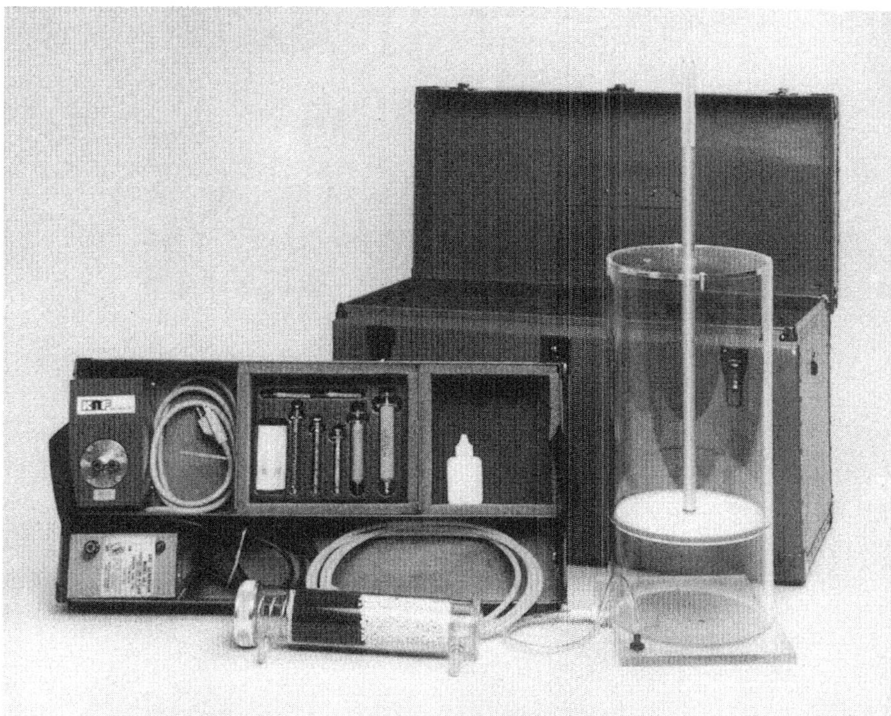

Figure 19-7. Gas Calibration Kit
(Courtesy of Houston Atlas)

Figure 19-8. Use of a Gas Calibration Kit
(Courtesy of Houston Atlas)

Rangeability

As indicated previously, the lead acetate detector operates within a range of H_2S concentrations. Increasing the upper detection for gas phase samples is accomplished by any one or a combination of the following methods:

1. Reducing the electronic gain
2. Making design changes to the sample contact area
3. Decreasing the sample flow rate
4. Combining the sample at a known rate with H_2S-free gas, using flow measurement devices (dilution)
5. Sampling volumetrically
6. Utilizing a diffusion barrier to control the rate of H_2S addition to a clean carrier gas
7. Decreasing the measurement cycle time

Implicitly assumed with each of the seven methods is that a steady-state H_2S gas phase concentration passes the lead acetate detector.

Reducing the electronic gain is done by adjusting a potentiometer. Instructions on how to do this are typically found in the operating manual supplied by the manufacturer.

Design changes to the sample contact area involve reducing the effective physical contact between the sample gas and the lead acetate detector. As the concentration of H_2S increases, the rate of PbS formation increases. The rate of PbS formation and, hence, the sensitivity of the detector can be reduced by restricting the ability of sample gas to come into contact with the lead acetate-impregnated tape.

Sample flow adjustments are easily made. When the sample flow rate is decreased, the residence time of the sample within the analyzer is increased. An increased residence time translates into a slower response time. Other limiting factors as to how low the sample flow can be decreased are the accuracy of the flow measurement device and the available sample pressure. Flow measurement accuracy and sample pressure also limit the available upper rangeability of the lead acetate detector when the sample is combined with a H_2S-free carrier gas. Volumetric sampling and gas transfer units are discussed in the section "The System."

As the concentration of H_2S increases, the volume of sample needed and the transfer area required decrease. Cascade arrangement of injection valves and gas transfer units is required to achieve 0 to 100 mol% H_2S full-scale calibrations.

Cascade arrangement of injection valves and gas transfer units is required to achieve 0 to 100 mol% H_2S full-scale calibrations.

Decreasing the sample measurement cycle increases the tape consumption rate.

Decreasing the lower detectable range of the lead acetate detector involves using any one or a combination of the following methods:

1. Increasing the electronic gain
2. Increasing the sample flow rate
3. Making design changes to the sample contact area
4. Increasing the measurement cycle time.

Electronic gain increases are possible to a point. Depending on the electronic design, at some level of amplification the noise levels will no longer be within acceptable levels. Typical noise levels for the lead acetate detector are ±2% of full scale. Power conditioners that stabilize input voltage fluctuations may be required to achieve acceptable responses at very high gain settings.

Increasing the sample flow rate increases the amount of H_2S that passes by the detector in a given period of time. An increased amount of H_2S per unit time increases the absolute signal strength from the detector. Flow rates exceeding about 400 cm^3/min can cause bubbler solution entrainment, which can lead to tape breakage. Very high sample flow rates can transport the sample past the detector so rapidly that the lead acetate/hydrogen sulfide reaction does not occur.

Design changes to the sample contact area center around increasing the effective contact between the sample gas and the lead acetate-impregnated tape. Increasing the effective contact between the sample gas and the tape causes more color change to occur, thus increasing the absolute signal strength from the detector.

Changing the sample measurement cycle time on lead acetate detectors using analog electronics requires adjusting the setting of a potentiometer. With microprocessors, this setting can be made from the user software interface. Cycle time adjustments directly affect the consumption rate of the lead acetate sensing tape.

Liquid Sample Measurement

To this point, the discussion of the lead acetate detector has dealt with gas phase samples. Hydrogen sulfide analyzer applications exist for liquid as well as gaseous samples. Two cases of liquid samples are of concern. The first is when the sample is normally a gas, but because of pressure exists as a liquid. Such samples are referred to as liquefied petroleum gases (LPGs) and natural gas liquids (NGLs). By reducing the pressure and adding heat, these samples are converted to gaseous samples. Heat addition is required so that a cold trap does not form when the pressure is reduced. Typical LPG and NGL samples are sufficiently volatile so that, after pressure reduction and heat addition, they remain in the vapor state. After sample conditioning, the analysis for H_2S proceeds as for a gas sample.

For the class of samples that retain their liquid character at room temperature and pressure, extending the rangeability of the lead acetate detector requires adding special sample conditioning. Examples of these applications include waste water and crude oil. Obtaining analytical results for these and other liquids involves transferring the H_2S from the liquid phase to the gas phase. To do this requires the use of a noncondensable carrier gas. Nitrogen is normally used; air can be used if the sample contains no combustible components and if sensitivity is not an issue.

Transferring H_2S from the liquid to the gas phase involves the vapor/liquid equilibrium relationship of hydrogen sulfide. For dilute solutions, the equilibrium relationship is linear and can be expressed mathematically by:

$$Y_{H_2S} = H \times X_{H_2S} \tag{19-5}$$

where:

Y_{H_2S} = mole fraction H_2S vapor phase

X_{H_2S} = mole fraction H_2S liquid phase

H = Henry's law constant

Transferring a chemical compound from the liquid phase to the vapor phase using a gas is called stripping.

Equation (19-5) is referred to as Henry's law. Vapor/liquid equilibrium relationships, such as Henry's law, are functions of both temperature and pressure. Temperature and pressure, therefore, must be controlled to obtain consistent analytical results. Transferring a chemical compound from the liquid phase to the vapor phase using a gas is called stripping. After the stripping operation is completed, the gas and liquid must be separated. Upon separation, the gas stream goes to the lead acetate H_2S detector for analysis. Excess sample is evacuated either manually or automatically.

Liquid standards containing H_2S can degrade due to outgassing of the liquid. Standards should be stored in a refrigerator to enhance long-term stability or should be prepared fresh when needed. Standards stored or prepared should have a minimum amount of head space.

Calibration issues for liquid samples are similar to those for gas samples. In both cases, a calibration standard is required. For LPG and NGL samples, a standard produced by a compressed gas manufacturer is recommended. For normal liquid samples, a calibration kit or a standard is normally available from the analyzer manufacturer. A typical liquid calibration kit is shown in Figure 19-9.

Figure 19-9. Liquid Calibration Kit
(Courtesy of Houston Atlas)

Standard Methods

Four ASTM methods presently exist using the rate-of-change of color lead acetate detection system. Two methods are for H_2S analysis and two are for total sulfur analysis:

Method Number	Title
D 4045-87	Sulfur in Petroleum Products by Hydrogenolysis and Ratiometric Colorimetry
D 4085-82	Analysis of Hydrogen Sulfide in Gaseous Fuels (Lead Acetate Reaction Rate Method)
D 4323-84	Hydrogen Sulfide in the Atmosphere by Rate-of-Change of Reflectance
D 4468-85	Total Sulfur in Gaseous Fuels by Hydrogenolysis and Ratiometric Colorimetry

Copies of these methods are available from ASTM (1916 Race Street, Philadelphia, PA 19103-1187).

Certifications

On-line analytical instrumentation in hydrocarbon service usually is installed in electrically-hazard-rated areas. This means that explosive mixtures can surround the analyzer continuously or intermittently. Since analyzers are electrical devices, the potential for an explosion is real. When followed, various electrical codes and design procedures yield instruments suitable for installation in hazardous environments. Numerous electrical codes and design practices exist worldwide. There also are a number of certifying authorities. Table 19-1 lists major electrical codes, certifying authorities, and the countries where the codes apply.

Table 19-1. Electrical Codes

Electrical Code	Country/Region	Certifying Authority
National Electric Code	United States	Underwriters Laboratory (UL) Factory Mutual (FM)
Canadian Electric Code	Canada	Canadian Standards Association (CSA)
Cenelec	Europe	
	United Kingdom	British Approvals Service for Electrical Equipment in Flammable Atmospheres (BASEEFA)
	France	Laboratoire Central Des Industries Electriques
	Belgium	Institut National Des Industries Extractives
	Germany	Physikalisch Technische Bundesanstalt (PTB)
Japanese Standards Association	Japan	Japanese Industrial Standard (JIS)
City of Los Angeles	Los Angeles, CA	City of Los Angeles
Standards Association of Australia	Australia	Standards Association of Australia

Design certifications cover the design of a specific piece of equipment.

Installation certifications are focused on a particular piece of equipment at a specific site.

Manufacturers indicating a code acceptance typically have had a unit built and sent to the certifying agency for testing. A certificate of conformity indicating compliance is normally provided to the manufacturer. A stamp indicating certification is usually affixed to equipment manufactured according to a given code. Obtaining the required electrical certification for a given installation can occur in two ways: (1) certification of the design and (2) certification of the installation. Design certifications cover the design of a specific piece of equipment. Manufacturing processes must adhere to the specifications for the certification to be valid. Installation certifications are focused on a particular piece of equipment at a specific site. A unit duplicating the design of an installation certification does not carry the electrical approval. Duplicate units based on design certifications do carry the electrical approval.

When instrumentation is placed in an electrically-hazard-rated area, the electrical classification must be determined. Without knowledge of the area electrical classification, it is impossible for a manufacturer to comply with the installation requirements. Electrical classifications take into consideration the chemical nature of potentially explosive mixtures. In general, the design considerations for an explosion involving hydrogen are more rigorous than those for an explosion involving methane. The energy released by a hydrogen explosion is much greater than for a methane explosion, and the small physical size of the hydrogen molecule allows it to get through gaps and seals with relative ease.

Bibliography

1. Cines, M. R., et al., "Molecular Sieves for Removing H2S from Natural Gas," *Chemical Engineering Progress*, p. 89-93, August 1976.
2. Clevett, Kenneth J., *Process Analyzer Technology*, New York: John Wiley & Sons, 1986.
3. "Flow of Fluids Through Valves, Fittings, and Pipe," Technical Paper No. 410, New York: Crane Co., 1982.
4. Leffler, William L., *Petroleum Refining for the Non-Technical Person*, Tulsa, OK: PennWell Publishing Co., 1985.
5. Moore, Patrick J.,"Recent Developments in H2S Analysis of Process Streams," *Advances in Instrumentation*, 43 (3): 959-962, 1988.
6. Moore, Patrick J., "The Use of a Gas Transfer Unit to Sample Fuel Gas Streams Containing Hydrogen Sulfide for EPA Compliance Applications," Research Triangle Park: Instrument Society of America, May 1990.
7. Moore, Patrick J., "Determination of the Concentration of Hydrogen Sulfide in Liquid Streams Using Lead Acetate Detection Technology," *Advances in Instrumentation and Control*, 45: 537-547, October 1990.
8. Moore, Patrick J., "Hydrogen Sulfide Detection and Determination," *Proceedings of the Twenty-Fifth Gulf Coast Measurement Short Course*, Houston, TX: American School of Gas Measurement Technology, 1990.
9. Moore, Patrick J., "Emissions Control Safeguards Environment," *Intech*, 38 (11): 20-22, November 1991.
10. Nelson, W.L., *Petroleum Refining Engineering*, New York: McGraw-Hill, 1941.
11. Perry, Robert H., and Chilton, Cecil H., *Chemical Engineers Handbook*, 5th ed., New York: McGraw-Hill, 1973.
12. Lide, David R., Ed., *Handbook of Chemistry and Physics*, 77th ed., Boca Raton, FL: CRC Press, 1996.
13. Wiseman, P., *Petrochemicals*, Chichester, U.K.: Ellis Horwood Ltd., 1986.

About the Author

Patrick J. Moore is a Chemical Engineering graduate of the University of Delaware. He has worked for the Olin Corporation, Lummus, Envirotech Controls Inc., and presently is the Manufacturing Manager for Houston Atlas Inc. Mr. Moore is a member of the American Institute of Chemical Engineers and the Instrument Society of America. He has published numerous articles in the field of analytical instrumentation and the psychological aspects of literature.

20

Total Sulfur Analyzers

With the introduction of bimetallic catalyst materials into refinery operations in the 1970's, reformer yields and selectivity were greatly enhanced. Improvements in yields and selectivity of desired hydrocarbons could only be maintained if the catalyst material was not exposed to significant amounts of sulfur. Sulfur reacts with the catalyst material, forming metal sulfides and rendering the catalyst nonfunctional.

Initial attempts to control sulfur exposure involved measuring hydrogen sulfide in the hydrogen recycle line. Unfortunately, this measurement provided data only after the catalyst was already exposed. True control could only be accomplished by measuring the sulfur in the feedstock material. Because of the reducing atmosphere in the catalytic reformer vessel, sulfur compounds are converted to hydrogen sulfide. In the feedstock, sulfur can exist in a wide variety of forms, including mercaptans, sulfides, disulfides, and thiophenes (see Table 20-1). A hydrogen sulfide analysis of the feedstock does not provide a complete picture of the potential for sulfur exposure. A total sulfur value of the feedstock, however, does provide the information needed to prevent sulfur exposure to the catalyst material.

Table 20-1. Compounds Containing Sulfur

NAME	FORMULA (a)
Carbonyl sulfide	COS
Carbon disulfide	CS_2
Hydrogen sulfide	H_2S
Mercaptan (thioether)	RSH
Sulfide	R'SR
Disulfide	R'SSR
Higher-order sulfide	$R'S_nR$ $n = 3, 4, ...$
Sulfur dioxide	SO_2
Sulfur trioxide	SO_3
Thiophene	C_4H_4S
(a) R' and R = C_1, C_2 . . .	

Total sulfur measurements are becoming increasingly important for environmental applications. In 1993, the allowable limit for sulfur in diesel fuel decreased from 0.5 to 0.05 wt%. California environmental regulations require a maximum of 40 ppm/wt (weight basis) of sulfur in gasoline. Limits for the sulfur content in gasoline nationwide are specified in the 1991 Clean Air Act Amendment. Fuel gas to combustion devices is presently limited in H_2S content. Like feedstocks, the sulfur in fuel gas exists in a variety of forms. Total sulfur in fuel gas is rapidly becoming a concern, with limits of 40 ppm/v (volume basis) on a calibration range of 100 ppm/v full scale under active consideration. Concern for the total sulfur content in fuels such as diesel, gasoline, and fuel gas centers on the emission of oxides into the atmosphere. Oxides of sulfur and nitrogen are the main contributors to acid rain. When such oxides combine with atmospheric water, sulfurous, sulfuric, nitrous, and nitric acids are formed.

Use of the lead acetate detector for total sulfur measurement began in the late 1960's. (See Chapter 19 for a discussion of lead acetate detectors.) Initially, systems employed a catalyst to convert the total sample sulfur to hydrogen sulfide. In 1974, Houston Atlas developed the noncatalytic hydrogenation technique for sulfur conversion. In the early 1980's, the oxyhydrogenation sulfur conversion technique was introduced. For on-line applications, the oxyhydrogenation method initially was limited to gases. In the laboratory, the method was applicable to gases and liquids. On-line oxyhydrogenation for liquids was achieved around 1990. Presently, all three approaches for sulfur conversion to hydrogen sulfide are in use with the lead acetate detector.

The Analysis System

Obtaining a total sulfur measurement using the lead acetate detector requires converting all of the sulfur in the sample to H_2S. Three approaches to achieving this conversion are presently in use:

1. Catalytic hydrogenation
2. Noncatalytic hydrogenation
3. Oxyhydrogenation

Catalytic conversion has appeal because of its lower pyrolyzer temperature requirements. Volatile sulfur compounds such as carbonyl sulfide and methyl mercaptan can convert to H_2S catalytically around 800°F (425°C). While catalysts can convert sulfur to H_2S at lower temperatures than noncatalytic systems, there are some important restrictions that need to be kept in mind. First, as the molecular weight of the sulfur compound increases, the conversion temperature also increases. Second, the conversion efficiency of individual heavier sulfur compounds is not a consistent function of temperature. Third, when the catalyst fails, the output reading goes to zero, provided that there is no H_2S in the sample. Fourth, some catalyst systems require a regeneration cycle. Catalyst regeneration is done in a hot air stream. While the catalyst is regenerating, the analyzer is off-line. A nitrogen purge cycle to eliminate explosive gas mixtures forming in the catalyst area is needed before introducing the air.

Noncatalytic hydrogenation sulfur conversion to H_2S occurs at temperatures from 1830 to 2200°F (1000°C to 1200°C). Temperatures of 2000 to 2200°F (1100 to 1200°C) are necessary to convert thiophene sulfur. Use of a high-temperature heating element permits pyrolyzer operating temperatures that are significantly below the 2900 to 3100°F (1600 to 1700°C) design maximum temperature. The upgrade in the heating element greatly enhances operational reliability.

Oxyhydrogenation (oxyhydro) pyrolysis achieves the same result as hydrogenation pyrolysis. Sulfur in the sample is converted to H_2S. Instead of a single-step conversion, oxyhydrogenation employs a two-step process. Sample sulfur is initially oxidized to SO_2 and SO_3 and is then reduced to H_2S. Oxyhydrogenation has the additional benefit of converting samples, such as benzene and gasoline, without carbon formation. A flow diagram of the oxyhydrogenation pyrolysis process is shown in Figure 20-1.

Achieving quantitative sulfur conversion to H_2S without carbon formation in the pyrolyzer is necessary when using the lead acetate detector in total sulfur analyzer service. Carbon formation creates a site for sulfur loss because it absorbs H_2S. The extent to which a hydrocarbon forms carbon appears to depend on the degree of its aromaticity and the length of the carbon chain. Highly aromatic and unsaturated hydrocarbons, such as benzene and butadiene, and long-chain hydrocarbons, such as cetane, form carbon easier than compounds like methane and ethane.

An easy way to check for carbon formation is to examine the filter element at the pyrolyzer exit. A black discoloration will be visible if carbon formation has occurred in the pyrolyzer.

Oxyhydrogenation pyrolysis combusts the hydrocarbons as well as the sulfur. Combusted hydrocarbons form carbon monoxide, to a lesser extent, and carbon dioxide, which do not easily further react to form carbon. Water is also formed from the combustion process. It is removed from the system by cooling the gas exiting the pyrolyzer. Expelling the water from the analyzer system after each analysis cycle is recommended. Coupled with the acidifying effect caused by the carbon dioxide, expelling water at the end of each cycle prevents H_2S loss in the water-removal system.

Oxyhydrogenation has the added benefit of combusting ring-bound thiophenic sulfur compounds. Conversion of thiophenic sulfur is easier in oxyhydrogenation pyrolysis compared to hydrogenation only. Rings broken by combustion reactions tend to be exothermic, whereas rings broken by hydrogenation reactions tend to be endothermic.

Chemically, the overall reactions that describe the conversion of sample sulfur to H_2S for straight reductive pyrolysis and oxyhydro pyrolysis are

Straight Reductive Pyrolysis

$$R - S + H_2 \rightarrow H_2S + R - H \qquad (20\text{-}1)$$

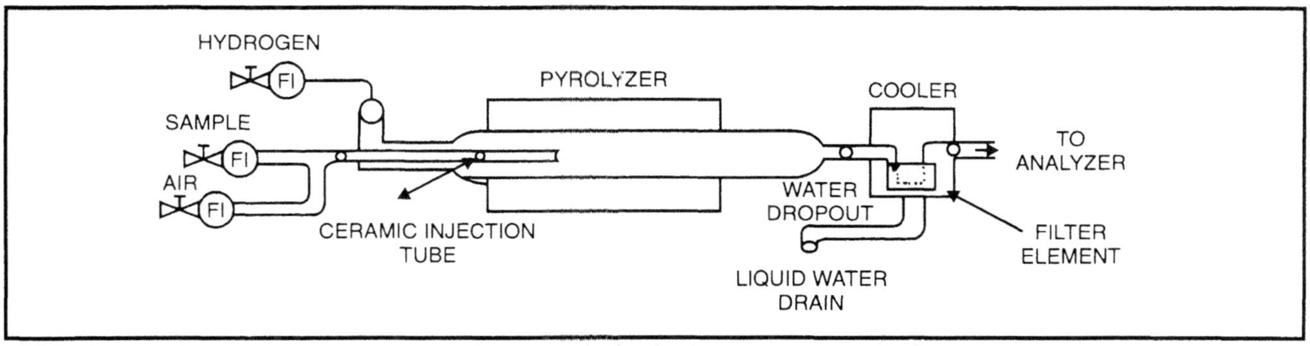

Figure 20-1. Flow diagram for the oxyhydrogenation pyrolysis of sulfur compounds to H_2S
(Courtesy of Houston Atlas)

Oxyhydro Pyrolysis

$$R - S + O_2 \rightarrow SO_2 + CO_2 + H_2O \qquad (20\text{-}2)$$

$$SO_2 + \frac{1}{2}O_2 \Leftrightarrow SO_3 \qquad (20\text{-}3)$$

$$SO_2 + 3H_2 \rightarrow H_2S + 2H_2O \qquad (20\text{-}4)$$

$$SO_3 + 4H_2 \rightarrow H_2S + 3H_2O \qquad (20\text{-}5)$$

Reaction (20-1) ignores the consumption of hydrogen used to saturate carbon-carbon bonds. The amount of hydrogen needed for carbon bonds will depend on the nature of the sample. In general, the oxyhydro pyrolysis approach is employed when the molar quantity of unsaturated hydrocarbons exceeds 10%. The extent to which the combustion products of CO_2 and H_2O form depends on the nature of the hydrocarbon being sampled. Water formed from the reduction of SO_2 and SO_3 is insignificant, because the sulfur concentrations are at a level of parts per million.

Detection sensitivity is governed by two broad concerns: (1) the electronics utilized to process the detector's signal and (2) the strength of the signal generated by the detector. Issues related to detector electronics are covered in Chapter 19. Increasing the signal strength involves a consideration of the carrier gas-to-sample ratio. Both air and hydrogen additions, in fact, represent a dilution of the sample sulfur concentration, even though they are carrier gases necessary for operation. Air contains 80% nitrogen, which is inert in the oxyhydro pyrolysis reactions. Maximizing the signal concentration to the detector means minimizing the amount of air and hydrogen consumption. In the oxyhydro pyrolysis system, minimizing the air and hydrogen consumption simultaneously minimizes the amount of water formation. Minimizing the amount of water formation translates into faster and easier water removal.

Two limits on the extent of carrier gas reduction must be observed in both the straight reductive and oxyhydro pyrolysis processes. One is the formation of carbon. The other is the quantitative conversion of sulfur. Exceeding either limit can significantly impair the quality of the analytical results. As part of the analyzer calibration procedure, the manufacturer determines the flow rate parameters suitable for a particular application.

It is important to note that the carrier gases must be free of sulfur. Sulfur that is present is converted to H_2S and is added to the sample sulfur. The net result is an artificially high sulfur measurement. From a practical standpoint, the limit of sulfur tolerable in the carrier gases is at a concentration well below the noise level. As detection levels decrease, the elimination of sulfur in the carrier gases becomes more critical, because the base sulfur concentration can exceed the noise limits of the detector. Using analog electronics, the lower limit of detection is approximately 30 ppb total sulfur. Issues related to removal of volatile sulfur compounds in carrier gases to sub-ppb purity levels is just now being addressed. Should removal methods limit the amount of sulfur purity to a level above the limit of detection, mathematical adjustment techniques may have to be employed. With the advent of microprocessor-based signal conditioning, mathematical adjustments become a matter of coupling software programming with analyzer computations.

Rangeability

On-line total sulfur analysis using the lead acetate detector is applicable to gaseous and liquid samples. As discussed in Chapter 19, liquid samples are classified into two categories: (1) liquefied gaseous samples and (2) stable liquids at atmospheric pressure. Obtaining an analysis of either form of liquid sample requires converting the sample state to a gas. With liquefied gas samples, the vaporization typically occurs prior to sample metering. For stable liquids, the vaporization occurs after metering. Because of the inherent variability in the vapor/liquid distribution in two-phase flow, accurate repeatable sample metering is infinitely easier to achieve when the sample exists in a single phase.

While, in principle, it is possible to sample liquefied gases in the liquid state, there are several benefits to vaporizing before metering:

Vaporize before metering.

1. Pressure and temperature control after vaporization typically are not critical.

2. Operating a liquid metering system at an elevated pressure increases the probability for maintenance.

3. Flashing inside the valve when the downstream pressure is reduced can cause a local "cold" spot to occur, which can trap sample sulfur.

Estimating the temperature and pressure conditions needed to maintain a single-phase sample requires knowledge of the sample stream composition and vapor pressure data of each sample component as a function of temperature. Coupled with the assumption of ideal gas and liquid phase solutions, the dew point and/or the bubbler point temperature at a given pressure is calculated. This allows use of appropriate heating or cooling methods to assure that the sample is maintained in a single phase.

Because of the uncertainty in compositional values and in physical property data, it is desirable to factor in a safety margin when deciding what temperature is needed to maintain a single-phase sample. As a rule of thumb, adding 10F° (5C°) to the dew point temperature and subtracting 10F° (5C°) from the bubble point temperature at a given pressure provides for reasonable variations in composition and for uncertainties in physical property data. Wide compositional changes can drastically alter the calculational results. Given the relatively small dimensions of analyzer sampling systems compared with the operating units in which they are utilized, high precision in design calculations is not necessarily economically critical. Computer-aided calculations provide a much greater degree of precision in the calculated results.

Because the lead acetate detector measures H_2S, when the detector is coupled to a reductive pyrolysis furnace it is possible to obtain results for both H_2S and total sulfur in gaseous and easily gasified samples. Implicitly assumed is that the H_2S concentration is above the noise level of the total sulfur measurement range. Normally, a minimum signal-to-noise ratio of 5 to 10 identifies the lowest statistically useful detection limit on a given calibration range. Data collected near the noise level can have absolute uncertainties as large as the measurement obtained. Achieving H_2S and total sulfur values from a total sulfur analyzer with a lead acetate detector simply involves diverting the combined carrier and samples flows around the pyrolyzer.

Assuring reliable results when the flow around the pyrolyzer is diverted requires consideration of fluid flow characteristics. For straight reductive pyrolysis systems, the sample stream is mixed with hydrogen alone. Using a mixed carrier and sample stream mimics the molecular weight of the pyrolyzer exit gas. This maintains a flow consistency between the H_2S analysis and the total sulfur analysis. Since sample addition is at a much lower rate than hydrogen addition, chemical structure changes undergone by the sample through the pyrolysis furnace represent a small perturbation to the bulk properties of the pyrolyzer exit gas. For oxyhydro pyrolysis, separate air and hydrogen flowmeters may be required to assure a reasonably close reproduction of the postcombustion molecular weight.

Extending the applicability of a lead-acetate-based total sulfur analyzer system to sulfur chromatography is straightforward. On-line sulfur chromatography is possible because it uses the lead acetate-based total sulfur analyzer as a gas chromatographic detector. Presently this extension is in its infancy. When employed in a sulfur chromatographic mode, sulfur species separation prior to the total sulfur analyzer is required. Separation prior to pyrolysis allows the conversion to H_2S to proceed one sulfur compound at a time instead of simultaneously (as during a total sulfur analysis). Sulfur species' concentrations are determined using integration, comparing the area counts of the unknown with the area counts generated by a known sulfur species' concentration injection. Identification is made on the basis of elution times.

Standard packed columns such as Porapak® QS and Carbopack™ B are used for separating sulfur species. In principle, megabore columns can be used; in practice, column flooding limits the quantity of sample that can be injected. Quantity restrictions constrain the lower limit of deductibility when using capillary columns.

Calibration

Calibration issues for total sulfur analyzers using the lead acetate detector are similar to H_2S analyzers using the same detector (see Chapter 19). Because the lead acetate detector is a H_2S analyzer, it is possible to use H_2S as the sulfur compound for calibration. The difficulty with using H_2S is that only detector response is verified.

Total sulfur analyzers using lead acetate detection depend on the conversion of sulfur compounds to H_2S to obtain a measurement. Verification of the pyrolysis furnace functionality requires the use of a sulfur compound other than H_2S. By using a different sulfur compound, the conversion to H_2S is forced to occur. This verifies operation of the pyrolysis furnace. For gaseous samples, carbonyl sulfide (COS) or methyl mercaptan (CH_3SH), can be used. Carbonyl sulfide blends are preferable to methyl mercaptan blends because of their longer-term stability. Use of gas calibration kits employing volumetric mixing are an excellent way to prepare fresh gaseous standards. For specific concerns related to cylinder blends of sulfur standards, a gas supply company should be consulted.

For liquid calibrations, two approaches are available. The first approach employs the use of a liquid calibration standard. Standard questions involve the volatility of the sulfur compound used, the temperature at which the standard is stored, the amount of headspace above the liquid surface and the reactivity of the sulfur compound. These issues relate to the equilibrium distribution of the sulfur compound between the liquid and vapor phases. In general, the stability of a liquid standard is enhanced when measures are taken to reduce or eliminate the possibility of sulfur in the gas phase.

Reactivity can be reduced by selecting a molecular arrangement that shields the two unshared electron pairs on the nucleus of the sulfur atom. Such isolation occurs by using an organic sulfide compound. An organic functional group of sufficient length in the organic sulfide can also mimic the properties of a typical hydrocarbon liquid. When a balanced organic sulfide is selected, the reaction propensity of one side of the molecule is no greater than the other. Heuristically, the selection of di-normal butyl sulfide satisfies the requirements outlined above.

Calibration blends are typically made in solutions of isooctane. Isooctane is selected because of its availability in a very sulfur-pure form. Concentration units of weight of sulfur per unit volume of standard are preferred. Ease of preparation governs the unit selection. Unit conversion to ppm/wt requires knowledge of the specific gravity of the bulk material. Standards can easily be made in solvents other than isooctane (e.g., naphtha, kerosene, and diesel fuel).

Use of a gaseous standard is another approach. As mentioned previously, it is preferable to use a sulfur compound other than H_2S in order to force H_2S conversion in the pyrolyzer. Using a gaseous standard to calibrate an analyzer designed for liquid service requires an accurate gas-phase delivery system. In order to correlate the response of a gaseous standard on an analyzer designed for liquid service with a liquid standard calibration, the mass flow rate of sulfur per unit time past the detector must be more or less the same in both cases. Consistency in the gaseous delivery system can be compromised by temperature variations. Gas samples, because of their compressibility, are much more susceptible to density variations due to temperature than are liquid samples. In low-ppm-detection liquid applications, the mass flow rate of sulfur across the detector represents very low concentrations of sulfur in gaseous standards. Adequate accuracy in the delivery system may be a problem. While it is possible in theory to use a gaseous standard on a liquid service total sulfur analyzer, this approach is not recommended for sub-ppm applications.

Standard Methods and Certifications

For information regarding ASTM standard methods, electrical certification codes and authorities, see Chapter 19.

Bibliography

1. Barnes, W.T. and Moore, Patrick J., "Odorant Analysis Using A Sulfur Specific Detector," *Advances in Instrumentation*, 42(1): 557-566, 1987.
2. Clevett, Kenneth J., *Process Analyzer Technology*, New York: John Wiley & Sons, 1986.
3. Crane Co., "Flow of Fluids Through Valves, Fittings, and Pipe," Technical Paper No. 410, New York: Crane Co., 1982.
4. Drushel, H.V., "Trace Sulfur Determination in Petroleum Fractions," *Analytical Chemistry*, 50(1): 76-81, 1978.
5. Kisa, Erik, "Determination of Trace Amounts of Sulfur in Organic Solvents," *Analytical Chemistry*, 54(8): 1450-1459, 1982.
6. Leffler, William L., *Petroleum Refining for the Non-Technical Person*, Tulsa, OK: Penn Well Publishing, 1985.
7. Moore, Patrick J., "The Importance of Sulfur Measurement on Synthesis Gas Streams," *Advances in Instrumentation*, 43(3): 963-966, 1988.
8. Moore, Patrick J., "Recent Advances in On-line Total Sulfur Measurement for Liquid Hydrocarbon Streams," *Advances in Instrumentation and Control*, 44(4): 1419-1426, 1989.
9. Moore, Patrick J., "Recent Advances in the Measurement of Total Sulfur in Diesel Fuels," *Proceedings of ISA/92* (Toronto), Research Triangle Park: Instrument Society of America, p. 63-71, 1992.

10. Nelson, W.L., *Petroleum Refining Engineering*, New York: McGraw-Hill, 1941.

11. Perry, Robert H. and Chilton, Cecil H., *Chemical Engineers Handbook*, 5th ed., New York: McGraw-Hill, 1973.

12. Lide, David R., Ed., *Handbook of Chemistry and Physics*, 77th ed., Boca Raton: CRC Press, 1996.

13. Wiseman, P., *Petrochemicals*, Chichester, U.K.: Ellis Harwood Ltd., 1986.

About the Author

Patrick J. Moore is a Chemical Engineering graduate of the University of Delaware. He has worked for the Olin Corporation, Lummus, Envirotech Controls Inc., and Houston Atlas Inc. He is presently Manufacturing Manager for Houston Atlas. Mr. Moore is a member of the American Institute of Chemical Engineers and the Instrument Society of America. He has published numerous articles in the field of analytical instrumentation and the psychological aspects of literature.

21

Octane and Cetane Analyzers

Octane Analyzers

The octane number of a gasoline is a measure of its combustion characteristics in an engine. Most gasoline engines operate on a four-cycle sequence (Figure 21-1). On the intake cycle, the connecting rod, which is operated by the crankshaft, pulls the piston downward. The intake valve opens and allows a mixture of gasoline vapor and air from the carburetor to enter the cylinder. The piston then moves up on the compression cycle, the intake valve closes, and the mixture is compressed to a high pressure in the cylinder. Shortly before the piston reaches the top of the cylinder, a spark is generated at the spark plug and the fuel mixture ignites. The combustion forces the piston downward and generates the engine power. Then the piston rises again on the exhaust stroke. The exhaust valve opens and the combustion products are forced out to the exhaust system.

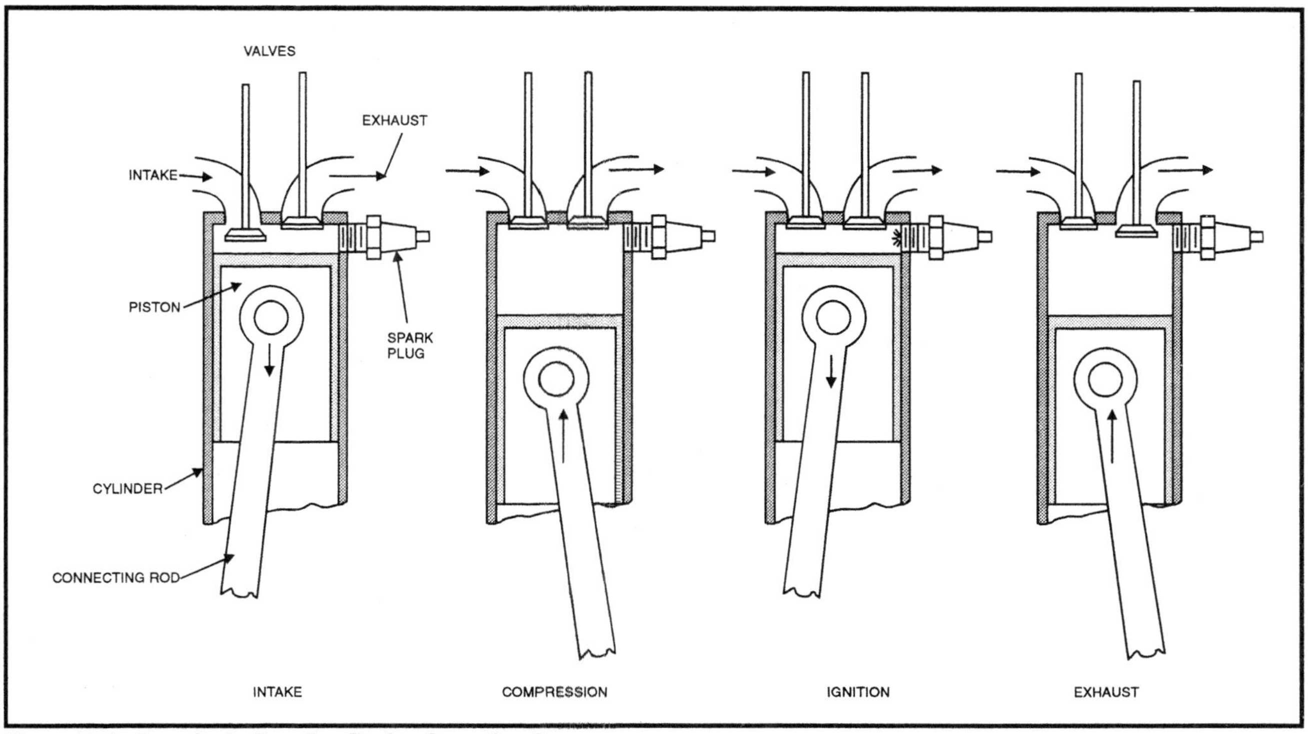

Figure 21-1. Four-Cycle Gasoline Engine Operating Sequence
(Courtesy of Sun Refining and Marketing Co.)

For smooth operation of the engine, the combustion should start at the spark plug and radiate outward from it until the entire volume of the fuel is burned. However, some fuels develop a "cool flame" and preignite. That is, portions of the fuel may start to burn before the flame wave moving through the cylinder reaches them. This will produce knocking, which usually occurs when the engine is operating under a heavy load. The knock is an audible metallic clacking noise that occurs when high-pressure, high-frequency waves bounce off the top of the piston and the cylinder walls.

Continuous "knock" can melt holes in the top of pistons!

The octane number of gasoline is based on two pure compounds. One, normal heptane, with a straight chain of seven carbon atoms, is designated as zero octane. The second, isooctane, with a chain of five carbon atoms and three methyl branches, is designated as 100 octane. Mixtures of these two pure compounds can produce any desired octane number between zero and 100 and are known as primary reference fuels (PRFs). An example of a PRF is the mixture of 85% isooctane and 15% *n*-heptane, which has an octane number of 85.

Some engines, particularly aircraft engines, require gasoline with an octane rating higher than 100. Primary reference fuels for these applications can be prepared by adding tetramethyl or tetraethyl lead to isooctane. Toluene, an aromatic hydrocarbon that has an octane rating above 100, is also mixed with isooctane and the mixture used as a reference fuel.

Laboratory Test Methods

Octane numbers of gasolines have been determined in the laboratory for many years with engines known as cooperative fuel research (CFR) engines. They were developed and are manufactured by Waukesha Engine Company (Waukesha, WI). The laboratory methods are substantially unchanged today.

Motor Octane Number (MON) correlates well with engine operation at high speeds and high temperatures.

ASTM Standard Test Method D-2700. The first method developed was the motor method test, which measured the motor octane number (MON). This test gives results that correlate well with the operation of gasoline engines at high speeds and high temperatures.

A single-cylinder engine with a variable-compression ratio mounted on a stationary-base is used for this test. The assorted controls and measuring instruments are mounted beside the engine. The carburetor is equipped with three fuel bowls. One is for the fuel to be analyzed. The other two are for reference fuels, one with an octane number lower than that expected for the test fuel and the other with a higher octane number. The fuel bowls are adjusted to feed their samples into the carburetor to provide a fuel-to-air ratio that produces the maximum knock intensity as measured by the knock meter.

Research Octane Numbers (RONs) correlate better with engine operation at low speeds and low temperatures.

The engine is run at 900 rpm and the spark plug is energized at 13° before the piston reaches top dead center (TDC). The compression ratio is adjusted to give a standard knock intensity with a specific reference fuel. The test fuel is then fed to the engine and the knock intensity compared with that of the reference fuel.

ASTM Standard Test Method D-2699. In the 1940s, improvements were made in the design of commercial engines, and compression ratios were increased. Improved gasolines with higher octane ratings became available, and a second test method was developed. This "research method" is run at 600 rpm on an engine almost identical with the one used for motor octane. Research octane numbers (RONs), which are obtained by this procedure, correlate better with engine operation at more mild conditions (i.e., low speeds and low temperatures).

The road octane number, which is the average of the research and the motor octane numbers, (RON + MON)/2, gives a good correlation with overall car performance on the road.

Road Octane Number (RON + MON)/2 or (R + M)/2 correlates best with overall engine operation at over-the-road conditions.

Other Laboratory Test Methods. Three other ASTM procedures are available for testing engine fuels.

The *D-2886 test* for measuring knock characteristics of motor fuels by the distribution octane number (DON) is claimed to provide an improved measure of the road antiknock ratings of motor fuels in passenger vehicles operating at high load and low engine speed. This condition may be encountered with vehicles equipped with manual transmissions. Since most vehicles now have automatic transmissions, the method is used very little.

The *D-2623 test* measures the motor octane number of liquefied petroleum gases that are used as engine fuels.

The *D-909 supercharge method* measures the knock characteristics of aviation fuels.

On-line Measurement of Octane Number

ASTM Standard Test Method D-2885 is available for monitoring RON and MON ratings using on-line analyzers.

When octane tests were first developed, refineries generally produced gasoline in batches. The products were stored in tanks and tested in the laboratory before being shipped to customers. As the demand for gasoline increased, refineries became more efficient and produced these products by in-line blending. The blended stocks were mixed and shipped to their final destinations with little delay. More rapid product analyses were necessary, and on-stream analyzers were developed.

The laboratory engine could not be used to monitor process samples continuously, because within a period of hours or days, the apparent octane rating of a known sample would decrease due to changes in engine characteristics. However, switching periodically from the process fuel to a calibration fuel with a known octane rating can compensate for any deviations in engine performance. The engines are programmed to measure the difference between the process fuel and the calibration fuel, known as the delta octane number (ΔON). From this difference, the actual octane number of the process fuel can be calculated from the known octane number of the calibration fuel. ASTM Standard Test Method D-2885 is available for monitoring RON and MON ratings using on-line analyzers.

Measurement Principles

A number of different systems are available to measure octane numbers on-line. Some of them use automated versions of the Waukesha engines. Two of the others measure the preignition reactions that cause engine knocking. More recently, new methods based on near-infrared (NIR) spectroscopy have been developed.

All of these methods can be used to monitor the individual components — such as light naphthas, reformates, and alkylates — that are blended into gasolines, as well as the finished gasolines themselves, the compositions of which can vary considerably with the seasons and with the operating conditions in the refinery. Other methods analyze the gasoline by gas chromatography (GC) or measure the refractive index (RI) of the sample and correlate it with the octane number. These correlation methods are limited to the analysis of gasoline components, such as reformates, that have relatively consistent compositions.

ETHYL CORPORATION MODEL 8021 OCTANE ANALYZER

This instrument compares the octane number of a process stream with that of a prototype fuel having a similar composition. Therefore, changes in engine performance affect both fuels equally. No primary fuels are used in this analyzer.

The system uses the conventional CFR engine with a comparator console. A fuel panel regulates conditions and the delivery of the sample and prototype fuel to the engine. A modified carburetor has been added to minimize sample dead time. A mechanical chiller maintains the components and both fuels at the proper temperature.

The engine operates 5 min on the prototype fuel and then switches automatically to the process fuel for 25 min. The prototype is fed to a conventional carburetor, and the process sample is fed through a flow-through carburetor assembly at a rate ten times that required by the engine. A suction tube removes the excess sample and returns it to the process. Sample dead time is minimized in this assembly.

Knock intensity of the sample is measured and compared to the desired intensity level every few seconds. If a deviation occurs, a control unit generates a signal to drive the compression motor to correct it. This system has a reproducibility better than 0.2 MON or RON in the range of 90 to 100 octane.

ETHYL CORPORATION COMP-CON III COMPUTER OCTANE ANALYSIS SYSTEM

This system can control as many as three research method engines and three motor method engines that are monitoring as many as three process samples simultaneously. The computer switches between process and prototype fuels, controls the knock intensity of each engine by adjusting the compression ratio, and computes the road octane number from the RON and MON. Each engine can be set up from the operator's panel. Panel settings adjust fuel-to-air ratios, change the detonation meter sensitivity, and select the proper compression ratios for the prototype fuels. When a process sample is not available, a standby mode can operate the engines on prototype fuel.

DUPONT MARK IV OCTANE COMPARATOR

Mark IV is the latest in a series of analyzers developed by the DuPont Company. It uses a Waukesha engine to measure the difference in knock intensity between prototype and production fuels at a constant compression ratio and calculates ΔON. The actual octane number of the prototype fuel is determined by comparing it to a pair of primary reference fuels. From that value the actual octane number of the product fuel is calculated. The engine runs for 4 min on the prototype fuel and then 4 min on the product fuel, thus minimizing the effects of engine drift.

The analyzer has an operator console mounted on the engine chassis, a fuel selection system, a cooling unit, and a special carburetor to minimize dead time in the product sample system. The detonation meter, the knock intensity recorder, and the ΔON meter are mounted on the upper panel of the comparator. The lower panel contains the cycle programmer, the octane number computer, and the necessary power supplies.

The computer takes the knock signal from each fuel at the end of each 4-min cycle and calculates the ΔON. The actual octane number of the product fuel is calculated from the known octane number of the prototype fuel. The programmer controls all functions of the system, switches fuels, and tells the computer when to sample the data and when to compute. The Mark IV system can handle gasolines with octane numbers from 70 to 100, with standard deviations of 0.06 for RON and 0.12 for MON.

Available accessories can calculate the road octane number from the RON and MON and the cumulative octane barrel indicator (COBI) of off-spec product from the number of barrels produced. They can also signal alarms for engine malfunctions and environmental problems.

DUPONT COMOC II COMPUTER OCTANE COMPARATOR SYSTEM

This analyzer can control four engines, two for RON and two for MON, operating on two blending systems. A COMOC console is mounted on each engine stand to interface the engines with the central computer. The computer calculates absolute and delta octane numbers for the product fuels and the total blend error (COBI) from the flow data of the blender. At the end of a blend, the system can calculate the average ΔON, the COBIs for research, motor, and road octanes, and the total volume of the blend. In addition, the computer monitors for engine malfunction, high or low knock intensity, changes in compression ratios, and variations in knock intensity in the prototype fuel.

ASSOCIATED OCTEL MARK 5 OCTANE COMPARATOR

This model is a microprocessor-based system, the latest in a line produced by Associated Octel, Ltd. in Great Britain. The comparator controls a CFR engine and feeds the prototype and product fuels to the engine for 3-min intervals. From the ΔONs it calculates the octane number of the product fuel. The system can provide analog signals to recorders, digital outputs to host computer/printer facilities, and status and alarm signals.

In addition to the Mark 5 Comparator, Octel can supply complete systems with engines for RON and MON, reference fuel storage tanks, and sampling systems. All units can be supplied in weatherproof shelters.

UNIVERSAL OIL PRODUCTS (UOP) MONIREX OCTANE ANALYZER

UOP's Monirex instrument measures the spontaneous preignition reactions that cause knocking in gasoline engines. The analyzer uses a cylindrical reactor into which the fuel and instrument air are introduced continuously. The reactor is maintained at a temperature high enough to generate preignition reactions. The heat generated by these reactions raises the temperature of the reaction products above that of the reactor vessel (Figure 21-2).

When the reactions are completed, the temperature will reach a maximum at a specific distance from the reactor inlet. If the octane number of the sample fuel increases, the rate of the reactions will decrease and the peak temperature will move farther down the length of the reactor. A change in air pressure also affects the rate of the reactions. If the air pressure in the reactor increases, the speed of the reactions will increase and the peak will move closer to the reactor inlet. By controlling the air pressure, the peak temperature point can be maintained at a specific location in the reactor.

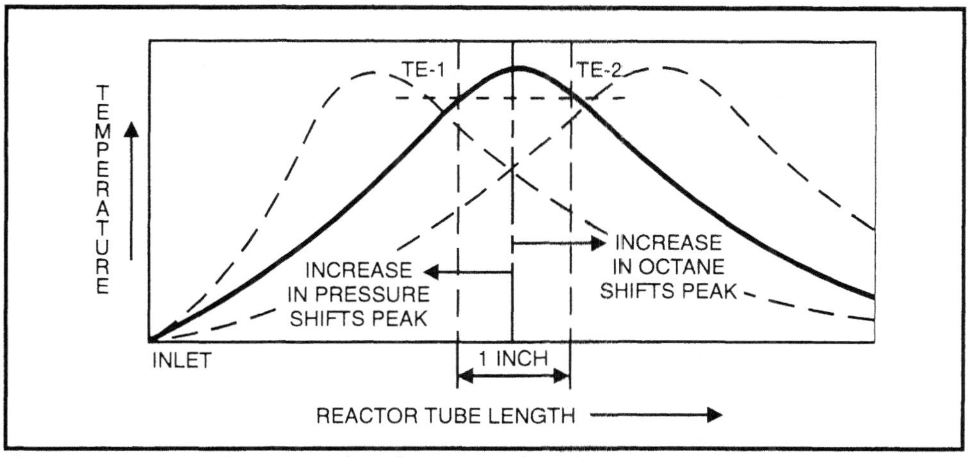

Figure 21-2. Monirex Octane Monitor: Operating Principle
(Courtesy of UOP/Monirex Systems)

The reactor contains two thermocouples located 1 in. (25 mm) apart. If the peak temperature occurs ahead of the thermocouples, TE-1 (see Figure 21-2) will indicate a higher temperature than TE-2. Conversely, if the peak occurs after the thermocouples, TE-1 will indicate a lower temperature than TE-2. If the peak is centered between the two thermocouples, they will both indicate the same temperature. Since the peak is maintained at a fixed point in the reactor by controlling the reactor pressure, the pressure correlates with the octane number of the sample.

A flow diagram is shown in Figure 21-3. The fuel sample enters the system at the lower left, and a 1 mL/min sample is pumped through the preheater and into the reactor by the metering pump. Excess incoming sample returns to the fast loop through the rotameter (FI). Instrument air enters the system at the lower right and is controlled at 45 psig by the pressure controller (PC). Part of the air stream flows through the flow indicator (FI) and the flow controller (FC). It passes through the preheater coil in the bath heater and joins the fuel sample in the preheater. This mixture enters the reactor tube and passes the differential thermocouple (TE-3), which senses the position of the reaction peak. The controller (TDC) adjusts the reaction pressure by operating the pressure control valve at the upper right of the unit. This pressure is recorded by the recorder (PR) and is correlated with the octane number of the sample. The waste sample/air mixture from the reactor tube passes through the afterburner with additional air, where the combustion is completed. The final products are vented to the atmosphere through a vent stack. High- and low-pressure switches, (PSH and PSL) shut off the sample pump if the sample pressure becomes too high or if the air pressure is too low.

The field unit, including a sampling system, is rated for installation in Class I, Group D, Division 1 hazardous areas. The operating panel is mounted in the control room with a strip-chart recorder for recording results.

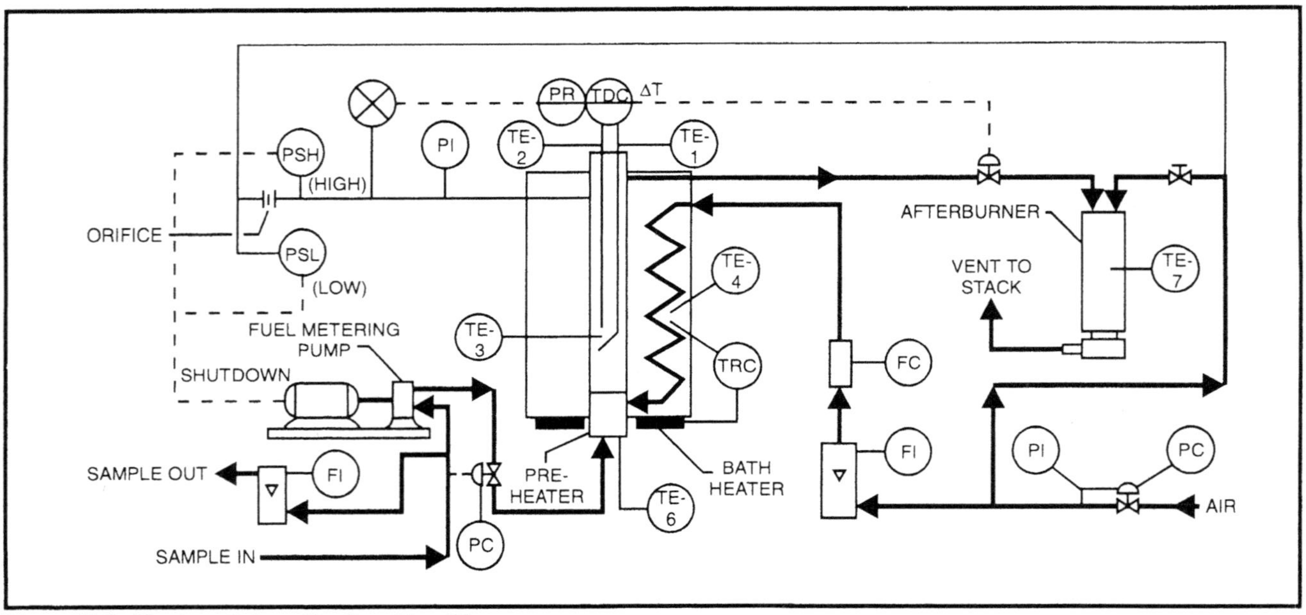

Figure 21-3. Monirex Octane Monitor: Flow Diagram
(Courtesy of UOP/Monirex Systems)

Three primary reference fuels and a prototype fuel similar to the process sample are required to calibrate the analyzer. Two of the primary reference fuels should have octane ratings that bracket the expected octane number of the process sample. The third primary rule and the prototype should have octane ratings equal to that of the process sample. The three primary reference fuels are then used to determine the slope of the response curve of the analyzer. The analyzer is calibrated with the third reference fuel and the prototype by varying the reactor temperature until the same response is obtained from each.

ARCAS MODEL 81-P PROCESS OCTANE ANALYZER

This analyzer is based on research by Gulf Research and Development and uses the cool flame reactions to determine octane numbers. Gulf licensed Anacon, Inc., to produce the Anacon Model 80-P instruments for Gulf refineries. Later, Anacon incorporated microcomputer technology into the system and produced the Arcas Model 81-P octane analyzer. This model is now marketed by Foxboro Analytical Division of the Foxboro Company.

The analyzer measures the magnitude of the exothermic reactions that occur when a gasoline sample is injected with air into a heated reaction chamber (Figure 21-4). After sample injection, the temperature of the fuel/air mixture remains constant during the induction period before the cool flame reactions begin. The temperature will then rise rapidly to a peak and decrease. As the octane number decreases, the reactions become more severe and the induction period decreases. Both the induction period and the peak temperature can be correlated with the octane number of the sample. Subsequent studies showed, however, that the peak temperature gives a better correlation and the analyzer measures only this variable. Figure 21-5 shows correlations between the peak temperatures (in millivolts) and the RONs. The four curves are for different reactor temperatures. Reactions have been observed at temperatures from 520 to 625°F (270 to 330°C), but the best correlations have been obtained in the range of 570 to 590°F (300 to 310°C).

The reactor assembly (Figure 21-6) consists of an insulated stainless steel cylinder with a spherical void in the center. It is mounted in an explosionproof housing suitable for installation in Class I, Group D, Division 1 hazardous areas. A proportional controller in the programmer signal-conditioning assembly, which is installed in the control room, heats the cylindrical block to a temperature near 570°F (300°C). The actual temperature depends on the octane range and the specific application. Air from an external flow controller enters the top of the reactor block and exits from the bottom to an atmospheric vent. Every 5 min, 12 µL of the gasoline sample is injected into the reactor by a chromatographic-type liquid injection valve. The explosionproof enclosure is heated to 160°F (70°C), with another proportional controller, to vaporize the sample before it is injected into the reactor and to preheat the incoming air.

A microthermocouple in the center of the reactor cavity detects the exothermic reaction when the sample ignites. The temperature of the reactor block is measured by the reference thermocouple. The difference between these two temperature readings is converted to a millivolt signal for transmission to the programmer. The programmer converts the signal to an octane number, which is recorded on a strip-chart recorder. The correlation between the temperature output in millivolts and the RON for ten different reformate samples is shown in Figure 21-7.

The analyzer unit consists of a freestanding frame with three explosionproof enclosures. The lower left enclosure contains the reactor unit, the upper left enclosure the microprocessor-based controller, and the upper right enclosure the digital display unit. The sample-conditioning equipment, the airflow control equipment, and the sample metering pump are also mounted on the frame. The analyzer should be installed indoors or in an analyzer shelter to protect it from the weather and to facilitate maintenance.

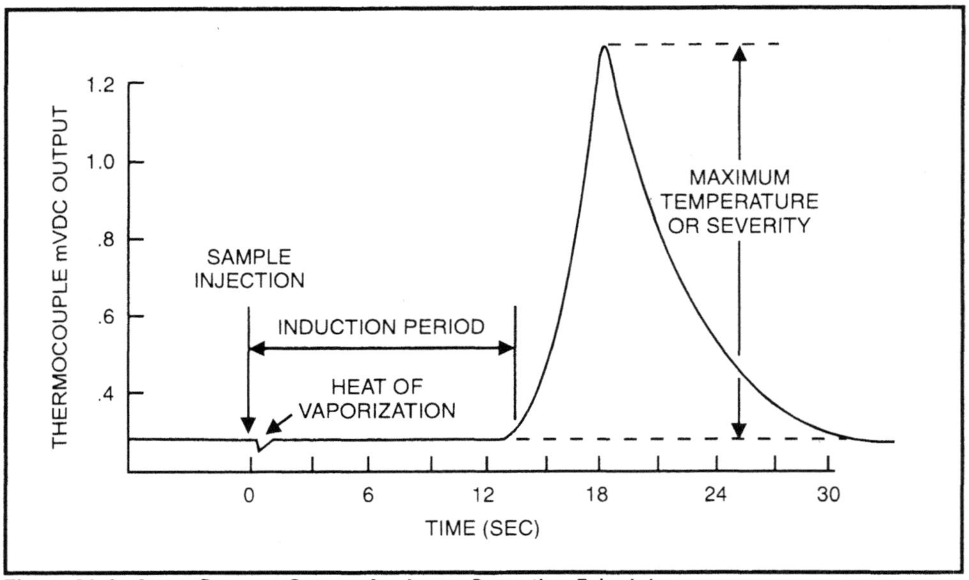

Figure 21-4. Arcas Process Octane Analyzer: Operating Principle
(From Process Analyzer Technology, *Kenneth J. Clevett, New York: John Wiley & Sons)*

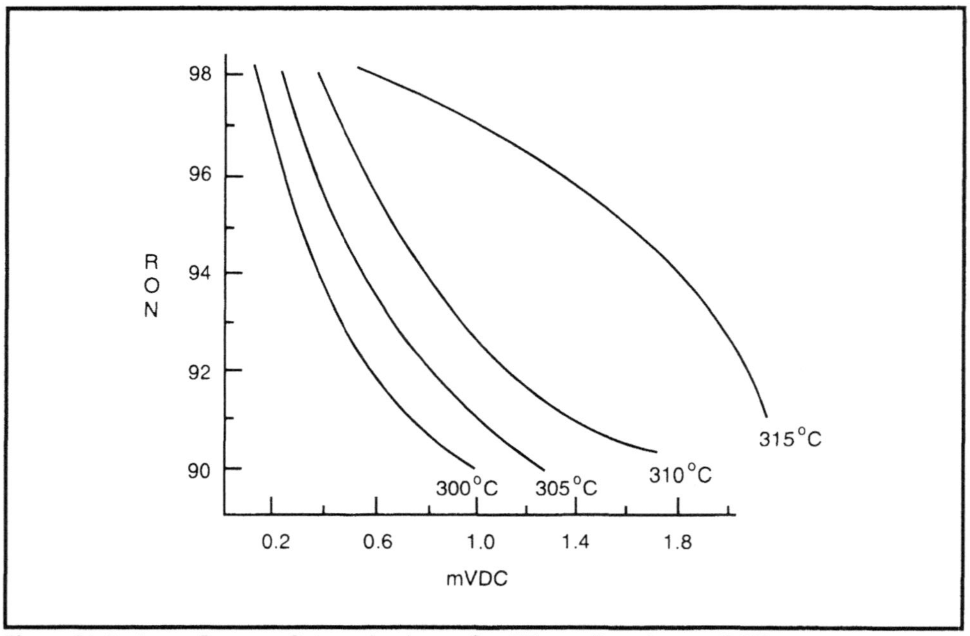

Figure 21-5. Arcas Process Octane Analyzer: Cool Flame Reaction vs. RON
(From Process Analyzer Technology, *Kenneth J. Clevett, New York: John Wiley & Sons)*

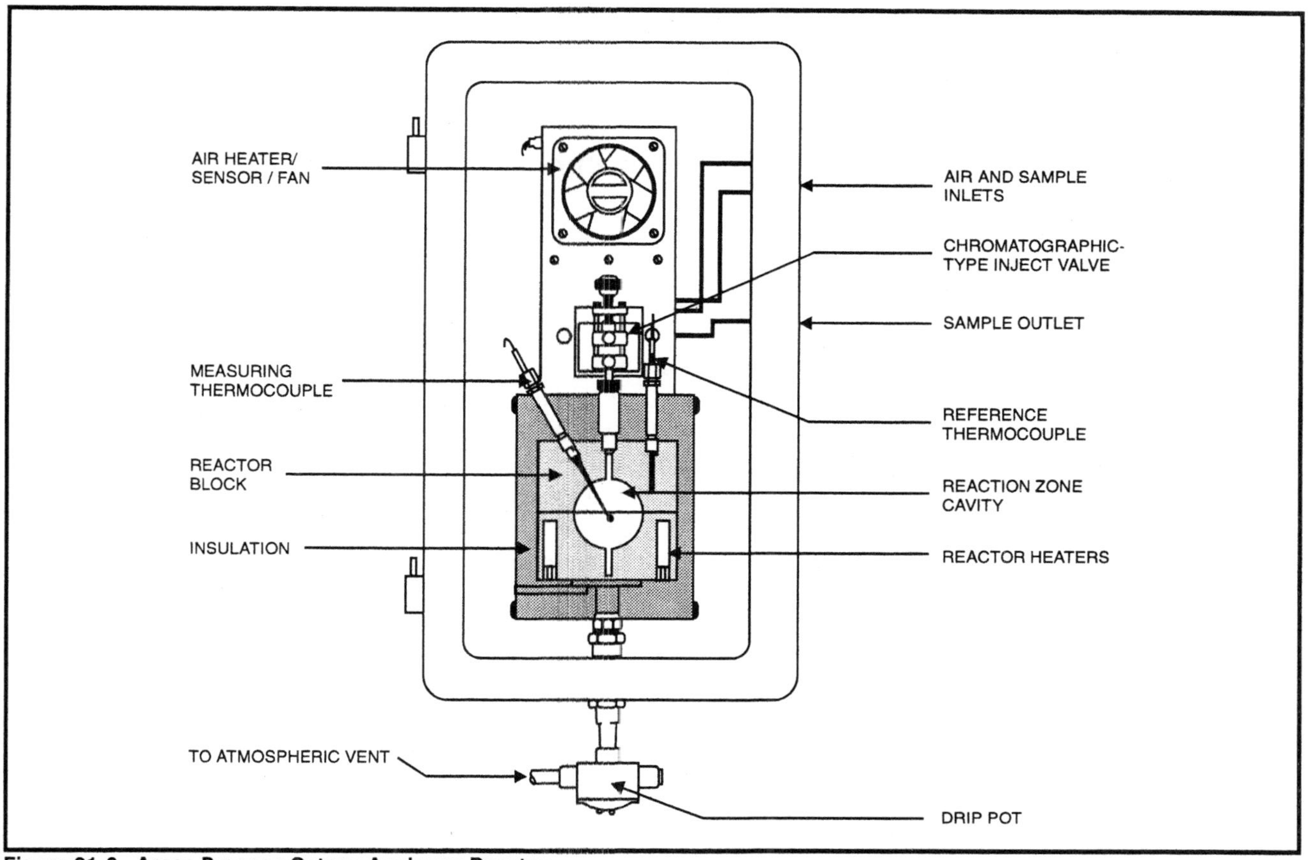

Figure 21-6. Arcas Process Octane Analyzer: Reactor
(From Process Analyzer Technology, *Kenneth J. Clevett, New York: John Wiley & Sons)*

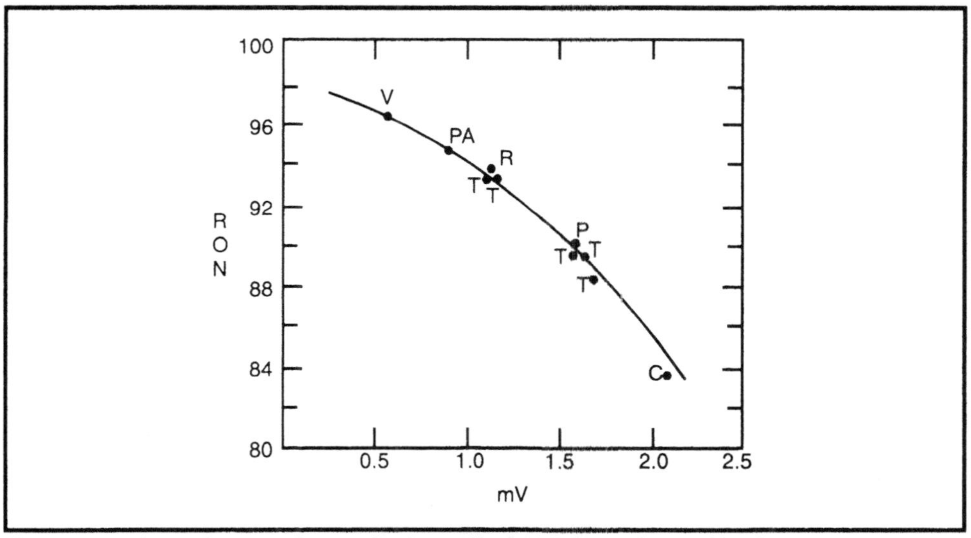

Figure 21-7. Arcas Process Octane Analyzer: Peak Height vs. Octane Number
(From Process Analyzer Technology, *Kenneth J. Clevett, Ed., New York: John Wiley & Sons)*

The analyzer can be field-adjusted over a 5 ON span. Repeatability is 0.1 ON. One SCFM of instrument air at 60 to 100 psig is required. Sample requirements are 300 mL/h at a maximum pressure of 300 psig and a temperature between 50 and 105°F (10 and 40°C).

ANACON MODEL 31S DIFFERENTIAL REFRACTOMETER

The reforming processes of catalytic reforming and hydroforming are used to produce components for high-octane gasolines. the major reactions involved are the conversion of paraffins and naphthenes to aromatics. The unsaturated aromatic compounds have much higher refractive indices than those of the saturated paraffins and naphthenes.

The octane number correlates very well with the refractive index of the reformed product from a particular feedstock. The correlation may be different for different feedstocks; however, by measuring the refractive index of the feedstocks and the products alternately, appropriate correlations have been obtained.

The Anacon refractometer has been used successfully to monitor reformates. The refractive index of the feedstock and its product are recorded in the control room, where the operators can maintain proper conditions for the reforming process.

APPLIED AUTOMATION CHROMATOGRAPH

In the early 1970s, Applied Automation found that the analysis of paraffins, naphthenes, and aromatics by multiple-column gas chromatography could be used to predict the octane number of reformates. A typical chromatogram of a reformate is shown in Figure 21-8. Low-resolution columns are used and individual components are not separated. Instead, a peak for the total saturates (i.e., the paraffins plus naphthenes) and a peak for the total aromatics are obtained. The percent of aromatic components is calculated from the areas under these peaks.

The measurement of total aromatics in the reformate produced from a specific feedstock permits the calculation of the octane number of the product with a reproducibility of 0.1 ON. A single laboratory octane engine test provides an adequate correlation for the total aromatic concentrations in the products from different feedstocks.

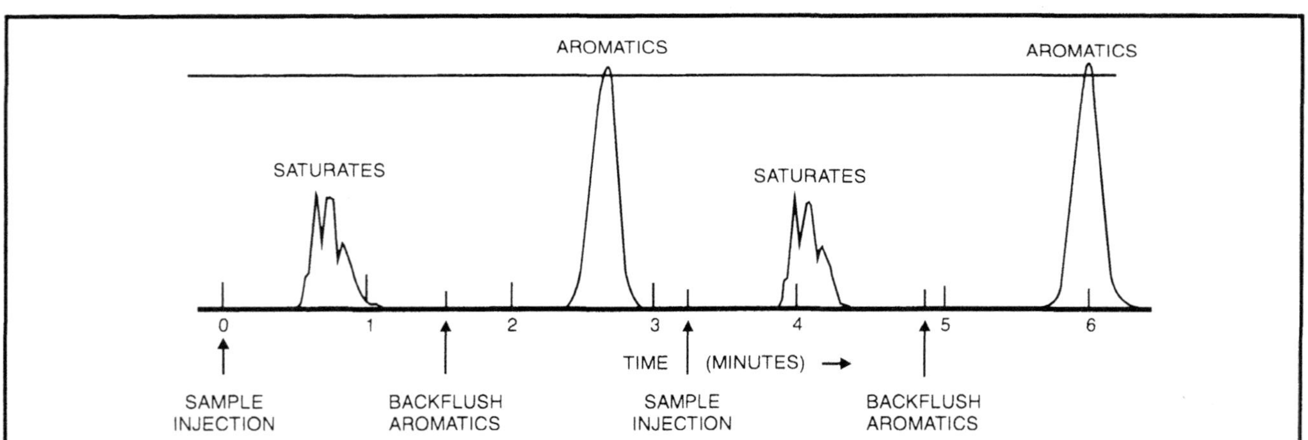

Figure 21-8. Applied Automation Chromatogram: Separation of Saturates and Aromatics
(From Process Analyzer Technology, *Kenneth J. Clevett, New York: John Wiley & Sons)*

NEAR-INFRARED SPECTROSCOPY

Several companies have developed octane analyzers based on NIR spectroscopy. Methyl, methylene, and aromatic components of molecules, which contribute significantly to the octane numbers of gasolines, generate relatively wide absorption bands in the 1100 to 1250 nm range of the NIR spectrum. Computer analysis of these absorption bands produces correlations with the research, motor, and road octane numbers of the gasolines, with reproducibilities two to three times better than the results from octane engines. The analyzers use sample probes that can be inserted directly into a gasoline blender product line, so complex sampling systems are not required. Fiber-optic cables transmit light to the sample probe and, from it, back to the spectrometer, which can be located a considerable distance from the blender. The spectrometer, therefore, can be located in a nonhazardous area, such as a control room, where it is available to operating personnel.

More recently, a version of the NIR spectrometer was developed to monitor the octane numbers of gasolines. Spectra of several gasoline samples are shown in Figure 21-9. The differences in the individual spectra are small, and no one region is specific for the octane rating. However, analysis of the data by the partial least-squares method showed good predictions of octane numbers from the absorption bands in the 1100 to 1250 nm region (Figures 21-9 and 21-10). The results from 32 samples from 11 different sources, 8 of which contained methyl tertiary butyl ether (MTBE), are shown in Table 21-1 and Figure 21-11. The standard error of prediction was ±0.20, compared with a standard deviation for an octane engine of ±0.60. An explosion-proof process-monitoring center, with an easy-to-use operator interface, including an operator keyboard and a liquid crystal display, can be installed next to a gasoline blender.

Probes for installation in the process stream consist of two transmission probes with sapphire windows, a coupler to maintain a consistent optical pathlength, and a process pipe flange with gland seals. The probes need not be taken out of the line for maintenance. The packing glands on the sides of the flange can be loosened and the probe slid sideways until the section between the windows is outside the process pipe, where it can be cleaned. The probes are connected to the monitoring center with single-strand fiber-optic waveguides, which can be hundreds of feet long.

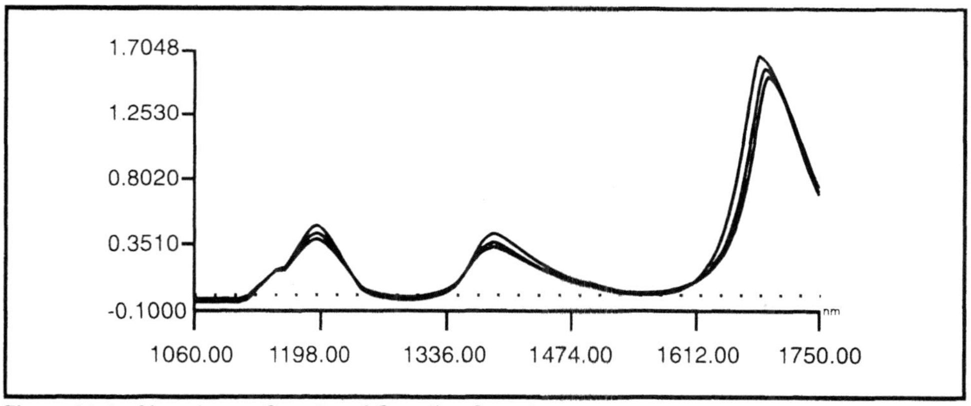

Figure 21-9. Absorbance Spectra of Gasoline Samples
(Courtesy of Guided Wave, Inc.)

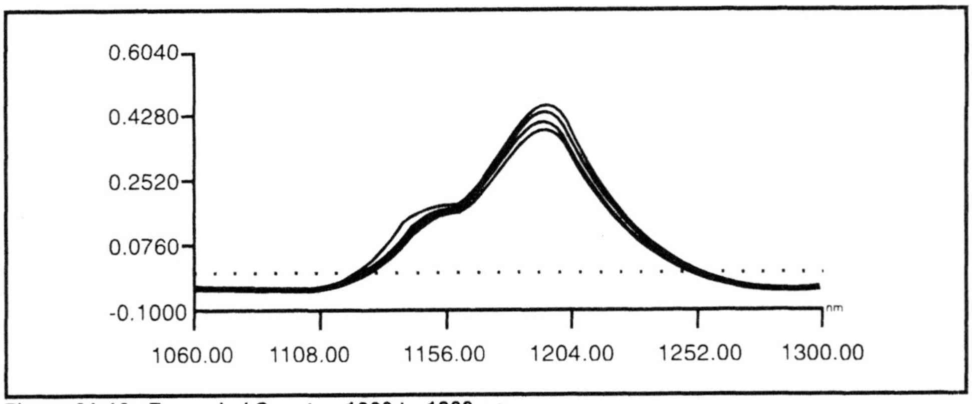

Figure 21-10. Expanded Spectra, 1060 to 1300 nm
(Courtesy of Guided Wave, Inc.)

Table 21-1. Results of NIR Spectral Determination of Octane Numbers
(Courtesy of Guided Wave, Inc.)

Octane Number Unknowns and Predicted Values		
(R + M)/2	Predicted	Error
86.700	86.590	–0.110
87.000	87.270	0.270
87.000	86.870	–0.130
87.000	86.750	–0.250
87.100	87.420	0.320
87.200	87.070	–0.130
87.400	87.770	0.370
88.600	88.410	–0.190
88.600	88.520	–0.080
88.600	88.790	0.190
88.800	88.740	–0.060
89.000	89.090	0.090
89.100	89.090	–0.010
89.100	89.130	0.030
89.200	89.121	–0.079
89.400	89.310	–0.090
89.900	90.002	0.102
90.100	90.324	0.224
90.500	90.435	–0.135
90.800	91.220	0.420
91.100	91.344	0.244
91.200	91.040	–0.160
91.200	91.260	0.060
91.400	91.080	–0.320
91.400	91.345	–0.055
91.600	91.560	–0.040
91.700	91.370	–0.330
91.700	91.450	–0.250
91.800	92.020	0.220
92.400	92.440	0.040
92.400	92.237	–0.163
93.200	92.992	–0.208

Multiple sample points can be monitored with a high-speed multiplexer, allowing the analysis of up to six separate feed streams into the blender or of several blenders side-by-side in a refinery. The system can also monitor a special prototype fuel of known octane number as a continuous check on system operation.

In addition to the research, motor, and road octane numbers, the system can be programmed to determine concentrations of MTBE and alcohol additives, percent aromatics, Reid vapor pressures, and flash points of the gasolines. Figure 21-12 compares the NIR flash point results for 86 samples of gasolines with the results of the ASTM test.

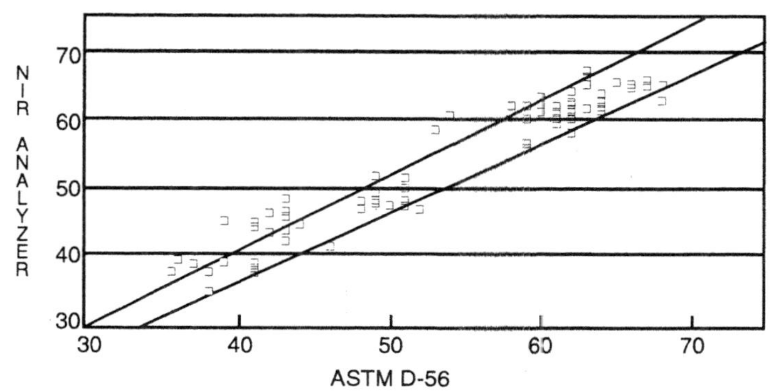

Figure 21-11. Flash Points of Gasoline Samples
(Courtesy of Guided Wave, Inc.)

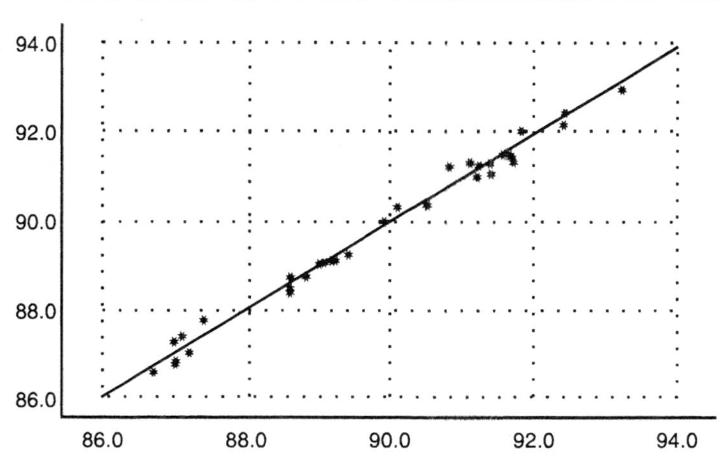

Figure 21-12. NIR Octane Numbers vs. Octane Engine Results
(Courtesy of Guided Wave, Inc.)

An NIR octane analyzer was developed by Ashland Petroleum Company (Ashland, KY). This NIR analyzer was used with a computer to process the data.

The NIR spectra of a group of gasoline samples is shown in Figure 21-13. Multiple linear regression was used to develop the calibration. A second derivative was used to eliminate baseline drifts, normalize the data, and increase the apparent degree of band resolution. The results are shown in Figure 21-14. The resulting absorption bands of aromatics and normal and isoparaffins are shown in Figure 21-15.

The system was installed in Ashland's Saint Paul Park refinery in Minnesota in late 1990. One month was required for calibration, and the system was operated for four more months to test its performance. Then it was interfaced with a Honeywell gasoline blending system and placed on closed-loop control. It was found to be extremely reliable and about two to three times more accurate than the knock engines. It is insensitive to changes in vapor pressure, which occur in summer- and winter-grade gasolines. One calibration equation for each octane — research, motor, and road — was suitable for all grades of gasolines and all volatility ranges.

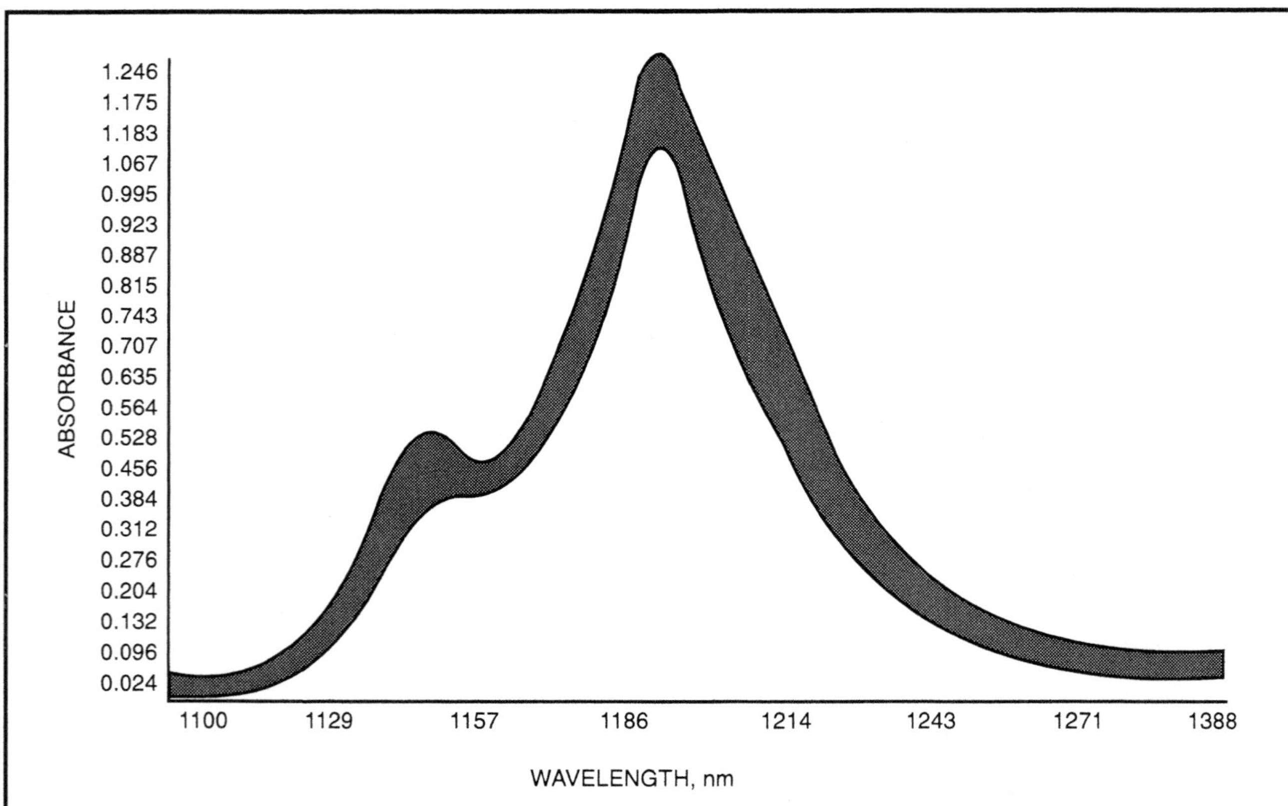

Figure 21-13. NIR Spectra of Gasoline Samples
(Courtesy of Fluid Data, Inc.)

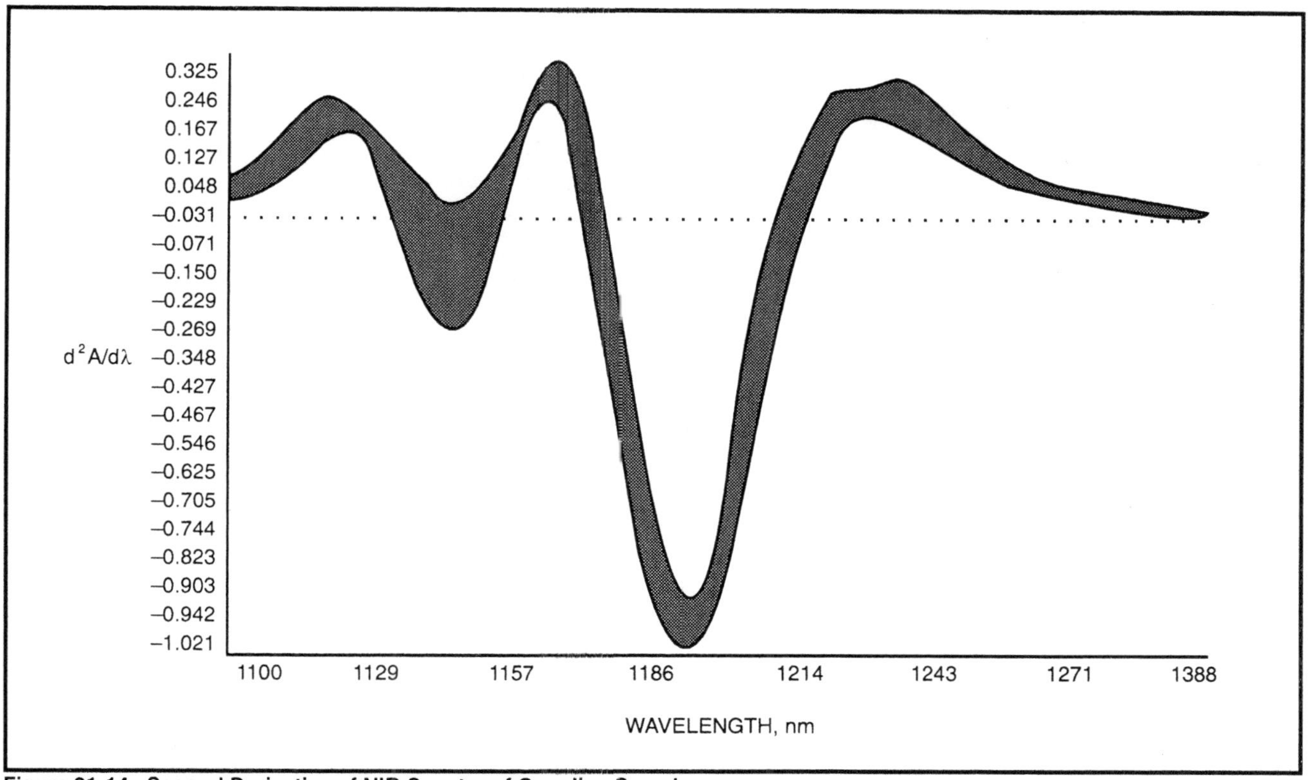

Figure 21-14. Second Derivative of NIR Spectra of Gasoline Samples
(Courtesy of Fluid Data, Inc.)

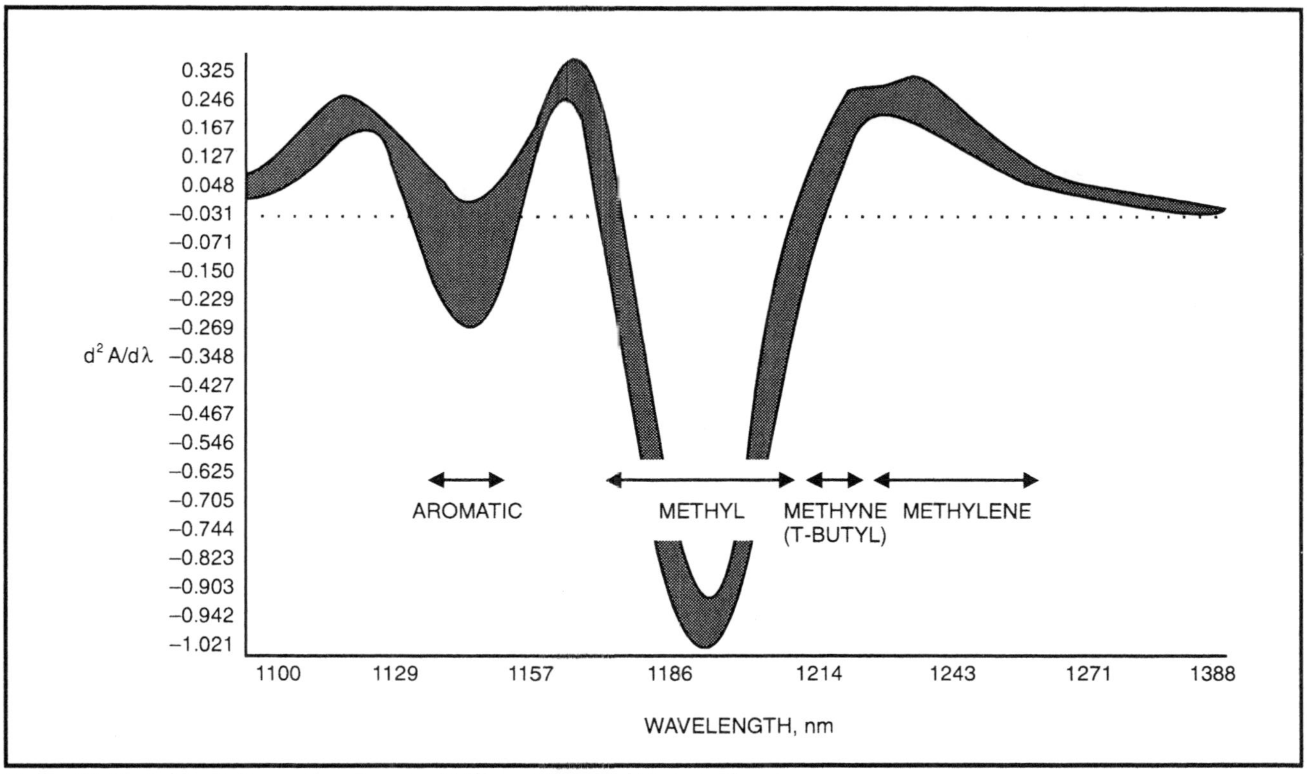

Figure 21-15. Infrared Absorption Bands of Different Types of Hydrocarbons
(Courtesy of Fluid Data, Inc.)

Figure 21-16 compares the NIR and knock engine results during a 18-h blending operation. Because of its consistency (Figure 21-17), which results in increased production of on-spec products, the installation can be paid for in a matter of months.

The InfraTane™ system (developed by Ashland Oil) has been licensed to Fluid Data for manufacture, marketing, and maintenance. The system is a complete, self-contained unit certified for installation in hazardous areas. It contains the spectrometer in the upper housing and the sampling system and electronics in the two lower housings, and also includes a switching system for the protofuel validation. A simpler version is available to monitor the components, such as reformate, that are blended into the finished gasolines.

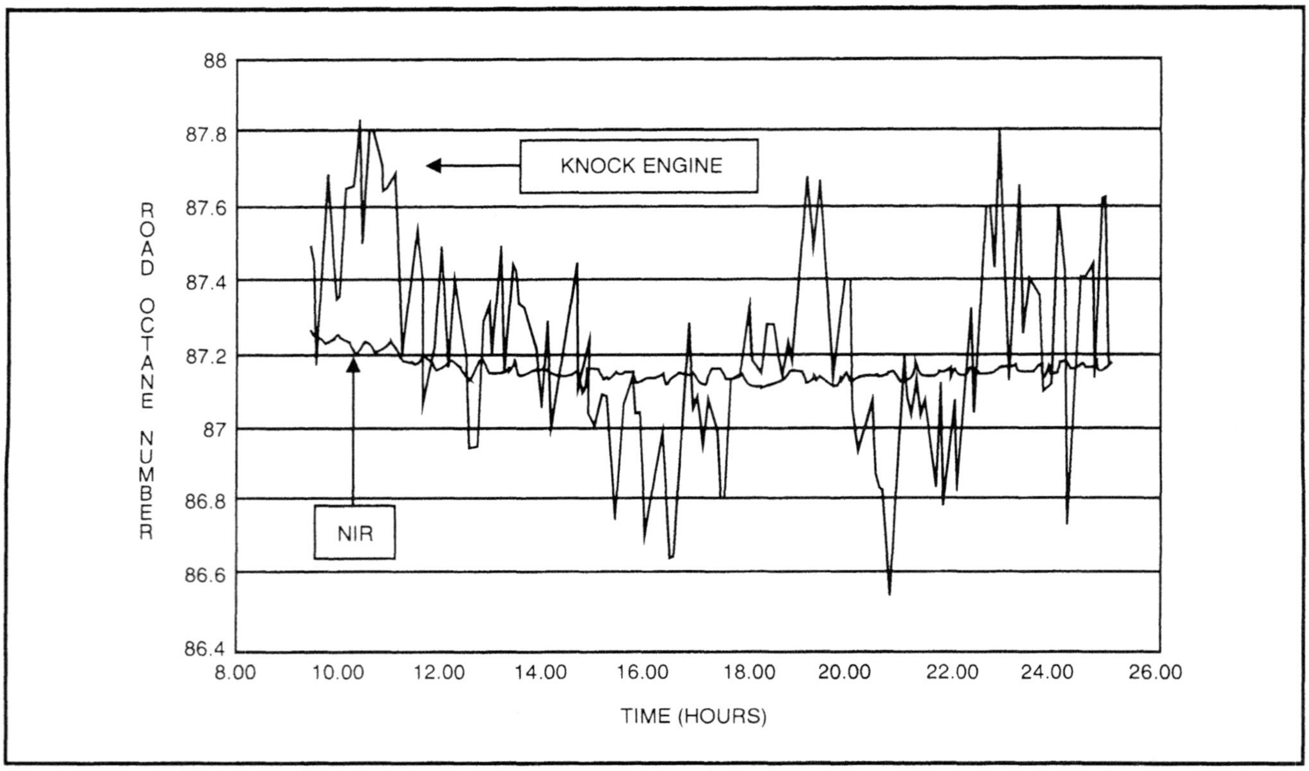

Figure 21-16. Comparison of Octane Numbers from NIR Spectra and Octane Engines
(Courtesy of Fluid Data, Inc.)

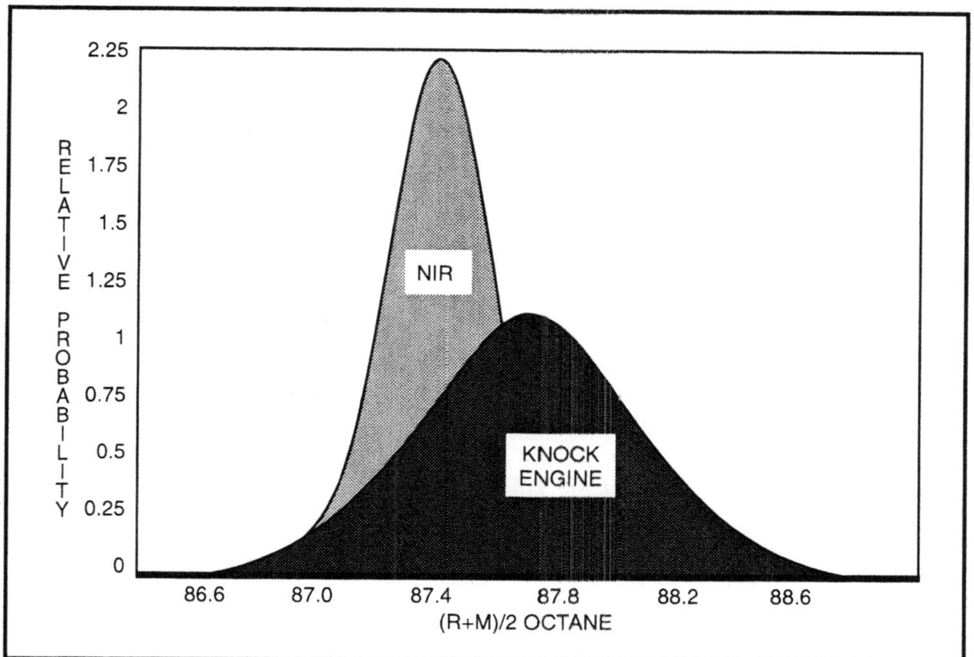

Figure 21-17. Relative Probability Distributions for Octane Numbers from NIR Spectra and Octane Engines *(Courtesy of Fluid Data, Inc.)*

Cetane Analyzers

Diesel engines are reciprocating engines similar to gasoline engines, except that spark plugs do not ignite the fuel. A diesel-fuel/air mixture is injected into the cylinders by a fuel pump and compressed to a much higher compression ratio than that used in gasoline engines. The high compression heats the mixture to a temperature at which it ignites spontaneously. Diesel fuels are comparable in price but safer to handle than gasolines. They are used mainly in trucks and heavy equipment that require long-term continuous operation.

Measurement Principles

Cetane number is a measure of the performance of diesel fuel, comparable to the octane number of gasoline. Cetane is a saturated hydrocarbon with 16 carbon atoms in a straight chain. It defines the high end of the cetane number range. Hepta-methyl nonane (HMN), a hydrocarbon with nine carbon atoms in a straight chain with seven branched methyl groups, defines the low end. Cetane number is calculated by:

$$\text{Cetane Number} = (\%n\text{-cetane} + 0.15) \times \%\text{HMN}$$

A reference fuel containing 100% cetane would have a cetane number of 100, while one containing 100% HMN would have a cetane number of 15.

In the laboratory, cetane numbers are measured in a one-cylinder diesel engine, which is manufactured by the Waukesha Motor Company (Waukesha, WI). It is similar to the octane engine described earlier. The performance of the test fuel is compared to mixtures of reference fuels that bracket the expected cetane number of the test fuel. The procedure is detailed in ASTM Standard Test Method D-613.

On-line cetane analyzers, using NIR analyses, can be used to monitor the cetane number of diesel fuels.

Common Applications of Octane Analyzers

The most extensive use of on-line octane analyzers is in the blending of finished gasolines. The components are mixed and delivered to product pipelines with little or no intermediate storage in tanks. The time required to sample the products and analyze them in the laboratory could result in the production of a considerable amount of off-spec product. If the octane numbers were ultimately found to be too low, the product would have to be sold as a lower grade of gasoline or, if too high, excessive, expensive high-octane components would be given away.

Since the compositions of gasolines can vary widely because of variations in refinery feedstocks and the seasonal variations in composition, the most popular on-stream octane analyzers for finished products are the automated CFR engines. For intermediate operations such as catalytic cracking, reforming, and alkylation, where the compositions are relatively uniform, the correlative (NIR, GC, RI and exothermic reaction) analyzers are adequate.

Ranges of Variables

Both the automated Waukesha engines and the correlative techniques are suitable for products ranging form regular-grade gasolines with octane numbers in the low 80s to aviation gasolines with numbers above 100.

Options and Accessories

Accessories are available for the automated engines to compute road octane numbers from the MON and RON and to compute the COBI, which is a measure of the total value of off-specification gasoline.

Limitations

The automated octane engines are very expensive to install and require frequent, extensive maintenance. However, because of the valuable data they produce, they can still pay out in a matter of months.

Installation Hints

The manufacturers of the automated engines can supply complete systems that include the engine, controls, and sampling system in shelters suitable for installations in electrically hazardous areas.

Specific Safety Hazards

Gasoline samples are volatile and the vapors are flammable. Care must be taken to assure safe handling. The vapors are also toxic and must be vented outside the analyzer building.

Unique Sampling System Requirements

The sampling systems should include fast loops to minimize the dead time in moving the samples to the analyzers. Also, disposition of the flammable waste samples requires special care.

Typical Calibration Requirements

Primary reference fuels with octane numbers close to those of the process samples must be blended or purchased. Prototype fuels similar to the process fuel must also be available.

Troubleshooting and Maintenance

Octane engines must be cleaned and calibrated periodically. The instruction manuals will detail the procedures. Octane simulators, such as the units that measure the preignition reactions and the gas chromatographs and the refractive index monitors that determine aromatic concentrations, must be serviced and calibrated according to procedures given in their manuals.

About the Author

Conard Donnell received a Bachelor's and a Masters degree in Chemistry and Physics from Temple University. His career with automatic process analyzers began and ended at Sunoco Refinery where he developed on-line instrumentation to measure the physical properties of process streams automatically and much faster than laboratory analysis. Since retirement, Mr. Donnell has worked as a consultant for Sunoco, Fluid Data, Precision Scientific Co., Lockwood and McLorie, and DuPont working on projects that included design and installation of process analyzer systems in refineries worldwide.

22

Vapor Pressure Analyzers

Vapor pressure is defined as the pressure of a confined vapor in equilibrium with its liquid phase at a specific temperature. If the temperature increases, more of the liquid will evaporate and the vapor pressure will increase. Vapor pressure is an important property of volatile petroleum products, such as automotive and aviation gasolines, because it affects start-up, warm-up, and the tendency to vapor lock at high operating temperatures or high altitudes (low pressures).

ASTM Standard Test Method D-323 for Reid Vapor Pressure. This is the universally accepted test method for the petroleum industry. It is used for volatile crude oils and volatile, nonviscous petroleum products (except for liquefied petroleum gas, which is covered by ASTM D-1267 and will not be reviewed here).

Two chambers are used for the test (Figure 22-1). One is a liquid chamber, the other a vapor chamber which has a volume four times that of the liquid chamber. The liquid chamber is chilled in an ice-and-water bath and then filled with cold sample and connected to the vapor chamber. Both chambers are immersed in a constant-temperature bath maintained at 100°F (40°C) and allowed to reach an equilibrium. As the cells warm, liquid evaporates until a vapor/liquid equilibrium is reached. At this point, the reading on the pressure gage, which is connected to the vapor chamber, is recorded as the Reid vapor pressure (RVP).

Two different liquid chambers are used. The lower one in Figure 22-1 has a single opening and is used for samples, such as most gasolines, whose vapor pressures are below 26 psig. The sample and the chamber are chilled in an ice-and-water bath separately before the chamber is filled and attached.

If the vapor pressure of a sample is above 26 psig, this procedure could allow light vapors to escape and the measured vapor pressure would be too low. For these samples the upper vessel in Figure 22-1, which has two openings, should be used. The sample container is connected to a cooling coil, which is immersed in an ice bath, and the cooling coil is connected to the bottom valve (J) on the liquid vessel. When the bottom valve is opened and the upper valve (I) is cracked to allow air to escape, cold sample fills the vessel completely. The liquid vessel, filled with sample, is then connected to the vapor vessel and both are immersed in the 100°F (40°C) bath as in the standard procedure.

Reproducible results require extreme care when carrying out the test. A small loss of vapor (most probably propane or propylene) from the sample can greatly affect the test results. The prescribed procedure for sample transfer, immersion of the test equipment into the constant-temperature bath, and measurement of the vapor pressure must be followed carefully.

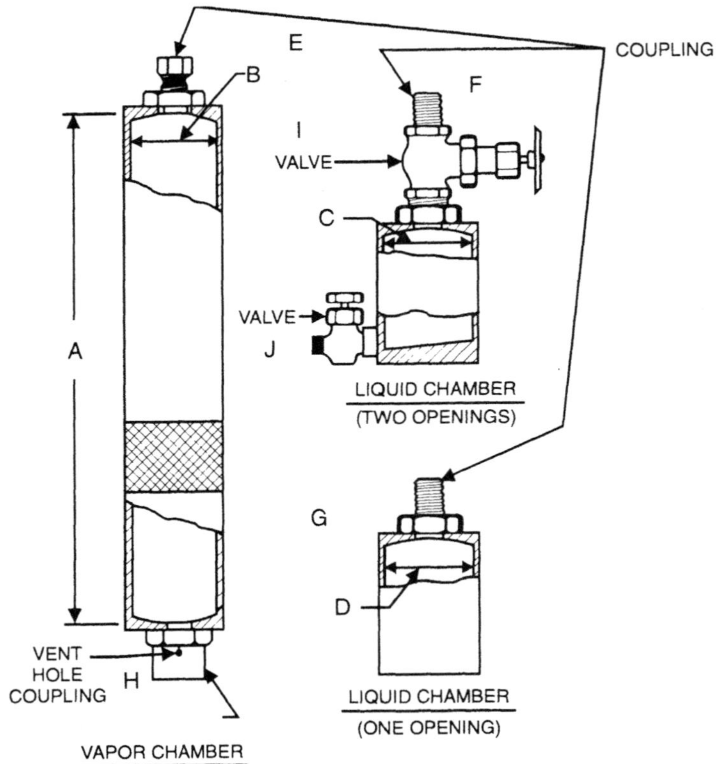

DIMENSIONS OF VAPOR PRESSURE APPARATUS			
KEY	DESCRIPTION	mm	in.
A	VAPOR CHAMBER	10 ± 1/8	254 ± 3
B,C,D	VAPOR AND LIQUID CHAMBER, ID	2 ± 1/8	51 ± 3
E	COUPLING, ID MIN	3/16	4.7
F,G	COUPLING, OD	1/2	12.7
H	COUPLING, ID	1/2	12.7
I	VALVE	1/2	12.7
J	VALVE	1/4	6.35

Figure 22-1. ASTM D-323 Reid Vapor Pressure (RVP) Test: Laboratory Test Cells
(Courtesy of ASTM)

On-line Process Analyzers

Precision Model 44770 Reid Monitor

This analyzer continuously measures and records the RVP of distillate fuels. It duplicates the measurement principles of the ASTM D-323 test, produces results in RVP units, and correlates directly with the ASTM values. The Reid monitor performance is unaffected by changes in the distillation range of the sample. It is well suited to applications such as gasoline blending, where frequent changes in formulations can occur.

A schematic flow diagram is shown in Figure 22-2. One head of a duplex pump sends sample at a rate of 100 mL/min through a heat-exchanger coil, which is immersed in an oil bath maintained at 100°F (40°C). The heated sample goes into the saturation chamber through a spray nozzle. Liquid level in the upper part of the saturation chamber is maintained by a float. An orifice controls liquid sample flow to the lower part of the saturation chamber and to the exhaust pump. As sample enters the chamber, it lifts the float and the needle valve from the orifice.

The design of the exhaust pump maintains a flow rate of 500 mL/min for the two-phase product. It creates a vacuum in the lower section of the chamber, which causes part of the sample to vaporize through the orifice, and the required 4:1 vapor/liquid ratio is established.

The pressure transmitter senses the vapor pressure in the lower section of the saturation chamber. It generates two outputs. One provides the saturation air that mixes with the sprayed sample entering the upper chamber; the other generates the 3 to 15 psig output signal proportional to the 2 to 19 lb RVP range.

Response time to a change in sample RVP is 60 s, and the repeatability is 0.1 RVP. This analyzer is rated for installation in Class I, Group D, Division 1 locations.

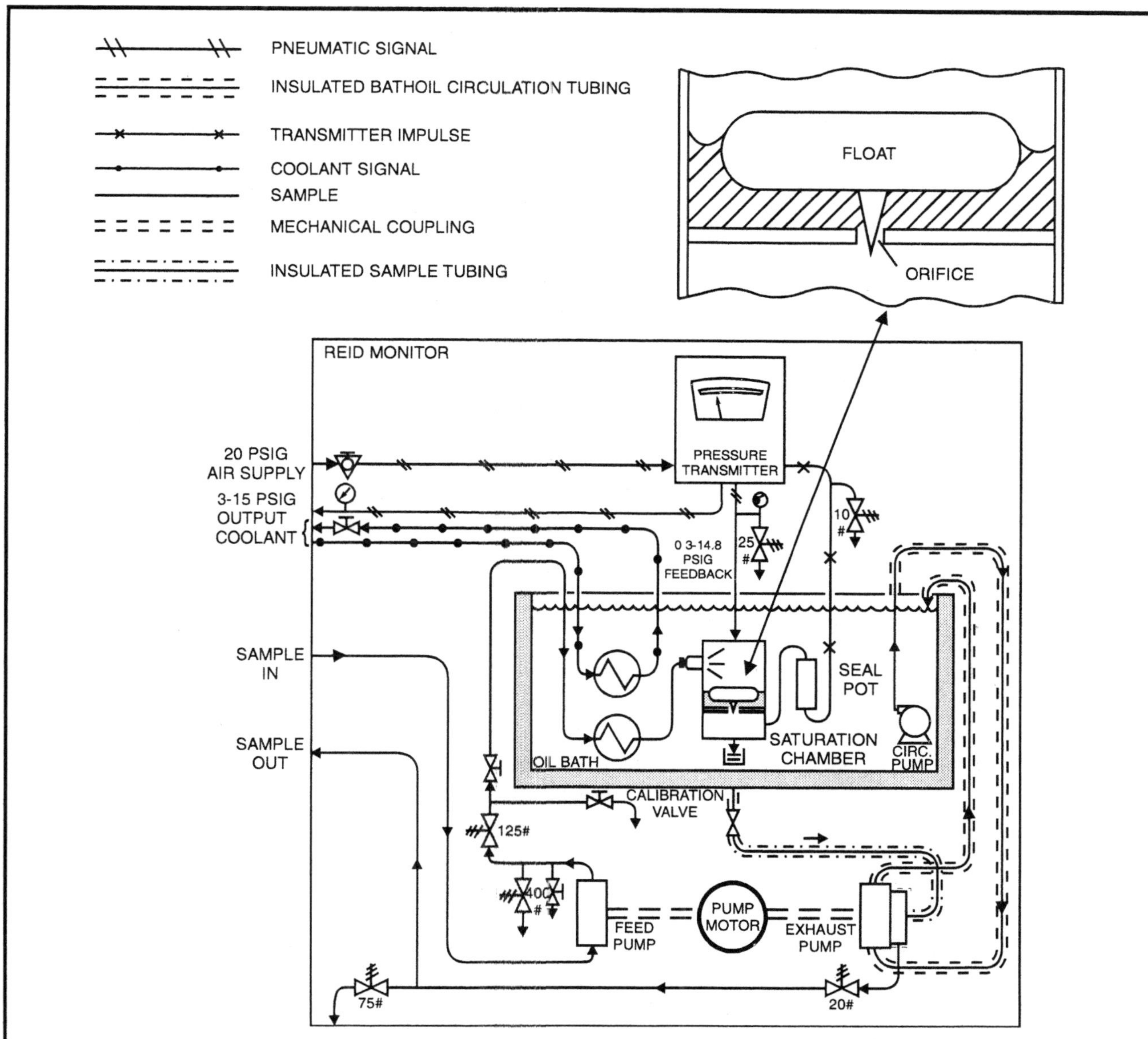

Figure 22-2. Precision Model 44770 Reid Monitor: Flow Diagram
(Courtesy of Precision Scientific Petroleum Instruments Co.)

Fluid Data Model 1354 Absolute Vapor Pressure Analyzer

The Absolute Vapor Pressure analyzer can be adjusted to correlate with RVP by adjusting the configuration of the nozzle.

The Fluid Data analyzer operates on a principle very similar to a common laboratory eductor pump. The sample, at 100°F (40°C), is pumped through a jet orifice (Figure 22-3). The increase in velocity of the fluid at this point creates a pressure drop sufficient to induce vaporization of the sample. When the static pressure of the liquid at the nozzle exit equals that of the vapor pressure, the pressure drop is at a minimum. The analyzer gives the absolute vapor pressure reading, which is defined as the pressure exerted when a liquid is in equilibrium with its own vapor. This is not the same as RVP, because the Reid test includes a factor depending on the expansion of a volume of added air. The analyzer can be adjusted to correlate with RVP by adjusting the configuration of the nozzle.

A schematic flow diagram of the analyzer is shown in Figure 22-4. Sample enters the analyzer at the upper left. Sample temperature must be between 75 and 120°F (25 and 50°C). It passes through the pressure regulator, which maintains a constant downstream pressure, and then through a heat exchanger in a constant-temperature bath maintained at 100°F (40°C). The sample passes through the nozzle assembly, a detailed diagram of which is shown in Figure 22-5, and out of the system through the back-pressure regulator. The two pressure regulators maintain a constant differential pressure across the orifice. The ASTM thermometer in the top of the nozzle assembly indicates the sample temperature. A two-bladed impeller provides complete mixing of the bath oil to assure even temperature distribution. Cooling water can be circulated through the stainless steel coil to maintain the 100°F (40°C) bath temperature for hot samples. The pressure transmitter, connected to the vapor chamber of the ejector, provides the output signal, which is normally 0 to 20 psig but which can be as high as 0 to 150 psig.

The main components are mounted on the top cover of the oil bath. The explosionproof housing on the left side contains the temperature controller for the oil bath and power switches for the oil bath heater system and the stirring motor. The oil bath is screwed to the top cover. It can be disconnected and lowered manually on supports between the legs when it requires maintenance. Sample and cooling water connections are on the vertical panel on the right side of the assembly.

The response time of the analyzer is 43 s, and the repeatability is 0.1 psig. The sample flow rate of 15 to 50 gal/h minimizes response time.

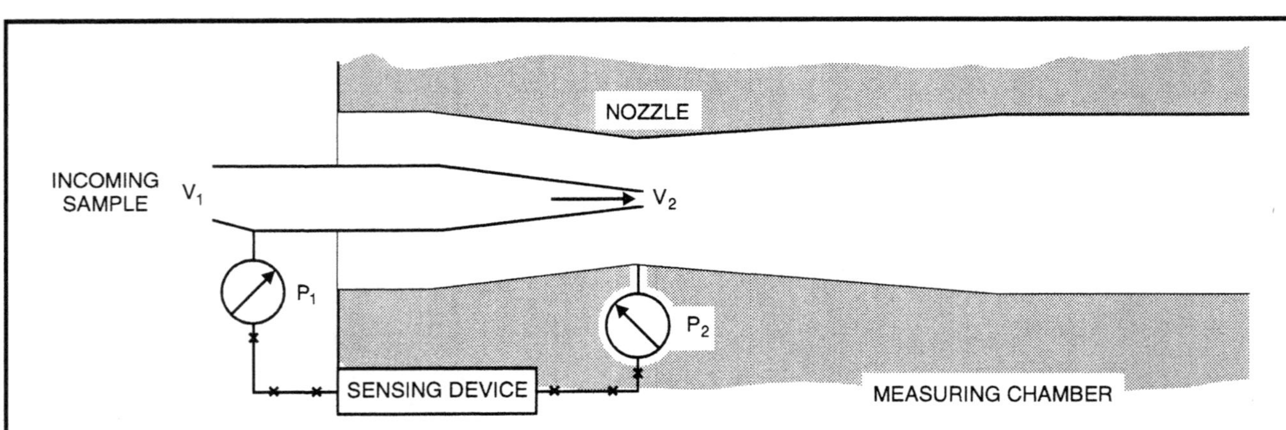

Figure 22-3. Fluid Data Absolute Vapor Pressure Analyzer: Operating Principle
(Courtesy of Fluid Data, Inc.)

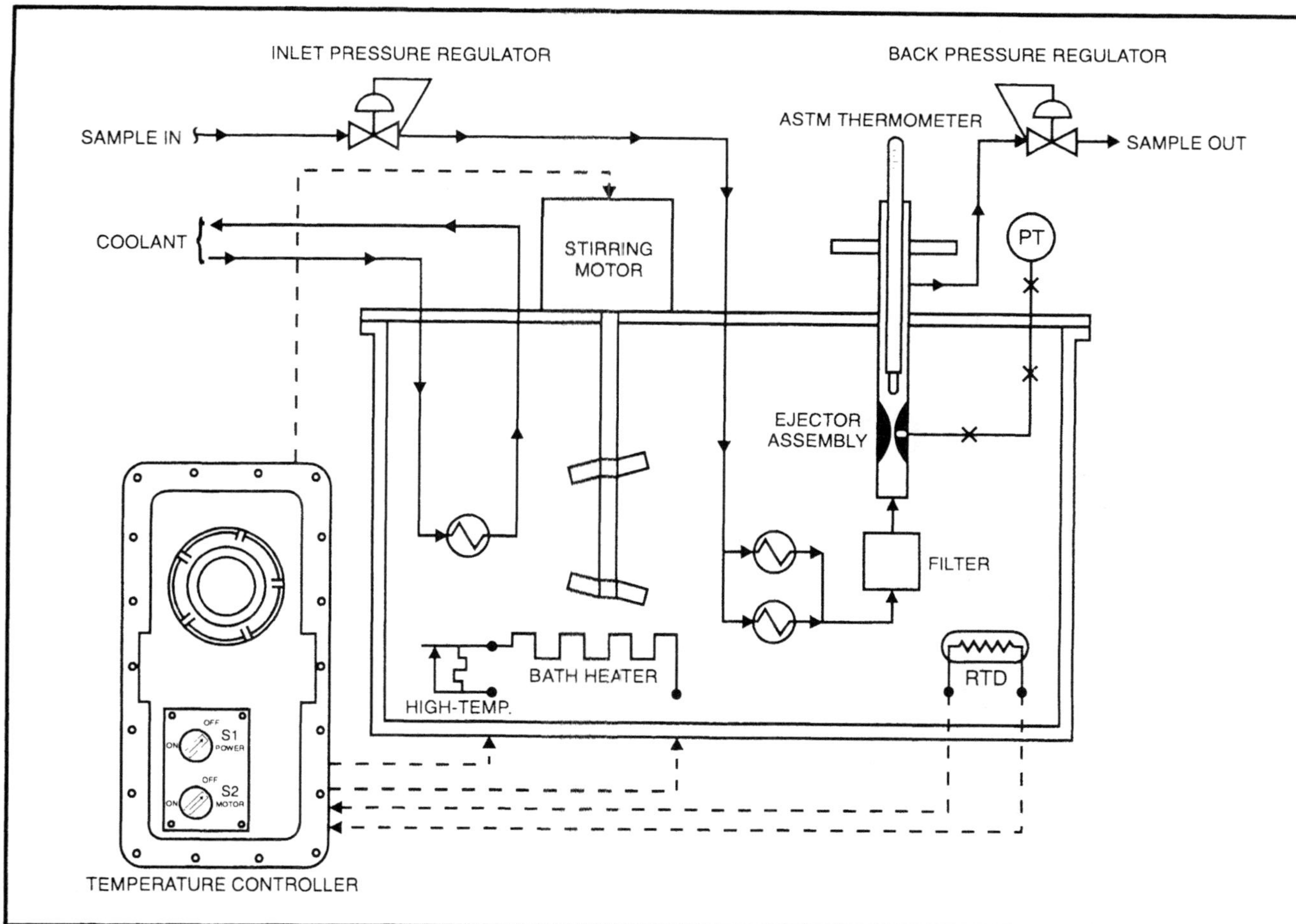

Figure 22-4. Fluid Data Absolute Vapor Pressure Analyzer: Flow Diagram
(Courtesy of Fluid Data, Inc.)

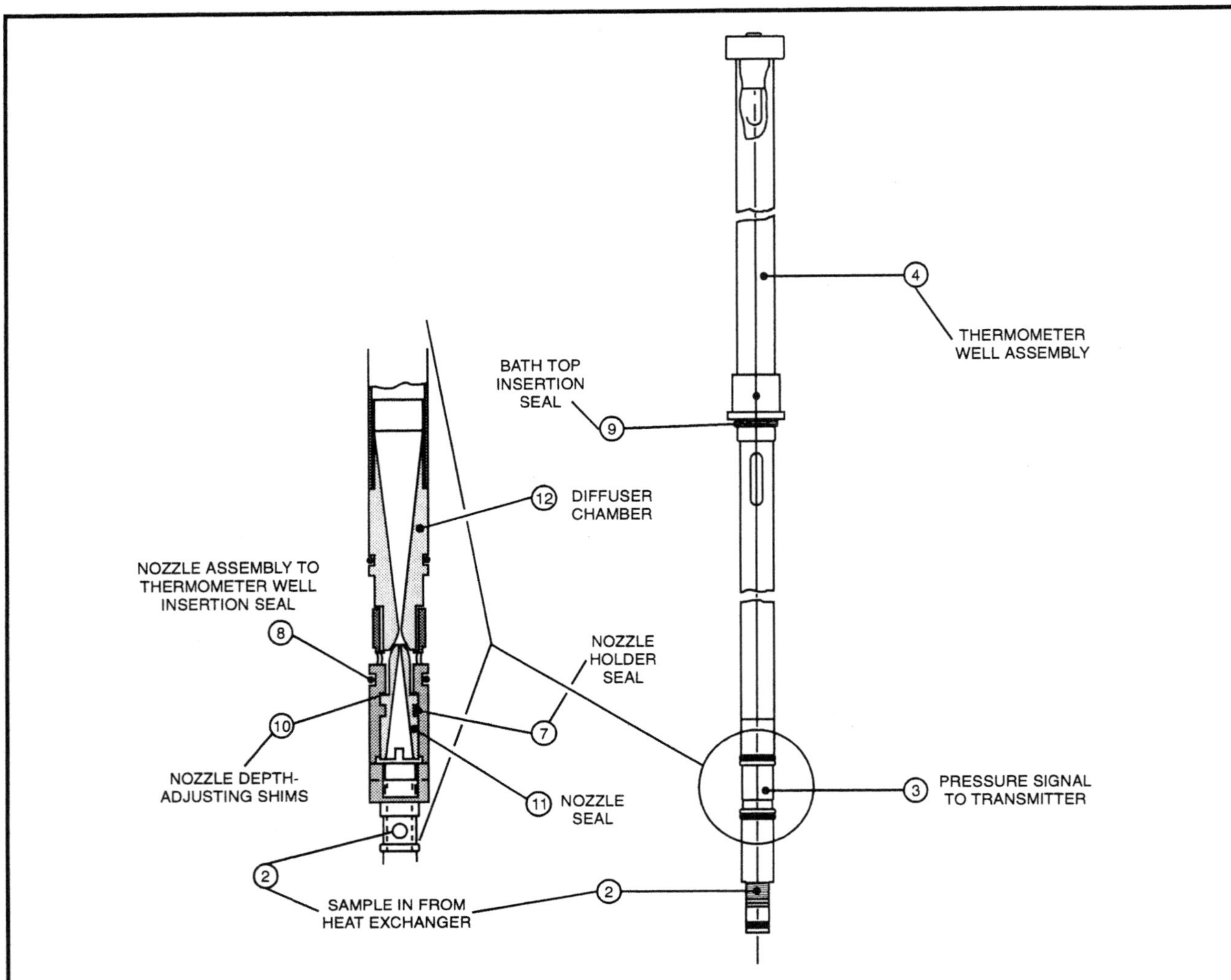

Figure 22-5. Fluid Data Absolute Vapor Pressure Analyzer: Thermometer Holder and Ejector Nozzle Assembly
(Courtesy of Fluid Data, Inc.)

ABB Vista Model 4100 RVP Analyzer

This analyzer uses a sample cell with a volume of 12.5 mL maintained at 100°F (40°C). Periodically, a 2.5-mL sample is introduced, leaving a vapor volume of 10 mL to give the required 4:1 vapor-to-liquid ratio. The system schematic is provided in Figure 22-6 and the operating cycle is detailed in Figure 22-7.

The analyzer operates on a 7.5-min cycle, which begins with a 20-s nitrogen purge (1, in Figure 22-7), followed by a "false" fill (2). The cell fills with fresh sample and immediately empties to purge the connecting tubing between the solenoid valves and the measuring cell. Nitrogen purges the cell three times (3, 4, and 5), with each purge lasting 20 s. Three purges, instead of one long one, allow time for vapor trapped in the cell to evaporate and be removed by the following purge.

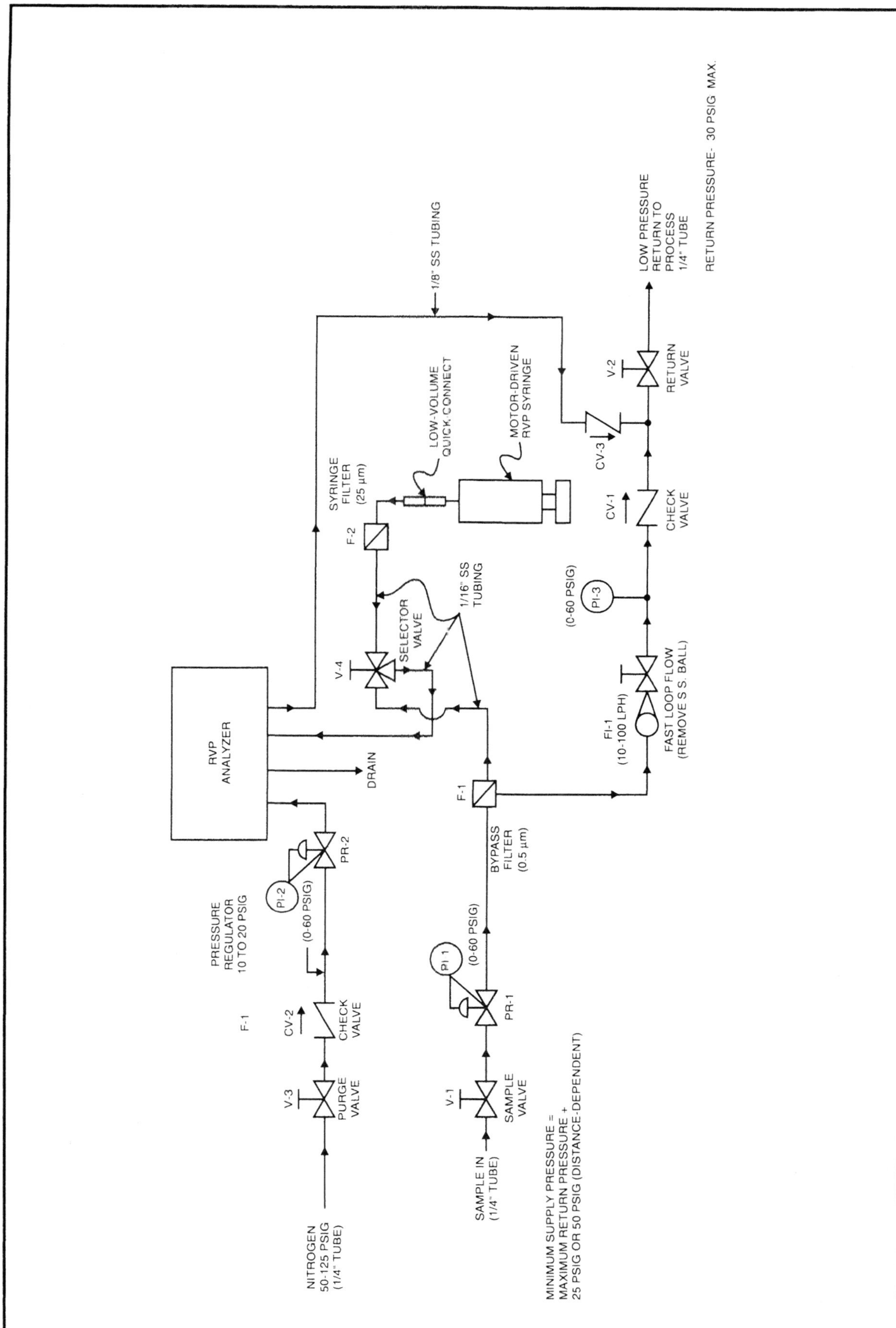

Figure 22-6. ABB Vista Model 4100 V/L Analyzer: Flow Diagram
(Courtesy of ABB)

The cell then stands for 50 s (6) to reach temperature equilibrium. At equilibrium, the nitrogen inside the cell is at 100°F (40°C) and at ambient pressure, since the drain solenoid is open. The pressure at this time is recorded by the controller as the zero pressure and then the drain valve closes. The 2.5-mL sample (7) is admitted to the cell, allowed to stand for 4 min to reach equilibrium (8), and the pressure is recorded again. From this reading, the zero pressure, and the atmospheric pressure, the controller calculates the RVP at 760 mm pressure and transmits the result to the recorder. The syringe shown in the diagram of the sample recovery system (Figure 22-8) is used to introduce prototype samples into the analyzer to check its performance. This unique design eliminates the need for calibration.

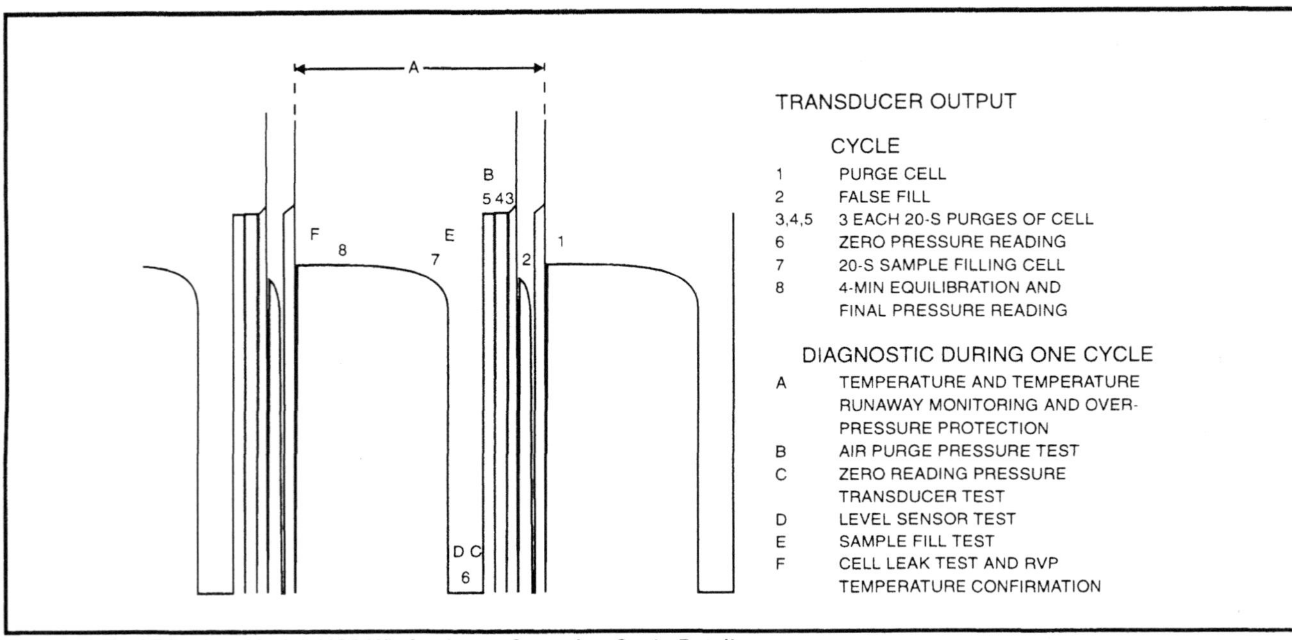

Figure 22-7. ABB Vista Model 4100 V/L Analyzer: Operating Cycle Detail
(Courtesy of ABB Vista)

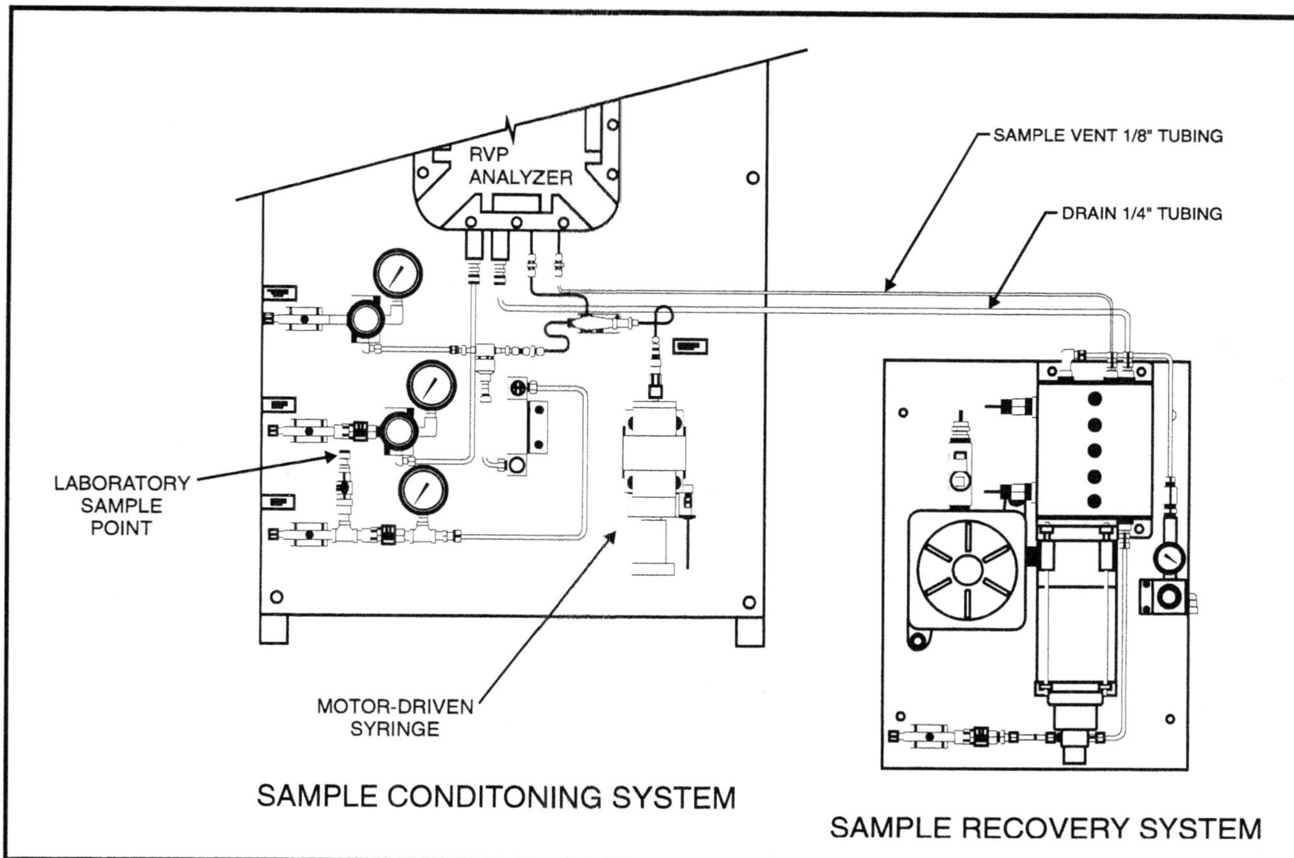

Figure 22-8. ABB Vista Model 4100 V/L Analyzer: Sample Conditioning System (L) and Sample Recovery System (R)
(Courtesy of ABB Vista)

Common Applications

Major applications of vapor pressure analyzers are for gasoline blenders to meet volatility requirements and to conform with government emission specifications. The response of the Reid Vapor Pressure monitor is not affected by changes in gasoline composition, which can occur frequently in blending operations. Many other products, such as kerosenes, naphthas, and liquefied petroleum fuels, have vapor pressure specifications; furthermore, the vapor pressures of crude oils in storage tanks and pipelines must be maintained within safe limits.

Many RVP analyzers are being used for in-line gasoline blending systems to maximize the addition of lower-cost butane without exceeding maximum vapor pressure limits. These analyzers minimize the need to reprocess off-spec blends and can pay for themselves in a very short time.

Ranges of Variables

The Precision Reid monitor covers the vapor pressure range of 0 to 20 psig, with a reproducibility of 0.1 psig. The usual range for the Fluid Data absolute vapor pressure analyzer is also 0 to 20 psig, but it can be used for samples with vapor pressures as high as 150 psig. The ABB Vista analyzer also covers the 0 to 20 psig range, with a reproducibility of 0.1 psig.

Options and Accessories

Precision Scientific can supply a sample-conditioning system, which filters the sample, removes water from it, and regulates its pressure. The company also supplies sample recovery systems and a prototype fuel calibration system. Fluid Data can provide sampling systems designed to meet the needs of their applications.

Waste sample from the ABB analyzer must be vented to atmospheric pressure, but the amount is only about 1,200 cm^3 per day. A sample recovery system can be supplied by ABB for this purpose (Figure 22-8).

Limitations

The Precision Scientific Reid monitor requires cooling water at 30 gal/h at a maximum temperature of 90°F (30°C). The Fluid Data analyzers require water at 12 gal/h at a maximum temperature of 35°F (2°C). For the operation of the ABB analyzers, the sample temperature must not exceed 90°F (30°C), and sample pressure must be at least 40 psig.

Installation Hints

Vapor pressure analyzers should be installed indoors or in an analyzer shelter to protect them from the weather and to facilitate maintenance.

Specific Safety Issues

All analyzers are rated for installation in Class I, Group D, Division 1 hazardous locations. Since the samples to be analyzed may be extremely flammable, the areas in which the analyzers are installed should be well ventilated and monitored by hydrocarbon leak detectors and flame recognition systems.

Unique Sampling System Requirements

The Precision Reid Vapor Pressure monitor requires clean, dry sample at 100 mL/min at a pressure between 10 and 100 psig and a temperature between 50 and 110°F (10 and 45°C). Maximum back pressure must not exceed 50 psig. For the ABB Vista analyzer, the sample temperature must be less than 90°F (30°C), and the sample pressure must be above 40 psig.

Typical Calibration Requirements

Calibration of vapor pressure analyzers with laboratory samples is difficult because of the possible loss of vapors from volatile samples — not only while taking samples from the process line, but also while handling them in the laboratory. Common calibration systems (Figure 22-9) can also introduce errors. Volatile vapor can be lost while filling the sample tank, or the nitrogen that is used to pressurize the sample into the analyzer can dissolve in the sample and increase the apparent vapor pressure.

A two-tank system (Figure 22-10) suitable for handling gasoline can minimize these problems. The right-hand tank contains the gasoline calibration sample floating on the water layer in the bottom of both tanks. Nitrogen pressure on top of the water in the left-hand tank forces water into the right tank and the gasoline into the analyzer. The interface between the gasoline and the water is visible in the sight glass beside the right tank. The water layer prevents the nitrogen from dissolving in the gasoline, and the system can be charged without the loss of light ends (high-vapor-pressure material).

The system is charged by first filling the two tanks slightly more than half full with water. Nitrogen is admitted to force the water into the right tank until it is completely full. Gasoline is then admitted to the top of the right tank until the interface between it and the water drops almost to the bottom of the sight glass. The water, which is forced into the left tank, nearly fills this tank. The nitrogen, of course, must be vented.

One detail is important. The connections at the bottom of the right tank should be constructed as shown in Detail A of Figure 22-11. If they are connected as shown in Detail B, the water from the left tank may tend to flow horizontally toward the level gage instead of upward into the right tank. The interface between the water and the gasoline can be forced upward into the sight glass, and the interface there will not correspond to the actual interface in the tank. If the connections are made as shown in Detail A, the main flow from the left tank will be directed upward into the right tank. Since the right tank has a much larger volume than the sight glass, the level of the two interfaces will remain the same.

The ABB analyzer requires no calibration, but provision is made for the introduction of prototype samples to monitor its operation and to detect malfunctions.

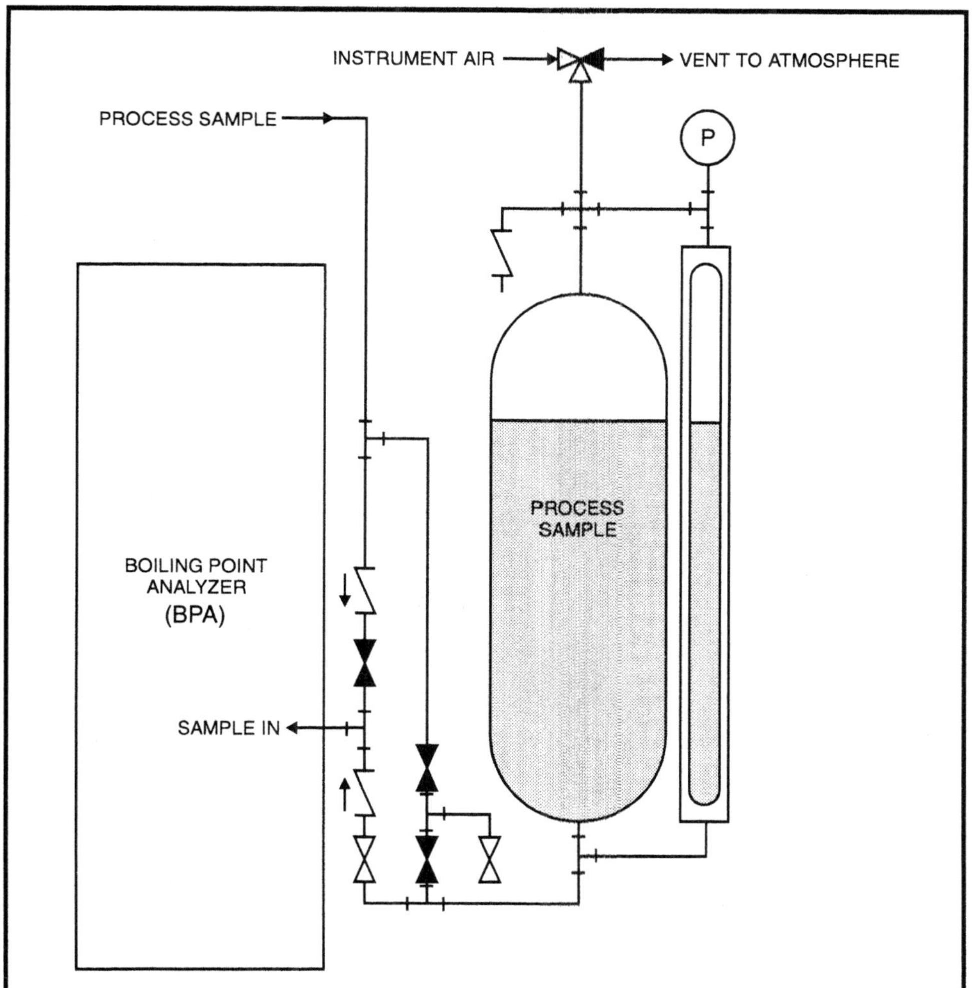

Figure 22-9. Typical Distillation/Boiling Point Analyzer Calibration System
(Courtesy of Sun Refining and Marketing Co.)

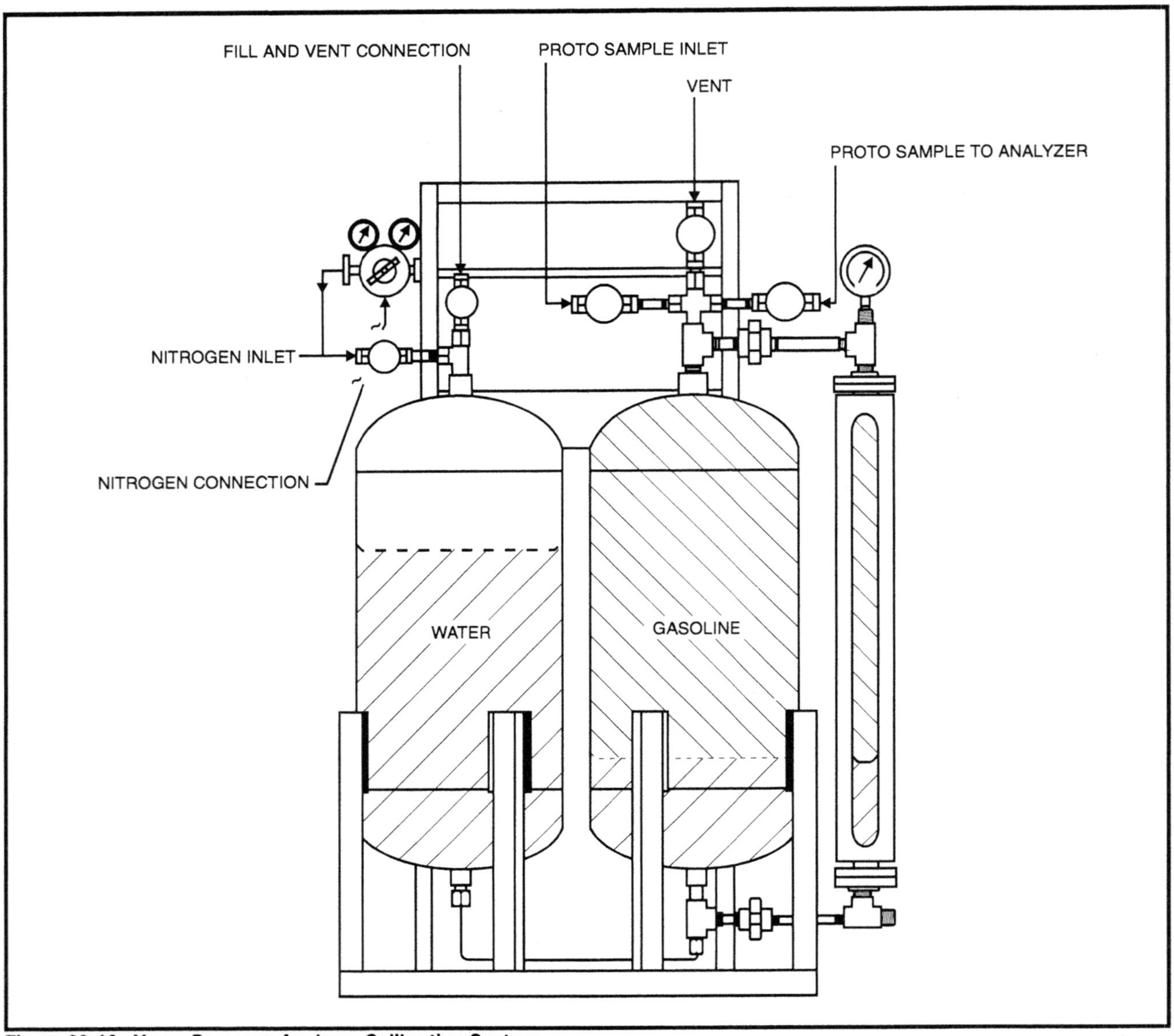

Figure 22-10. Vapor Pressure Analyzer Calibration System
(Courtesy of Sun Refining and Marketing Co.)

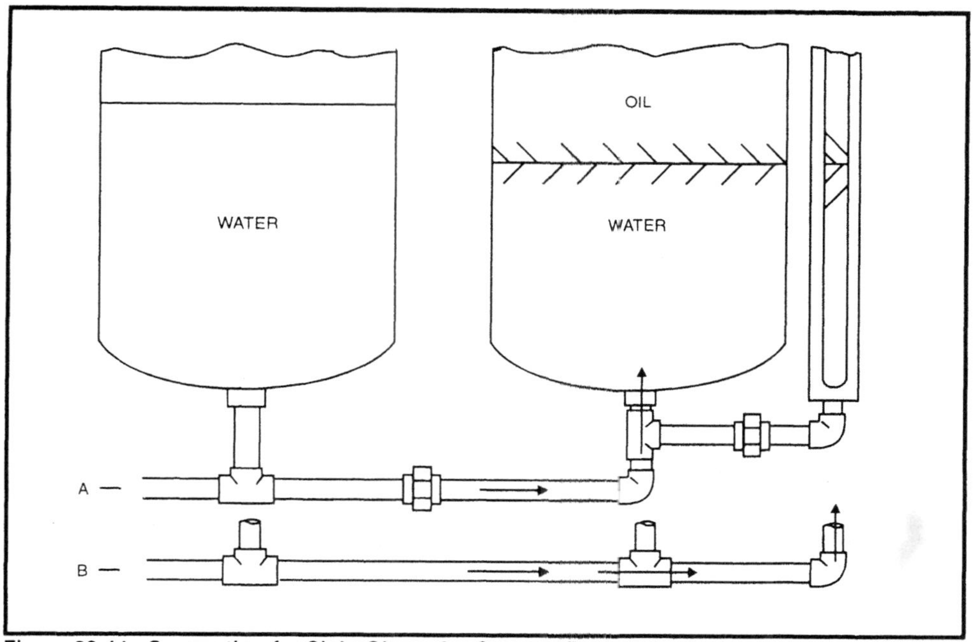

Figure 22-11. Connection for Sight Glass: A – Correct; B – Incorrect
(Courtesy of Sun Refining and Marketing Co.)

Troubleshooting and Maintenance

A typical maintenance procedure for the Fluid Data absolute vapor pressure analyzer includes the following:

Daily Checks:

1. Check the temperature indicated on the mercury thermometer. It should be exactly 100°F (40°C).
2. Check the signal lamp on the temperature controller. It should be flashing on and off at about 2-s intervals, indicating that the controller is functioning.
3. Check the oil level in the bath and add oil if needed.

Monthly Checks

1. Remove the nozzle from the eductor and clean it.
2. Remove the inlet filter and clean it.

Recommended checking sequences and actions to be taken are detailed in the manufacturer's operation and maintenance instruction manuals.

The ABB Vista analyzer has a built-in diagnostic system that can detect malfunctions. If one occurs, the analyzer will cease to operate and an alarm will sound for operator assistance. A message at the instrument identifies the source of the problem.

About the Author

Conard Donnell received a Bachelor's and a Masters degree in Chemistry and Physics from Temple University. His career with automatic process analyzers began and ended at Sunoco Refinery where he developed on-line instrumentation to measure the physical properties of process streams automatically and much faster than laboratory analysis. Since retirement, Mr. Donnell has worked as a consultant for Sunoco, Fluid Data, Precision Scientific Co., Lockwood and McLorie, and DuPont working on projects that included design and installation of process analyzer systems in refineries worldwide.

23

Vapor-to-Liquid Ratio (V/L) Analyzers

The vapor-to-liquid ratio, *V/L*, is a measure of the volume of vapor formed over a specific volume of liquid at a specific temperature. Like the Reid vapor pressure and the front-end distillation data, it is a measure of the volatility of gasolines and is related to the performance of the fuel in cold starting of engines, vapor locking at high operating temperatures, and emission of pollutants into the air. Exact correlations between these measurements are difficult to obtain. An ASTM method for *V/L* was published in 1966, but an on-stream analyzer to measure the ratio has become available only recently.

ASTM Standard Test Method D-2533 for Vapor/Liquid Ratio of Spark-Ignition Engine Fuels. In this test, a measured volume of cold 32°F (0°C) sample is introduced from a syringe into a burette (Figure 23-1), which has been filled with glycerol from a leveling bulb (Figure 23-2). The burette is placed in a constant-temperature bath maintained between 90 and 180°F (30 and 80°C). Part of the sample vaporizes and forces some glycerol back into the leveling bulb. The level in the bulb is adjusted to equal the level of the liquid in the burette, except for a small difference that may be necessary to correct the reading to 760 mmHg (atmospheric pressure). Refer to the ASTM method for detailed instructions for this correction.

When the sample reaches bath temperature, the volume of vapor in the burette is noted as *V*. The *V/L* ratio is calculated from the known volume of liquid sample injected from the syringe. The temperature for a specific *V/L* can be determined by measuring the ratio at several temperatures, plotting the results, and reading the temperature at the desired value from the plotted line.

Process Analyzers

Ethyl Corporation VOLRAC *V/L* Analyzer

The VOLRAC *V/L* analyzer measures the vapor-to-liquid ratio of gasoline products from on-line blending operations. The results correlate very well with ASTM D-2533 laboratory tests. This is the only on-line analyzer available for *V/L* measurements. The flow diagram is shown in Figure 23-3.

Process sample is filtered, coalesced to remove solids and moisture, and cooled by a chiller. A fuel metering pump feeds the sample into the vapor-liquid separator, which is maintained at a selected temperature between 100 and 150°F (40 and 65°C). The gasoline enters the separator through an orifice that produces a spray and allows the vapor to separate from the liquid quickly. The vapor flow rate is measured by the vapor flowmeter, which is maintained at the same temperature as the separator chamber. Liquid sample flows from the bottom of the separator to waste. The *V/L* ratio is calculated from the measured vapor flow rate and the fixed liquid flow rate from the metering pump. The results can be compensated for variations in temperature and barometric pressure.

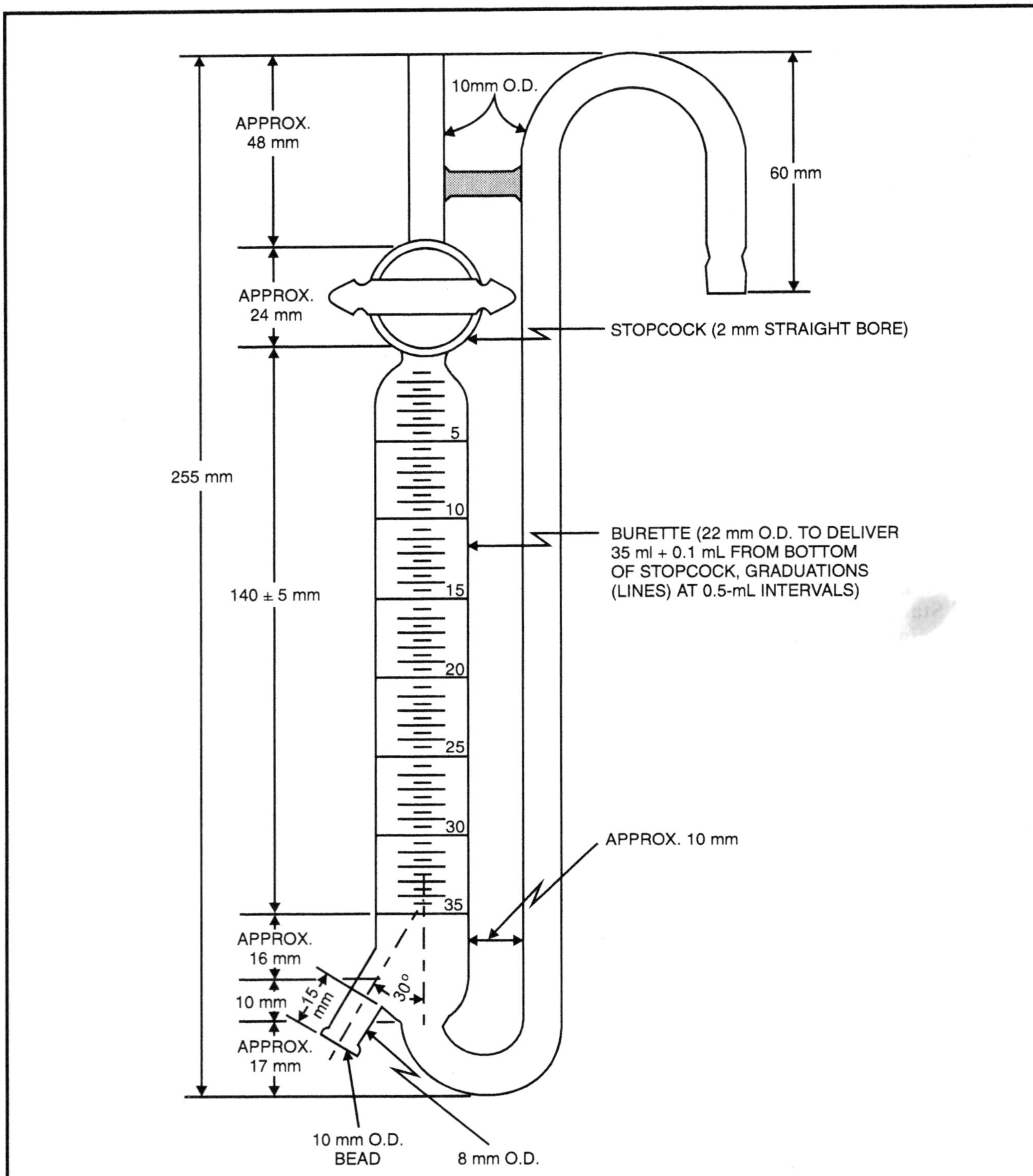

Figure 23-1. ASTM D-2533 Test Apparatus
(Courtesy of ASTM)

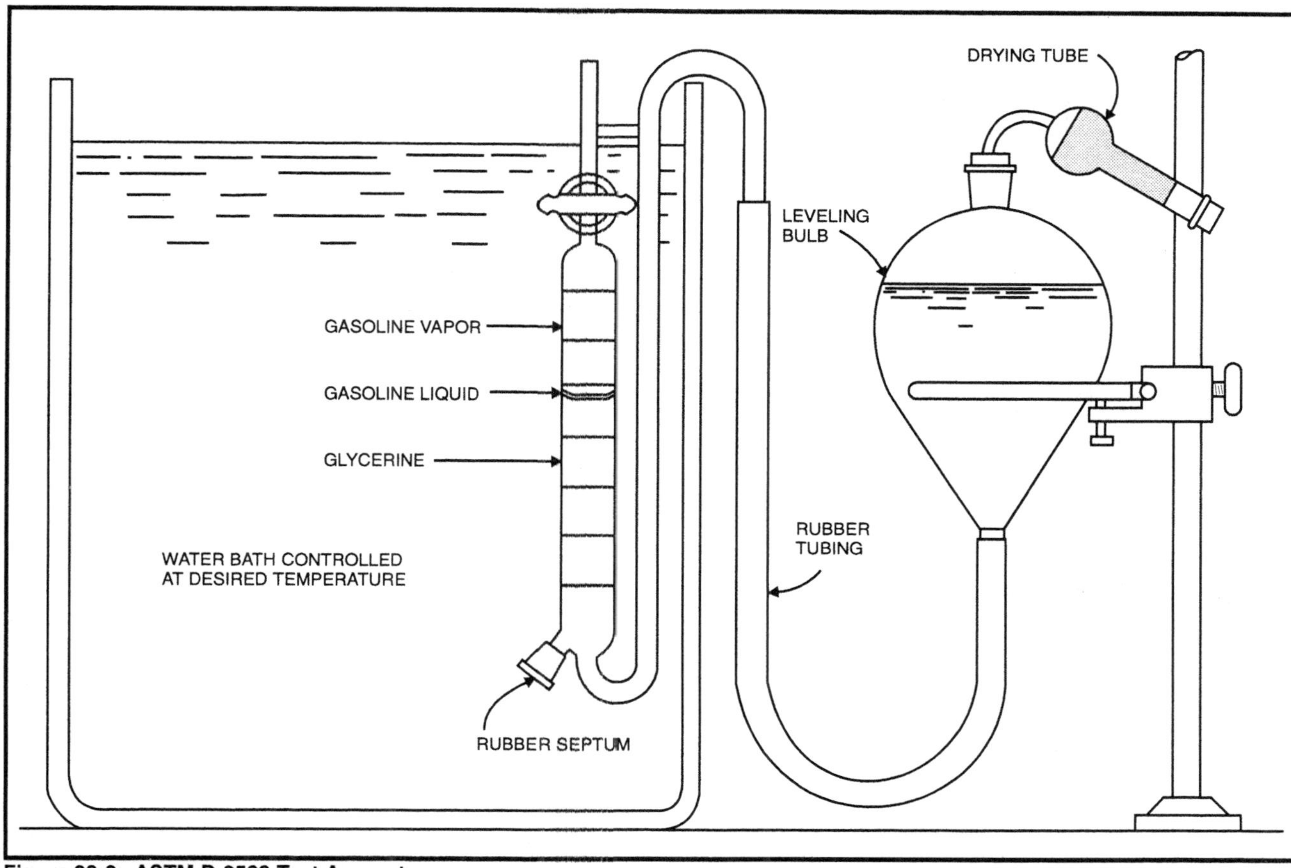

Figure 23-2. ASTM D-2533 Test Apparatus
(Courtesy of ASTM)

Common Applications

The VOLRAC V/L analyzer is currently in use in refineries for in-line gasoline blenders to control the volatility of the products while adding the maximum possible amount of butane to the gasoline. Many states have passed laws limiting allowable *V/L* ratios, especially in the summertime, to minimize the emission of vapor pollutants into the atmosphere. Control of the *V/L* ratio by direct measurement is profitable.

Ranges of Variables

The VOLRAC V/L analyzer can measure *V/L* ratios from 10 to 40 at temperatures from 100 to 150°F (40 to 65°C) with a response time of 60 s. Accuracy is 0.5 *V/L*. The analyzer output is a Δ*V/L* from –4 to +4 units from the desired value, and outputs are provided in 4-20 and 10-50 mADC signals.

Options and Accessories

A sample recovery system is available from the manufacturer.

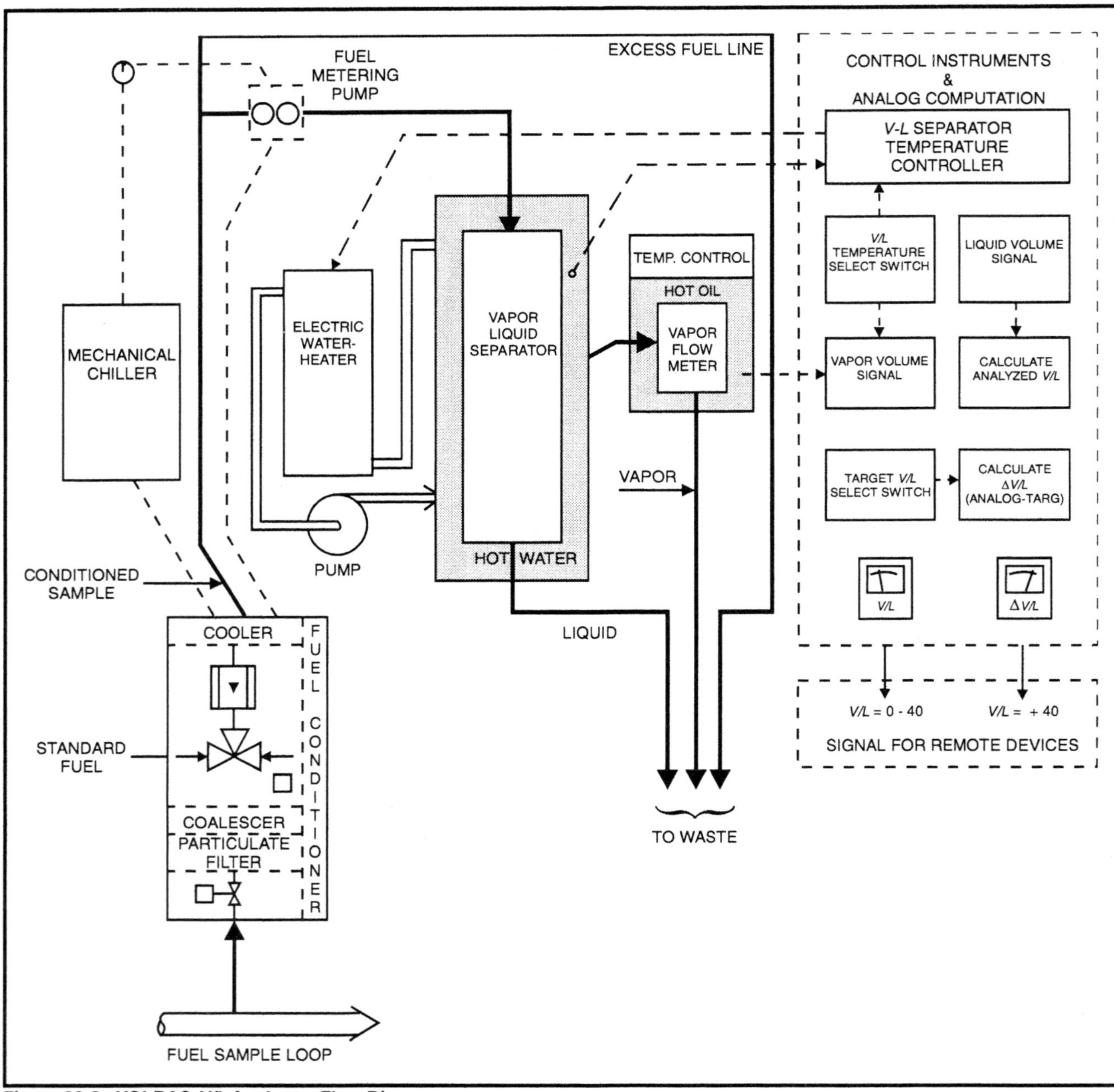

Figure 23-3. VOLRAC V/L Analyzer: Flow Diagram
(Adapted from Process Analyzer Technology, Kenneth J. Clevett, New York: John Wiley & Sons)

Limitations

Since the effluent liquid and vapor samples are discharged at atmospheric pressure, a sample recovery system will usually be required to dispose of the waste material.

Installation Hints

The analyzer is electrical-hazard rated for installation in NEC Class I, Group C and D, Division 2 (IC/D2) locations with the electrical enclosures purged by instrument air. It should be installed indoors or in an analyzer enclosure with protection from the weather and for ease of maintenance.

Specific Safety Issues

Since gasolines are hazardous substances, proper provision must be made for disposing of the waste samples and for venting vapors to safe areas. The analyzer should be installed in a climate-controlled building, vented with a positive pressure fan exhausting out of the upper wall area of the building and have hydrocarbon detectors/flame detectors monitoring the analyzer area.

Unique Sampling System Requirements

The analyzer requires sample flowing at a rate of 6 to 8 gal/h at a pressure from 50 to 150 psig. Only a small portion of this is fed into the analyzer, but the sample recovery system must be capable of disposing of all of it safely.

Typical Calibration Requirements

The VOLRAC analyzer must be calibrated with prototype gasolines that have been analyzed in the laboratory enough times to give a value with a reproducibility equal to or better than that of the analyzer. Reproducibility of the analyzer can be as small as 0.2 *V/L*. A calibration system, like that described in Chapter 23 for vapor pressure analyzers, would be desirable.

Troubleshooting and Maintenance

VOLRAC analyzers require regular attention to maintain the correct flow rates of samples and correct operating temperatures. Regular inspection of the particulate filters, coalescers, and mechanical chillers is necessary. Refer to the analyzer instruction manual for specific details.

About the Author

Conard Donnell received a Bachelor's and a Masters degree in Chemistry and Physics from Temple University. His career with automatic process analyzers began and ended at Sunoco Refinery where he developed on-line instrumentation to measure the physical properties of process streams automatically and much faster than laboratory analysis. Since retirement, Mr. Donnell has worked as a consultant for Sunoco, Fluid Data, Precision Scientific Co., Lockwood and McLorie, and DuPont working on projects that included design and installation of process analyzer systems in refineries worldwide.

Electrochemical Process Analyzer Systems

Electrochemical analyzers are found in a variety of applications in industry. Virtually all manufacturing sites produce waste streams that must be cleaned prior to discharge into public waterways. Invariably, the acidity (or pH) and the purity (or conductivity) of the stream must be monitored and controlled according to limits set by regulatory agencies in the site's operation permit. Additionally, many chemical processes are pH dependent and must be monitored and controlled to maximize profitability. Electrochemical analyzers are also found in boiler feedwater and recycle applications to monitor blowdown and prevent corrosion and boiler fouling.

Many ion-selective electrodes have been developed to measure various ionic species in aqueous solutions. A key to successful application of ion-selective electrode applications is an appropriate sample conditioning system. Few electrochemical analyzers may be operated in-situ due to the possibility of fouling of the sensor portion of the electrode and potential operator exposure to process when the probe must be removed from the stream and calibrated. The need to extract a process sample, prepare it for continuous contact with the sensing element and provide for calibration often turns a rather simple measurement into a complicated system application.

For example, the periodic measurement of ammonia in the laboratory involves minimal sample preparation and relatively infrequent calibrations of the ammonia electrode. However, in a process application, the sample must be extracted from the process line and filtered to remove particles that would eventually coat the surface of the sensor. Often the pH must be adjusted to assure the ammonia is in a form that will respond to the sensing element. The sample must then be returned to a different location at a lower pressure. This same sample preparation system must have provisions built in to allow the sensor to be calibrated with known concentrations of ammonia, while minimizing potential exposure of the mechanic to the process stream. As a result, most electrochemical analyzer applications are "engineered systems" rather than simple in-situ (insertion) measurements.

Many oxygen analyzers are also based on electrochemical measurement techniques. For example, zirconium oxide oxygen sensors detect the current generated by the difference between the number of ionic species created on the surface of a reference porous platinum electrode exposed to oxygen in clean instrument air and the number of ions produced on a measuring electrode by any oxygen present in a boiler or furnace stack. Galvanic cells are also commonly used as oxygen analyzers. These cells generate current (mADC) when oxygen permeates through a special membrane and reacts with a cathode. The cathode typically is made from gold-coated silver or platinum. The cathode is immersed in a cell containing a bath solution of potassium hydroxide and an anode made of some reactive metal, like lead or zinc. The oxygen molecules produce hydroxyl anions and the atoms of the anode surface are consumed by oxidation to their respective oxide species. The current produced is proportional to the concentration of oxygen permeating through the membrane.

Chapters 24 through 27 will cover the more common types of electrochemical analyzers, as follows:

Chapter 24	Conductivity Analyzers
Chapter 25	pH Analyzers
Chapter 26	ORP (Redox) Analyzers
Chapter 27	Trace Oxygen Analyzers

24

Conductivity Analyzers

Analysis of solutions employing conductance measurements evolved with development of solution and electrolyte theory in the early 1900s, though prior to this, chemists had been measuring the resistance of solutions since the early 1800s. Today, conductance and resistance measurements are made with conductivity analyzers to measure total ionic concentration of electrolyte samples. Measurements are nonspecific; they cannot be used to distinguish one ion from another but can be used effectively, for example, in measuring water purity and analyzing binary mixtures of electrolytes. A variety of instrumental techniques have been developed, some using electrodes in contact with the sample, and some not. The newest methods incorporate modern microelectronics and integrated software programs.

Cells and instrumentation are designed to measure electrical resistance (or its reciprocal, conductance) in electrolyte samples and to convert these measurements to units of concentration for use in monitoring and control.

Conductance measures the flow of charge when an electric field or potential is applied across a conductor. In metals the charge carrier is the electron, as it is for *n*-type semiconductor materials. In *p*-type semiconductors, positive charge or electron holes carry charge across electric fields. A semiconductor material can therefore support the flow of either positive or negative charge, but generally not both. In solutions, however, equivalent quantities of positive and negative charge result from the dissolution of electrolyte. Both move, although in opposite directions, when an electric potential is imposed across their solution.

Conductance (the inverse of resistance), flow of charge (current), and force or electric field (voltage), both for electronic and ionic charge flow, are exactly expressed by Ohm's law. In electrolytes, however, well-defined practices must be employed to produce accurate measurements. Ions have ten thousand times the mass of electrons; therefore, electrolytic conduction is accompanied by transfer of mass as well as charge. Also, ions drift through their solutions under the influence of the imposed field at velocities approximately five times less than do electrons in metals. These factors generally make the magnitude of ion conduction much smaller than that observed for the electron process.

An important distinction arises experimentally when an electronic instrument is employed to measure ionic conductance, since electrodes must be employed to connect it to the solution. Ions do not flow in wires, nor do electrons in solution; consequently, charge or current crosses interfacial boundaries (metal electrodes in contact with the solution) and must be accompanied by formation or discharge of ions. This results in voltage drops, which must be separated from that in the sample electrolyte to allow Ohm's law to accurately reflect solution conductivity measurement.

Conductivity Analysis Advantages: Nearly linear output over the full range of conductivities presented by the aqueous process world. Wide selection of corrosion-resistant materials. High measurement accuracy available at high process temperatures and pressures.

This chapter will discuss the properties of solutions, processes at electrodes, and common conductance measurement techniques. Conductivity cells, instrumentation, and several applications also will be discussed, as will an important branch of conductivity called resistivity measurement. A list of the major suppliers of conductivity analyzers in the United States is provided in Table 24-1.

Table 24-1. U. S. Manufacturers of Conductivity Analyzers

The Foxboro Co. 38 Neponset Ave. Foxboro, MA 02035
Great Lakes Instruments 9020 West Dean Rd. Milwaukee, WI 53224
Johnson Yokogawa 4 Dart Rd. Newnan, GA 30265
Honeywell Inc. 795 Horsham Rd. Horsham, PA 19044
Rosemount Analytical 2400 Baranca Parkway Irvine, CA 92714
TBI–Bailey Controls 2105 Lockheed Way Carson City, NV 89709

Conductance Properties of Electrolytes

Materials exhibiting electrical conductance (1/resistance) have charged particles that are free to move. Charge moves in response to a force in the form of an electric field. The particles move with velocities related to their chemical and physical properties, and they move in the direction of the field opposite to the sign of their charge. Electrolytes provide these charged particles. When they dissolve, they dissociate to form mixtures of positively and negatively-charged ions. Because positive ions migrate toward the negative electrode (cathode) they are called cations. Likewise, ions attracted toward the positive electrode (anode) are negatively-charged and are called anions.

Movement of ions in a solution with an imposed electric field constitutes electric current, the magnitude of which is fundamentally related to the concentration as well as the velocity and the charge of those ions. We are interested here primarily in the relationship of conductance with concentration of ions. However, because the analytical utility of this relationship depends so strongly on the properties of ions in solution, we will first discuss some properties of electrolyte solutions.

Dissociation of Electrolytes

If a material completely dissociates in water into its component ions, it is termed a strong electrolyte and is said to be 100% dissociated. Examples of strong electrolytes are salts such as $NaCl$ and $BaCl_2$, acids such as HCl and H_2SO_4, and bases such as $NaOH$ and $Ba(OH)_2$.

If a material only partially dissociates when placed in water, it is termed a weak electrolyte, and if it does not dissociate at all, it is termed a nonelectrolyte. Examples of weak electrolytes are acids such as HF and acetic, bases such as ammonium hydroxide, and salts of weak acids and bases such as sodium acetate and ammonium chloride, respectively. Examples of common nonelectrolytes are sugars, oils, and large numbers of hydrocarbon-based liquids and solvents.

The conductance of strong electrolytes is composed of contributions from all of its ions, whereas the conductance of weak electrolytes is composed of contributions from only those that dissociate. The conductance of nonelectrolytes is nil. Conductance analyzers are therefore of greatest value when measuring strong electrolyte samples, but can also be applied to weak electrolyte samples. They are of no value in measurement of concentration of nonelectrolytes.

Conductance analyzers are therefore of greatest value when measuring strong electrolyte samples, but can also be applied to weak electrolyte samples. They are of no value in measurement of concentration of nonelectrolytes.

Molar and Equivalent Conductivity of Electrolytes

The familiar equation for Ohm's law is

$$I = E/R \tag{24-1}$$

where I is the current (flow) in amperes, E is the applied potential (force) in volts, and R is the resistance of the solution expressed in ohms. When two electrodes are used to measure the resistance of a solution, the measurement is related to the area of the electrodes, a, and the distance between them, l; that is:

$$R = r\,(l/a) \tag{24-2}$$

where R is in ohms; l and a are in centimeters; and r, termed the specific resistivity with the units $\Omega \cdot \text{cm}$, is the resistance of 1 cm^3 (equivalent to 1 mL) of solution.

For solutions, it is often more convenient to speak of conductance, L (the reciprocal of resistance), and conductivity, k (the reciprocal of resistivity). From Equation (24-2):

$$L = 1/R = k\,(a/l) \tag{24-3}$$

where L is the conductance in siemens (S) and k is the specific conductivity in $S \cdot \text{cm}^{-1}$.

The terms resistance and resistivity and conductance and conductivity are often incorrectly used. Resistivity and conductivity include a geometric term (cm), which we will see later is called the cell factor. Resistance and conductance are electrical measurements and do not contain this unit.

If a gram-mole of a substance is dissolved in 1 cm^3 of solution, its conductivity is termed molar conductivity. Likewise, a gram-equivalent dissolved in 1 cm^3 of solution has a conductivity called equivalent conductivity. Note that this definition of equivalent conductivity creates an inconvenience. If we want to measure the molar conductivity of a 1 M solution of electrolyte, we would need to observe the conductance of its entire volume placed between two electrodes 1 cm apart, and this would require that the electrode areas be 1,000 cm^2. To avoid this, we use the simple expression:

$$\Gamma = 1000k/C \tag{24-4}$$

where k is the specific conductivity, C is the concentration of the solution, and 1,000 comes from the identity, 1 L = 1,000 cm^3. If C is in moles/liter, Γ is the molar conductivity in $\text{S} \cdot \text{cm}^2/\text{mole}$, and if C is in equivalents/liter, Γ is the equivalent conductivity with the units $\text{S} \cdot \text{cm}^2/\text{equivalent}$.

Since charge flow in solutions is by both anions and cations, the conductance of an electrolyte is the sum of the equivalent ionic conductances of all of its ions. Therefore, using the relationships $C = \Sigma c_i$, $\Gamma = \Sigma\Gamma_i$, and Equations (24-3) and (24-4), we can write:

$$L = 10^{-3}\left(\frac{a}{l}\right)\sum_i z_i \cdot c_i \cdot \Gamma_i \tag{24-5}$$

where:

L = conductance in Ω^{-1} or siemens
a = area of electrodes in cm²
l = distance between electrodes in cm
c_i = concentrations of the participating ions in mole/cm^3
Γ_i = equivalent conductivity of the participating ion in $S \cdot cm^2$/equivalent
z_i = charge on the ion in equivalents/mole

The product $z_i \cdot \Gamma_i$ is the molar conductivity and has the units S·cm^2/mole.

Physical and Chemical Effects on Conductivity

Since conductance depends on the movement of ions and different ions have dissimilar velocities, the faster-moving ions will carry a greater proportion of the current through the solution than the slower ones. For example, in dilute HCl, H^+ which moves much faster than Cl^-, carries about 82% of the current and Cl^- only 18%.

The mobility of an ion is the equivalent conductivity of that ion, and the equivalent conductivity of an electrolyte is the sum of the mobilities of its ions. Thus, in using Equation (24-5) to estimate the conductivity of a solution, the mobilities of all ions present must be included.

Values of Γ are not exactly proportional to the concentration of all ions in solution. If Γ (or Γ_i for ions) is plotted against C, as the solution is diluted, and finally extrapolated to zero concentration, the Γ or Γ_i intercept, termed the equivalent conductivity at infinite dilution, is obtained. This is the conductivity where the ions are so far apart that they do not interact or retard one another's movement. It is a convenient reference point and is commonly employed for tabulation of numerical values of Γ_i. Table 24-2 gives the equivalent conductivities, often called mobilities, of several common ions at infinite dilution.

Table 24-2. Equivalent Conductivity and Temperature Coefficients for Several Ions at Infinite Dilution at 25°C [(a)] (S·cm^2/equivalent)

Cations	$\Gamma°$	Temperature Coefficient [(b)]	Anions	$\Gamma°$	Temperature Coefficient [(b)]
H^+	349.8	0.0139	OH^-	198.6	0.018
K^+	73.5	0.0193	Cl^-	76.4	0.0202
Na^+	50.11	0.0220	NO_3^-	71.42	0.020
Ca^{+2}	59.50	0.0230	SO_4^{-2}	80.0	0.022
Mg^{+2}	53.06	0.022	CO_3^{-2}	69.3	0.02
Cu^{+2}	53.6	0.02	HCO_3^-	44.5	—
$(n-Bu)_4N^+$	19.5	0.02	$Picrate^-$	30.4	0.025

(a) Data from Ref. 1. (b) Temperature coefficient = $(1/\Gamma°)(d\Gamma°/dT)$

Example: Using the values in Table 24-2 and Equation (24-5), estimate the conductances of (a) 0.01 *M* KCl, (b) 0.05 *M* $CaCl_2$, and (c) 0.02 *N* H_2SO_4. Assume that $a/l = 1$ cm.

(a) $L = 10^{-3} \times 1\ \{1 \times 0.01 \times 73.5 + 1 \times 0.01 \times 76.4\} = 0.0015$ **S, or 1.5 mS**

(b) $L = 10^{-3} \times 1\ \{2 \times 0.05 \times 59.5 + 1 \times 0.10 \times 76.4\} = 0.014$ **S, or 14 mS**

(c) 0.02 *N* H_2SO_4 is 0.02 *M* H^+ and 0.01 *M* SO_4^{-2}

$L = 10^{-3} \times 1\ \{1 \times 0.02 \times 349.8 + 2 \times 0.01 \times 80.0\} = 0.0086$ **S, or 8.6 mS**

Remember, the values calculated here are only approximations; limiting ionic conductivities were used and the actual measured conductances for these solutions will be somewhat smaller than those calculated. The error resulting will increase as the concentration of the solution whose conductivity is being estimated increases.

Sources of Conductivity Data

Volume VI of the *International Critical Tables* (1929) provides data for a large number of electrolytes, up to 1 *M* concentration and at several temperatures. Some higher concentrations are also given. Although all of the data were compiled before 1929, it is quite accurate and a valuable resource.

For industrial process streams, complete data on conductivity versus concentration and temperature for common materials such as strong acids, bases, and salts are scattered throughout the literature. These data are usually available from the manufacturers of conductivity analyzers.

A fairly comprehensive listing of the conductivities of common electrolytes at various concentrations is presented in Table 24-3.

Conductivity and Concentration

Units of concentration vary among the many users of conductivity measuring equipment. Units commonly employed for solutions with high concentrations of electrolytes are percent by weight — for example, grams of electrolyte per 100 g of solution. For less concentrated solutions, parts per million (ppm) as NaCl (i.e., mg/l) and sometimes grains per gallon (gpg) as NaCl or calcium carbonate may be used. Table 24-4 shows the relationships between these units.

Table 24-3. Conductivity vs. Concentration for Common Solutions
(From Foxboro TI612-112, with permission.)

Weight Percent	ppm mg/L	NaCl	NaOH	NH_4OH	NH_3	HCl	H_2SO_4	HNO_3	HF	SO_2	Acetic Acid (a)
		Conductivity, $\mu S \cdot cm^{-1}$ ($\mu\Omega \cdot cm^{-1}$) at 25°C									
0.0001	1	2.2	6.2	4.1	6.6	11.7	8.8	6.8	10	6.4	4.2
0.0003	3	6.5	18.4	8.3	14	35.0	26.1	20	30	18	7.4
0.001	10	21.4	61.1	17	27	116	85.6	67	99	54	15.5
0.003	30	64	182	31	49	340	251	199	290	150	30.6
0.01	100	210	603	58	84	1140	805	657	630	450	63
0.03	300	617	1,780	102	150	3,390	2,180	1,950	1,490	1,200	114
0.1	1,000	1,990	5,820	189	275	11,100	6,350	6,380	2,420	3,600	209
0.3	3,000	5,690	16,900	329	465	32,200	15,800	18,900	5,100	7,900	368
1.0	10,000	17,600	53,200	490	810	103,000	48,500	60,000	11,700	17,200	640
3.0	Rarely used	48,600	144,000	790	1,110	283,000	141,000	172,000	34,700	32,700	1,120
5.0		78,300	223,000	958	1,115	432,000	237,000	275,000	62,000	42,000	1,230
10.0		140,000	358,000	1,115	1,120	709,000	427,000	498,000	118,000	61,000	1,530
20.0		226,000	414,000	968	4,351	850,000	709,000	763,000	232,300	Sat	1,600
30.0		Sat	292,000	725	Sat	732,000	828,000	861,000	390,000	Sat	1,405
40.0		Sat	191,000	460	Sat	Sat	770,000	820,000	NA	Sat	1,080
50.0		Sat	150,000	285	Sat	Sat	620,000	717,000	NA	Sat	740
75.0		Sat	Sat	Sat	Sat	Sat	182,000	340,000	7.8 (0°C)	Sat	168
100.0		Sat	Sat	—	<1	Sat	10,000	50,000	4 (0°C)	<1	<1
Point of Maximum Solubility		26%	~ 50%	13.6% (1 atm)	28% (1 atm)	37%	—	—	—	11.7% (1 atm)	—
Point(s) of Maximum Conductivity		26%	16%	2.67%	5.5%	18.5%	31% 92.5%	31%	~ 35%	11.7%	19%
Maximum Conductivity		244,000	412,000	1,120 (18°C)	1,120 (18°C)	852,000	830,000 139,000	862,000	NA	66,000	1,600

(a) HAc data at 18°C.

Table 24-4. Conversion Chart

Conductivity µS/cm, 25°C	Resistivity at 25°C	Dissolved Solids, ppm (a)	Grains/U.S. Gallon, gpg (a)
0.055	18.3 MΩ · cm	0	0
0.063	16 MΩ · cm	0.00363	0.0002
0.071	14 MΩ · cm	0.00776	0.00045
0.083	12 MΩ · cm	0.0133	0.00078
0.1	10 MΩ · cm	0.0210	0.0012
0.5	2 MΩ · cm	0.206	0.012
1.0	1 MΩ · cm	0.44	0.026
10.0	0.1 MΩ · cm	4.6	0.27
100	0.01 MΩ · cm	47	2.7
200	5,000 Ω · cm	91	5.3
1,000	1,000 Ω · cm	495	29
10,000	100 Ω · cm	5,400	315
20,000	50 Ω · cm	10,800	631

(a) ppm and gpg data are expressed as NaCl; corresponding figures for calcium carbonate are obtained by multiplying these NaCl values by 0.856, i.e., the factor (MW $CaCO_3$/2 MW NaCl).

Conductivity measurements above 20,000 μS · cm^{-1} are commonly applied in chemical streams, gas scrubbers and waste stream measurements having high concentrations of dissolved materials. The graphs shown in Figure 24-1 for sulfuric acid and sodium hydroxide are typical of the conductivity concentration relationships that are encountered; they are nearly linear at low concentrations, but at high concentrations the rate of change of conductivity with concentration falls off, leading to nonlinearity.

Conductivity and Temperature

Tables 24-2 and 24-3 and Figure 24-1 show that temperature has an effect on the conductivities of ions. Temperature coefficients are always positive. This is because the velocity of ions increases with temperature, increasing their mobilities. At the same time, solution viscosity decreases, resulting in less resistance or drag on movement of ions, again increasing their mobilities. Temperature dependence is usually expressed as a relative change per degree Celsius at a particular temperature, commonly as %/°C at 25°C. Ultrapure water, whose conducting ions are contributed by the temperature-dependent ionization constant of water, has the largest temperature slope, 5.2%/°C at 25°C, whereas weak electrolytes have values of about 2%/°C at 25°C and strong electrolytes are typically somewhat lower.

When conductivity measurements are made at temperatures other than the reference temperature, the data are usually normalized back to what they would be at the reference temperature, often (but not necessarily) 25°C. To do this, temperature sensors mounted in the measuring probes, used with resistance networks and electronic circuity that are configured to have temperature slope characteristics similar to those of solutions, are used to control gain in the analyzer circuitry. Modern instruments that contain a microprocessor may use files of temperature data stored within the analyzer. They measure the temperature and conductivity of the sample and deliver a temperature-compensated value that is limited in accuracy only by the number and accuracy of the data points in the file. In either case, accurate measurements require knowledge of the identity of the electrolyte and the relationship of its conductivity with temperature.

... accurate measurements require knowledge of the identity of the electrolyte and the relationship of its conductivity with temperature.

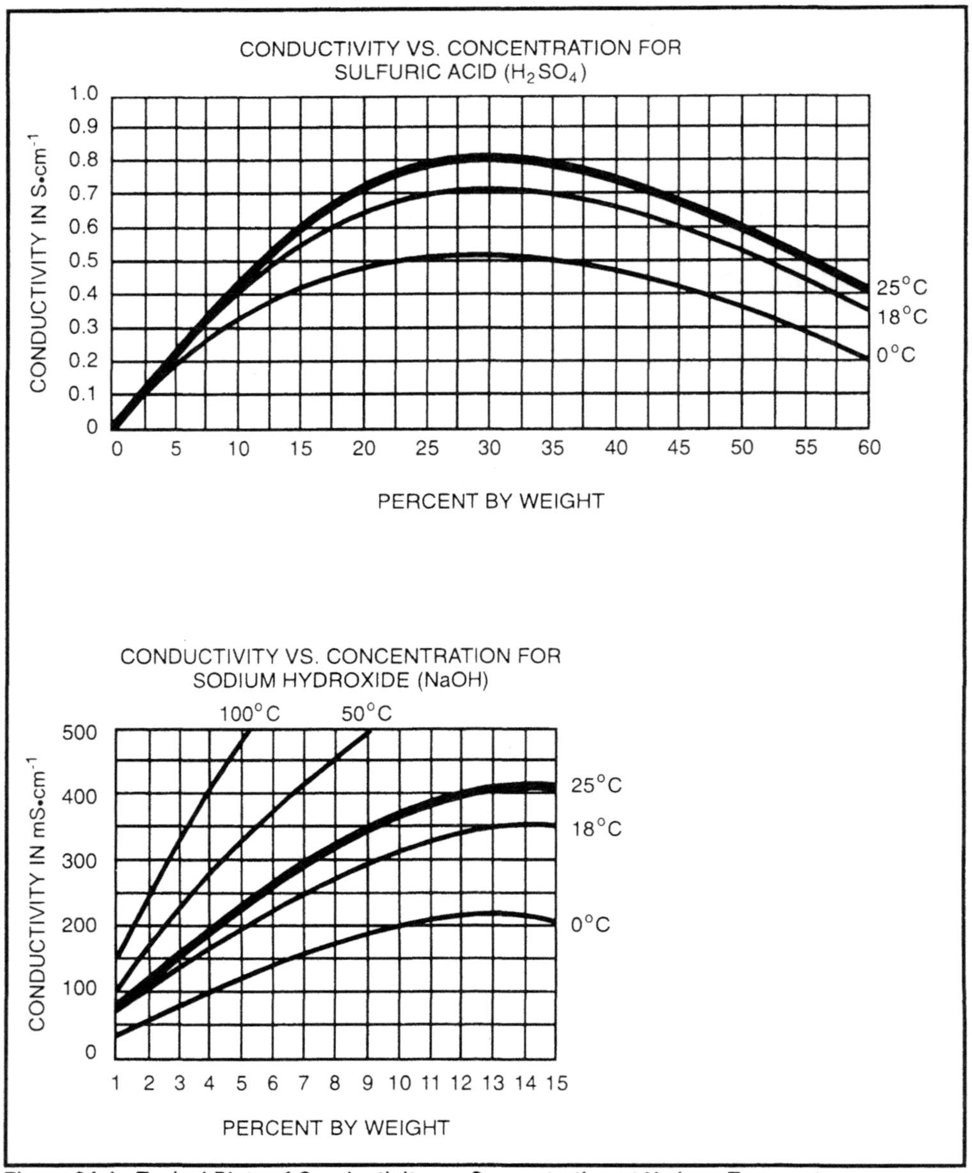

Figure 24-1. Typical Plots of Conductivity vs. Concentration at Various Temperatures

The Cell Constant

Measured conductance, L, in Equation (24-5) may be related to the concentration of participating ions if the factor (a/l) is known. Transposing (a/l) to the left side of this equation and multiplying by L converts conductance in the units S to conductivity in the units $S \cdot cm^{-1}$. The ratio l/a, given the symbol Θ, is the cell constant with the units cm^{-1}.

Values of Θ are furnished by the manufacturers of conductivity cells, with specified accuracies that range from ±5% to ±1%. For precise and accurate work, however, the user may wish to determine cell constant experimentally. This is done with accurately known solutions of potassium chloride, whose conductivities, in $S \cdot cm^{-1}$, have been determined precisely using cells with known geometry. These reference solutions have been adopted as conductivity standards by ASTM and are shown in Table 24-5. The cell constant is readily determined using the expression $\Theta = k/L$, where k is taken from the tabulated values and L is measured as the cell is being calibrated, with careful attention to the measurement temperature.

Table 24-5. Conductivity of Potassium Chloride Solutions Used for Determination of Cell Constants

Approximate Normality	Weight of KCl in g/L solution	Temperature, °C	*k* (a), µS · cm^{-1}
1.0	72.2460	0	65,176
		18	97,838
		25	111,342
0.1	7.4365	0	7,138
		18	11,167
		25	12,856
0.01	0.7440	0	773.6
		18	1,220,5
		25	1,408.8
0.001	Dilution of 0.01 M KCl to 1 L at 25°C	0	77.69
		18	127.54
		25	146.93

* The conductivity of the water used to make these solutions must be measured and the conductivities corrected as necessary. Data from Reference 1.

As the dimensions of the cell are changed, the cell constant varies as the ratio of *l* to *a*. For measurement of solutions having low conductivity (about 0.05 to 200 µS · cm^{-1}), the electrodes should be close together and have cell constants in the range of 0.1 to 0.01 cm^{-1}. For more conductive solutions (about 10 to 20,000 µS · cm^{-1}), electrode separations should be increased to give cell constants of 1, 10, or sometimes 100 cm^{-1}. Selection of the appropriate cell constant for the application is necessary to allow the actual conductance read by the instrument to be in an accurately measurable range.

Selection of the appropriate cell constant for the application is necessary to allow the actual conductance read by the instrument to be in an accurately measurable range.

Signal to noise considerations limit resistance measurements to less than about 2 MΩ (corresponding to 0.5 µS); at the other end, high signal level magnitudes limit conductance measurements to less than about 0.2 mS (corresponding to 5 kΩ). (Note: MΩ = 1×10^6 Ω = 1,000,000 Ω; kΩ = 1×10^3 Ω = 1,000 Ω.) Table 24-6 shows cell constants for various solution conductivity ranges as recommended by ASTM D1125-82. Cells with a cell constant greater than 10 cm^{-1} are large in size, so most cells used in industrial measurement and control applications fall in the range of 0.01 to 10 cm^{-1}.

Table 24-6. Recommended Cell Constants for Various Conductivity Ranges

Conductivity Range, µS · cm^{-1}	Cell Constant, cm^{-1}
0.05-20	0.01
1-200	0.1
10-2,000	1.0
100-20,000	10.0
1,000-200,000	50.0

Data from Ref. 2.

Instrumentation

A schematic diagram of the conductance cell and its equivalent electronic circuit is shown in Figure 24-2. Resistance and conductance analyzers measure the electrical impedance of the total cell. This measurement includes four components: electrolysis impedance; double layer capacitance impedance at the electrode contacts with the solution; parallel capacitance impedance across the cell; and finally the desired measurement, electrolyte resistance or conductance. A voltage is applied across the cell to drive the ionic current. For charge to pass from the electrode into the solution, two processes must occur. The first involves the double-layer. Here, a positively-charged electrode attracts a layer of negatively-charged ions as does the negative electrode a layer of positively-charged ions. This stored charge constitutes capacitance. When the voltage applied is very small, no current flows beyond that necessary to charge this capacitor and a voltage drop is created. If the applied voltage is further increased to force charge through this double-layer, then electrolysis must occur to form or discharge ions. This process shunts the double-layer and behaves electrically like a nonlinear resistor, increasing significantly as the potential applied across it is increased. This too gives a voltage drop. Both effects are minimized by using a small-amplitude sine wave having a frequency large enough to limit capacitance impedance in the double-layer but with small enough amplitude to limit impedance due to electrolysis.

Typical double-layer capacitance for clean electrodes is 10 to 100 $\mu F/cm^2$. For a 1000-hz sine wave, this results in an impedance range ($1/2\pi fC$) of approximately 1.5 to 15 Ω, where f (frequency) is expressed in hertz and C (capacitance) is expressed in farads. This impedance decreases with increasing frequency and capacitance. Platinized electrodes, used in high-precision laboratory work, have capacitance significantly greater than "bare" metal electrodes and are used for this reason. Because of their cost, however, platinum electrodes are not practical for industrial applications and are seldom used.

The parallel capacitance across the cell and signal lead capacitance limits the usable frequency of the applied sine wave to about 1,000 Hz. Thus, most commercially available analyzers are operated in the range 60 to 1,000 Hz with cells having cell constants giving measured resistance values in the 10^3 to 10^5 Ω range where these parasitic impedances are negligible.

Composition measurement using conductivity is popular for industrial process measurement and control because of the inherent simplicity and reliability of the technique when it is applicable.

Composition measurement using conductivity is popular for industrial process measurement and control because of the inherent simplicity and reliability of the technique when it is applicable. Cells are available that cover the 10^8 range of conductance or resistance encountered in the aqueous process world. Three types of cells are widely used: two-electrode, four-electrode, and electrodeless.

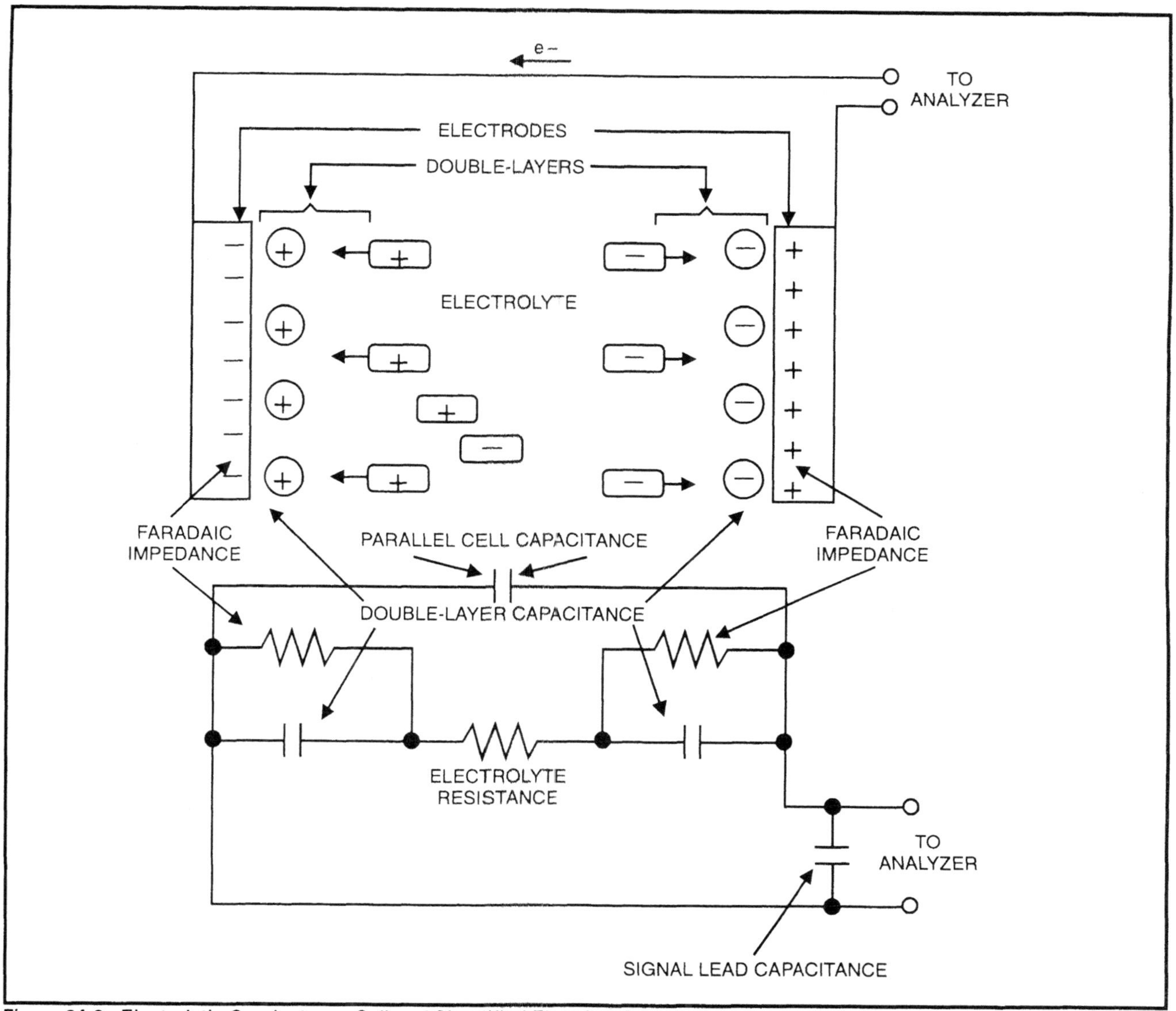

Figure 24-2. Electrolytic Conductance Cell and Simplified Electrical Schematic

Two-Electrode Measurements

Two-electrode cells are best suited for measurements in clean solutions to avoid coatings and films that may form on electrodes and produce ohmic contact error. Selection of the largest practical cell constant helps in this regard. Two-electrode cells, and their associated instrumentation, are the least expensive type to purchase and are commonly employed for applications where high accuracy is not critical — for example, leak detection, level limit detection of solutions in tanks, aqueous/organic solvent phase detection, and concentration monitoring and control in plating and rinsing baths.

Two-electrode cells are best suited for measurements in clean solutions to avoid coatings and films that may form on electrodes and produce ohmic contact error.

A simplified electrical schematic diagram of a typical two-electrode conductivity analyzer is shown in Figure 24-3. The cell is driven through an amplifier using a sine wave derived from an oscillator with 0° and 180° output phases. The drive amplifier prevents loading of the oscillator by the cell when high-conductivity solutions are being measured.

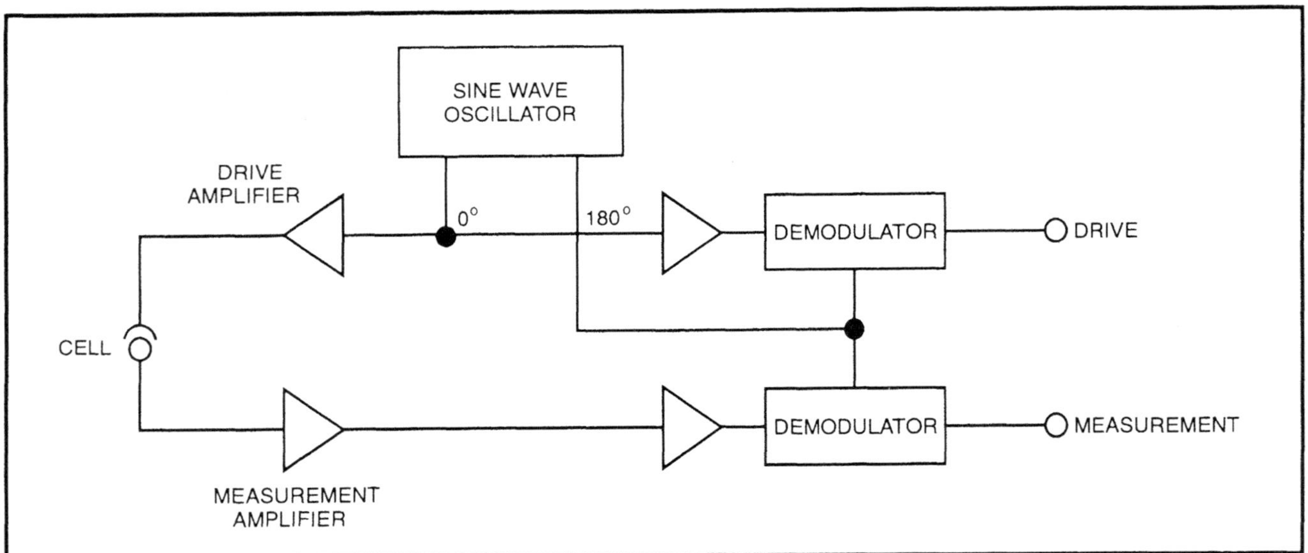

Figure 24-3. Simple Electrical Schematic of a Two-Electrode Conductivity Analyzer

Cells of 0.01 and 0.1 cm^{-1} are usually made with concentric cylinders of inert metals such as stainless steel, titanium, or monel. Their configuration provides a well-defined electric field distribution between the electrodes, and their cell constants remain constant over wide concentration ranges. Larger cells, those with cell constants in the range of 1 to 10 cm^{-1}, have electrodes of inert metal, or sometimes carbon, mounted in chemically-resistant polymeric materials. The geometry imposed by their size generally requires that the electrodes be enclosed within an electrically-inert solution shield to control the shape of the electric field and maintain the cell constant at its desired value.

The cell conductance, which is the input resistance of the measuring amplifier, is output as a voltage proportional to the conductivity of the solution being measured. Synchronous demodulators remove noise and other interferences from the measurement signals and provide dc voltage level signals proportional to solution conductivity.

RESISTIVITY ANALYZERS

Resistivity analyzers are widely used for measurement of trace ionic contamination in the highly pure water used in semiconductor manufacturing, in steam for steam turbine power generation, and other critical areas.

Resistivity analyzers are an important outgrowth of two-electrode conductivity analyzers. They are widely used for measurement of trace ionic contamination in the highly pure water used in semiconductor manufacturing, in steam for steam turbine power generation, and other critical areas.

The theoretical conductivity of pure water is 0.055 $\mu S \cdot cm^{-1}$ or 18.18 $M\Omega \cdot cm$ at 25°C [Ref. 3], and the limit of ionic impurity detection is about 0.1 ppb. Over the temperature range from 0 to 100°C, the ultimate resistivity of water ranges from about 84 $M\Omega \cdot cm$ to 1.3 $M\Omega \cdot cm$, corresponding to 0.13 to 8.4 $M\Omega$ measured resistance using a 0.1 cm^{-1} cell. At these levels, capacitance effects in the leads and parallel cell capacitance can produce significant error. The resistivity analyzer minimizes these errors using a circuit of the type shown in Figure 24-4.

Parallel cell and lead capacitance (Figure 24-2) have values in the range 10 to 100 pf for the typical 0.01 or 0.1 cm^{-1} cell. This represents AC impedance levels, $X_C = 1/2\pi fC$, in the range of 1.5 to 0.15 $M\Omega$ at 1000 Hz. These values are roughly equivalent to the resistance of the sample in the cell and thus must be eliminated. Their magnitude is first reduced by reducing the drive frequency to as low a value as possible, typically 100 Hz, but some type of phase-sensitive detection must also be employed to distinguish reactive impedance from the desired resistivity measurement.

A way of doing this is shown in Figure 24-4. As with the conductivity cell, the resistivity cell is driven with a sine wave through an amplifier to prevent loading of the oscillator. The oscillator also provides sine waves 90° and 180° out of phase from the drive signal. Since the cell is in the feedback loop of the measurement amplifier, its output is directly proportional to the resistance of the sample being measured. Synchronous demodulators provide DC voltage levels for drive, quadrature, and measurement signals that are used in the calculation of resistivity. Demodulated drive and measurement signals are used to reduce noise, and demodulated quadrature is used to eliminate the contribution of impedance resulting from capacitance in the cell.

Cells commonly used in resistivity measurements have cell constants in the range 0.01 to 0.1, and their calibration presents some difficulty. The least conductive ASTM calibrating solution shown in Table 24-5 has a conductivity of 146.93 $\mu S \cdot cm^{-1}$ or 6,806 $\Omega \cdot cm$ at 25°C. This gives resistance of 680.6 ohms with a 0.1 cm^{-1} cell and only 68.06 Ω with a 0.01 cm^{-1} cell. A few ohms of resistance due to oxide films or coatings on the electrodes, or contact and lead resistance, can result in significant error when cell constants are determined using these conductive solutions. This consideration leads to selection of the 0.1 cm^{-1} cell as the more accurate choice, but careful attention should be given to maintaining cleanliness of cells, and particularly to prevention of their being exposed to dirty or oily solutions.

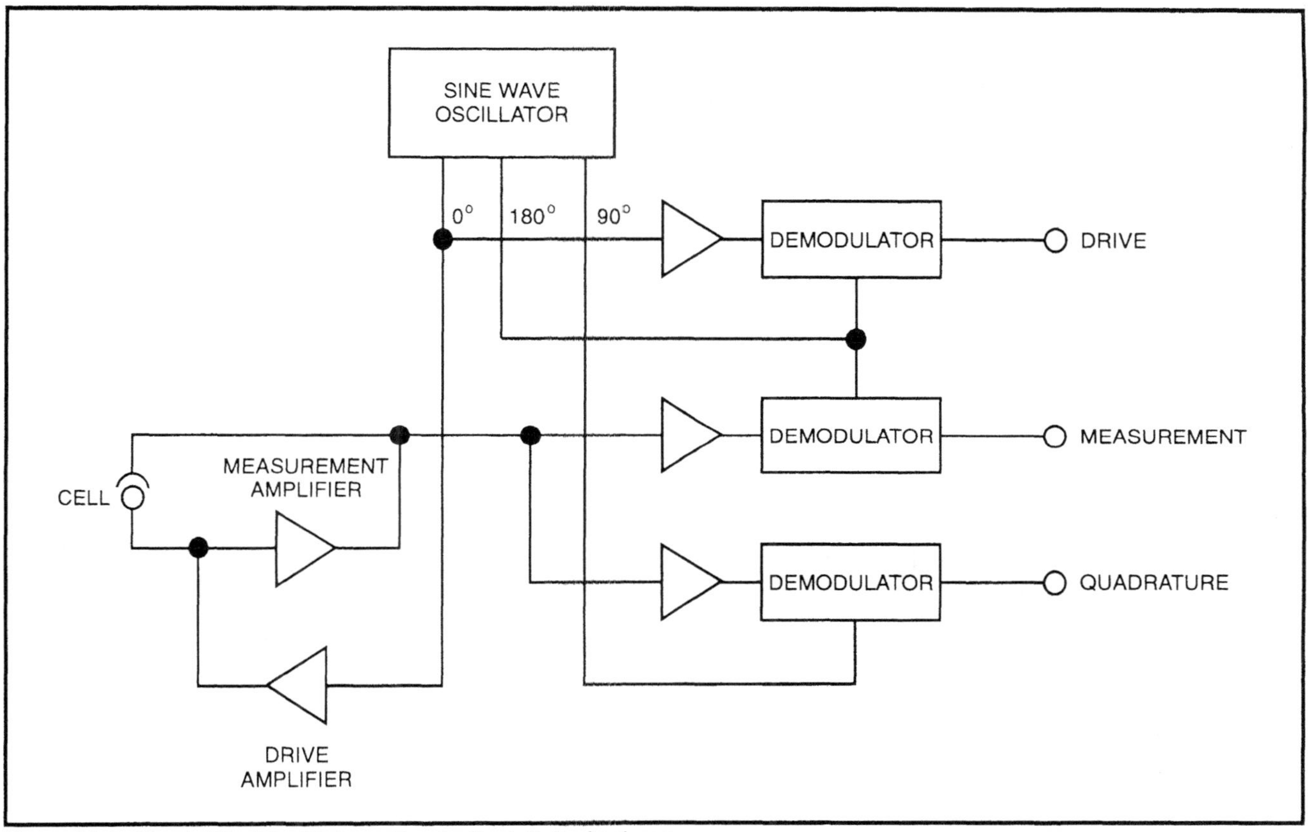

Figure 24-4. Simple Electrical Schematic of a Resistivity Analyzer

Four-Electrode Measurements

Four-electrode conductivity is useful when coating and fouling of electrodes occurs.

Four-electrode conductivity is useful when coating and fouling of electrodes occurs. Current is imposed across two drive-electrodes, and the potential drop through the electrolyte is detected using two sense-electrodes. A typical arrangement is shown in Figure 24-5. The sense-electrodes are monitored with a high-input-impedance voltage-measuring amplifier to minimize the current drawn through them. In this way, even if they become coated, the potential drop across the ohmic resistance of the coating is small and contributes no appreciable error to the measured potential drop in the solution between them. These contacts may be reference electrodes joined to the process sample via salt bridges and liquid junctions, but simple metal contacts are usually practical since redox potentials at the two should be equal and will cancel.

Interfacial effects at the drive-electrodes, their double-layer capacitance, faradaic impedance, concentration polarization, and buildup of coatings, all of which create voltage drops at their contacts to the sample, will have no bearing on the voltage drop measured by the sense-electrodes as long as sufficient voltage is available to drive the control current. As coating builds up on the drive-electrodes, the voltage necessary to maintain the control current increases. This can be measured and used as a diagnostic tool to signal the user that coatings are present and to indicate that cleaning is required.

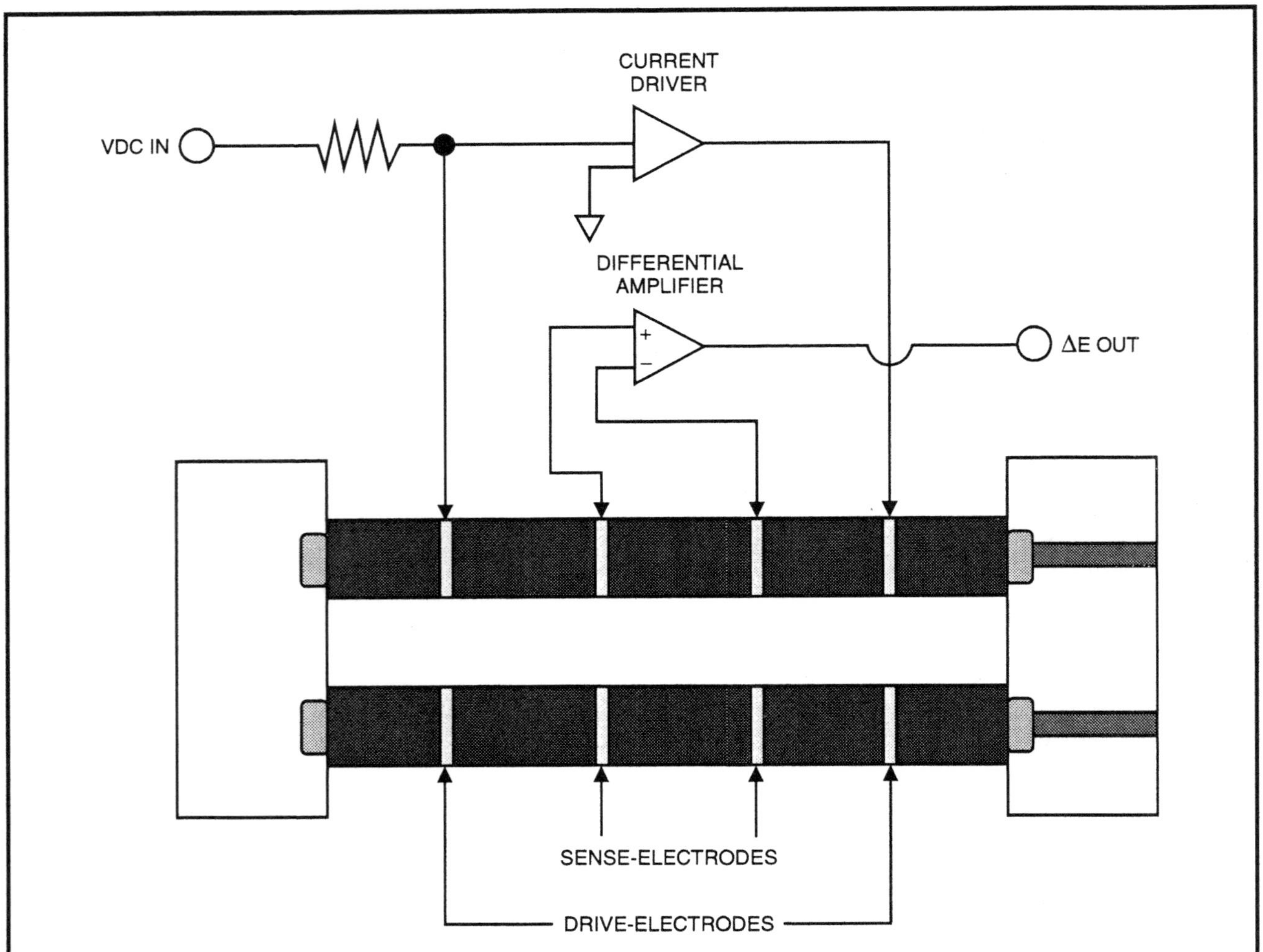

Figure 24-5. Four-Electrode Cell and Conductivity Analyzer

Geometrical considerations usually render four-electrode techniques inapplicable to probe configurations because it is usually not feasible to design probes where the field across the sense-electrodes can be made uniform. Thus, four-electrode probes give a very nonlinear relationship between conductivity and concentration, but can be used to advantage when conductivity set-point control is the desired result.

Four-electrode techniques are better suited to cylindrical flow-through cells, such as pipes, that allow linear distribution of current along the axis of their diameter. A typical configuration uses drive and sense-electrodes mounted in the walls of an otherwise nonmetallic pipe and provides a totally obstructionless measurement cell.

Four-electrode techniques are better suited to cylindrical flow-through cells, such as pipes, that allow linear distribution of current along the axis of their diameter.

Cell constants are easily calculated using the expression a/l, where a is the area (in cm^2) determined by the internal diameter of the pipe and l is the distance (in cm) between the sense-electrodes. A few cell constants for 1-, 2-, and 4-in. internal-diameter cells with a few convenient sense-electrode spacings are shown in Table 24-7.

Typical applications include measurement of concentrated and dirt laden salts, acids, and alkalis in chemical processes in the mining, metallurgy, pulp and paper, and aluminum industries where samples often contain solids, oils, or other materials that form insulating coatings on the electrodes.

Table 24-7. Typical Cell Constants for 1-, 2-, and 4-in. Diameter Four-Electrode Flow-Through Cells

Sense-Electrode Spacing, in.	Cell Constant, cm^{-1}, for Pipe Diameter, inch		
	1	2	4
1/2	0.25	0.06	0.01
1	0.5	0.13	0.03
2	1.0	0.26	0.06
4	2.0	0.50	0.12

Electrodeless Conductivity Measurements

One way to eliminate electrode polarization effects is to eliminate the electrodes. Techniques for this are called electrodeless conductivity [Ref. 4]. Instrumentation for this is shown schematically in Figure 24-6(a). Operationally, it resembles a transformer. Two encapsulated toroids, when immersed in electrolyte, form a conductive loop of solution shared by both. One toroid radiates an electric field in this loop, and the other detects a small (induced) electric current. Practically speaking, the two toroids are the coils of a transformer interconnected by the resistance of the electrolyte (Figure 24-6b). The radiated field is typically 20 kHz, and the induced current, which is proportional to the conductivity, is amplified, rectified, and displayed.

Practically speaking, the two toroids are the coils of a transformer interconnected by the resistance of the electrolyte (Figure 24-6b).

The available ranges in commercially-available electrodeless conductivity instruments are 0 to 100 $\mu S \cdot cm^{-1}$ at the sensitive end to 0 to 2 $S \cdot cm^{-1}$ at the least sensitive end. Accuracies are within a few tenths of a percent when careful attention is paid to temperature.

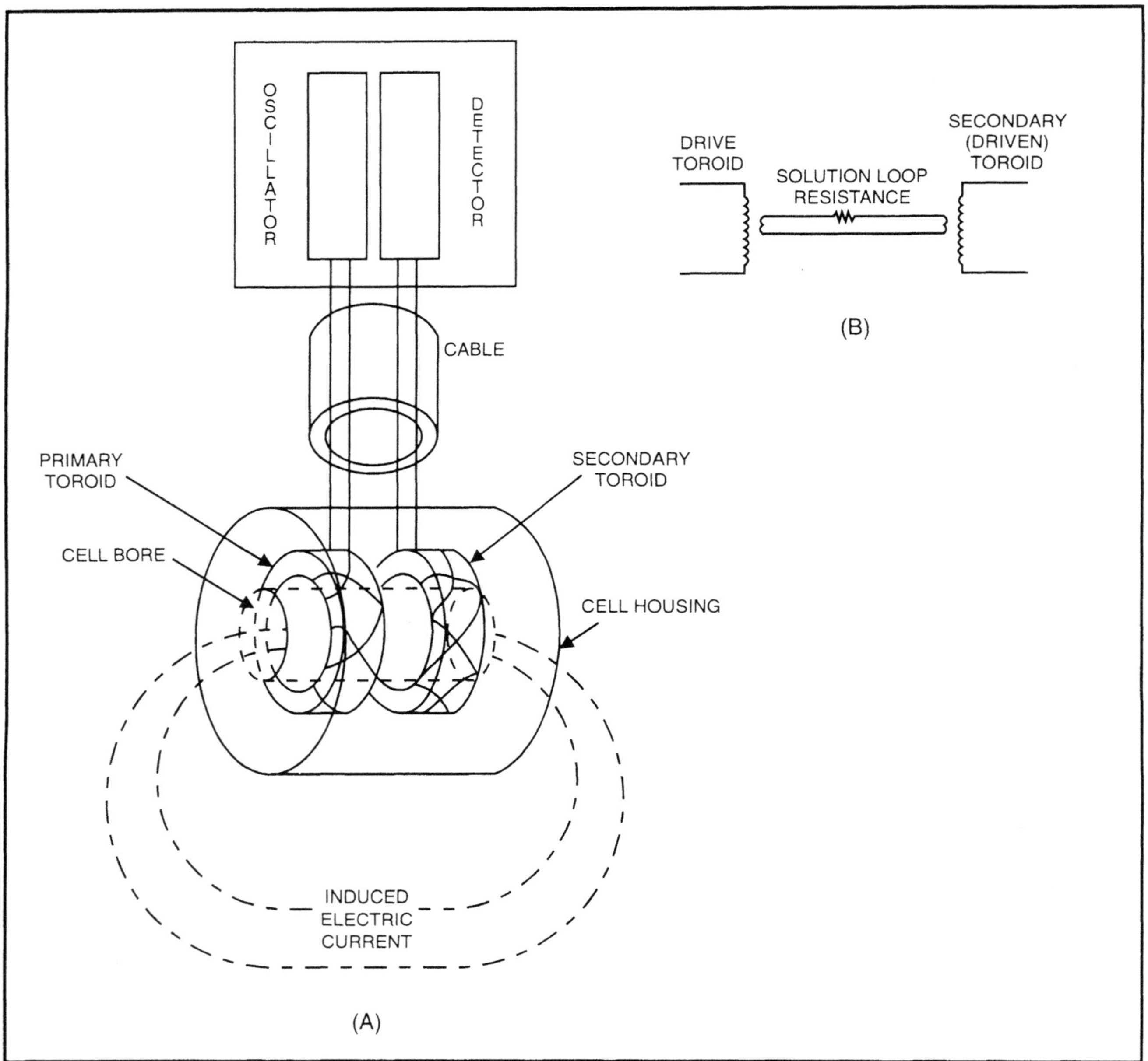

Figure 24-6. (a) Electrodeless Conductivity Analyzer; (b) Transformer Analog Schematic

ELECTRODELESS PROBES AND CELLS

Various configurations of electrodeless cells are available. Some designed for immersion are shown in Figure 24-7. The cores are covered with an encapsulant of a nonconductive, temperature-stable, and chemically-resistant material such as epoxy, Noryl, PFA, PEEK, or polypropylene. Cells are available for use at temperatures and pressures as high as 200°C and 500 psig.

A particular advantage of these cells is their relative immunity to fouling.

A particular advantage of these cells is their relative immunity to fouling. The effect of coatings, as long as they are nonconductive, is purely geometric; they affect cell constant only to the extent that they displace solution from the surface of the probe, primarily on the internal diameter of the core. Large-bore sensors are usually selected to minimize this effect when coating buildup is a significant problem.

Figure 24-8 shows the electrodeless technique applied to a flow-through cell configuration. Since the operational requirement of this measurement is the generation of current in a conductive loop intercepted by the sensing coil, it is necessary to configure the hardware into a flow loop using an arrangement as shown. The problem encountered is that cell constants become somewhat large due to the path length created by the solution loop, and this restricts their application to samples having fairly high conductivities. Because of cost and the variability of line size, electrodeless flow-through cells are not generally commercially available. Four-electrode contacting techniques are probably the better option in most applications where an obstructionless flow-through cell is required.

Since the operational requirement of this measurement is the generation of current in a conductive loop intercepted by the sensing coil, it is necessary to configure the hardware into a flow loop using an arrangement as shown.

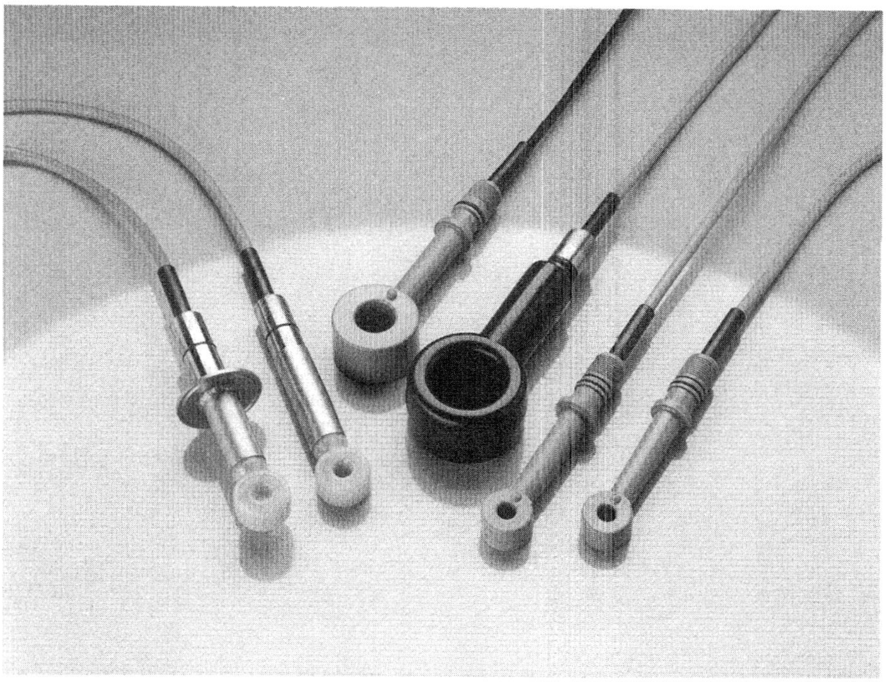

Figure 24-7. Immersion-type Electrodeless Cells

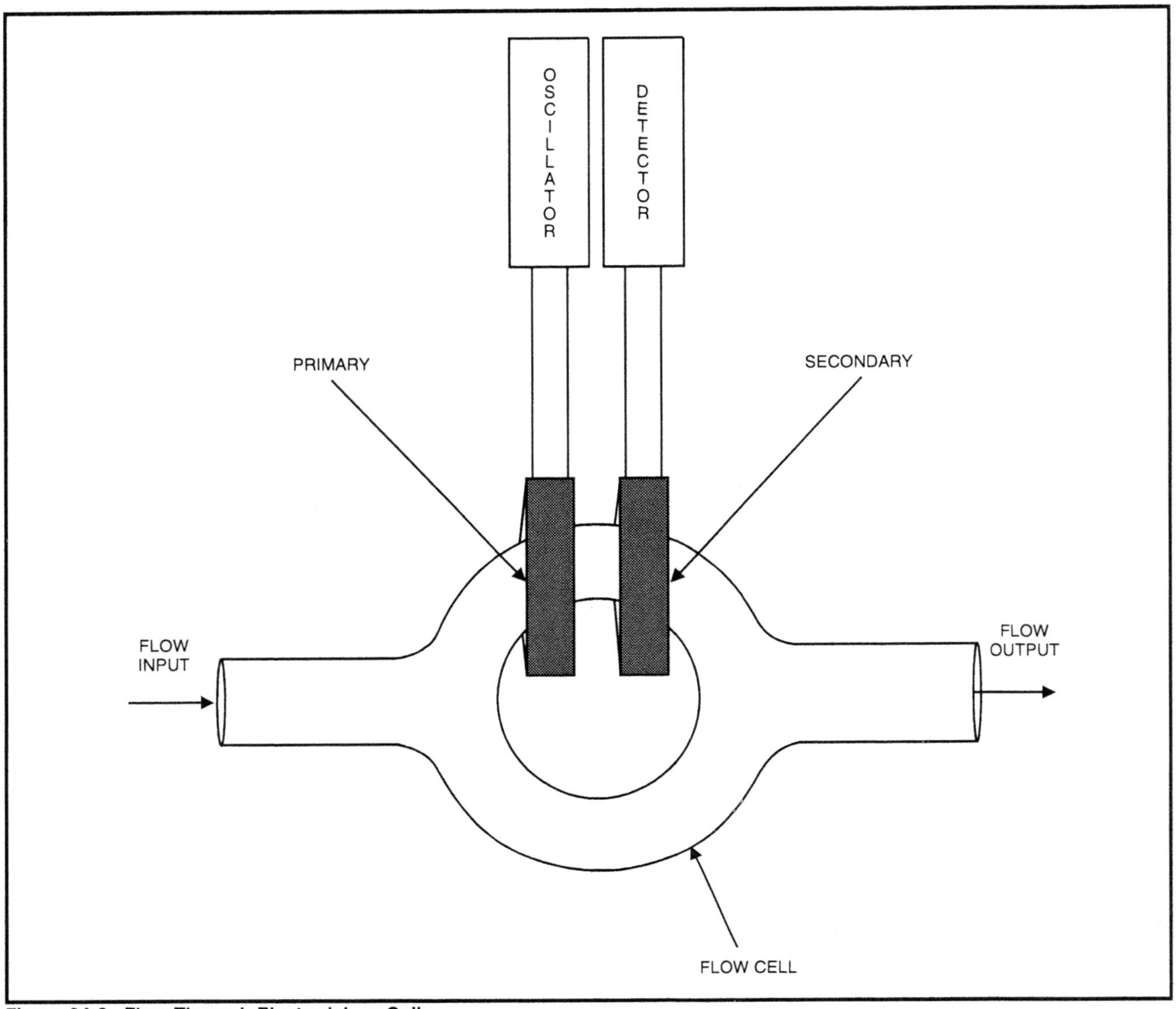

Figure 24-8. Flow-Through Electrodeless Cell

Equation (24-6) assumes infinite distance of the probe from container walls. Practically speaking, the electrodeless probe should be mounted so that the distance of the active cell, the doughnut structure, is at least three diameters from the container or process vessel wall.

ELECTRODELESS CELL CONSTANTS

Cell constants of electrodeless probes are determined by the ratio of the current path length to the cross-sectional area of the hole in the doughnut-like sensor. It is possible to calculate this to a moderate degree of accuracy using the expression [Ref. 5]:

$$\Theta = (c/\pi a^2) + \left(\frac{1}{2}a\right) \tag{24-6}$$

where a is the internal radius of the bore (in cm), and c is the thickness of the core along the internal axis (in cm).

The dimensions c and a establish the range of cell constants available for practical cell sizes. For a small cell having $a = 0.792$ cm and $c = 4.20$ cm, the typical dimensions of a commercially available small-bore cell, the cell constant calculated is 2.76 cm^{-1}. For a large bore cell having $a = 2.47$ cm and $c = 5.32$ cm, the calculated cell constant is 0.48 cm^{-1}. Commercially available cells therefore have cell constants in the range of about 0.5 to 3.0 cm^{-1}.

The analyzer and cell can be calibrated as a system using either a precision resistance decade box or known solutions. In either case, for a lower-range value of 0 $\mu S \cdot cm^{-1}$ or 0 $mS \cdot cm^{-1}$, the clean and dry sensor is suspended in air and the zero is adjusted. The upper-range value is then calibrated with a resistance box or a known solution. For the resistance procedure, a wire is looped through the probe bore and connected across the resistance box. The resistance value corresponding to the upper-range conductivity value desired is calculated for the cell constant of the cell using the expression:

$$R = \Theta \cdot 1000 / F_{sc} \qquad (24\text{-}7)$$

where:

R = resistance required in ohms
Q = cell constant in cm^{-1}
F_{sc} = full-scale value desired

For the solution calibration, the probe is immersed in the known solution and the analyzer is adjusted to output the solution's known conductivity.

CALIBRATION OF ELECTRODELESS CELL CONSTANTS

Many commercial instruments do not output conductance (in units of the siemens), but rather conductivity (in units of $S \cdot cm^{-1}$). The cell constant is built into the instrument, and the accuracy of the measurement is dependent on the builder's manufacturing tolerances. A simple procedure to check accuracy is to immerse the probe in a solution of known conductivity (e.g., a KCl solution selected from Table 24-5) and note the conductivity reading. The probe is then removed from the solution; a wire is passed through the probe bore, and the ends are connected across a precision decade box. Resistors are dialed in until the indicated value of conductivity is the same as that noted for the solution. The cell constant is then the simple product of the known solution conductivity and the resistance on the decade box.

A simple procedure to check accuracy is to immerse the probe in a solution of known conductivity (e.g., a KCl solution selected from Table 24-5) and note the conductivity reading.

Intelligent Conductivity Analyzers

A microprocessor-based conductivity analyzer is a powerful analytical tool. It can measure solution properties, such as conductivity and temperature. It can combine these measurements with compensation data (for temperature, either user or manufacturer-supplied). It can output analyses in selected concentration units suitable both for simple monitoring and for concentration control.

Modern conductivity analyzers, with on-board computers, provide essential features such as temperature-correction to reference values, digital display of concentration data from measured conductivity values, controller functionality, self-diagnostics, and calculations specific to the user's needs. Temperature compensation and concentration computation data for common acids, bases, or salts can be embedded in the instrument software; data for less common materials can sometimes be loaded by the end-user.

Microprocessor-based conductivity analyzers go beyond simple conductivity measurement; they interpret measurements, diagnose themselves and their cells for faults, and are able to output both measurement and process control action.

To construct these data tables, the user can measure conductivities of known concentrations of process solution at the selected reference temperature, and input these data to the analyzer. The user can then determine conductivities at various temperatures and input these data. In this way, the analyzer gives the highest possible temperature-compensated accuracy. Also, relay contacts within the analyzer, under control of the microprocessor, can be programmed to open or close for selected time intervals at process setpoints. This is useful for control of processes in tanks, such as those used in gas scrubbing where chemical reagent addition may have delays associated with mixing and dissolution of solid reactants.

Conductivity and resistivity analyzers are available with the capability for measuring two cells and the ratio of the conductivities or resistivities measured with the cells can be computed and output as a signal for use in applications where treatment or processing is measured, monitored, and controlled by comparing the process input and output measurements. Typical applications are ion exchange and reverse osmosis operations monitored by resistivity cells mounted at the inputs and outputs. The ratio R_{in}/R_{out} indicates exchange efficiency; R_{in} can be used with a flow measurement to predict ion-exchange capacity, and R_{out} to signal breakthrough and resin bed exhaustion. Other applications include parts rinsing to control efficiency of cleaning and heat-exchanger water monitoring to detect process breakthrough and contamination.

Typical applications are ion exchange and reverse osmosis operations monitored by resistivity cells mounted at the inputs and outputs. Other applications include parts rinsing to control efficiency of cleaning and heat-exchanger water monitoring to detect process breakthrough and contamination.

Installation and Maintenance

Choosing the correct sensor and mounting it so that it has a representative sample of solution are probably the most important considerations for proper installation. The following criteria are paramount:

1. Select the proper cell constant (see Table 24-6).
2. Select and calibrate the proper range on the analyzer.
3. Select corrosion-resistant probe materials.
4. Select suitable mounting hardware for installation.
5. Select a suitable process location for installation.

For proper sensor mounting, the following conditions should be observed:

1. The solution between the electrodes or around the probe should be representative of the solution as a whole.
2. A moderate flow across the cell should be assured to provide an up-to-date solution sample.
3. Mount contacting cells in attitudes that prevent accumulation of air bubbles between the electrodes (for example, in a pipe tee with the sample directed into the cell chamber rather than perpendicular to it). Likewise, for electrodeless cells, mount the probe so that air bubbles do not collect in the bore.
4. Mount electrodeless cells where tank and pipe walls are at least three probe diameters distant from the cell.
5. For both contacting and electrodeless cells, mount where sludge or sediment cannot collect around the cells or electrodes.

Probes should be placed where samples are well mixed, representative, and free of bubbles, and where solids do not accumulate.

For two-electrode contacting cells, the mounting arrangements shown in Figure 24-9 provide excellent service if plumbing is convenient; for larger pipes, weldments allow insertion that provides adequate sampling with good immunity from air bubbles. Other common mounting configurations include ball valves that allow insertion or retraction of cells and probes from the sample without interruption of the process. Sample cells may also be used if the sample is to be withdrawn from the process through a bypass line.

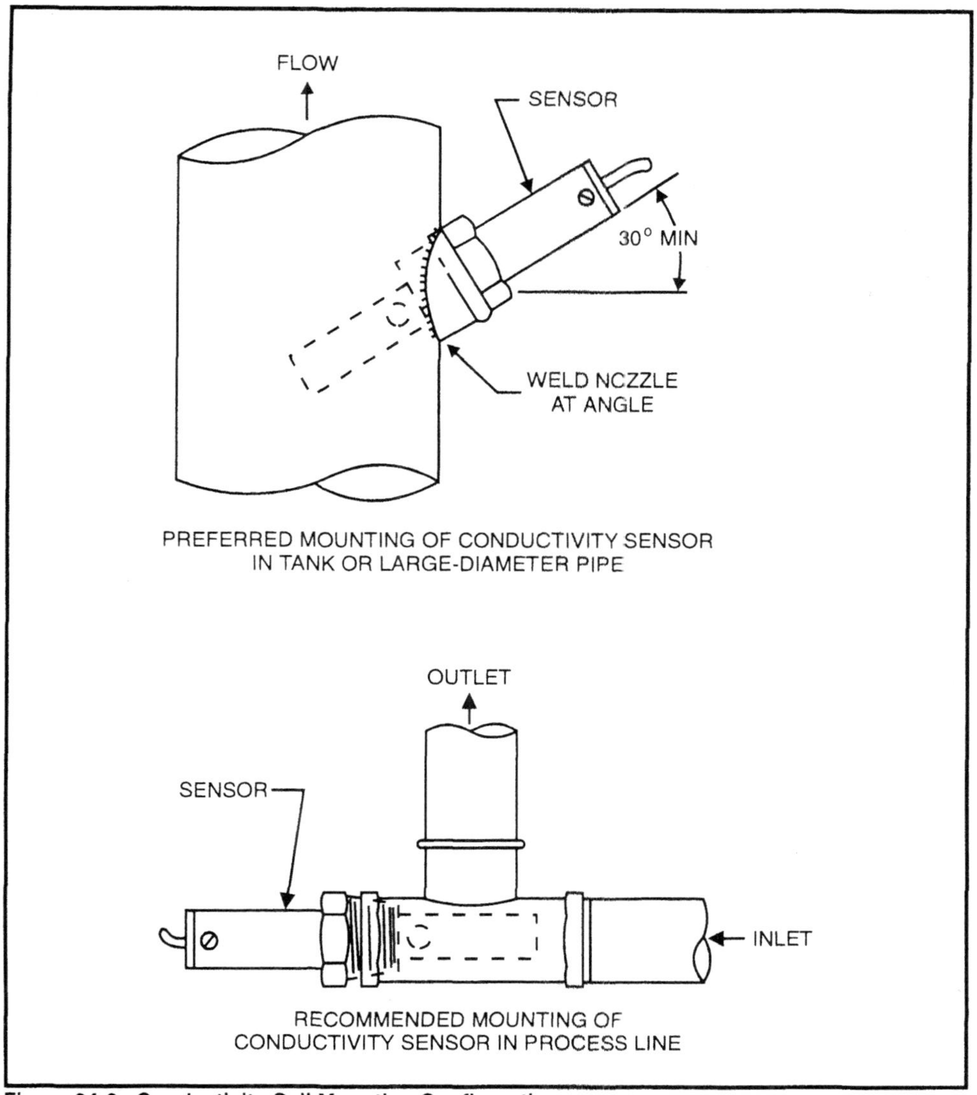

Figure 24-9. Conductivity Cell Mounting Configurations

Acknowledgments

The author would like to thank Phil Reese and Joe Downey of The Foxboro Company for technical help in the preparation of this chapter.

References

1. "The Measurement of Electrolytic Conductance," *Handbook of Analytical Instrumentation*, G.W. Ewing, Ed., New York: Marcel Dekker, 1989.

2. ASTM D1125-82, "Standard Test Methods for Electrical Conductivity and Resistivity of Water," *Annual Book of ASTM Standards*, Vol. 11.01, Philadelphia: ASTM, 1983, p. 149-156.

3. Thornton, R.D., and Light, T.S., "A New Approach to Accurate Resistivity Measurements of High Purity Water," *Ultrapure Water*, 6(5): 14-26, 1989.

4. Light, T.S., "Conduct Electricity without Electrodes," *Chemtech*, p. 496-501, August 1990.

5. Light, T.S., McHale, E.J., and Fletcher, K.S., "Electrodeless Conductivity," *Talanta*, 36(1/2): 235-241, 1989.

About the Author

Kenneth S. Fletcher received his B.S. Degree in Chemistry from Trinity College, Hartford, Connecticut in 1963 and a Ph.D. in Electroanalytical Chemistry from The University of Massachusetts at Amherst in 1967. His entire working career has been spent at The Foxboro Company working, initially as a researcher, and later as a group leader, developing Electrochemical Sensors for Process Monitoring Applications. These activities have led to new products and product enhancements in pH, Contacting and Electrodeless Conductivity, and Dissolved Oxygen measurements. Many publications in the scientific and trade literature have resulted from these activities as well as patents.

25

pH Analyzers

The concentration of hydrogen ion in aqueous solutions, or pH, is measured using pH analyzers (pH meters) in conjunction with electrochemical cells. Electronic pH measurement, by direct potentiometry, became established in the 1930s with the introduction of the glass electrode and high-input impedance electronic amplification equipment based on the vacuum tube. Complementary techniques, based on metal oxide electrodes such as antimony/antimony oxide and colorimetric techniques that rely on dissolved pH-sensitive indicators, were available prior to this time and are still used in special situations. However, specially formulated glasses have been developed (primarily in the 1950s and '60s) that show remarkable selectivity to aqueous hydrogen ion; great inertness to chemical, physical, and environmental interferences; and wide rangeability, covering the approximate 14 orders of magnitude in hydrogen ion concentration — from strong acid to strong base.

Advantages: Dip-type or flow-through designs are available, with high selectivity, sensitivity, and environmental immunity.

Advantages: Dip-type or flow-through designs are available, with high selectivity, sensitivity, and environmental immunity.

Commercially available probes for measurement of pH in industrial processes contain three essential integral components: (1) the pH-sensitive probe (electrode) with its internal filling solution and reference element, (2) a reference probe (electrode) with a salt bridge that contacts the sample by a restrained diffusion barrier, and (3) a thermistor or resistance temperature detector (RTD) to measure process temperature and compensate for its effect on measured pH. The electrode probes constitute the two elements of a galvanic cell, which develops a millivolt signal that is proportional to the pH of the solution. Because this millivolt signal is developed across the high electrical resistance of the pH-sensitive glass, these probes often contain a preamplifier to allow transmission of the pH signal over appreciable distance to the pH analyzer (up to 1,000 ft [600 m]).

The probes are available in chemically resistant materials with diverse process mounting configurations that allow them to be dipped into weirs, flumes, ponds, or tanks. They can also be placed in-line by means of standard pipe fittings. Probes are also available with optionally-fitted mechanical cleaners, such as electrically-actuated brushes, to provide automatic cleaning of interfering coatings and deposits. This is perhaps the most significant maintenance requirement of these probes in continuous monitoring and control applications.

This chapter will provide an introduction to pH theory of aqueous electrolytes as it relates to the operational function of the pH electrode. The pH meter equation will be introduced, and operational parameters that need to be controlled for accurate pH measurement will be discussed in detail. Finally, instrumental aspects of measurement amplifiers and intelligent pH analyzers will be presented. Table 25-1 provides a list of manufacturers of industrial pH analyzers.

Table 25-1. Major U.S. Manufacturers of pH Analyzers

Manufacturer
The Foxboro Company 600 N. Bedford St. E. Bridgewater, MA 02333
Great Lakes Instruments 9020 W. Dean Rd. Milwaukee, WI 53224
Honeywell Inc. 795 Horsham Rd. Horsham, PA 19044
Johnson-Yokogawa 2 Dart Rd. Shenandoah, GA 30267
Rosemount Analytical 2400 Barranca Pky. Irvine, CA 92714
Signet Scientific 3401 Aerojet Ave. El Monte, CA 91734
TBI-Bailey Controls 2175 Lockheed Way Carson City, NV 89709

The pH Scale

Aqueous solutions are governed by an ionic equilibrium relationship, Equation (25-1), which states that the mathematical product of the total concentration of hydrogen ion and hydroxide ion at a given temperature is a constant:

$$K_W = [H^+][OH^-] \tag{25-1}$$

This constant, K_W, the ion product constant for water, has a value of 1.008×10^{-14} at 25°C. Values at other temperatures are shown in Table 25-2.

The well-known pH scale is derived from this relationship. By definition, pH = – log $[H^+]$. Using similar notation for the other terms in Equation (25-1), we write:

$$pK_W = pH + pOH, \text{ where } pK_W = 13.996 \text{ at } 25°C \tag{25-2}$$

For neutral solution, the concentration of hydrogen ion, chemically an acid, and the concentration of hydroxide ion, chemically a base, are equal. From Equation (25-1):

$$[H^+] = [OH^-] = (K_W)^{1/2} = 1.008 \times 10^{-7} \text{ at } 25\ °C \tag{25-3}$$

or equivalently, from Equation (25-2):

$$pH = pOH = 1/2pK_W = 6.998 \text{ at } 25°C \tag{25-4}$$

Other definitions are apparent from these relationships. For an acid solution:

$$[H^+] > [OH^-], \text{ or } pH < pOH \tag{25-5}$$

For a basic solution:

$$[OH^-] > [H^+], \text{ or } pH > pOH \tag{25-6}$$

The following example calculations show the interconversion of $[H^+]$ and pH.

(a) $[H^+] = 2.0 \times 10^{-5}\ M$; calculate the pH.

$$pH = -\log 2.0 \times 10^{-5} = \log (1/2.0 \times 10^{-5}) = \log(1/0.20 \times 10^{-4})$$

$$= \log(5.0 \times 10^{4}) = \log 10^{4} + \log 5.0 = 4.00 + 0.70 = 4.70$$

(b) pH is 5.92; calculate $[H^+]$

$$[H^+] = 10^{-5.92} = 10^{0.08} + 10^{-6} = 1.2 \times 10^{-6} M$$

(c) A solution 10 *M* in acid would have a pH of –1 and a pOH of 15; likewise, 10 *M* base would have a pH of 15 and a pOH of –1. In reality, the usable range of electrometric pH measurement is somewhat less than this — usually 2 to 12 pH, depending on the accuracy required.

It should be noted that the pH of the neutral point is not defined as a solution having pH = 7.00, but as a solution having equal concentrations of hydrogen and hydroxide ion. Neutral pH at other temperatures, calculated using corresponding values for the ion product of water, varies from about 7.5 to 6.1 over the temperature range of 0 to 100°C. Values are shown in Table 25-2.

Solution Principles

Chemical materials that may be present in process samples may make the concentration of hydrogen ion observed with the pH electrode less than actually present. Examples are weak acids, weak bases, and, sometimes, complex ions. An example is acetic acid, which only partially dissociates:

$$HA = H^+ + A^- \qquad (25\text{-}7)$$

This equilibrium lies predominantly to the left. Free hydrogen ion, that which is measurable using the pH electrode, is much less than the total concentration of acid, *HA*, in the sample.

An equilibrium expression can be written to quantitatively express this process:

$$K_a = [H^+][A^-]/[HA] \qquad (25\text{-}8)$$

K_a, the acid dissociation constant for various weak acids and bases, is tabulated in various sources (e.g., [Ref. 1]).

Solving Equation (25-8) for $[H^+]$ gives:

$$[H^+] = K_a\,[HA]/[A^-] \qquad (25\text{-}9)$$

This expression shows that, in the presence of weak acid, pH or $[H^+]$ depends on the value of K_a as well as the ratio $[HA]/[A^-]$. Similar considerations apply to samples having weak bases, such as ammonium hydroxide or sodium carbonate, for which a similar equilibrium expression can be written in terms of the basic dissociation constant K_b and $[B]/[HB^+]$:

$$[OH^-] = K_b\,[B]/[HB^+] \qquad (25\text{-}10)$$

The following example calculations show the effects of a weak acid (acetic acid) on measurement of a strong base, and a weak base (ammonium hydroxide) on strong acid.

Calculate the pH of a 50-mL sample of 0.1 *M* acetic acid after addition of 10 mL of 0.1 *M* NaOH; $K_a = 1.75 \times 10^{-5}$.

One-fifth of the acetic acid is converted to sodium acetate; therefore, from Equation (25-9):

$$[H^+] = 1.75 \times 10^{-5} \times (0.4/0.1) = 7.00 \times 10^{-5}$$
$$\text{pH} = 4.15$$

If this same quantity of strong base had been added to 50 mL of pure water:

$$[OH^-] = (10\ \text{mL})(0.1\ M)/60\ \text{mL} = 0.02 = 2.0 \times 10^{-2}$$
$$\text{pOH} = 1.70,\ \text{pH} = 12.3$$

This shows that the presence of the weak acid prevented the change of pH that would be expected if this quantity of strong base was added to a water solution without it.

Calculate the pH of a 50-mL sample of 0.1 *M* ammonium hydroxide after addition of 25 mL of 0.1 *M* HCl; $K_b = 1.75 \times 10^{-5}$.

One-half of the ammonium hydroxide is converted to ammonium chloride; therefore, from Equation (25-10):

$$[OH^-] = K_b = 1.75 \times 10^{-5}$$
$$\text{pOH} = 4.75,\ \text{pH} = 9.25$$

If this quantity of strong acid had been added to 50 mL of pure water:

$$[H^+] = (25\ \text{mL})(0.1\ M)/(75\ \text{mL}) = 0.033 = 3.3 \times 10^{-2}$$
$$\text{pH} = 1.48$$

As above, weak base present in the solution prevented the pH change that would otherwise have been expected.

Reference 2 provides a comprehensive mathematical treatment of ionic equilibria.

Activity and Concentration

Rigorous quantitative descriptions of chemical equilibria and electrode measurement of dissolved ions require introduction of the concept of ionic activity. Electrodes measure activity, not concentration. So far in this discussion we have used concentrations in the equations describing chemical equilibria. Concentration of solutions, expressed in molar units (i.e., moles per liter), represent weights of pure materials (e.g., grams/gram molecular weight, expressed in grams/mole) dissolved in 1 L of water. This is the true analytical concentration of material in the sample, denoted by c_i. Temperature is defined usually at 25°C.

Activity is commonly called "effective" concentration. When salt or electrolyte is put into solution, the chemistry of its ions is affected by the presence of all other ions that may be present, through ion-ion interaction, hydrolysis, ion complexation, or whatever. In very dilute solutions these interactions are less important, and activity and concentration are virtually equivalent. Activity and concentration are related by:

$$a_i = c_i f_i \tag{25-11}$$

where f_i, a dimensionless quantity called the activity coefficient, measures the ratio of ion activity, a_i, to ion concentration, c_i. Activity coefficients are primarily determined by the total ionic strength of the solutions (I), which is defined by:

$$I = 1/2 \sum c_i z_i^2 \tag{25-12}$$

where c_i is the concentration and z_i is the charge of each ion.

Values of activity coefficient for acids and bases have been tabulated (e.g., [Ref. 1]). They vary from unity for very dilute solutions to about 0.9 at 0.005 M and about 0.8 at 0.1 M ionic strength. At higher ionic strengths, they may decrease significantly and may even progress to values greater than unity for very high ionic-strength solutions.

The effect of activity coefficient on measurement accuracy is minimized by performing pH probe standardization (described later in this chapter) using buffer solutions having ionic strength similar to that of the sample.

Table 25-2. Ion Product Constant, K_W and pK_W, and pH of the Neutral Point for Water at Various Temperatures [Ref. 1]

Temperature, °C	Ion Product Constants		pH of the Neutral Point
	K_W	pK_W	
0	0.114×10^{-14}	14.943	7.472
5	0.185	14.734	7.367
10	0.292	14.535	7.268
15	0.451	14.346	7.173
20	0.681	14.167	7.084
25	1.008	13.996	6.998
30	1.469	13.833	6.917
35	2.089	13.680	6.840
40	2.919	13.535	6.768
45	4.018	13.396	6.698
50	5.474	13.262	6.631
55	7.297	13.137	6.569
60	9.614	13.017	6.509
65	12.6	12.90	6.45
70	15.9	12.80	6.40
75	20.4	12.69	6.35
80	25.1	12.60	6.30
85	30.9	12.51	6.26
90	38.0	12.42	6.21
95	46.8	12.34	6.17
100	55.0	12.26	6.13

Typical pH Ranges

Approximate pH values of some common materials are shown in Figure 25-1.

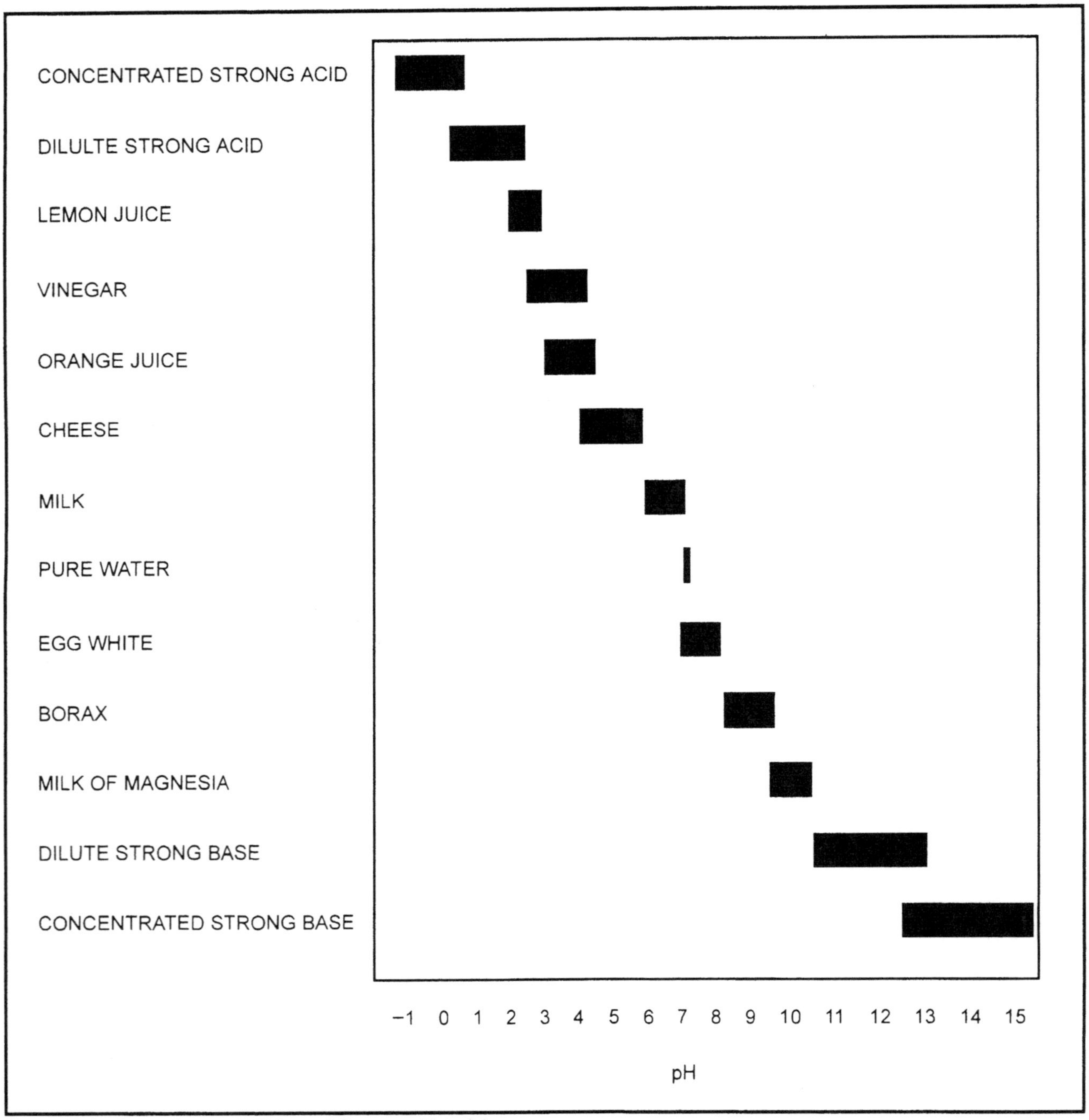

Figure 25-1. Approximate pH Values of Common Materials

Buffer Solutions

Equation (25-9) describes the behavior of a pH buffer solution. Buffer solutions, made by combination of weak acids and their salts, or weak bases and their salts, have the property that they resist change in pH.

Inspection of Equation (25-9) shows that when $[HA] = [A^-]$, their ratio is 1 and $[H^+] = K_a$, or $pH = pK_a$. If the total concentration of weak acid (i.e., $[HA] + [A^-]$) is large, then their ratio is little affected as acid or base is added to the system. Such solutions, called buffer solutions, have stable pH and are widely used for standardization of pH analyzers.

Buffer solutions, as certified reference materials (except $Ca(OH)_2$), are available from the National Institute of Science and Technology (NIST) or as reagent-grade materials from various laboratory supply houses and many pH analyzer manufacturers. Table 25-3 lists pH values over the temperature range of 0 to 90°C for the most commonly employed buffer solutions.

Table 25-3. pH of Commonly Employed Buffer Solutions at Various Temperatures [Ref. 1]

Temperature, °C	Potassium Oxalate	Potassium Tartrate	Potassium Phthalate	Potassium Phosphate	Borax	Calcium Hydroxide
0	1.67	...	4.01	6.98	9.46	13.43
5	1.67	...	4.01	6.95	9.39	13.21
10	1.67	...	4.00	6.92	9.33	13.00
15	1.67	...	4.00	6.90	9.27	12.81
20	1.68	...	4.00	6.88	9.22	12.63
25	1.68	3.56	4.01	6.86	9.18	12.45
30	1.69	3.55	4.01	6.85	9.14	12.30
35	1.69	3.55	4.02	6.84	9.10	12.14
40	1.70	3.54	4.03	6.84	9.07	11.99
45	1.70	3.55	4.04	6.83	9.04	11.84
50	1.71	3.55	4.06	6.83	9.01	11.70
55	1.72	3.56	4.07	6.84	8.99	11.58
60	1.72	3.56	4.09	6.84	8.96	11.45
70	1.74	3.58	4.12	6.85	8.93	...
80	1.77	3.61	4.16	6.86	8.89	...
90	1.80	3.65	4.20	6.88	8.85	...

For most applications, the following buffers can be made with reagent-grade chemicals.

Potassium Oxalate (0.050 *M*). Dissolve 12.70 g of potassium tetroxalate dihydrate, $KH_3(C_2O_4)_2 \cdot 2H_2O$, in distilled water and dilute to 1 L at 25°C.

Potassium Acid Tartrate (Saturated at 25°C). Saturate distilled water with excess salt at 25 ± 3°C and remove excess solid prior to measurement. This solution is approximately 0.034 *M*.

Potassium Acid Phthalate (0.050 *M*). Dissolve 10.21 g of potassium acid phthalate, $KHC_8H_4O_4$, in distilled water and dilute to 1 L at 25°C.

Potassium Phosphate Mixture. Dissolve 3.40 g of potassium dihydrogen phosphate, KH_2PO_4, and 3.55 g of sodium monohydrogen phosphate, Na_2HPO_4, in distilled water and dilute to 1 L at 25°C. Note: Na_2HPO_4 absorbs water and thus should be dried at 110°C for several hours and cooled in a desiccator prior to weighing out the quantity required.

Borax (0.010 *M*). Dissolve 3.81 g of sodium tetraborate decahydrate, $Na_2B_4O_7 \cdot 10H_2O$, in distilled water and dilute to 1 L at 25°C. This solution is susceptible to contamination by CO_2 in the atmosphere. It should be kept tightly sealed when not in use.

Calcium Hydroxide (Saturated at 25°C). Add a considerable excess of freshly opened, finely granular, reagent-quality calcium hydroxide to distilled water and shake or stir vigorously. After adjustment of temperature to 25°C ± 2C°, filter suspended material with a medium-porosity sintered glass filter. The solution must be protected from CO_2 and should be replaced when turbidity develops.

The pH Measuring Cell

A schematic diagram of a typical pH measuring cell is shown in Figure 25-2. Essential components are two reference electrodes, a pH-responsive glass membrane, a salt bridge, and a liquid junction. When this cell is connected as shown to a suitable voltmeter, an electrical potential can be measured. A cell of this type is a galvanic cell; the potential developed depends on the concentrations of substances involved in electrode reactions. Contributions to the total measured potential come from the several liquid junctions (solution interfaces), denoted by bold vertical bars in Figure 25-2. One interface resides on each side of the glass membrane, at each reference electrode, and at the liquid junction. Each of these elements generates a potential that contributes to the total measured cell potential. Each will be discussed to provide an understanding of the analytical application of this cell to pH measurement.

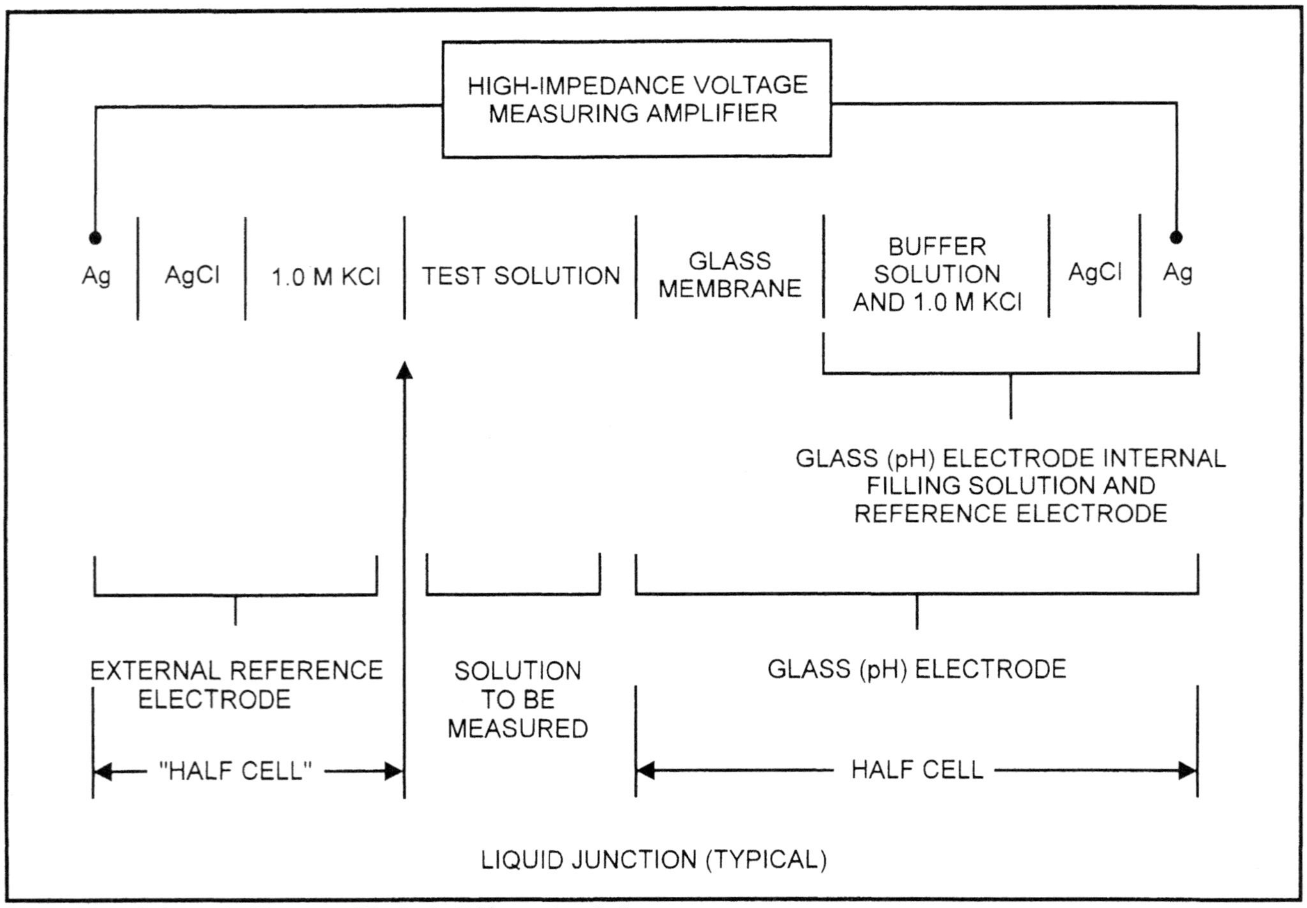

Figure 25-2. Schematic Diagram of Typical pH Measuring Cell

The pH-Sensitive Membrane

The pH response is provided by the unique property of various glass compositions to develop potentials related to the concentration or activity of hydrogen ion (H^+) in the solution to which they are exposed. By forming the glass as a thin membrane interposed between solutions of dissimilar pH, the potentials developed on each surface combine to generate a potential gradient across the membrane which is exactly described by a form of the Nernst equation. The operational form of this equation is simply:

Nernst Equation

$$E_g = 0.19841T \log\{[H^+]_1/[H^+]_2\} \quad (25\text{-}13)$$

where E_g is the potential generated across the membrane, T is the temperature in degrees Kelvin, and $[H^+]_1$ and $[H^+]_2$ are the molar concentrations of hydrogen ion in the two solutions.

Equation (25-13), which is fundamental to analytical potentiometry with ion-selective electrodes in general, simply states that the measured potential varies logarithmically with the ratio of the concentrations of the measured ion and has a slope, termed the Nernst coefficient, determined solely by temperature. Values of this slope, precisely calculated at various temperatures, are given in Table 25-4.

Glass compositions yielding potentiometric response to hydrogen ion are generally mixtures of oxides of elements of oxidation states 3 or greater (e.g., silicon or aluminum) with elements of oxidation states 1 or 2 (e.g., lithium, sodium, or calcium). Other elements may be present to effect physical properties of the resulting glass structure. In the solid state, the monovalent ions are the electrically-mobile species. Their presence is required to allow the minute flow of current required for the measurement of potential. Conventional electrodes generally present an exposed area of about 1 cm^2 to the solution, are about 1 mm thick, and present typical resistance in the range of 10 to 100 mΩ to the voltmeter.

Table 25-4. Values of the Nernst Coefficient, 0.19841 $T^{(a)}$, from 0 to 100°C

Temperature, °C	Nernst Coefficient (Slope), mV/decade pH
0	54.197
5	55.189
10	56.181
15	57.173
20	58.165
25	59.157
30	60.149
35	61.141
40	62.133
45	63.126
50	64.118
55	65.110
60	66.102
65	67.094
70	68.086
75	69.078
80	70.070
85	71.062
90	72.054
95	73.046
100	74.038

(a) $T = t\,(°C) + 273.15$

These membranes provide fixed anionic sites within the glass, facilitating electrolytic transport of cationic charge (i.e., Li^+ or Na^+) through the membrane. Dissimilar concentrations of hydrogen at the two faces, in equilibrium with these anionic sites, are the source of the observed potential. The degree to which this ion-exchange reaction is specific to the ion of interest determines the selectivity of the membrane. Theoretical descriptions of the pH function of glass suppose the presence of a hydrated gel layer on the glass surface in which this ion-exchange process occurs. Accurate and precise response requires that this surface be rigorously maintained. Many compositions are available with high selectivity for the hydrogen ion, and this forms the basis for the large commercial availability of glass pH probes.

Reference Electrodes

Reference electrodes at either side of the pH membrane (Figure 25-2) provide the electrolytic contacts necessary to allow measurement of the potential gradient across the membrane. They must meet three interrelated criteria: reversibility, reproducibility, and stability. Reversibility measures their ability to support current flow without change of potential. Reproducibility measures their ability to track changes in temperature as predicted by the Nernst equation without measurement lag (hysteresis). Finally, stability measures the useful life of the electrodes as reflected in their ability to maintain stable potentials over the life of the cell.

Common reference electrodes, silver/silver chloride (Ag/AgCl) and mercury/mercurous chloride (Hg/Hg_2Cl_2 [calomel]), are bathed in solutions having fixed concentrations of chloride, usually as the potassium salt. Calomel, because of the toxicity of mercury, is becoming less popular for use in pH measuring cells, even though it is somewhat the more stable of the two. Electrode reactions associated with these half-cells are

$$2Ag + 2Cl^- = 2AgCl + 2e^- \quad (25\text{-}14)$$

$$2Hg + 2Cl^- = Hg_2Cl_2 + 2e^- \quad (25\text{-}15)$$

These reactions allow electron flow to the voltmeter while their chemical equilibria resist change of potential when current is drawn through the cell. Electrodes with this property are termed well-poised or nonpolarizable.

If identical electrodes, in thermal equilibrium, are used in the pH cell, then the potential and temperature coefficient developed by each is canceled in the differential measurement of cell potential. It is difficult to design probes where close thermal equilibrium between the two reference elements is maintained, particularly in processes that undergo rapid temperature cycling. This can lead to errors and instability in measurements.

The internal filling solution of the pH half-cell (right side of Figure 25-2) has a fixed concentration of hydrogen ion, $[H^+]_1$, and a fixed concentration of electrolyte salt (e.g., a pH buffer solution containing KCl and saturated with AgCl in contact with Ag wire). The reference probe half-cell (left side of Figure 25-2) has similar construction. It too uses a filling solution containing the chloride electrolyte saturated with AgCl. Here, however, the electrolyte must also contact the sample. This is done by means of a salt bridge containing a relatively large concentration of electrolyte and a constricted diffusion liquid junction, such as a porous ceramic, cloth wick, or small hole.

It is worth noting that AgCl or Hg_2Cl_2, which have finite solubility in the KCl electrolyte must be excluded from this salt bridge to prevent them from contaminating the liquid junction. Therefore, the reference electrode often has a second junction, in close proximity to the reference element, that defines a chamber containing KCl and the Ag or Hg metal salt. Outside this chamber is the salt bridge, usually containing KCl but possibly other electrolytes, if KCl at the process junction is undesirable.

In some commercially available salt bridges, the electrolyte is gelled to help inhibit ingress of process sample and loss of internal electrolyte.

In some commercially available salt bridges, the electrolyte is gelled to help inhibit ingress of process sample and loss of internal electrolyte.

Reference Probe Liquid Junction

The liquid junction between the salt bridge and the process sample is required to define an ionic diffusion barrier between these dissimilar solutions and is a source of an electrical potential that gives an unavoidable pH measurement error. The magnitude of this error can be small and stable or large and erratic.

The dissimilar ionic concentrations across this barrier establish a nonequilibrium flux of ions due to their concentration gradient. They move at different velocities, or mobilities, from the concentrated to the dilute phase. If the diffusing salt is an equitransferrent electrolyte — that is, one whose co-ions have equivalent ionic mobilities — then the potential developed is small. Typical salts of this type are KCl and NH_4NO_3. However, if the electrolyte contains ions of nonequivalent mobilities, such as HCl, then significant potential may occur across the junction. The mobility of H^+ is significantly greater than Cl^-, so when HCl diffuses across a concentration gradient, H^+ moves much faster than Cl^-, giving a negative charge to the concentrated phase and a positive charge to the dilute phase. Likewise, for NaOH, OH^- leaves the concentrated phase positively charged. Because of its origin, this interfacial potential is sometimes called a diffusion potential. Approximate values of liquid junction potentials for several solutions against a saturated KCl salt bridge are shown in Table 25-5.

Ceramic or other types of constrained barriers used as liquid junctions in commercially available probes may entrap materials such as HCl or NaOH. Its presence there can create unpredictable and unstable junction potentials as the composition of the process solution changes.

Table 25-5. Approximate Liquid Junction Potential, E_j, at 25°C (a) [Ref.1]

Solution X	E_j, mV	pH Error
HCl, 1 *M*	–14.1	+0.24
HCl, 0.1 *M*	–4.6	+0.08
HCl, 0.01 *M*	–3.0	+0.05
K acid phthalate, 0.05 *M*	–2.6	+0.05
KCl, 0.1 *M*	–1.8	+0.03
KCl, 1 *M*	–0.7	+0.01
KCl, saturated	0.0	0.00
$NaHCO_3/Na_2CO_3$, 0.025 *M*	–1.8	+0.03
NaOH, 0.01 *M*	–2.3	+0.04
NaOH, 0.1 *M*	0.4	–0.01
NaOH, 1 *M*	8.6	–0.15

(a) Junction: Saturated KCl/Solution X

The magnitude of this potential is fixed by the identity of the diffusing ions. Its stability, however, is determined by the method of forming the barrier. To establish a fixed and stable diffusion gradient, the solutions to either side of the barrier should be quiet and unstirred. Further, the ions should move along a linear concentration gradient; that is, they should be shielded from sample solution flow and turbulence. These are very difficult criteria to realize in pH cell designs suitable for industrial applications. The type of junction most commonly found in commercial reference probes, the porous frit, is a compromise of ruggedness and utility with junction stability.

Temperature cycling of the process may produce process solution thermal pumping into, and electrolyte solution thermal pumping out of, the reference solution chamber. This phenomenon may shorten useful cell life by creating unstable junction potentials and contamination of the reference electrolyte. The effect may be reduced by using long salt-bridge path lengths and junctions with higher flow restrictions.

Salt bridges with pressurized reservoirs of electrolyte offer some advantages. They are used in difficult applications to continuously push fresh electrolyte through porous junctions or small holes designed to allow flow rates in the range of 2 to 5 mL/day. They may provide the best solution in critical measurement applications requiring high accuracy, stability, and immunity from fouling. Unfortunately, they are seldom applied due to their relative complexity, the difficulty of maintaining isothermal conditions between the reference and pH elements of the cell, and these elements' need for frequent maintenance.

Potentiometric Equation for the pH Cell

E_g glass membrane potential
E_j liquid membrane potential

Based on the above, the total potential measured using the cell shown in Figure 25-2 is given by:

$$E_{cell} = E_g + E_j = 0.19841T \log \{[H^+]_1/[H^+]_2\} + E_j \qquad (25\text{-}16)$$

where E_g and E_j are the potentials across the glass membrane and liquid junction, respectively. We have assumed identical reference electrodes so their potentials cancel and need not be included.

The pH probe is constructed by filling the pH half-cell with a solution having reference electrolyte, KCl saturated with AgCl for the reference electrode, and a pH buffer to fix $[H^+]_1$ within the glass electrode. Equation (25-16) can be rewritten to show that the measured potential is now a function only of $[H^+]_2$ and E_j, the other term being constant. Thus:

$$E_{cell} = \text{const} + 0.19841T \log[H^+]_2 + E_j \qquad (25\text{-}17)$$

The adherence of the potential generated by the glass pH electrode to that predicted by Equation (25-17) over the full range of pH values exhibited in the aqueous world is nothing short of remarkable, although some conditions do produce error. Sodium error, or, more accurately, alkaline error, represents a degree of membrane response to large concentrations of alkali ions in solutions of high pH. Sodium is most common, hence sodium error, although lithium and potassium exert the same effect at degrees related to their ionic sizes. This apparent loss of selectivity results from competition with hydrogen ions, in short supply at high pH, with the chemically similar alkali ions for the ion-exchange sites in the glass. The error increases significantly with temperature. Special compositions of glass, called low-sodium-error or high-pH-range compositions, limit this error in applications requiring its management; trade-offs include lack of ruggedness and higher resistance. Manufacturer's literature should be consulted for quantitative application data.

A second error is noted at the other extreme of pH. The acid error, in solutions of strong acids (pH 1 or less), causes measured pH to be somewhat greater than true pH. Its magnitude increases with concentration of acid and often with time of exposure. The source of this error has received considerable study, but a definitive conclusion is not yet available. In measurement applications where this error is important, consideration should be given to alternate measurement techniques, such as conductivity or perhaps automated titration.

The pH Meter Equation

Potentiometric measurement of the pH of sample solutions is conveniently accomplished by a comparative procedure; the pH measuring cell with associated analyzer is first standardized or calibrated with buffer solutions of known pH. Then the potential of the cell in contact with the sample is measured. The methodology for this is based on the pH meter equation, which is derived from Equation (25-17), now rewritten as:

$$E = \text{const} - 0.19841T\text{pH} + E_j \qquad (25\text{-}18)$$

To measure pH of an unknown solution, the potential of the cell, E_S, is first determined using a buffer of known pH_S. The cell, or probe, is then placed in the unknown solution and pH_X is determined after measurement of the new cell potential, E_X. Writing Equation (2-18) in terms of pH_s and E_S, and pH_X and E_X, and subtracting gives:

$$pH_x = pH_s + \frac{1}{0.19841T}(E_x - E_s) + (E_{js} - E_{jx}) \qquad (25\text{-}19)$$

This is the pH meter equation. After standardizing with buffer, the observed ΔE (or $E_X - E_S$) is divided by the known slope factor, mV/decade pH, at the measurement temperature (Table 25-4), to indicate the sample pH. Plots of the pH meter equation, E versus pH, at several temperatures are shown in Figure 25-3. Note that the curves intersect at an isopotential point and show increased slope with temperature.

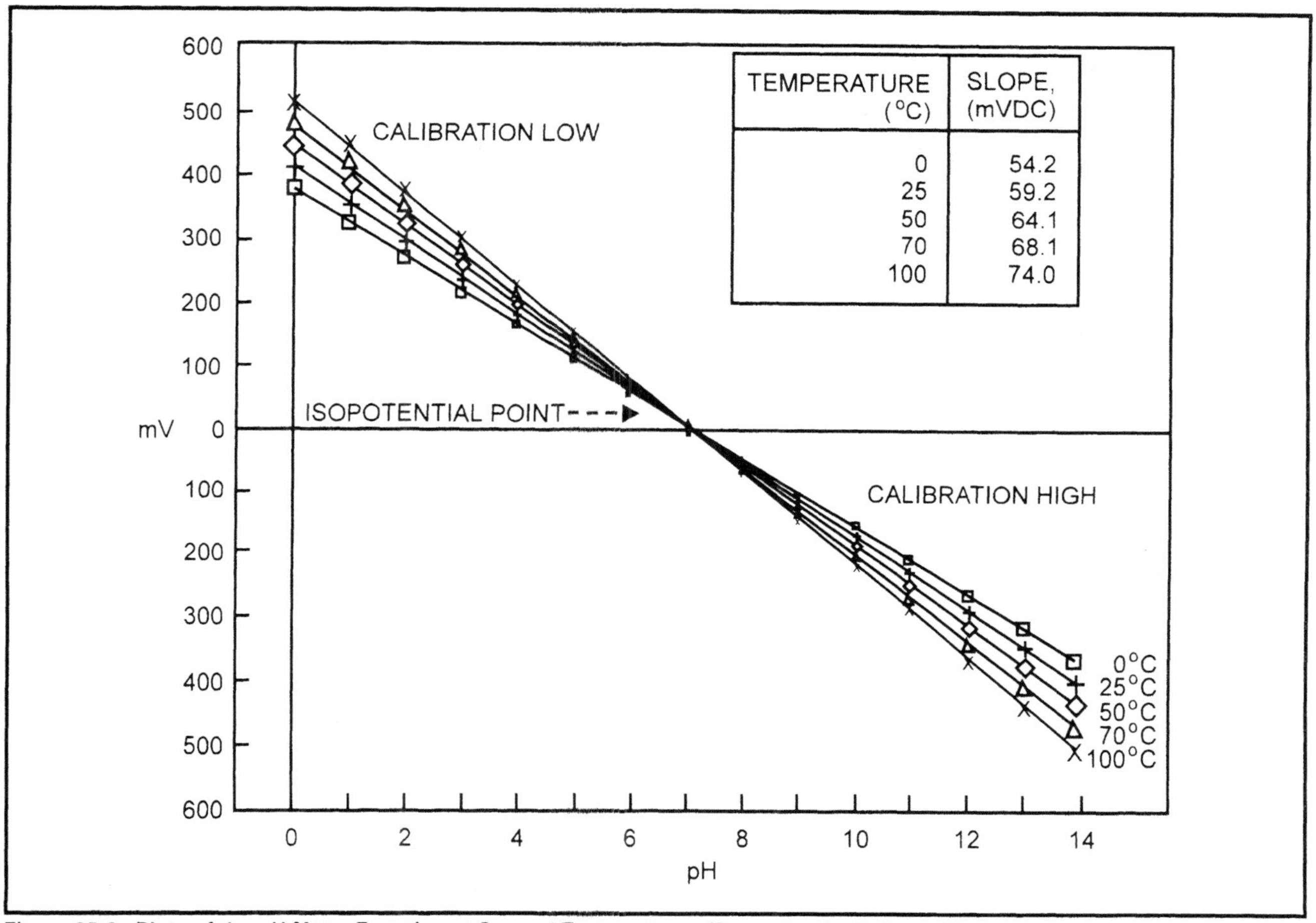

Figure 25-3. Plots of the pH Meter Equation at Several Temperatures (For a standard glass pH sensor and Ag/AgCl reference electrode)

The term $(E_{js} - E_{jx})$, the residual junction potential, is an unavoidable consequence of measuring pH with electrochemical cells of the type we are discussing. For liquid junctions with saturated KCl, its value may be as high as 14 mV (0.24 pH) against 1 *M* HCl and 0.9 mV (0.15 pH) against NaOH. Higher errors result when using salt bridge electrolytes having lower salt concentrations. At the other extreme, for low ionic strength solutions such as boiler feedwater and soft fresh water, considerable, unpredictable, and erratic residual liquid junctions may result when using the ceramic frit junctions commonly incorporated in measuring cells.

The consequence of the residual junction on the isopotential point is to move it along the abscissa, to the right or left in Figure 25-3, depending on its sign and magnitude.

pH Measurement Temperature Dependency

Temperature affects electrometric pH measurement in two ways — through its effect on the electrodes and on the solutions.

pH CELL TEMPERATURE DEPENDENCY

Only one temperature effect is normally compensated in modern pH analyzers: the Nernst slope term (Table 25-4). The other lies in the $(E_x - E_s)$ term of the pH meter equation (Equation [25-19]). The slope term is controlled either manually by the operator or automatically using electronic circuitry within the analyzer from measured values supplied by a temperature resistor or thermistor mounted in the probe.

Note in Equation (25-16) that when $[H^+]_{ext} = [H^+]_{int}$, the log term goes to zero and the potential generated by the membrane is zero and independent of temperature. The pH of the internal electrolyte defines this isopotential point (see Figure 25-3). This pH is chosen, by convention, usually as neutral (pH 7.00) in commercially-available electrodes. If the probe is immersed in a solution having the pH of its internal filling solution, a nonzero potential will manifest as a temperature variation in the terms E_x and E_s of the pH meter equation. Manual adjustment of the meter to 0 mV output may be provided with an adjustment on many pH analyzers. This nonzero output at the isopotential point rises from nonideality of the glass membrane (termed its asymmetry potential), differences in composition of internal and external solutions, nonideality of the reference electrodes, and the residual liquid junction potential.

Temperature compensation for sample solutions at or near the isopotential point, pH 7.00, is not important since the temperature coefficient here is 0.00 mV/C°. However, for this same electrode, the temperature coefficient is 1.39 mV/C° (i.e., 0.024 pH/C°) in samples at pH 0 or 14. As the sample pH moves away from the isopotential pH, accuracy of temperature measurement and compensation for it becomes significant.

SOLUTION TEMPERATURE DEPENDENCY

Two dependencies are noted for solutions. The first rises from their volumetric expansion with temperature. The result is that concentration, expressed in molar units, decreases by approximately 0.04%/deg as the temperature is increased.

The second dependence rises from the effect of temperature on ionic equilibria and their associated equilibrium constants and activity coefficients. For strong acids and bases, the known temperature dependence of the ion product of water (Table 25-4) can be used to construct these temperature corrections. For some weak acids and bases, particularly those used as standard buffer solutions and listed in Table 25-3, values have been determined experimentally. Data of this type are specific to the sample being measured and can be entered into modern microprocessor-based pH analyzers. This function provides additional compensation when required.

The best practice for reducing temperature errors from all sources is to standardize and measure solutions at the same temperature.

A Few Words About pH Control

pH control, usually applied in neutralization during waste treatment, can be a difficult problem due to the nonlinear response of the pH cell to concentration of acid or base and also to dead time associated with mixing (and often dissolving) of reagents that are added to adjust pH toward preselected set points.

We have already seen from Equations (25-9) and (25-10) that weak acids and bases resist change in pH. The value of pH is little affected when strong acid or base is added to solutions where the ratios $[HA]/[A^-]$ and $[B]/[BH^+]$ are in the range of about 0.1 to 0.9. However, as these ratios move out of this "buffering" range during neutralization, the pH will change at an increasing rate with reagent flow. To control the pH of the process at a set point in the range where this pH changes so rapidly requires extremely efficient mixing, rapid pH probe response, a large range of control of the reagent metering valve, and special attention to the controller and its tuning. A comprehensive treatment of the topic is available in Ref. 3.

Commercially-Available pH Probes

Diagrams of typical commercially-available pH probes and cells are shown in Figure 25-4. Cells using discrete and separate pH probes with reference probes (Figure 25-4a) grew out of their laboratory counterparts and are still used. Today, many of these cells, both for laboratory and process, are manufactured as combination probes with various configurations (Figures 25-4b and 25-4c).

In both figures 25-4(b) and 25-4(c), the glass electrode has a sealed inner chamber that contains a buffered filling solution and a reference electrode. Because of the high electrical impedance of the glass, this internal reference electrode is connected to the outside world by shielded cable (Figure 25-4a), or may be connected directly to the input of an integral high-input impedance amplifier, commonly available within cells of the type shown in Figures 25-4(b) and 25-4(c). Likewise, the reference probes consist of common elements, the liquid junction, salt bridge, and reference electrode element connected to a simple insulated wire leading to the pH analyzer.

Fermentation Application

Probes of the type shown in Figure 25-4(a) are particularly suited to fermentation, where periodic high-temperature sterilization is required. Glass probes with membrane compositions able to withstand these temperature extremes without loss of pH function are available. The reference probe, with its small ceramic junction, may be filled with sterile electrolyte and then pressurized to maintain clean electrolyte in the salt bridge and stable flow into the fermentation medium. Isolation of the reference electrode from the liquid junction during the heat sterilization phase helps extend the useful life of this probe.

The combination probe shown in Figure 25-4(b) is designed for ruggedness. Its flat surface contains a flat glass membrane with significantly greater thickness than the bulb types shown in Figures 25-4(a) and 25-4(c). The ceramic junction is also on this surface. This design offers resistance to breakage and reduced fouling because the sensitive membrane is not exposed to entrained solids in the process. There is a trade-off, however. Increased thickness of the glass requires compositions having low intrinsic resistivity, and these turn out to have high sodium ion errors. Such electrodes are therefore restricted to use in pH ranges from 2 to 12. The combination probe shown in Figure 25-4(c) is suitable for wide-range pH measurement with low sodium error. The sensing electrode is a thin membrane of conventional spherical design. The sensing and reference electrode with integral filling solutions and ceramic junction are field replaceable. The pH glass-sensing electrode is sealed within a chemically inert plastic cartridge that plugs into the process probe. Sealing is effected using O-rings, and electrical connection is made with an integral contact. Likewise, the ceramic junction is molded into a chamber that seals into the probe salt bridge, again using an O-ring. This greatly facilitates membrane and liquid junction replacement in the field.

High-Temperature (Fermentation) pH Electrode

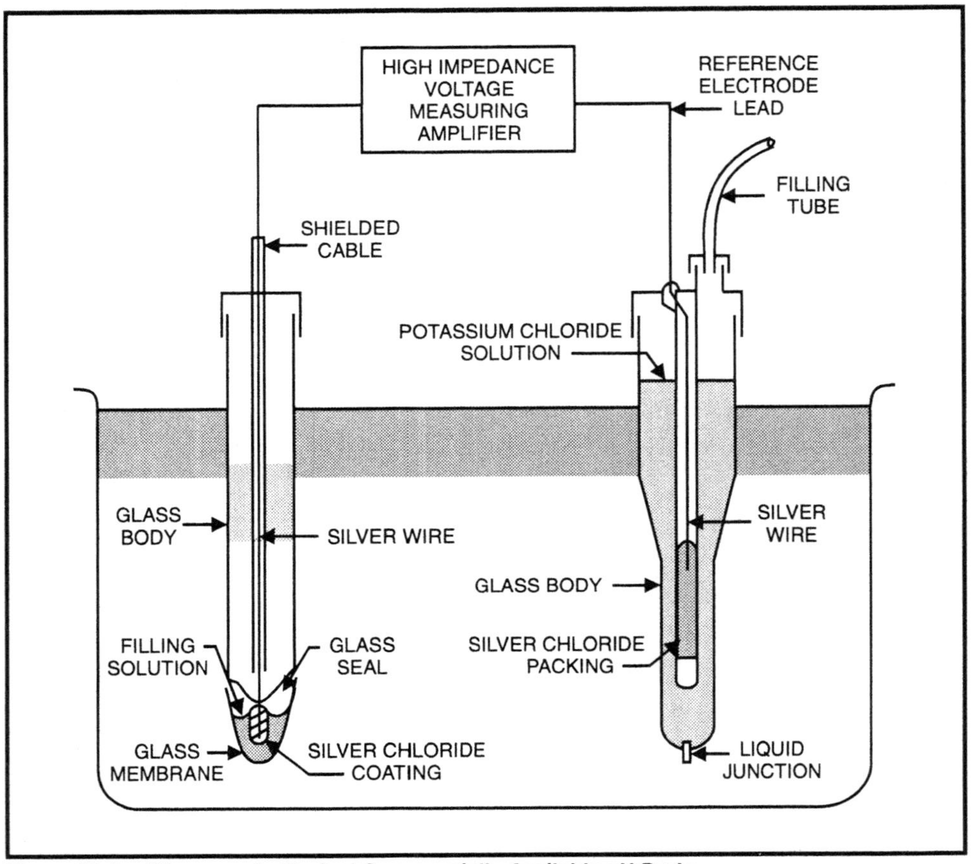

Figure 25-4a. Diagram of a Typical Commercially-Available pH Probe

Rugged pH Electrode

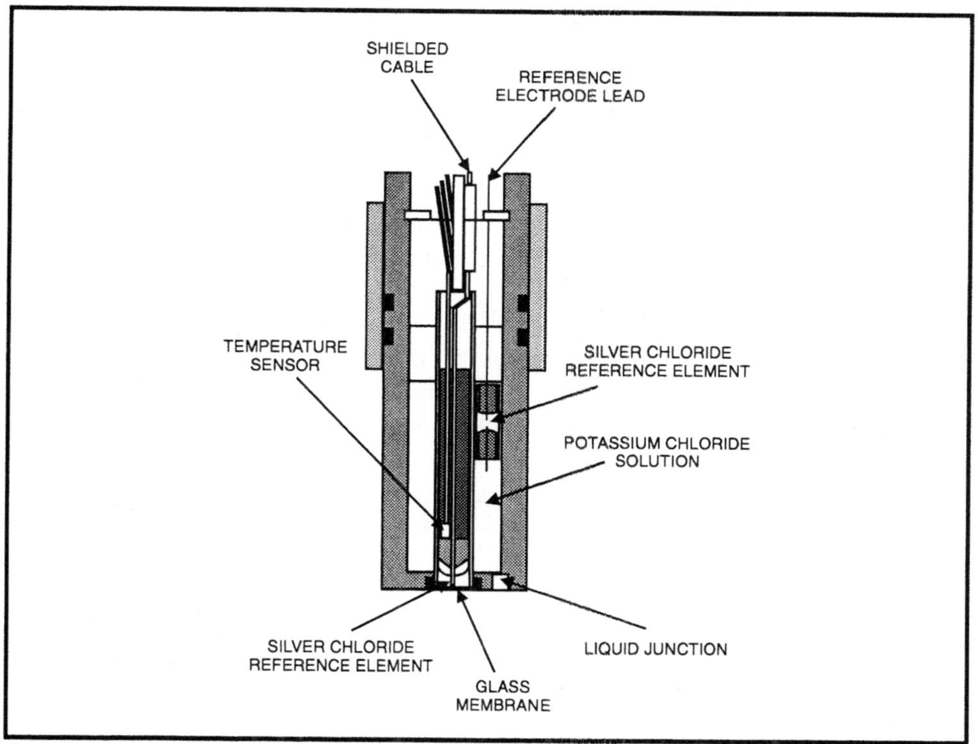

Figure 25-24b. Diagram of a Typical Commercially-Available pH Probe

Wide-Range pH Electrode

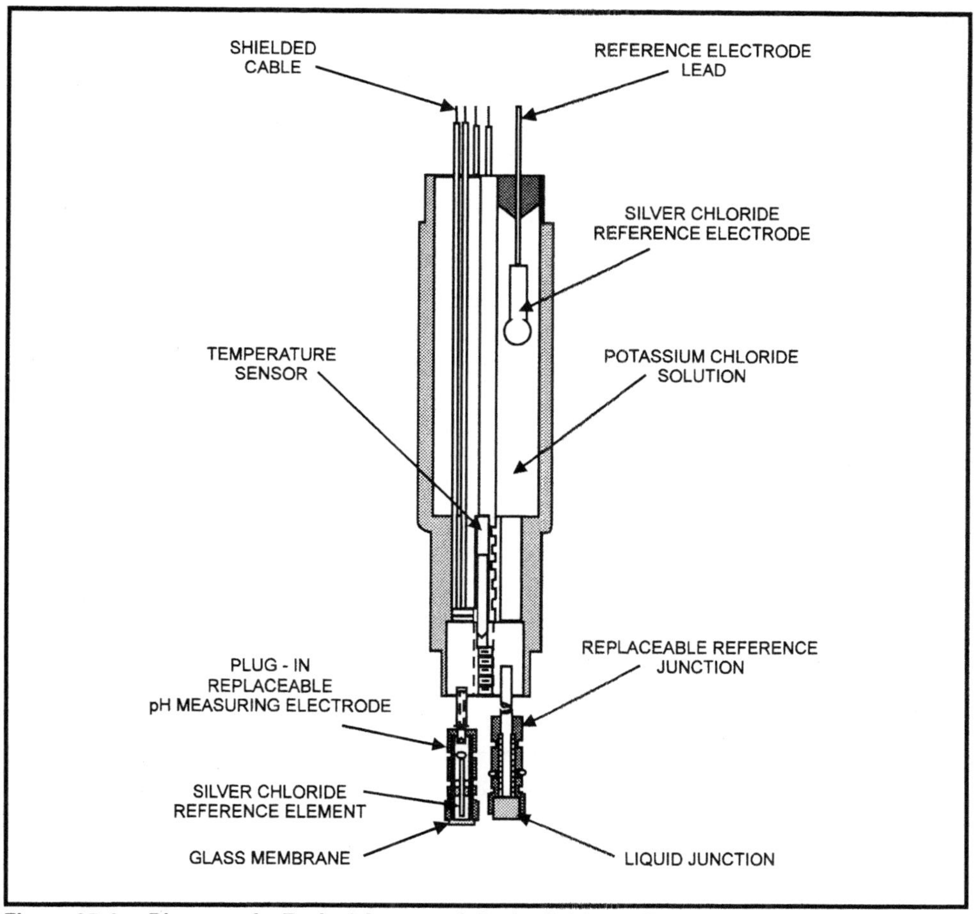

Figure 25-4c. Diagram of a Typical Commercially-Available pH Probe

Calibration and Standardization

pH meters are CALIBRATED.
Ph Cells are STANDARDIZED.

Calibration and standardization are distinct operations, although the terms often are used interchangeably. pH meters are calibrated using accurate millivoltage sources. pH cells are standardized using known buffer solutions and pH temperature relationships. Calibrated pH meters, with known compensation, may be used with pH cells to standardize the pH analyzer system (the cell [probe] and the meter).

pH cells, operated isothermally without liquid junctions and with sample solutions at constant ionic strength, are capable of virtually perfect pH function over wide ranges of pH [Ref. 4]. Perfect pH function is the slope factor in mV/decade pH, as predicted by Equation (25-13) and shown for various temperatures in Table 25-4. The degree to which a cell exhibits this property is conveniently expressed by its electromotive efficiency. This value, B_e, is defined as the ratio of the observed slope to the theoretical slope at a given temperature:

$$B_e = \frac{\text{slope}_{(\text{experimental})}}{\text{slope}_{(\text{theoretical})}} \tag{25-20}$$

Under laboratory conditions, general-purpose and wide-range pH electrodes have been shown to have $B_e > 0.995$. Under process conditions, B_e values in the range of 0.90 to 0.95 are not uncommon.

It is thus critically important that two-point calibration, using buffer solutions that bracket the expected range of pH for the sample, be employed.

Lower- and upper-range values can then be adjusted to compensate for nonideal B_e. Commercial manufacturers of pH measuring electrodes and cells do not publish values of B_e for their cells. Rather, they take the practical approach, using two-point standardization and suitably designed instrumentation to compensate for it. By determining and tracking B_e values at suitable time intervals and under reference conditions of temperature and standard buffers, users can increase the reliability of their analyzer systems.

pH Probe Maintenance

In clean solutions of nonaggressive chemicals at moderate temperature and pressure, pH probe reliability is measured in months. In aggressive media, such as high concentrations of acids or bases or samples with entrained solids, or at extremes of temperature and pressure, cell life may be measured in hours.

In clean solutions of nonaggressive chemicals at moderate temperature and pressure, pH cell reliability is measured in months. In aggressive media, such as high concentrations of acids or bases or samples with entrained solids, or at extremes of temperature and pressure, cell life may be measured in hours.

Probe problems, if not obvious on-line, are best detected during probe standardization and by observation of probe B_e. Problems in the probe most commonly occur at the glass membrane or the liquid junction.

The glass electrode probe may be broken, fouled, or shorted with a path of sample solution to the high-impedance electrical lead leading to the analyzer. Aside from physical inspection, these problems may be apparent if pH readings from the cell seem sluggish, nonvariant, or outside the range of normal operating conditions. Broken or shorted electrodes may be detected without removal by making a resistance measurement.

A simple method for doing this, using the analyzer in its mV output mode, employs a resistance shunt applied across the pH and reference electrode terminals of the cell. In the absence of the shunt, the millivolt reading, E_1, is noted. A shunt resistance of appropriate size to reduce the cell potential by a readable amount, say half the unshunted value, is placed across the cell.

The resistance of the cell, from Ohm's law of parallel resistors, is given by:

$$R_e = R_{shunt}\left[\frac{(E_1 - E_2)}{E_2}\right] \tag{25-21}$$

where:

R_e = cell resistance
R_s = shunt resistance
E_1 = cell voltage without shunt
E_2 = cell voltage with shunt

If values of R_s, equivalent to values of R_e (say 100 MΩ), do not cause E_2 to vary from E_1, it is reasonable to expect that the glass membrane is broken or otherwise shorted.

Coatings on electrodes are not easily detected while they are installed in the process, unless they are installed in a sample flow loop that allows injection of reagents enabling measurement of response time. The response of coated electrodes to process changes is sluggish or nonexistent; operator familiarity with the process may allow for inference of this condition. The electrode must be removed for cleaning or be replaced to correct this situation. The nature of the process should serve as a guide for the type of cleaning (mechanical or chemical) best suited to restore pH function.

pH electrodes that are operated for long periods of time at elevated temperature may exhibit sluggish response, develop high resistance, and become noisy or erratic. They can sometimes be rejuvenated by brief immersion in 2% ammonium bifluoride solution. If the reference electrode junction is clogged or contaminated, it can sometimes be cleared by grinding lightly with a fine abrasive or often by immersing the tip in a solution of KCl and then heating the solution.

Keep a laboratory-type pH probe, soaking in a buffer solution, and a reference electrode, with its reference junction soaking in a moderately concentrated solution of KCl, in a clean and secure area of the maintenance shop. This, along with a high-quality pH meter to allow checks of the individual components of troubled process cells, will greatly facilitate troubleshooting.

Keep a laboratory-type pH probe, soaking in a buffer solution, and a reference electrode, with its reference junction soaking in a moderately concentrated solution of KCl, in a clean and secure area of the maintenance shop. This, along with a high-quality pH meter to allow checks of the individual components of troubled process cells, will greatly facilitate troubleshooting.

Instrumentation

The salient electronic features of a typical process pH analyzer are shown in Figure 25-5. The boxed portion of this figure is the measurement preamplifier, which may be integrated into the probe. The circuit shown will accept signals from the pH cell, whether or not they are preamplified. Because of the high resistance of the pH probe (10^8 to 10^9 Ω), electrical isolation of its signal lead from ground and the input impedance of the measuring amplifier must be at least 10^{12} to limit measurement error to less than 0.1%.

Both glass electrode and reference leads are connected through series resistors to unity-gain buffer amplifiers A1 and A2. This converts the measured potentials to low-impedance signals, suitable for transmission to the remote analyzer, where they are applied to the inputs of a unity-gain differential amplifier, A3. The pH probe measuring amplifier is also used to drive the shield of signal lead from the glass probe. This guard shield, driven with a lower-impedance source but at the same potential as the signal lead, provides an effective barrier from electronic noise and the effect of cable capacitance. For transmission of signals over large distances, say 500 to 2,000 ft (150 to 600 m), without impedance conversion using a preamplifier in the probe, the phenomenon of lead capacitance in the signal lead, in combination with the impedance level of the probe, would provide time constants so large that signal changes would be seriously delayed or possibly not transmitted at all. With properly guarded shields, the remarkably fast response times exhibited by pH probes may be transmitted to their remote analyzers with virtually no degradation.

The output of amplifier A3 is connected to the input of a second differential amplifier A4 having offset and gain, which may be used for isopotential and temperature slope adjustment, depending on the design of the compensation circuitry. Leads from the temperature-sensitive resistor, found in most process probes, are not shown; they are monitored with additional circuitry and may be used to control gain at amplifier A4.

Also shown in Figure 25-5 is solution ground. Process cells utilize a metallic or electrically-conductive contact, mounted in the probe, connected to analog ground in the analyzer. Glass and reference electrode probes are measured differentially against this contact, thereby providing common mode rejection from noise generated by spurious electric fields in the process. To prevent ground paths through the analyzer, its output must be isolated by design or by using an isolation module between it and devices to which it is connected. The green wire from the power line is not connected to the analog ground of the analyzer power supply or to solution ground. It is provided solely for safety and shielding.

When using a pH analyzer designed to operate with solution ground, with a cell not having provision for this ground (e.g., a laboratory cell operated in a glass beaker), a wire should contact the solution and be connected to the solution ground terminal on the analyzer. Alternatively, a jumper may be temporarily installed between the reference and solution ground terminals to operate the cell since the glass beaker provides the necessary isolation between the cell and the external world.

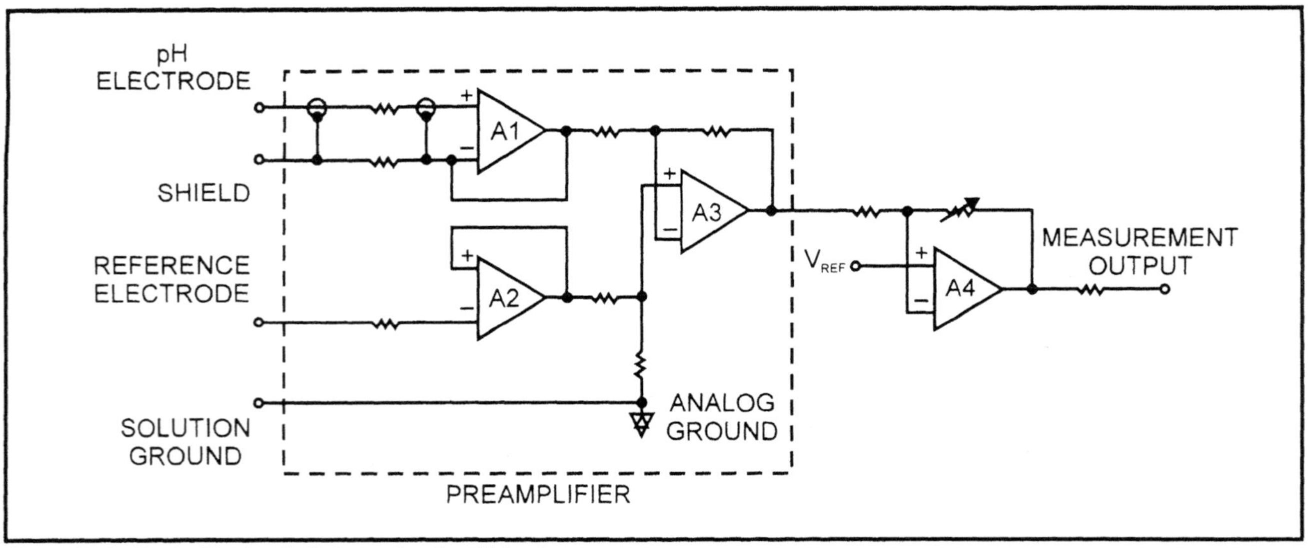

Figure 25-5. Basic Electronic Schematic of a pH Analyzer

Intelligent pH Analyzers

Digital electronics, microprocessors, menu-driven software, data-table storage, and convenient keyboard-user interfacing have greatly expanded the capability and utility of pH analyzers. The jobs of interpreting, compensating, calibrating, maintaining, and troubleshooting pH cells are greatly facilitated by this equipment. The components of a typical intelligent instrument are shown in Figure 25-6.

Signal interfacing accepts analog signals from the pH cell — its measurement, its temperature, and possibly diagnostic information. The multiplexer directs signals selected by the CPU (central processing unit, or microprocessor) to an integrating A/D converter, where they are converted to digital format; a clock starts in the CPU at the start of the conversion and stops when a signal from the A/D tells it that the conversion is complete. The time of the conversion is inversely related to the amplitude of the selected input voltage, and these inverse time relationships are used by the CPU with on-board software for data processing.

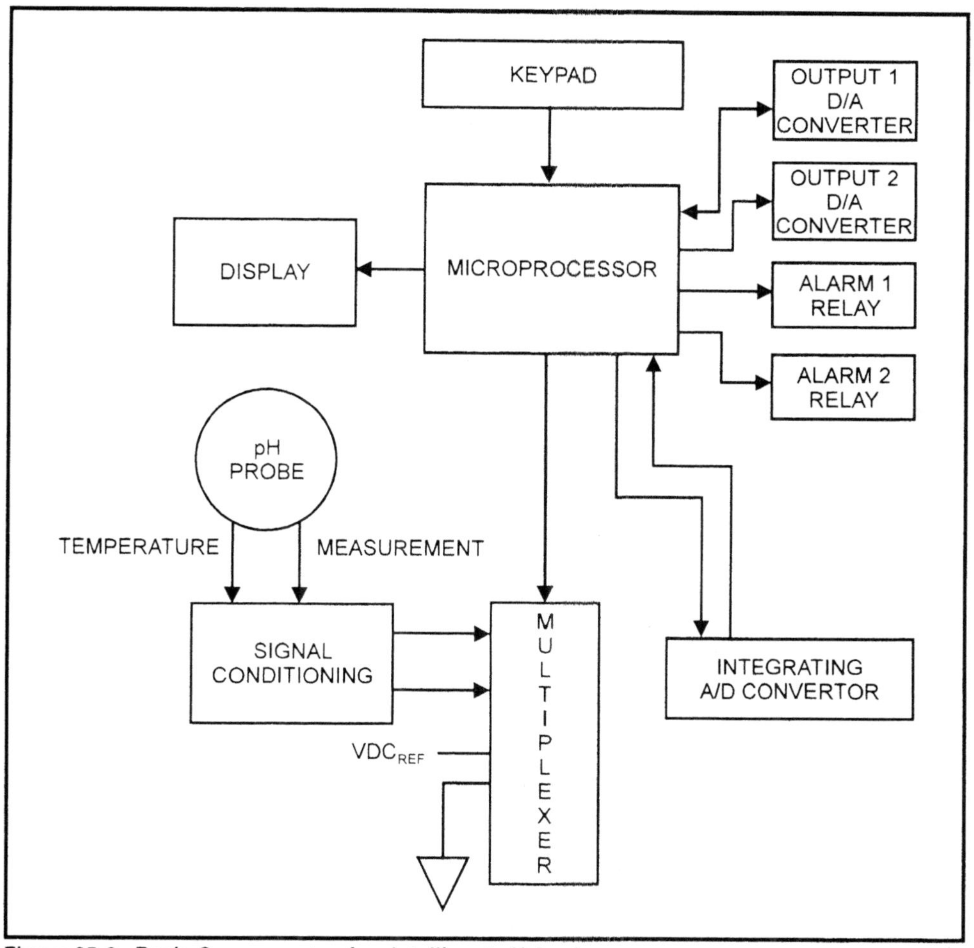

Figure 25-6. Basic Components of an Intelligent pH Analyzer

The keypad, together with the display, provides operator interface with the CPU. It allows user selection of measurement display and its units, data entry such as cell standardization values, range selection, alarm setups, and custom data files like those used for pH compensation of specific materials for solution temperature dependency. It may also be used for data storage and retrieval from previous cell standardizations, such as cell slope, to aid the user in cell maintenance.

The alarms and outputs of the analyzer, also under control of the user by means of the keypad and CPU, may provide selection of outputs, isolated voltage or current, and alarm programming, such as timed alarm activation switching (on or off). This latter feature is of value in processes where reagent addition is employed to control pH and measurement delay, or dead time, due to mixing or reagent solubility, must be compensated to effect proper process control dynamics.

Analyzer Safety and Environmental Protection

pH analyzers are available in packaging suitable for the full range of environmental exposure — from the general-purpose NEMA 1 rating (i.e., no environmental protection but suitable for the laboratory and control room) to the weatherproof, corrosion-resistant protection afforded by the NEMA 4X rating. They are available in a variety of mounting configurations, such as wall, pipe, and DIN-size panel mounting (usually for 1/4 and 1/2 DIN panel cutouts), and are suitably packaged for radio frequency interference/electromagnetic interference (RFI/EMI) immunity. Additionally, 100/110 and 220/240 VAC two-wire models, 24 VAC models, and a variety of agency certifications (such as FM, CSA, and UL for the U.S. and Canada, and CENELEC or BASEEFA for Europe; see Table 25-6) are readily available from most analyzer manufacturers.

Table 25-6. Common Product Safety Specifications and Their Industrial Applications

Testing Laboratory, Types of Protection, and Area Classification	Area of Application
BASEEFA(a) certified intrinsically safe EEx, Ia, IIC (CENELEC). Suitable for Zone 0 or 1, or in Zone 1 only, depending on whether the power source is certified 1a or 1b, respectively.	Zone 0: Explosive air/gas mixture always or likely present Zone 1: Explosive air/gas mixture likely to occur under operations-upset conditions
BASEEFA certified nonincendive ExN, II, for use in Zone 2 hazardous locations	Zone 2: Explosive air/gas mixture not likely to occur
CSA(b) certified for use in general-purpose, ordinary locations	Explosive conditions never present
FM certified intrinsically safe for Class I, Groups A, B, C, and D, Division 1, and Class II, Groups E and G, Division 1	Division 1: Similar to BASEEFA Zone 0 Class I: Flammable gases Class II: Flammable dusts
FM(c) certified nonincendive for Class I, Groups A, B, C, and D, Division 2 hazardous locations	No sparks or hot surfaces generated by equipment; similar to BASEEFA Zone 2

(a) British Approvals Service for Electrical Equipment in Flammable Atmospheres.
(b) Canadian Standards Association. (c) Factory Mutual Engineering and Research Organization.

Many manufacturers supply two-wire pH analyzers, specifically designed for remote installations where alternating-current power is not available, or for hazardous areas where intrinsic safety requires limits on power levels to field-mounted analyzers. With this instrumentation, signal outputs are transmitted on power inputs (hence two-wire) which are held at intrinsically safe power levels. Local indication may be supplied by an analog or digital display sealed behind a window. Instruments of this type are usually mounted in heavy die-cast metal housings with gasketed access covers that conform to stringent environmental, electrical, and safety requirements. A typical two-wire pH analyzer is shown in Figure 25-7.

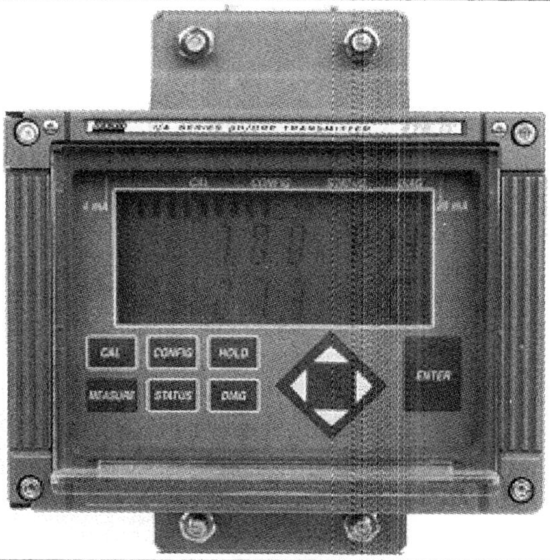

Figure 25-7. Typical Two-Wire pH Analyzer and Probe

Installation

Cell Mounting in the Process Plant

A variety of configurations are available for mounting pH cell assemblies in-line in tanks or flowing streams or off-line in sample loops. Several of these configurations are shown in Figure 25-8.

Ball valve assemblies (Figure 25-8a) permit the sensor to be inserted or removed from process streams or tanks under rated temperature or pressure without draining the system or resorting to bypass arrangements. O-ring seals on the probe assembly seal against the inner diameter of the ball and shaft assemblies. These valves are available with tube fittings, pipe fittings, or weldments to allow purging and in-situ cleaning.

Flow cell chambers (Figure 25-8b) allow installation on off-line sample loops. A major advantage of this type of installation is the ability to easily admit cleaning and standardizing solutions through the cell for maintenance and calibration.

Figure 25-8(c) shows a probe with an integral pipe bushing for mounting into tanks or pipelines with integral weldments or into flow-through tees (Figure 25-8d). Finally, for use in tanks or troughs, protection sleeves are available for mechanical protection of the glass membrane (Figure 25-8e).

Flow-through and submersible electrode cleaning systems are available from some manufacturers. These use electrically or pneumatically driven brushes or wipers. They are not universally effective, but can be of some value for probes in locations that are difficult to access.

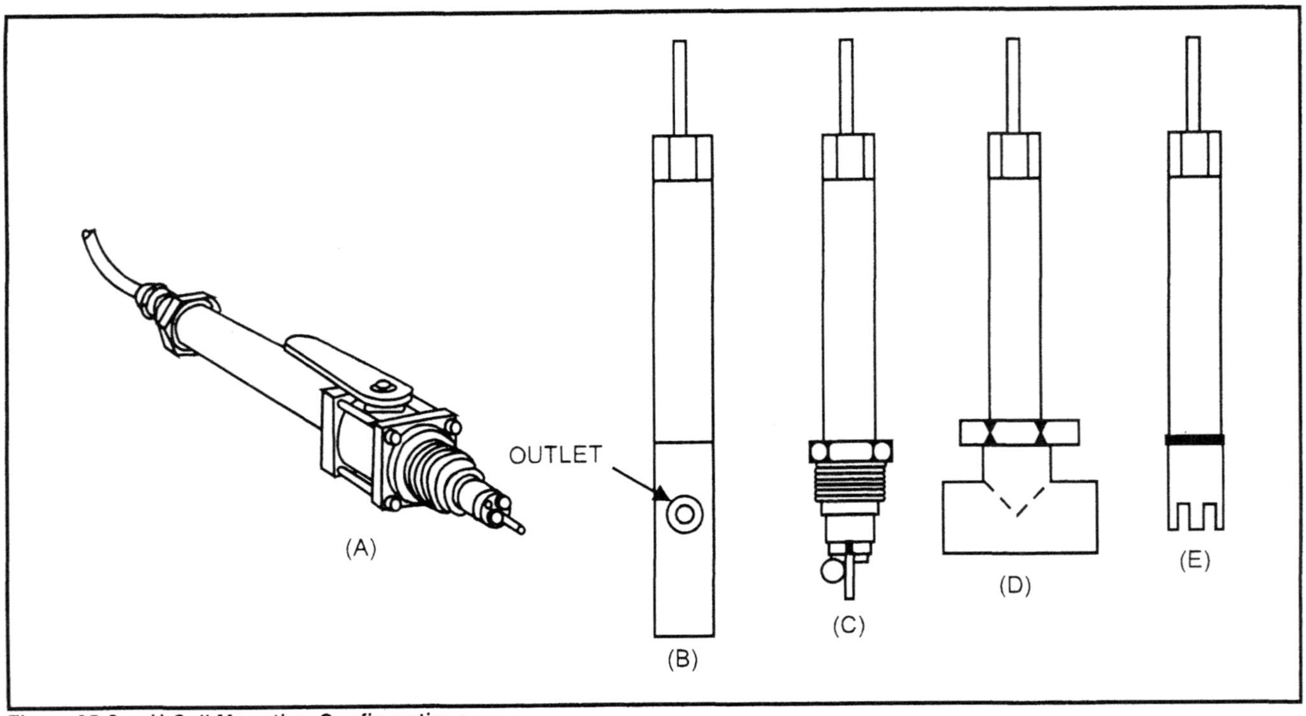

Figure 25-8. pH Cell Mounting Configurations

The following general installation procedure applies to all pH analyzer systems:

1. Use an EYS (explosionproof, Y-shaped seal) fitting in permanent wiring near the probe to prevent leakage of process liquid along wiring from the probe to other parts of the system.
2. Assure that the probe is firmly fitted into the process with a leaktight seal.
3. Install the probe where solution is representative of the process. In flow-through cells, proper piping can often minimize effects of trapped air, abrasives, accumulation of material, and coating of electrodes.
4. Have access to a spare probe and a complete set of proper range standard solutions or buffers.
5. Install probes and analyzers where they will be accessible for maintenance.
6. When wiring to high-impedance input, follow these experienced instructions and suggestions:
   a. Avoid spliced connections unless absolutely necessary; if necessary, use only high-impedance insulation and terminal strips and put connections in grounded enclosure to minimize electrical pickup.
   b. Use purge system, instead of desiccant, when protection from moisture is required.
   c. Connect shields only as specified by the manufacturer. Driven shields, often used to reduce capacitance of sensor leads, must not be connected to ground.
7. Before start-up, check operation of system with a simulated input signal. Most pH analyzers read 7 pH with 0 mV (shorted) input.

8. Standardize electrodes in buffer and check calibration at second point. If difference in temperature between process and buffer is greater than about 10°C (20°F), allow time for temperature of the probe to stabilize.

Troubleshooting

The best way to test sensor performance is to perform a two-point standardization with fresh buffer solutions. The probe is immersed in each buffer with sufficient time allowed for full thermal and chemical equilibrium to occur. Equilibration times greater than a minute or so indicate a problem. Depending on the analyzer being serviced, the electromotive efficiency, or the change in millivolts observed for the known pH difference between the buffers, is determined. Ninety-nine percent of theoretical (i.e., $B_e = 0.99$) or greater at the temperature of the standardization indicates that the cell is functioning well; B_e in the range of 0.95 to 0.99 indicates satisfactory operation; and $B_e < 0.95$ indicates trouble. Table 25-7 lists solutions to commonly encountered problems.

Do not use a grab sample unless the pH of the process is known to be stable. The pH of the sample can change, often due to exposure to CO2 in the air, during the time required to make the measurement.

Table 25-7. Troubleshooting Guide

Problem	Possible Causes	Solution
None or only slight pH response	Broken pH electrode	Replace
	Heavily coated electrode	Clean/replace
	Broken lead to probe	Replace
Elongated span	Calibration incorrect; manual	Recalibrate
	Temperature compensator set too low, automatic temperature compensator in error	Correct temperature compensation
pH Offset	Reference junction fouled	Clean/replace
	Reference solution contaminated	Replace
Noisy	Solution ground open	Repair
	Intermittent cell contact to sample	Assure cell immersion and absence of air bubbles
	Improper reference junction	Replace
	Reference junction fouled	Clean/replace
	Air bubble in salt bridge	Remove bubble
	Broken or shorted lead	Replace
Slow response	Dehydrated glass membrane	Soak in weak salt solution for 2-3 h
	Coated or dirty membrane	Clean/replace
	Improper cell placement	Place cell in area of uninterrupted flow

Acknowledgments

The author would like to thank Joe Downey, Annette Nowiszewski, Norm Gilmartin, and Phil Reese of The Foxboro Company for their technical assistance.

References

1. Bates, R.G., *Determination of pH, Theory and Practice*, New York: McGraw-Hill, 1964.
2. Butler, J.N., *Ionic Equilibrium: A Mathematical Approach*, Reading, MA: Addison-Wesley, 1964.
3. Shinskey, F.G., *pH and pION Control in Process and Waste Streams*, New York: John Wiley & Sons, 1973.
4. Light, T.S., and Fletcher, K.S., "Accurate Evaluation of pH Glass Electrodes in Solutions of High Ionic Strength," *Analytical Chemistry*, 39: 70, 1967.

About the Author

Kenneth S. Fletcher received his B.S. Degree in Chemistry from Trinity College, Hartford, Connecticut, in 1963 and Ph.D. in Electroanalytical Chemistry from The University of Massachusetts at Amherst in 1967. His entire working career has been spent at The Foxboro Company working, initially as a researcher, and later as a group leader, developing Electrochemical Sensors for Process Monitoring Applications. These activities have lead to new products and product enhancements in pH, Contacting and Electrodeless Conductivity, and Dissolved Oxygen measurements. Many patents as well as publications in the scientific and trade literature have resulted from these activities.

26

ORP (RedOx) Analyzers

ORP analyzers are versatile monitoring devices that have long been used for scientific investigations in laboratories throughout the world. The ORP measurement has been most widely used for end-point detection of potentiometric redox titrations in laboratories, but has also been successfully applied in the process environment.

The terms ORP (oxidation-reduction potential) and redox (reduction/oxidation) are used synonymously to describe the ratio of the activities of oxidized to reduced species in solution. ORP measurements are nonspecific by nature; any species that is capable of contributing electrons will affect the ORP. A redox potential forms in solution as a result of the potential of one chemical species to transfer electrons to another. Actual redox potentials are observed only in truly reversible systems; however, measurements in nonreversible reactions, such as natural waters and fermentations, can still yield useful data.

For simplicity's sake, we'll say that *electron activity* is responsible for the oxidation-reduction potential observed in a solution (although free electrons do not actually exist in an aqueous solution).

Measurement Principles

The Basic Framework

Oxidation and reduction reactions always occur simultaneously as the reactants undergo a change in oxidation state and electrons are transferred.

Consider the reaction:

$$A_{ox} + B_{re} \Leftrightarrow A_{re} + B_{ox} \tag{26-1}$$

The ox and re subscripts refer to the oxidized and reduced forms of A and B. The reaction in Equation (26-1) is actually the sum of two half-cell reactions:

$$A_{ox} + ne^{-} \Leftrightarrow A_{re} \tag{26-2}$$

$$B_{re} \Leftrightarrow B_{ox} + ne^{-} \tag{26-3}$$

Equation (26-2) represents the reduction of A_{ox} to A_{re}, and Equation (26-3) represents the oxidation of B_{re} to B_{ox}. An electron has a negative charge and is written as e^- in the half-cell equations. An electrochemical potential is assigned to each half-cell reaction.

Historically, oxidation was a term used to describe a reaction with oxygen. It has since been generalized to refer to any reaction in which an atom or ion loses electrons, resulting algebraically in an increase of oxidation state. A substance that can accept electrons and undergo a decrease in oxidation state is referred to as an *oxidizing agent*. Conversely, reduction refers to a reaction in which an atom or ion gains electrons, resulting algebraically in a decrease of oxidation state. A substance that gives up electrons and undergoes an increase in oxidation state is referred to as a *reducing agent*.

Certain electrodes, called ORP electrodes, develop potentials in contact with these solutions and develop a scale of potentials, called redox potentials, which forms the basis of this analytical technique.

A method exists for determining the potential of oxidizing agents to oxidize and of reducing agents to reduce. Certain electrodes, called ORP electrodes, develop potentials in contact with these solutions and develop a scale of potentials, called redox potentials, which forms the basis of this analytical technique.

In solutions, oxidation-reduction potentials are determined by using a potentiometric electrochemical technique quite similar to that used in pH measurements. However, instead of using a measurement system directly analogous to pH units, ORP analysis directly measures the solution potential, usually in volts or millivolts.

ORP is measured using an inert-metal sensor known as an ORP or redox electrode, a reference electrode, and a specially-designed voltmeter that limits current flow through the solution. These specific components will be discussed in some detail. The installation, maintenance, and use of the measurement system is simple, but interpretation of the potentials and electrochemical reactions occurring in the measured solution requires consideration and discussion.

The Measurement System: Ideal Versus Reality

The redox potential of the system is the potential achieved by the chemical equilibria of the reversible redox couples present. Equations (26-2) and (26-3) show an equilibrium relationship between chemical activity (or concentration) and electrical activity (or potential). We will see that these are related quantitatively by the Nernst equation (which will be defined later). In these equilibria, electron flow requires that oxidized forms be converted to reduced forms and reduced forms be converted to oxidized forms. If these processes occur easily, the system is called reversible or well-poised and will have all of the redox couples in equilibrium at the same potential. An ORP sensor and reference electrode form a galvanic cell — that is, one that develops a potential in response to the chemical activities or concentrations of a redox couple. The function of the ORP sensor, or indicator electrode, is to develop a potential in response to the redox potential of the solution to which it is exposed. The reference electrode is isolated from the measured solution and provides the constant reference potential against which the redox potential is measured.

Simply placing a metal sensor into the solution containing the redox couple will not necessarily produce the appropriate Nernst equation response (measured against some reference electrode), except perhaps under ideal reversible conditions. Any application of ORP requires an understanding of the basic principles involved when an electrode develops a potential. The ORP sensor itself, the reference electrode, and the redox reaction(s) can all contribute to nonideal behavior in the measurement system. In the following sections, these three components of the ORP measurement system, as well as the ORP analyzer, will be discussed along with minimal applicable theory.

ORP SENSOR

An ORP sensor develops a potential in response to the redox potential of the solution in which it is immersed. The metal electrode material must be chemically inert to all of the species present in the sample; in practice, platinum is the commercial ORP sensor material of choice.

The ideal ORP sensor possesses these traits:

- Inert
- Noble metal
- Electroactive to both oxidized and reduced species
- High exchange-current density

Inert metal electrodes do not take part in the solution redox chemistry. They serve simply as sites for electron transfer between oxidized and reduced species in the solution. Their ability to do this is reflected by their adherence to the Nernst equation (discussed later). This is a property of noble metals and other materials that are inert and have high exchange-current densities for the redox system being measured.

All metals are made up structurally of ion lattices with electrons moving through them. Some metal electrodes, such as mercury and silver, respond to the presence of their own ions in solution. Under ideal conditions, these metals form stable potentials when concentrations of their ions are fixed, and these metals are used as reference electrodes. The noble metals, however, do not contribute their own ions to solution and may serve as exchange sites for soluble redox couples, developing potentials specific to that chemistry. Platinum, iridium, gold, or palladium are common choices for reference electrodes. Graphite and boron carbide have been used to a lesser extent [Ref. 1, 2].

Under ideal conditions, these metals form stable potentials when concentrations of their ions are fixed, and these metals are used as reference electrodes.

Consider an electrode in a system containing the redox couple:

$$A_{ox} + ne^- \Leftrightarrow A_{re} \qquad (26\text{-}2)$$

where n is the number of electrons exchanged. When the rates of the oxidation and reduction reaction are equivalent, the reaction is said to be at equilibrium. During equilibrium, the net current flow from the oxidation and reduction is zero, and the potential of the reaction is fixed. Although net current flow is zero, the current flow associated with either the oxidation or the reduction is not. These opposing current flows are called exchange currents.

Exchange current, and exchange-current density (i.e., current per unit electrode area) are expressions used to characterize the reversibility of ORP sensors in particular redox solutions. A current flows any time that an exchange takes place at the electrode surface between ions in solution. If the exchange is fast, we have a high exchange current, and the reaction is said to be reversible. In this case, the potential generated from the ORP sensor at 25°C is accurately given by the Nernst equation.

$$E = E^o + \frac{0.05916}{n}\log\frac{[\text{oxidized}]}{[\text{reduced}]} \qquad (26\text{-}4)$$

where:

E = the measured potential of the solution as measured by the electrode

E^o = the standard redox potential of the solution

n = the number of electrons that transfer in the reaction

[oxidized] = the concentration of the oxidized species in the reaction

[reduced] = the concentration of the reduced species in the reaction

Note that when [oxidized] = [reduced], then $E = E^o$ (the standard potential). Standard potentials will be covered later in this chapter.

Under standard conditions (1 cm^2 sensor surface area and 1 M [mole/liter] solutions), exchange-current densities for different metals have been found to range between 10^0 and 10^{-25} A/cm^2. These currents are dependent on electrode material, the redox system employed, and its concentration in the sample solution. This measurement is not commonly used to evaluate redox systems because it is very time consuming. The measurement is, however, of fundamental interest in describing the reversibility of redox systems and is of practical value in corrosion studies for nonnoble metal/electrolyte systems.

Reactions with high exchange-current density have stable, accurate, and reproducible ORP potentials. Platinum ORP sensors have gained widespread acceptance due to the high exchange-current density they show in many applications. Additionally, the catalytic properties of platinum facilitate measurements in nonreversible redox systems, since microequilibria are sometimes established on the platinum surface.

A sluggish response can be expected if the Pt/PtO system potential and the sample potential are close together.

Platinum sensors are known to operate with or without oxidized surfaces (Figure 26-1). When a platinum sensor is in an oxidizing solution, the surface will become oxidized if the potential is greater than the Pt/PtO redox system. The potential of this system is dependent on pH. At elevated pH, the electrode functions with an oxidized Pt surface; and at low pH, with a reduced Pt surface. A sluggish response can be expected if the Pt/PtO system potential and the sample potential are close together. In either case, an electron exchange takes place with the solution, but at dissimilar rates.

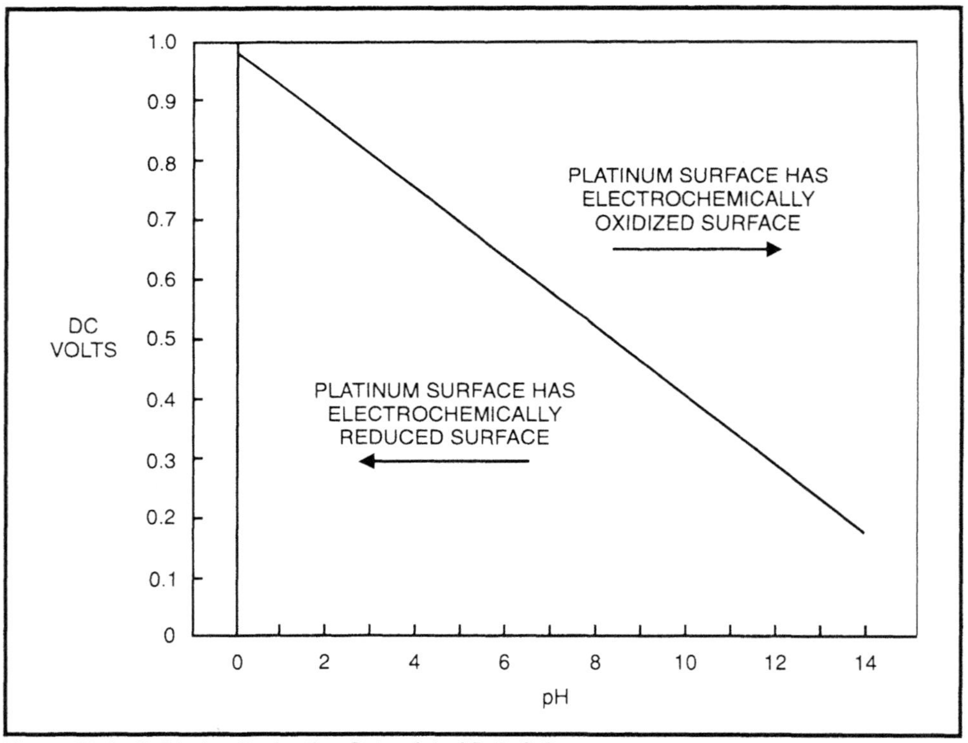

Figure 26-1. Oxidation-Reduction Potential of Pt/PtO System

Voltage differences of 200 mVDC have been observed between ORP sensor materials when immersed in the same solutions. Systems that contain or dissociate to strongly oxidizing ions (Cl_2, MnO_4^-, or $Cr_2O_7^{-2}$) usually demonstrate wide voltage differences, depending on the metal electrode used. Extensive electrochemical oxidation of the electrode surface can lead to a decrease in the exchange-current density, resulting in a sluggish response, erroneous voltage, and poor precision.

Gold ORP electrodes have been reported to offer an advantage over platinum when used in strong oxidizing agents.

Gold ORP electrodes have been reported to offer an advantage over platinum electrodes when used in strong oxidizing agents.

REFERENCE ELECTRODE

Redox measurements must be made by measuring the potential of the ORP electrode against a suitable reference electrode. These electrodes are typically combined for process measurements. Figure 26-2 shows a schematic drawing of a combined ORP sensor.

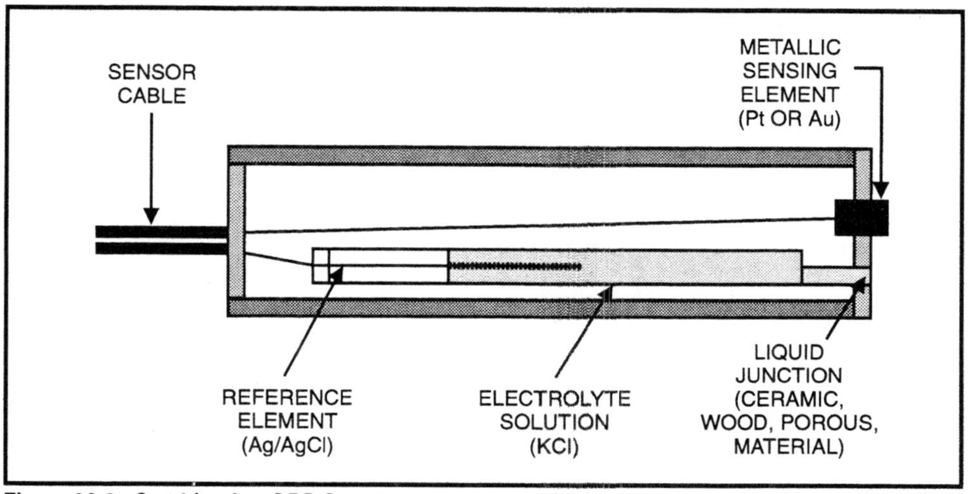

Figure 26-2. Combination ORP Sensor

The ultimate reference electrode is the standard hydrogen electrode (SHE). By convention, the potential of the SHE is given as 0.00 V at all temperatures. This reference electrode responds to the oxidation of hydrogen, given by:

$$H^+ + e^- \Leftrightarrow \frac{1}{2}H_{2(g)} \qquad (26\text{-}5)$$

It is made of platinum foil immersed in a 1 *M* acid solution. The potential of the hydrogen electrode varies with the partial pressure of hydrogen gas in equilibrium with the platinum surface. Voltages are corrected to a partial pressure of hydrogen at 760 mm Hg. The potential is given by:

$$E = E^o + 0.059 \log \frac{H^+}{\sqrt{p_{H_{2(g)}}}} \qquad \text{where } E^o = 0.0 \text{ V} \qquad (26\text{-}6)$$

The potential of the SHE is defined for 1 M acid and H_2 pressure of 1 atm. Although this is a reference electrode in the classical sense, a fixed partial pressure of hydrogen is required for its use, making it impractical for use outside the laboratory. For process applications, a more rugged type of reference electrode is desired.

The ideal reference electrode possesses these characteristics:

- Stable, rapidly attained potential
- Electrolyte nonreactive with sample
- Low junction potentials
- Low maintenance, easy to handle
- No thermal hysteresis
- *dV/dT* known and reproducible
- Low polarizability
- Rugged, compact, and convenient

The reference electrodes used for ORP measurements are the same as are used with pH measurements.

The reference electrodes used for ORP measurements are the same as are used with pH measurements. In the laboratory, Hg/Hg_2Cl_2 (calomel) and Ag/AgCl types are used as reference electrodes, but Ag/AgCl types are used almost exclusively in process. The Ag/AgCl electrode fulfills most of the above-mentioned reference electrode criteria.

The electrode is made by immersing a silver (Ag) wire, usually plated with silver chloride (AgCl), into an electrolyte solution of potassium chloride (KCl) saturated with AgCl. The electrolyte contacts the sample through a porous barrier constructed of ceramic, wood, fiberglass, porous plastic, or other materials. The equivalent ion mobility of potassium and chloride results in lower liquid junction potentials, making potassium chloride the preferred *salt bridge* electrolyte. The Nernst expression for this electrode is

$$E = E^{o}_{\mathrm{AgCl}} + 0.059 \log \frac{1}{[\mathrm{Cl}^-]} \tag{26-7}$$

The concentration of chloride ions dictates the electrode potential. High concentrations of potassium chloride are used in commercial preparations because evaporation will have minimal effect on the concentration and resulting potential. Additionally, diffusion of process into the reference cavity will be minimized. Table 26-1 lists potentials for this reference electrode with different chloride concentrations. It is important to note that many electrolytes supplied by manufacturers contain additives to prevent freezing or boiling, and may have Cl^- activities different from their actual concentrations.

Table 26-1. Potentials for Ag/AgCl Reference Electrodes

Electrolyte	DC Voltage, VDC
Saturated KCl	+0.199
3.0 *M* KCl	+0.209
1.0 *M* KCl	+0.237
0.10 *M* KCl	+0.290

The terms *activity* and *concentration* have both been used to describe chloride ions. In very dilute solutions, the activity and concentration of the ions are approximately equal, but in more concentrated solutions they differ. The two are related mathematically by a term called the activity coefficient (*f*):

$$a = c \times f \tag{26-8}$$

Therefore, the activity will be less than the actual concentration in concentrated solutions.

Silver/silver chloride reference electrodes change voltage as a function of temperature (and the solubility of AgCl in the KCl). A known, predictable dV/dT is known for the Ag/AgCl reference system and is tabulated for two potassium chloride concentrations in Table 26-2. When the reference electrode returns back to room temperature after being exposed to elevated temperatures, it should return to its original potential. When it doesn't, the electrode is said to exhibit thermal hysteresis. The Ag/AgCl electrode does not exhibit this property.

Table 26-2. Standard Potentials of Ag/AgCl Electrodes as a Function of Temperature and KCl Concentration

Temperature, °C	3 *N* KCl	Saturated KCl
0	0.255	0.2228
5	0.222	0.2186
10	0.2197	0.2138
15	0.2164	0.2089
20	0.2129	0.2039
25	0.2090	0.1985
30	0.2054	0.1935
35	0.2017	0.1884
35	0.1978	0.1831
40	0.1978	0.1831
45	0.193	0.1780
50	0.190	0.1733

It is important to know what reference electrode system is employed when comparing results or trying to reproduce a process measurement indicated by an ORP sensor.

It is important to know what reference electrode system is employed when comparing results or trying to reproduce a process measurement indicated by an ORP sensor.

When comparing results, potentials should be referred to the same reference system. This is also important when using calibrating solutions. Publications often use a value referenced to the SHE; the values can be converted by:

$$E_{SHE} = E_{observed} + E_{ref} \quad (26\text{-}9)$$

where:

E_{SHE} = the potential of the redox solution referenced to the SHE
$E_{observed}$ = the potential observed utilizing the reference electrode
E_{ref} = the potential of the reference electrode

Issues relating to the maintenance and troubleshooting of reference electrodes are covered later in this chapter.

ELECTRONIC ANALYZER/CONTROLLER

The ORP analyzer/controller is similar to that used in pH measurements. A high-input impedance is required in the circuit to limit current flow through the solution (sometimes the sensor contains the integral preamplifier). Most modern instrumentation employs some form of digital display and solid-state electronics for accurate, sensitive, and reliable measurement. Many ORP analyzer/controller units are integral with basic pH analyzers. By either flipping a few switches or depressing a couple of keypads, the monitor is converted from pH to ORP. The basic electronic schematic for an ORP analyzer is shown in Figure 26-3.

Although some ORP analyzers use the resistance signal to monitor temperature, it is not utilized to correct the mVDC values displayed.

The displayed measurement typically is millivolts. The voltage range required by pH analyzer/controllers is insufficient for many redox systems, so an expanded voltage range (typically ±1500 mVDC) is provided. An ORP analyzer/controller is designed to read absolute mVDC. ORP analyzers do not utilize temperature compensation. Changes in the Nernst slope factor are not presently compensated, perhaps because a single isothermal point is not found (as with pH measurements; see Chapter 25). Additionally, the n term in the Nernst slope factor (indicating the number of electrons that were transferred in the reaction) would have to be known in each reaction. Note: Some ORP analyzers use the resistance signal from the temperature transducer in the sensor to monitor temperature, but this measurement is not utilized to correct the mVDC values displayed.

Figure 26-4 illustrates a rugged process-style combination ORP sensor and analyzer.

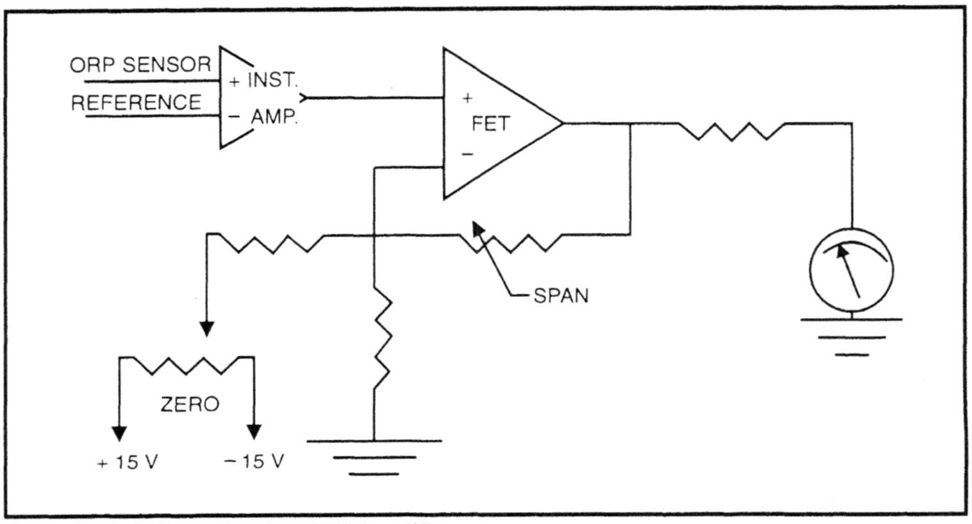

Figure 26-3. Schematic of an ORP Amplifier

Figure 26-4. Typical Process ORP Analyzer and Sensor
(Courtesy of The Foxboro Co.)

REDOX SOLUTION

The final component of the measurement system is the redox solution itself. The redox solution is a combination of two or more redox couples that form a chemical reaction which can be monitored and controlled by utilizing the output of the ORP analyzer/controller. Analogous to pH, a parallel between the two measurements may be made as follows:

	pH	Redox
Species transferred	Proton; H^+	Electron; e^-
Donor	Acid	Reducing agent
Acceptor	Base	Oxidizing agent

Characteristics of an ideal ORP solution include:

- Rapid redox equilibria
- No competition with atmospheric oxygen
- No mixed potentials
- No components that can poison sensor
- Constant temperature
- pH-controlled

The redox equilibria in solution must be rapid to allow a Nernstian response to occur. Also, the species must be in the true equilibrium concentration. A slowly drifting voltage is observed if the kinetics of the reaction are slow. Reactions which result in components leaving as gases will never reach a true equilibrium state. Drifting voltages can also occur if a counter-oxidation process competes with the redox system of interest. Oxygen from the air may also compete with a dilute redox system, depending on kinetics. However, the system can be purged with an inert gas before measurements are taken to reduce interferences from atmospheric oxygen. Unfortunately, this procedure may drive volatile chemicals (whose activity could influence the redox couples) out of solution and should be checked before this procedure is applied.

The system can be purged with an inert gas before measurements are taken to reduce interferences from atmospheric oxygen.

In systems with multiple redox couples, a mixed potential can occur. The redox couple with the highest exchange current will make the predominant contribution to the measured potential, producing a measurement error.

The potential developed by the sensor should be a function of the solution in which it is immersed, and not an artifact caused by a faulty electrode. Organic coatings and chemical/physical contaminants from the solution can cause deterioration of the ORP electrode response. The surface of the electrode should be kept as clean as possible. Cleaning suggestions are found in the section on "Maintenance" in this chapter.

Redox solutions may have high temperature coefficients. Changes in ion dissociation, activity coefficients, and interactions of ions are functions of temperature. The temperature should be noted when trying to reproduce a reaction while monitoring ORP.

When the oxidizing or reducing power is the main interest, it is possible to control to a fixed voltage and allow temperature to fluctuate. Pilot studies should be conducted to test the feasibility of this approach. A savings in materials and thermal control could result [Ref. 3].

When the oxidizing or reducing power is the main interest, it is possible to control to a fixed voltage and allow temperature to fluctuate. Pilot studies should be conducted to test the feasibility of this approach.

pH should be monitored and controlled in many ORP processes. Many redox reactions involve hydrogen ion as a reactant or product, and control of pH may have a direct effect on reaction rates. An example of this occurs in reduction of iodine with sodium sulfite:

It is a good idea to have a second ORP sensor available to referee a questionable value. If two sensors agree within ±10 mV of each other, it will verify the potential developed by the original sensor.

$$Na_2SO_3 + I_2 + H_2O \Leftrightarrow Na_2SO_4 + 2HI \qquad (26\text{-}10)$$

The two half-cell reactions (only reactive ions shown) are

$$I_2 + 2e^- \Leftrightarrow 2I^- \quad \text{(reduction)} \qquad (26\text{-}11)$$

$$SO_3^{-2} + H_2O \Leftrightarrow SO_4^{-2} + 2H^+ + 2e^- \quad \text{(oxidation)} \qquad (26\text{-}12)$$

When an excess of sodium sulfite exists in the solution, the potential observed will be dependent on Equation (26-13). However, because the term [H^+] appears in the equation, the potential depends on the hydrogen ion activity or the pH:

$$E = E^o + \frac{0.05916}{n} \log \frac{[SO_3^{-2}]}{[SO_4^{-2}]\,[2H^+]^2} \qquad (26\text{-}13)$$

When iodine (the oxidant) is in excess, the hydrogen ion concentration does not come into play. The observed potential is given by:

$$E = E^o + 0.05916 \log [I^-]^2 \qquad (26\text{-}14)$$

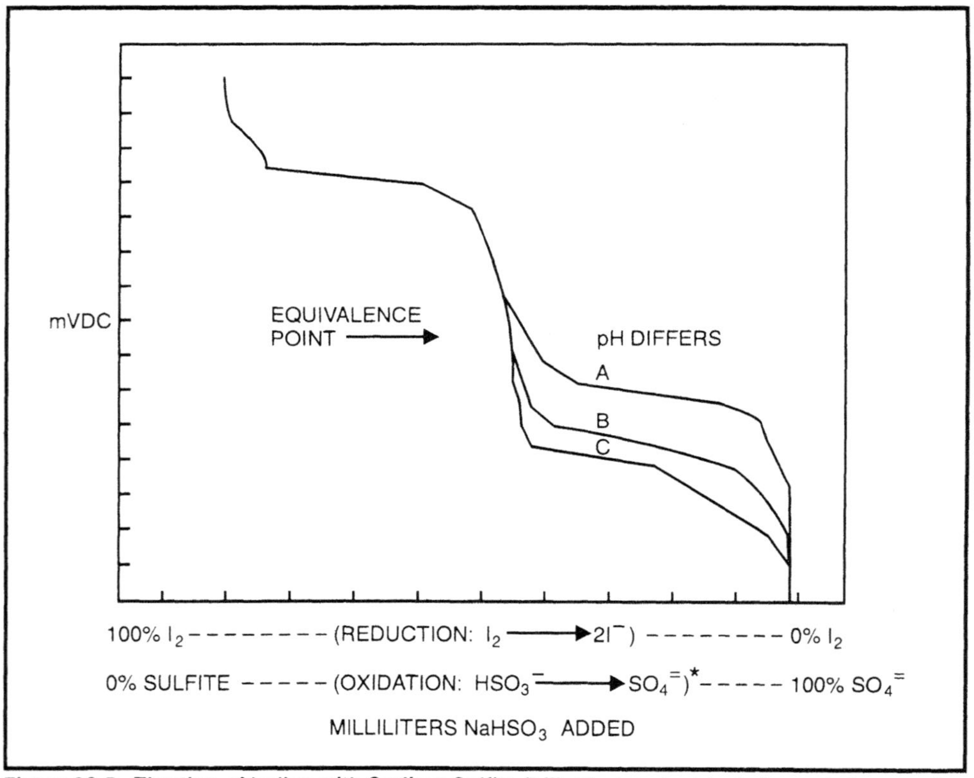

Figure 26-5. Titration of Iodine with Sodium Sulfite (* Illustrates electrical-charge balance only; not chemically-balanced)

Figure 26-5 illustrates this system. Although oxidation and reduction occur simultaneously, only one reagent is in excess at any one time, and the potential developed at the ORP sensor is dependent on the reagent that is in excess, not that which has been consumed. When sodium sulfite is in excess, the potential is dependent on Equation (26-12) and on the pH. This type of plot is considered a titration curve. The *y*-axis contains the observed potential in millivolts (mVDC), and the *x*-axis contains the volume (milliliters) of sodium sulfite ($NaHSO_3$) added to the iodine.

Calculating an ORP Potential

The methods and calculations used as examples in this section are based on pure ORP situations. Although the calculations may seem tedious, and not appropriate in a practical guidebook, they do provide a practical understanding of the chemistry that leads to the measured potential.

In practice, it is not possible to calculate the concentration of a single redox component from an ORP potential, since the potential is always determined as a ratio of oxidized-to-reduced species. A combination of redox, pH, and activity of dissolved species can affect solution equilibria. Extensive data relating pH and ORP potentials are available in more advanced texts (e.g., [Ref. 4]).

It is not possible to calculate the concentration of a single redox component from an ORP potential, since the potential is always determined as a ratio of oxidized-to-reduced species.

Most of the chemical properties of redox reactions can be predicted or calculated. It is possible to calculate concentrations of all the species at equilibrium, equilibrium constants, and cell potentials that are produced; to predict whether the reaction will occur spontaneously; and to balance redox equations. Unfortunately, the kinetics of the reactions are not predictable from these data, and reactions that theoretically should occur on thermodynamic grounds may not occur at all!

Recall the reaction presented in the opening paragraphs of this chapter. It was stated that every oxidation reaction was accompanied by a corresponding reduction reaction and vice versa:

$$A_{ox} + B_{re} \Leftrightarrow A_{re} + B_{ox} \qquad (26\text{-}1)$$

and that the reaction in Equation (26-1) was the sum of two half-cell reactions:

$$A_{ox} + ne^- \Leftrightarrow A_{re} \qquad (26\text{-}2)$$

$$B_{re} \Leftrightarrow B_{ox} + ne^- \qquad (26\text{-}3)$$

Standard half-cell potentials have been determined for many reactions and are tabulated as reduction potentials in accordance with the International Union of Pure and Applied Chemistry (IUPAC) convention. The reduction potential has the same sign (– or +) as the log of the equilibrium constant for the reduction reaction. Table 26-3 presents a partial list of standard reduction potentials. The half-cell potentials represent the *free energy* per electron transferred. The potentials listed may be thought of as a *scale of power*, where measurements made in strongly reducing systems are, by convention, large negative voltages, and those made in strongly oxidizing systems are usually large positive voltages. The H_2/H^+ half-couple occupies the middle of the scale and has the potential of 0.000 VDC.

Table 26-3. Partial List of Standard Reduction Potentials

Half-Cell Reaction	$E°$, emf, VDC
$O_3 + 2H^+ + 2e^- \leftrightarrow O_2 + H_2O_2$	+2.07
$S_2O_8^{-2} + 2e^- \leftrightarrow 2SO_4^{-2}$	+2.00
$Co^{+3} + 1e^- \leftrightarrow Co^{+2}$	+1.842
$H_2O_2 + 2H^+ + 2e^- \leftrightarrow 2H_2O$	+1.776
$Au^+ + 1e^- \leftrightarrow Au$	+1.68
$MnO_4^- + 4H^+ + 3e^- \leftrightarrow MnO_2 + 2H_2O$	+1.679
$HClO + H^+ + e^- \leftrightarrow 1/2Cl_2 + H_2O$	+1.63
$MnO_4^- + 8H^+ + 5e^- \leftrightarrow Mn^{+2} + 4H_2O$	+1.491
$Au^{+3} + 3e^- \leftrightarrow Au$	+1.42
$Cl_2 + 2e^- \leftrightarrow 2Cl^-$	+1.3584
$O_2 + 4H^+ + 4e^- \leftrightarrow 2H_2O$	+1.229
$Pt^{+2} + 2e^- \leftrightarrow Pt$	~1.2
$2IO_3^- + 12H^+ + 10e^- \leftrightarrow I_2 + 6H_2O$	+1.19
$ClO^- + 2H_2O + 2e^- \leftrightarrow Cl^- + 2OH^-$	+0.90
$Ag^+ + 1e^- \leftrightarrow Ag$	+0.7996
$Fe^{+3} + 1e^- \leftrightarrow Fe^{+2}$	+0.770
Quinhydrone electrode, H^+, $a = 1$	+0.6995
$O_2 + 2H^+ + 2e^- \leftrightarrow 2H_2O_2$	+0.682
$ClO_3^- + 3H_2O + 6e^- \leftrightarrow Cl^- + 6OH^-$	+0.62
$I_2 + 2e^- \leftrightarrow 2I^-$	+0.535
$Cu^{+2} + 2e^- \leftrightarrow Cu$	+0.522
$CNO^- + 2H^+ + 2e^- \leftrightarrow 2HCN$	+0.33
$AgCl(s) + 1e^- \leftrightarrow Ag + Cl^-$	+0.2223
$Cu^{+2} + 1e^- \leftrightarrow Cu^+$	+0.158
$2H^+ + 2e^- \leftrightarrow H_2$	+0.000
$O_2 + H_2O + 2e^- \leftrightarrow HO_2^- + OH^-$	–0.076
$Pb^{+2} + 2e^- \leftrightarrow Pb$	–0.1263
$Ni^{+2} + 2e^- \leftrightarrow Ni$	–0.230
$Co^{+2} + 2e^- \leftrightarrow Co$	–0.28
$Cr^{+3} + 1e^- \leftrightarrow Cr^{+2}$	–0.4026
$Ag_2S + 2e^- \leftrightarrow 2Ag + S^{-2}$	–0.7051
$Zn^{+2} + 2e^- \leftrightarrow Zn$	–0.7628
$2H_2O + 2e^- \leftrightarrow H_2 + 2OH^-$	–0.8277
$Na^+ + 1e^- \leftrightarrow Na(s)$	–2.7109
$Li^+ + 1e^- \leftrightarrow Li(s)$	–3.045

BALANCING EQUATIONS BY THE ION-ELECTRON METHOD

An electrochemical cell reaction is written so that all atoms appearing on one side of the equation are balanced in type and number by all atoms appearing in the other side of the equation. The charge total on each side must also be equal.

The rules used to balance an equation by the ion-electron method follow:

1. First break down the redox reaction into half-reactions; one half representing the oxidation step, the other half the reduction step.
2. Balance each individual half-cell in terms of ion charges and electrons. Water molecules (H_2O), protons (H^+), and hydroxyl ions (OH^-) can be added as required.
3. Multiply either/or both equations by a factor so that the number of electrons lost in the oxidation is equal to the number of electrons gained in the reduction.
4. Add the half-cell reactions together.
5. Simplify the result; cancel electrons, water, and common factors on each side of the equation.

When redox equations are balanced by this method, it is important that the correct half-cell be selected from Table 26-3.

BALANCING OXIDATION-REDUCTION EQUATIONS — AN EXAMPLE

Dichromate ion will oxidize ferrous ion to ferric ion in an acid solution. Balance the equation:

$$Cr_2O_7^{-2} + H^+ + Fe^{+2} \rightarrow Cr^{+3} + Fe^{+3} + H_2O \quad \textbf{(26-15)}$$

The oxidation half-reaction ($Fe^{+2} \rightarrow Fe^{+3}$) is balanced by adding an electron to the right side of the equation:

$$Fe^{+2} \rightarrow Fe^{+3} + e^- \quad \textbf{(26-16)}$$

The reduction half-reaction ($Cr_2O_7^{-2} + ?H^+ \rightarrow ?Cr^{+3} + ?H_2O$) is balanced by inspection. Seven atoms of oxygen from $Cr_2O_7^{-2}$ ion require 14 H^+ ions to produce seven molecules of water. Two Cr^{+3} ions are formed in the reaction:

$$Cr_2O_7^{-2} + 14H^+ \rightarrow 2Cr^{+3} + 7H_2O \quad \textbf{(26-17)}$$

In terms of electrical charge, six electrons must be added to the left side of the equation:

$$Cr_2O_7^{-2} + 14H^+ + 6e^- \rightarrow 2Cr^{+3} + 7H_2O \quad \textbf{(26-18)}$$

−2 +14 −6 +6

(continued)

(continued)

Before adding both half-reactions, the same number of electrons must be exchanged. The oxidation half-reaction must be multiplied by six:

$$6Fe^{+2} \rightarrow 6Fe^{+3} + 6e^{-} \quad \textbf{(26-19)}$$

Upon addition of the balanced half-cells, the six electrons on either side of the equation will cancel and the reaction becomes:

$$\textbf{OXIDATION:} \quad 6Fe^{+2} \rightarrow 6Fe^{+3} + 6e^{-} \quad \textbf{(26-20)}$$

$$\textbf{REDUCTION:} \quad + Cr_2O_7^{-2} + 14H^{+} + 6e^{-} \rightarrow 2Cr^{+3} + 7H_2O \quad \textbf{(26-21)}$$

$$\textbf{OVERALL:} \quad Cr_2O_7^{-2} + 6Fe^{+2} + 14H^{+} \rightarrow 6Fe^{+3} + 2Cr^{+3} + 7H_2O \quad \textbf{(26-22)}$$

When redox equations are balanced by this method, it is important that the correct half-cell be selected from Table 26-3.

CALCULATING E^0 AND AN EQUILIBRIUM CONSTANT K

The standard electrode potentials and the equilibrium constant for a reaction can also easily be determined. The overall standard potential, E^0, is determined by following these steps:

1. Write the balanced equation for the overall equilibrium reaction.
2. Express the equation in terms of two suitable half-reactions.
3. Find the standard electrode (reduction) potentials for each half-reaction.
4. If the half-cell is a reduction (electrons on the left), the half-reaction potential is as given. If the half-reaction is written as an oxidation (electrons on the right), change the sign of the half-reaction potential.
5. Add the half-reaction potentials to get the overall standard potential, E^0, for the reaction.

CALCULATING THE EQUILIBRIUM CONSTANT, K, AND E°

Consider the balanced equation shown below:

$$5Cr^{+2} + MnO_4^{-} + 8H^{+} \rightarrow Cr^{+3} + Mn^{+2} + 4H_2O \quad \textbf{(26-23)}$$

In this example, we will determine what the standard potential will be when Cr^{+2} reacts with permanganate ions. The balanced equation for the reaction is as shown in Equation (26-17).

The oxidation reaction where Cr^{+2}, the reducing agent, is converted to Cr^{+3} is shown below along with its half-cell potential (shown as a reduction):

$$Cr^{+3} + 1e^{-} \rightarrow Cr^{+2} \qquad \textbf{–0.41 V} \quad \textbf{(26-24)}$$

(continued)

(continued)

The reduction reaction where permanganate ion is reduced to Mn^{++} and the half-cell potential follow:

$$MnO_4^- + 8H^+ + 5e^- \rightarrow Mn^{+2} + 4H_2O \qquad \mathbf{1.491\ V} \qquad \mathbf{(26\text{-}25)}$$

In order to add these two equations together, the same number of electrons must be transferred; this can be done by multiplying each equation by the number of electrons appearing in the other. The oxidation reaction is rewritten to reflect this (Equation [26-18] is multiplied by a factor of five) and is now written as an oxidation, and a sign change has been indicated in accordance with step 4 above. The reduction equation will remain the same (multiplication by one):

$$5Cr^{+2} \rightarrow 5Cr^{+3} + 5e^- \qquad \mathbf{0.41\ V} \qquad \mathbf{(26\text{-}26)}$$

Since five electrons appear on both sides of the equation, it is balanced and the number of electrons involved in the oxidation of Cr^{+2} are the same as the number involved in the reduction of MnO_4^-.

The overall standard potential for the reaction is:

$$\boldsymbol{E^o = 1.491 + 0.41 = 1.901\ V} \qquad \mathbf{(26\text{-}27)}$$

The equilibrium constant for the reaction at 25°C can be calculated using Equation (26-28):

$$\boldsymbol{\log K = 16.9\ nE^o} \qquad \mathbf{(26\text{-}28)}$$

For the reaction above, $E^o = 1.901$ V and $n = 5$. The equilibrium constant K would be equal to:

$$\boldsymbol{\log K = 16.9\ (5)\ 1.901 = 160.6345} \qquad \mathbf{(26\text{-}29)}$$

$$\boldsymbol{K = 4.310 \times 10^{+160}} \qquad \mathbf{(26\text{-}30)}$$

SPONTANEITY AND FREE ENERGY (G°)

The equilibrium constant just calculated allows us to predict whether or not a given mixture of reactants and products will proceed to form more products. The relationship between free energy (G°) and the equilibrium constant is given by:

$$G^o = -2.303\ RT \log K \qquad (26\text{-}31)$$

When K is a value larger than 1, the calculated value of G^o is negative, and the reaction is considered spontaneous and is strongly favored. When K is less than 1, the calculated value of G^o is positive and the reaction is not spontaneous.

When K is a value larger than 1, the calculated value of G^o is negative, and the reaction is considered spontaneous and is strongly favored. When K is less than 1, the calculated value of G^o is positive and the reaction is not spontaneous.

The mathematical approach to chemical reactions will be summarized before introducing actual applications that provide guidance in predicting the feasibility of using ORP measurements to monitor or control processes.

First, balance the reaction that is occurring in the process. In certain cases, more than one reaction may occur.

The measurable redox potential as a function of concentration is given by the Nernst equation:

$$E = E^o + \frac{0.05916}{n}\log\frac{[\text{oxidized}]}{[\text{reduced}]} \qquad (26\text{-}4)$$

When [oxidized] = [reduced], at the equivalence point, $E = E^o$.

To convert potentials referenced from one reference system to another use:

$$E_{observed} = E_{ORP\ sensor} - E_{ref} \tag{26-9}$$

where $E_{ORP\ sensor}$ is the calculated potential versus the SHE, E_{ref} is the standard potential of the reference electrode you will be using (Tables 26-1 and 26-2), and $E_{observed}$ is the potential you should expect for your system in the solution.

From the equilibrium constant calculation, the likelihood that the reaction will occur can be predicted. The kinetics (actual occurrence) of the reaction cannot be predicted from the equations reviewed.

Common Applications

Successful applications of ORP in process environments include chemical reduction of chromate wastes (see below) and the chemical oxidation of cyanide waste with chlorine.

Enzymes, yeasts, amino acids, anaerobic and aerobic microorganisms, and other biological organisms are highly dependent on the oxidation-reduction medium in which they exist. ORP sensors are used to monitor the redox potential in bioprocesses. Although mixed parameters are measured, in some applications the redox potential is a valuable empirical parameter. For example, one advantage of the redox electrode is an indication of an abrupt potential drop when changing from aerobic to anaerobic conditions [Ref. 5].

The most common use of ORP consists of maintaining the ORP at or above the practical value necessary to complete the reaction of one of the reactants [Ref. 3]. Figure 26-6 shows a combination ORP sensor used in fermentation processes. Other applications depend only upon the imaginations of the plant engineers or chemists.

Pilot runs of the process, with comparisons to the amount of oxidant or reductant used, will demonstrate whether ORP is the proper technology for the control that is required.

Pilot runs of the process, with comparisons to the amount of oxidant or reductant used, will demonstrate whether ORP is the proper technology for the control that is required.

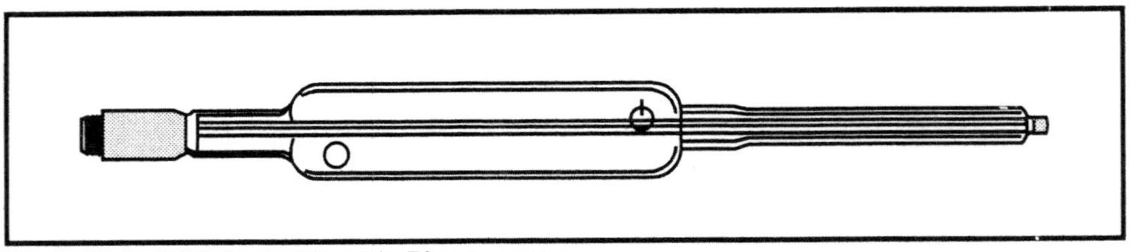

Figure 26-6. ORP Sensor Used in Bioprocesses
(Courtesy of Mettler Toledo Process Analytical, Inc.)

Chromate Waste Treatment

Highly toxic hexavalent chromate solutions are produced from plating and rinsing procedures used in the plating industry. These waste products require proper treatment to neutralize the highly toxic hexavalent chromate into the less toxic insoluble trivalent form, which is capable of being precipitated as a hydroxide at a pH of about 8.5. Although industrial processes vary between facilities, ORP analysis has been used with success to monitor the reduction of chromium to the trivalent form.

Three reagents are commonly used to reduce the chromate waste: ferrous sulfate, sodium metabisulfite, and sulfur dioxide. Ferrous sulfate and sodium metabisulfite require pH adjustment to approximately 2.5 for rapid reduction to occur. Sulfur dioxide gas forms sulfurous acid in solution and typically does not require additional acid additions for the reduction to occur. The chemical reactions that occur with each of the reagents are given in Table 26-4.

The reduction of hexavalent chromate in the waste stream is monitored using an ORP analyzer. The relationship between ORP potentials and units of reagent added is shown in Figure 26-7. Recall that ORP is a measure of the ratio of the concentration of oxidized to reduced species in the waste stream, and not a single concentration, so the relationship is not linear. Control for the reaction is a single potential that represents the conversion of all of the chromium to the trivalent form. This potential can be calculated or derived experimentally in a laboratory setting.

Remember to correct the calculated potential for the reference electrode you will be using.

Outputs from the ORP analyzer can signal a process controller to add the reducing reagent until the tank reaches the appropriate redox potential. Nominal potentials (versus the SHE) for reactions with the reagents discussed also are included in Table 26-4.

Once the chromium has been converted to the trivalent state, it is in the appropriate form to complex with hydroxide. Either lime or caustic is added to the stream to bring the pH near 8.5. The chrome precipitate is then allowed to settle out and can be removed as a sludge. Figure 26-8 illustrates the relationship between pH and trivalent chromium solubility.

This application illustrates some of the principles discussed earlier. A generalized schematic of the instrumentation is shown in Figure 26-9.

Table 26-4. Neutralization Reactions of Chromium to the Trivalent Form Using Different Reagents

Reagent and Reduction Potential versus SHE	Type of Reaction	Equation
Ferrous sulfate, +500 mV at pH 2	Hydrolysis	$CrO_3 + H_2O \rightarrow H_2CrO_4$
	Reduction	$H_2CrO_4 + 6FeSO_4 + H_2SO_4 \rightarrow$
		$Cr_2(SO_4)_3 + 3Fe_2(SO_4)_3 + 8H_2O$
	Precipitation	$Cr_2(SO_4)_3 + 3Ca(OH)_2 \rightarrow 2Cr(OH)_3 + 3CaSO_4$
Sodium metabisulfite, +380 mV at pH 2.5	Hydrolysis	$Na_2S_2O_5 + H_2O \rightarrow 2NaHSO_3$
	Reduction	$H_2CrO_4 + 3NaHSO_3 + 3H_2SO_4 \rightarrow$
		$Cr_2(SO_4)_3 + 3NaHSO_4 + 5H_2O$
	Precipitation	$Cr_2(SO_4)_3 + 3Ca(OH)_2 \rightarrow 2Cr(OH)_3 + 3CaSO_4$
Sulfur dioxide, +165 mV at pH 2.9	Hydrolysis	$SO_2 + H_2O \rightarrow H_2SO_3$
	Reduction	$H_2CrO_4 + H_2SO_3 \rightarrow Cr_2(SO_4)_3 + 5H_2O$
	Precipitation	$Cr_2(SO_4)_3 + 3Ca(OH)_2 \rightarrow 2Cr(OH)_3 + 3CaSO_4$

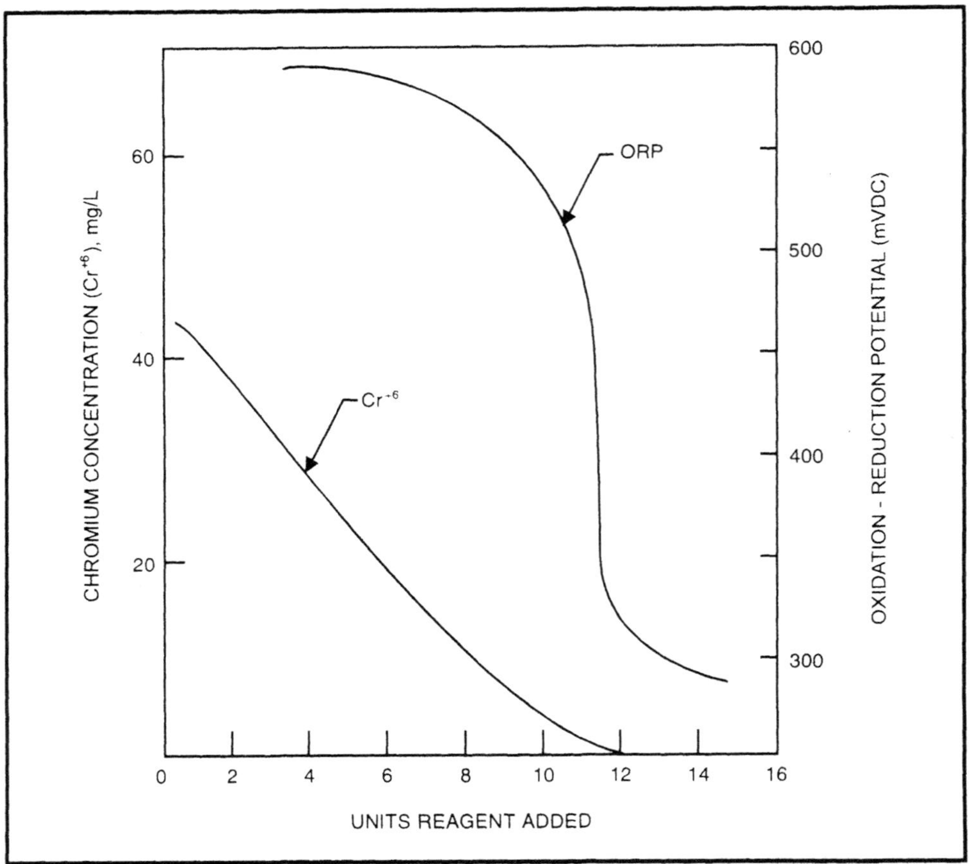

Figure 26-7. ORP Potential Observed during Chromium Reduction with Sodium Metabisulfite
(SHE reference)

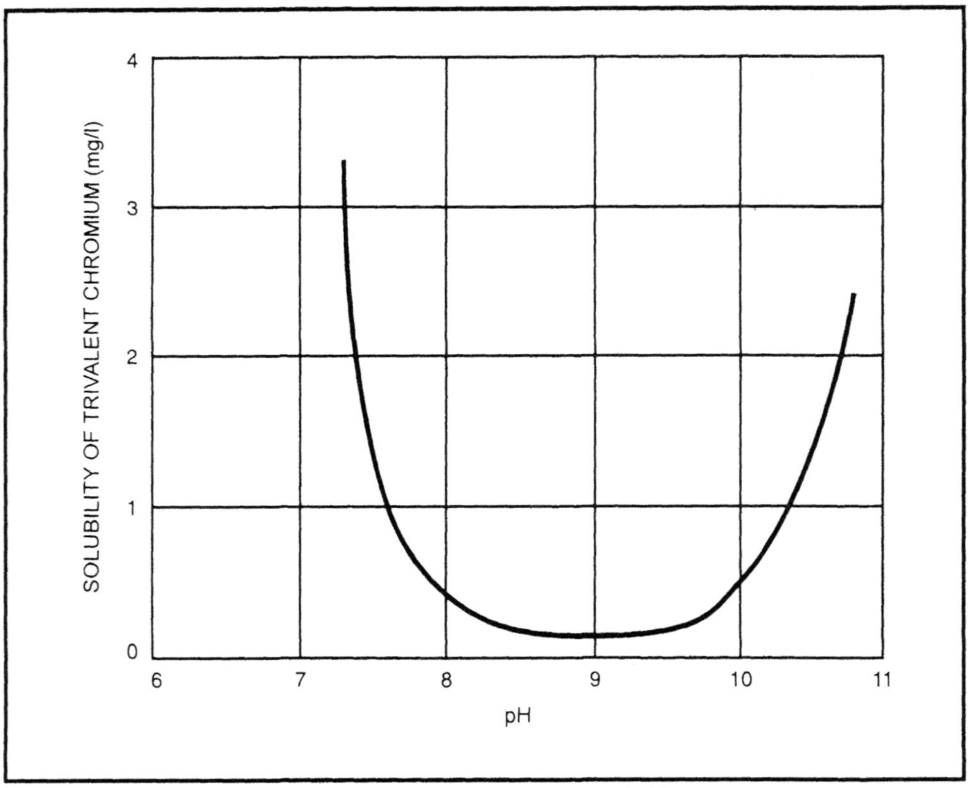

Figure 26-8. Solubility of Trivalent Chromium as a Function of pH
(Chrome precipitates out as solid sludge.)

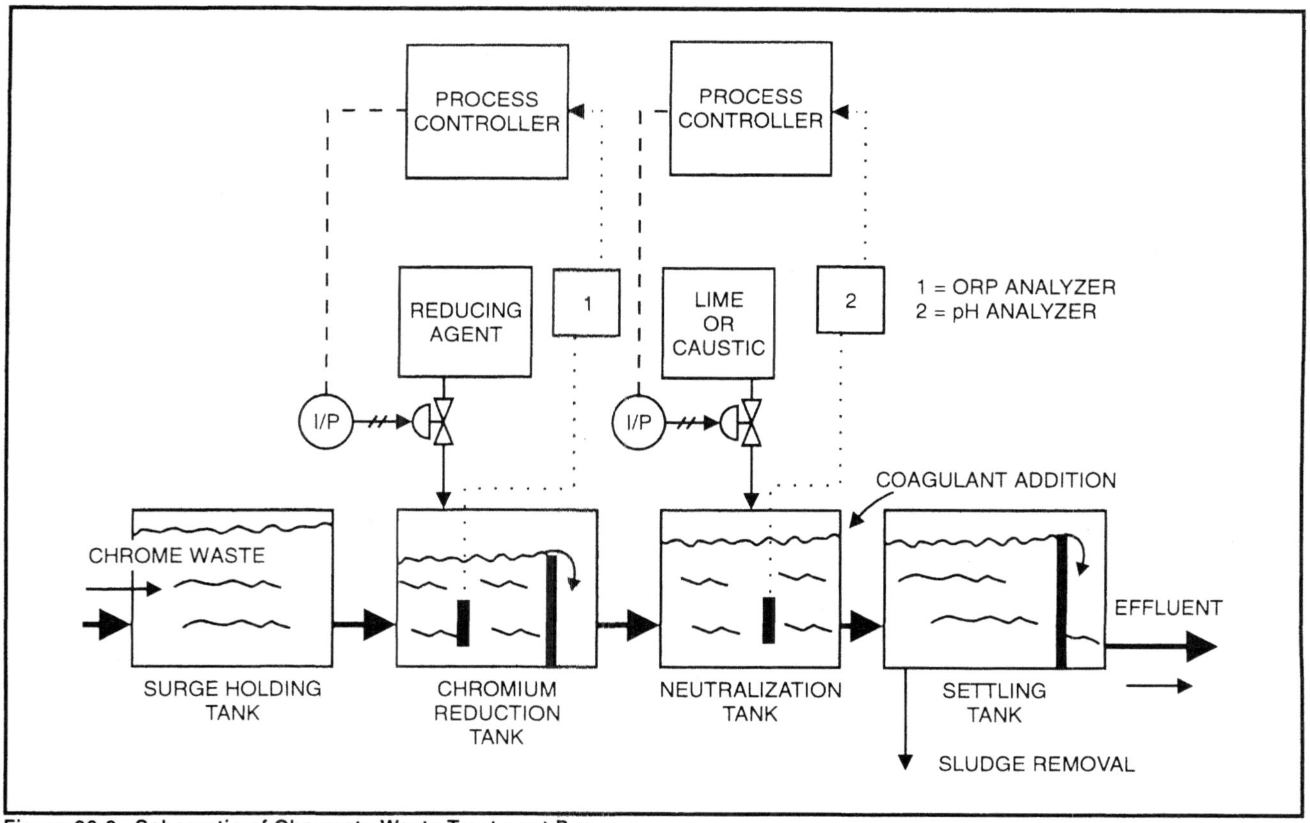

Figure 26-9. Schematic of Chromate Waste Treatment Process

Options and Accessories

ORP sensors and analyzers utilize approximately the same options and accessories as pH sensors. The following list includes many of the options available from various manufacturers:

ORP Sensor Options

- Integral preamplifier
- Flowing or gelled reference electrode
- Bridged electrolyte (double junction)
- Platinum, gold, or other noble metal reference electrode
- Billet, disk, or metal band
- Plastic or glass body
- Detachable cables/various connectors
- Various mounting configurations for pipe or tank
- Electrode cleaning systems
- Electrode calibration systems
- Steam sterilizable

ORP Analyzer Options

- Operating voltage
- Current or voltage outputs
- Various mounting options, including panel, surface, and pipe

- Alarms with dead-bands or timers
- Analog or digital displays
- Electrical classifications

Table 26-5 includes many of the major manufacturers of ORP equipment. Check the *ISA Directory of Instrumentation* under "pH Analyzers" for a more complete listing of ORP sensor and analyzer distributors.

Table 26-5. Major U.S. Manufacturers of ORP Analyzers

The Foxboro Company 600 North Bedford St. East Bridgewater, MA 02333
Great Lakes Instruments 9020 W. Dean Road Milwaukee, WI 53224
Honeywell Inc. 795 Horsham Rd. Horsham, PA 19044
Johnson-Yokogawa Corporation 4 Dart Rd. Newnan, GA 30265-1040
Rosemount Analytical Inc. 2400 Barranca Parkway Irvine, CA 92714
TBI-Bailey Controls 2175 Lockheed Way Carson City, NV 89701

Limitations

The measurement of ORP potentials is limited to systems in which a change in oxidation or reduction occurs. Not all redox reactions are easily measured using an ORP sensor or analyzer. Problems with the ORP sensor, reference electrode, or the chemistry of the system can contribute to drifting or erratic voltages. Oxygen from the air or bubbles from agitation have been known to interfere with the measurement. In summary, it is up to the user to determine whether a measurement is sufficiently stable to control a process.

Problems with the ORP sensor, reference electrode, or the chemistry of the system can contribute to drifting or erratic voltages.

Selection and Installation Hints

Utilize the following criteria when choosing and installing an ORP sensor:

1. Choose sensors with the correct temperature and pressure limits and chemical inertness.
2. Mount the sensor suitably for the process of interest (see Figure 26-10).
3. A properly-installed sensor responds to a representative sample of the process.
4. The sensor should see only a moderate flow rate. Excessive flow past the sensor could result in an incorrect reading, and premature failure.
5. Position the sensor so that the sensing surface faces toward the flow. This will minimize air bubbles and fluid cavitation.

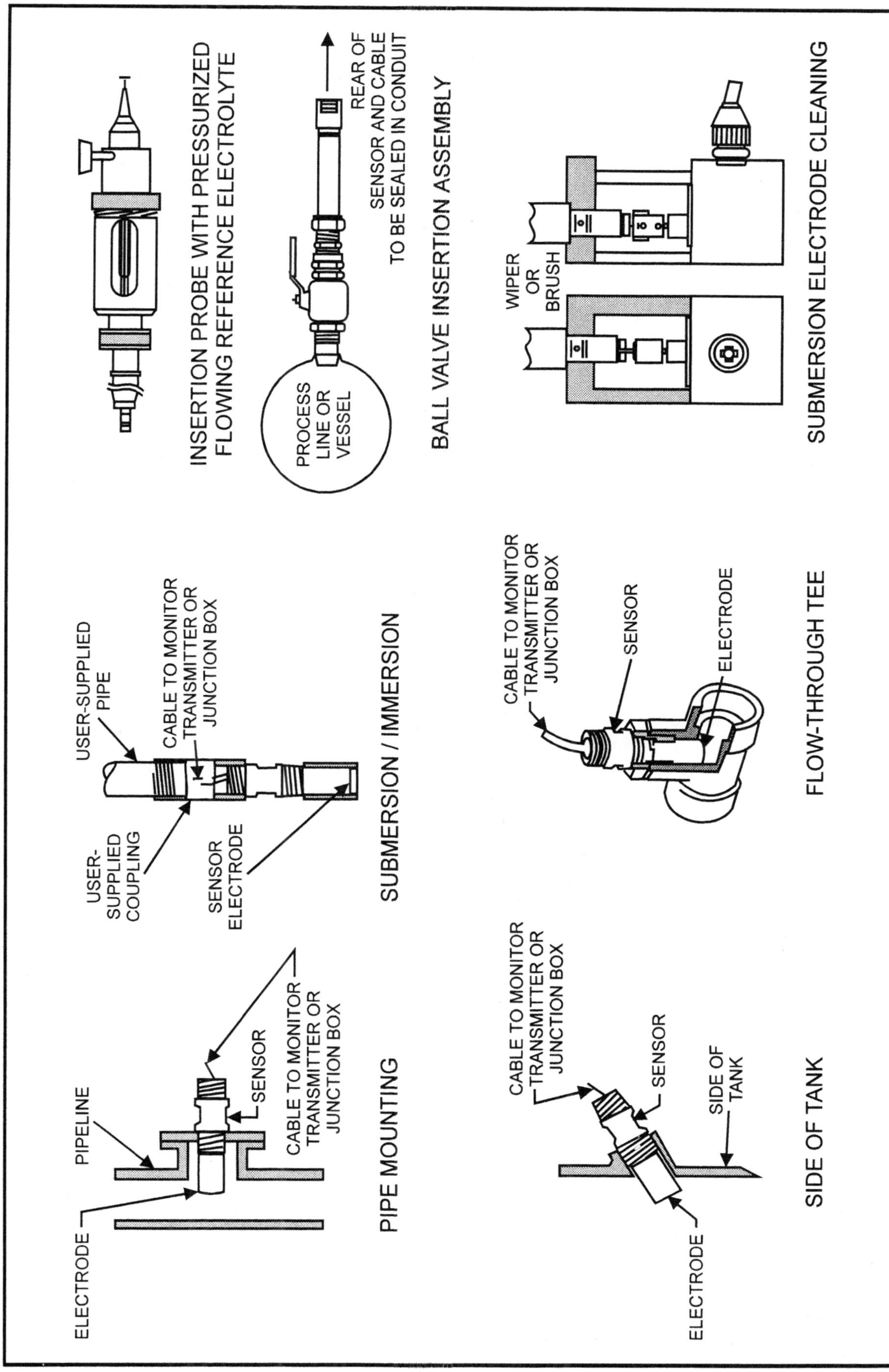

Figure 26-10. Various Sensor Mounting Configurations
(Hardware may vary with manufacturer.)

6. Mount sensors within a 0° to 45° angle from the vertical. The reference electrode may become electrically discontinuous (open circuit) if sensors are operated inverted due to electrolyte flow away from internal components.
7. Sensor cables should be placed in conduit or protected from the environment.
8. Analyzers must have the correct electrical-hazard classification for the area in which the monitor will be used.
9. Both sensor and analyzer should be in an area accessible for maintenance.

Typical Calibration Requirements

The Analyzer

An ORP analyzer is a measuring device for electrode potential. A factory-calibrated analyzer should not, in theory, require recalibration.

Since the analyzer measures an absolute voltage, it can be checked by simply shorting the reference and sensor inputs together. When the unit is powered, the display should read 0 mV. If this value is not correct, the monitor should be calibrated using a voltage standard (with <1% accuracy). An electronic calibration procedure should be included in the manufacturer's instruction manual. This is typically a two-point calibration performed on both extremes of the voltage range that can be displayed on the instrument.

The Sensor

To assure that the sensor has continuity, a resistance measurement is taken at the ORP indicating sensor surface and the cable lead (or connector) is used to carry the signal. This measurement is used only on sensors without integral preamplifiers. A sensor that has electrical continuity will read an electrical short or very low resistance.

To assure that the sensor is functioning properly once it is connected to the analyzer, it should be placed into an ORP solution of known value.

To assure that the sensor is functioning properly once it is connected to the analyzer, it should be placed into an ORP solution of known value. The solution should be stable in air (well-poised), and chosen to correspond to a value close to the potential being monitored in the process. Adjustment of the analyzer is typically not necessary if the sensor does not produce *exact* agreement with the solution's known value. The sensor develops voltages corresponding to the ratio of total oxidized species to total reduced species in the solution, minus the reference electrode's potential. The ORP test solutions recommended should produce rapid and reproducible voltage values. The exchange density in the solutions recommended is high and well-poised. The formulations for a few of these are found in Table 26-6. Be aware that a sensor may still be unstable in a process even though it was very stable in the test solutions. Additionally, two sensors may indicate the same potentials in these well-poised solutions, but differ significantly in poorly-poised solutions.

A redox sensor can also be tested in pH buffers (0 to 7.00 range only) if the buffers are saturated with quinhydrone prior to measurement. The lifetime of these solutions is very short (approximately 1 h).

The potentials of common pH buffers saturated with quinhydrone versus the SHE at 25°C are listed below:

pH 7 buffer; phosphate buffer	285 mV
pH 4 buffer; phthalate buffer	462 mV
pH 1.68 buffer; potassium tetroxalate	600 mV

Redox standards having negative potentials do not exist. This is because solutions with negative potentials would be rapidly oxidized with oxygen from the air. This is also true for pH buffers above pH 7 saturated with quinhydrone.

Table 26-6. Formulations for ORP Test Solutions

Formulation 1 [Ref. 6]: +675 mVDC versus SHE
1 *M* H_2SO_4; sulfuric acid 0.100 *M* $Fe(NH_4)_2(SO_4)_2 \cdot 6H_2O$; ferrous ammonium sulfate 0.100 *M* $Fe(NH_4)(SO_4)_2 \cdot 12H_2O$; ferric ammonium sulfate See Table 26-7 for potentials of a platinum ORP sensor in this solution versus various reference electrodes.
Formulation 2: +503 mVDC versus SHE
0.01 *M* $K_4Fe(Cn)_6$ 0.05 *M* $K_3Fe(Cn)_6$
Formulation 3: +437 mVDC versus SHE
0.10 M $K_4Fe(Cn)_6$ 0.05 M K_3 $FE(Cn)_6$

Table 26-7. Potentials of the Platinum Electrode Versus Various Reference Electrodes at 25°C in Standard Ferrous-Ferric Solution (Formulation 1 from Table 26-6) [Ref.6]

Reference Electrode	emf, mVDC
Hg, Hg_2Cl_2, saturated KCl (SCE)	+430
Ag, AgCl, 1.00 *M* KCl	+439
Ag, AgCl, 4.00 *M* KCl	+475
Ag, AgCl, saturated KCl	+476
Pt, H_2 (p = 1), H^+ (a = 1) (SHE)	+675

Troubleshooting and Maintenance

An electrode should be tested in a standard solution to verify that it is operational. A list of possible symptoms and solutions is presented in Table 26-8.

Table 26-8. Troubleshooting Guide

Symptom	Solution
The magnitude of the voltage is correct, but the polarity is incorrect.	Check to assure that sensor and reference connections are made properly. Verify that polarity switch (if present) is in correct position.
The electrode is very unstable in the test solution, but is reading near the expected voltage value.	This may be the symptom of a contaminated ORP surface or a contaminated reference junction. Clean ORP surface. Clean or replace reference junction. Replace sensor.
The electrode is very unstable in the test solution and is not reading anywhere near the expected voltage value.	See immediately above. Check to verify that the ORP sensor is continuous. Substitute known working reference electrode into circuit and remove reference connection. Replace sensor.
The electrode has the correct polarity, but is reading a large voltage difference from the published specification.	Verify that the correct reference electrode was used in calculations for expected voltage.
The electrode has the correct polarity, but is reading a different voltage from the published specification.	See immediately above. ORP surface is contaminated. Clean. Reference junction is contaminated. Clean.
The electrode has the correct polarity, but is reading a large voltage difference from the published specification.	Verify that the correct reference electrode was used in calculations for expected voltage.

Maintenance

ORP sensors may require periodic cleanings to remove deposits that form. Generally speaking, choose a solvent system capable of dissolving the film.

Remember, like dissolves like. *For example, if the sensor is coated with an oily film, use water and a household detergent (such as Joy or Dawn) capable of solubilizing the oil to remove it. A thorough rinsing in clean water will often improve the response of an ORP sensor.*

Other treatments using chemicals are listed in Table 26-9. Care should be taken in using these chemicals. The optimum treatment for the ORP surface often can damage the reference junction if the ORP surface and the reference junction are combined in one body.

For mechanical cleaning, lightly polish the ORP surface with an extremely fine-grained emery cloth or powdered alumina. A smooth surface will reach equilibrium faster than a roughened one.

Table 26-9. Chemical Treatments for ORP Sensor

(1) Rinse electrode in concentrated sulfuric acid (H_2SO_4). Soak for 10 min in 50°C solution of 3% potassium dichromate in 10% sulfuric acid. Soak in water for several hours, followed by conditioning in 10% ascorbic acid dissolved in 0.1 *M* HCl.
(2) Soak in alcoholic KOH. Rinse in water.
(3) Clean in aqua regia. Rinse in water.
(4) To prepare an oxidized coating on a platinum electrode: Adjust 10 to 15% NaOCl solution to pH 7 with 1 *M* HCl. Soak electrode for a few minutes. Rinse in water.
(5) To prepare a reduced surface on a platinum electrode: Soak electrode in a ~0.1 *M* $FeSO_4$ solution for a few minutes. Rinse in water.

Reference Electrode Problems

A well-maintained laboratory-style reference electrode, stored with its junction soaking in KCl solution, can aid in troubleshooting a suspect reference electrode. The two potentials can be compared to one another on a regular basis to determine when fill solutions, junctions, or total electrodes should be replaced.

A well-maintained laboratory-style reference electrode, stored with its junction soaking in KCl solution, can aid in troubleshooting a suspect reference electrode. The two potentials can be compared to one another on a regular basis to determine when fill solutions, junctions, or total electrodes should be replaced.

KCl electrolytes used in Ag/AgCl reference electrodes require saturation with AgCl to prevent dissolution of the internal silver wire. The electrolyte also contacts the process sample through a liquid junction and reacts with constituents of the sample. Sulfides (or other compounds capable of forming sparingly-soluble silver precipitates) have caused problems with measurements. Darkening of the junction as well as drifting or erratic potentials are indications of a troubled reference junction. Reference cleaning solutions are available from some manufacturers that are quite effective in dissolving sparingly-soluble silver precipitates from the junctions. Some electrode designs offer an external AgCl-free salt bridge chamber to contact the sample. These are sometimes called bridged or double-junction reference electrodes. This style of reference electrode is available as a separate electrode and in combination electrode designs. When the electrolyte is replaceable, and is replaced frequently, junction problems are greatly reduced. In cells having fixed, nonreplaceable bridge electrolytes, diffusion between chambers masks any benefits of this design, except perhaps when both chambers utilize gel.

Some manufacturers and users have switched to reference electrodes containing gel electrolytes. These sensors may become dehydrated from storage in dry conditions. A new installation that involves an electrode such as this might appear as an open circuit or one with an extremely high resistance and may alarm intelligent equipment, indicating that the reference connection is electrically open.

This problem can be fixed by allowing the sensor to stand in a container of warmed KCl solution for at least 15 min, then letting the reference cool to ambient temperature while still immersed in the solution. The expansion during heating and vacuum formed during cooling normally will allow rehydration of the gel to occur. Because little else can be done to revive a troubled gel-filled electrode, these are simply thrown away when problems occur.

Another approach that is used to "fix" some electrodes is to replace the junction with a new one when erratic readings occur. An example of this style was illustrated earlier in Figure 26-4. Both ORP sensor and reference junction can be replaced on this electrode without disposing of the entire sensor body when replacement is indicated.

Acknowledgments

The author would like to thank Joseph Downey, Kenneth Fletcher, and Annette Nowiszewski of The Foxboro Company and James Sonia of Mettler-Toledo Electrodes for assistance in preparing some of this material.

References

1. Ives, D.J., and Janz, G.J., *Reference Electrodes*, New York: Academic Press, 1961.
2. Clark, W.M., *Oxidation-Reduction Potentials of Organic Systems*, Baltimore, MD: Williams & Wilkens, 1960.
3. "Theory and Application of Oxidation-Reduction Potentials," Foxboro Technical Information Sheet, TI1-61A, Foxboro, MA: The Foxboro Co.
4. Pourbaix, M.J.N., *Atlas of Electrochemical Equilibriums in Aqueous Solutions*, Translation by J.A. Franklin, Centre Belge d'Etude de la Corrosion, Pergamon, Oxford, 1966.
5. Twork, J.V., and Yacynych, A.M., *Sensors in Bioprocess Control*, New York: Marcel Dekker, 1990.
6. Light, T.S., "Standard Solution for Redox Potential Measurements," *Analytical Chemistry*, 44(6): 1038, 1972.

About the Author

Jane Konieczny-Burdick received a B.S. degree from the University of Massachusetts-Amherst, and a Master's degree in chemistry from Northeastern University in Boston, MA. She has 20 years of experience in the field of electrochemistry holding positions in research and development, engineering, applications, marketing and sales support.

27

Trace Oxygen Analyzers

Oxygen in the air we breathe or in the inert protective gases consumed by industry can be monitored using an electrochemical oxygen transducer instrument system. This chapter discusses the "micro-fuel cell" electrochemical oxygen transducer and how it performs while used as the sensor in an oxygen analyzer. One instrument supplier (Meeco) markets the trace-oxygen instrument system combined with a trace-moisture instrument system. Confidence in the instrument's performance should be acquired with practice well in advance of the day when a confined space must be entered or when trace-oxygen information is needed.

Before and during every entry into a confined space, the percent-oxygen of the breathing air should be monitored. Many lives have been lost in industry by inadequate preparation, lack of ventilation (of hydrocarbon vapors or nitrogen gas), or lack of fresh-air breathing equipment. Trace amounts of oxygen contamination can reduce yields, adversely affect product quality, waste valuable resources and present an explosion hazard. Oxygen can be the cause of embrittlement during the heat treating of many metals. In inert gas welding, excessive oxygen contamination can result in inferior welds that may fail prematurely [Ref. 1]. Air pumped into an automotive exhaust system (providing oxygen for burning hydrocarbons) can be quantified. Knowing the oxygen content after the catalytic converter permits calculation of the dilution (from nitrogen in air) and provides catalyst performance information.

When air (having 21% oxygen and 78% nitrogen) dilutes a sample gas because of an air pump in the exhaust system or because of excess air in the combustion process, the oxygen information can be used to evaluate emission values (i.e., compensate the reported values for the air dilution in the sample gas being analyzed).

Monitoring oxygen contamination of inert gases is commonplace in the production of chemicals, petrochemicals, semiconductors, heat-treated metals, and in virtually every industry where inert gases are used in a manufacturing process.

Monitoring oxygen contamination of inert gases is commonplace in the production of chemicals, petrochemicals, semiconductors, heat-treated metals, and in virtually every industry where inert gases are used in a manufacturing process.

Principle of Operation

The micro-fuel cell electrochemical transducer (Figure 27-1) converts oxygen energy into electrical energy. We are more familiar with how a telephone (transducer) converts sound (mechanical) energy into electrical energy. The most common electrochemical transducer is a battery, which transforms chemical energy into electrical energy. With a fuel cell, the chemical energy (oxygen) is stored externally; the rate at which the oxygen (chemical) is fed into the fuel cell determines the amount of energy (electrical current) generated. Oxygen from the flowing sample diffuses through both the membrane and the thin film of electrolyte to reach the surface of the cathode, creating a linear electrical output (proportional to oxygen's partial pressure) [Ref. 2].

The fuel cell has lead stored within. When oxygen from outside the device (sensor) enters the cell, a chemical reaction provides electrical power in the 1 to 200 μW range, hence the name "micro-fuel cell." The lead (anode) is oxidized to lead oxide; two electrons are transferred for each atom of lead oxidized, resulting in the overall reaction:

$$O_2 + 2Pb^{++} \rightarrow 2PbO + 2e^-$$

This "fuel cell" electrochemical transducer is specific for oxygen, except when iodine, bromine, chlorine, or fluorine gases are in the sample stream. These acid gases also oxidize lead.

The fuel cell is designed as a disposable unit (1-1/4 in. [32 mm] in diameter and 3/4 in. [20 mm] thick). The rear of the cell is fitted with two metal-ring contacts, electrically common to the anode and cathode (respectively) inside the cell. The anode (proprietary) maximizes the reaction. The cathode is a convex metal perforated disk wetted with electrolyte (an aqueous solution of constant pH). The external surfaces of the cathode are plated with gold, silver, rhodium, or platinum depending on the application. A flexible membrane seals the "front" of the cell, providing internal-volume-compensating performance during the life of the cell (due to moisture loss). The thickness of the protective membrane covering the cathode determines the speed of response and controls the moisture-loss rate when monitoring a very dry sample.

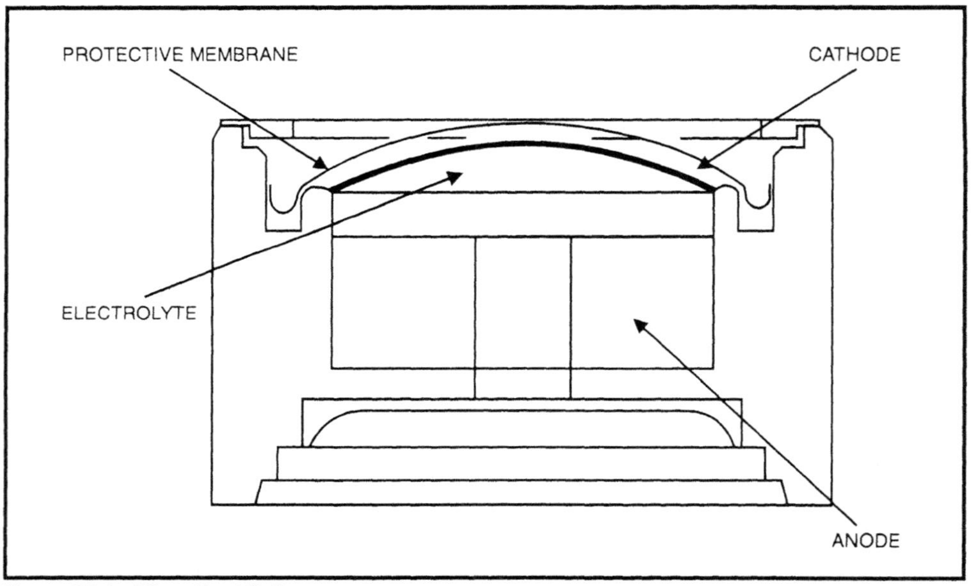

Figure 27-1. Micro-Fuel Cell

The membrane material is made of plastic whose thickness is accurately selected to control the speed of response and the life of the cell (by retaining water). The cathode material in the cell is the sensing electrode. The site where oxygen that has diffused through the membrane from the sample gas reacts chemically while the cathode is "bathed" in a common pool of electrolyte.

The cathode is a convex metal perforated disk. Its external surface is plated with gold for percent-oxygen sensors. For trace-oxygen sensors, the cathode is plated with silver, rhodium, or platinum. The anode material is composed of lead configured in a proprietary manner to maximize the amount of metal available for reaction. The anode material is oxidized (in a basic media) to lead oxide as the sensor performs (converting oxygen energy to electrical energy). Two electrical metallic-ring contacts are located concentrically on the outside flat base of the main body of the cell which is fabricated from high-density polyethylene.

While in storage, the cell contacts remain shorted. When the cell has been removed from the instrument for any period of time, the electrical contacts need to be shorted (with a metal clip) to discharge the electrical current. This reduces "run-in" time after the cell has been returned into the instrument system.

While in storage, the cell contacts remain shorted. When the cell has been removed from the instrument for any period of time, the electrical contacts need to be shorted (with a metal clip) to discharge the electrical current. This reduces "run-in" time after the cell has been returned into the instrument system.

Applications

Every oxygen electrochemical sensor is constructed for a particular application. The questionnaire application needs to be written to represent the actual chemical composition of the sample gas. Because most applications are common ones, instrument systems may be manufactured and stocked ready to ship. For most combustion-gas continuous-monitoring applications, a sample system may be required to scrub out acid gases and particulates.

Engineering of an instrument system is the business of a supplier. Traditionally, the responsibility ended with the electrical output signal (e.g., 4-20 mADC) when it represented, specifically, the oxygen component-of-interest in the sample. Today, many instruments are "packaged" into a system, mounted in a panel with a common sampling system. A computer recalls and graphically reports statistical data on the gas-quality profile during the past month to document for the customer that the process has been in control.

U.S. manufacturers of oxygen instruments include:

- Baldwin Environmental, Reno, NV
- Teledyne Brown Engineering, City of Industry, CA
- Meeco, Inc., Warrington, PA
- AMETEK Process and Analytical Instrument Division, Newark, DE

Percent-Oxygen in Breathing Air

Before and during entry into a confined space (e.g., a tank from which gasoline or inert-gas vapors have been vented), the atmosphere needs to be confirmed and monitored to ensure that the breathing air in the confined space has an oxygen content of 20.9%. The performance of each percent-oxygen analyzer and sample pump must be confirmed as normal prior to the test. With nitrogen gas, the percent-oxygen instrument should go to zero (within 2 min). To check percent-oxygen instrument performance, blow exhaled human breath into the sensor to verify that the reading does respond to 13 to 15% oxygen (within 30 s). Select and use the instrument designed and engineered for your application.

To check percent-oxygen instrument performance, blow exhaled human breath into the sensor to verify that the reading does respond to 13 to 15% oxygen (within 30 s).

Review, understand, and follow the written standard operating procedures (SOP) for your company prior to doing an oxygen-in-breathing-air test (and prior to any entry). Normally, an entry permit will be provided that has been signed by a qualified, authorized person. Expect to find, for example, all product lines and vent lines connecting to the confined space modified (by inserting isolation blinds). The tank will be well-vented and declared safe from flammable and toxic gases. Inert or flammable vapors or flammable liquids must not enter the confined space during the entry. Use the sampling pump (built into the instrument) and a probe to withdraw the initial-gas sample from the confined space to a flow-through adapter (covering the sensor) on the percent-oxygen instrument.

Process-Monitoring Applications

Process-monitoring instrument systems serve the medical, underwater diving, and industrial markets. Following are several common applications where the electrochemical oxygen transducer provides valid information:

Percent-Oxygen Applications (and Typical Ranges)

- Blends of breathing gases (1.0 to 99.0% oxygen)
- Breathing air in a confined space (20.9%)
- Combustion oxygen in a furnace (1-10%)
- Combustion oxygen from a lime kiln (1-10%)
- Automotive exhaust air quality (1-5%)
- Incinerator emission gases air quality (1-10%)

Trace-Oxygen Applications (and Typical Ranges)

- Inert (protective) atmosphere from an annealing oven (1-100 ppm oxygen)
- Inert (protective) atmosphere in metal bonding (1-100 ppm)
- Inert protective atmosphere in oil refining (1-1,000 ppm)
- Industrial gas quality (1-2,000 ppm)
- Oxygen in hydrogen, methane, ethylene, or natural gas (1-10,000 ppm)

Annealing

For heat-treating of metallic parts, a protective gas atmosphere of hydrogen, argon, nitrogen, or helium is used to prevent oxidation on the hot surface. Annealing furnace leaks can be located by monitoring the gas from the furnace for trace-oxygen. The sampling systems used must be of metal tubing to prevent air contamination of the sample during monitoring. Generally, there is a positive sample pressure to permit sampling and analysis. When a sample pump is used, a special sampling system needs to be designed to prevent air contamination of the sample. Generally, moisture content (e.g., frost point) as well as trace-oxygen is quantified to define the inert-gas quality.

Combustion Gases and Auto Exhaust

By using a pump and a metallic sample probe, samples of combustion gases (or auto exhaust) can be withdrawn from the hot source (e.g., furnace stack or tailpipe) to a flow-through adapter covering the sensor. When water-soluble gas components-of-interest are to be quantified, the sample gas will be kept hot (using a heat-traced metallic sampling line) to prevent moisture condensation. A flash water-chiller quickly removes the water without removing the water-soluble components-of-interest (e.g., SO_2, SO_3, NO_2). When percent-oxygen and combustibles only are needed for continuous combustion control of the process, water can be introduced at the probe (or in the sample line) to help flush the sample toward the vacuum pump, the water separator, and on to the oxygen analyzer. Particulates in the sample normally will not interfere with the analysis.

Condensed water can be separated by a trap-and-drain filter. A filter in the line is not required; but if a filter is used, it will not change the percent-oxygen results. The sample pump normally should be located between the sample source and the sensor. For valid information, the sample must be vented to atmosphere with no back-pressure when it leaves the sensor.

The cell is always oriented either membrane-side down or so that the membrane is vertical.

The cell is always oriented either membrane-side down or so that the membrane is vertical.

Any moisture condensing on the membrane of the sensor must flow down and drain out of the flow-through adapter. If water was to collect and completely cover the sensor, the speed of response would be reduced. For this reason, sensors are always oriented so that liquid water will not collect on the sensor membrane.

For every 10 min of monitoring a combustion gas sample, the sensor should be exposed for 20 min to ambient air. This is called "intermittent" operation, and will allow the cell to perform for months.

For every 10 min of monitoring a combustion gas sample, the sensor should be exposed for 20 min to ambient air. This is called "intermittent" operation, and will allow the cell to perform for months.

Special cells are available that have been chemically-buffered during manufacturing for continuous use in combustion gas service. The speed of response depends on the thickness of the membrane. Generally, the thicker the cell membrane, the longer the cell life.

Referee Parameter

Oxygen can be used as a "referee" parameter for calculating the amount of sample dilution (by ambient air) that has taken place within the sampling/analysis system. For example, auto exhaust has about 1% oxygen when sampled directly from the tailpipe with no hole in the muffler and without dilution air pumped into the system. This oxygen content can change when the sample goes through the analysis system. This increase in oxygen indicates that air has diluted the sample (i.e., the sample has been diluted by the oxygen and the nitrogen present in the air). Since air is about 80% nitrogen, the total dilution within the sample-analysis system has actually been five times the measured increase in oxygen. Therefore, the oxygen results can be utilized as the "referee" parameter to document the lack of dilution or the amount of dilution that has occurred within the sampling-analysis system. The numbers (for the components-of-interest) can be adjusted to 3, 5, 10, or 15% oxygen, depending on the application (e.g., auto exhaust, mobile-combustion sources, stationary-combustion sources) [Ref. 3].

Blending Gases

Gas standards also can be contaminated by air within a gas-blending system. For example, when a 100 ppm (of component X) gas standard with 10 ppm trace-oxygen (when tested directly from the outlet of the cylinder regulator) is being blended with a diluting gas (having 10 ppm oxygen), the gas mixture should also have 10 ppm oxygen. When the blend of the two gases test at 1,000 ppm of trace-oxygen (not the expected 10 ppm oxygen), this indicates air dilution has occurred during the blending operation. The actual (true) amount of the X component-of-interest in the mix has been reduced because of air dilution. The "true" number of the X component-of-interest resulting from the blend will be less than the expected number by a factor of five times the change (Δ) in oxygen value. Trace-oxygen values can be utilized as a referee parameter. Comparing the actual trace-oxygen content of each gas component being combined with the trace-oxygen value in the blended gas allows the air dilution to be calculated.

Leaks in the Distribution System

When there is a leak in the gas-handling system, the sensor appears to be flow sensitive (i.e., producing higher than true results, especially at the lower gas-flow rates of 1 to 100 cm^3/min). With no leaks in the sampling system (and no plastic tubing in the sampling system), the sensor will not react as a flow-sensitive detector. The trace-oxygen value will be stable as the sample flow rate is reduced from 1,000 to 10 cm^3/min. The trace-oxygen instrument system thus can be used to check a sampling system (or a gas distribution system) for leaks. The trace-oxygen reading at a "low" flow rate will be the same (trace-oxygen reading) as at all "higher" flow rates in the distribution system.

Do not think that gas pressure on the inside of a gas-distribution system (blowing gas out through a small hole) will protect the gas inside from air contamination. Because the partial pressure of oxygen in the air (e.g., 10,000 × 20.9 = 2,090,000 ppm) is much greater than the partial pressure of oxygen in the sample gas, oxygen from the air will push oxygen (upstream through the hole) into the higher-pressure inert gas, notwithstanding the pressure drop across the hole.

Rangeability

10,000 PPM = 1.0000 PCT

Oxygen gas in the 0 to 99% range can be analyzed quantitatively using a percent analyzer system. Various ranges, as needed for the application, can be obtained when specified. Oxygen measurements in the 0 to 10,000 ppm range can be done with a trace-oxygen instrument system. Most trace-oxygen instruments have 0-10, 0-100, 0-1,000, and 0-10,000 ppm ranges and one percent-oxygen as a calibration range.

Percent-oxygen analyzers are limited at the low end by the design of instrumentation amplifier and sampling systems. One instrument with a single range of 0 to 100% oxygen requires no amplification. Percent-oxygen instruments can also have three ranges (e.g., 0-5%, 0-10%, and 0-25%; or 0-10%, 0-25%, and 0-100%).

Trace-oxygen instruments are limited in the 0 to 10 ppm oxygen range, where they have accuracy of ±1 ppm; state-of-the-art electronic amplification systems are required on all ranges. Each trace-oxygen instrument has multiple ranges (e.g., 0-10, 0-100, 0-1,000, 0-10,000 ppm and 0-25% oxygen).

For accuracy in the 0-1,000 ppm ranges, all-metallic sampling tubing must be used because oxygen from air diffuses through most plastic and rubber tubing, resulting in higher-than-true oxygen data.

Interferences

In the absence of oxygen, there is a zero-electrical output. When the analyzer is calibrated using ambient air at 1 atm pressure (i.e., 0.1 MPa, or at the altitude where the instrument will be used), the instrument provides valid oxygen data. Normally, gas samples under pressure will be reduced by the use of a pressure regulator. Because the cell has a 10% potassium hydroxide electrolyte, the chemistry has a strong affinity for acid gases (e.g., HCl, SO_2, SO_3, SO_4, H_2S, CO_2, etc.). These acid gases pass through the membrane and interfere with the cell life. Therefore, in continuous service, these acid gases must be scrubbed from the sample gas. For continuous combustion gas service where carbon dioxide is constantly present, a special chemically-buffered cell should be used (e.g., A-2 for <20% CO_2; or A-5 for >20% CO_2). For intermittent combustion gas service, normal sensors must be exposed to ambient air every 10 min (for at least 20 min duration); the sensor will offgas CO_2 from short exposure to 0.1 to 20% CO_2 in the sample (while exposed to a gas sample with zero CO_2).

Calibration

Percent-oxygen and trace-oxygen analyzer systems should be calibrated to 20.9% oxygen while the sensor is exposed to air at local ambient-air pressure. The oxygen content in the earth's atmosphere is generally regarded as a gas standard (i.e., 20.9% oxygen). Even at 15,000 ft (4,570 m) above sea level, we have 20.9% oxygen (Figure 27-2).

Adjusting the span setting with a so-called gas standard below 209,000 ppm oxygen (i.e., 20.9% oxygen) decreases the reliability and accuracy of the calibration.

Calibration must be performed at whatever elevation (above or below sea level) that the analyzer system will be used (Figure 27-3). Calibration once every month is recommended by manufacturers to monitor the performance and the cell-life expectancy and to calibrate analyzer response. Only during air calibration should the span be adjusted (Figure 27-4).

Exposing the sensor to a "known" gas standard every day (or every hour) can be reassuring to verify (as frequently as desired) the speed-of-response of the trace-oxygen instrument system; however, avoid changing the span setting during these frequent checks.

Exposing the sensor to a "known" gas standard every day (or every hour) can be reassuring to verify (as frequently as desired) the speed-of-performance of the trace-oxygen instrument system; however, avoid changing the span setting during these frequent checks.

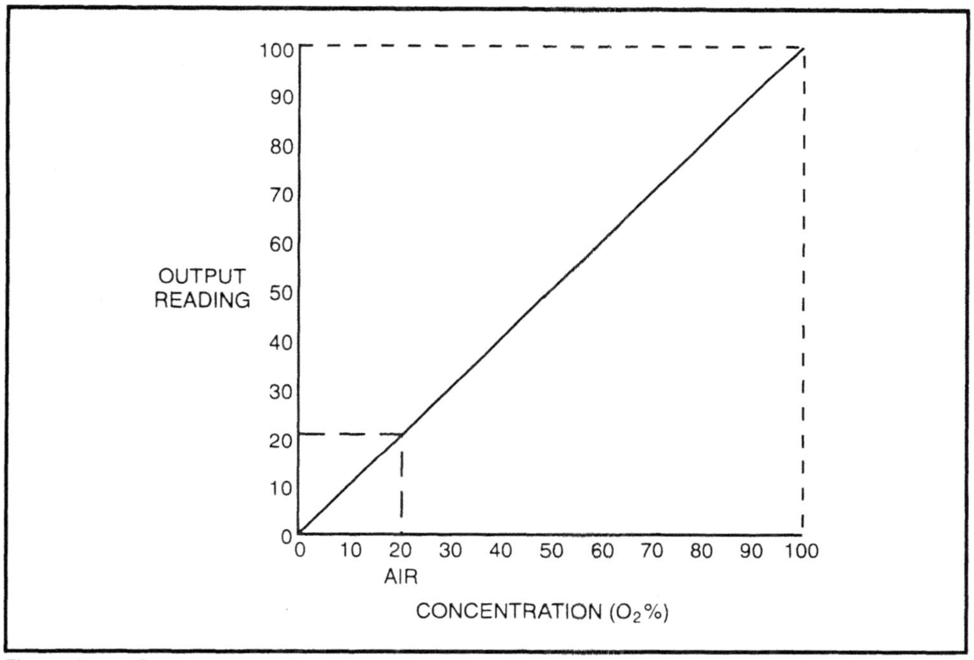

Figure 27-2. Output versus Concentration

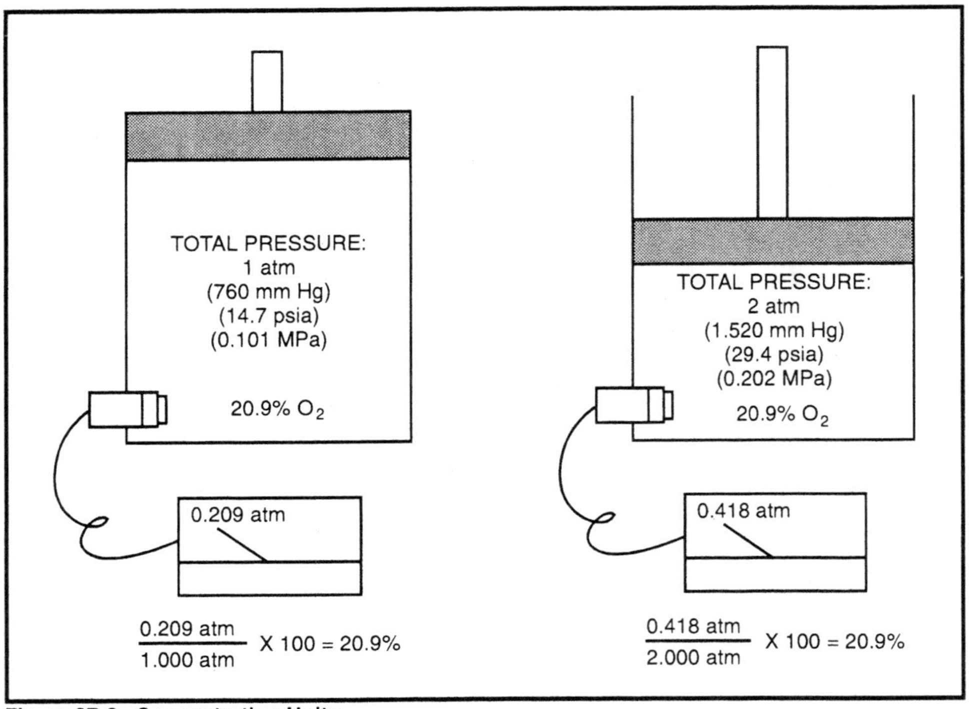

Figure 27-3. Concentration Units

Mechanical Zero

Mechanical zero of an instrument system (with a meter readout) needs to be checked and adjusted (if required) before a calibration can be started. Check that the analog readout device reads zero under the following conditions: power off, and cell contacts shorted (or cell out of the analytical instrument). Set the meter to read zero using the small screw built into the analog meter face. Set any recorders to zero also.

Electronic Zero

Electronic zero of a trace-oxygen instrument system needs to be checked and fixed (if required) before a calibration can be started. Check that the analog readout signal and recording devices electronically read zero under the following conditions: power on, cell in the instrument with inert gas flowing (for more than 1 h) while the instrument is in the calibration position. When the electronic zero does not seem normal (i.e., the same 0.0 as mechanical zero), call the manufacturer for service or advice.

For portable instruments, check that the batteries are charged. A charge takes a minimum of 24 h with the instrument in the 'OFF' position.

For portable instruments, check that the batteries are charged. A charge takes a minimum of 24 h with the instrument in the 'OFF' position.

A written log tells that the instrument has been recharged and calibrated. The log, signed by the person performing the calibration, indicates when the instrument should next be calibrated.

When placing a cell in the instrument, remove the shorting clip, but do not puncture the plastic membrane sealing the cell. Put the cell into the instrument (so the two metal rings on the cell make contact with the two metal contacts in the instrument). Start purging the instrument with nitrogen. After at least 30 min of nitrogen purge, switch the sample gas to compressed air (ambient air through the cell chamber) at a flow rate of from 10 to 100 cm^3/min with the instrument in the calibration (CAL) position. Unlock and adjust the span knob until the readout device indicates at the CAL mark, then lock the span-pot knob. Confirm that at a 0 cm^3/min airflow rate through the instrument the reading still reads at the CAL mark. This checks that the air is not pressurizing the sensor (during calibration). Minimize the time that the cell is exposed to ambient air (5 or 10 min calibration time). Purge the air from the instrument for a few hours using a gas with low trace-oxygen (e.g., argon, helium, or nitrogen).

Do not purge with carbon dioxide gas or any gas containing carbon dioxide. An acid gas will react with the electrolyte in the sensor and cause the cell to become permanently dead.

Do not purge with carbon dioxide gas or any gas containing carbon dioxide. An acid gas will react with the electrolyte in the sensor and cause the cell to become permanently dead.

Normally, the instrument purges down smoothly to below 100 ppm in 1 h. However, the cell may take 24 h (or more) before it will purge down to below 5 ppm after calibration-on-air. Plan so that one day each month (following calibration) the trace-oxygen instrument system will be purging down to the lowest readings.

Checking that the cell does go to the CAL mark (when exposed to air) is the best way to confirm that the cell is alive and that it has at least another month of life. When the cell does not rise to the CAL mark (when calibrated on air with 20.9% O_2), the cell has lost moisture and needs to be replaced with a new cell.

Checking that the cell does go to the CAL mark (when exposed to air) is the best way to confirm that the cell is alive and that it has at least another month of life. When the cell does not rise to the CAL mark (when calibrated on air with 20.9% O_2), the cell has lost moisture and needs to be replaced with a new cell.

Effects of Air Pressure

If the air calibration is done on a "high"-pressure day (e.g., 30 in. Hg), what is the error the next week during a "low"-pressure day (of 27 in. Hg)? The 3-in. Hg change in ambient pressure would create a 10% error in the reading. This would represent a 0.6% of oxygen lower deviation. This is true for all oxygen sensors in any application. The accuracy statement of "±1 ppm in the 0 to 10 ppm range" takes this normal pressure deviation into consideration (assuming constant-temperature conditions).

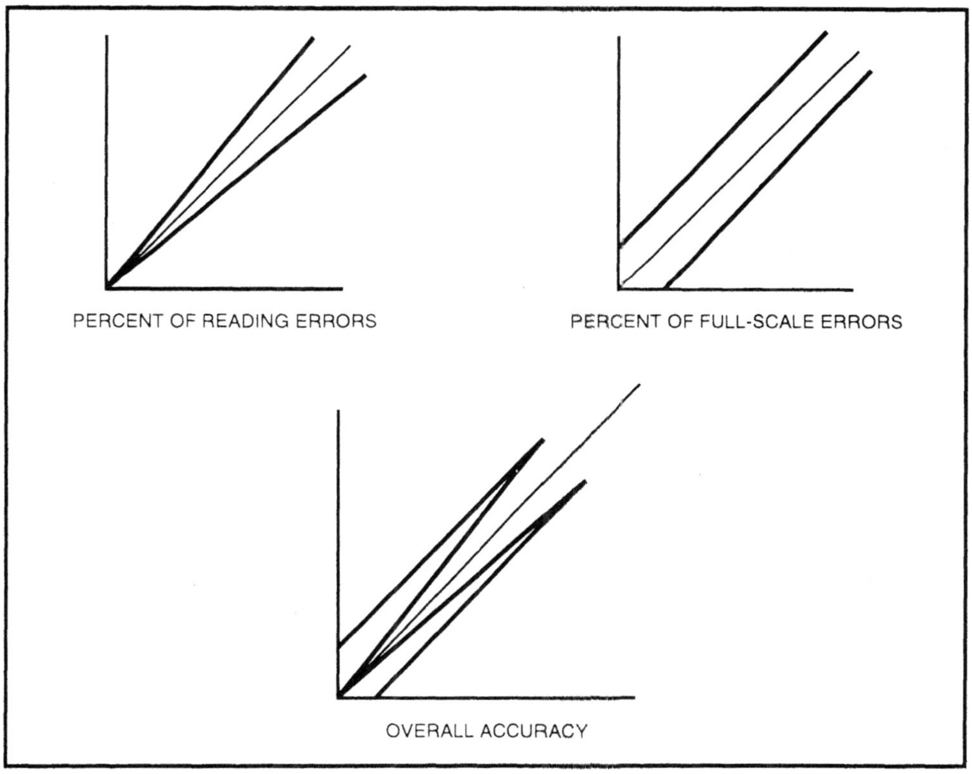

Figure 27-4. Analyzer Performance Envelope

Temperature Variation Compensation

The micro-fuel cell electrochemical oxygen transducer has a positive temperature reaction: When the temperature increases, the cell's output increases. Therefore, an electrical compensation signal is built into the feedback circuit between 0 and 50°C. The accuracy statement of "±5% of reading throughout the usable temperature range" takes this normal temperature deviation into consideration (assuming constant atmospheric pressure conditions). The chemical in the cell can freeze at temperatures below 24°F (–4°C). Internal heaters may be used to keep the cell temperature above the freezing point for cold ambient conditions. Freezing of the cell does not harm the cell, although it will stop the chemical reaction and lower the electrical output signal to near zero. When the cell is warmed, it will again become active. Freezing the cell does prolong storage life (as does keeping the cell enclosed in the sealed plastic shipping container).

Cleaning

Cleaning of the fuel cell is easy. Wash it with soap and water, taking care not to rupture the thin plastic membrane sealing the cell. Cleaning the electrical contacts on the rear side of the cell can be done using a rubber eraser and fine steel wool.

Storage and Replacement Cells

Keep only one extra cell for each instrument in service; do not store cells for a long time.

Each new cell is packaged in a plastic bag during shipment to decrease air exposure. Some cells will leak chemicals into the plastic bag during storage. The shorting clip may turn black, but the cell will still work. Each cell is guaranteed for 6 months. They generally last in-service for 6 to 12 months, depending on the water-loss rate, which in turn depends on the dryness of the sample gas and the thickness of the protective membrane. After 6 months of continuous service in a very dry sample gas, some decrease in response speed can be expected; the cell will perform well, however, for another 6 months before failing to span to the CAL mark when exposed to ambient air. Keep only one extra cell for each instrument in service; do not store cells for a long time.

References

1. Brown, Jeffrey A., "Monitoring Trace Oxygen Contamination in Process Gases," *Process Engineering*, January 1987.
2. Lauer, Jay, "Sensors for Gas Phase Oxygen Analysis," City of Industry, CA: Teledyne Brown Engineering, Analytical Instruments Division, unpublished paper.
3. Federal Regulations (FR) 120:1286 (Part 50, App. A, Method 20); 120:0619, Stationary Sources, Sec. 60:332 (a)(2); 120:0621 Stationary Sources (gas turbines) Sec. 60:335 (f)(1) 10-15-92.

About the Author

Byrd F. Parmelee is an independent sales representative with Memeco Sales and Service, Inc., a firm specializing in process analyzer system applications located near Chicago, Illinois. Mr. Parmelee is a past Quality Assurance Manager for Air Liquide America and a past sales-service representative with DuPont Petroleum Chemicals, Teledyne Analytical, and Technicon Industrial. He has nearly 30 years of experience in process analyzers and environmental monitoring and is the author of a Laboratory Technical Training Program. Mr. Parmelee holds a Bachelor of Arts degree in Chemistry from The Colorado College, Colorado Springs, Colorado.

Spectrophotometric Process Analyzer Systems

The fundamental principle of spectrophotometric analyzers is based on the fact that molecules in a process stream absorb light (energy) at specific frequencies. The energy absorbed may be released in the form of heat or by emission of light (at a wavelength of light characteristic to the molecule). In a stream of known composition, it is often possible to select one or more wavelengths of light that are unique to a component within the stream. An appropriate technique can then be employed in which either the absorption of light at the specific frequency or the emission of light from the component of interest is measured.

In the first instance, a broad-band light source is passed through a cell containing a small flowing sample of the process stream. A detector capable of selectively measuring the wavelength of interest is utilized to measure the energy absorbance. The majority of instruments in use today employ this technique. Instruments have been developed over the entire electromagnetic spectrum, from high energy x-rays to low energy infrared radiation.

In the second technique, a broad-band of light is used to excite the component(s) of interest, and the resultant light emitted is measured at a wavelength characteristic of the component(s) of interest. Techniques used to excite the sample include reactions with ozone, the use of electrical discharges and low-level radioactive sources.

Recent advances include the development of multiple wavelength detectors such as Photodiode Array (PDA) detectors. This has enabled analyzers with PDAs to measure and report several stream components simultaneously. A common application of a PDA detector includes the measurement of several gases, such as H_2O, CO_2, and CO in process heater stacks (a measurement typically required as a part of the regulatory permit to operate the heater).

Fourier Transform Infrared (FTIR) techniques have historically been utilized in the laboratory. Robust hardware is now available to be applied for continuous analyses. These instruments are now starting to replace other equipment due to speed of analysis (less than 10 seconds) and the development of models that can predict several stream properties simultaneously (based on data accumulated from thousands of samples at multiple wavelengths, correlated to measurements previously obtained by laboratory techniques). After the correlation is developed, these models are used to predict the physical properties of unknown process streams based solely on the FTIR measurements. This correlation technique is often referred to as chemometrics.

Typical uses for Spectrophotometric Analyzers at different portions of the electromagnetic spectrum include:

- X-ray absorption — measurement of low levels of water in solids or asphalt

- X-ray fluorescence — measurement of low ppm metals in waste streams
- Ultraviolet — measurement of H_2S in process streams
- Visible — measurement of the color or process streams
- Infrared — measurement of CO or CO_2 in heater stack emissions

Chapters 28 through 32 will cover the more common types of spectrophotometric analyzers, as follows:

Chapter 28	Infrared (IR and NDIR) Absorption Analyzers
Chapter 29	Fourier Transform Infrared (FTIR) Analyzers
Chapter 30	Near Infrared (NIR) Analyzers
Chapter 31	Ultraviolet/Visible (UV/VIS) Absorption Analyzers
Chapter 32	Chemiluminescence (CHEMIL) Analyzers

28

Infrared (IR and NDIR) Absorption Analyzers

The history of spectroscopic-type analyzers [Ref. 1] can be traced to 1666, when Sir Isaac Newton passed sunlight through a prism to display an array of colors. Subsequently, in 1814, Fraunhofer's studies of the sun's spectrum revealed a large number of fine, dark lines (discontinuities) in the spectrum. In 1848, Foucault observed that a flame containing sodium ions would absorb the yellow light from a strong arc placed behind it. These observations were correlated by Kirchoff in 1859 when he explained the Fraunhofer lines by interaction of elements in the sun's cooler atmosphere absorbing the continuous spectrum emitted by the hot interior of the sun—qualitative analysis of the sun's atmosphere!

The discovery of the infrared region occurred in 1800, when Herschel, using a simple thermometer, found that the spectrum's radiant heat was greatest outside the visible spectrum near the red end. Tyndall, studying the heat absorption properties of various gases, developed an apparatus very similar to conventional nondispersive infrared (NDIR) analyzers, with a source of infrared energy, an absorption cell, and a differential heat (infrared) detector.

The discovery of the infrared region occurred in 1800, when Herschel, using a simple thermometer, found that the spectrum's radiant heat was greatest outside the visible spectrum near the red end.

For about 50 years, little was done in the field of infrared instrumentation development. Progress is rare in one field of technology without advancements in others. Infrared analytical instrumentation in the 1930s and beyond was spurred by developments in the field of electronics. It was during this period that the highly-sensitive and selective Luft-type detector was developed. By the end of World War II, more than 400 process infrared analyzers were operating in German synthetic gasoline plants [Ref. 2]. Recent infrared sensor development efforts have focused on military applications utilizing high-sensitivity fast-response infrared detectors in missile guidance, missile detection, and fire-control systems. Infrared detectors also are used for intruder alarms.

Typical infrared detectors include thermopiles, photovoltaic cells, photoconductive cells, Golay cells, solid-state infrared sensors, and the sealed pneumatic detector (Luft and Microflow-sensor types). The discussion in this chapter will be limited to those types used in process analysis instrumentation for quantitative analysis purposes.

Measurement Principle

For a compound to be infrared-active — that is, to absorb infrared energy — it must undergo a change in dipole moment during one of its vibrational modes. For example, consider two simple diatomic molecules: carbon monoxide (CO) and oxygen (O_2) (Figure 28-1). CO is somewhat polar, with the carbon atom side being slightly positive and the oxygen atom side slightly negative. Consequently, the CO molecule has a small dipole moment. Imagine that the C and O atoms are connected by a spring (the molecular bond) and that the molecule vibrates in a simple harmonic motion based on the masses of the two atoms and the tension of the bond. Light (infrared) energy at this frequency will be absorbed by the CO molecule, increasing the amplitude of vibration. Thus, the CO molecule is defined as infrared-active.

Conversely, the oxygen molecule does not have a dipole moment and thus is not infrared-active. An explanation of this can be found in quantum mechanics, which is beyond the scope of this article. Elemental diatomic and noble gases such as nitrogen, hydrogen, argon, and halogens are not infrared-active; all other molecules exhibit infrared activity.

Plots of energy transmission versus wavelength between 2 and 15 µm (5,000 and 666 cm^{-1}) are known as infrared absorption spectra (Figure 28-2). Such spectra are generated by spectrometers that use prisms or gratings to disperse the infrared energy into various wavelengths. The prism or grating is rotated so that the frequency of light passing through the measurement cell is continuously changed and the resulting absorptions are recorded as spectra. Thousands of spectra are catalogued for pure compounds [Ref. 3]. These are used for molecular structure studies and qualitative/quantitative analysis of unknowns. Near-infrared (NIR) absorption occurs at wavelengths between the visible and infrared region 0.8µ - 2µ. NIR spectra are the result of harmonics and combinations of the fundamental molecular vibration frequencies.

The number of infrared absorption bands and the intensity of absorption are directly related to molecular structure—a fundamental property of the compound. However, there are molecular-specific absorptions exhibited within a series of compounds. For example, all organic compounds tend to exhibit infrared absorption at about 3.4 µm, the group frequency of the C–H bond vibration. Thus, it may not be possible to discriminate among a mix of hydrocarbons at this wavelength, and other absorption bands should be utilized.

Process infrared analyzers are dedicated quantitative analysis devices designed to monitor a single "key" component in a multicomponent stream. A wavelength of sensitization is selected for which the component of interest (analyte) strongly absorbs infrared and background components are transparent. Process infrared analyzers do not use prisms or gratings and thus are defined as nondispersive infrared (NDIR) analyzers.

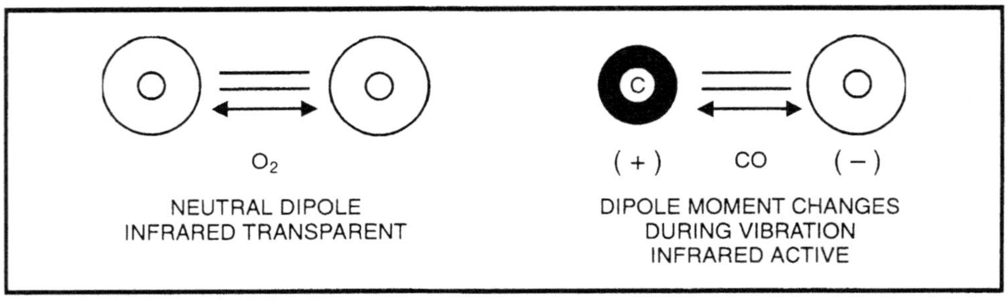

Figure 28-1. Molecular Vibration

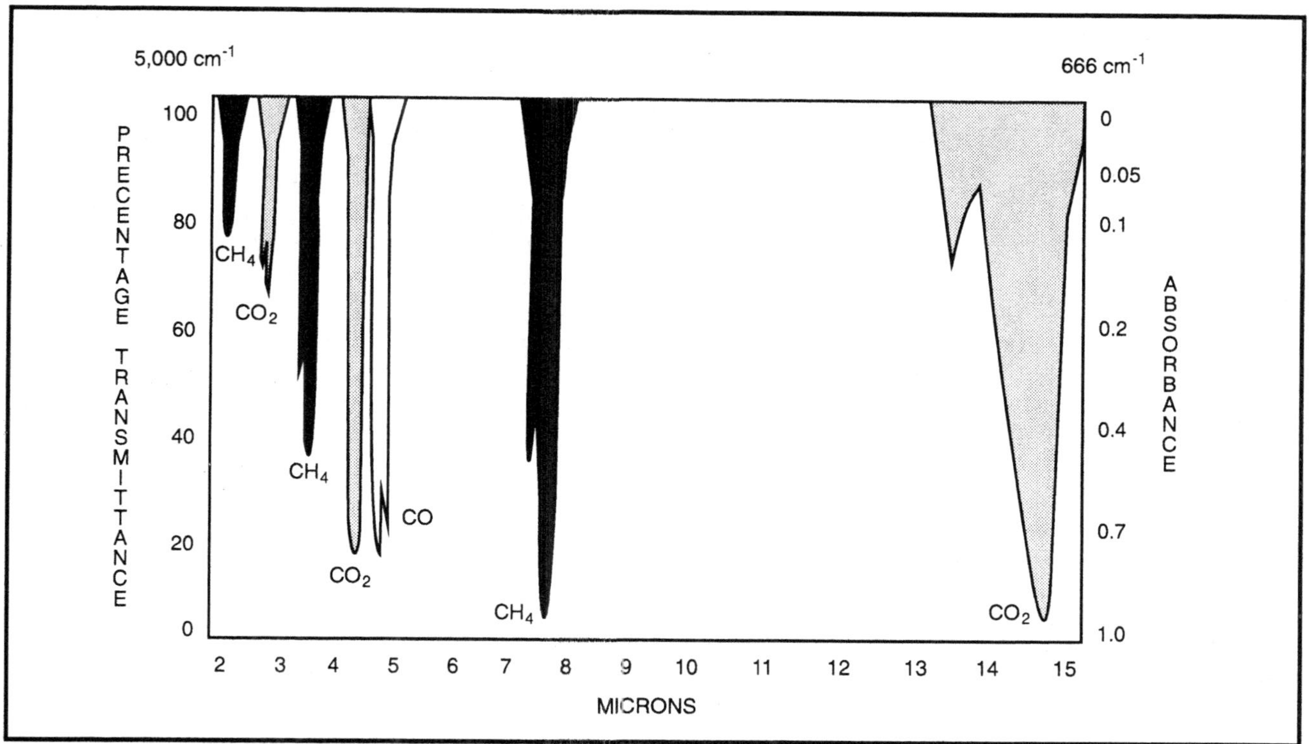

Figure 28-2. Infrared Spectra of CO, CO_2, and CH_4

Virtually all spectrographic systems are based upon the Beer-Lambert law [Ref. 4] of energy absorption (Figure 28-3). The law is stated as:

$$A = \text{Log } I_o/I = abc$$

where

I_o = energy at λ_x entering an absorption cell
I = energy at λ_x exiting the absorption cell
a = absorption coefficient of compound X at wavelength λ_x.
b = length of absorption cell
c = concentration of compound X
A = absorbance

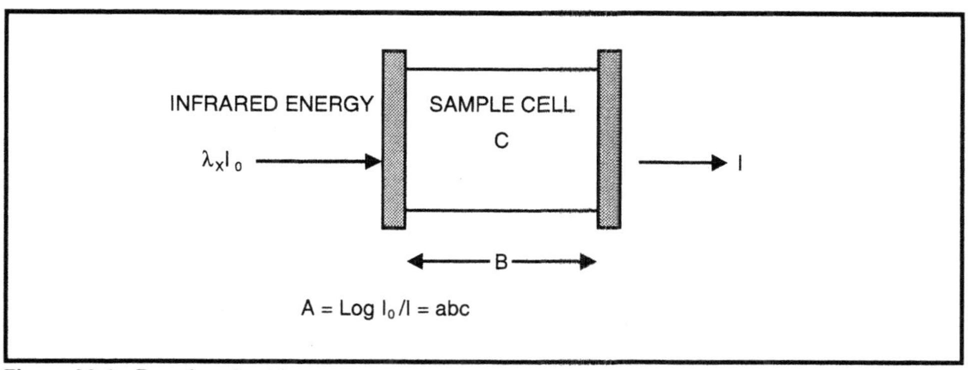

Figure 28-3. Beer-Lambert Law

The Beer-Lambert law simply states that if the cell length b is fixed and a is a function of the analyte's molecular structure, changes in absorbance (A) are proportional to concentration (c). Because process infrared analyzers typically operate at absorbances of less than 0.2 units, their output can be considered linear.

Instruments based on this principle can be found in many configurations of optical systems: dual beam, single beam, selective detectors, broadband detectors, optical filters, gas-filled detectors, rotating optical filters, rotating gas filters, multiple analytical/reference wavelength systems, open-path, and so on. For illustrative purposes, simple dual-beam and a single-beam systems will be described.

As illustrated in Figure 28-4, the typical dual-beam infrared analyzer optical system consists of a polychromatic infrared source, an absorption cell, a reference cell, optical filters, and a differential infrared detector. Infrared energy passes through the optical filter, which limits the radiation to the analytical wavelength(s). The radiation then passes through the sample (absorption) cell to the detector. The reference-beam radiation, also limited by an optical filter, then passes through the reference cell to the detector. The detector compares the infrared energy between the two beams.

When the analyte is in the sample cell, infrared energy is absorbed (at the analytical wavelength) proportional to the analyte's concentration (Beer-Lambert). The energy imbalance between the sample and reference beams is measured, electronically-conditioned, and refined to a quantitative analytical signal.

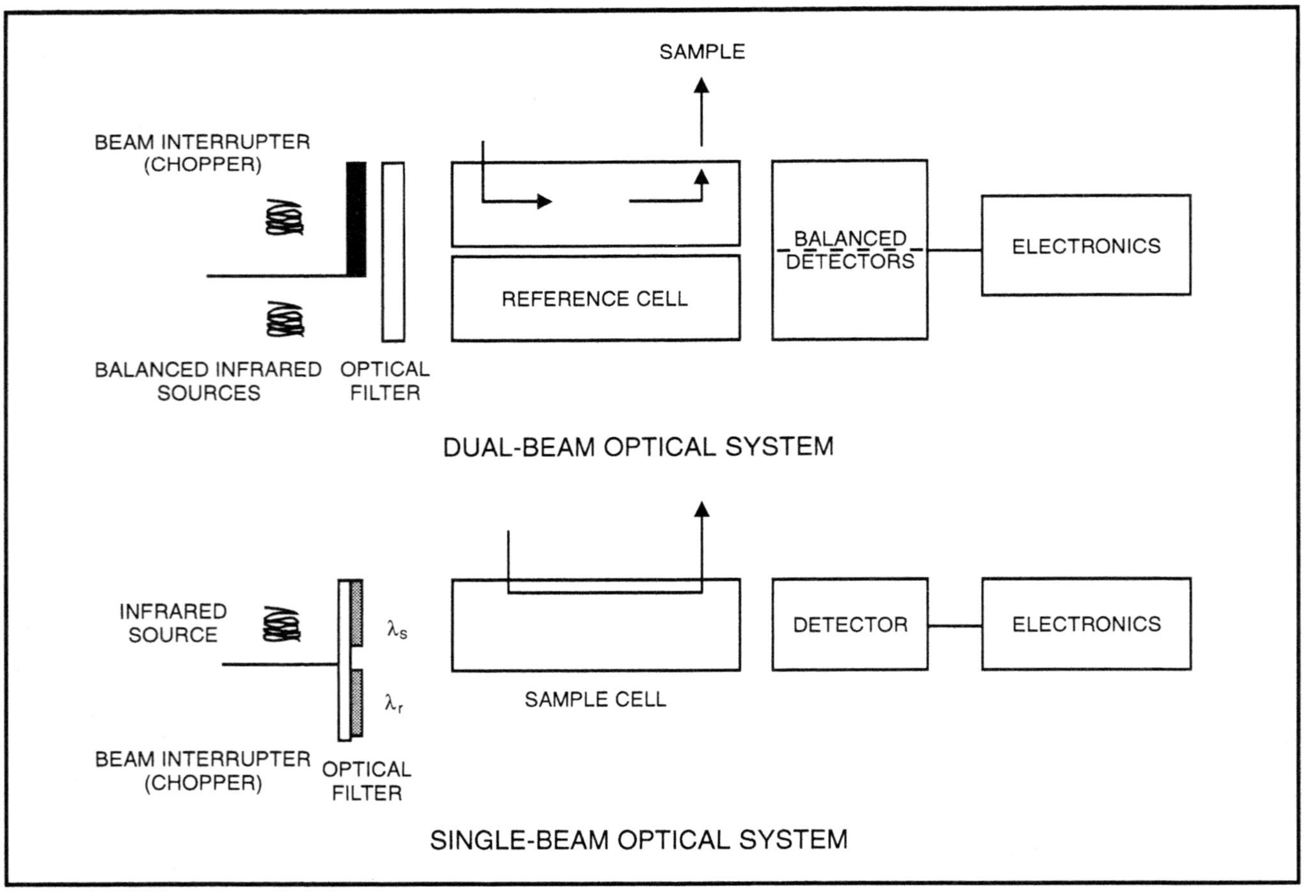

Figure 28-4. Typical NDIR Optical Systems

Some manufacturers use a single-beam optical system (Figure 28-4). The beam is mechanically chopped by two rotating optical filters: one with maximum transmission at the analytical wavelength and the other at a reference wavelength. The detector has broadband response and compares infrared energy transmissions at the selected wavelengths.

Sources and Cells

Infrared sources are usually nichrome elements, heated to a dull red color, emitting energy in the 2 to 15 μm region. Radiation is limited by optical materials to the infrared region of interest.

Cell designs vary widely — from 0.025 mm path length and multiple internal reflection (MIR) cells for liquid service to 1 m single and folded-path length designs for high-sensitivity gas applications. The materials of construction vary with the application. Aluminum is a common cell material, as are stainless steel and Hastelloy. The material will vary with the application and the corrosivity of the sample. Cell liners can be polished gold-plated stainless steel tubing to enhance energy transmission through the tubular optics.

Detectors

Pneumatic (Luft-type) detectors (Figure 28-5) consist of two sealed chambers, one each for the sample and reference beams. Both are connected to a third chamber containing a condenser microphone. These chambers are filled with a sensitizing gas and hermetically-sealed. When the interrupter exposes the infrared source, the energy traverses the optics through the detector window. The sensitizing gas absorbs infrared energy (at its infrared absorption spectrum wavelengths), increases its temperature, and generates a small pressure pulse that moves the membrane, changing the capacitance of the microphone proportional to the incident energy. Electronically, the pressure pulses are compared from the reference and analytical beams and imbalances are proportional to concentration.

Response to the wavelength of interest depends on the detector gas. To measure CO_2, the detector gas is CO_2; to measure methane, methane is the detector gas. By changing the sensitizing gas, spectral response can be "built-in" to the wavelength of interest.

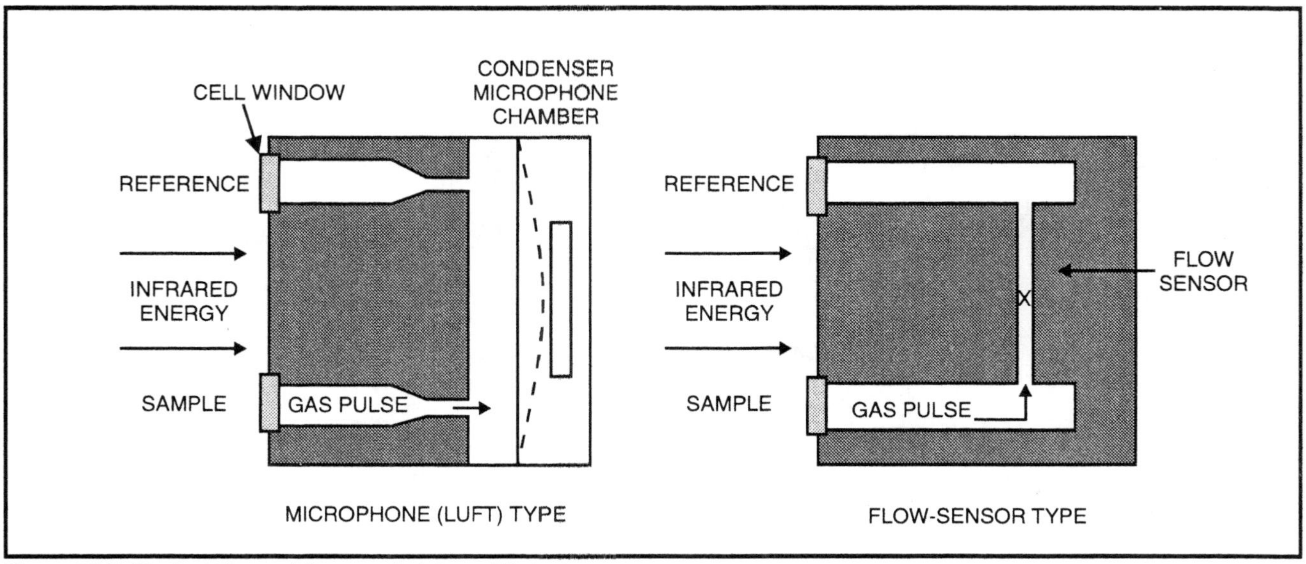

Figure 28-5. Typical Thermal-Pneumatic Infrared Detectors

On the other hand, broadband-response infrared detectors (semiconductor photodetectors) exhibit high response over a wide spectral range. Selectivity to the analyte depends on limiting the source radiation to the wavelengths of interest with optical filters (Figure 28-6).

A variation of the Luft concept employs a slightly different detector design with a thermal flow-sensor (Microflow) in place of the microphone. The flow sensor measures detector gas movement within a sealed chamber as a result of infrared absorption between the sample and reference radiation.

The Luft-type and Microflow thermal pneumatic detectors are more sensitive than semiconductor photodetectors. Additionally, pneumatic detectors have more flexibility since response to the wavelength of interest can easily be varied by selection of the detector fill gas.

The Luft-type and Microflow thermal pneumatic detectors are more sensitive than semiconductor photodetectors. Additionally, pneumatic detectors have more flexibility since response to the wavelength of interest can easily be varied by selection of the detector gas.

Photoacoustic Optical Systems [Ref. 5]

A recent addition to NDIR-type optics design is the photoacoustic method (Figure 28-7). In this technique, polychromatic infrared energy, limited to the wavelength of interest by an optical filter, is chopped at an acoustic frequency. The energy then passes into an absorption cell that contains a microphone.

When the trapped analyte absorbs infrared energy, it increases its temperature and pressure, producing an acoustic wave at the chopping frequency. This wave is detected by the microphone. The microphone response is electronically-processed to produce a quantitative analysis signal: The stronger the sound wave (audio pulse), the higher the concentration. Such a system inherently has little zero-drift, since in the absence of the analyte there is no signal.

Photoacoustic principles lend themselves to multicomponent analysis with individual "optics" per component. In addition, microprocessor electronics can be used to compensate for background responses (cross-sensitivities) among the various analytes.

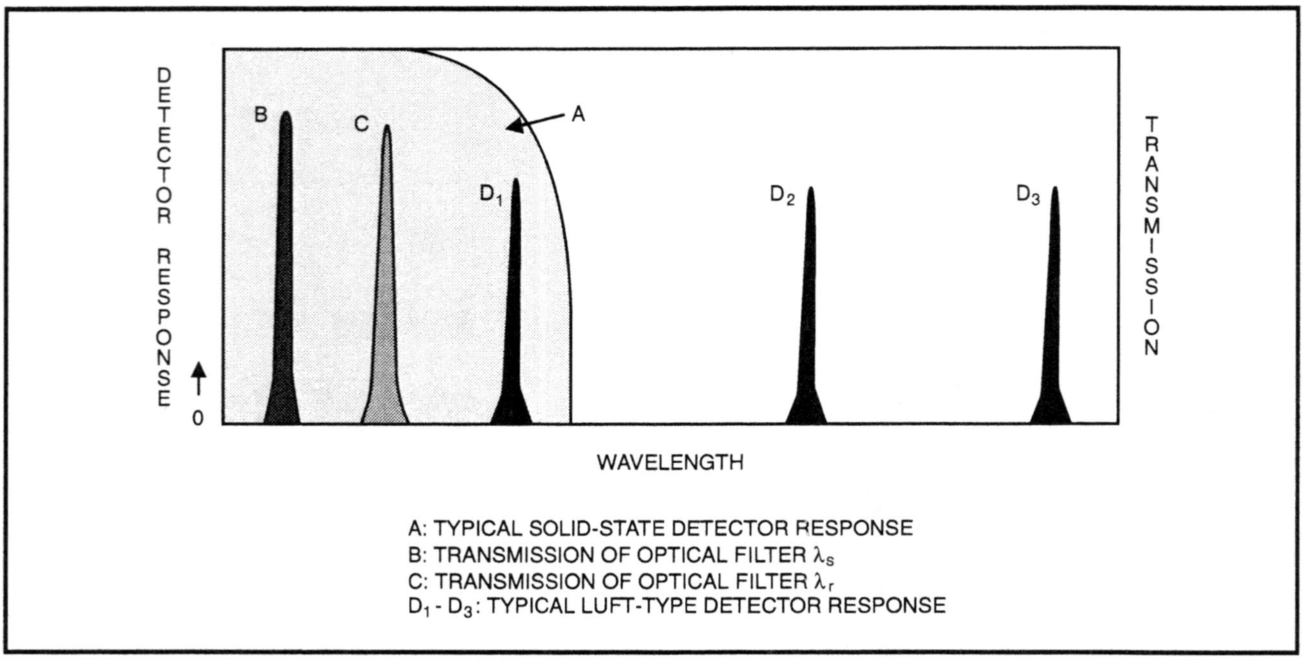

Figure 28-6. Detector Response Versus Optical Filter Transmission

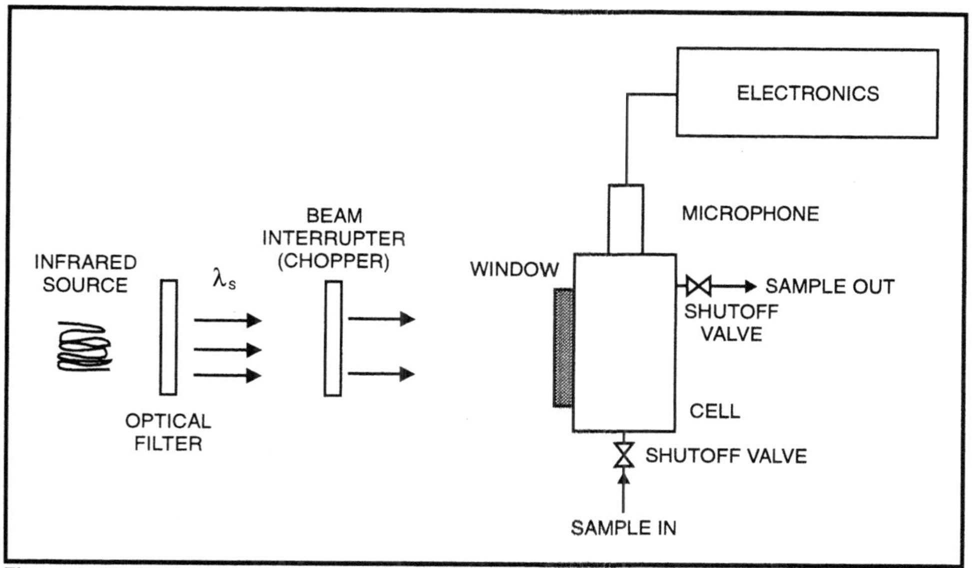

Figure 28-7. Photoacoustic Infrared Detector

Multicomponent Analysis

NDIR analyzers are dedicated spectrometers designed to monitor a single component in a multicomponent stream. To monitor more than one component, several NDIR analyzers must be used or another analytical technique considered.

NDIR analyzers are dedicated spectrometers designed to monitor a single component in a multicomponent stream.

Since the need to monitor "key" process compounds was the driving force for NDIR advancements, the need for more process analytical data has influenced development of multi-component NIR and Fourier transform infrared (FTIR) analyzers (see Chapters 29 and 30).

One alternative is to consider a spectrometer to scan the infrared region. Conventional dispersion-type devices have limitations of speed and sensitivity. However, FTIR spectrometers [Ref. 6] are fast, with the ability to monitor the complete spectrum and provide multicomponent analysis.

Most such devices are based on the Michelson interferometer, which uses a moving mirror to produce an interferogram (Figure 28-8). If the source of energy is monochromatic, the interferogram will be a sine wave of intensity versus mirror position. If the source of energy is polychromatic, the interferogram will be complex, requiring digital computers to calculate the transmission spectra. Once that is done, the computer can then perform a quantitative analysis calculation for the multiple components of interest.

NIR systems use scanning spectrophotometers in the NIR-visible-UV regions and are useful for multicomponent on-line process analysis. NIR analyzers use gratings to generate the wavelengths of interest and fiber optics (Figure 28-9) to peer into the sample stream and interface with the detector. Digital computers are used for data collection, manipulation, and concentration calculations.

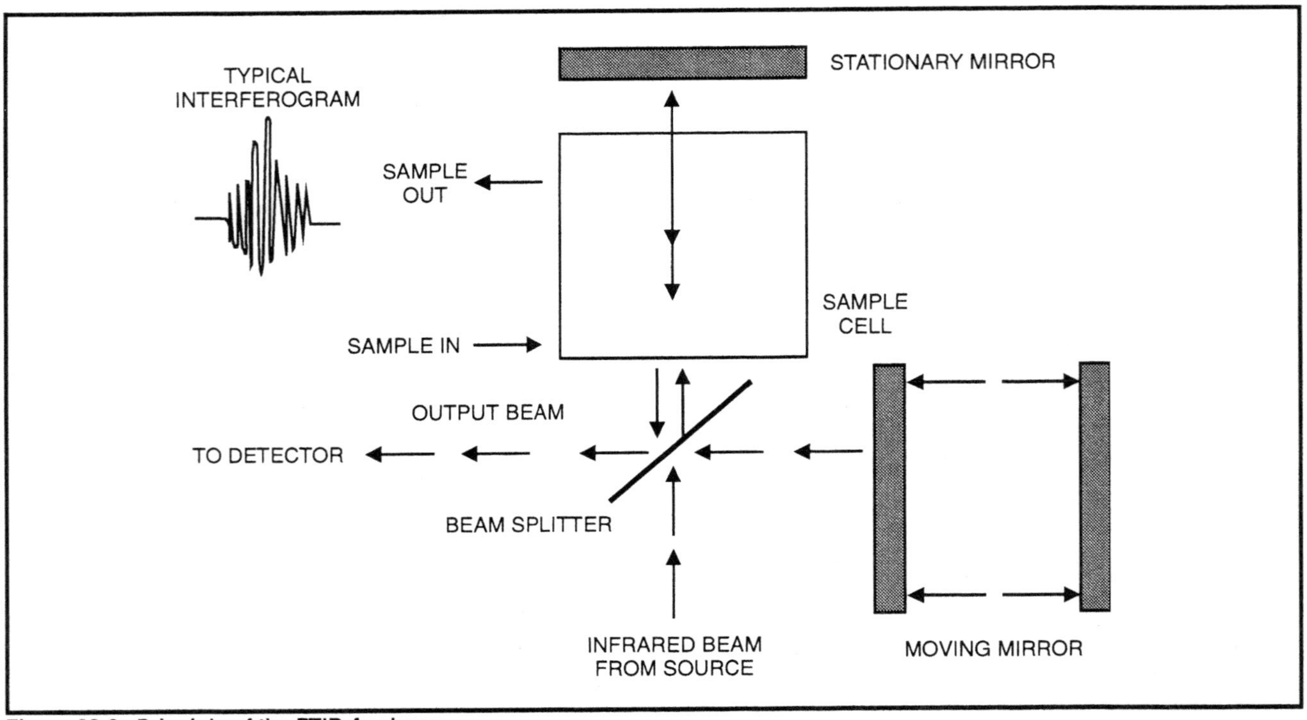

Figure 28-8. Principle of the FTIR Analyzer

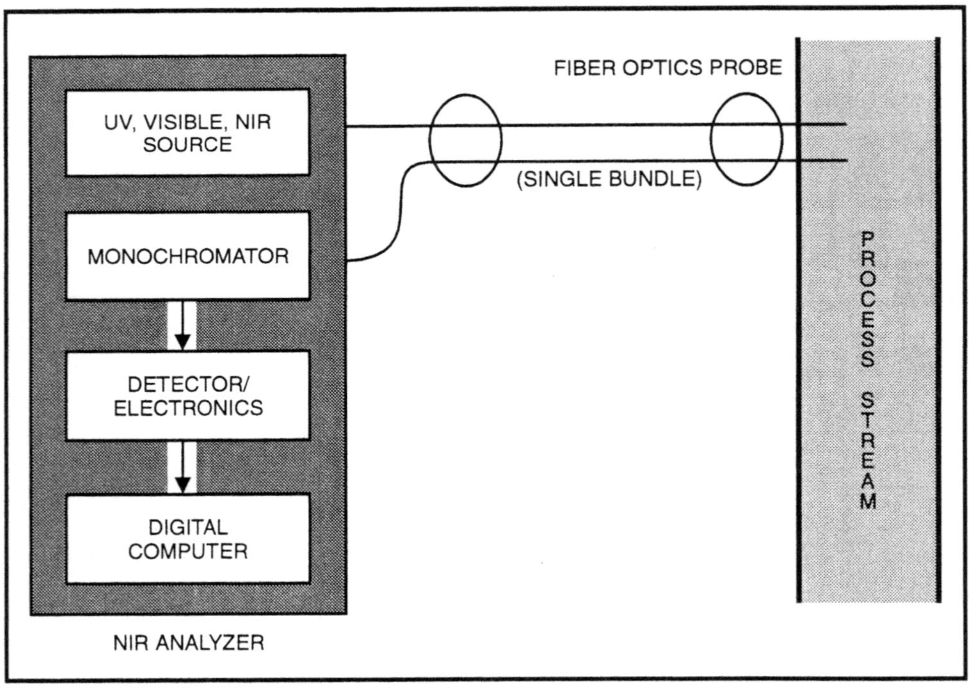

Figure 28-9. NIR Analyzer

Common Applications

Process infrared analyzers are used for process control, ambient air monitoring (personnel/plant protection), water-quality systems, air-quality measurements, energy conservation (combustion control), and so on. Infrared analyzers typically are utilized to monitor a single "key" component in a multicomponent stream. The analytical signals are used for process control. Due to the multipurpose capacity of infrared analyzers, applications range from gaseous and liquid applications to monitoring solids and films. Typical process examples include ammonia synthesis, ethylene, ethylene oxide, synthetic rubber, chlorinated hydrocarbons, and ferrous and nonferrous metals.

Due to the multipurpose capacity of infrared analyzers, applications range from gaseous and liquid applications to monitoring solids and films.

A typical process application is the measurement of 0 to 1% methane from primary and secondary reformers in an ammonia synthesis plant. The stream(s) contains hydrogen, nitrogen, CO_2, and up to 50% water vapor at high temperatures and pressures. The sample is cooled and condensables removed. The analysis is performed at 3.3 μm (3030 cm^{-1}), where methane absorbs infrared energy and background compounds (CO_2, H_2O, CO, N_2) are transparent (Figure 28-2). Analysis signals are used to monitor/control reformer operating conditions.

Hydrogen plants typically monitor process streams for total carbon oxides (0 to 20 ppm) for catalyst protection. Such applications may use sample cells up to 1 m in length.

In an ethylene oxide plant, NDIR analyzers have been used to monitor ethylene feed gas purity (methane and ethane contaminants), ethylene in the reactor feed and recycle streams, CO_2 in the reactor inlet and outlet streams, ethylene oxide product in reactor outlet, and CO in the reactor outlet for detection of side reactions.

The measurement of carbon monoxide-in-air to meet air-quality standards mandated by the Environmental Protection Agency (EPA) is often performed by NDIR analyzers. Many designs have met the stringent EPA stability/selectivity requirements, earning the designation of EPA analytical method. EPA-mandated continuous emission monitoring systems (CEMS) are used on stacks for measurement of SO_2, NO_x, and other pollutants. A carbon dioxide (or oxygen) analysis is required to correct for flue-gas air dilution. NDIR analyzers are used for the CO_2 analysis as well as measurement of the pollutants.

Sensors for water-quality analyzers, such as total carbon (TC) and total organic carbon (TOC) systems, are infrared analyzers.

Loss-prevention applications involve sensing process leaks in flammable and toxic ranges; again, infrared absorption principles are applicable. Open-path optical systems to 200 m pathlength are applicable for path integrated (ppm meter) hydrocarbons-in-air measurements. Smaller systems are used to detect flammable concentrations.

Oxygen analyzers are used to control a combustion process at or near the stoichiometric point for energy conservation purposes. Signals from such analyzers are used to control air/fuel ratios. Infrared CO analyzers are also used to detect the point at which CO (combustibles) appears. This avoids creation of reducing conditions to below the stoichiometric point and the possibility of an explosion or fire in the duct.

Greenhouses use NDIR analyzers for maintaining high levels of ambient CO_2 to stimulate plant growth.

Liquid applications involve the measurement of water in petrochemical liquids, such as 0 to 200 ppm water in ethylene dichloride.

NIR and FTIR analyzers are more sophisticated applications of infrared spectroscopy to process analysis applications.

NIR and FTIR analyzers are more sophisticated applications of infrared spectroscopy to process analysis applications. Using digital computing technology (chemometrics) to process infrared detector signals, such analyzers are applicable to multicomponent analysis. The computer, using chemometric data-handling techniques, is able to calculate stream compositions as well as certain physical properties such as cloud point, boiling point, and vapor pressure.

Another application of process infrared technology has been open-path and reflection-type analyzers (Figure 28-10). These open-path [Ref. 7] instruments electronically ratio the reflected energy of two beams, sample and reference, to obtain analytical data. Applications range from ambient air monitoring to measuring film thicknesses (Figure 28-11) to monitoring water in synthetic rubber crumb or dog food.

Ranges of Variables

NDIR analyzers can be used at full-scale ranges from 0 to 10 ppm to 0 to 100% (v/v). Suppressed ranges such as 90 to 100% are also possible. However, for most compounds, the lower full-scale range may be at the 250 to 500 ppm level.

As previously mentioned, ranges depend on cell lengths, with short cells being used for high concentrations and long cells for low ranges. NDIR analyzers can support field range changes by a factor of 2 to 20 and still have a linear response. Full-scale ranges are also dependent on background compounds, selectivity requirements, and so on. Typical stream conditions and compositions (maximum, minimum, and average levels) must be determined prior to application confirmation. Suppliers must be informed of this information prior to bid. Such information is required to properly design the optics to be responsive to the analyte and insensitive to background compounds. The vendor should be contacted concerning any change of service, as new optics may be required. Figure 28-12 shows a typical form outlining the required information.

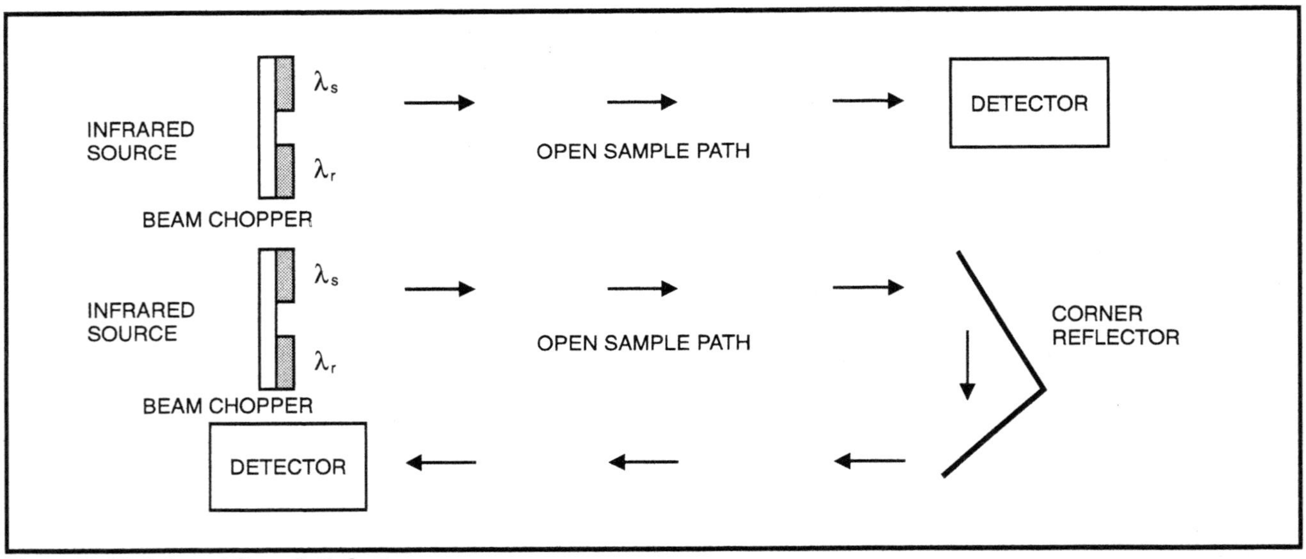

Figure 28-10. Infrared Open-Path Systems

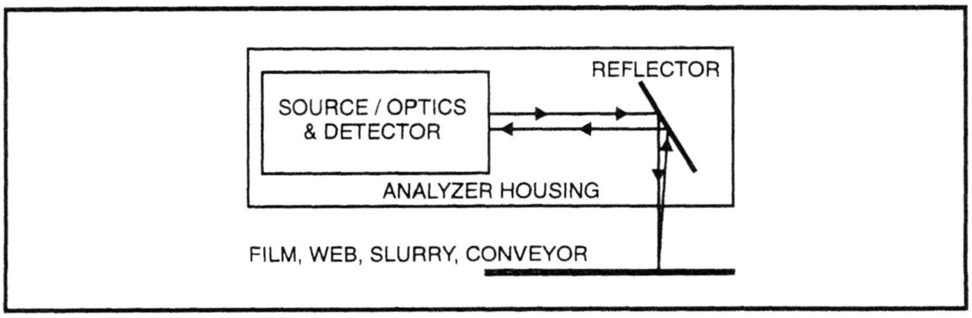

Figure 28-11. Infrared Reflection System

PROCESS ANALYZER SYSTEM DATA

Ref: Spec. No:
Contract No:
Req. No:
By: Chk'd: Appr:

GENERAL

Tag Number: Service:
Flow Sheet/Dwg. Ref: Loop No:
Analyzer Type:
Manufacturer & Model:

SAMPLE SOURCE COMPOSITION DATA

Phase: Vapor Liquid Mixed — Describe:

COMPOSITION (Concentration Units ____)

Component	Minimum	Average	Maximum	Measured Range

Contaminants: Liq: Solid:
Describe:

STREAM CONDITIONS

Sample Pressure MIN NORM MAX Temp. MIN NORM MAX Location Flow Rate
Return Pressure MIN NORM MAX Temp. MIN NORM MAX Location Flow Rate

PIPING INFORMATION

Sample Material Size Tracing Insulation
Source Line: Length Transport Time (est.)

Sample Tap: Flange Thread Horiz Vert Other

Sample Material Size Tracing Insulation
Return Line: Length Transport Time (est.)

Return Tap: Flange Thread Horiz Vert Other

ANALYZER INFORMATION (Operating)

Flow Rate Pressure Temperature
Other

SPECIAL NOTES AND COMMENTS

Figure 28-12. Sample Form Showing Information Required by Analyzer Vendor

Options and Accessories

NDIR analyzers are available in general-purpose and NEMA enclosures as well as housings meeting NEC hazardous-area classifications. Many analyzer designs can be used with ISA purge systems: Types X, Y, or Z. Additionally, optics can be remote from the analyzer control section (zero and span). Accessories include dual-range capabilities, autostandardization units, linearization circuits, and alarms.

Typical performance specifications are:

Response Time: Step change to 90% with final reading in 2-5 s
Repeatability: ±1/2%
Zero Drift: ±1%/24 h
Span Drift: ±1%/24 h
Linearity: 2-5%
Output Signal: mVDC, mADC, VDC, as required
Flow Rate: 500-2,000 cm^3/min

Limitations

Most spectroscopic analyzers perform a direct analysis of a conditioned process sample. In spectroscopic terms, the sensitivity limit is about 0.01 to 0.05 absorbency units and is application-dependent. In practical terms, this full-scale limit is in the range of 10 to 100 ppm with sensitivities of 0.1 to 1 ppm.

As previously indicated, any compound that is infrared active can, in theory, be measured by an NDIR analyzer. Elemental diatomic and noble gases are not infrared active.

It is necessary to select an analytical wavelength where the component of interest absorbs infrared energy and background components are transparent.

Many compounds have similar molecular structures (the hydrocarbon homologs), and thus the absorption bands may overlap. Selectivity limitations are usually evidenced as a positive bias by interference(s) from a background component(s). Thus, it is necessary to select an analytical wavelength where the component of interest absorbs infrared energy and background components are transparent. Sometimes this is not possible because a minor overlap of absorption bands may affect the analysis signal.

It is possible to estimate this response from the Beer-Lambert law. If the component absorptivities at a given wavelength are known, or can be estimated from infrared spectra, the absorbencies for the analyte and background component can be ratioed, thus providing an estimate of the bias.

Although this technique is suitable for bias estimating, experimental data usually exhibit better selectivity than the calculation. Manufacturers of process infrared analyzers can provide discrimination ratios (DR) [Ref. 8] for various sensitizations. For example, typical DR values for a carbon monoxide analyzer are: CO_2, 5,000:1; H_2O, 2,500:1. In other words, 5,000 ppm of CO_2 would be equivalent to 1 ppm CO, or on a typical flue gas analysis problem a change from 9.5 to 10.0% CO_2 would be equivalent to 1 ppm CO — a negligible interference. Discrimination ratios are discussed in greater detail in Appendix 4.

When the bias is intolerable for a successful application, alternative sensitization methods can be used (dual detectors). Additionally, the bias can be "compensated for" during calibration, or another analyzer can be used to monitor the background and correct the analytical signal. Alternate technology can also be employed, such as flow injection analysis (FIA), ultraviolet (UV) spectroscopy, gas chromatography (GC), and mass spectrometry (MS).

Installation Hints

Installation requirements will vary among applications. However, analyzers must be easily accessible for normal maintenance. Provision should be made in the sample-conditioning system to extract samples for backup laboratory analysis when analyzers are taken from service for maintenance or repair. The design of the sample system should also allow laboratory samples to be taken without interrupting the operation of the on-line analyzer while it is in process service. Although most NDIR analyzers are designed for outdoor installation, weather protection must be provided. For best performance, all types of process analyzers should be installed in an analyzer shelter or comparable controlled environment.

The failure of infrared analyzers to perform properly can usually be related to the sampling system. Extreme care must be used during sample-conditioning system design and installation to assure a representative sample, proper removal of particulate and condensates, and a means to calibrate the analyzer. See Chapter 3 for additional information on sample-conditioning systems.

Specific Safety Issues

NDIR analyzers are available in NEC Class I and II, Division 1 and 2, and CSA designs. Additionally, many process infrared analyzers are designed to accept ISA-Types X, Y, and Z purge systems.

Note, there are hazards associated with sample streams. Many process fluids are flammable, toxic, and/or corrosive. Thus, proper venting and disposal of exhaust fluids and the use of corrosion-resistant system components must be considered. Material safety data sheets should be reviewed by maintenance or engineering personnel.

All sample-handling systems should undergo a safety review, paying specific attention to possible exposure of the process to maintenance personnel, especially those systems which have provision for taking laboratory samples. The design of these systems is also critical from the standpoint of routine maintenance and operating personnel safety.

Unique Sampling System Requirements

As NDIR analyzers are basically molecule counters, care must be taken to assure that the sample cell is operated at constant pressure. For most gaseous analysis applications, this means that the sample should vent to a safe exhaust at atmospheric pressure. Restrictions (flowmeters and valves on the outlet) should not be permitted. A 1% change in cell pressure will result in a direct analytical error of 1%.

Sampling systems for liquid applications should have a pneumatic accumulator to prevent hydraulic pulses (from process pumps and valves) doing damage to optics. For samples with high dew points, it may be necessary to heat the sample transport line, sample-conditioning system, sample cell, and fast-loop return.

Sampling systems frequently require particulate filters and coalescers/condensers for removal of particulates and condensate. Typical additional components include traps, automatic drains, and so forth. For such systems, parallel installation of these types of components (one in a standby mode) should be considered to expedite changeover when maintenance is required.

Typical Calibration Requirements

As with most instrumentation, calibration is required at scheduled intervals to assure that the analytical signal is valid and that the instrument has not drifted because of environmental conditions, component aging, or other factors. The calibration of an infrared analyzer is a two-point operation: zero and span. Calibration methods are well covered in manufacturers' instruction manuals and generally involve introducing known samples and performing the necessary analyzer adjustments. Because such analyzers are related to the Beer-Lambert law, a plot of signal versus concentration will be slightly nonlinear. Manufacturers typically select cell lengths to limit this nonlinearity to less than 5% of the full-scale measurement value. Alternately, the manufacturer may provide a "linearizer" board that compensates the output signal at the high concentration portion of the span range.

The calibration of an infrared analyzer is a two-point operation: zero and span.

Calibration samples should be validated (analyzed) by the plant's laboratory prior to use by maintenance personnel. Process samples can be used provided that they are stable and have been preanalyzed in the laboratory. However, a mix of pure components is preferred because the mixture can be replicated. Care must be taken to assure that gas samples will not condense inside the container due to cylinder pressure and temperature conditions, and that the calibration samples have not stratified. Where interferences (matrix effects or cross-sensitivities) exist, calibration and analytical data can be affected. The bias should be determined, and, if significant, interfering compounds should be included in the calibration samples.

Zero calibration samples should be equivalent to the average process stream less the analyte. The span sample should be a concentration of the analyte (about 80% of the full-scale range) in the zero material. In the previous example of CO in air, the zero gas would be nitrogen and the span gas CO in nitrogen (as oxygen is infrared-transparent).

It should not be forgotten that the object of on-line analysis instrumentation is to provide quality analytical data to the client (operations, regulatory agencies, etc.). Regular calibration efforts will assure this. Calibration data should be recorded in the form of a chart (SPC control chart) to verify that the analyzer is providing quality analytical data.

To confirm selectivity, response to background compounds should be checked on a scheduled basis. Note: This is comparable to running a chromatogram for confirming the peak gating. Such selectivity checks should also be done if the feedstock changes radically. Interferences usually result in a positive data bias.

Troubleshooting and Maintenance

Process infrared analyzers are quite rugged, frequently exhibiting lifetimes in excess of 10 years. However, when failures occur, they can generally be field corrected. Detector failure will be exhibited by a decay of sensitivity and selectivity. Beam interrupter (chopper) motors and temperature-control systems exhibit operational lifetimes greater than 18 months. Current designs include microprocessor-based maintenance diagnostics to expedite troubleshooting. Additionally, most manufacturers provide comprehensive Troubleshooting and Maintenance sections in their instruction manuals. These sections should be reviewed periodically to avoid stressful situations when the inevitable failure occurs.

Most analyzer problem areas lie in the sample-handling systems.

Nevertheless, most analyzer problem areas lie in the sample-handling systems. Preventive maintenance of filters, traps, and condensers will help to avoid analyzer downtime.

To troubleshoot NDIR analyzers, the instrument should be operated on calibration gases. If the analyzer runs stable on zero and span gases and unstable on-line, the problem may reside in the sampling system. If the analysis has a positive bias as compared to laboratory data, the optics may have a selectivity problem. Response to various process stream components (pure) should be measured and, if unacceptable, the analyzer manufacturer should be contacted.

References

1. Kendall, *Applied Infrared Spectroscopy*, Reinhold.
2. "Instrumentation and Control in the German Chemical Industry," British Intelligence Operations Sub-committee Report #1007, 1946.
3. Pierson, R.H., Fletcher, A.N., and Glantz, E.S., "Catalog of Infrared Spectra for Qualitative Analysis of Gases," *Analytical Chemistry*, 28: 1218-1239, August, 1956.
4. Willard, Merritt, *Instrumental Methods of Analysis*, Van Nostrand.
5. Rosenweig, A., *Photoacoustic Spectroscopy*, New York: Wiley & Sons, 1980.
6. Boyle, W.M., "Principles and Applications of Fourier Transform Infrared (FTIR) Process Analysis," *Process Control and Quality*, 2: 11-41, 1992.
7. Dailey, W.V., "The Determination and Use of Discrimination Ratios for Non -Dispersive Infrared Analyzers," presented at ISA/AID National Conference (Montreal), 1965.
8. Dailey, W., Sonner, R., and Wright, D., "An Open Path Hydrocarbon Detector," presented at ISA/AID Symposium (Chicago), 1984.

Bibliography

1. Clevett, K.J., *Process Analyzer Technology*, New York: John Wiley & Sons, 1986.
2. Nichols, G.D., *On-Line Process Analyzers*, New York: John Wiley & Sons, 1988.
3. Sall, A.O., "Selectivity and Sensitivity of Infrared Gas Analyzers," *Priborostroenic*, No. 9, 1956.
4. Sall, A.O., "The Theory of the Microphone Chamber in a Veingerov Spectrophone and of an Optical-Acoustical Gas Analyzer," *Soviet Physics*, 26: 152-169, 1956.

About the Author

William V. Dailey has made a career of applying analytical technology to the process field while with Mine Safety Appliances Company and Exxon Research and Engineering Company. He is currently President of Process Analyzer Resources, Inc. (a consulting firm), and an active speaker on topics in the process analytical technology field. He has served as speaker, program chairman or session developer for over 30 technical programs. He is a prolific author as well with 27 papers or other published works to his credit. Some of his works include "Infrared Analyzers for On-Stream Analysis of Ethylene Oxide Process Streams," "A Novel NDIR Analyzer for Nitric Oxide, Sulphur Dioxide and Carbon Monoxide Analysis," and "The Whys, Whats, Whos and Hows of Process Analyzer Maintenance." Mr. Dailey has received the Gilmer-Thomason-Fowler Award from the Instrument Society of America's Analysis Division (1984), the ISA Excellence-in-Documentation Award (1985) for his paper on "An Open Path Infrared Hydrocarbon Detector," and is an ISA Fellow. He holds a B.S. from the University of Pittsburgh and a Bachelor of Law degree from La Salle Extension University. He has been an instructor on process analyzers for ISA and at the Centre for Applied Process Technology at Lambton College in Ontario.

29

Fourier Transform Infrared (FTIR) Analyzers

Fourier transform infrared (FTIR) analyzers measure the chemical composition of raw materials, intermediate, and final products in a wide variety of process industries. As laboratory instruments, they have been available since about 1960. However, analyzers based on this technique have been used in process industries only since the mid-1980s. The earliest continuous-monitoring FTIR applications involved the measurement of toxic gases in ambient air. Today, the range of applications is extremely broad—from analyzing parts per billion level components in gases and sparged water streams to measuring the major component percentages in solids, liquids, and gases. Measurement speed, one of the most important advantages of these analyzers, typically is about 30 s but can range from one measurement per second to one every 5 min. Because of this speed and other advantages, FTIR analyzers are used to measure continuous and batch processes at plants operated by many of the world's major chemical manufacturers.

Measurement speed, one of the most important advantages of these analyzers, typically is about 30 s but can range from one measurement per second to one every 5 min.

The basic principles of operation, range of applications, and installation considerations will be discussed in this chapter. Because of the broad range of suitable applications for the Fourier transform infrared technique, you probably will need the assistance of an analyzer vendor in developing a specific application.

Measurement Principle

Fourier transform infrared analyzers are fundamentally based on the same physical measurement principle as nondispersive infrared analyzers (NDIR) (see Chapter 28). This principle is that virtually all chemicals of importance to manufacturing absorb infrared radiation in a unique and reproducible manner. The specific pattern of absorption for each chemical is caused by the excitation of vibration modes in the bonds between atoms in the sample. This absorption pattern is collected as the spectra measured by the infrared analyzer. The fact that these patterns are linear and additive is the basis of the infrared analyzer's ability to simultaneously and quantitatively determine the composition of complex mixtures of chemicals.

This principle is that virtually all chemicals of importance to manufacturing absorb infrared radiation in a unique and reproducible manner.

Figure 29-1 is an infrared absorption spectrum of dichloromethane (methylene chloride) annotated to illustrate the assignment of some characteristic bands. In general, weaker bonds linking heavier atoms are found at lower frequencies, shown at the right of the spectrum display, while stronger bonds and lighter atoms appear at the upper end of the spectrum. In this example, the weak CH band appears at the left, with the bands containing chlorine to the right. Interpreting distinct spectroscopic band assignments as shown here is a useful tool for chemists developing calibrations, but is not necessary for the engineer applying the system in the field.

In general, weaker bonds linking heavier atoms are found at lower frequencies, shown at the right of the spectrum display, while stronger bonds and lighter atoms appear at the upper end of the spectrum.

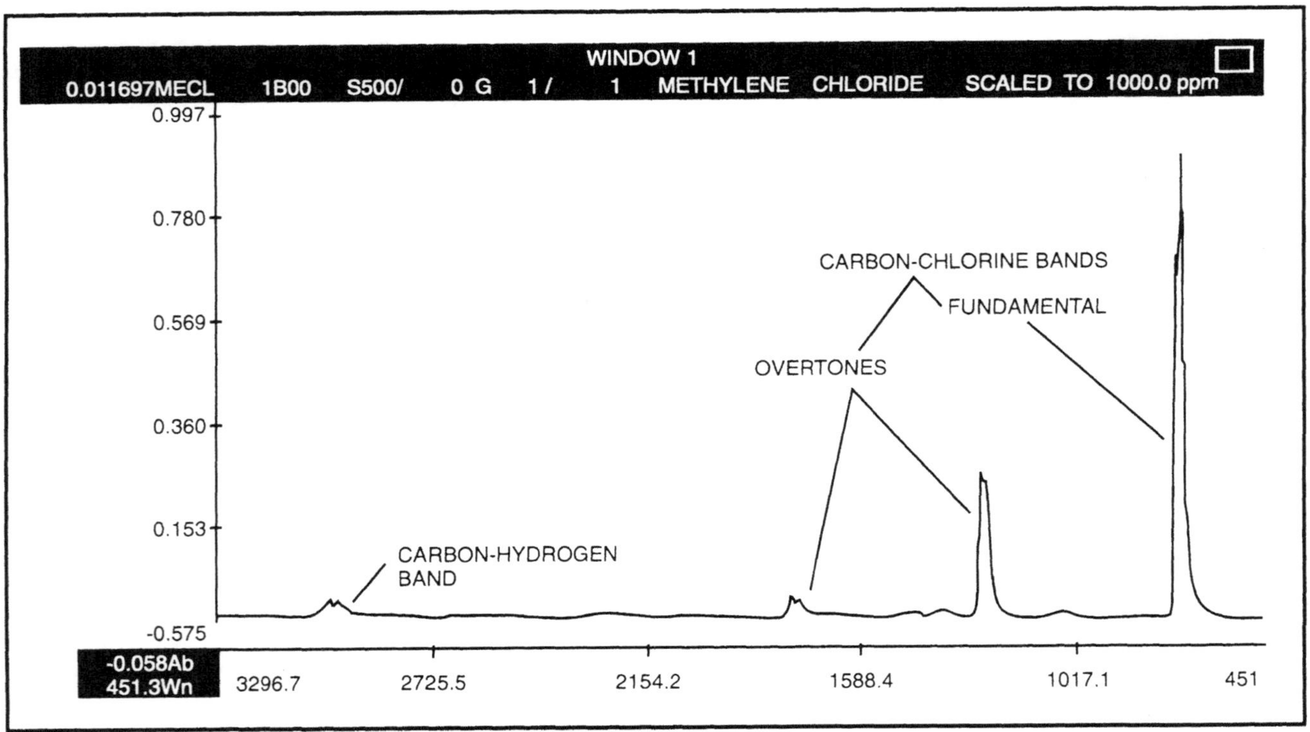

Figure 29-1. Dichloromethane Infrared Spectrum with Band Assignments

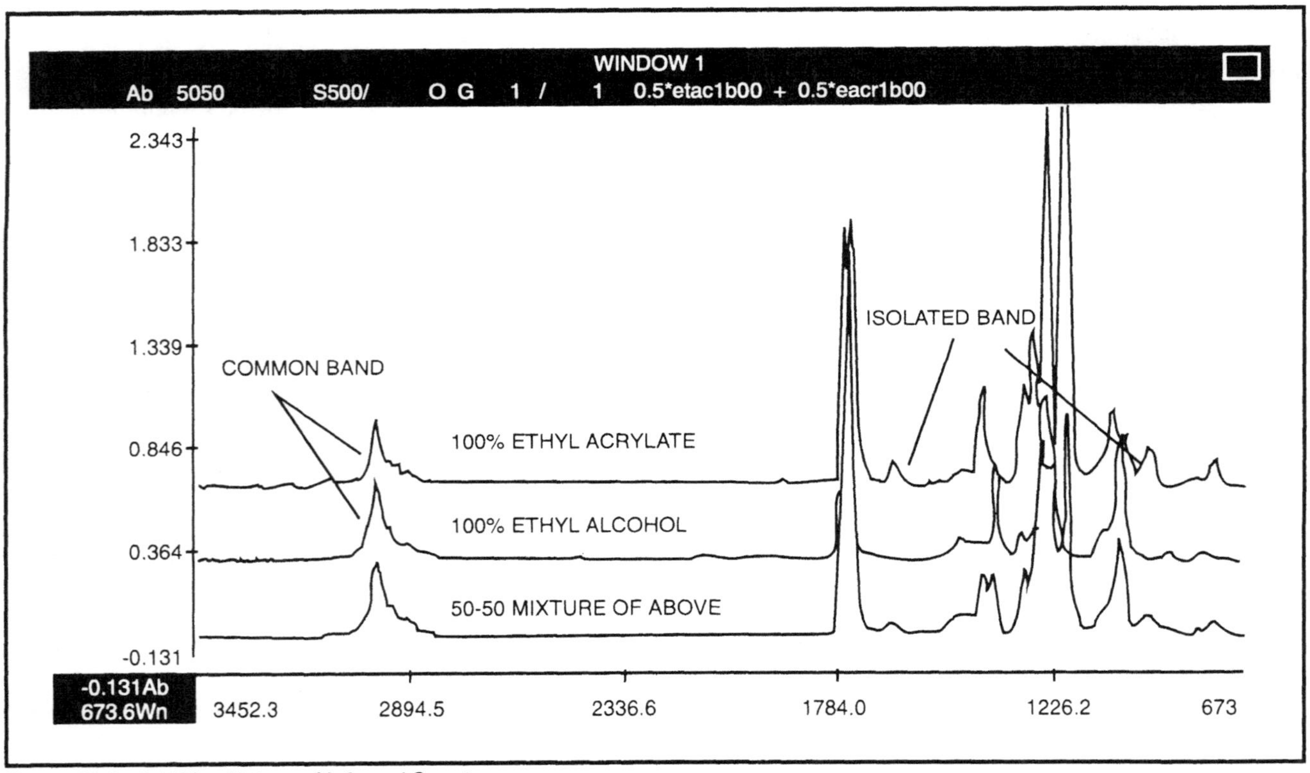

Figure 29-2. Additive Nature of Infrared Spectra

Figure 29-2 illustrates the additive nature of infrared spectra. This feature makes calibrations possible for this class of analyzer. The Beer-Lambert law, often shortened to "Beer's law," states that absorbance *A* measured by an infrared spectrometer is proportional to the product of the intrinsic absorptivity *a* of the band being measured, the path length *b* the radiation passes through, and the concentration *c* of the species being measured.

The three spectra shown in Figure 29-2 are ethyl alcohol, ethyl acrylate, and a 50:50 mixture of the two. The molecular structures of these materials have both common and distinct bands. Note that where the bands match, the 50:50 mixture spectrum band is approximately the same height as each of the pure spectrum bands. Where there are isolated bands, the mixture bands are half the height of the pure component bands.

A number of mathematical techniques, commonly grouped under the term "chemometrics," is used to produce a statistically-optimum calibration from a set of known concentration spectra. All of these techniques make use of this linear additive behavior of spectra.

Fourier Transform

The term "Fourier transform" applies to the method in which the spectrum is obtained, illustrated schematically in Figure 29-3. This method is fundamentally a process of modulation or encoding followed by demodulation or decoding. A source of infrared radiation, generally from a heated coil of wire, is collimated and passed through an interferometer. Once in the interferometer, the radiation is split into separate transmitted and reflected beams that are bounced off moving and fixed mirrors, respectively. The beams are then recombined at the beam splitter, where they produce an interference pattern. This pattern varies with time as the moving mirror travels back and forth.

This method is fundamentally a process of modulation or encoding followed by demodulation or decoding.

The result of this process is that the longest radiation waves are modulated at low frequencies while the shortest waves are modulated at high frequencies. The composite pattern of all these frequencies, when sampled, is called an interferogram and is the Fourier transform (a mathematically-defined operation) of the spectrum of the infrared (IR) source.

Interferometers built in this form can be found in many modern laboratory spectrometers. However, due to the sensitive equilibrium required to maintain the alignment of the mirrors, they are quite susceptible to vibration and temperature changes and thus are often not suitable for use in a process manufacturing environment.

Practical Process Interferometers

Although an interferometer of a geometry very similar to that shown in Figure 3-30 could in principle be used in a process analyzer, this is not practical because of its extreme sensitivity to vibration and temperature changes. At the time of this writing, two companies have developed and fielded interferometers much more suitable for use in a process environment.

At the time of this writing, two companies have developed and fielded interferometers much more suitable for use in a process environment.

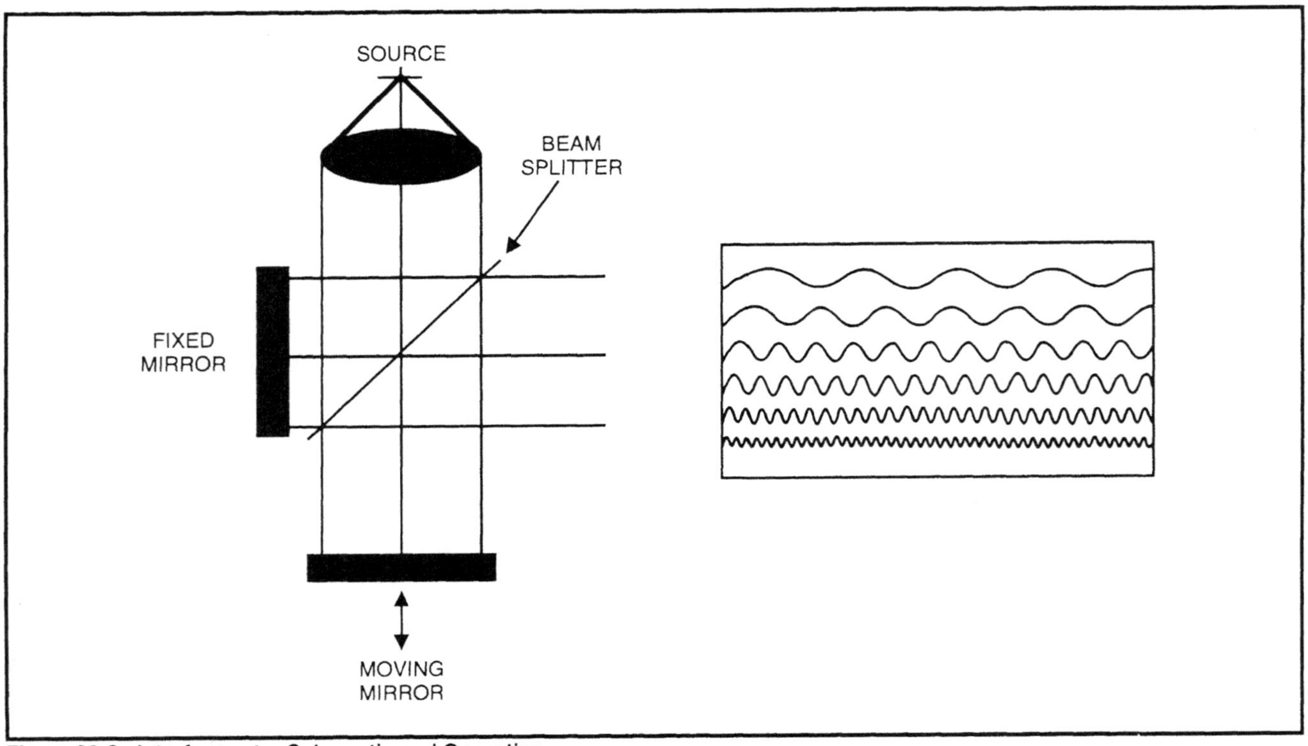

Figure 29-3. Interferometer Schematic and Operation

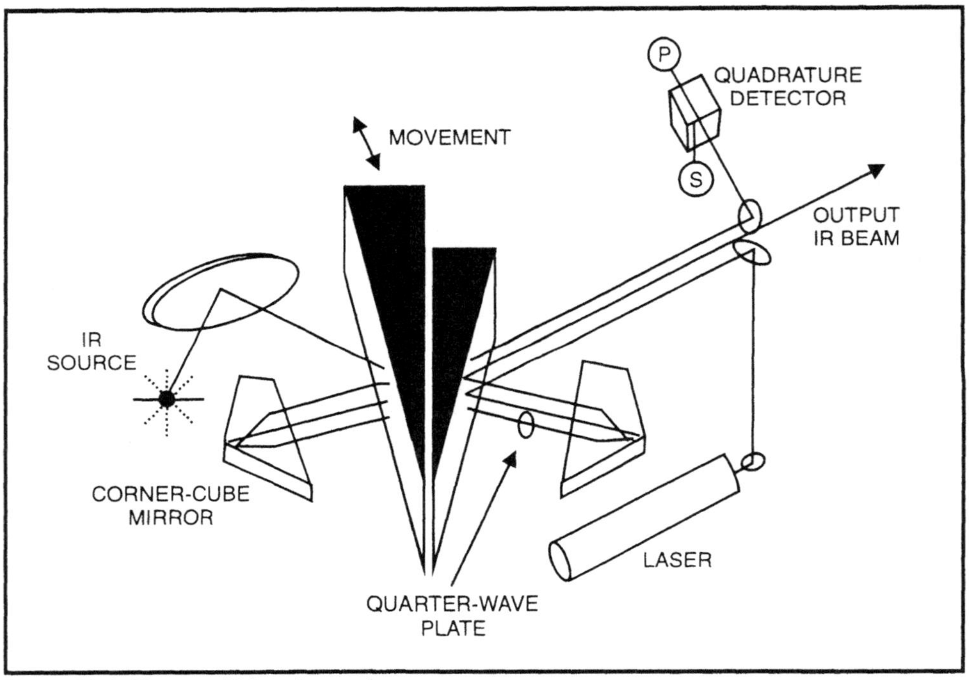

Figure 29-4. Transept™ Interferometer

KVB/ANALECT TRANSEPT™ INTERFEROMETER

The Transept interferometer (Figure 29-4) was designed from the start for harsh duty. The fixed and moving mirrors of the instrument shown in Figure 29-3 are replaced by two permanently-fixed corner-cube mirrors. The beam-splitter is fixed on the inside surface of the smaller of the two wedges of refractive material (potassium bromide or calcium fluoride). The path-length modulation is accomplished by translating the larger compensator wedge across one beam of the interferometer. As it travels, it displaces air with an index of refraction of 1.0 by the refractive material with an index of 1.4 to 1.5. The modulations produced by this design are fully equivalent to those of the standard design. A quadrature laser position-detection scheme allows precise acquisition of interferogram data and a highly-stable frequency scale.

The main benefits of this geometry are permanent-alignment and insensitivity to vibration and temperature changes achieved by permanently mounting the three critical optical surfaces (2 corner-cube mirrors and smaller refractive material wedge) of the interferometer.

BOMEM/MICHAELSON WISHBONE INTERFEROMETER

The wishbone-design Michaelson interferometer used in the Advance FTIR and Advance FTNIR from Bomem/Applied Automation, Inc./Hartmann & Braun/ Elsag Bailey (at the time of this writing) utilizes dual moving corner reflectors. This design features permanent alignment of the interferometer optics. As can be seen in Figure 29-5, the relative alignment of the two reflectors throughout the scan motion results in negligible effects due to tilt and shear.

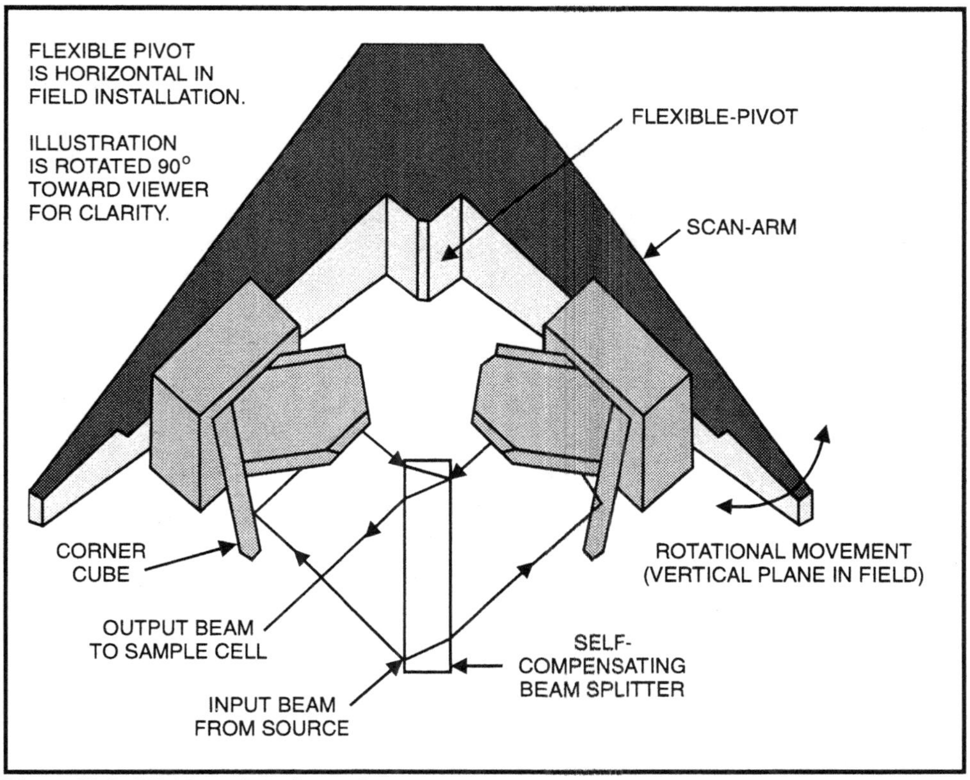

Figure 29-5. Wishbone Interferometer

The design is also insensitive to vibration, because the moving mechanism is supported by a laterally-stiff flexible pivot. It is center-balanced so that lateral impulses do not affect the rotational movement. Thermal stability is assured by taking advantage of the high thermal conductivity of aluminum in constructing both the scan arm and the beam splitter support.

Other unique features include (1) the use of a self-compensating beam splitter, (2) spatial separation of incoming and outgoing beams, allowing increased versatility without moving components, and (3) the use of quadrature detection for determining the direction of scan-arm movement.

Basic Operating Procedures

Even though interferometer designs may differ, the operating procedures for both laboratory and process systems are essentially the same. The modulated radiation from the interferometer is measured twice. The first time it is collected without the sample present, the second time with the material to be measured in the beam's path. Each of these interferograms is analyzed by the same IR detector and transformed using a fast-Fourier transform (FFT) algorithm applied by the analyzer's data system. This mathematical process is the inverse of the one used by the interferometer. Thus, the interferograms are demodulated or decoded.

The negative log of the transmission spectrum is called an absorbance spectrum. In this format, the band heights obey Beer's law.

The two spectra produced have all the spectral contributions of the source, interferometer, sample cell, and detector. The second spectrum has all of these and, in addition, the contributions of the sample to be measured. The ratio of these two therefore consists of the transmission spectrum of the sample plus some noise.

The negative log of the transmission spectrum is called an absorbance spectrum. In this format, the band heights obey Beer's law. This format is used in the quantitative computations of composition. This may appear as an overly complex method of obtaining the spectrum of a material. There are, however, three important reasons to apply this technique. We will state them here without proof in the interest of brevity. The multiplex or Fellegits-advantage is that all frequencies of light are incident on the detector at all times, shortening the time needed to produce an accurate spectrum. The throughput or Jacquinot-advantage is that, compared to a dispersive spectrometer, no angle-limiting slits are required to obtain a spectrum with a good resolution. This again reduces the time required for a measurement. Finally, the frequency scale or Connes-advantage is that a helium-neon-laser is used to determine the frequency scale of the instrument in all modern FTIR analyzers. The use of this laser is critical in maintaining quantitative stability in the face of process environmental changes.

Common Applications

All that is required is that the sample be positioned so that some part of the interferometer beam can be absorbed as it makes its way to the detector.

One of the most important characteristics of FTIR analyzers is the range of applications for which they are suited. All that is required is that the sample be positioned so that some part of the interferometer beam can be absorbed as it makes its way to the detector. The sample phase need not be changed in order to make the measurement, as is the case with gas chromatography (GC) analyzers. Gases and liquids are generally measured while constrained to a fixed-path-length in some type of sample cell with IR transparent optics. On the other hand, solids can be measured by either of two methods. If they are thin enough, they can be measured by direct transmission, as in the case of moving films. Thicker solids can be measured by reflection from the surface of bulk materials.

Table 29-1 presents a brief list of some of the process measurements that have been performed successfully by FTIR analyzers. The scope of materials measured has been limited more by the small number of vendors pioneering this field than by the constraints of the technique. New applications are going on-line every month.

Table 29-1. FTIR Analyzer Applications

Phase	Major Constituents	Trace Constituents
Gas	Composition of polymer feed streams Methanol and methylene chloride in nitrogen	Trace toxins in ambient air Flammable gases in ambient air Methane and HCl in process hydrogen Contaminants in chlorine, bromine
Liquid	Molten polymer composition Batch reaction end-point determination Continuous synthesis composition	Contaminants in liquid chlorine Stabilizers in reactive liquids Composition of supercritical-phase ethylene

Table 29-2. FTIR Analyzer Measurement Ranges

Sample Type	Units	Minimum Detection	Full Scale	Notes
Trace gases in air	ppm	0.1-10	100-10000	10-20 m multipass cell
Process gases	%	0.01-1.0	100	0.5-20 cm single-pass cell
Stack gases CO, NO_X, SO_2	ppm	1.0	3000	1-5 m cells
Process liquids	ppm/%	100-1000	10-100%	0.1-0.5 mm transmission cells
Process liquids	%	0.1	100	2 to 11 bounce ATR* cells
Volatile organics in wastewater	ppb/ppm	1-500	1-100	Sparging + long-path gas cells

* Attenuated Total Reflectance

Measurement Ranges and Accuracy

It is difficult to establish a hard and fast measurement range and accuracy specification for an analyzer technology with such a wide scope of possible applications. The information presented in Table 29-2 and in the remainder of this section is therefore offered simply as a guideline in reviewing the feasibility of using this technology for a specific application. In the end, the best resource for determining whether an application is a candidate measurement with an FTIR analyzer is an FTIR vendor with experience in developing process systems. The applications engineers at process FTIR vendors constantly are evaluating new applications and may have previously developed a related system or, almost as important, determined that FTIR is not an appropriate technique for the application. Numerous good process FTIR applications have already been proven, but putting even one of these systems on-line is a relatively complex and expensive (hardware, calibration, and commissioning – $100,000 US ± $25,000 US) process. Avoiding failed applications is essential in this as in any emerging analyzer technology industry.

The best resource for determining whether an application is a candidate measurement with an FTIR analyzer is an FTIR vendor with experience in developing process systems.

Accuracy

Accuracy is a complex issue in this class of analyzers. It is affected by noise and nonlinearity in the basic instrument, but may be even more limited by the difficulty of obtaining precise calibration standards and the need to account for potentially-compound interferences. If the calibration standards can be considered absolute, accuracy of ±1% of range can be expected.

An additional error term proportional to the minimum detectable quantity of material becomes important at the low end of each component range. There also may be an increase in the relative error due to the fact that the majority of the IR energy in the measurement bands is absorbed at the high end of the measurement ranges.

Precision or relative accuracy of ±1% of reading or ±0.1% of full scale can often be obtained.

Wavelength Coverage

The majority of applications to date for process FTIR have used the traditional mid-IR region of the spectrum. This region normally runs from 400 to 4400 cm^{-1} (25 to 2.27 μm). Recently, extended mid-IR ranges from 450 to 7,000 cm^{-1} or near-IR ranges from 12,000 to 3,000 cm^{-1} also are being used. These newer ranges access the overtone and combination bands of the fundamental bands in the mid-IR. The near-IR region is particularly useful because it can be carried over the fiber-optic cables that are discussed later in this chapter. The selection of a specific region for a given application is based on chemistry, sample conditions, and spectroscopy.

Alignment Considerations

An interferometer is the critical component in the measurement process. Alignment of this component must be extremely accurate and stable in order that the process analyzer continues to deliver accurate measurements. This requirement was a limitation with some early applications of the technology, but most current process FTIR analyzers are permanently aligned. Field alignment is not required and in most cases is not even practical due to the requirement for specialized tools. Some lower-cost analyzers from other than the manufacturers mentioned in this chapter require periodic field alignment. The support cost for providing this service should be factored into any procurement decision.

Options and Accessories

FTIR systems have become more capable in recent years due to the availability and integration of a number of options and accessories. These accessories allow the system to be more efficiently customized to specific applications.

Measurements Cells

The most important option to be considered when configuring an FTIR analyzer is the sample interaction cell that presents a representative sample to the analyzer for absorption of the infrared energy. The cells for process FTIR analyzers all have some common requirements, which will be discussed here before reviewing the specific properties a few of the most widely-used sample cells:

1. They must contain the sample to a stable fixed-path-length; any variation in path length will affect the readings.
2. They must be chemically-compatible with the sample. This consideration must take into account the full range of temperature and composition variations over an extended period of time. A cell that works fine in the lab for a few measurements may rapidly fail on-line.
3. The optical components must transmit infrared light of all wavelengths needed for the analyses. Often, there may be a trade-off here as the best IR transmitting materials often are relatively weak and may be sensitive to chemical attack.
4. All components of the cell must be capable of keeping the sample safely contained. Again, the full range of process variables over an extended period of time must be considered.

SHORT-PATH GAS CELLS

These cells fundamentally consist of a hollow tube with gas fittings in the sides at each end and infrared transparent windows on each end. The sample flows through the tube while the infrared energy passes axially through the gas volume. Figure 29-6 shows a typical gas cell. The spacing between the windows for this type of cell ranges from a few millimeters to about 20 cm.

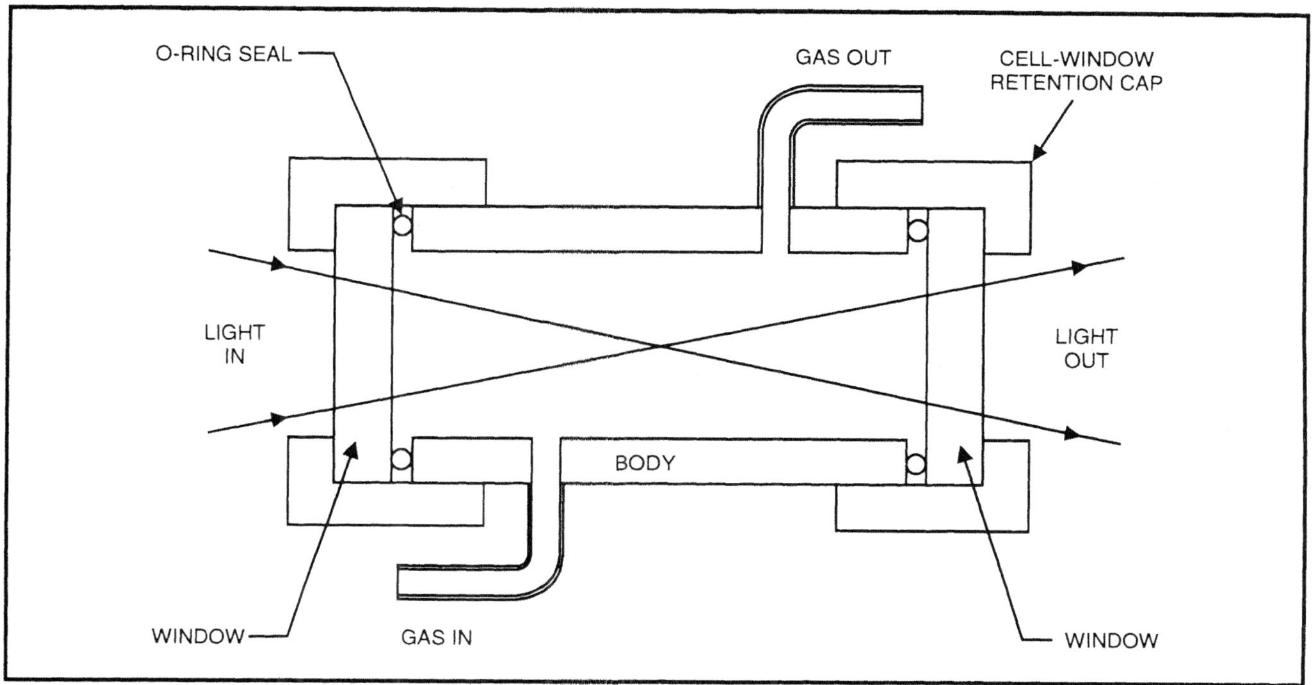

Figure 29-6. Typical Short-path Gas Cell

LONG-PATH GAS CELLS

Long-path gas cells present a large number of molecules to the analyzer beam for measurement. The majority of long-path gas cells used today are of the "white cell" design, which folds a number of reflected beams back and forth in a sample volume. Figure 29-7 shows the optical schematic of such a cell. Effective path lengths of these cells generally run in the range of 2 to 40 m.

The majority of long-path gas cells used today are of the "white cell" design, which folds a number of reflected beams back and forth in a sample volume.

These white cells are excellent for clean gases and can be made of materials that stand up to many corrosive chemicals. However, since the beam bounces off the mirrors many times, their use is limited when sample components reduce the reflectivity of the mirrors. Consider that a 20-pass cell has 40 reflections. The throughput limited by the mirrors is r^n, where r is the reflectivity and n is the number of bounces. For a system with 98% reflectivity at 40 reflections, that is 45% throughput. A reflectivity of 95% yields only 13% throughput.

For applications not suitable for multiple reflection, a new class of cells has recently become available. These tubular cells send the beam through one or more sections of straight reflective tubing. Multiple-tube versions use pairs of 45° mirrors to fold the light from one tube to a second parallel tube. Figure 29-8 shows such a cell. These cells have two major advantages. First, they have greater tolerance for dirty or chemically aggressive samples; second, they exchange samples faster than in an equivalent white cell. This second benefit comes from the plug flow of sample through the long, slim tube. It can be important in measuring rapidly changing concentrations or when switching between streams. Disadvantages include lower path-length-to-volume ratio and higher cost.

Many long-path gas measurements are made on streams containing a significant amount of water vapor. For example, ambient air has between 1 and 4% water vapor based on temperature and relative humidity. Combustion stack gases have up to 20% water vapor. In a long-path cell, these relatively high levels of water vapor blank out some regions of the spectrum. To reduce this problem, some systems employ water-removal equipment. Two types of such systems are often used. Condenser-type gas chillers reduce the water level to approximately 0.6%, whereas permeation-type systems can lower it into the low parts per million range. Both of these techniques tend to remove certain chemicals along with the water vapor, so they cannot be used for all applications.

Particulate matter and any heavy stream component that would tend to form a coating on mirror surfaces must be excluded from the cell.

Another issue that arises with long-path cells using multiple mirror reflections is stream cleanliness. Particulate matter and any heavy stream component that would tend to form a coating on mirror surfaces must be excluded from the cell. The former is done by including particulate filters upstream of the cell, while the latter is often done by operating the cell at elevated temperatures to prevent condensation of any stream components.

The process versions of both classes of long-path cells allow for disassembly and mirror cleaning, but this is done only after extended operation or due to a major process or sample system upset. Background measurements for these measurements are routinely done after flushing out the previous sample with nitrogen or instrument air.

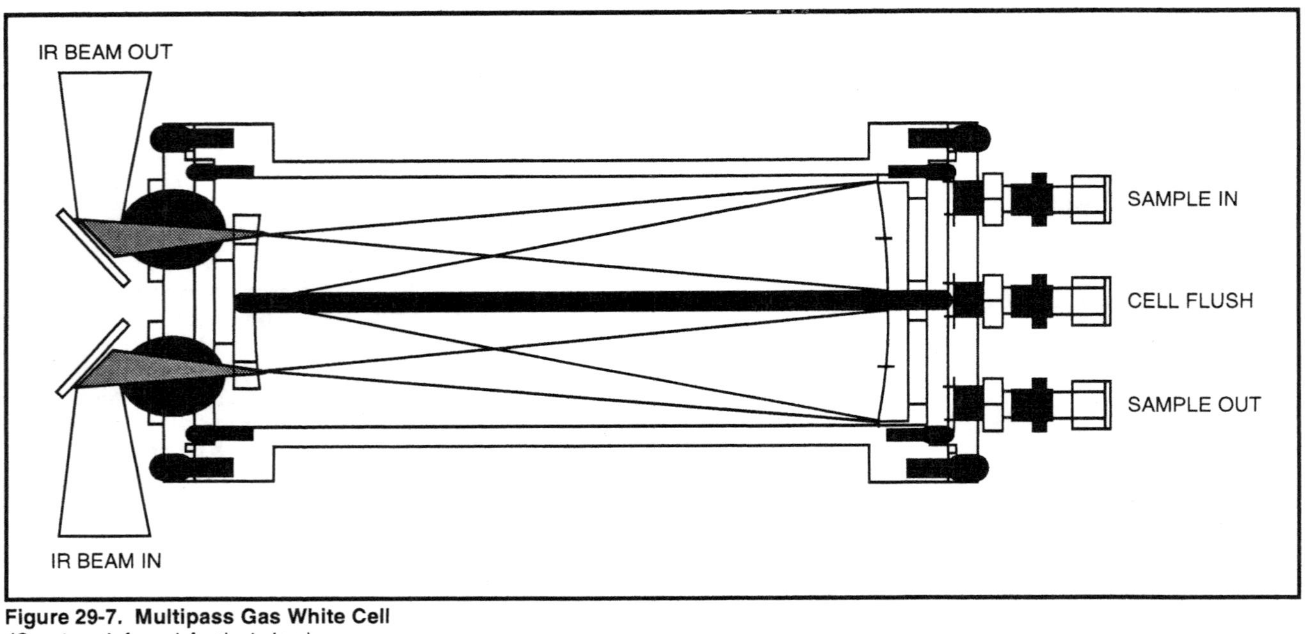

Figure 29-7. Multipass Gas White Cell
(Courtesy Infrared Analysis Inc.)

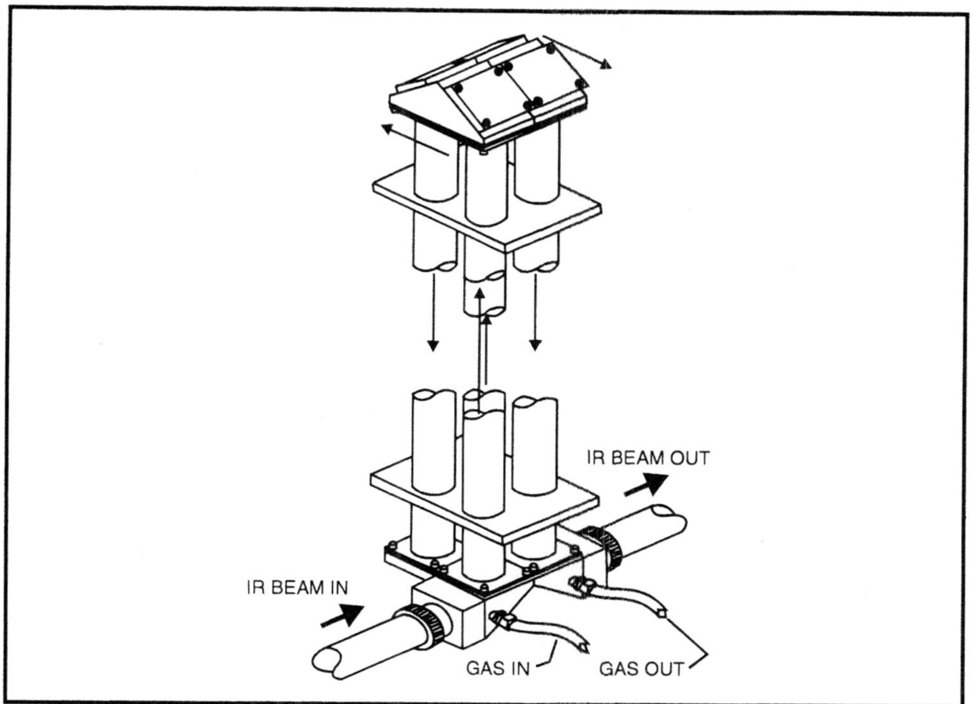

Figure 29-8. Laminar-Flow Gas Cell
(Courtesy Axiom Analytical Inc.)

TRANSMISSION LIQUID CELLS

Transmission cells similar to those described above often are used for liquid measurements. Consider, however, that the scaling of the spectrum is proportional to the number of molecules in the path. Liquids are typically approximately 1,000 times more dense than gases, so the corresponding path lengths are proportionally smaller. A 10-cm gas cell will have approximately the same number of molecules as a 0.1-mm liquid cell. These small gaps at times present difficulties, but many important applications use this class of cell. Transmission liquid cells for FTNIR are easier to use, as the intrinsic band strengths in this region of the spectrum are 10 to 1,000 times weaker.

ATR LIQUID CELLS

A technique called attenuated total reflectance (ATR) is used to avoid flow problems through very small gaps. In this technique, the light is passed through a transparent crystal in such a way that it reflects off the inner surface. If the angle of reflection is appropriately chosen, then all the light is reflected. Figure 29-9 shows the beam geometry of a simple ATR measurement; Figure 29-10 shows a modern ATR Tunnel™ cell. While the light is totally reflected, a small part of the electric vector of the electromagnetic field of the light penetrates through the surface. If an infrared-absorbing material is brought in contact with the outer surface, then this field can electric-vector-couple into the sample and be absorbed. This is the attenuation in attenuated total reflectance.

The length of this electric-vector-field penetration is on the order of one-fourth to one times the wavelength of the IR light, or from 0.5 to 20 μm for mid-IR radiation. By choosing the number of bounces, the effective path length can be adjusted for a particular application. The flow path outside the crystal can be any appropriate size. Some ATR cells actually are placed inside a reaction vessel.

ATR cells have some limitations. The crystal used must maintain a good surface quality for an extended period when exposed to the sample materials.

Cleaning normally is done by adding a stream to the sample system that provides a solvent appropriate to remove all traces of the sample material.

Samples that coat the crystal will exclude new sample and thus only measure the coating material. Finally, ATR is generally not used in the NIR region.

Instrument background measurements for liquid applications are generally done through the measurement cell. This requires that the cell be automatically cleaned as part of the background measurement process. Cleaning normally is done by adding a stream to the sample system that provides a solvent appropriate to remove all traces of the sample material. The subsequent background is either taken through this solvent or is taken after the cell is dried with nitrogen or air.

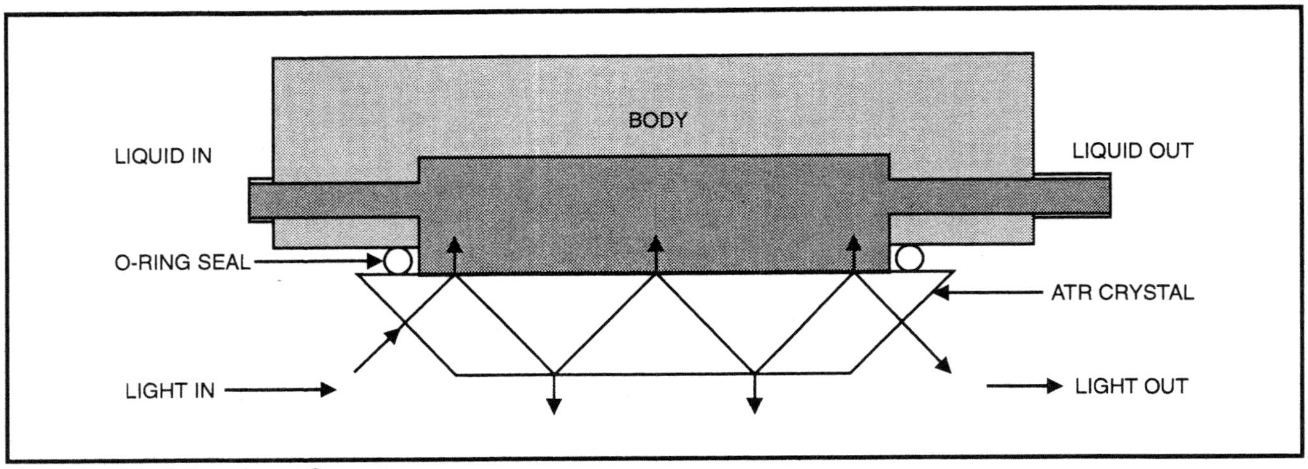

Figure 29-9. ATR Experiment Geometry

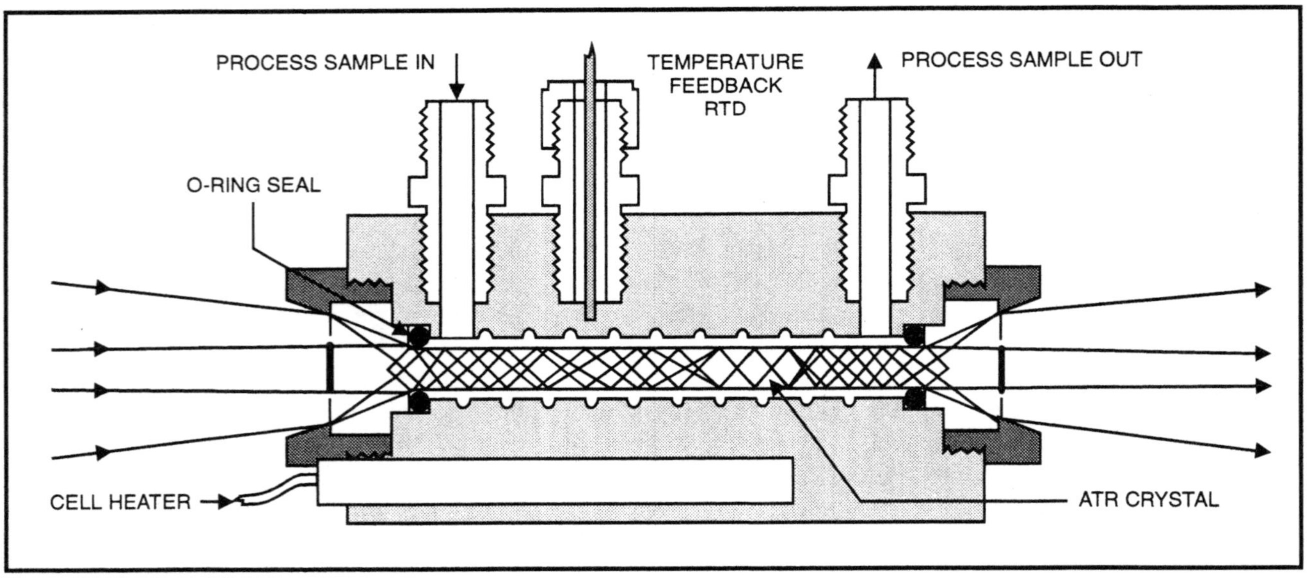

Figure 29-10. Tunnel™ Cell
(Courtesy Axiom Analytical)

Sample Systems

Most process FTIR applications involve extractive or on-line sampling. A sample is removed from the process, measured, and then either returned to the process at a different point or sent to waste. The sample systems to accomplish this range from trivially simple to extremely complex. Generally, the valves, pumps, and flow monitoring associated with the sample system is under the control of the same software that is making the measurements. Sample system maintenance is one area where service can generally be easily done by plant personnel. Most process FTIR sample systems use the same type of valves, fittings, and other components as found in many types of on-line analyzers (see Chapter 3).

FIBER-OPTIC MEASUREMENT OPTIONS

In an increasing number of applications, particularly in the NIR region, it is appropriate to send the light to the sample rather than bringing the sample to the analyzer. This includes applications where the sample is under extreme pressure, is of an extremely high temperature, or presents an unacceptable risk in terms of the possibility of leaks in the sample system.

In an increasing number of applications, particularly in the NIR region, it is appropriate to send the light to the sample rather than bringing the sample to the analyzer.

Fiber-optic cables, consisting of extremely pure glasses or other materials, can send light over long distances with little attenuation. For silica fibers, spectral ranges run from the visible to approximately 2 μm. Fiber runs longer than 100 m have been achieved. For heavy metal fluoride glasses, the long-wave cutoff is extended to about 4.5 μm, including the fundamental C–H, N–H, and O–H vibrations. These fibers can be obtained in lengths up to about 50 m. Fibers that extend into the mid-IR region are becoming available for specialized short-run (1 to 3 m) applications.

COMMUNICATIONS OPTIONS

The information generated by a process FTIR system generally is communicated to a process control system. Most of the present applications use multiple 4-20 mADC loops for outputs. More effective communication can be achieved using Modbus communication, custom RS-232-C protocols, or ethernet connections to conventional local area networks (LANs). The restrictions in using these advanced capabilities generally are attributed to the cost and complexity of programming the current generations of distributed control system (DCS) units rather than any limitations in the analyzer.

Limitations

The most important limitation of process FTIR is that the individual concentration data must be mathematically extracted from a single complex spectrum. This imposes two major restrictions.

Lack of Stream Knowledge

The mathematics that extracts the concentrations must be taught the spectral signatures of all the materials that will be in the FTIR beam. The chemist or process engineer requesting the analyzer must be able to identify these materials, and the chemist who builds the application model must obtain spectra of each. The system engineer or chemist cannot say, "I am not interested in measuring component C, so don't bother to calibrate for it." A corollary to this rule is that after the application goes on-line, changes in the stream composition generally will have a negative impact on the measurements and a calibration update will be needed. Statistical measures of model fit are often performed along with each analysis to warn of the presence of unknown components.

The mathematics that extracts the concentrations must be taught the spectral signatures of all the materials that will be in the FTIR beam.

Stream Complexity

Since mathematics are used to extract the individual concentrations, there is a limit to the number of components that can be simultaneously monitored. Routinely, 6 to 15 components are often measured at one time, but any more than this is rare.

One class of applications can be an exception to this general rule of thumb. In these applications, only the maximum current concentration possible must be known, such as in hazard-detection systems. While the actual concentrations are not determined, this "not-to-exceed" value for a large number of materials can be useful.

Another way to get around this limitation is to model functionally-similar components together. For example, it is more accurate to model the weight percent of oxygen in oxygenated fuels by modeling ethers as one group and alcohols as another group than by modeling every possible ether and alcohol, then adding their oxygen contributions.

Case Histories

The following application case histories are intended to provide an overview of the process of bringing one of these analyzers to routine operation.

Industrial Hygiene Analysis of Ambient Air

A manufacturer of copier equipment uses a number of volatile solvents in the process of coating optical drums for copiers. At some level, all of these materials are toxic; therefore, the manufacturer wants to protect workers from harm should any of the normally-sealed equipment leak.

The manufacturer sent the FTIR analyzer vendor a list of the materials to be monitored, along with the alarm levels required for adequate protection. Most of the materials were common industrial chemicals, and the FTIR vendor had existing spectral libraries showing that the application was feasible for all of the compounds found. The manufacturer sent a small sample of the one compound not found in the libraries to be measured in the vendor's laboratory.

Since the application was spectroscopically feasible, the next step was to define the hardware and software configuration options, such as the number of measurement points, measurement rate, and alarm outputs. A final quotation was then presented and accepted.

The completed analyzer was calibrated by collecting spectra of varying concentrations of each species to be measured and of expected nonmeasured interferants, such as CO_2 and H_2O. Finally, a chemometric model that allowed the prediction of all desired species was built and installed in the environmental monitoring automation program.

The instrument was delivered and operated for about a year and a half with no false alarms. A few real alarms were traced to leaks in the company's equipment. Suddenly, however, a large number of false alarms started to occur at two sample points. Discussions between the manufacturer's industrial hygiene and production groups brought to light that a new solvent had been added to the process. In response to the change, the FTIR vendor sent a service technician experienced in collecting gas bag samples to measure several quantitative samples of the new material on the analyzer, then forward the results on floppy disk to the vendor. These new measurements were added to the calibration matrix, and the vendor provided a new program of setup files to be installed by the customer. The problem was solved.

Polymer Processing from Input Streams to Final Product

This discussion is not a single application but an illustration of the way process FTIR can be applied throughout an industry to solve related problems. In the late 1980s, several manufacturers came to a process FTIR vendor for help with the control of polyethylene reactors. Each manufacturer had different processes and different ideas about the best way to make the measurements, so consequently the vendor produced different instrument configurations for each.

The solvent-based polymer reaction was measured by liquid transmission. The high-pressure gas-phase processes were measured by looking at the ethylene and additives at 2,000 to 4,000 psig, again in short-path transmission cells. These were measurements of supercritical fluids midway between liquids and gases. Finally, in the early 1990s, a technique was developed to let down the high pressure of the gas-phase processes to near-ambient pressure without some of the additives coming out as liquids. New analyzer systems were then configured to measure the low-pressure gases in short-path transmission gas cells. These cells and all stream-switching valves had to be kept at high temperature to prevent any liquid dropout.

Other customers wanted to directly measure the properties of the final polymer. A polymer equipment manufacturer developed a sample system that pumped molten polymer through a very-high-pressure liquid cell. A standard process FTIR spectrometer was then configured to make the measurements.

Even this advanced application has evolved in its configuration. Near-IR probes now are available that can be placed directly on polymer processing equipment to make measurements without the added step of pumping the samples. This new approach allows the analyzer to be situated away from the processing equipment, and, with optical multiplexers, more than one extruder can be monitored with the same analyzer.

Gasoline Composition and Physical Properties Measurement

Gasoline is one of the most important industrial products in our society, and its manufacture is an immense business. Extracting a few more liters of salable product from a barrel of crude oil can make an enormous difference in a refiner's bottom line. For this reason, refiners and instrument vendors have been working for years on spectroscopic methods to measure octane number and compositional properties of blended gasoline and gasoline components.

This research has resulted in several complete spectroscopic systems that provide automated on-line prediction of gasoline properties. These instruments are probably the closest thing to standard products from any of the process FTIR vendors. Even so, to do the best predictions, sample sets of carefully characterized samples must be measured on the analyzer or on an optically-equivalent laboratory unit. These spectra are used to produce a chemometric model matching the specific conditions at the refinery. The most advanced of these systems have provisions for correcting normal variations in instrument output. They also can detect and capture gasoline samples that are not accurately being measured, based on statistical tests. These captured samples are then used to improve the model accuracy.

Installation Hints

Process FTIR is a rugged technique. Perhaps this is the reason that many of the earliest analyzers were placed directly in process plant areas exposed to the elements. Although these analyzers have operated reliably compared to other process analyzers, analysis of the service histories indicates that a more controlled environment is probably warranted. Consequently, newer analyzers normally are located in instrument shelters. At installations where there is significant heat buildup in purged enclosures, instruments are cooled by vortex coolers or small air conditioners.

Specific Safety Issues

FTIR analyzers, like most complex composition-monitoring equipment, employ electronics with power levels above intrinsic safety levels. In addition, the infrared source is a potential source of ignition. Both problems are normally addressed by purging the analyzer with air or nitrogen and, depending on the area classification, monitoring the purge conditions. The energy in the infrared signal itself is not hazardous. The laser used for frequency reference is kept inside the interferometer to prevent eye hazards.

Unique Sample System Requirements

The sample systems for FTIR analyzers are similar to those for other more familiar technologies (see Chapter 3). A representative sample must be safely transferred from the process to the analyzer. Often the amount of sample is larger than actually required, allowing a better averaging of composition variables. Since the measurement is nondestructive, the sample often is maintained at the same process temperature and pressure, so it can be reintroduced to the process.

Typical Calibration Requirements

One of the most important advantages of using process FTIR techniques is that the final measurement is a fundamental property of nature.

One of the most important advantages of using process FTIR techniques is that the final measurement is a fundamental property of nature. Thus, in a very real sense, FTIR calibrations are permanent and transferable. As noted earlier, provisions must be made for detecting and re-calibrating for components not in the original calibration, but automatic on-line calibration systems are seldom used except where required for legal reasons.

FTIR calibrations are complex and require the skills of an experienced spectroscopist or chemometrician to be done correctly, except in the simplest of applications.

Along with the benefit comes a cost. FTIR calibrations are complex and require the skills of an experienced spectroscopist or chemometrician to be done correctly, except in the simplest of applications. Chemists with these skills may be accessible in some large companies, but are becoming less so with corporate cost cutting and downsizing. Outside calibration services are available from the FTIR vendors as well as from independent consultants.

Troubleshooting and Maintenance

Properly-designed applications software takes advantage of the volume of internal instrument information available from the spectrum collected for each measurement. This information is made available to the on-site personnel to indicate when service may be required and, in many cases, what the nature of the problem is. Some of the required service may be done by on-site module-level replacement. More complex problems usually must be dealt with by the vendor's service staff.

As for periodic maintenance, process FTIR analyzers generally require a minimum of service. The infrared source is the only element with a predictable wear-out mode and needs to be replaced approximately once a year. FTIR sample systems have the same maintenance requirements as other analyzer systems. Filters must be kept clean, valves checked for leaks, and all components subject to attack by the sample should be checked for corrosion.

As for periodic maintenance, process FTIR analyzers generally require a minimum of service.

About the Author

Bruce C. McIntosh received a B.S. in Electrical Engineering and a B.A. in Humanities from Union College in Schenectady, NY, in 1966. He worked for GE Radio Receiver Department from 1966 to 1972, finishing up by heading a multi-department project to develop the first electronic digital clock radio. He moved to Laser Precision Corp. which introduced the first microprocessor-based FTIR in 1980 under the Analect brand name. In the late 1980s, he developed relationships with several large chemical companies to field FTIR process analyzers. This experience led to the development of the turn-key FTIR process analyzers now provided by Analect.

30

Near Infrared (NIR) Analyzers

Near-infrared (NIR) photometry is widely used in industry for on-line process measurements. Properly applied, NIR instruments provide accurate, fast, reliable, noncontact measurements of critical process stream composition parameters. NIR is one of few reliable on-line technologies that can make a nondestructive measurement of individual chemical constituents of a process sample. These on-line measurements can be useful or essential for real-time process and quality-control systems.

NIR instruments can analyze most solids using NIR reflectance and most liquids and gases using NIR transmission. The most common and easiest NIR measurement is moisture. Depending on the application, the NIR moisture measurement accuracy can be as good as 0.01% in solids or a few parts per million in liquids. Other standard measurements include oil and fat content, coating weight and thickness, resin content, and chemical material composition.

NIR instruments can analyze most solids using NIR reflectance and most liquids and gases using NIR transmission. The most common and easiest NIR measurement is moisture.

For process measurements, a NIR sensor is installed on-line in a survivable location where it sees a representative sample of the process stream. The sensor must be calibrated by measuring several samples with both the NIR measurement and an independent reference method, comparing the results, and installing correction factors in the NIR instrument to make it agree with the reference method. A properly installed and calibrated NIR sensor provides a constantly updated process measurement to the operators and control systems without disturbing the process.

Theory of NIR Analysis

Near-infrared instrumentation makes measurements by illuminating the sample with NIR light and measuring the amount of light that the sample absorbs. Different materials absorb NIR light at different wavelengths, and this absorbance depends on the concentration of the material. By measuring the light absorbance at a characteristic wavelength of a specific material, an NIR instrument can measure that material's concentration.

Physics of NIR Light

Near-infrared falls between the visible and infrared wavelength ranges of the spectrum and includes light with wavelengths in the range of 1100 to 2500 nm. NIR light can be reflected, absorbed, or transmitted when it strikes a sample, as shown in Figure 30-1. Reflectance can be diffuse, from materials that scatter light, or specular. Typical light-scattering materials include powders, plant matter, and matte paper and ceramics. The diffuse reflection and transmission caused by light scattering is described by the Kubelka-Munk theory. Transparent materials or glossy surfaces do not scatter light and instead produce specular reflection and direct transmission. Typical nonscattering materials include clear liquids, thin plastic films, and polished metal surfaces. Light absorption in these materials is described by the Beer-Lambert equation.

Both the Kubelka-Munk and the Beer-Lambert theories are discussed in most NIR textbooks. The physical and chemical composition of the sample determines the character of these interactions. For transparent samples such as clear liquids and thin films, absorbance and transmission are the primary interactions. For opaque samples, including most solids, the NIR light is either reflected or absorbed.

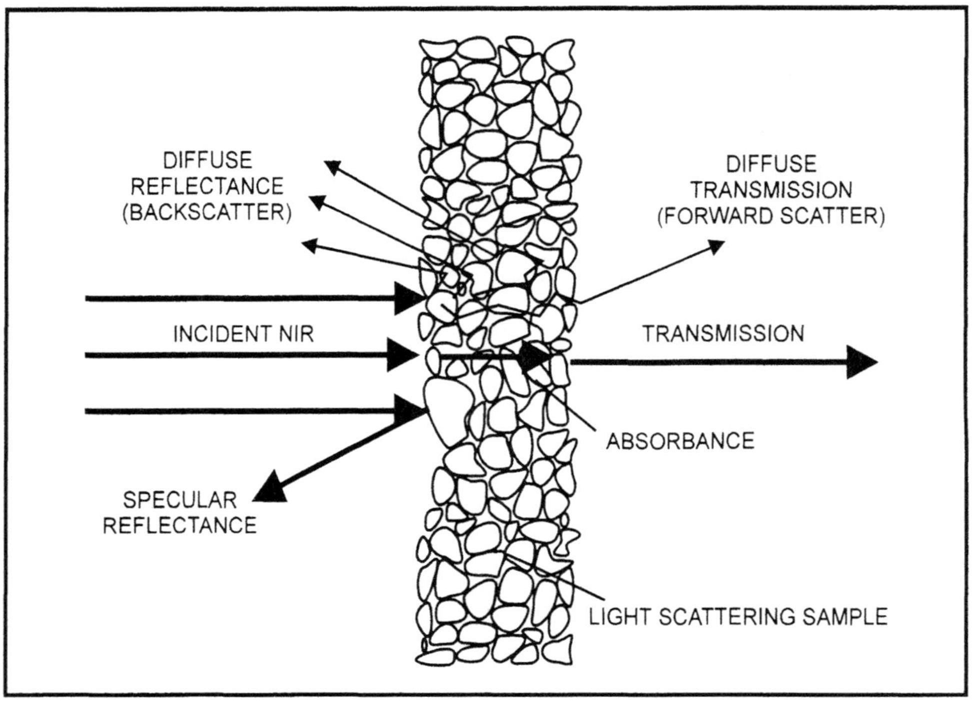

Figure 30-1. NIR Interactions with Solids

Light absorption in the NIR region is due to overtone and combination-band absorbances, which occur at higher energies and shorter wavelengths than the fundamental vibrations which absorb in the mid-infrared region. Due to quantum mechanical selection rules, only vibrations involving hydrogen (i.e., C–H, O–H, and N–H bonds) show significant absorbance in the NIR region, and these absorbances are much weaker than mid-infrared absorbance bands from the same vibrations. A detailed discussion of the theory is beyond the scope of this work and can be found in standard textbooks.

Measurements with NIR Light

The NIR absorbance of a material depends primarily on its chemical composition. Different compounds absorb at different NIR wavelengths, depending on their chemical structure. Typical absorption wavelengths for a variety of materials and chemical structures are shown in Table 30-1.

In general, only materials containing hydrogen, such as water, organic chemicals, and most plastics, have significant NIR absorbances.

Figures 30-2 and 30-3 show NIR spectra of several selected materials. Because absorbance is approximately proportional to concentration, a NIR instrument can determine concentrations of NIR absorbers by measuring the absorbance at the appropriate wavelengths.

Table 30-1. Typical NIR Absorbance Wavelengths

Chemical Structure	Typical Materials	Absorbance Wavelengths, nm
O–H (water)	Water	1,450 - 1,940
O–H (alcohols)	Alcohols, glycols, carbohydrates	2,100
C–H (aliphatic)	Coatings, plastics, solvents	1,720 - 2,300
C–H (aromatic)	Coatings, plastics, solvents	2,150

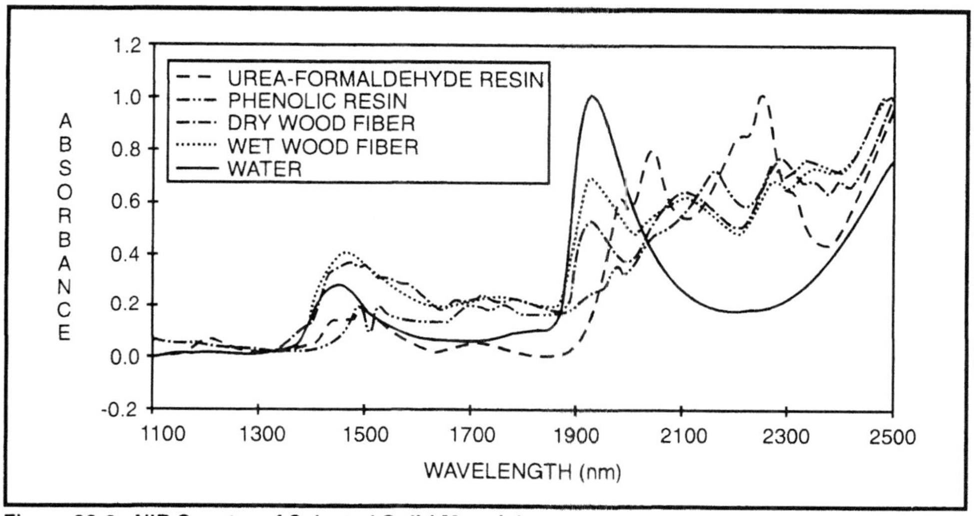

Figure 30-2. NIR Spectra of Selected Solid Materials

NIR instruments typically make measurements by shining NIR light on the sample and measuring the reflected light at several wavelengths. At least one wavelength is absorbed by the material being measured, and the others are chosen as appropriate to give the best stability and accuracy. The instrument uses the amount of reflected light to calculate the concentration of the material being measured. The exact calculation used depends on the instrument and the measurement, and is determined as part of setting up and calibrating the instrument.

The calculations for each measurement are determined in advance by comparing the NIR instrument response with a reference laboratory measurement on a variety of calibration samples. There is no a priori way to determine what calculations are necessary. Thus, NIR is a secondary method. Calibration is necessary to any NIR measurement. The accuracy of any NIR measurement is limited by the reference method accuracy.

Calibration is necessary to any NIR measurement. The accuracy of any NIR measurement is limited by the reference method accuracy.

Water is the most common and easiest NIR measurement. This is because water is an extremely strong NIR absorber at its two absorption wavelengths of 1,940 and 1,430 nm, as shown in Figure 30-3. Also, no other common materials absorb strongly at the 1,940-nm wavelength, making it easy for NIR techniques to distinguish water from the other constituents of the sample.

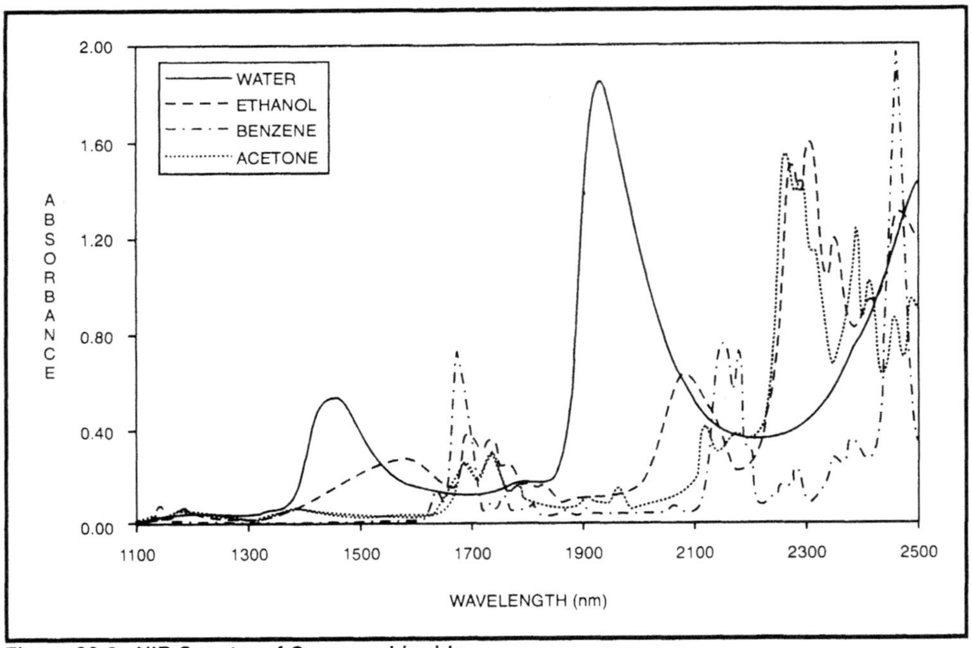

Figure 30-3. NIR Spectra of Common Liquids

On-Line NIR Instrument Design

Process NIR instrumentation must make reliable, accurate, fast measurements in an on-line environment. The on-line environment presents a variety of difficulties, including dust and dirt, high temperatures, hazardous atmospheres, changing sample composition and presentation, and untrained operators. Several types of commercial NIR instruments are available for process measurements. The most commonly used are fixed-wavelength photometers, which make measurements using a few fixed wavelengths of NIR light. Spectrophotometers can use the entire NIR spectrum to make measurements. Photometers are generally less expensive, easier to use, and better suited to process conditions than spectrophotometers, but have less versatility and power for making difficult measurements in complex systems.

Photometers are generally less expensive, easier to use, and better suited to process conditions than spectrophotometers, but have less versatility and power for making difficult measurements in complex systems.

NIR instruments typically use a transmission methodology for transparent samples, including most liquids and gases. In transmission, the NIR light passes through the sample before it is detected. For opaque samples, including most solids, a reflectance geometry is used. In this geometry, the instrument detects the light reflected from the sample.

PHOTOMETERS

Most NIR process instruments meet the requirements of the process environment using a fixed filter design. Figure 30-4 shows generic designs for transmission and reflectance process NIR instruments. The analyzer produces NIR light with a quartz halogen bulb. Light from the bulb is nonhazardous and does not heat the sample. The light passes through two fixed-wavelength optical interference filters mounted on a rotating filter wheel. The filters are chosen to pass NIR light at the specific wavelengths required for the measurement. The rotation of the filter wheel chops the light into pulses at each of the wavelengths selected by the interference filters. Each rotation of the wheel yields one measurement. The wheel typically rotates 10 to 30 times per second, so the measurement is updated rapidly. From the filter wheel, the light passes through the sample cell to the detector. The instrument electronics amplify the detector signal and send it on to a processor, which uses the signal from each of the NIR wavelengths to calculate and display the concentration. The processor can be located remotely for greater convenience in monitoring the NIR measurement, or to avoid harsh process conditions.

Process NIR instruments use the transmittance or reflectance of at least two NIR wavelengths to make a measurement. In the simplest measurement technique, one wavelength, referred to as the measure, is chosen at an absorbance peak of the constituent being measured. The other wavelength, referred to as the reference, is chosen away from any of these absorbances. The instrument uses the ratio of the transmittance at the reference and measure wavelengths as the basis for the measurement. The wavelengths are chosen to make the reference/measure transmittance ratio linear with the concentration of the constituent being measured.

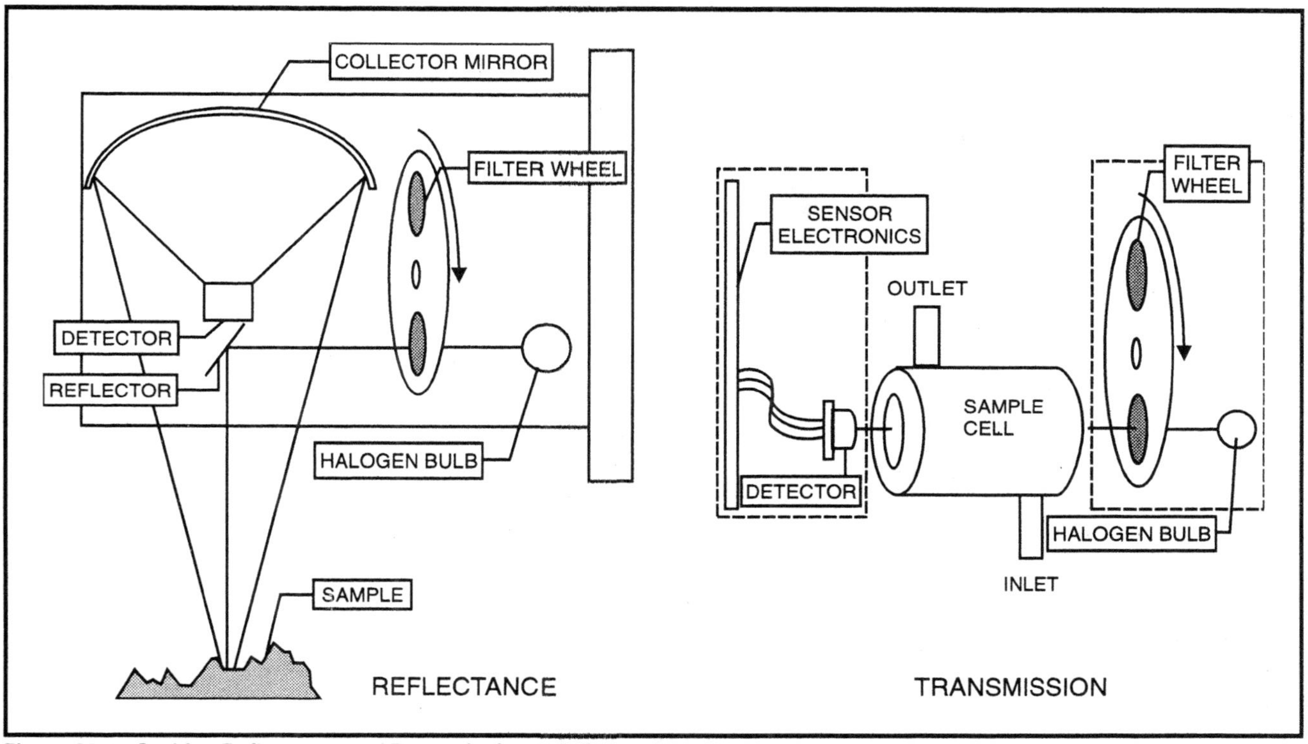

Figure 30-4. On-Line Reflectance and Transmission NIR Photometers

SPECTROPHOTOMETERS

Spectrophotometers use a variety of methods to produce a spectrum of NIR light, including diffraction gratings, diode array detectors, infrared emitting diode (IRED) light sources, and variable-wavelength optical interference filters. Grating-based instruments can use the full NIR spectrum, giving them immense versatility and power for difficult measurements. However, they tend to be expensive, relatively fragile, and less tolerant of variations in sample presentation. Diode array and IRED-based instruments are just coming into use. They have no moving parts, but are limited at present to wavelengths below 1,600 to 1,800 nm by the available sources and detectors.

Full-spectrum instruments typically use the transmittance or reflectance of many wavelengths to calculate the desired measurement. The appropriate calculation and wavelengths are determined by any of a wide variety of mathematical methods known as chemometrics. Chemometrics develops an optimal measurement using sophisticated statistical techniques to analyze spectra of many samples. Standard chemometric methods include multiple linear

Chemometrics develops an optimal measurement using sophisticated statistical techniques to analyze spectra of many samples.

By using chemometric techniques and many wavelengths, a full-spectrum instrument can often make difficult measurements in complex process streams and applications.

regression (MLR), principal component regression (PCR), and partial least squares (PLS). The details of chemometric analysis are beyond the scope of this work and can be found in standard texts. By using chemometric techniques and many wavelengths, a full-spectrum instrument can often make difficult measurements in complex process streams and applications.

FIBER OPTICS

In transmission measurements, fiber optics are often used to transmit NIR light to and from the sample, allowing the NIR instrument to be located remotely in a mild environment. A fiber optic multiplexer can be used to transmit light from many locations to one instrument, allowing a single instrument to be used for several different measurements. Fiber optics are seldom used in reflectance geometries, because the fiber is too small to capture more than a small fraction of the reflected light as it spreads out from the sample.

Many full-spectrum instruments use fiber optics so the relatively expensive and fragile instrument can be located in a mild environment. Fiber-optic multiplexers are often used to minimize the number of expensive full-spectrum instruments required in an installation. However, in the absence of special considerations, an instrument is generally slightly less accurate and stable with optical fibers than without, because of the smaller amount of light transmitted through the fibers. Photometers seldom use fiber-optic systems, because photometers are generally robust enough to operate on-line, and because the price of a fiber optic system is comparable to that of a photometer.

Strengths of the NIR Region for Process Analysis

ROBUST INSTRUMENTATION

Spectroscopically, the NIR spectral region is very convenient for on-line instrumentation design. NIR light works well with conventional optical and optoelectronic materials. Halogen bulbs are very bright NIR sources, optical glass is transparent to NIR light, and lead sulfide is a sensitive, stable, and inexpensive NIR detector. This combination gives NIR instruments very good signal/noise ratios and long-term stability. Because these critical components are durable and relatively inexpensive, process NIR instrumentation can be affordable, accurate, stable, and reliable.

EASY SAMPLE HANDLING

Process NIR instrumentation is noncontact and nondestructive. No sample preparation is required. With many instruments and applications, it is not necessary to maintain an exactly repeatable sample presentation; that is, the position of the reflectance sample can vary by several inches without causing significant error in the measurement.

For transmission measurements, NIR instruments typically use a relatively long path length to maintain good sensitivity. The long path length makes NIR sample handling relatively easy, because there is little impedance to the sample flow.

WATER MEASUREMENTS

Water has a very intense, isolated NIR absorption peak at 1,940 nm. This makes water measurements relatively easy for NIR instrumentation. NIR is usually the best method available for on-line water measurements, unless one of the limitations discussed below makes NIR measurements difficult or impossible.

NIR is usually the best method available for on-line water measurements, unless one of the limitations discussed below makes NIR measurements difficult or impossible.

Limitations of NIR for Process Measurements

The physics of NIR spectroscopy impose several limitations on NIR measurements. Many of these limitations can be avoided by appropriate sample handling and wavelength selection. Tables 30-2 and 30-3 list the primary factors, discussed in more detail below, that can adversely affect NIR measurements.

TRANSMISSION MEASUREMENTS

Table 30-2. Potential Limitations of NIR Transmission Measurements:

Potential Limitation	Measurement Solution
Opaque samples	Short path length and/or short wave-length, sensitive optics
Particulates and bubbles	Filter the stream
Temperature variations	Control or compensate for temperature
NIR interference	Measure and compensate for interference

Opaque Samples

A NIR measurement will only work if a significant amount of NIR light returns to the sensor after reaching the sample. For transmission measurements, this implies that opaque samples cannot be measured by NIR, because little or no NIR light can transmit through them. Sensitivity can be increased by using a sensitive detector or an intense light source. Transmission can be increased by using a shorter path length or by moving to shorter light wavelengths, which are less strongly absorbed. However, both of these options decrease the sensitivity of the NIR measurement.

Particulates, Bubbles, and Two-Phase Streams

Any light scattering in the stream will affect the stream's background NIR absorbance and thus influence any NIR measurement. Light scattering can be caused by particulates, bubbles, or additional liquid phases. These interferences should be removed by filtration or some other method.

Temperature Variations

Stream temperature changes will affect the NIR absorbance of the stream by changing the stream density or the hydrogen-bonding characteristics. Almost any measurement involving water, alcohols, or amines will be significantly affected by changes in the stream temperature. These effects can be eliminated by controlling the stream temperature or by measuring the temperature and compensating for its effects on the measurement. Temperature control and compensation are discussed in more detail later in this chapter.

NIR Interference

If other materials in the stream are chemically similar to the constituent being analyzed, they will absorb at the same NIR wavelengths and interfere with the measurement. For instance, it is usually difficult to measure one alcohol in the presence of other alcohols. In many applications, the interfering component can be measured at another wavelength, and this measurement can be used to compensate for the interference on the primary measurement.

REFLECTANCE MEASUREMENTS

Table 30-3. Potential Limitations of NIR Reflectance Measurements:

Potential Limitation	Measurement Solution
Strongly absorbing samples	Short wavelengths, sensitive detectors
Surface sensitivity	Short wavelengths, equilibrated samples
Surface texture and composition	Choose compensating wavelengths

Strongly Absorbing Samples

For a NIR reflectance measurement, the sample must reflect a significant amount of NIR light. NIR measurements of extremely dark samples such as carbon black are difficult because such materials reflect little or no light. The measurement can be improved by using extremely sensitive detectors and intense light sources.

Surface Sensitivity

In a reflectance measurement, the NIR light penetrates only a limited distance into the sample, and makes a measurement only within that distance. The penetration distance depends on the sample and NIR wavelength and can range from almost nothing for dark, opaque samples to very deep for transparent materials. In most samples the penetration depth is less than 1 mm. The problem is most pronounced in measuring water in thick materials such as whole grains or nuts. A thick sample that has just been dried typically has much more moisture in the core than on the surface where an NIR instrument can detect it, possibly causing errors in the NIR measurement. Surface sensitivity effects can be reduced by using shorter NIR wavelengths, which penetrate farther into the sample, or by choosing a measurement location where the surface and core are not different.

Surface Texture and Composition

Anything that affects the NIR reflectance from a sample can potentially cause errors in a NIR reflectance measurement. Factors that can influence the reflectance include particle size, surface texture, color, moisture content, sample composition, and sample thickness. Because water is a very strong NIR absorber, changes in the water content of a sample are likely to interfere with other NIR measurements. Interferences due to these variations can often be compensated out by choosing the appropriate measurement wavelengths.

Installation Hints

The great majority of process NIR instruments consist of a processor and one or more sensors. The sensors are installed on-line and measure the NIR absorbance of the sample. The processor is installed off-line. It uses absorbance information from the sensor to calculate the desired measurements and displays the measurement results. Figure 30-5 shows typical NIR instrument installations for transmission and reflectance measurements.

The sensor installation is critical to the success of an NIR measurement. The sensor should be mounted in a survivable environment where it can see a consistent, representative, real-time sample of the process with the correct sample presentation. The details of the processor installation are less important, since processors can be installed remotely and avoid the full rigor of process conditions.

Sample Presentation

Good sample presentation is critical to a successful NIR measurement. The NIR sensor must see a consistent, representative sample of the process. For solids measurements using reflectance, the product should be positioned at an appropriate distance and angle from the sensor. In transmission measurements, the product must be free of bubbles and particulates and should be controlled within an appropriate temperature range. With both reflectance and transmission measurements, it should be possible to collect grab samples from the process at the sensor's measurement location, and these samples should be representative of the product measured by the sensor.

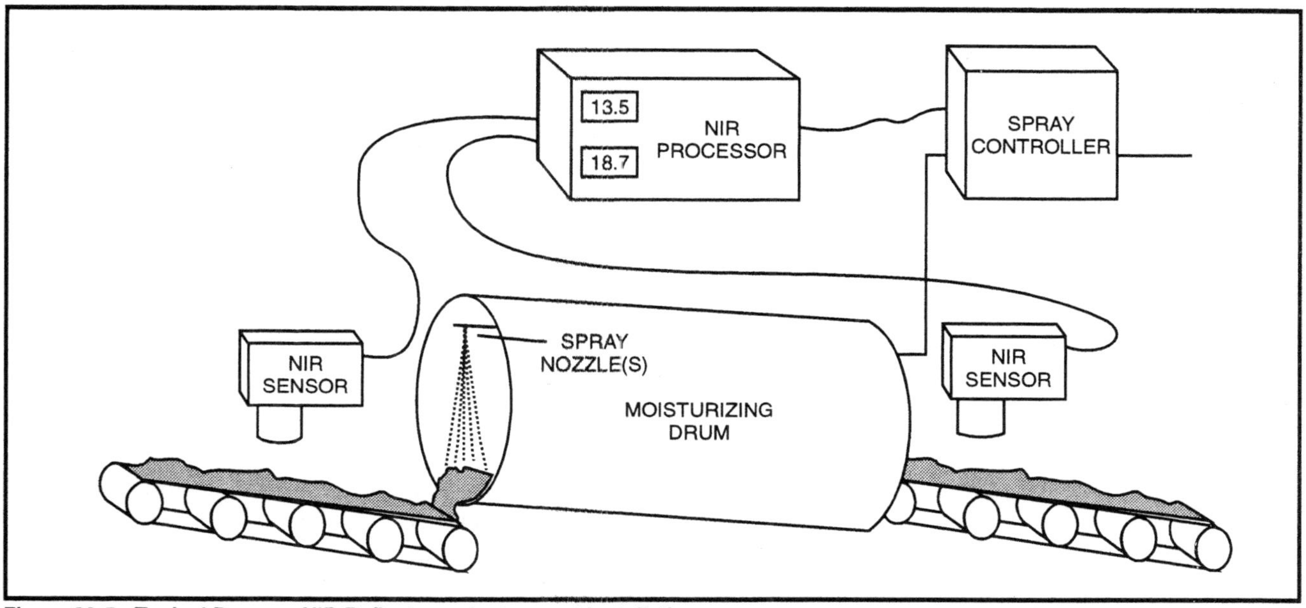

Figure 30-5. Typical Process NIR Reflectance Instrument Installation

REPRESENTATIVE SAMPLES

NIR sensors usually measure only a portion of the process stream. For the NIR measurement to be useful in control or monitoring, that portion must be representative of the process. Transmission sensors usually measure a relatively small sidestream, while reflectance sensors usually see a strip 1 to 2 in. wide running along the process. A well-mixed process stream will ensure that the NIR measurement is representative.

PROCESS VARIABILITY

If possible, a NIR sensor should be located at a point where the variability in the process stream is relatively small or relatively slow. Large, rapid process variation makes control and calibration difficult. In calibrating the instrument, large point-to-point variation and process instability cause large sampling errors, as discussed in the section "Typical Calibration Requirements." Large sampling errors decrease the accuracy and reliability of the NIR instrument calibration.

CONTINUOUS PROCESS STREAM

Voids or gaps in the process stream can cause large changes in the reading of an NIR sensor, because a void has different NIR absorbance properties than the process. A void may cause the NIR instrument to send erroneous signals to the control system. If gaps in the process are unavoidable, they should be detected by a photoeye system or some other means so that the NIR measurement can be frozen (sample-and-hold system) while the gap is present.

ACCESS TO CALIBRATION SAMPLES

A NIR mounting location should be chosen to assure that calibration is possible. Most NIR instruments are calibrated using grab samples taken from the process. In this case, a sampling point immediately downstream from the sensor should be accessible to allow the collection of grab samples. The calibration procedure should be considered in any NIR instrument installation.

PRESENTATION GEOMETRY

Reflectance NIR sensors have a range of acceptable presentation geometries. The geometry includes the distance between the NIR sensor and the sample, and the sample presentation angle. The range of acceptable geometries depends on the instrument. In most applications, typical process NIR photometers will handle changes of a few inches in the sensor-sample distance and several degrees in the presentation angle while maintaining reasonable accuracy. However, the presentation angle is critical when measuring shiny or glossy surfaces.

Presentation geometry is seldom a concern in transmission measurements, because the process usually flows through an immobile sample cell. This locks the sample into the correct geometry.

SAMPLE CONDITIONING

Stream conditioning is necessary to obtain the best possible accuracy in a liquid- or gas-phase process NIR measurement. It is generally less critical for reflectance measurements on solids. The instrument must see a clean, single-phase stream at a reasonably constant temperature and pressure. Measurement on an uncontrolled stream will usually result in a less accurate and less reliable measurement value. Stream conditioning for liquid and gas measurements is discussed in more detail later in this chapter.

Environmental Conditions

Typical process NIR sensors can operate in a variety of locations. The primary considerations are the sample presentation, the environmental temperature, and the cleanliness of the environment. Other potential difficulties include hazardous environments, intense ambient lighting, vibration, and exposure to the elements. These potential problems can almost always be solved using the appropriate sensors and accessories.

ENVIRONMENTAL TEMPERATURE

NIR sensors are generally limited to operation in ambient environmental temperatures of less than 120 to 140°F (50 to 60°C) by their internal electronics. Extended operation at excessively high temperatures will cause instrument failure. The working temperature range can be extended with cooling accessories. Extremely hot locations, such as the exit of a process dryer, should be avoided. As a rule of thumb, locations that are too hot for people are also too hot for NIR instruments. Fiber-optic probes can be used in transmission measurements to completely remove the sensor from a hot area.

As a rule of thumb, locations that are too hot for people are also too hot for NIR instruments.

The calibration of all NIR sensors is slightly sensitive to ambient temperature. For typical process NIR sensors, the temperature changes occurring indoors have only a negligible effect on the measurement.

CLEANLINESS

Except for the sensor windows, process NIR sensors are generally sealed against water and dust, and are not affected by dirty environments. Corrosive materials and constantly wet (condensing humidity) conditions should be avoided.

The windows on all process NIR instruments must be kept clean to assure accurate measurements, since dirt on the window blocks the passage of NIR light. The preventive maintenance program for any NIR instrument should include regular checks of window cleanliness. For reflectance measurements in dirty environments, the window should face down to prevent dust from falling onto it. Most instruments use a curtain of air blown across the windows to keep away dust, dirt, and moisture. The air must be clean and dry [–40°F (–40°C) dewpoint]. Standard plant air generally blows oil and water onto the window, making the problem worse.

HAZARDOUS ENVIRONMENTS

NIR sensors can be installed in hazardous atmospheres by mounting the sensor in an approved explosion-proof housing or a housing equipped with a suitable, certified purge package.

FOOD PROCESSING INSTALLATIONS:

NIR sensors are available in materials compatible for food processing. Typical materials are a stainless steel sensor housing and a Kel-F window. Kel-F is a polymer chemically related to Teflon; it is transparent to NIR light and will not shatter.

AMBIENT LIGHT

Intense NIR sources, such as direct sunlight, infrared dryers, some vapor lamps, and bright incandescent lamps, may interfere with the NIR measurement. The sensor and the product it sees should be shielded from these strong sources.

VIBRATION

The NIR measurement itself is not affected by vibration, although excessive vibration may damage the instrument. NIR reflectance sensors are in widespread use for measurements on vibrating conveyors.

EXPOSURE TO THE ELEMENTS

An NIR sensor installed outdoors must be shielded from exposure to precipitation, large temperature swings, and direct sunlight. Precipitation can damage the electronics and will certainly influence any water measurements. Very large temperature changes, such as those experienced outdoors in changeable weather, can cause noticeable calibration shifts. To shield the sensor from these problems, outdoor NIR sensors are typically enclosed in an insulated or temperature-controlled housing.

An NIR sensor installed outdoors must be shielded from exposure to precipitation, large temperature swings, and direct sunlight.

Developing and Evaluating NIR Measurements

Before an NIR instrument can provide a useful process measurement of some constituent of a process stream, it is necessary to find a way to calculate the constituent concentration from the NIR properties of the process stream. The wavelengths and calculations used to determine the concentration are known as the measurement algorithm or model. Developing an appropriate model is essential to any NIR measurement. In practice, NIR instrument manufacturers have prior experience with most applications, and may already have a good model.

Both the end user and the instrument manufacturer are critical to developing a new model, if one is necessary. The end user usually supplies samples from the process, which the manufacturer uses to develop the model. Several (sometimes many) samples are necessary. The NIR instrument is used to measure the NIR properties of the samples, and the constituent concentrations of those samples are measured with an independent reference method. A variety of mathematical and statistical tools and chemometric techniques can then be used to search for a satisfactory measurement model. The process is similar to calibration, described later. However, model development is more complicated because of the need to choose wavelengths and calculations in a measurement model, as opposed to only the instrument sensitivity and offset in a calibration. Familiarity with NIR spectroscopy and the appropriate mathematical tools is essential in developing a good measurement model. For this reason, NIR measurement models are almost always developed by the instrument manufacturer's applications specialists before the instrument is installed.

NIR measurement models are almost always developed by the instrument manufacturer's applications specialists before the instrument is installed.

To give a good NIR measurement, the measurement model must be sensitive to the constituent being measured. It must also be insensitive to other process changes that could introduce errors in the NIR measurement by changing the NIR properties of the process stream. To assure that the model meets these criteria, it must be developed with samples in which the measured constituent concentration varies over the full range of interest. The samples should also include examples of any other process variations that might affect the measurement. The end-user needs to work with the model developer to ensure that the developer knows the full range of variation expected in the process and has development samples which cover that range.

The most important types of interfering factors are the chemical composition and the physical characteristics of the process stream. If any chemical constituents of the stream absorb at a wavelength used for the measurement, changes in the stream composition may interfere with the measurement, causing error. Potentially important physical characteristics include the stream density, refractive index, and light-scattering properties. These errors can often be removed by measuring the interfering constituent using other wavelengths and using the measurement to compensate for the interference. The compensation wavelengths and method are usually determined using statistical and chemometric techniques discussed earlier.

Typical Calibration Requirements

All on-line NIR techniques are a secondary measurement method. Like any secondary method, NIR instruments must be calibrated by comparison to a primary reference method to provide a useful engineering measurement. Calibration finds the relationship between a sample's NIR properties to the desired measurement. Because the instrument is calibrated by comparison to a reference method, the NIR measurement accuracy can be no better than the reference method accuracy. Calibration of an on-line instrument involves special considerations to minimize errors in sampling from the process and comparing the on-line sensor to the reference method.

Basic Calibration Procedure

For on-line NIR instruments, calibration consists of determining the sensitivity and offset that make the NIR measurement agree with the reference method. Calibration involves several steps. Before calibrating, the instrument must be correctly configured and installed for the desired measurement. Calibration data are collected by measuring a number of samples with both the NIR instrument and the reference method. The data are analyzed by linear regression, and the regression statistics are evaluated to determine the quality and reliability of the calibration. If the calibration is statistically reliable, the new sensitivity and offset are installed in the instrument.

As with any other secondary measurement, an NIR instrument cannot be proved better than the reference method, and the accuracy of the calibration will be no better than the accuracy of the comparison between the NIR and the reference methods.

Errors can arise in the reference method, the sampling procedure, and the NIR instrument. All three error sources should be considered and minimized to the extent feasible in order to get the best possible calibration.

The samples should be representative of the product being measured. They should cover the widest feasible range of variation among samples, but have the smallest feasible point-to-point variation within a sample. The instrument location for the calibration should be chosen to give the best compromise among these concerns. In some applications, this requires calibrating the instrument near-line or in the laboratory, and moving it on-line after the calibration is complete.

NIR AND REFERENCE METHOD MEASUREMENT

The data for a calibration are acquired by measuring a number of samples with both the NIR instrument and the reference method. Table 30-4 shows a sample calibration data set.

Table 30-4. Calibration Data from an NIR Measurement of Moisture in Potato Chips.

Regression Statistics			
Correlation Coefficient		0.982	
Standard Error of Calibration		0.183	
Slope of Regression Line		1.472	
Y-Intercept of Regression Line		2.249	
	% Moisture Measurement		
Sample No.	Reference Method	Actual NIR	Corrected NIR
1	3.2	7.2	3.3
2	1.6	4.4	1.5
3	1.7	4.5	1.6
4	2.2	5.7	2.3
5	2.8	6.1	2.6
6	0.8	3.6	0.9

Gathering the calibration data is the critical step in a calibration. All unavoidable errors in a calibration occur during data gathering. Therefore, the sampling method, NIR measurement, reference method, and the measurement location should be chosen to minimize the errors and give the best possible sample characteristics.

Most applications should use three to seven samples for a calibration. The number of samples needed for a calibration depends on the accuracy of the measurement, including reference method, sampling, and NIR instrument error. Applications with large error require larger numbers of samples. Even in extremely good applications with very little error, at least three data points should be used. Calibrations using many samples give more reliable values of the sensitivity and offset, better outlier detection, and better estimates of the measurement accuracy.

LINEAR REGRESSION AND EVALUATION

The data are analyzed by linear regression to determine the sensitivity and offset that will make the instrument accurately measure the desired constituent concentration. Linear regression is a statistical technique that determines the equation of the best-fit line between the reference method and NIR method measurements.

Linear regression supplies two statistics that are basic measures of calibration quality: the correlation coefficient and the standard error of calibration (SEC). The correlation coefficient is a measure of the linearity of the data, ranging from +1.0 to −1.0. A correlation of +1.0 or −1.0 is perfect. Correlations greater than 0.95 generally indicate a good calibration, whereas correlations less than 0.85 generally indicate a problem. The SEC is a good measure of the error to be expected from a calibrated instrument. A large standard error indicates an inaccurate measurement. For an on-line measurement to be useful, the SEC should be better than the required accuracy. However, the SEC includes any error contributed by the reference method, sampling, and the NIR instrument, and will therefore never be better than the reference method. Advanced regression methods and statistics are available to improve the quality and help determine the reliability of a calibration. These advanced methods are discussed in a variety of statistics and regression texts.

The correlation coefficient is a measure of the linearity of the data, ranging from +1.0 to −1.0. A correlation of +1.0 or −1.0 is perfect. Correlations greater than 0.95 generally indicate a good calibration, whereas correlations less than 0.85 generally indicate a problem.

The standard error of calibration (SEC) is a good measure of the error to be expected from a calibrated instrument. A large standard error indicates an inaccurate measurement. For an on-line measurement to be useful, the SEC should be better than the required accuracy.

In a poor calibration, the sensitivity determined by the calibration is unreliable and may be dramatically incorrect. Poor calibrations occur when the calibration error is greater than or comparable to the range of variation in the samples. If a high standard error or low correlation coefficient shows that a calibration is poor, the instrument sensitivity should be left unchanged, and the instrument offset should be corrected to make the instrument and reference measurements agree when the process is on target. This procedure will make the instrument read well when the process is on or near the target. The accuracy will decrease as the process gets farther from target.

Figure 30-6 shows a scatter plot of the actual and corrected NIR data from Table 30-4. The corrected values give a good estimate of the performance to be expected from the instrument after calibration.

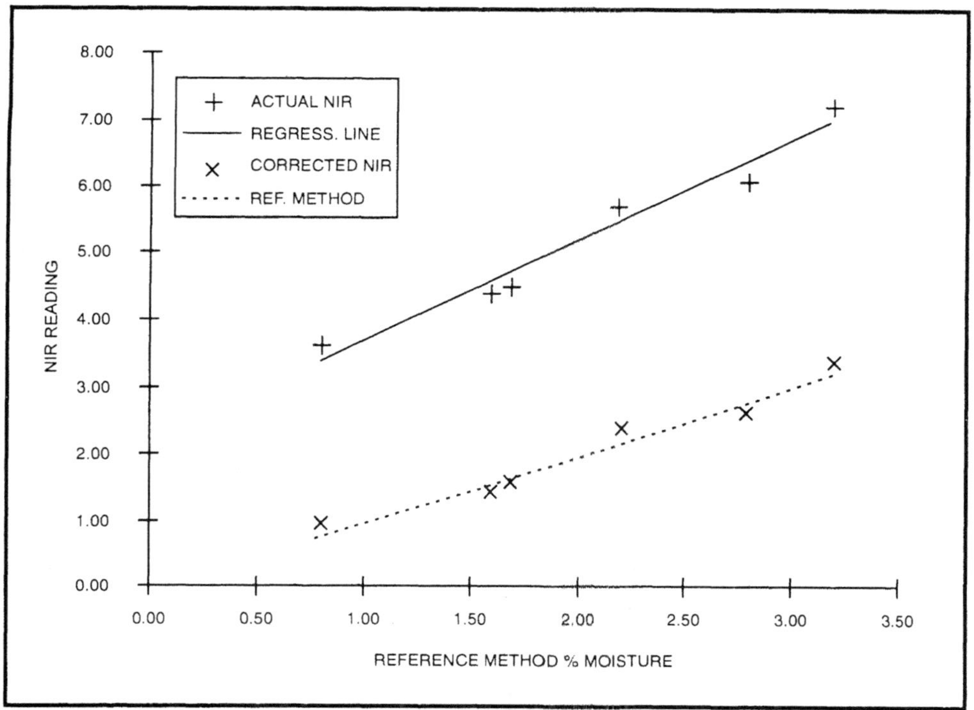

Figure 30-6. Original and Corrected Regression Plots from a NIR Measurement of Moisture in Potato Chips (See Table 31-4 for raw data)

Obtaining Accurate Calibrations

SOURCES OF CALIBRATION ERROR

The three sources of calibration error are the reference laboratory, the sampling procedure, and the NIR instrument. In any difficult calibration, all three error sources should be evaluated to determine which are important and correctable. Work toward improving the calibration should concentrate on reducing or eliminating the important error sources.

The three sources of calibration error are the reference laboratory, the sampling procedure, and the NIR instrument.

Reference Method Error

The reference method is the standard of comparison for any NIR calibrator. However, like any measurement, it has some error. Causes for error in the reference method include the inherent error of instruments used in the reference method (scales, volumetric measurement, chromatographs, etc.), variations in the operator's technique, and systematic limitations in the method. To avoid error due to operator technique, the reference method should follow documented procedure.

In difficult calibrations, the reference method error should be evaluated to determine whether it is a significant factor in the calibration error. This evaluation can be based on experience with the reference method, comparison to some other method, or tests of the reference method repeatability. Repeatability tests can be difficult if the reference method is destructive, because it is not possible to test the same sample more than once. In this situation, repeatability is estimated by making measurements of several nominally identical subsamples. However, if there is significant sampling error (i.e., variation between the subsamples), the reference method error and sampling error are indistinguishable.

Moisture measurements present special problems for the reference method, because moisture changes rapidly in any sample exposed to air. Moisture samples should be handled and stored as little and as quickly as possible to minimize this exposure. The sample should be kept in an airtight container for transport and storage, even for short times or distances.

Sampling Error

Sampling error occurs when the nominally identical samples measured by the NIR instrument and the reference method actually have different values of the constituent being measured. Sampling error occurs most frequently when there is large point-to-point variation in the sample. In this situation, if the NIR instrument measures a sample at one point, and the reference method sample is taken from a slightly different point, the difference between samples at the two points will appear as an error in the calibration. This problem is often present in on-line calibrations, where the NIR measurement averages a long strip of sample and the reference method uses a small grab sample. If the reference error is known to be small, the sampling error can be evaluated by repeated measurements on different subsamples from the same sample.

Figure 30-7 compares calibrations with large and small sampling errors. In this application, sampling was the only significant error source. Simply eliminating the sampling error improved the calibration quality from abysmal to very good. The sampling error was eliminated by comparing the average of several NIR measurements with the average of several reference measurements from the same sample. This assured that both methods saw an accurate average of each sample.

NIR Instrument Error

The NIR instrument is an obvious potential source of calibration error. In the initial calibration, confidence in the NIR instrument is usually low, and errors resulting from the reference and sampling methods are frequently ascribed to the instrument. These other error sources should be considered before too much effort is expended in searching for errors in the NIR instrument.

Error from the NIR instrument can take several forms, including noise, nonlinearity, changes in the NIR reading caused by routine variations in the sample composition, and electronic failure. Electronic failures can generally be dealt with on-site. Most other problems are related to the instrument design or the measurement algorithm, and should be referred to the manufacturer.

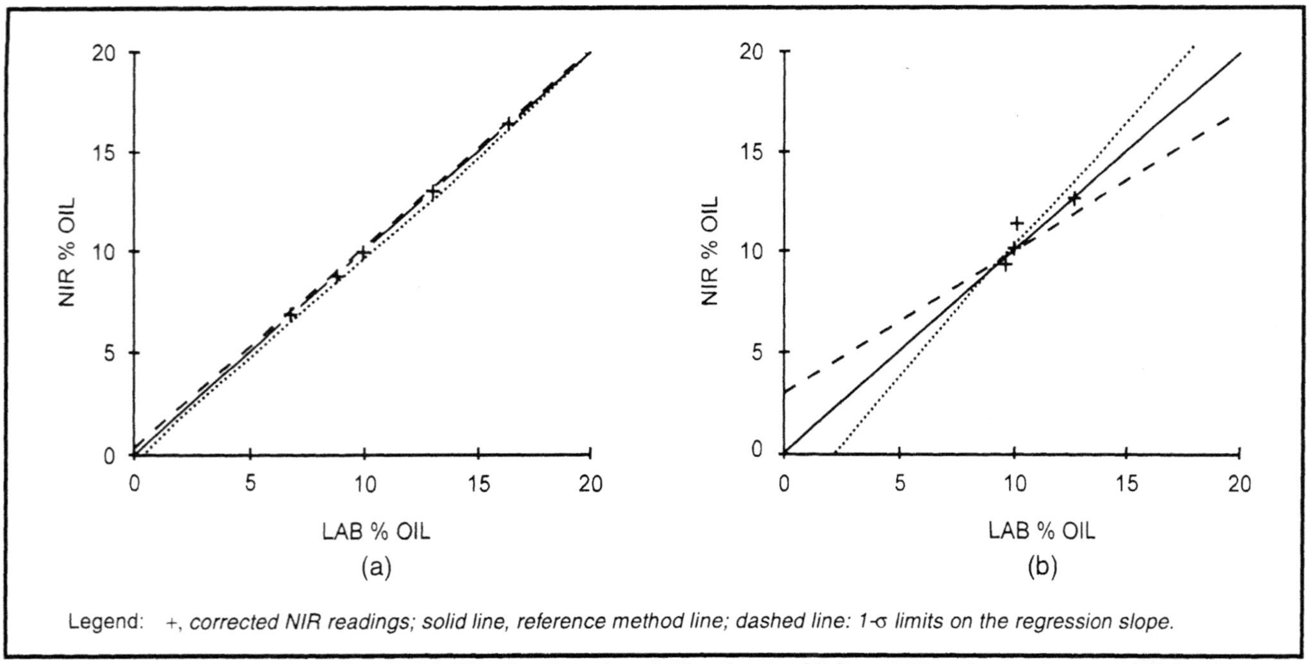

Figure 30-7. The Effects of Sampling and Calibration Error in a Calibration for Oil-Based Flavoring on Gourmet Cheese Popcorn
(a) Good calibration, with a correlation coefficient of 0.998 and a standard error of 0.03% oil. (b) Bad calibration, with a correlation coefficient of 0.557 and a standard error of 1.35% oil. Except for the sampling procedure, the two data sets were collected under identical conditions.

SAMPLING AND CALIBRATION METHODS

The samples and sampling method used in a calibration can be critical to the calibration's success. The sampling method should be chosen so that the samples vary over a wide range, have little point-to-point variation, and are representative of on-line production samples. A large range of variation in the sample assures that the NIR instrument has the correct sensitivity and is not affected by variations in the product composition or presentation. Small point-to-point variation minimizes the sampling error in the calibration. Use of representative samples assures that the NIR instrument is calibrated on the same material it will measure and eliminates systematic errors in the calibration.

Wide Range of Variation

The samples should cover the widest feasible range of variation to assure that the calibration is accurate and that the instrument will respond correctly in all process conditions. The constituent (component) being measured should vary over a range at least five times larger than the expected error in the calibration. This increases the accuracy in determining the sensitivity value for the NIR instrument. If the range of sample concentrations is too small, the instrument sensitivity determined by the calibration will be unreliable. If the process cannot produce a sufficiently wide range of variation on-line, the instrument should be calibrated near-line or in the laboratory.

The sample set should include the full range of variation that the process can produce. There should be at least three samples, near the minimum, target, and maximum concentrations of the constituent being measured. Any other important process variations should also be included.

Minimum Point-to-Point Variation

To minimize sampling error, the samples should have the smallest possible range of point-to-point variation. Ideally, the concentration of the constituent being measured will change only slightly over a short distance, so that samples grabbed for the laboratory will be essentially identical to the samples viewed by the NIR instrument. Unfortunately, many applications are not ideal, and near-line or laboratory sampling or special sampling techniques are required.

If point-to-point variation is a problem, special care is necessary to assure that the NIR instrument and the reference method measure the same samples. For off-line calibrations, point-to-point variation can be eliminated with a measurement procedure that assures that both NIR and the reference method measure the entire sample. This may involve averaging several NIR measurements and dividing the sample into several subsamples for the reference method. In an on-line measurement, the NIR instrument sees a long, narrow, shallow strip of product on top of the process. Grab samples for reference method measurements should be taken from this strip, so that they are the closest feasible match to the sample seen by the NIR instrument. A good technique for grabbing samples is to take many small, shallow pecks from the top of the product bed, immediately downstream of the NIR instrument. Samples should be grabbed over a time comparable to the response time of the instrument, and the NIR instrument reading noted at the end of the sampling period.

A good technique for grabbing samples is to take many small, shallow pecks from the top of the product bed, immediately downstream of the NIR instrument.

Representative of Production

The samples for a calibration should be representative of actual on-line production. If there are significant differences between the calibration samples and actual on-line samples, the on-line NIR measurement will be inaccurate, even though it was correct on the calibration samples. The easiest way to avoid this problem is to calibrate the instrument on-line, in which case the calibration and production samples are identical. If the calibration must use nonrepresentative samples, it is usually possible to correct the measurement by adjusting the instrument offset so that the instrument reads correctly when the process is running on target.

Off-line calibrations for moisture are usually forced to use nonrepresentative samples; the total moisture of a sample and the moisture distribution within it can change rapidly. For instance, the surface of a dough sheet will be relatively dry after it leaves a dryer. If the sheet is stored off-line for a few minutes, however, moisture will move to the surface from the wet center of the dough sheet until the surface and center have approximately equal moistures. An instrument calibrated on these samples would read erroneously low when used on-line.

SENSOR LOCATIONS: ON-LINE, NEAR-LINE, AND LABORATORY

An instrument can be calibrated with the sensor installed in its ultimate location viewing the process directly (on-line), near the process but viewing grab samples (near-line), or away from the process, viewing artificially made samples (laboratory) (Table 30-5). The optimal calibration method varies depending on the application. The method should be chosen to achieve the best possible combination of representative sampling, a wide range of variation, and minimal sampling error, given the constraints of the process and application.

On-Line Calibration

This is the method of choice in applications where point-to-point variation is small and the process can produce a wide range of variation. In an on-line calibration, the sensor sees the product directly as it moves through the process. Calibration samples are taken by grabbing a sample from the line while simultaneously noting the instrument reading. Unrepresentative sampling is not a problem, since the sensor inherently sees the same samples it will see in routine measurements.

However, the process may have large point-to-point variation, causing significant sampling error. Also, acquiring samples that cover a wide enough range of variation will be a problem if the product specifications are tighter than the necessary range. In this situation, an on-line calibration requires the manufacture of bad product. Most production facilities are naturally reluctant to do this!

In this situation, an on-line calibration requires the manufacture of bad product. Most production facilities are naturally reluctant to do this!

Near-Line Calibration

This is the method of choice when the process has substantial routine point-to-point noise and the process can produce a wide range of variation. In a near-line calibration, the instrument is installed off-line, close to its ultimate on-line location. Samples are grabbed from the line, immediately presented to the sensor, and then sent to the laboratory for the reference method measurement.

Near-line sampling eliminates sampling error if both the NIR sensor and the reference method analyze the entire grab sample. Also, near-line samples are very representative because the sample has little time to change between the grab and the NIR analysis. However, near-line calibration relies on the process to provide a sufficient range of variation in the sample.

Laboratory Calibration

This type of calibration should be used when creating a large range of variation by changing the process is not feasible. Laboratory calibration can eliminate sampling error, but samples prepared in the laboratory may not be representative of the on-line product.

In a laboratory calibration, the instrument is installed in a laboratory, and samples are measured there by both the NIR instrument and the laboratory reference method. Samples for the calibration are made artificially in the laboratory. This gives great control over sampling and the range of sample variation. However, samples made in a laboratory are frequently not perfectly representative of actual process samples, and the laboratory calibration may be less accurate on-line. Therefore, calibrations using grab samples are usually better near-line than in the laboratory.

If a calibration developed in the laboratory is in error when applied to an on-line measurement, the error can often be corrected by adjusting the instrument offset to make the instrument agree with the actual value.

If a calibration developed in the laboratory is in error when applied to an on-line measurement, the error can often be corrected by adjusting the instrument offset to make the instrument agree with the actual value.

Table 30-5. Relationships between Sensor Location and Sample Selection

	Sensor Location		
Sample Selection	On-Line	Near-Line	Laboratory
Representative samples	Best	Good	Poor (process dependent)
Range of variation	Fair (process dependent)	Fair (process dependent)	Best
Sampling error	Fair (process dependent)	Good	Good

Troubleshooting and Maintenance

Once an NIR instrument is calibrated, it should be checked at intervals to assure that the instrument is stable and maintaining its accuracy. This verification is required for programs such as pharmaceutical validation and ISO 9000 certification. Today's growing documentation requirements make a formal calibration check procedure increasingly necessary.

Several verification methods are in use, all of which compare the NIR instrument reading with some reference method. A significant difference between the methods indicates questionable calibration stability. Maintaining control charts of the difference between NIR and reference measurements can be very effective, because it can detect subtle trends in the calibration that could be masked by the comparison errors. Several verification methods are in common use, including comparison with operator expectations, comparison with NIR reflectance standards, comparison with individual grab samples, and full calibration.

In all of these methods, the total expected error of the comparison, including reference, sampling, and NIR instrument errors, should be used to set a trigger level. Disagreement between the NIR and reference measurements indicates a problem if it is greater than the trigger level. If the trigger level is set too small, normal random errors will trigger indications of calibration problems. Recalibration due to normal errors in the measurement amounts to "chasing your tail," wasting time and effort in frequent, useless recalibrations.

Recalibration due to normal errors in the measurement amounts to "chasing your tail," wasting time and effort in frequent, useless recalibrations.

The easiest and least rigorous check method involves the operator, who frequently makes informal comparisons between his or her estimate of the process conditions and the NIR instrument's measurement. Operator checks are only as good as the operator's judgment and are impossible to document. They can serve as an alert that something is wrong, but should be backed up by a more rigorous test at regular intervals and if problems are suspected.

NIR instrument stability can be checked by using stability check standards. Check standards for on-line use provide a constant NIR absorbance. The materials used for the standards are thermally stable, do not degrade with time, and do not absorb water. For reflectance measurements, check standards typically consist of a ceramic reflector and a glass cover, enclosing a rare-earth doped-glass or a polymer chosen to simulate the product. For transmission measurements, standards typically consist of a rare-earth glass window, possibly with a neutral-density filter. They are supplied in pairs, to check the instrument sensitivity and offset. Because the check standards are intended to simulate a specific product, they do not have an absolute reading. Instead, a reference reading is determined by measuring the check standards on an instrument known to be well-calibrated. The calibration can then be verified as needed by measuring the standards and comparing the results to the reference reading. Stability check standards provide a highly repeatable, simple, fast test of sensor stability.

Stability check standards provide a highly repeatable, simple, fast test of sensor stability.

A calibration can be verified directly by comparing NIR and laboratory measurements of grab samples. Many on-line installations carry out these checks at regular intervals. The verification samples are generally clustered around the process target, with no significant range of variation. Therefore, they do not provide a good check on the instrument sensitivity. Sampling and reference method errors make the direct comparison less repeatable than reflectance standards.

A full recalibration of the instrument is the most difficult and rigorous verification method. It can require some expense, and significant effort by plant personnel. It is usually easier to verify the validity of an existing calibration using one of the simpler methods than it is to generate a new calibration from scratch. Most plants avoid full calibration whenever possible to minimize this effort.

Applications: NIR Transmission Analysis

NIR transmission instruments are currently in on-line use for a variety of chemical process measurements. One common liquid-phase application is parts per million water measurement in streams containing reactive chlorine, where maintenance of a low water level is necessary to prevent corrosion by eliminating HCl formation. In other applications, water measurement at parts per million levels may be required to meet specifications for water content. NIR can be used to control distillation operations for solvent recovery and solvent stripping, measuring both water in solvent solutions and solvents in water solution, and to monitor chemical component concentrations in reaction mixtures and product streams. Gas-phase applications include corrosion prevention, reaction mixture control, and reaction monitoring.

Tables 30-6 and 30-7 show a selection of transmission NIR measurement applications. These lists are only a sampling of the possibilities for process NIR measurements and are by no means complete. Many other types of measurements are possible. The detection limits and accuracies depend strongly on the application. In most applications, the detection limit for water ranges from 1 to 20 ppm, while organics can be detected at 0.01 to 1.0% levels.

In most applications, the detection limit for water ranges from 1 to 20 ppm, while organics can be detected at 0.01 to 1.0% levels.

Liquid-Phase Applications

Water is the most commonly analyzed compound when applying NIR technology to the measurement of interest. The unique absorption characteristics of water in this region of the spectrum allow for an accurate and repeatable measurement of water concentrations ranging from low parts per million to high percentages.

For comparison purposes, Figure 30-3 shows NIR absorption spectra of various liquid compounds. Please note that none of the compounds exhibits any absorbance in the region where water absorbs most strongly, and these spectra were taken using a 1-in. path length cell—approximately 250 times the path length used for obtaining a water spectrum (1-mm path length), while obtaining similar strength of absorption. These points are further evidence of the unique characteristics of water absorption and an indication of its ease to isolate for an accurate on-line analysis.

In addition, the spectra indicate the characteristic absorption bands of a sampling of compounds exhibiting O–H (nonwater), and C–H bonding. NIR transmission analysis is appropriate for the direct constituent measurement of these and other organic compounds that contain hydrogen.

Table 30-6. Selected Liquid-Phase Applications of Transmission NIR Photometry

Measurement	Application	Minimum Range
Water	Ethylene dichloride (EDC)	0-100 ppm
	Refrigerants	0-50 ppm
	Glycols	0-400 ppm
	Liquid halogens (Cl_2, Br_2)	0-100 ppm
	Heat exchange fluids	0-100 ppm
	Alcohols	0-400 ppm
	Organic acids	0-1%
	Hydrocarbons	0-100 ppm
	Amines	0-500 ppm
	Mineral acids	0-5%
	Solvent recovery	0-100 ppm
Sodium hydroxide	Caustic scrubbers	0-20%
Metaphenylene diamine	Dimethylacetamide	5-15%
Ethyl hexanol	Diisooctylphthalate	0-1,000 ppm
Polymerized material	Polymerization reaction mixtures	5-30%
Acetic acid	Water	0-15%
Dimethylacetamide	Water	0-15%
Organic solvents	Solvent stripping	0-15%

Gas-Phase Applications

The relative density of gases is much less than liquids. For on-line analyzers designed with fixed-volume sample cells, this factor results in the need for a sample cell of much larger volume in order to achieve resolution and accuracy of the constituents being analyzed.

With regard to gaseous applications using NIR transmission technology, rules similar to those for liquids apply. Table 30-7 lists common applications.

In addition, some process NIR analyzers can be configured to operate in the mid-IR region of the spectrum. For gaseous applications, this region offers the strength of fundamental absorption bands and expands the constituents available for analysis to include: CO and CO_2, acids, ammonia, water at ppm levels, as well as low-ppm concentrations of other compounds containing hydrogen.

Table 30-7. Selected Gas-Phase Applications of Transmission NIR Photometry

Measurement	Application	Minimum Range
Water	Halogens (Cl_2, Br_2)	0-100 ppm
	Natural gas	0-100 ppm
	Phosgene	0-100 ppm
	Air, inert gases	0-100 ppm
	Vinyl chloride monomer	0-100 ppm
Ammonia	Polymerization feedstock stream	0-1%
Methane, ethane	Reaction gas mixture	0-1%
Hydrogen cyanide	Reaction gas stream	0-5%
Ethylene	Flue gas stream	0-1%
CO, CO_2	Flue gas stream	0-5%

Sample Presentation Considerations for Transmission Analysis

The primary elements to consider in control of a liquid or gaseous process stream for NIR measurements are discussed in the following paragraphs. See the section "Limitations of NIR for Process Measurements" for solutions to some of the problems listed here.

PHASE

It is essential to introduce the liquid or gas sample to the analyzer in a single phase. The stream must be properly conditioned in order to avoid the possibility of the sample boiling or condensing within the sample cell. Bubbles or particulates in the process stream must be removed upstream of the sample cell inlet. Immiscible and opaque solutions are unacceptable for NIR transmission analysis as well.

TEMPERATURE

Stream temperature control is the most critical element in on-line NIR transmission analyses. In an uncompensated measurement, temperature variations of as little as 2F° (1C°) can result in error on the order of 10 to 20% of the full-scale measurement range. Measurements involving water, alcohols, or amines are particularly sensitive to temperature variations, because temperature affects the hydrogen bonding characteristics of these materials. Typical installations control the temperature to within at least ±9°F (±5°C) and rely on temperature compensation techniques to eliminate residual temperature effects.

Temperature control/compensation is achieved in one or in a combination of the following ways:

- *Heat-tracing* of the sample cell and a portion of the incoming sample line to a constant temperature of a few degrees higher than the maximum operating temperature of the process (individual process conditions may alter this general rule).
- *RTD:* Installation of a resistance temperature device (RTD) at the outlet of the sample cell. By establishing and incorporating into the measurement algorithm the appropriate temperature coefficient, and then applying the signal from the RTD to the algorithm, accurate temperature compensation at the measurement point can be achieved, often to ±0.2F° (±0.1C°).
- *Optical correction*: Establish and incorporate the appropriate reference optical wavelengths to compensate/cancel the effects that temperature-based spectral variations may have on the component(s) being analyzed.

PRESSURE

The sample pressure deserves critical consideration when gaseous processes are being analyzed because pressure is directly proportional to concentration. Since the sample cell is a fixed volume, control of pressure prior to sample cell inlet is essential to an accurate, repeatable measurement. The sample takeoff and return pressures assist the supplier in determining the need for installation of a process pump.

The sample pressure deserves critical consideration when gaseous processes are being analyzed because pressure is directly proportional to concentration.

COMPOSITION

Identification of the composition and concentration range of all the components of the sample stream, including background components, is essential. Knowledge of sample stream composition is necessary in order to (1) verify feasibility and desired range of measurement, (2) compensate for any background variations that may affect the absorption (and thus the measurement) of the component(s) to be analyzed, and (3) compensate for any possible interfering compounds, such as chemical compounds whose spectral absorption overlaps the component(s) to be analyzed.

MATERIALS OF CONSTRUCTION

Choose compatible materials of construction for all wetted parts to be in contact with the sample cell and sample conditioning system in order to avoid cell window coating and /or corrosion of the cell and its parts.

OTHER PROCESS VARIATIONS

Additional consideration must be given to the presence of any condition/substance that may alter and/or prevent the passage of light to and through the sample cell. Examples of such light blockers include undissolved particulates, bubbles, black or opaque samples, coated windows, and so forth.

Installation Configurations for Transmission NIR Measurements

Typical sampling configurations for process NIR instruments include sidestream (near-line), in-line, and fiber-optic probe installations. All spectrophotometric analyzers perform nondestructive measurements, thereby avoiding loss of product and/or chemical waste formation. In most instances, the analyzed sample is fed directly back into the process line. It is essential to consider all process specifications relating to the measurement of interest in order to determine the best configuration for a particular application. Table 30-8 summarizes the advantages and disadvantages of the various installation configurations.

Table 30-8. Advantages and Disadvantages of Transmission NIR Installation Configurations

Installation Configuration	Advantages	Disadvantages
Sidestream	Control stream conditions at point of measurement: temperature, pressure, filtering Isolate analyzer for maintenance without interrupting the process. Sensor can switch between multiple streams	Requires a sample conditioning system Possible long delay time
In-line	Instant response Representative sampling No sample conditioning required	Cell maintenance is only possible when the process is not operating Process stream must be clean, with constant temperature and pressure
Fiber-optic probe	No sample conditioning required Sensor can be installed in safe areas With a multiplexer, one sensor can be used with multiple streams	Only short pathlengths available Low light intensity Process stream must be clean, with constant temperature and pressure Fiber optics are expensive.

SIDESTREAM ANALYSIS

A sidestream analysis (Figure 30-8) involves extraction from the main process line of a small volume of the process liquid or gas stream to be analyzed. The slip stream of process fluid is continuously fed through the process sample cell. Thus, the analyzer is constantly measuring new product as it is flowing past the takeoff point.

Other advantages of sidestream analysis include the ability to:

- Filter "dirty" streams
- Condition incoming sample to constant conditions
- Maintain the sample cell and sample system without interfering with the main process
- Feed the sample back into the main process stream without creating any waste product

Sidestream analysis is convenient in that it allows for control of all process conditions via sample conditioning, including temperature control, pressure regulation, and filtering.

Sidestream analysis is convenient in that it allows for control of all process conditions via sample conditioning, including temperature control, pressure regulation, and filtering. The NIR absorption characteristics of liquids and gases are both temperature and pressure dependent. Thus, it is imperative to properly condition the incoming sidestream to a constant temperature and pressure (control of pressure is much more important for gas-phase than for liquid-phase analysis). Additional temperature and pressure control measures for fine control are then taken within the sample system.

Finally, analyzer location must be considered in relation to response time and flow requirements. The sensor should be located as close to the process takeoff point as possible to minimize the measurement delay caused by the time required for the sample to flow through the sample system. The delay time can be calculated by:

$$\text{Delay time} = \frac{\text{Volume of sample cell} + \text{Volume of sample system inlet plumbing}}{\text{Sample system flow rate}}$$

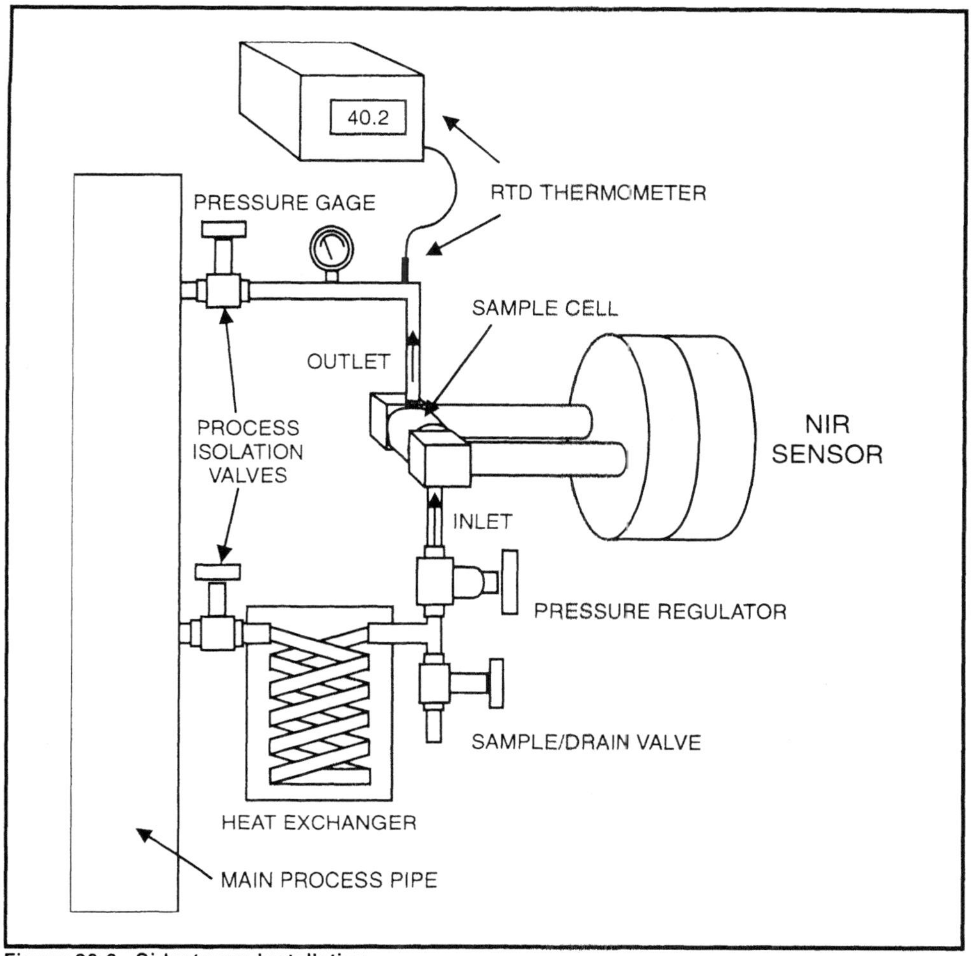

Figure 30-8. Sidestream Installation

NIR measurements are generally flow independent. Therefore, to minimize the delay time, the flow rate should be as fast as possible without creating significant turbulence in the sample cell. The flow rate through a sample system depends on the sample viscosity, the pressure drop between the takeoff and return points, and the sample system's impedance to flow. If flow is poor, it may be necessary to use a fast-loop bypass system or to include a pump in the sample system to assure adequate flow and minimal delay time.

The flow rate through a sample system depends on the sample viscosity, the pressure drop between the takeoff and return points, and the sample system's impedance to flow.

IN-LINE INSTALLATION

In an in-line installation (Figure 30-9), the NIR sensor looks directly through the main process pipe. This typically involves modifying the process pipe to install windows on each side, around which the sensor assembly is fitted. The windows can be mounted on a flanged pipe section that replaces part of the process pipe. In-line installations eliminate the need for a sample-conditioning system. However, in-line installations are suitable only if the stream is clean and particulate free and the process temperature is constant to within ±9F° (±5C°). Also, in-line installations require the manufacturing process to be shut down in order to clean the cell windows or perform maintenance on the analyzer.

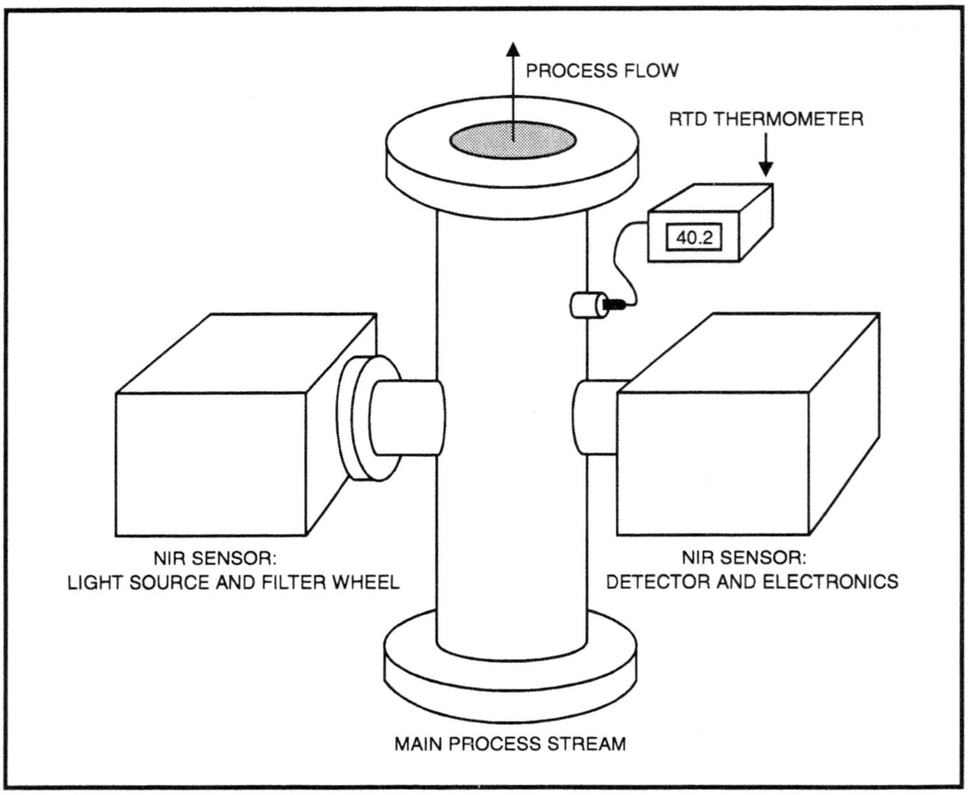

Figure 30-9. In-Line Installation

FIBER-OPTIC PROBE INSTALLATION

Another common installation configuration is a fiber-optic probe (Figure 30-10) installed in a process line and connected via fiber-optic cable to a remotely-located sensor assembly. Probes are constructed of industrial-grade materials and are capable of handling high process temperatures and pressures. Major advantages of this type of installation include elimination of a sample-conditioning system and location of the sensor in a nonhazardous, controlled environment.

Probes are constructed of industrial-grade materials and are capable of handling high process temperatures and pressures.

However, it is important to consider a few important factors when deciding on a fiber-optic probe installation for a specific application. Unlike sidestream analyses where the measurement is being made at the sensor assembly, fiber-optic cable loses signal strength over distance. In addition, sample path length is limited in comparison to a sidestream configuration. Therefore, applications involving low-ppm measurements, low signal strength (i.e., weak absorption) of the constituent(s) being analyzed, and/or long probe-to-sensor distances may not be suited for a fiber-optic probe installation.

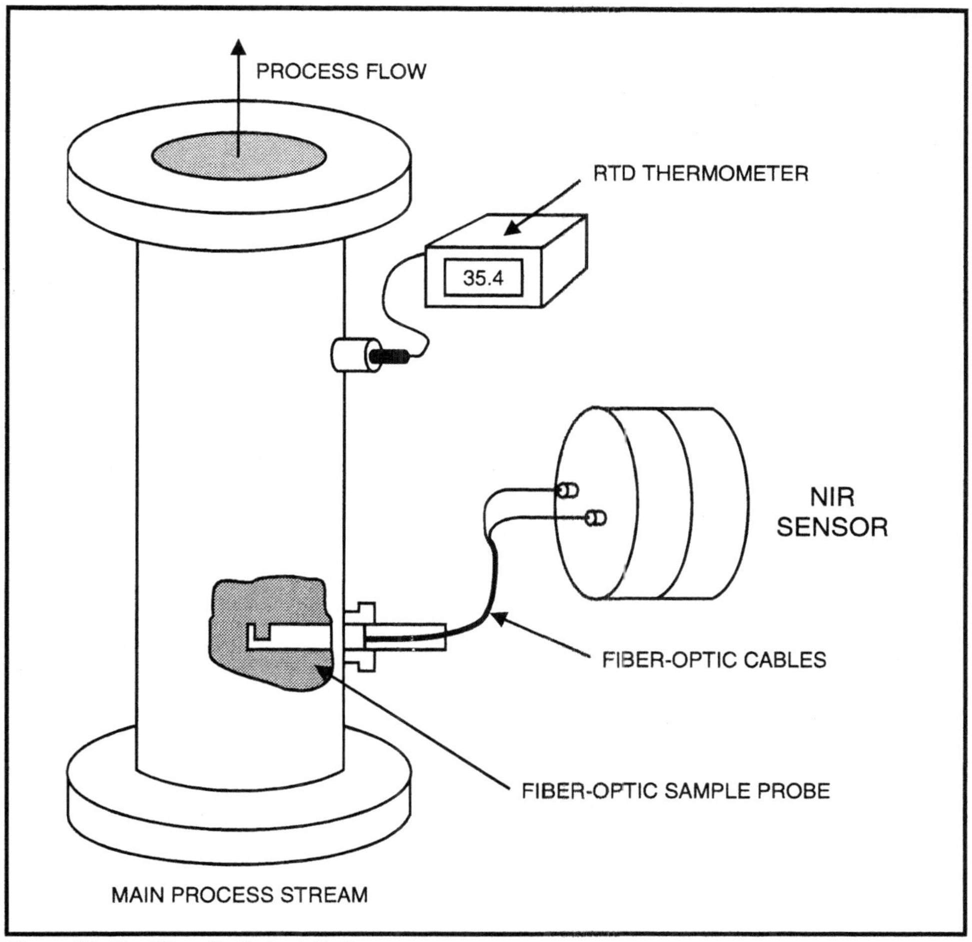

Figure 30-10. Fiber-Optic Installation

Liquid/Gas System Calibration Methods

All on-line analyzers provide a secondary measurement and thus must be calibrated against some reference or laboratory method for comparison purposes. Several factors must be considered when calibrating an on-line NIR transmission analyzer:

1. A minimum of three calibration samples representative of the range of analysis (i.e. low, middle, and high concentrations of the constituent being analyzed) must be included in the calibration group.
2. Calibration must be conducted with samples that represent the process operating conditions.
3. The reference analysis method must be repeatable and accurate.

General methods for process liquid and process gas calibrations are listed below, assuming that the on-line analyzer has been installed at or in the process. Depending on the specific application, operating conditions, hazardous nature of the product, and so forth, some of the methods listed are more or less attractive. The most favored method of calibration must be chosen individually.

GRAB SAMPLES

Samples are grabbed and analyzed via the lab or reference method. Results are compared with the corresponding analyzer values obtained when the samples were grabbed.

KNOWN-STANDARD SAMPLES

Pre-made standards of known concentrations are analyzed off-line. A zero adjustment is performed when the analyzer is installed on the process line to compensate for any variations between the calibrating conditions and the process conditions. It is necessary to analyze a minimum of two standards (preferably three), representative of at least a low- and a high-concentration sample.

CREATE STANDARD SAMPLES IN THE SYSTEM

The sample-conditioning system for the process stream may be capable of being controlled so as to allow the creation of calibration samples at or within the process. Startup or shutdown of the process is sometimes amenable to the variation desired for calibration. Use of a sample-conditioning system to isolate a portion of the process and subsequent control of constituent concentrations also result in an accurate on-line calibration with minimum sample handling.

Case Studies

0 TO 100 PPM H_2O IN ETHYLENE DICHLORIDE

Ethylene dichloride (EDC) is a precursor to the formation of vinyl chloride, trichloroethylene, vinylidene chloride, and trichloroethane. The presence of water in EDC results in the unwanted formation of hydrogen chloride in the process stream and the subsequent corrosion of process piping and other wetted process components.

The on-line continuous analysis of moisture in EDC is accomplished using a NIR transmission analyzer. The most critical installation locations include: (1) after completion of the reaction process, (2) just prior to storage, and (3) at the outlet of the storage tank as the material is used or transported.

Installation of an analyzer system for this measurement results in both economic savings and improved quality control. The economic benefits include protection of plant process piping via prevention of unnecessary pipe leaks/ replacement and avoidance of unwanted plant shutdown in the event of a break in the process piping. Quality of final product is also maintained or improved via continuous checks for moisture contamination.

Measurement is accomplished using an optical filter combination of 1.89 μm for the water absorption and 1.82 μm for the reference. The analysis is temperature sensitive, and thus coarse and fine temperature control is necessary.

For a sidestream installation, coarse temperature control is achieved via heat-tracing of a portion of the incoming sample line and the sample cell to a constant temperature. For either an in-line and/or a fiber-optic probe installation, it is necessary for the process stream at the point of analysis to be maintained to a constant temperature range of ±5F° (±3C°).

Regardless of installation configuration, fine temperature control is essential and is achieved via use of an RTD coupled with an associated temperature transmitter. The RTD is installed either at the outlet of the sample cell or in-stream with the fiber-optic probe. The RTD feeds the temperature signal directly into the microprocessor, which contains the algorithm that has been modified to include a temperature coefficient to automatically compensate for temperature variations of ±2 to 9F° (±1 to 5C°). In the event that even small amounts of contaminants (such as chlorine) exist in the process stream, special materials may be required for the sample cell (i.e., Kynar or Monel), sample cell windows (i.e., sapphire or calcium fluoride), and O-rings (Kalrez or Teflon).

On-line system calibration is accomplished by introducing to the analyzer at least three process EDC samples of increasing moisture concentrations (i.e. low, mid, and high H_2O concentrations corresponding to the desired range of measurement) and comparing the precalibrated analyzer values to the values obtained with the same samples, using the laboratory/reference method at the process operating temperature. The analyzer computes the span and zero values for the application via linear regression analysis. Periodic grab sample comparison is recommended to check the long-term integrity of the calibration.

MOISTURE IN ETHYLENE GLYCOL AND GLYCOL MIXTURES

Ethylene glycol is used primarily as a coolant and as antifreeze, but also as an intermediate compound in various chemical and manufacturing processes, a solvent, and an ingredient in airplane deicing fluid. The control of moisture is essential to the production of high-quality, consistent product.

The on-line continuous analysis of moisture in ethylene glycol is accomplished using a NIR transmission analyzer. The most critical installation locations include: after completion of the reaction process, just prior to storage, and at the outlet of the storage tank, as the material is used or transported.

Due to the strong –OH bonding effects that occur between ethylene glycol and water, this measurement is extremely temperature sensitive. Without process temperature compensation, a 1.0F° (0.56C°) change in temperature results in an average of 40 ppm variation in the analyzer reading. Two compensation methods are available to account for process temperature variations: use of an optical filter in combination with an associated algorithm (1 to 4 ppm/C° sensitivity), and use of an RTD, applying the temperature coefficient over the process temperature range to the algorithm as a correction factor. By combining these two methods, the measurement sensitivity is reduced to less than 1 ppm/C°.

Glycol is not corrosive; thus, the sample cell can be constructed of type 316 stainless steel, with glass windows and Teflon O-rings. However, glycol does have the potential to coat windows. Sidestream analysis with a sample cell provides a relatively large surface area and increased sample volume analyzed (as opposed to a fiber-optic probe). These factors assist in maintaining a relatively high signal-to-noise ratio for this measurement. Individual analyzer performance should be considered when choosing an installation configuration for applications of less than 1% moisture in ethylene glycol. Calibration procedure is similar to the procedure outlined for the measurement of moisture in EDC.

Applications: NIR Reflectance Analysis

NIR reflectance analyzers are currently in widespread on-line use for a variety of process applications. Because water is a critical parameter in many processes, and an easy measurement for NIR technology, process NIR reflectance sensors are primarily used to measure water. Many other measurements are possible, including oils and fats, organic coatings, nicotine, protein, caffeine, sugars, and solvents. Table 30-9 lists several common applications.

Sample Presentation for Reflectance Analysis

Both the distance between the sample and the NIR sensor and the orientation of the sample with respect to the sensor affect the NIR reflectance of the sensor. Typical on-line NIR photometers in most applications will accept variations of a few inches in distance and 10° to 20° in angle without introducing significant measurement error. Most full-spectrum NIR instruments are more sensitive to sample positioning. For any instrument, variations in the sample presentation should be minimized to the extent allowed by the process and should be held within the instrument manufacturer's specifications.

Typical on-line NIR photometers in most applications will accept variations of a few inches in distance and 10° to 20° in angle without introducing significant measurement error.

An NIR instrument measures the reflectance of whatever is in view. A gap in the process stream lets the sensor view the conveyor or something other than the product. The conveyor's reflectance is typically unrelated to the product's reflectance, so gaps can cause erroneous measurements. Even momentary voids in the process stream can produce errors, unless the gap is detected by some other method and the NIR measurement is frozen (sample-and-hold) while the gap is in view. Gaps are a particular problem when they reveal a highly reflective metal surface. This is often the case with vibrating conveyors or metal mesh conveyor belts. It may be possible to remove the gaps by installing a dam in the process or by relocating the sensor.

Table 30-9. Selected Solid-Phase Applications of Reflectance NIR Photometry

Application	Measurement	Range
Tobacco	Water Nicotine Total reducing sugars	0-50% 0-10% 0-30%
Fried snack foods	Water Oils/fats	0-5% 25-40%
Particle board	Water Resin	0-10% 0-20%
Paper (finished)	Water Coating thickness Coatweight	0-10% 0-10 mils 0-250 g/m^2
Paper (in-process)	Water	30-50%
Dried pet foods (finished)	Water Protein Oils/fats	0-5% 0-20% 0-5%
Dried pet foods (in-process)	Water	40-60%
Breakfast cereals (finished)	Water Sugar coating	0-5% 0-20%
Breakfast cereals (in-process)	Water	40-60%
Pharmaceuticals	Water Other solvents	0-30% Varies
Pharmaceutical tablets	Tablet coating thickness	
Coal	Water	0-15%
Metal ores	Water	0-10%
Polymer fibers	Water	0-10%
Polymer pellets	Water	0-0.5% 0-2% (nylon)
Ceramic powders	Water	0-5%
Coated-metal sheets	Coating thickness or weight	0-20 μm
Polymer films	Water Film thickness	0-5% 0-40 mils

Installations for Reflectance NIR Measurements

The most common on-line reflectance NIR installation has the NIR sensors mounted over a conveyor belt where it can view the product moving on the belt, as shown in Figure 30-5. Other installation methods are available for special process situations, as shown in Figure 30-11. Some of these situations include:

- *Web products*: In many web processes, such as paper and film manufacture or coating, it is useful to obtain a profile of the moisture, thickness, or coatweight across the width of the web. To do this, NIR sensors can be mounted on traversing frames that move the sensor back and forth.
- *Falling product*: In measuring falling products, it is necessary to assure that the product constantly fills the NIR sensor's field of view. With a sufficiently dense process stream, this can be done simply by installing a window in the process pipe. It may be necessary to slant the window into the stream to retard the flow and produce a reliable product buildup on the window. Alternatively, the window can be combined with a cup to capture and hold the product. The cup is periodically blown clean with compressed air or nitrogen.
- *Thin film*: Measurements of thin films on reflective surfaces must be made using polarized light incident on the surface at approximately 58° angle of incidence. These installations are critically sensitive to small changes in the sample presentation.
- *Screw conveyors*: In a screw conveyor, the screw passes through the NIR instrument's sample spot once per revolution of the conveyor. Because the screw is typically a highly reflective metal, this disturbs the NIR measurement. To avoid this, a small section of the screw is removed and the sensor is installed there, or the NIR sensor is installed just past the end of the screw, as the product falls into the next processing step.
- *Transflectance measurements*: In some web processes, it is desirable to measure a moisture distribution through the web. This can be done by combining reflectance and transmission measurements. The reflectance measurement is most sensitive to the web surface, whereas the transmission measurement is equally sensitive to the whole thickness of the web. Comparing the two measurements gives some information on the moisture distribution.

Most NIR photometers used in dirty environments use an air purge—a curtain of air blown across the sensors' window to keep away dust, dirt, and moisture.

Many NIR solids measurements are made in dusty environments. Keeping the sensor window clean is particularly critical in these applications. Most NIR photometers used in dirty environments use an air purge—a curtain of air blown across the sensors' window to keep away dust, dirt, and moisture. The purge air must be clean and dry or the purge will blow oil and water onto the window, making the problem worse. Oil and water filters should usually be installed in the air supply line immediately before the sensor. Facing the sensor window downward will also help by preventing dust from falling onto the window.

When measuring very thin, translucent, or transparent samples it is useful to have a reflector mounted behind the sample. As a rule of thumb, a reflector is probably needed if a person can see light through the sample.

When measuring very thin, translucent, or transparent samples it is useful to have a reflector mounted behind the sample. These samples may reflect little of the NIR light from the sensor. A reflector increases the sensitivity of the measurement by assuring that a much larger fraction of the NIR light is reflected back to the sensor. As a rule of thumb, a reflector is probably needed if a person can see light through the sample. The reflector should have a brightly reflecting matte surface to cause light diffusion. The reflector material should have no strong NIR absorbances; Teflon and ceramics are suitable materials. The reflector must be kept as clean as the sensor window.

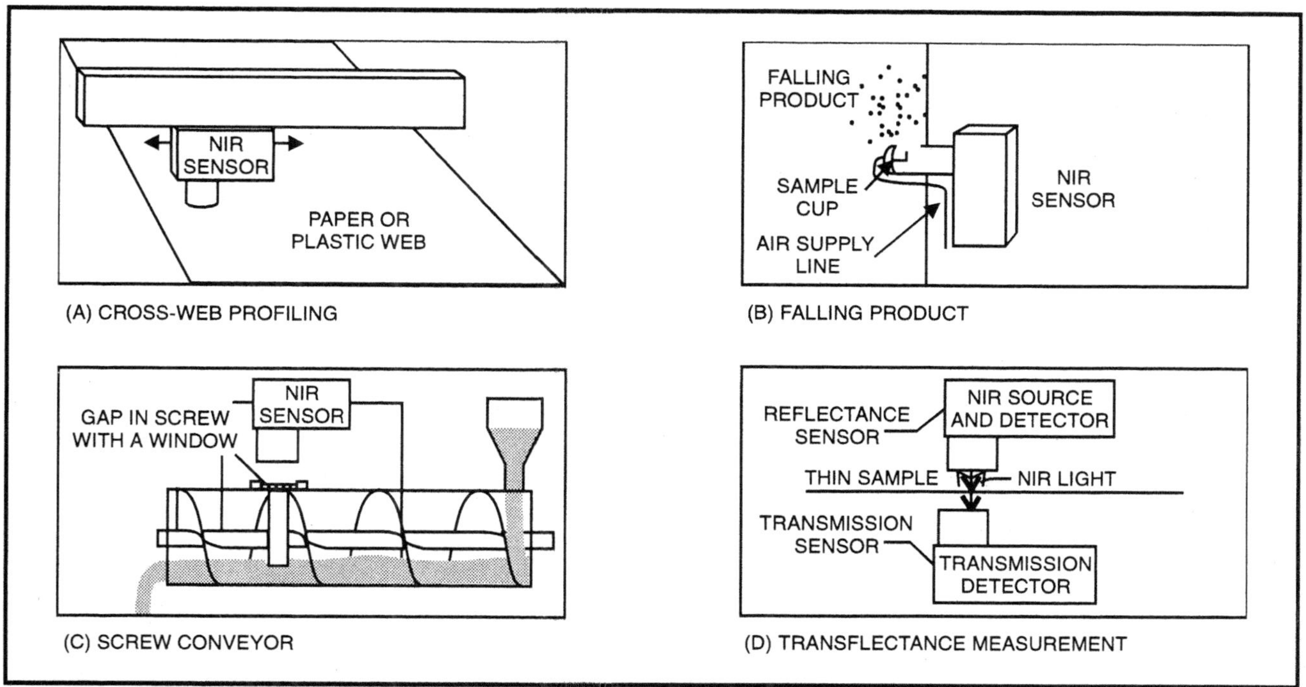

Figure 30-11. Selected Installations for Process NIR Reflectance Photometers

Reflectance Calibration Methods

On-line calibrations are usually preferred for on-line reflectance NIR sensors for solids. Laboratory and off-line calibrations are usually not as good, because it is often difficult to reproduce a dynamic on-line sample with static laboratory samples. However, on-line calibration for a solid can involve significant difficulties in sampling and handling errors. If the process is not homogeneous, sampling errors will be large. The samples frequently are changing rapidly as they pass the sensor, and rapid sample handling is required. Samples should be grabbed just downstream of the sensor, from the top of the product bed. Sampling error can be minimized by grabbing samples over several seconds and averaging the NIR measurement over the same period.

CALIBRATION OF MOISTURE MEASUREMENTS

In some applications, calibration of a moisture measurement may be difficult because of problems in sampling and the reference measurement method. Sampling is difficult if the product dries out or picks up moisture very rapidly when exposed to air. Moisture calibration samples should always be handled as little and as quickly as possible to minimize this error source. If the moisture in the product changes rapidly from point to point, calibration grab samples may not have the same moisture as the process sample seen by the NIR sensor.

For solid materials, weight loss on drying (LOD) is the most common moisture reference method. It is inexpensive, simple, and reliable, but has an important limitation in that LOD detects whatever materials evaporate from the sample, and only those materials. Typical drying methods do not remove all the water, and therefore do not detect it. They do detect other volatile materials. Variations in the drying conditions, sample composition, or volatiles cause error in the water measurement. Few LOD methods attain accuracies better than ±0.1% water, and accuracies of ±0.2 to 0.5% are more common. If better accuracy is required, wet chemical methods such as Karl Fischer analysis should be used.

Special Considerations for NIR Moisture Analysis in Solids

A NIR water measurement can be very accurate and is often believed to be more reliable than the reference method used to calibrate it. On the other hand, calibrating NIR instrumentation for on-line water measurements can present several difficulties.

A NIR water measurement can be very accurate and is often believed to be more reliable than the reference method used to calibrate it. On the other hand, calibrating NIR instrumentation for on-line water measurements can present several difficulties. The problems can affect both calibration and routine operation.

One difficulty is the volatility of water: A process sample exposed to the air will gain or lose water very rapidly. In routine operation, a process sample with rapidly changing moisture levels will produce errors when the process line speed changes. For instance, imagine a product that is drying quickly as it leaves an oven. If the line speed drops, the product will have more time to dry between the oven and the sensor, and the sensor will measure a lower moisture. Changes in the air temperature, relative humidity, or air currents over the process can also affect the moisture reading by changing the drying rate.

Also, solid process streams are often inhomogeneous, with the moisture changing rapidly from point to point. For instance, tobacco leaves contain widely varying amounts of moisture, depending on the type of tobacco and its growth and storage conditions. A NIR instrument measuring tobacco leaves sees a different moisture level for each leaf it measures. Unless the instrument is used with a control system intended to control moisture on this small scale, the instrument should be set with a slow response time to average over a large amount of the stream. This is particularly important in closed-loop control systems, where the control system should not try to respond to changes in the measurement caused by normal point-to-point variation.

About the Author

Dr. Edward K. Baldwin earned his B.S. in Chemistry at Furman University in 1981, and Ph.D. in Physical Chemistry at Harvard University in 1986. His thesis research was a study of the ultra-high vacuum surface chemistry of nitrogen and nitrogen oxides on tungsten surfaces. Since 1987, he has been employed by Moisture Systems, a world leader in process NIR photometer instrumentation, as applications manager and technical director. At Moisture Systems, his primary responsibilities are development of new NIR-based process measurements, continual improvement of existing measurements, and development of new instrumentation and measurement techniques. He has several publications on NIR measurements and ultra-high vacuum research, and one patent pending for a NIR measurement of plasticizer in cigarette filters.

31

Ultraviolet/Visible (UV/VIS) Absorption Analyzers

During the past half century, photometric analyzers based on ultraviolet/visible (UV/Vis) absorption have found increasingly wide application throughout the process industries. UV/Vis photometric analyzers were among the first plant stream analyzers used for direct monitoring of process stream composition and for closed-loop control.

Until the late 1950s, most UV/Vis analyzers were developed within petroleum and chemical companies for proprietary use; a few were designed at university research facilities. Many of these early instruments either used the mercury 253.7 nm resonance line from a germicide lamp as the measuring wavelength or adapted the Beckman DU spectrophotometer.

In pre-1930 Germany, Dr. Kurt Mueller developed one of the first photometric analyzers to monitor elemental mercury vapor, using the 253.7 nm line as the measuring wavelength [Ref. 1]. He established the mercury 253.7 nm molar extinction coefficient at 4.66×10^6. More than 25 years later, D.J. Troy of DuPont, unaware at the time of Mueller's work, precisely determined the molecular extinction coefficient to be 4.53×10^6, within 3% of Mueller's figure [Ref. 2]. In 1931, Professor Farington Daniels at the University of Wisconsin developed a photometric analyzer to measure the ratio of nitrogen dioxide/nitrogen tetroxide (NO_2/N_2O_4) in a laboratory pilot plant [Ref. 3]. Eight years later, T.T. Woodson of General Electric described a sensitive mercury analyzer using a G4T4 low-pressure mercury discharge germicidal lamp as the source for the 253.7 nm line [Ref. 4].

In 1940, the DuPont Company was producing as much chlorinated hydrocarbon solvents as possible from 29 batch reactors at its Niagara Falls plant, mostly for the British armed forces. Process control was poor at best, depending on operators to shut off chlorine feed upon observing the chlorine gas color through sight glasses in the reactor off-gas line. V.J. Hanson, at the time a young DuPont physicist, placed a tungsten lamp at one sight glass and a blue optical filter before a barrier-layer photocell at the opposite sight glass. At the initial chlorine release, this simple system triggered an alarm to shut off the chlorine feed. Placing these systems on all 29 reactors resulted in savings of millions of 1940 dollars!

After World War II, a number of petroleum and chemical companies — including Gulf Oil, Shell Development, Sun Oil, Hercules, as well as DuPont — developed proprietary UV/Vis process analyzers. Several of these analyzers were commercialized and are still used throughout the process industries.

In recent years, the elements used in UV/Vis analyzers have been greatly improved. Advances in thin-film technology have resulted in more stable interference filters with greater resistance to high temperatures and improved transmittance with more narrow bandpass. "Quieter" discharge lamps with longer life expectancies are now available, and silicon photodiodes have been improved for use in the UV region with greatly improved signal-to-noise ratios and temperature resistance equivalent to that of vacuum photodiodes. Highly pure quartz with high transmittance throughout the UV/Vis region is now available for windows, lenses, and fiber optics; other optical glasses are being developed.

The development of linear photodiode array detectors (LPDADs), holographic gratings, and relatively inexpensive yet powerful industrial computers has resulted in commercially available process photodiode array (PDA) analyzers with multiwavelength outputs for multicomponent analyses [Ref. 5].

Measurement Principles

The Nature of Electromagnetic Radiation in the UV/Vis Region

The electromagnetic spectrum covers radiation from radio waves to gamma waves, and in the midst of this broad spectrum are the UV and visible regions. The near-UV wavelengths covered in this discussion extend from 190 nm (10^{-9} m) to about 380 nm, the deep violet end of the visible region which extends to the deep red wavelength of about 750 nm. (The human eye responds to the "visible" region with maximum sensitivity at about 525 nm in the green region of the spectrum. "Light" literally refers only to electromagnetic radiation in the visible region; however, since "light" is a simple and short word, in this chapter it will also be used in place of "electromagnetic radiation" in the UV region.)

The energy attached to electromagnetic waves is directly proportional to the frequency of this radiation ($E = h\nu$) and inversely with wavelength. Hence, UV has higher energy than the visible and orders of magnitude greater than the mid-infrared (IR). This high energy means that highly intense light sources and sensitive detectors are available for application in the UV region.

Remember the man that keeps the spectrum — ROY G BIV!

Red	700 nm
Orange	625 nm
Yellow	575 nm
Green	525 nm
Blue	475 nm
Indigo	450 nm
Violet	400 nm

The Nature of UV/Visible Absorption

Absorption of light is observed as a decrease in light intensity resulting from the interaction of the light energy with the absorbing material. The observed loss of radiant energy usually results from conversion to heat energy, although some energy may be used in producing fluorescence (emission of light at a lower frequency) and chemical reactions.

In the UV/Vis region, the absorption is due to excitation of more loosely bound valence electrons, including the "unsaturation" (π) electrons of multiple bonds and unpaired electrons of the free radicals. This corresponds to photoelectron energies between 3 and 6 eV. The UV/Vis spectra of molecules generally are composed of broad bands when compared with the line spectra of atoms. Each electronic state has associated with it numerous vibrational and rotational energy levels. The vibrational-rotational fine structure of the bands is usually not resolved except for low-pressure gas-phase spectra of low molecular weight, simple molecules. For example, sulfur dioxide and chlorine dioxide show fine structures when studied at high resolution. Also, the absorption spectra of the simpler aromatic compounds (e.g., benzene and toluene) show distinct fine structures even in the liquid phase. In general, however, pressure broadening and solvent interaction make a spectra appear as the envelope of the otherwise-expected fine-structured spectrum.

Generalized UV/Vis Absorption Analyzer

Ultraviolet/visible analyzers, as well as all other absorption analyzers, have the same functional elements, although specific building blocks may vary considerably from instrument to instrument. Figure 31-1 shows a simplified block diagram of an absorption analyzer.

The analyzer must have a *source* (light source) of the radiant energy. Sources, of course, will differ depending on the instrument design and may be tungsten, metallic discharge, deuterium, or xenon discharge lamps — or even sunlight for special applications.

Wavelength isolation (optical filters) may be adequately provided by the selective emission of the source and response of the detector in some analyzer systems. However, for most UV/Vis absorption analyzers, wavelength isolation must be provided by special filtering systems. The filtering can be in the form of gratings, prisms, or interference filters located before or after the sample.

The *cell* is a device to contain the process sample, usually on a flow-through basis. The cell must be chemically and mechanically compatible with the process. A cell typically is cylindrical, with ends capped by optical windows sealed to the body by gaskets or O-rings and having sidearm ports for process sample flow.

The *detector* is a device to convert radiation intensity to an electrical (or mechanical) signal that is a function of the intensity. Basic detectors range from photomultipliers or silicon photodiodes to photodiode arrays.

Readout devices can be indicating or recording instruments. In certain applications, alarm units (audio and/or visual) may meet the requirements of the systems.

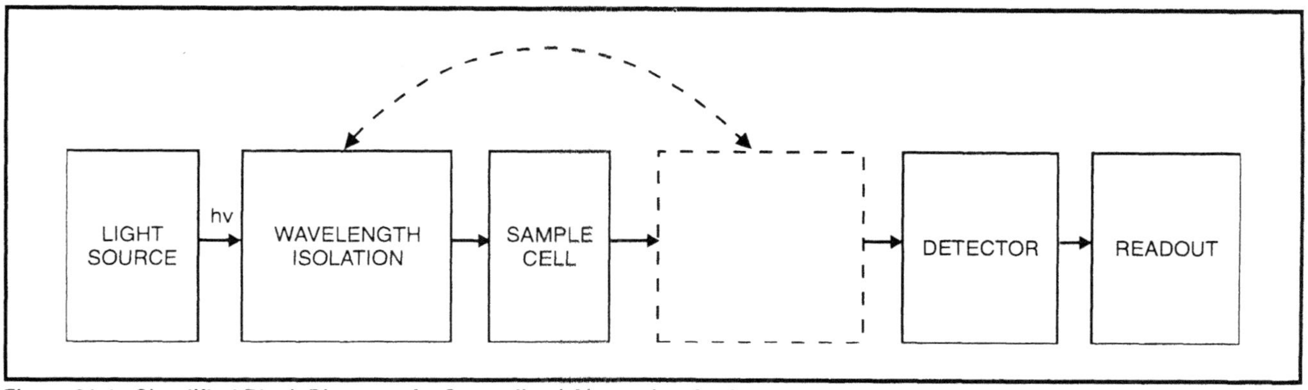

Figure 31-1. Simplified Block Diagram of a Generalized Absorption Analyzer

Beer's Law (Lambert-Beer Law)

Beer's law of light absorption is essential in the application of process UV/Vis absorption analyzers. The following review may be helpful:

$$A = \log\frac{I_0}{I} = \log\frac{1}{T} = abc \text{ (basic equation of Beer's law)}$$

where:

A = absorbance (a dimensionless function, linear with concentration and path length)
a = molar absorptivity in liters/mol · cm (also called molecular extinction coefficient)
b = path length in centimeters
c = concentration in moles/liter
I_0 = intensity of light striking the detector with nonabsorbing sample in light path
I = intensity of light striking the detector with concentration c of absorbing sample in light path b
T = transmittance = I/I_0

For the vapor phase:

$$A = \frac{abc'}{2,450} \text{ at } 25^{\circ}\text{C and 14.7 psia (atmospheric pressure)}$$

where c' is concentration in volume percent or mole percent. Or:

There are 24.5 liters of gas per mole at 25°C and 14.7 psia.

$$A = \frac{abc'}{2,450} \times \frac{P + 14.7}{14.7} \times \frac{298}{t + 273}$$

at any temperature or pressure, where P is pressure (psig) and t is temperature (°C).

For the liquid phase:

$$A = \frac{10abc''d}{\text{M.W.}}$$

where:

c'' = concentration in weight percent
M.W. = molecular weight of the absorbing material
d = density of the liquid solution

$\frac{10c''d}{\text{M.W.}} = c$, **concentration in moles per liter.**

Design Criteria for Process UV/Vis Analyzers

The problems associated with continuous sampling in a plant environment are usually formidable, and design criteria for process UV/Vis analyzers go well beyond those for laboratory spectrophotometers. The following basic design requirements are in part derived from those suggested by Glasser [Ref. 6].

Some of these criteria may not be applicable where fiber optics are used or when the analyzer is located in a non-plant atmosphere.

RUGGEDNESS

Plant analyzers generally are not handled with tender loving care; they are exposed to plant vibrations and fumes, and it is not unusual to see an analyzer used as a "stepstool" by plant operators and mechanics. Associated with the need for sturdy construction is the basic requirement that the field units meet the electrical hazard classification of the plant.

SERVICEABILITY

To gain operator confidence, an analyzer should have an "on-line service factor" approaching 100%. Only routine checks and minor adjustments (such as cleaning of sample cells) should be required over a period of months. The design should take into account the instrument technician who must maintain it. Simplicity, accessibility of components, and ease of troubleshooting will help to give a technician a positive attitude toward the analyzer, a factor which often may mean the difference between a highly successful and a merely adequate application.

ACCURACY, REPRODUCIBILITY, AND STABILITY

Reliable analyzer operation usually will require instrument stability far exceeding that of a laboratory instrument. The ambient temperature and power-supply variations in the plant often place demands on analyzer performance never experienced in a laboratory environment. The accuracy for gas analyses may be limited by the pressure control (±2% from atmospheric pressure variations). In an automatic control system, high reproducibility (precision) with a minimum of "dead band" is often essential.

LINEAR READOUT

Plant operators are partial to straight-line calibrations. (However, microprocessor circuits can now be used to linearize a nonlinear signal.) In an automatic control system, an output signal linear with concentration also is desirable from a control dynamics standpoint.

FAST RESPONSE

The response of an analyzer system is usually limited by the sample lag — the time it takes for a sample to reach the analyzer from the process line. For stable control in certain automatic control systems, the maximum time for an analyzer to respond to a concentration change in the process may be a few seconds. In the control of film thickness, a response on the order of milliseconds may be required.

SIMPLIFIED SAMPLING REQUIREMENTS

Some 80% of the problems associated with analyzer installations are in the sampling systems. The design of the analyzer should be such that sample-handling problems are minimized. The ability to handle hot and high-pressure samples and to compensate for dirt and bubbles is valuable in this respect.

The design of the analyzer should be such that sample-handling problems are minimized.

Common Applications

As described earlier, UV/visible light energy (190 to 380 nm) is in the region of the differences in the vibrational and rotational energy states of loosely bound valence electrons, unsaturation (π) electrons, and so on. Needless to say, many compounds have selective absorption in the ultraviolet and visible region of the spectrum, particularly in the short-wavelength region (200 to 250 nm). Fortunately, several important classes of compounds are free from ultraviolet and visible absorption or have broad "windows" in this region. Table 31-1 lists many of these compounds. Note the lack of UV/visible absorption by the atmospheric gases, water, aliphatic alcohols, ether, and saturated aliphatic hydrocarbons. A strong UV-absorbing material could, of course, be sensitively and selectively analyzed in the presence of only these compounds. These are the simple UV/Vis analyzer applications; however, such applications represent only a relatively small percentage of the thousands of analyses by ultraviolet and visible light absorption. Some specific applications are described below.

Table 31-1. Representative Gases, Vapors, and Liquids That Do Not Absorb Visible or UV Radiation

> 210 nm	≥ 254 nm
Inorganics	*Chlorides*
Argon	Ethyl chloride
Carbon dioxide	Ethylene dichloride
Carbon monoxide	Methyl chloride
Helium	Methylene dichloride
Hydrochloric acid	Vinyl chloride
Hydrogen	*Esters*
Krypton	Butyl acetate
Neon	Cellosolve
Nitrogen	Dimethyl sulfate
Oxygen	Ethyl acetate
Water	2-Ethyl butyl acetate
Xenon	Methyl cellosolve
Saturated Straight-Chain Hydrocarbons	Vinyl acetate
Butane	*Acids*
Ethane	Acetic
Methane	Butyric
Propane	Propionic
Unsaturated Straight-Chain Hydrocarbons	Sulfuric
Acetylene	*All Ethers*
Ethylene	*Lower Alcohols*
Propylene	Ethylene glycol
Lower Alcohols	Isobutanol
Ethanol	Isopropanol
Methanol	*Inorganic Salts*
n-Butanol	Potassium chloride
n-Propanol	Potassium sulfate
	Sodium chloride
	Sodium sulfate

Elemental Halogens

Elemental halogens, particularly chlorine, with ranges of 0 to <50 ppm all the way up to 100%, both in gaseous and liquid phases, are monitored by UV/Vis analyzers. All of the elemental halogens absorb broadly in the UV and/or visible region. Both iodine and bromine have strong absorption in both the visible and the UV region. Iodine shows the strongest absorption, which varies significantly with the solvent in which it is dissolved. Chlorine has broad absorption over a wavelength span (220 to 430 nm) covered by few other materials.

Because of the high corrosiveness of the halogens, materials of construction in contact with the sample must be carefully selected. Fluorinated hydrocarbon resins such as Teflon and special metals such as Hastelloy-C, Monel, and Inconel have been successfully used in halogen applications. Synthetic sapphire and magnesium fluoride cell windows are used where fluorine or hydrogen fluoride is present, since quartz is attacked by these compounds.

Precise control of many chlorination processes has been possible through the accurate measurement and control of chlorine concentration.

Oxidizing Agents

In general, oxidizing compounds have strong absorption bands. Besides the elemental halogens, other oxidizing materials such as sodium hypochlorite (NaOCl), chlorine dioxide (ClO_2), hydrogen peroxide (H_2O_2), ozone (O_3), and potassium permanganate ($KMnO_4$) are examples of compounds whose concentrations are being monitored with UV analyzers. "Bleaching" steps can often be controlled by monitoring the oxidizing agent concentration.

Aromatic Compounds

All aromatic compounds (benzene and its derivatives) are strong UV light absorbers. Concentrations of aromatics in aliphatic hydrocarbons are monitored in very low concentrations, and unit processes for removal of the aromatics are often controlled through this sensitive measurement. Aromatics, such as phenol (C_6H_6OH), in waste streams are continuously monitored with UV analyzers. In processes where an aromatic is a reactant, precise control may be provided through monitoring and controlling a slight excess of the aromatic in the reaction mass.

Color Measurement

Color is often a primary quality specification monitored with a visible analyzer. The special case of color measurement is covered later in this chapter.

Sulfur Compounds

Sulfur- containing compounds often strongly absorb UV radiation. Low concentrations of hydrogen sulfide (H_2S), carbonyl sulfide (COS), carbon disulfide (CS_2), and sulfur dioxide (SO_2) are monitored and controlled for safety, environmental compliance, and process efficiency. The measurement of SO_2 is probably the largest application of UV analyzers. Applications for monitoring low-ppm concentrations of H_2S in hydrocarbon streams have been successful, but at times background absorbance (often from trace aromatics) interferes with the measurement. With process spectrophotometers (i.e., process photodiode array [PDA] analyzers), compensation for interferences can be provided by a "chemometrics" software program in the analyzer computer.

Inorganics

Salts of transition metals such as iron, copper, chromium, manganese, and nickel are often strong absorbers of UV and/or visible light. Typical applications include measurement and control of ferrous ion concentrations in steel pickling baths, cupric ions in nitric acid production, Cr(VI) and Cr(III) in aluminum plating processes, and a number of metallic ions in plating baths.

Antibiotic Compounds

Many antibiotic compounds have strong UV absorption. During their production, several process steps can be controlled by measurement of the antibiotic concentrations with UV analyzers.

Elemental Mercury

The monitoring of elemental mercury vapor in the parts per billion range is actually an "atomic absorption" measurement, where the 253.7-nm resonance line of mercury from a low-pressure Hg discharge lamp is the measuring wavelength. The molar extinction coefficient of elemental Hg at its 253.7-nm resonance line is 4.53×10^6 L/mol · cm, some two orders of magnitude greater than any other material at this wavelength.

Monitoring Liquid Color

... there is a science of color. "Color *is* something that can be measured and expressed in numbers"...

A multitude of products throughout the process industries are graded on the basis of color; for many (lube oils, resins, and solvents) color is the primary quality specification. Color, per se, is a complex subject. The fact that color involves both physiology and psychology adds a subjective aspect to the field of color specification. Nonetheless, there is a science of color. "Color *is* something that can be measured and expressed in numbers" [Ref. 7], and colorimetric standards have been established based on firm scientific data.

The colorimetric system most universally recognized is the International Commission on Illumination (ICI) Tristimulus System of Color Specification [Ref. 8]. The official abbreviation for this commission is CIE, for the French name "Commission Internationale de l'Eclaiage." A complete discussion of the CIE system is beyond the scope of this chapter. Briefly, however, it involves the basic concept that, with a given light source, the color of an uniform liquid sample can be completely described by three numbers of a consistent color coordinate system, such as the CIE X, Y, and Z tristimulus values.

Although a three-number system provides a complete definition of the color of a product, this type of specification is difficult to obtain. Visualization of the color from a three-number specification is virtually impossible for the average person. Hence, for most products, single-number or single-letter color grading systems have been developed. These standards have simplified color grading and have been valuable in making color a practical quality specification.

Figure 31-2 shows an isometric plot of typical single number system color standards as the coordinates in color space, derived from the CIE tristimulus coefficients, with the x and y trichromatic coefficients and Y, the luminous transmittance represented on the "z" axis. If these color standards are used on a liquid product, it is necessary that as the product changes color, the colors, when plotted in color space, either lie on or be within specified limits of the curve AB (Figure 31-2). In fact, relatively broad limits can be placed on variations of Y, the luminous transmittance, based on results from extensive experimental data [Ref. 9]. These studies show that, in visual color grading of transparent materials, the average viewer will tend to neglect reasonably large differences in luminous transmittance when compared to variations in chromaticity (relative values of x and y).

More specifically, Figure 31-2 is actually a plot of the ASTM D-1500 color scale. Limits for the x and y trichromatic coefficients average approximately ±1%. For Y, the allowable deviation averages approximately ±30%.

When only small differences in luminous transmittance, Y, exist for samples of the chromaticity (x versus y), the single-number standard can be plotted on a simple two-dimensional chromaticity diagram. Examples are the American Public Health Association (APHA) and Saybolt standards.

The best color-match judgment of one analyst is likely to differ from that of another.

The basic premise for a single-number or single-letter color grading system for a product is that the colors of the product lie on, or be within specified limits of, the curve of color standards in color space. As a result, one-dimensional color grading systems have definite limitations and potential pitfalls. If a product discolors and the resulting color lies beyond the specified limits of the standard color curve, the color grading system can no longer be satisfactorily used. For example, the APHA, Hazen, and Saybolt color scales basically grade the yellowness (ratio of blue to green absorbance) of products; if contamination causes the product color to become "greenish," none of these scales will be fully suitable. The best color-match judgment of one analyst is likely to differ from that of another.

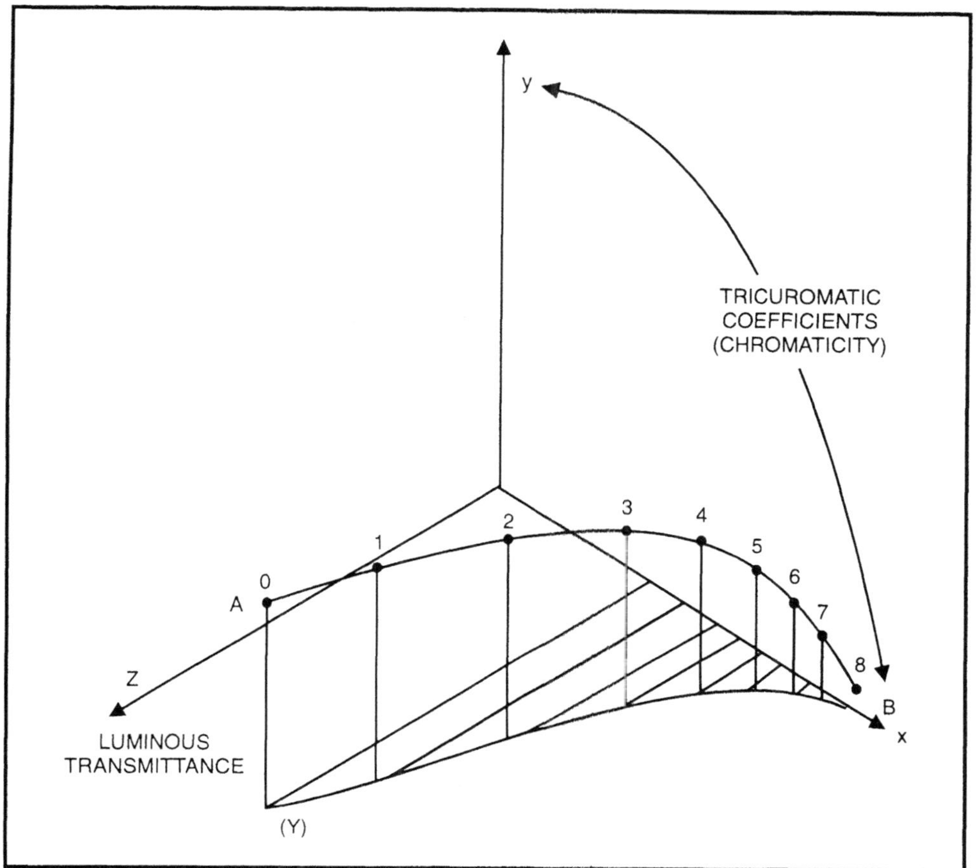

Figure 31-2. Three-Dimensional Plot of Color Standards

"Yellowness" Color Grading Scales

The most widely used color scales for yellowness are APHA and Saybolt. Both scales are based on the relative blue versus green absorbance; however, only the APHA scale, based on linear concentration increases of platinum-cobalt solution standards, is linear with absorbance. Saybolt color, used primarily in the petroleum industry, is measured by the height of a column of product adjusted to color-match that of a standard color-controlled disk; hence, the lighter the product color the greater the Saybolt color is, and the scale is inherently nonlinear with absorbance. Hazen color is similar to the APHA color scale, which has generally displaced it. The International Commission for Uniform Methods of Sugar Analyses (ICUMSA) color scale is used for yellowness color measurements in the sugar industry, and the corn syrup industry also has a yellowness color scale.

Color Scales Covering a Broad Band of Visible Wavelengths

Several materials are very dark brown in the raw state. As these products are refined, they become increasingly lighter, and there is usually a good correlation between the extent of refining and the color grade. In this category falls the refining of crude oil to lubricating or fuel oils, which are graded using the ASTM D-1500 color scale. To provide a linear response to the ASTM color standards, an analyzer must have a measuring channel with a response approximating that of the human eye or "standard-observer." Other similar scales are the Rosin Color (U.S. Official Rosin Color Scale) and the Maple Syrup Color Scale.

Reflected Color Measurement

The on-line measurement of reflected color, used primarily for textile and paper quality control, is a specialized field and will be only briefly be mentioned here. Except for a yellowness measurement, reflected color monitoring almost always requires a tristimulus system (CIE) measurement. It is essential that the light from the source match that of a CIE standard source [Ref. 9] to avoid erroneous results due to factors such as fluorescence.

Software is available for computers which control process spectrophotometers (both scanning and PDA types) to provide tristimulus coefficients from visible wavelength scans.

Ranges of Variables

Path-Length Ranges

Sample cells with path lengths as narrow as 0.001 in. (0.0254 mm) and as long as 6 ft. 6 in. (2.0 m) have been used with UV/Vis process analyzers. The high-intensity light sources and sensitive detectors used in UV/Vis analyzers allow for the use of open path "virtual cells" of many meters in length.

Absorbance Ranges

With the "quiet" light sources and improved power supplies and detector circuits now available in UV/Vis analyzers, full-scale absorbance ranges of less than 0.005 AU are feasible in special cases. Also, with high-intensity UV/Vis light sources and detector circuits with linear response over many decades of light intensity, full-scale absorbance ranges of greater than 4.0 AU (99.99% loss of light intensity) are available from UV/Vis analyzers.

Concentration Ranges

The molecular extinction coefficient (a in the Beer's law formula) of UV and visible absorbing compounds is often several orders of magnitude greater than in the infrared region. Hence, very low concentrations (in the low parts per million and even parts per billion ranges) can be measured with UV analyzers with long sample cells and low absorbance ranges. Even strongly absorbing compounds usually have no high concentration limit. Often, a compound may be measured at a wavelength on the tail of an absorption peak to allow for a reasonably-sized sample cell without an excessively high absorbance range. In a liquid solution, however, the measurement of the concentration may not follow Beer's law because the solution conditions change. For example, when acetaldehyde in water is measured over a range of 0 to 100%, the dilute solution of acetaldehyde is in an almost pure water solvent. The solvent becomes essentially pure acetaldehyde as the concentration approaches 100%, with changing solvent conditions in between.

Temperature Range

High-intensity light sources and sensitive detectors with relatively low temperature coefficients can be used in the ultraviolet and visible region. This allows for the light source, the detector circuitry, and the sample cell to be widely separated. Therefore, the temperature range is limited only by the materials of construction in contact with the sample Limits for both the high and the low temperature usually depend on the elastomeric material used to seal cell windows to the cell. Fluorocarbon elastomers are now available for continuous use at temperatures up to 580°F (320°C). With low sample temperatures, it is often necessary to purge the sample cell compartment with dry air or nitrogen. In some cryogenic applications, UV/Vis process analyzers are used where sample temperatures are as low as –70°F (–40°C).

Pressure Range

The high-pressure limits for samples being monitored with UV/Vis analyzers depend on the pressure ratings of the sample cells and associated equipment. UV/Vis process analyzers have been used to measure samples at pressures greater than 2,000 psig (142 bar). Samples under vacuum conditions are also being monitored with UV/Vis analyzers.

Elements of UV/Vis Absorption Analyzers

LIGHT SOURCES

The light sources used in UV/Vis absorption analyzers can be divided into two broad categories: (1) broadband emission sources and (2) discrete line emission sources. Broadband sources are almost always used in process UV/Vis spectrophotometers (dispersive analyzers) and increasingly in nondispersive UV/Vis analyzers with narrow band interference filters. Long-life gas discharge lamps emitting discrete lines (fixed wavelengths) are the most common light sources for ultraviolet analyzers.

BROADBAND EMISSION SOURCES

Deuterium discharge lamps are the most commonly used broadband sources in UV process analyzers and have almost completely displaced the less expensive but shorter-life *hydrogen discharge lamps*. The deuterium lamp allows any wavelength in the UV region to be selected but has a relatively short life, relatively low emission at any given wavelength, and requires an expensive power supply. Situations requiring relatively narrow bandpasses of wavelengths usually call for a highly sensitive detector, such as a photomultiplier or a charge-coupled photodiode array (CCD), as well as extraordinary precaution against stray light.

Pulsed xenon lamps are increasingly used as the broadband light source of choice in process UV analyzers. They have the same advantage as deuterium lamps, but with a relatively long life and high-intensity (short-time) pulses. Averaging of pulse intensities is often required for high-sensitivity (low-noise) operation. Because of the large differences in intensities at various wavelengths within the UV and visible region from the xenon lamp (as well as from the deuterium lamp), the detectors used with these lamps must have a linear response over many decades of intensity.

Tungsten lamps have broad emission in the visible and near-IR region, but the emission drops rapidly from the blue into the near-UV region. Nonetheless, since this lamp is inexpensive, has a very long life, and uses an inexpensive power supply, it is a popular choice for visible and very-near-IR analyzer applications.

For near-UV (wavelengths greater than 340 nm) as well as the visible and near-IR regions, *tungsten-halogen lamps* are often used as light sources for analyzer applications. This type of lamp has a quartz envelope and an incandescent tungsten filament with iodine doping (cycle). It generally has a long life, but for appreciable emission in the near-UV region, the lamp must be operated at or above rated voltage, and lamp life is greatly shortened.

GAS DISCHARGE LAMPS FOR DISCRETE LINE EMISSIONS

Gas discharge lamps with extremely narrow "spikes" or lines of emission are favored for process ultraviolet analyzers. For the most part, these lamps have long life expectancies, and the discharge lines are highly intense, usually three to six orders of magnitude above the background emission. Each discharge emission line is fixed and unchanging and is the exact wavelength, regardless of any shifts in the isolation filter(s). "Half-band widths" for discharge lines are meaningless since even the one-tenth bandwidth of a discharge line is usually less than 0.01 nm!

Mercury discharge lamps are favorite sources because of their long life and ready availability. Low-pressure mercury lamps have a highly intense and extremely narrow emission line at the mercury resonance wavelength of 253.7 nm. This lamp is used for monitoring not only mercury but also ozone, aromatic compounds, and others.

Medium-pressure mercury discharge lamps permit high-intensity emission of several mercury lines, including 265.2, 275, 280, 289, 297, 302, 313, 334, 365, 405, 436, 546, and 578 nm. The life expectancy of this lamp is exceptionally long; some have operated continuously for more than 7 years.

Zinc discharge lamps are used primarily where the 214 nm zinc line is the measuring wavelength.

Cadmium discharge lamps find application where the 228-nm discharge line is used as the measuring wavelength.

Techniques for Isolating Wavelengths

In process UV/Vis spectrophotometers, gratings are used almost universally for wavelength isolation, although sometimes prisms are also used. Interference filters and combinations of blocking and/or bandpass filters may be used to isolate discrete wavelengths in filter photometric analyzers.

GRATINGS

Most gratings now used in process spectrophotometers are holographically formed. These holographic gratings have this advantage: when light focused on a narrow slit passes through and strikes the grating, the dispersed light from the grating is linear with respect to wavelength.

INTERFERENCE FILTERS

Advances in thin-film technology have resulted in the availability of narrow-band, high-transmitting, excellent blocking, and relatively inexpensive interference filters in the visible and near-IR region. The shorter the wavelength, the thinner the dielectric layers required to make the filter; hence, UV interference filters are the most difficult to fabricate. Also, UV transmitting and stable dielectric materials are not readily available. Therefore, UV interference filters (particularly at the shorter wavelengths) are comparatively expensive. Stable interference filters with half-band widths of less than 2 nm are available and are being used in analyzers throughout the UV region. Since the transmittance peak of an interference filter will shift with the angle of incidence of the light striking it, the light striking the filter should be collimated at a zero angle of incidence (perpendicular to the face of the filter) to meet the filter specifications.

COMBINATIONS OF BANDPASS AND BLOCKING FILTERS

For isolation of spectral lines from a gas discharge lamp or for a relatively broad band of wavelengths, combinations of bandpass and/or blocking filters often provide the least expensive and most stable means of wavelength isolation. These filters may be in the form of special glass, special films, or sealed quartz bottles of gas or liquid. There must be no "windows" of transmittance with the filter combination where the detector responds.

Detectors

In general, UV/Vis detectors have high sensitivity, linear responses to light intensity over several decades, and comparatively low temperature coefficients. For process scanning spectrophotometers, photomultiplier tubes are universally used, and process photodiode array (PDA) analyzers use linear photodiode array detectors (LPDADs). Silicon photodiodes have become the detector of choice in most filter photometric analyzers, although a large number of analyzers in the field still utilize vacuum phototubes as detectors.

SILICON PHOTODIODES

The silicon photodiode is a light-sensitive semiconductor diode with a reverse bias applied, making it a photoconductive detector with a very fast response, high sensitivity, and good stability. The UV-enhanced silicon photodiodes can be used down to 200 nm, but are not as sensitive or as stable in the UV as in the visible region. The response extends into the near-IR, cutting off at about 1,100 nm. With the proper circuitry, this detector is highly linear with light intensity and can receive high light intensity levels without damage.

VACUUM PHOTOTUBES

The vacuum phototube uses the photoelectric effect to provide a current proportional to the light intensity (at a given wavelength) striking the tube's cathode. If the light intensity is high enough so that the current generated by the light is at least two orders of magnitude greater than the phototube dark current, the phototube current will be highly linear with intensity. As with the silicon photodiodes, very high levels of light can strike the phototube without causing damage or "fatigue." In general, the phototube with UV response (S-5 cathode surface) has an indefinitely long life, relatively low temperature coefficient, and high stability.

PHOTOMULTIPLIER TUBES (PMTs)

The photomultiplier tube offers the most sensitive detection means for UV and visible light (although charge-coupled LPDADs are now approaching the same sensitivity). However, precautions must be taken to avoid high light levels, such as room light, from striking the photomultiplier when voltage is applied. Excessive light can cause severe fatigue or even permanent damage to the photomultiplier. As with the phototube, the photomultiplier uses the photoelectric effect to provide electron flow from the tube cathode to an electrode at higher voltage. The impact of electrons causes the electrode to emit a much greater number of electrons to a similar electrode at an even higher voltage. This process of electron flow multiplication goes through several such stages, resulting in a final electron flow (current) that is orders of magnitude greater than the first stage. The photomultiplier requires a high multistage voltage supply and is sensitive to voltage variations. It also has a relatively high temperature coefficient.

BARRIER LAYER (PHOTOVOLTAIC) CELLS

The barrier layer cell is a semiconductor-diode light detector that produces a relatively high current signal as a function of light intensity. Because a barrier layer cell does not require a voltage supply, and its signal can be fed directly to an output meter, it provides one of the least expensive detection systems for UV and visible monitoring. Unfortunately, the barrier layer cell is temperature sensitive, nonlinear, and slow in response.

LINEAR PHOTODIODE ARRAY DETECTORS (LPDADs)

The linear photodiode array detector has become popular for process spectrophotometers because it allows a full spectrum to be provided without mechanical scanning. The LPDAD consists of a monolithic semiconductor chip into which is cast a linearly-spaced series of semiconductor photodiodes. Each photodiode produces a current output as a function of the light striking it. As configured in a process analyzer, each photodiode in the array sees a separate wavelength and provides an output for that wavelength. Commercial LPDADs are available with 320, 512, and 1,024 diodes in the array. Arrays with even greater numbers of diodes will be available when this handbook is published. It should be noted, however, that the greater the number of diodes, the more complex will be the electronic board(s) required to provide the raw transmittance and/or absorbance spectra which must be interfaced with a more powerful computer. The charge-coupled CCLPDAD has sensitivity orders of magnitude greater than the standard LPDAD and can be used at light levels almost as low as for a photomultiplier. Both the discrete LPDAD and the discrete CCD have low temperature coefficients and provide highly linear outputs with light intensities.

Types of Process UV/Vis Analyzers

Several process analyzers based on ultraviolet and visible absorption are commercially available. These instruments range from very simple to highly sophisticated designs.

Single-Beam Filter Photometric Analyzers

The simplest design is the single-beam single-wavelength filter photometric analyzer; its functional diagram is essentially the same as the general analyzer shown in Figure 31-1. The output of this type of analyzer will be affected by fluctuations and drift of the light source, dirt or bubbles in the sample cell, and any drift in the detector or detector circuit. As a result, single-beam instruments must operate on relatively low-sensitivity levels to provide reasonably stable analyses. Stabilized light sources and improved detector circuits have improved stability to some extent, and these simplified units have found extensive use in applications requiring only a "go/no-go" or broad-range measurement. As a detector for colorimetric reaction system analyses, a single-beam unit may prove adequate because of frequent automatic "zero" settings.

Dual Wavelength Filter Photometric Analyzers

There are two basic types of dual-wavelength analyzers: (1) the split beam photometric analyzer and (2) the filter-wheel chopper photometric analyzer. Both designs are based on a differential measurement at two wavelengths and have several distinct advantages, including compensation for dirt and bubbles in the sample cell. A glass filter with absorption characteristics equivalent to a selected concentration of the component to be measured can be placed in the common beam for a calibration check.

SPLIT-BEAM PHOTOMETRIC ANALYZERS

A functional diagram of a typical split-beam analyzer is shown in Figure 31-3. Light from the source is partially absorbed in passing through the sample. The light leaving the sample is divided into two beams by a semitransparent mirror. Each beam passes through an optical filter to a detector, usually a phototube or a photodiode. The filter removes all wavelengths except those to be measured. The ratio of the transmittance-to-reflectance of the beam splitter is selected so that the light intensities striking each detector will be reasonably close during normal operation of the analyzer.

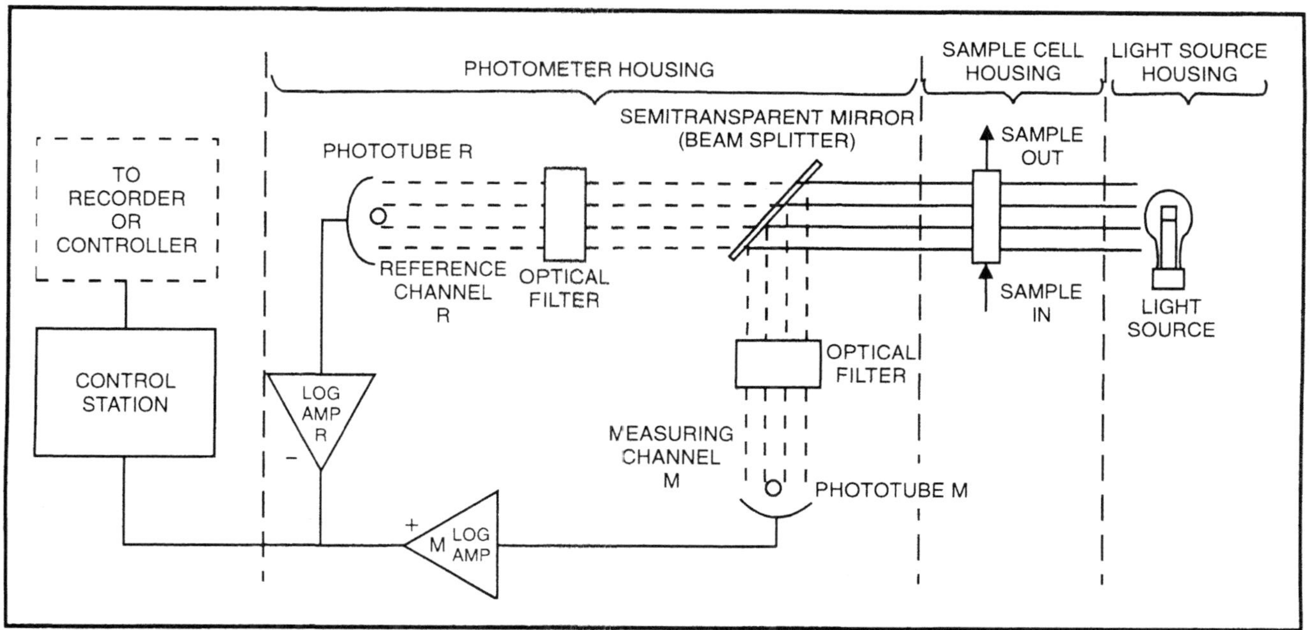

Figure 31-3. Functional Diagram of a Split-Beam Photometric Analyzer

Light at the analytical wavelength striking one of the detectors is absorbed comparatively strongly by the component whose concentration is being measured. Light at the reference wavelength, directed to the second detector, is absorbed weakly or not at all by the measured component. Each detector develops a current directly proportional to the light intensity striking it. This current is converted to a voltage proportional to the negative logarithm of the detector current by a logarithmic amplifier. The output voltages from the analytical and reference amplifiers are subtracted to produce a final output voltage proportional to the difference in the logarithm of the detector current. Going back one step, the output will also be proportional to the difference between the logarithms of the intensities of the reference and analytical beams. Based on Beer's law, the difference in the logarithms of the intensities is directly proportional to the concentration of the measured component. Hence, the analyzer output will vary linearly with the concentration.

Based on Beer's law, the difference in the logarithms of the intensities is directly proportional to the concentration of the measured component. Hence, the analyzer output will vary linearly with the concentration.

FILTER-WHEEL CHOPPER PHOTOMETRIC ANALYZERS

A functional diagram of a typical filter-wheel chopper photometric analyzer is shown in Figure 31-4. The chopper motor rotates the filter-wheel reference and analytical filters alternately and continuously into and out of the optical path. The source lenses (L_1, L_2, L_3) collimate the source light through the sample cell, and the detector lens (L_4) focuses the transmitted light to the detector. As with the split-beam analyzer, the light at the analytical wavelength is absorbed strongly by the component to be measured and the light at the reference wavelength is weakly absorbed or not at all. The detector provides two electrical signals that are proportional to the intensities at each wavelength and go through electronic circuitry with logarithmic conversion to linearize the output based on Beer's law.

Dual-Beam Filter Photometric Analyzers

The dual-beam analyzer using only one wavelength is fundamentally a more stable design than a single-beam dual-wavelength analyzer from an instrumental standpoint, provided that it has equivalent stability of optical and electronic components. However, it does not compensate for dirt or bubbles in the sample, nor can it compensate for potentially-interfering components in the sample.

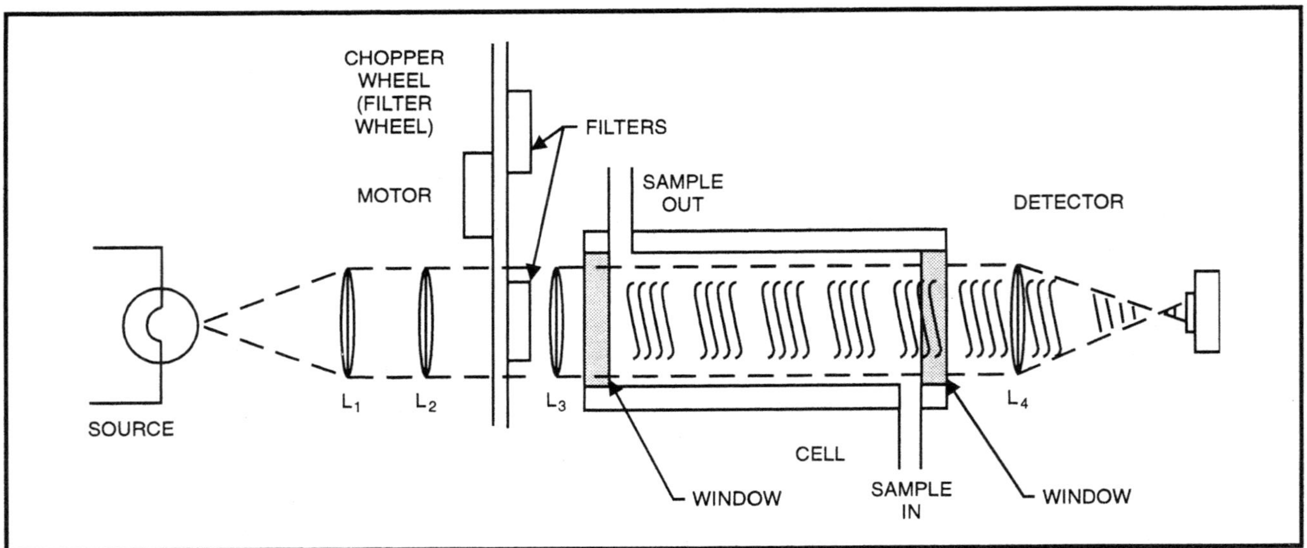

Figure 31-4. Functional Diagram of a Dual-Wavelength Analyzer with Chopper

Figure 31-5 is a functional diagram of a dual-beam analyzer. The optical and electronic components are basically the same as for the dual-wavelength analyzers, with the electronic circuitry performing the same functions. In the dual-beam analyzer, the measuring wavelength is usually isolated with an optical filter after the source light is collimated, and the selected wavelength of light is split into two beams before the sample, either by using a mirror and an open hole in place of filters in the equivalent of a filter wheel or with a beam splitter if two detectors are used. Mirrors are also used to direct the reference beam to the detector.

One commercially available analyzer has a *combination dual beam with split-beam dual-wavelength detection of each beam.* Theoretically, the combination will provide the photometric stability advantage of the dual-beam analyzer with the compensation features of the dual-wavelength measurement. However, four well-matched detectors are required.

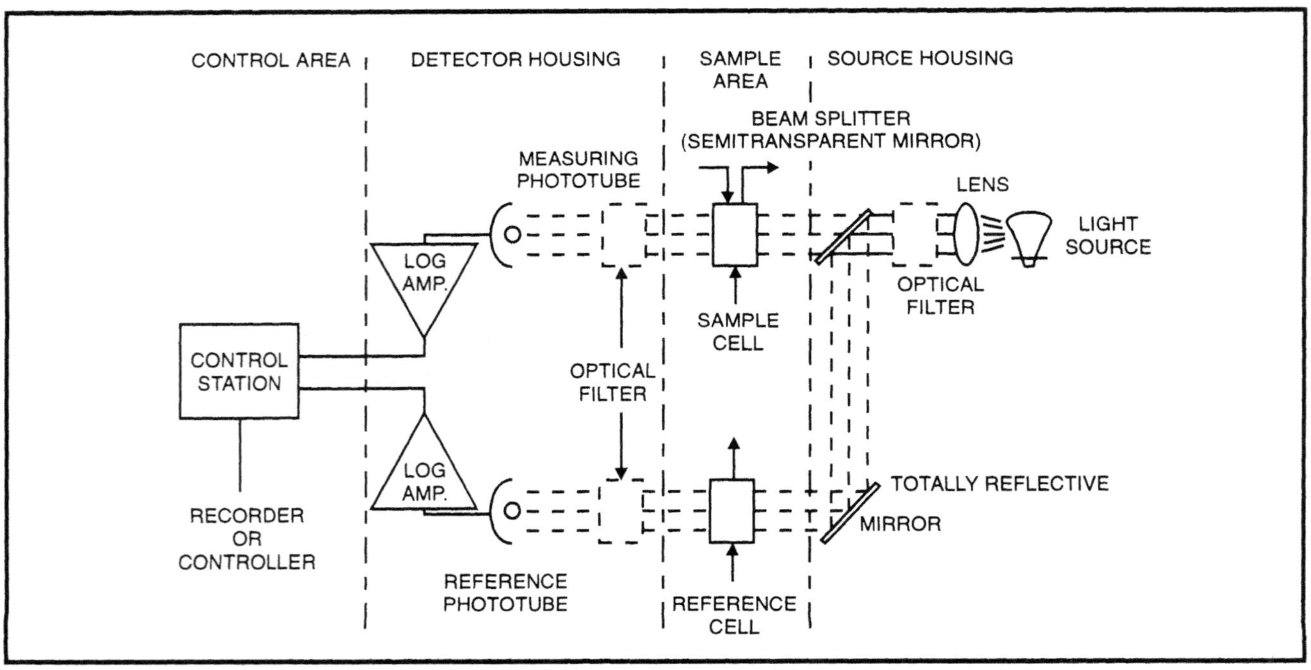

Figure 31-5. Functional Diagram of a Dual-Beam Analyzer

Process Spectrophotometers

There are two main types of process spectrophotometers: (1) the process scanning spectrophotometer and (2) the process photodiode array (PDA) analyzer. The scanning unit is almost always located in a safe, clean area and uses fiber optics to couple the analyzer to the sample. The PDA analyzer is usually rugged enough to mount in the plant atmosphere and can be built to meet the electrical classification of the area; moreover, it too can be provided with fiber optics. All process spectrophotometers (PDA and scanning) are capable of multicomponent analyses, a distinct advantage.

PROCESS SCANNING SPECTROPHOTOMETERS

A functional diagram of a typical process scanning spectrophotometer for UV and visible absorption is shown in Figure 31-6. The light is almost always a deuterium lamp and may have a tungsten-halogen lamp as a second source. A photomultiplier tube is usually the detector. Light from the source is focused on the entrance slit and directed to the diffraction grating by the combination of a flat mirror and a parabolic mirror. The grating disperses the light with respect to wavelength, the relative position of the grating determining the wavelength passing through the exit slit. (The width of the exit slit governs, for the most part, the resolution, or bandpass, of the instrument.) Light through the exit slit is focused on to a fiber-optic cable that transmits the light to the sample cell. A collimating lens terminating the cable directs the light through the sample cell to a second cable, also with a terminal lens, which focuses the light transmitted through the cell to the fiber-optic cable to pass the light on to the photomultiplier detector. Wavelength scanning is achieved by mechanically moving one of the elements, usually the grating, in the monochrometer. Although the scanning is very rapid in some of the process scanning spectrophotometers, time-related changes in a flowing sample may introduce errors.

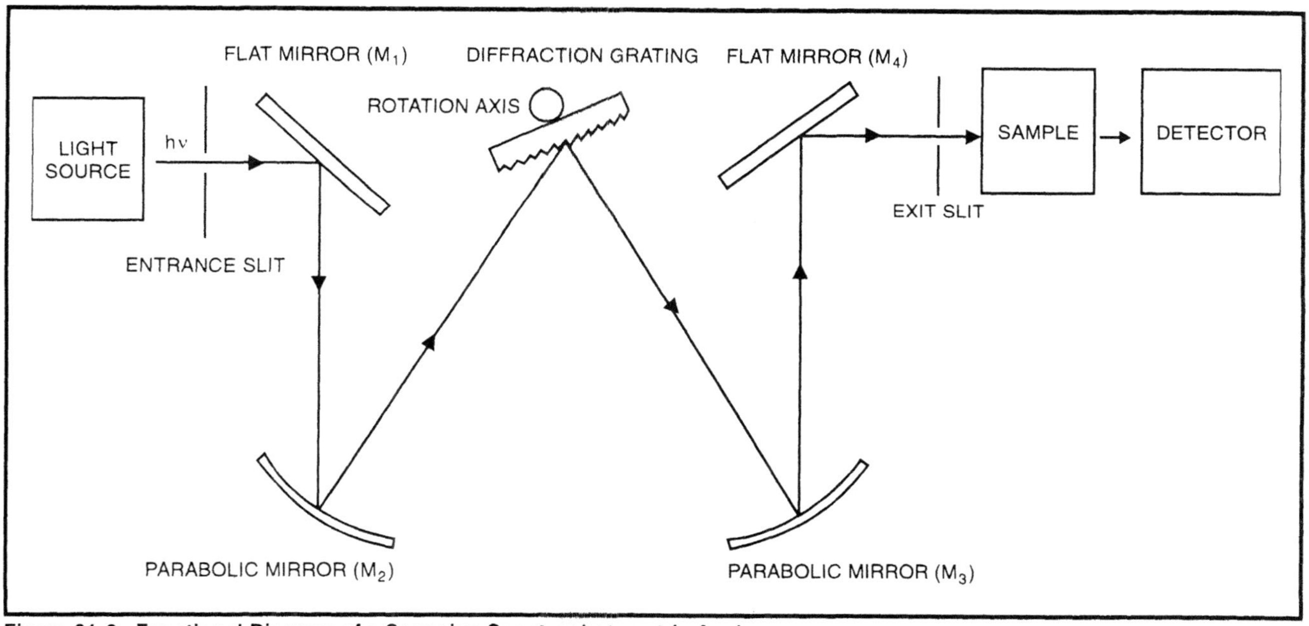

Figure 31-6. Functional Diagram of a Scanning Spectrophotometric Analyzer

PROCESS PHOTODIODE ARRAY (PDA) SPECTROPHOTOMETERS

A functional diagram of a process photodiode array (PDA) spectrophotometer is shown in Figure 31-7. Usually either a deuterium or a pulsed-xenon lamp, each of which emits a broad spectrum of light in the UV and visible regions, is used as the light source. The light is focused with lenses and/or mirrors directly or through fiber optics to the sample cell. The light emerging from the cell either directly or through fiber optics passes into the detector assembly, where the light is focused onto the slit and strikes a holographic grating that disperses the light onto the LPDAD.

The LPDAD is precisely positioned with respect to the grating so that each diode sees light *x* nm away from adjacent diodes, where *x* is usually selected to be between 0.2 and 2.0 nm. The higher the resolution (the smaller the *x*), the more applicable the instrument is for analyses of compounds with fine-structured absorption characteristics, as in vapor phase and simple aromatics. However, high resolution limits the wavelength span of the analyzer. Hence, in applying a process photodiode array (PDA) analyzer, a compromise may be required to attain adequate resolution over a broad enough spectral region for the application.

The photodiode array (PDA) spectrophotometer has no moving parts (except for a chopper when a deuterium lamp is used) and sees all wavelengths at essentially the same time. Thus, it is well-suited for process analyses.

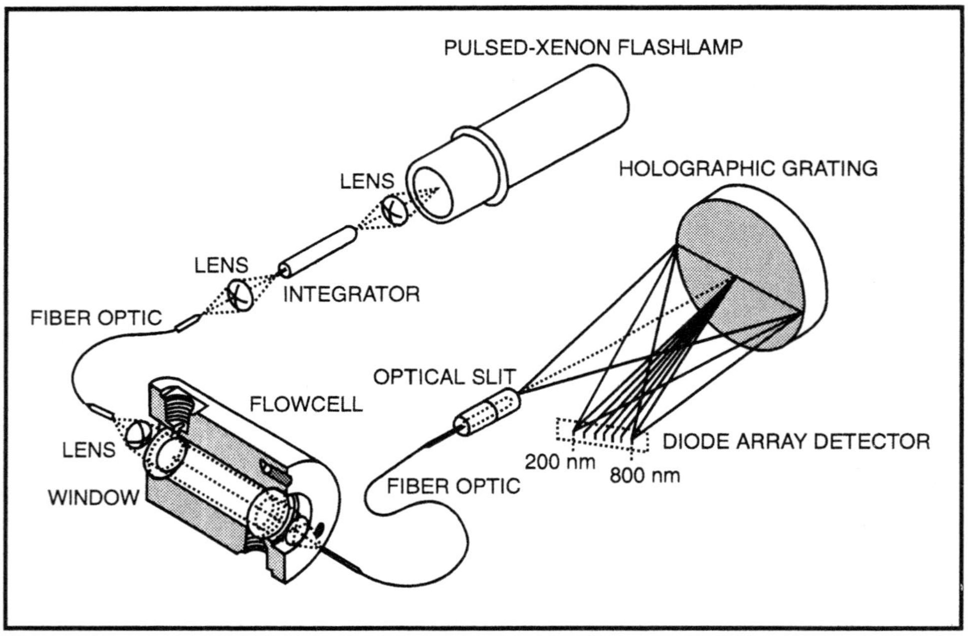

Figure 31-7. Process Diode Array Analyzer with Fiber Optics

Limitations of Process UV/Vis Analyzers

Because of the relatively high photoelectron energy associated with UV and visible regions of the electromagnetic spectrum and because ordinary atmospheric gases and water do not absorb in this region, many of the limitations for infrared analyzers do not pertain to the UV/Vis analyzers. However, there are a few specific limitations:

1. Potential process sample photochemical reactions activated by the light source
2. Excessively high absorbance
3. Window coating
4. Limited UV/Vis absorbance
5. "Solarization" of quartz windows and fiber optics

Potential Photochemical Reactions Activated by the Light Source

The UV and short-wavelength visible photoelectron energies may be at levels that initiate photochemical reactions. Some examples:

1. *The classic chain reaction of chlorine reacting with hydrogen to form hydrogen chloride may initiate when chlorine is activated by light in the 400-nm region. Similarly, chlorine may be activated to react with carbon monoxide to form phosgene ($COCl_2$). With samples containing relatively high concentrations of chlorine, hydrogen, and carbon monoxide, wavelengths in the 400-nm region should be blocked from striking the sample.*
2. *Wavelengths below 430 nm can initiate the decomposition of compounds such as chlorine dioxide (ClO_2) and nitrogen dioxide (NO_2). The decomposition of ClO_2 can be explosive!*
3. *UV-absorbing monomers can polymerize when exposed to UV light. An example is the polymerization of styrene to polystyrene. Polymer coatings on sample lines and on cell windows can be a maintenance headache.*

The decomposition of ClO_2 can be EXPLOSIVE!

Excessively High UV Absorption

In liquid streams, the measurement of high concentrations of strongly absorbing compounds in the UV region with a reasonable path-length sample cell (greater than 0.05 mm) may not be feasible since the absorbance may exceed 6 AU or higher (99.9999% of the light absorbed). Using a measuring wavelength on the tail of an absorption peak may in some cases provide a solution to this problem.

Sample Cell Window Coatings

Quartz is usually the preferred material for sample cell windows in the UV region because of its high UV transmittance and relatively low cost compared to other UV-transmitting materials, such as synthetic sapphire and magnesium fluoride. However, quartz, an excellent insulator, can have a static electric charge buildup on its surface, attracting deposits that coat the surface. These coatings can result in analyzer zero shifts and can scatter light enough to decrease the light intensity to an unusable low level. Where a spectrophotometer is used for multicomponent analyses, even a slight coating can ruin a calibration based on chemometrics or some other correlation technique. "Purged" windows, discussed later in this chapter, may sometimes be used to eliminate a coating problem.

Limited Number of UV/Vis Absorbing Compounds

In contrast to the ubiquitous infrared-absorbing compounds, the number of UV/Vis-absorbing compounds is quite limited. Saturated hydrocarbons, simple straight-chain alcohols, as well as the atmospheric gases and water do not have UV or visible absorption. Also, most UV-absorbing compounds have relatively broad absorption peaks (particularly in liquid phase) and do not have the "fingerprint" characteristic of infrared absorption.

"Solarization" of Quartz Windows and Fiber Optic Cable

When exposed to UV light over a period of time, quartz with only the slightest impurities will "solarize," a phenomenon observed as a decrease in UV transmission, particularly at the short wavelengths. This problem can usually be solved by periodically "zeroing" the analyzer. However, fiber optics near the light source may need to be replaced after an extended exposure if wavelengths below 230 nm are used for the analyses.

Case Histories

Thousands of UV/Vis analyzers are applied throughout the process industries. This section will review case histories taken from only a few of the most common of these applications:

1. Monitoring SO_2 and/or NO_x concentrations in stack gases for environmental control
2. Monitoring H_2S, SO_2, CS_2, and COS in sulfur recovery tail gas for precise control of the recovery process
3. Monitoring the color of light hydrocarbon products in petroleum refineries
4. Monitoring chlorine concentrations in off-gas from a chlorination process
5. Monitoring the ClO_2 concentration in water for bleaching of pulp

Monitoring SO_2 and/or NO_X in Stack Gases

The reduction of SO_2 and/or NO_x emitted from stacks in steam-generating plants, refineries, chemical plants, cement plants, steel mills, and smelters is a major objective of environmental legislation. UV absorption is the most common detection technique for monitoring SO_2 concentrations at stack gas levels and is also used extensively to monitor NO_x levels in stack gases. Levels over the range of 0 to 100 ppm and higher are usually required for both pollutants.

In the early 1970s, copper smelters were among the first to install UV analyzers to monitor stack gas SO_2 levels, in the range of 0 to 20%. One of the original installations in a copper smelting plant and subsequent modifications as SO_2 levels were lowered will be used for the case history presented here.

When the first UV analyzer was installed in an Arizona copper smelting plant, SO_2 concentrations in the stack were in the 10 to 20% range, with heavy loading of particulate matter, and essentially no vegetation grew in the vicinity of the plant. The UV analysis system was selected because there was no interference from the high moisture levels, the sample gas could be maintained at an elevated temperature (above 212°F [100°C]) and the analysis technique could withstand a limited amount of particulate matter in the sample gas.

(continued)

(continued)

Because of the high SO_2 concentration, the usual measuring wavelength of the 280-nm Hg line was abandoned in favor of the 313-nm line, where SO_2 has a very strong peak absorption. A short sample cell path length (1.5 in.) was selected to give an absorbance range of close to 0 to 1, a highly stable range with excellent linearity. The probe initially was a 1/2-in. NPT stainless steel pipe with a large internal filter element of sintered carborundum, positioned about halfway between the wall and center of the stack. The sample gas transfer line was maintained at 220°F (105°C), and the cell was pressure-regulated using a vacuum break/air aspiration system at 11.7 psia (at the altitude of this installation, the atmospheric pressure averages 13.2 psia). A hot-air blowback was automatically applied through a pneumatically-operated ball valve, not only to blow any accumulation of particulate matter off the filter element but also to provide a "zero" gas for automatic zero adjustment of the analyzer.

Despite the harsh plant conditions, the original photometric analysis system, with relatively minor changes (i.e., changing the measuring wavelength and installing a longer sample cell to accommodate lower levels of SO_2), operated essentially maintenance-free for more than 20 years. Sample cell windows were cleaned occasionally, and only three replacement lamps were required over two decades of operation. A new system, essentially the same as the original but with updated electronics, a microprocessor, and improved ball valves, was purchased as part of a large environmental improvement project in the early 1990s.

Monitoring of H_2S, SO_2, CS_2, and COS in Sulfur Recovery Tail Gas

One of the largest applications of UV analyzers is the monitoring of H_2S and SO_2 in a Claus process tail gas line for precise process control [Ref. 10]. In the Claus process, air is ratio-flow-controlled with the process feed gas, rich with H_2S, into a furnace to oxidize one-third of the H_2S to SO_2 to attain a 2:1 H_2S/SO_2 ratio to feed to the Claus catalyst (usually a form of aluminum oxide), where the H_2S reacts with the SO_2 to yield sulfur in accordance with following reaction:

$$2H_2S + SO_2 \rightarrow 3S\downarrow + 2H_2O$$

For maximum yield of sulfur and minimum sulfur compounds in the process tail gas going to an incinerator or to a cleanup process, the stoichiometric 2:1 H_2S/SO_2 ratio must be closely-controlled. Rapid, accurate, and reliable analysis for the H_2S and SO_2 (and in certain new processes, COS and CS_2) concentrations offers the best tool for precise control. UV analyzer systems are highly successful in this application.

(continued)

(continued)

Sampling of the tail gas presents several difficult problems, since the gas always contains sulfur vapor and often liquid sulfur carryover. Sulfur solidifies and will deposit on surfaces at temperatures below about 250°F (120°C). Sulfur also polymerizes at approximately 350°F (177°C), so sample system components must be maintained within a relatively narrow temperature range. Also, sulfur vapor has a strong UV absorption; hence, sampling systems for tail gas analyzers will almost always have either a demister or condenser operated at about 260°F (125°C) to drop the sulfur vapor level without sulfur solidification. The sample is generally heated to 300°F (150°C) for the UV analysis and is pulled through with a steam jet. Periodically, the sample flow is blocked to allow any accumulated sulfur liquid in the demister or condenser to drain and to air purge the sample cell for window cleaning and "zero" checking and adjustment if required. Often, in refineries, where part of the feed is "sour water stripping gas," ammonium-sulfur compounds with sublimation temperatures close to 300°F (150°C) are in the tail gas. In such cases, steam blowback may be required to remove these compounds.

When filter photometric analyzers are used for the tail gas analyses, several wavelengths are necessary to provide measurements for H_2S and SO_2 and to compensate for sulfur vapor interference. With process PDA analyzer systems providing a full spectrum, low levels of COS and CS_2 can also be measured using special chemometrics software.

Although sulfur-recovery plants have been using UV tail gas analysis systems since the early 1970s, some refineries resisted installing these units until required in the early '90s. When one of these plants finally installed a UV tail gas analysis system with closed-loop control, the increased sulfur yields were so great that a new tank car had to be purchased!

Monitoring the Color of Light Hydrocarbons in Petroleum Refineries

Monitoring the Saybolt color (pale yellow) of light hydrocarbons is an important quality measurement and usually requires a simple sample system. Within a week after the installation of such a system in one refinery, the analyzer output pegged full scale. At first the operators thought that the analyzer was defective, but then decided to doublecheck the product with a manually drawn sample, which was dark brown! They quickly found a heavy hydrocarbon valve accidentally opened to the product line. The savings resulting in this rapid correction of the problem were estimated to be greater than $300,000, about ten times the cost of the installation.

Monitoring of Chlorine Concentrations in Off-Gas of a Chlorination Process

In the manufacture of certain fluorochlorocarbon compounds, the concentration of chlorine in the off-gas is a direct measure of the catalyst status, a key for process control. The off-gas consists of HCl, HF, fluorochlorocarbon compounds, and Cl_2 at a pressure of 240 psig. The stream is highly corrosive, and the only metal acceptable for the wetted surfaces of the Cl_2 analysis sampling system is Hastelloy C. Since HF attacks quartz, sample cell windows are synthetic sapphire with Teflon gaskets. Pressure reduction is extremely difficult and sometimes unreliable. The solution to sampling problems is to operate at the controlled process pressure. The sample cell is made of heavy Hastelloy C with 3/8-in.-thick sapphire windows, pressure tested to 500 psig. At the 240 psig operating pressure, HF is partially dimerized to H_2F_2, which decreases the total molar concentration. Nonetheless, this analyzer system has been highly successful in closed-loop control and is considered essential for successful plant operation.

Monitoring ClO_2 in Water for Bleaching of Pulp

A highly successful application of UV/Vis photometric analyzers is the monitoring and control of ClO_2 concentration in water from ClO_2 generators to feed pulp bleach lines in paper plants. ClO_2 is highly corrosive, and sample wetted materials include titanium, fluorocarbons, and ceramics. Because of the strong absorbance of ClO_2, a short-pathlength sample cell is used. The usual measuring wavelength is the intense 436-nm Hg line, which allows for a high absorbance range with good linearity. One of the first of these installations successfully operated for more than 25 years even though the analyzer was installed on an outside wall of the ClO_2 plant.

Installation Hints

In general, UV/Vis analyzers are ruggedly constructed and can withstand hostile environments; however, they should be installed in housings or at least be shielded from direct sunlight and from rain. An inside installation will be particularly appreciated by the technician responsible for successful analyzer operation and maintenance. Other installation suggestions are discussed in the following paragraphs.

In a plant where vibration from large pumps or motors is noticeable, the analyzer and sampling system should be mounted on shock mounts. Reduction of vibration not only will result in more stable analyzer performance, but also will prolong the life of the light source and prevent tube connections from loosening in the sampling system.

If the analyzer is installed in an enclosed area and the sample is corrosive, toxic, and/or explosive, the sample flow to the analyzer should be limited and any high bypass flow should be kept outside the enclosed area. As an additional safety measure, the cell pressure should be regulated below atmospheric if possible.

Because analyzers with a long-path-length sample cell often occupy substantial horizontal wall space, the impulse may be to mount the analyzer vertically for space efficiency. However, some lamps must be mounted in only one direction for long life and low noise. Also, dirt will tend to deposit on the lower cell window in a vertical orientation. Thus, vertical mounting is unacceptable.

If the control station of the analyzer is mounted in a location remote from the field unit, a local indicator at the field unit would be very helpful to the instrument technician. Extra electrical leads in the connecting cable can be provided at a small incremental cost during initial installation and can prove useful for future needs.

Since UV/Vis analyzers often handle corrosive streams, a sample cell constructed of the best corrosion-resistant materials within reasonable cost differentials is generally found to be cost-effective in the long term. Process equipment may be made with inexpensive materials, such as black iron, that will almost always corrode, albeit at an acceptable rate. Such rates may be highly damaging for the analysis system, particularly in applications that involve measurement of low concentrations.

For liquid streams, the preferred sample cell flow pattern is to enter at the bottom and exit from the top of the cell to allow gas bubbles to be carried away with liquid sample. Flow in the opposite direction can cause gas not only to be trapped in the top of the cell but also to accumulate to such an extent that the gas phase intercepts part of the light beam. For gas streams, the flow should exit from the bottom of the cell to allow any solids or liquids to be carried away with the sample gas. The gas inlet can be either top or bottom. In installations where the gas sample outlet is on the top of the sample cell, it is not uncommon to eventually find the cell filled with liquid.

For sub-ambient temperature liquid samples, it often necessary either to heat the sample or to provide a dry air or N_2 purge for the sample cell compartment to prevent moisture in the atmosphere from condensing on the outsides of the windows. The analyzer system for ClO_2 is an example of such a condition.

Teflon tubing is excellent for sample lines because of its high-temperature rating and its excellent corrosion resistance. However, it is translucent enough to allow room light into the sample. Photochemical reactions activated by room light can occur, such as the potentially explosive H_2 and Cl_2 reaction to form HCl or the decomposition of ClO_2 to Cl_2 and O_2. Where there is any possibility of a photochemical reaction, the Teflon tube should be shielded from light by placing it within an opaque tube or even wrapping it with black vinyl tape.

Photochemical reactions activated by room light can occur, such as the potentially explosive H_2 and Cl_2 reaction to form HCl or the decomposition of ClO_2 to Cl_2 and O_2.

Heating of a gas sample may be necessary to prevent condensation, and the temperature, as well as the pressure, must be regulated since the moles of gas per unit volume is directly proportional to the absolute pressure and inversely proportional to absolute temperature (see the earlier discussion of Beer's law). If the gas cannot be heated because of its thermal instability (i.e., ClO_2), special traps may be required in the sample system.

Since many UV/Vis analyzers are used for closed-loop control, fast response is often essential. The largest single contributor to delay in analyzer response is sample lag, i.e., the time it takes for the sample to reach the sample cell from the sample tap. Ideally, an in-situ measurement, with the analyzer viewing the sample directly in the process pipe, eliminates sample lag, but in-situ analyzers pose other problems and are not nearly as successful as the concept promises. With an extractive sampling system, the analyzer should be installed as close as possible to the sample tap to minimize sample lag. In any case, a high-flow sample-bypass loop may be incorporated in the sampling system to decrease the effective sample lag.

Specific Safety Issues for UV/Vis Analyzers

Many of the safety issues unique to UV/Vis analyzers are associated with the intense light sources that emit UV light which can damage the eyes if looked at directly without protective eyewear. The low-pressure mercury discharge lamp can be deceivingly dangerous, since it emits relatively little visible light but has an extremely intense 253.7-nm discharge line; this is not visible to the human eye but can cause severe UV burns if not blocked by proper glasses. The most damaging UV light from a medium-pressure mercury lamp is blocked by safety glasses, but near-UV and visible light from this type of lamp can cause the equivalent of welder's burn to sensitive eyes and "sunburn" to skin exposed for a length of time (1-2 hrs).

As mentioned in the previous section, UV and visible light can activate explosive photochemical reactions if precautions are not taken. Because high levels of toxic and explosive compounds are often present in sample streams monitored by UV/Vis analyzers, analyzer shelters should be well-ventilated and equipped with toxic and flammable gas alarms to protect personnel from exposure to possible process sample leaks.

Samples also may be at highly elevated temperatures. Precautions must be taken by personnel working on the analyzer systems to avoid contact with hot surfaces that can cause severe burns. As with any analyzer in an area with an electrical hazard classification of Division 1 or 2, clearance from Operations must be obtained before power is applied to an analyzer when its explosionproof housings are open or when the purge system has been bypassed.

In general, an analyzer sampling system (including the sample cell) is an extension of the process equipment and plant safety precautions should be carried over to this system. "Booby traps" can be easily overlooked in a sampling system, particularly on dangerous corrosive streams. The design should be reviewed specifically for potential safety hazards.

Unique Sample System Requirements

The fact that moisture need not be removed from a gas stream for a UV or visible analysis means that the sampling system for UV/Vis analyzers can be less complex than many other types of analyzers. The sample can be maintained wholly in the vapor phase and special condensers are not required. It also means that the sample usually must be heated and regulated at an elevated temperature.

Many UV/Vis analyzers are used for closed-loop control, and thus speed of response is often essential. If a gas at a high process pressure is monitored at a reduced pressure by a UV/Vis analyzer, pressure reduction should be made at the sample tap so that the sample line will transport the gas at the reduced pressure, decreasing the sample line lag (based on the ideal gas law).

In liquid hydrocarbon streams being monitored for color, excessive water in the stream often comes out of solution and becomes dispersed in the sample as a second phase. The light can be scattered to a degree that none is transmitted through the sample cell. In such a case, a bypass coalescing filter is required to remove the water. Cooling the sample before the coalescer and then heating it up before it flows through the sample cell helps to keep the sample clear in the sample cell.

In UV/Vis photometric analyzers using compensating wavelengths, a small amount of particulate matter in the sample gas can flow through the sample cell without causing analysis errors. However, it is always advisable to have at least a coarse filter at the sample tap to prevent large particles from plugging and damaging sample system components.

In applications where trace amounts of a constituent are to be monitored, the proper materials of construction for sample wetted parts in the sample system are essential for a successful installation, particularly on a corrosive stream. It is also essential that the inside surface of a sample tube be free of oils; otherwise, the tube may act as a chromatographic column and hold up the trace material to be analyzed. A sample line with a corroded inner surface may act similarly, with corrosion product "holding onto" the trace material. An example of this problem was seen when construction personnel decided to install a black iron pipe sample line for a UV photometric analyzer to monitor trace Cl_2 in a gas stream with high levels of HCl. (The process pipe also was black iron.) Plant operators became upset when the analyzer failed to read any Cl_2! Installing a Teflon tube inside the sample line pipe immediately corrected the problem.

Flexible tubing with plasticizers usually should be avoided in UV analyzer systems because the plasticizer, often a phthalate that strongly absorbs in the UV, may leach out of the tube and cause erroneous analyses, or it may adsorb the material to be monitored. Similarly, polyolefin (polypropylene or polyethylene) tube absorbs NO_2 and should be avoided for samples with NO_2.

A common application of a UV analyzer is to monitor Cl_2 in a stream containing titanium tetrachloride ($TiCl_4$) as part of the manufacturing process for titanium oxide (TiO_2). Any moisture will immediately react with the $TiCl_4$ to form TiO_2, the best light-scattering material known. Hence, the sampling system should be designed to keep moisture out of the sample system. Included as part of the system are automatic dry N_2 purges into the analyzer discharge and inlet lines when the system is shut down.

No matter how reliable a sampling system may be, sooner or later maintenance will be required. The layout of the analyzer sample system should allow for ease of maintenance. Handling parts that have been exposed to corrosive and/or toxic streams does not appeal to any technician, and if the parts are not easily accessible and easily removed, a negative attitude toward maintaining the system can quickly develop. Using a heated enclosure for installation of components in a heated system rather than tracing and lagging the components is an example of an installation designed for maintenance.

Typical Calibration Requirements

UV/Vis analyzers are almost always initially calibrated with prepared concentrations of the material to be analyzed. For gas-phase applications, standard gases may be used, or pure gases may be precisely blended to provide the desired concentrations. For liquid applications, samples are prepared volumetrically and/ or gravimetrically. Concentrations may be confirmed by standard chemical tests, including titrations. In the case of process spectrophotometers using correlation analyses such as chemometrics, all potentially-interfering components may need to be blended to assure the validity of the chemometrics package.

UV/Vis analyzers generally have very stable spans, but may require periodic zero adjustments due to deposits on cell windows, aging of the light sources, filters, detectors, and solarization of windows and other optics. The zero is generally checked and adjusted if required by periodically purging the sample cell with a nonabsorbing gas such as air or N_2 on vapor-phase applications, or with a nonabsorbing liquid such as water in analyzers with water-soluble samples. In applications where there is no satisfactory zero liquid, the sample cell may be drained and purged with air or N_2 to check the zero of the analyzer.

With dual-wavelength filter photometers, spans can often be checked with a zero standard in the sample cell by inserting into the common light beam a glass filter with a differential absorbance at the measuring and reference wavelengths, equal to that of a given concentration of the material being measured. However, to comply with governmental regulations when monitoring stack emissions, standard gases must be injected at the sample tap to check calibrations.

Standard holmium oxide and didymium glass filters are available for wavelength checks above 270 nm for analyzers using broad sources. Discharge lines from mercury lamps are also used for more precise confirmation of wavelengths.

Troubleshooting and Maintenance

In general, 85% of the problems encountered in UV/Vis analyzer applications are associated with the sampling system and 5% with light source failures. The other 10% are distributed among the electronics, detectors, and other optical elements and changing process parameters that alter the basis of the measurement. (Deposits on sample cell windows are considered sampling system problems.)

Deposits on sample cell windows are considered sampling system problems.

The following procedure is recommended for maintenance and troubleshooting:

1. Read the manufacturer's operation and maintenance manuals.
2. Do not assume that the analyzer is functioning improperly just because operations personnel believe that it is not providing the analysis expected; the analyzer may be reading exactly what it sees. The process may be having problems, or unexpected compounds may be in the stream and interfere with the analysis.
3. Check the sample system to see whether flows, temperatures, and pressures are within specified limits. Check for the possibility of leaky switching valves by manually closing off each supply stream to see whether any flow is still indicated.
4. If possible, isolate the sampling system from the analyzer, fill the sample cell with the "zero standard" (air, N_2, water, etc.), and zero the analyzer. Follow the manufacturer's manual to check light source levels.
5. If light levels are within specified values, check the response of the analyzer by inserting an upscale standard. If the analyzer reads within 5% of the specified concentration of the standard, you can assume that the analyzer is operating properly.
6. If low light levels are indicated, check for deposits on the cell windows and to see whether the source is emitting light.

   Do not look directly at any UV lamp without proper eye protection!

   Replace the lamp and clean the windows if necessary. Note that UV-absorbing deposits may not be visible, and the sample cell may need to be removed to determine whether the windows are blocking UV light.
7. If the light source and windows are not the problem, refer to the manufacturer's manual for the procedure to isolate a defective optical or electronic component.
8. If the analysis system is functioning to specifications, operations and the quality control laboratory should be notified for further process evaluation. (In the case of process spectrophotometers, there may be an alarm to indicate that an unknown interference is in the sample. A spectrum of the sample usually can be provided by the analyzer to help evaluate the problem.)

References

1. Mueller, Kurt, *Zeitschrift für Physik,* 65:739-758, 1930.
2. Saltzman, R.S., et al., in *ISA Proceedings, 1962 Annual Conference,* Research Triangle Park, NC: Instrument Society of America, 1962, p. 40.1.62-1 to 40.1.62-8.
3. Verhock, F.H., and Daniels, F., *Journal of the American Chemical Society,* 53:1250, 1931.
4. Woodson, T.T., et al., *Industrial and Engineering Chemistry, Analytical Edition,* 11:83, 1939.
5. Saltzman, R.S., Small, J.R., and Steichen, in *1989 ISA Analysis Division Annual Symposium,* Paper #88-0107, "New Process UV/VIS Diode Array Analyzer Development," 0882-5785/89/069-083, Research Triangle Park, NC: Instrument Society of America, 1988.
6. Glasser, L.G., *Process Instruments and Controls Handbook,* 1st ed., New York: McGraw-Hill, 1957, p. 6-46 to 6-66.
7. Sears, F.W., *Principles of Physics III, Optics,* Cambridge, MA: Addison-Wesley, 1947, p. 301.
8. Judd, D.B., and Wyszecki, G., *Color in Business and Industry,* New York: John Wiley & Sons, 1963.
9. Osborn, R.H., and Kenyon, W.C., *Industrial and Engineering Chemistry, Analytical Edition,* 18:523, 1946.
10. Saltzman, R.S., and Hunt, E.B., Jr., *Analysis Instrumentation,* 10:55, 1972.

About the Author

Robert S. (Bob) Saltzman is a consultant on process stream analyzers, sampling systems, and on applying analyzers for process and environmental control. He earned his B.S. in engineering physics from the University of Maine, and early in his career, first at Hercules, Inc. (1950-1955) and then with E.I. DuPont de Nemours and Company, he was responsible for the design and application of various process stream analyzers.

From 1962 until he formed his own consultant company in 1994, his technical focus has been development and application of photometric analyzers. He has held various positions with the DuPont Instrument Products Division, where he was promoted to Research Fellow in 1987, and then with the Ametek Process and Analytical Instruments Division when Ametek purchased the Process Instruments Group from DuPont in 1990. A recognized innovator, Mr. Saltzman holds six patents and is author and co-author of 57 technical articles and six professional course presentations.

Mr. Saltzman was named an ISA Fellow in 1974 and has served as director of the Analysis Division, program chairman, session developer, and has taught numerous short courses. He received the ISA Kermit Fischer Environmental Award in 1993. He also received the Analysis Division's Gilmer-Thomason-Fowler Award and the Wilmington ISA Section's Charles R. Otto Award. In 1991, he received the Technical Excellence Award from the Diagnostics and Biotechnology Division of E.I. DuPont de Nemours.

32

Chemiluminescence (CHEMIL) Analyzers

Chemiluminescence, the emission of light by the product(s) of a chemical reaction, is today one of the most widely-used phenomena in stack gas and ambient air-monitoring instrumentation. Ozone, ammonia, peroxy acetyl nitrate (PAN), and a variety of reduced-sulfur compounds represent just a few of the many species detected using this technique. Oxides of nitrogen — in particular, nitric oxide (NO) and nitrogen dioxide (NO_2) — remain the compounds most frequently monitored via chemiluminescence. Therefore, this chapter will concentrate on nitrogen oxide analyzers.

Prior to the use of gas-phase spectroscopic analyzers, air-monitoring instrumentation employed primarily wet chemical techniques. In 1966 [Ref. 1], the mechanism for the chemiluminescent reaction between nitric oxide and ozone was clarified and quickly became the basis for the development of commercial NO monitoring instrumentation [Ref. 2-4]. Almost as quickly, the perceived advantages of such analyzers led to the development of methods to detect NO_2 utilizing the same basic mechanism [Ref. 5, 6].

The major advantages of chemiluminescence methods include:

- Increased sensitivity (detection limit)
- Improved specificity (accuracy)
- Rapid response time (control)
- Linearity over a wide dynamic range (precision)
- Continuous monitoring (control and reporting)
- Simplicity of design (maintenance)

Although these advantages were indeed realized in early commercial analyzers, much remained unknown and uncertain. Over the past 25 years, these uncertainties have led to fruitful study and research and will continue to do so because of the importance of oxides of nitrogen in atmospheric and air pollution chemistry. Commercial development has followed the research community's efforts, and today's chemiluminescent nitrogen oxide analyzer is much improved over its early predecessor, although the basic principles of operation are the same.

Commercial development has followed the research community's efforts, and today's chemiluminescent nitrogen oxide analyzer is much improved over its early predecessor, although the basic principles of operation are the same.

Measurement Principles

The gas-phase reaction of nitric oxide and ozone (O_3) produces a characteristic luminescence via the following basic chemistry:

$$NO + xsO_3 \rightarrow NO_2^* + O_2 \qquad (32\text{-}1)$$

$$NO + xsO_3 \rightarrow NO_2 + O_2 \qquad (32\text{-}2)$$

$$NO_2^* \rightarrow NO_2 + h\nu \qquad (\sim 600 \text{ to } 3{,}000\text{nm}) \qquad (32\text{-}3)$$

$$NO_2^* + M \rightarrow NO_2 + M \qquad (32\text{-}4)$$

Equations (32-1) and (32-3) essentially describe the method used in commercial instrumentation to measure nitric oxide. When NO reacts with O_3, some electronically-excited NO_2^* molecules are produced (Equation [32-1]). These molecules give off energy in the form of light emission (*hv*, Equation [32-3]). When the emitted radiation is monitored, it becomes a measure of the concentration of the NO in the reacting sample. The light emission occurs between 600 and 3,000 nm, with a peak at about 1,200 nm.

Equations (32-2) and (32-4) describe limitations (although not the only ones) to the measurement technique. Only a fraction (about 10%) of the NO that reacts forms NO_2^*, the excited state. The large majority of the NO molecules react to form NO_2, a nondetected spectroscopic state (Equation [32-2]). Of the approximately 10% excited NO_2^* molecules, only a fraction deliver their energy in the form of light emission. The remainder are "robbed" of that energy by the process of collisional deactivation or quenching (Equation [32-4]).

Despite these limitations, by maintaining the O_3 concentration at a large excess to the NO concentration and with a detection system designed for optimum light-collection efficiency, analyzers have been developed with linear dynamic ranges over 10,000 and with detection limits at the parts per trillion level.

Common air-pollution monitoring practice and instrumentation utilize fractional units, such as parts per billion or parts per million (by volume) as a measure of concentration. To convert parts per million to milligrams per cubic meter (mg/m^3), the following conversion equation is used:

$$C\,(\text{mg/m}^3) = \frac{C\,(\text{ppm}) \times \text{MW} \times 298 \times P}{24.45 \times T \times 760}$$

For example, to convert a 500-ppm NO sample at 760 torr and 298°K to mg/m^3, use the molecular weight (MW) of NO = 30 and then obtain:

$$C\,(\text{mg/m}^3) = \frac{500 \times 30 \times 298 \times 760}{24.45 \times 298 \times 760} = 613.5 \text{ mg/m}^3$$

Note that temperature is in units of degrees Kelvin (°C + 273) and pressure is in torr or millimeters of mercury.

In order to measure NO_2 using the same basic mechanism outlined by Equations (32-1) through (32-4), it is first converted to NO, after which the converted molecules react with ozone along with the original NO molecules in the sample. This results in a signal that approximates the sum of NO and NO_2. By taking the difference between the converted and nonconverted modes, a measure of the NO_2 concentration is obtained.

Basic Instrumentation

Figure 32-1 presents a simplified diagram of a chemiluminescence analyzer using the chemistry described above. To measure NO concentrations, the sample is blended with ozone in a flow reactor. The ozone required for the reaction is produced within the instrument from dry air or oxygen. The luminescence that results from the reaction of NO with O_3 is monitored through an optical filter by a high-sensitivity detector positioned at one end of the reactor. The filter/detector combination responds to the light emission in a narrow-wavelength band.

The luminescence that results from the reaction of NO with O_3 is monitored through an optical filter by a high-sensitivity detector positioned at one end of the reactor.

Figure 32-2 shows the spectral output of the chemiluminescent reaction (Trace 1) along with optical filter (Trace 2) and typical detector (photomultiplier tube [PMT]) response (Trace 3) characteristics. As can be seen, only a small portion of the spectrum is utilized for detection. This aids in gaining specificity for the analyzer. An electronics package takes the detector output and processes it to provide a workable signal. This can be in the form of meter movement, analog voltage, current level, or digital signal.

The sample inlet generally has two flow modes. The first (NO mode) is a direct path of sample to the reaction chamber. Ideally, only the NO in the sample reacts with the ozone to produce light emission (chemiluminescence). In the second mode (NO_x mode), the sample is routed through a converter where all the NO_x theoretically exits as NO.

The reacted sample mixture is drawn (or pushed) from the reaction chamber by a vacuum pump (or blower). The mixture may be scrubbed of its high ozone concentration and NO_2 prior to passing through the vacuum pump, or after passing through the pump to the exhaust.

The following sections discuss in greater detail the various analyzer components.

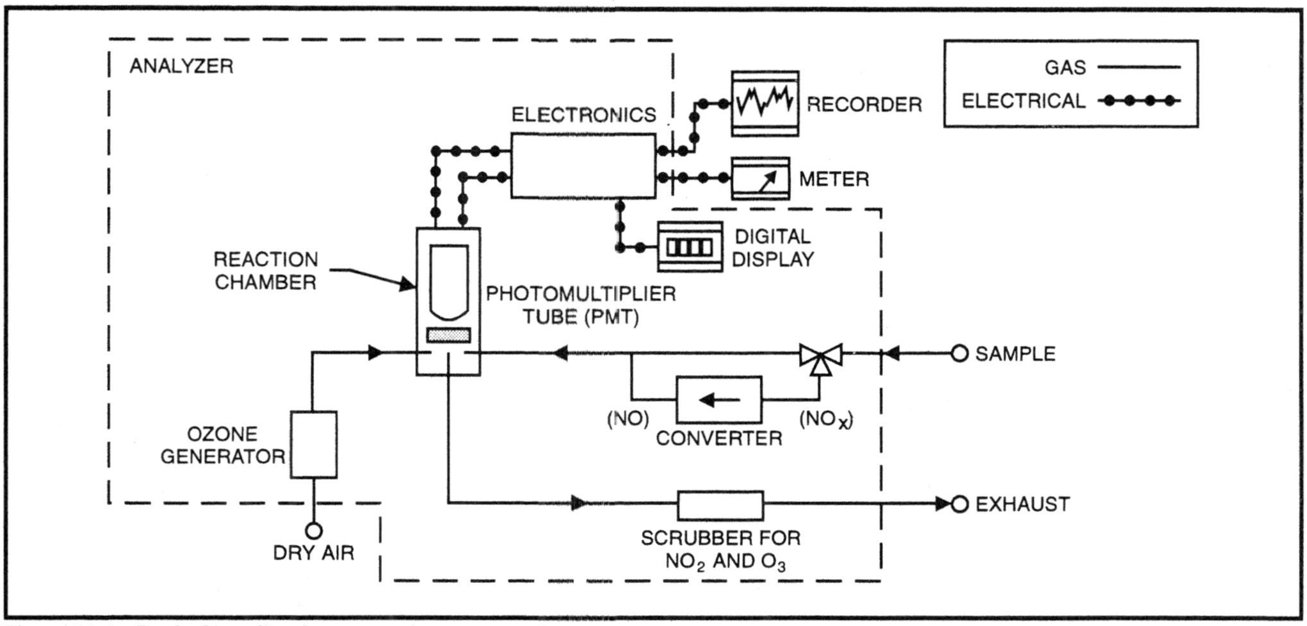

Figure 32-1. Basic Chemiluminescence Instrumentation

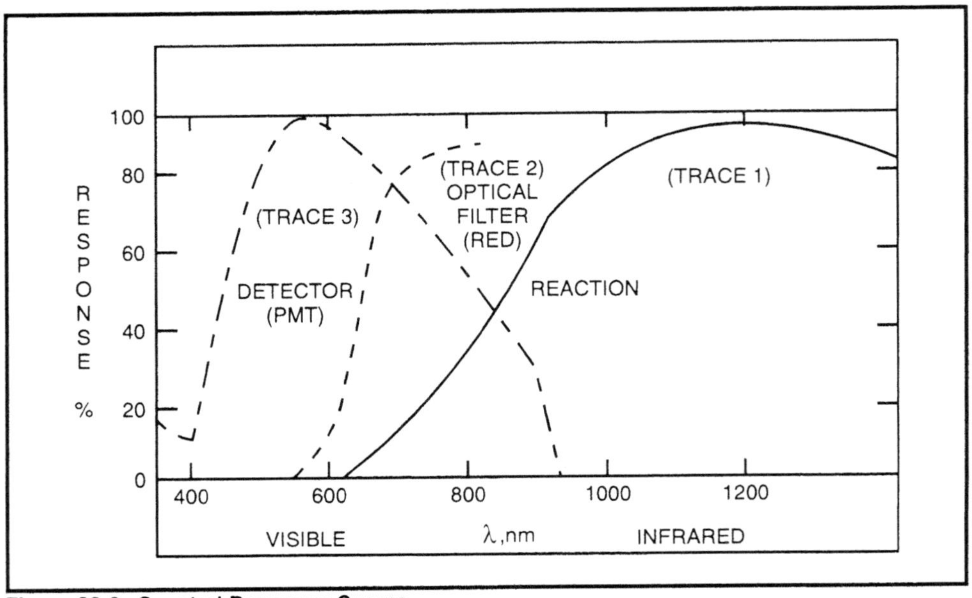

Figure 32-2. Spectral Response Curves

Gas Flow System

Figure 32-3 shows a detailed flow diagram of a typical analyzer. The chemiluminescent method relies on the measurement of the number of NO molecules entering the reaction chamber per unit time. In order to maintain instrument stability, two gas flow systems are used to regulate a constant flow of ozone and sample gas to the reaction chamber.

In Figure 32-3, the ozone flow system consists of a regulator and a single capillary. The function of the capillary is to restrict the flow in the line such that a controllable back-pressure is maintained. The pressure upstream of the capillary and the capillary diameter determine the flow. Other flow regulation schemes are possible. Mass flow controllers, critical orifices, and the regulation system described above have been used. All regulation systems are designed for the same purpose: to maintain a constant air or oxygen flow through the ozonator.

Since the chemiluminescent reaction is extremely flow sensitive, the sample flow system must assure a constant flow to the reaction chamber.

The prime requirement of the sample flow system is that the instrument output be insensitive to the flow of sample gas to the analyzer. Since the chemiluminescent reaction is extremely flow sensitive, the sample flow system must assure a constant flow to the reaction chamber. Figure 32-3 shows one standard industrial approach to fulfill these conditions.

The total sample passes into the analyzer through the inlet capillary restrictor. A fraction (about 5%) is bled off through a second smaller restrictor and directed to a mode valve. The portion of the sample bypassing the mode valve (and reaction chamber) exits the instrument through a bypass pump. The sample regulator functions to maintain a constant pressure drop across the smaller capillary. This method of flow control maintains a reaction chamber sample flow rate that is insensitive to the sample inlet pressure and total sample flow rate. This type of bypass flow system has the additional advantage of bringing the sample to the analyzer rapidly, reducing the lag time of the analyzer. It has the disadvantage of using a fairly large sample flow, which can be a limitation in smog chamber or bag sample studies. For typical process applications, however, there is no sample limitation.

Sample flow remains one of the most critical parameters to control.

Not all chemiluminescent analyzers use such a bypass-flow technique. Most ambient-air analyzers use only one sample restrictor and no bypass flow. Nevertheless, sample flow remains one of the most critical parameters to control. Mass flow controllers, critical orifices, and/or a regulator and capillary system using a pressurized sample have also been used.

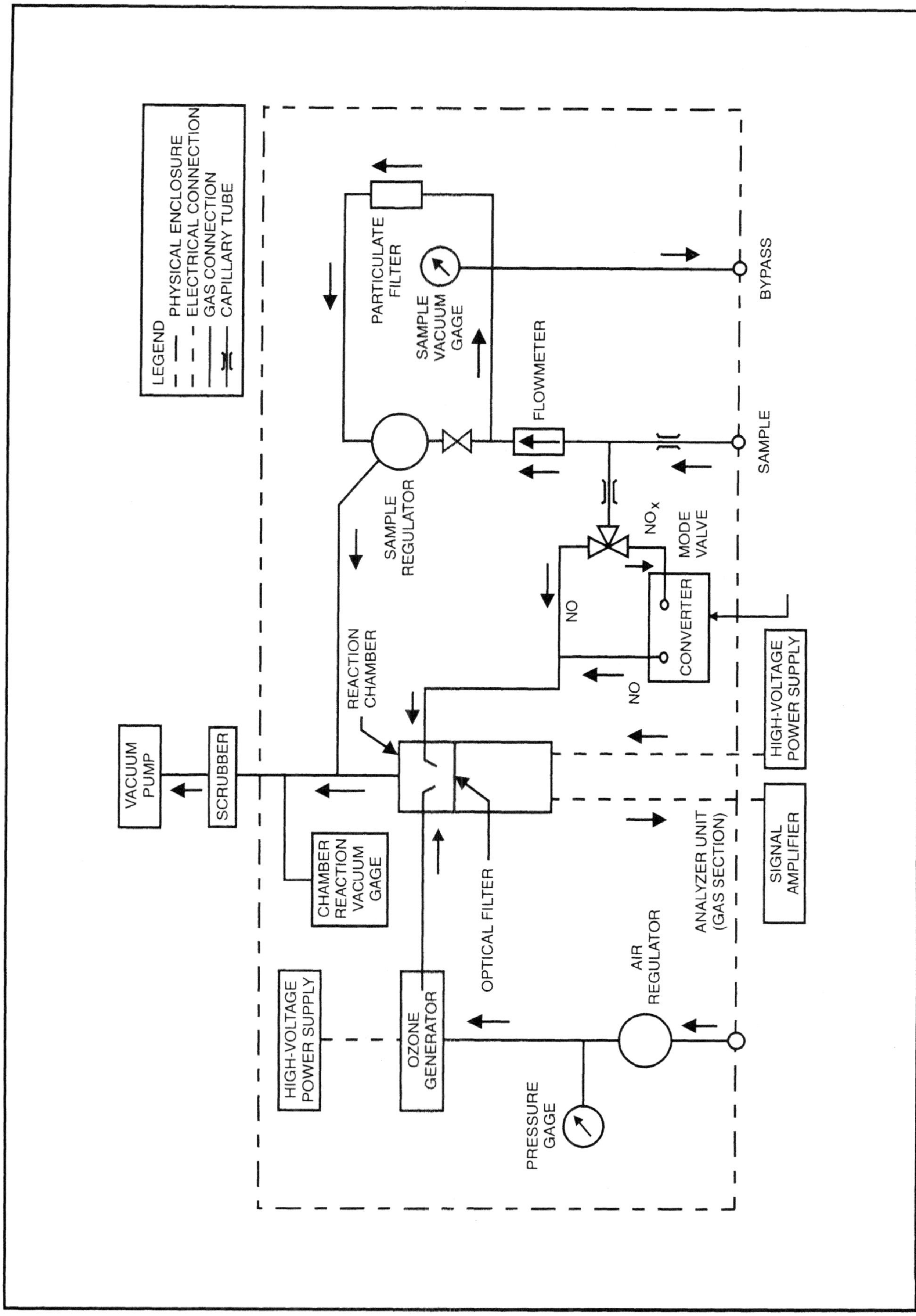

Figure 32-3. Chemiluminescence Analyzer Flow Diagram

Ozone Generation System

Although very early prototype analyzers used photolytic sources to generate O_3, this approach has never become prevalent in commercial instrumentation. The standard approach is the use of the silent-electric or corona discharge. A common design utilizes a condenser-style arrangement with a central electrode, air gap, and outer electrode. Air or oxygen passes over the surfaces of the inner electrode, which is inside a glass envelope, and then passes out to join the sample stream in the reaction chamber. By applying a high alternating or pulsing voltage (about 10,000 VAC) across the electrodes, a fraction of the oxygen in the air or oxygen stream is broken down into free radicals or single oxygen atoms. The majority of these single atoms combine with molecular O_2 to form ozone (O_3).

The fraction or percent of ozone produced depends on a variety of different parameters. The level of high voltage, the temperature of the air or oxygen stream, the flow rate through the ozonator, the stability of the voltage source, the gas composition of the ozonator feed, and the physical size of the ozonator all affect the O_3 output. Typical flow rates through commercial ozonators are on the order of 100 cm^3/min. At this flow rate, with a dry air feed, an ozone percentage of about 0.5% can be achieved. Using oxygen as the ozonator feed, ozone levels of 1 to 2% that can be achieved.

Using oxygen as the ozonator feed, ozone levels of 1 to 2% that can be achieved.

In most cases, where the sensitivity of NO analyzers drops off (becomes sublinear) at high concentrations (>2,000 ppm), the O_3 concentration is no longer in great excess to the NO concentration, making the kinetic assumptions necessary for extended linearity invalid. In order to maintain linearity, it is necessary to increase the relative O_3 concentration. This is usually done in either of two ways. First, the sample flow relative to the O_3 flow can be decreased; second, the O_3 flow can be increased, often by using oxygen as the ozonator feed rather than air.

There are certain critical precautions to take when operating a silent discharge ozonator. First and foremost, the user should be aware of the high voltages used to generate the ozone and take proper precautions working around or servicing the ozonator. The voltage source to the ozonator should always be turned off and disconnected, if possible, before servicing the analyzer. Only an individual who has been properly trained and has the appropriate test equipment (e.g., high-voltage probe) should attempt any servicing or analysis while the ozonator is turned on. Even then, the service department of the commercial analyzer supplier should be consulted prior to any service or maintenance.

The voltage source to the ozonator should always be turned off and disconnected, if possible, before servicing the analyzer.

Second, the ozonator should always be turned off if there is no gas flow through the ozonator. If high voltage is applied with no gas flow, the ozonator heats up rapidly. If gas flow is suddenly introduced, damage to either the electrode or the glass envelope often occurs due to the extreme temperature gradient that has developed. If pure oxygen is being used as the gas feed, a potential fire hazard also exists. In general, air is the preferred gas feed for the ozonator.

If high voltage is applied with no gas flow, the ozonator heats up rapidly.

Third, the air or oxygen used as the ozonator feed should be dry. The presence of water vapor with either feed reduces the production of ozone by reacting with the oxygen atom radicals before they can react with O_2 to form ozone. With air, where nitrogen is present, water vapor within the silent discharge produces what has been termed "ozone oil." This ozone oil, which consists primarily of nitric and nitrous acids, can plug restrictors, altering the ozone flow and causing corrosion within the chemiluminescent analyzer's plumbing and reaction chamber.

In commercial instruments, the gas feed to the ozonator is dried by a variety of methods. Indicating-calcium-sulfate or silica gel both have the advantage that, as the drying agent is consumed, the indicator turns color (usually from blue to pink). When the agent is completely consumed, it can be regenerated by heating to 230°F (110°C) for several hours.

Permeation dryers, operating on the principle of selective permeation of water through a suitable membrane, are also used. These have the advantage of being theoretically nonconsumable. Their disadvantage is that they require a more complicated plumbing arrangement, including additional capillaries or restrictors, as well as a pump to create the necessary water vapor concentration differential.

Heatless air dryers, molecular sieves, or refrigerated coolers have also been used to generate the dry air source for the ozonator within the chemiluminescent analyzer.

Permeation dryers, heatless air dryers, molecular sieves, or refrigerated coolers have also been used to generate the dry air source for the ozonator within the chemiluminescent analyzer.

Reaction Chamber

The reaction chamber is an enclosed vessel where the ozone and sample streams are mixed and luminescence is produced. The chamber material should be inert to the high concentration of ozone and be noncorrosive. Reaction chambers of polyvinyl chloride, Teflon, aluminum, glass, stainless steel, and gold have been used.

The reaction chamber is designed for rapid mixing of the two reagent flows so that as much of the chemiluminescence occurs in the view of the detector as possible. Depending on the pressure at which the chamber is operated and the flow rates of the mixing gases, the design volume of the chamber is determined. In general, the volume is chosen so that the majority of excited NO_2 molecules emit light while in the view of the detector rather than as they exit or are outside the chamber.

Usually included as part of the reaction chamber is the optical filter used to block wavelengths not associated with the $NO\text{-}O_3$ luminescence. Figure 32-2 shows the optical filter characteristics. The red filter cuts off virtually all wavelengths below 600 nm. Many unsaturated hydrocarbons (e.g., ethylene) and carbonyl compounds (e.g., nickel or iron carbonyl) react with ozone and chemiluminesce. However, the luminescence occurs in the visible or blue wavelength regions. By using an appropriate red filter, spectral interference from these types of compounds is virtually eliminated.

The reaction chamber is made leaktight via mechanical design (gaskets or O-rings) in order to prevent loss of the reacting mixture to the atmosphere, leak of the atmosphere into the chamber, and/or light leaks that the detector may see.

In some commercial analyzers, the reaction chamber is heated and controlled above ambient temperature (usually to about 120°F [50°C]). Since reaction (32-1) has a temperature dependence of about 2%/°C, stabilization and raising of the chamber temperature produces increased signal stabilization and luminescence. Heating also offers the advantage of minimizing surface adsorption of other gases on the reaction chamber walls.

One disadvantage of a heated reaction chamber is the increased complexity in circuitry and added cost. Heating the chamber can also raise the temperature of the detector. If the thermal coupling between the reaction chamber and detector is not properly maintained, the background signal due to the detector will rise, causing the detection limit to suffer.

One disadvantage of a heated reaction chamber is the increased complexity in circuitry and added cost. Heating the chamber can also raise the temperature of the detector. If the thermal coupling between the reaction chamber and detector is not properly maintained, the background signal due to the detector will rise, causing the detection limit to suffer.

The Detector

The detector most commonly used in chemiluminescence analyzers is a photomultiplier tube (PMT). PMTs convert incident radiation into electrical signals by use of photoemission, and then amplify those signals by means of secondary emission. Figure 32-4 shows a basic schematic arrangement of a typical PMT.

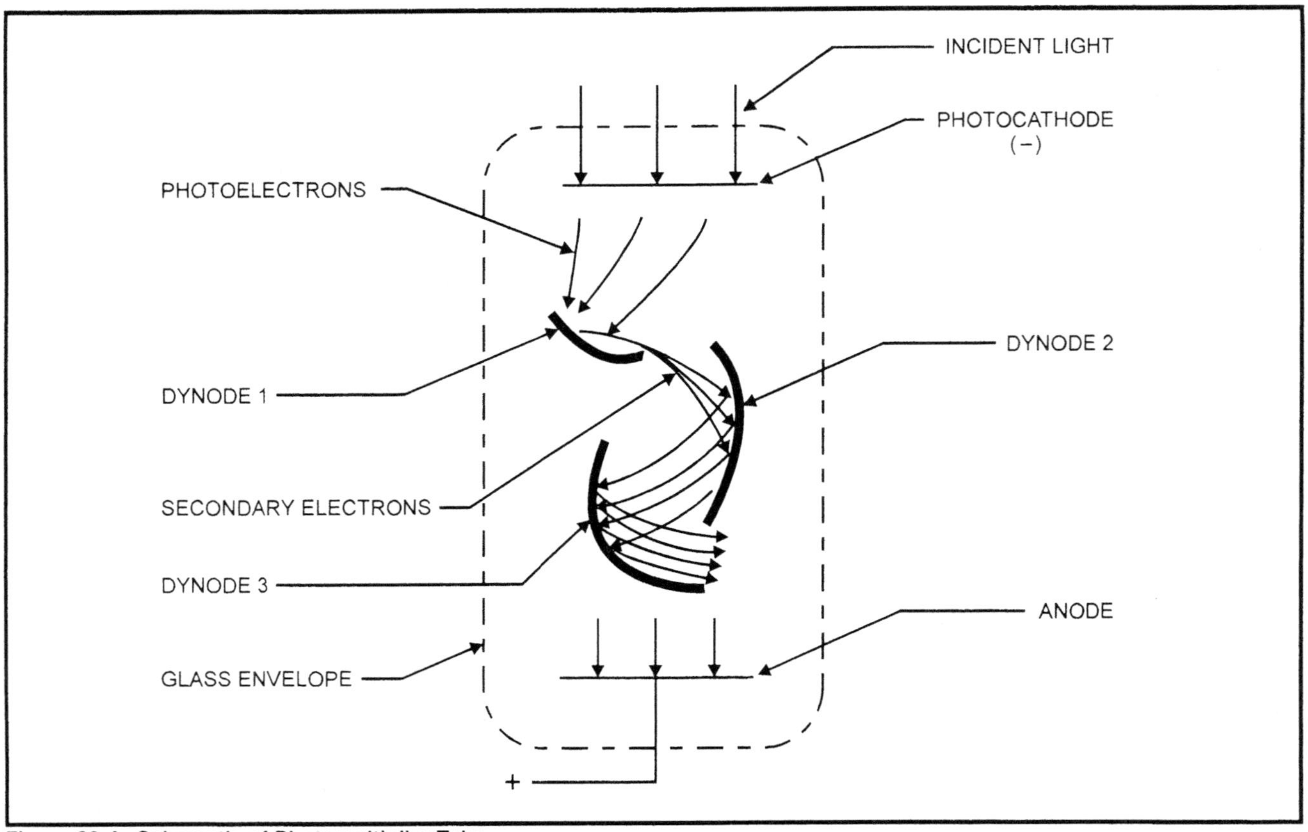

Figure 32-4. Schematic of Photomultiplier Tube

The photocathode, by absorbing photons from the $NO\text{-}O_3$ luminescence, emits electrons which are accelerated and directed by an electric field to the first dynode. When electrons strike the surface of the dynode, secondary electrons are emitted. These secondary electrons are themselves directed to a series of dynodes, eventually resulting in a huge number of emitted secondary electrons. The anode, situated at the end of the dynode string, collects these secondary electrons. The flow of electrons per second represents an electrical current output signal. Typical electrical current levels of 10^{-6} to 10^{-12} ADC are generated.

The PMTs most often used are of the multialkali or extended-red variety in order to have the infrared sensitivity required to detect NO_2 luminescence. Even these types of tubes respond only to wavelengths up to about 900 nm, as shown in Figure 32-2. Although at first glance it appears that the PMT is not a good choice as a detector, because of minimal overlap of the luminescence with the detector sensitivity, this type of system has proved successful in detecting concentrations in the parts per trillion range. In fact, the minimal overlap provides increased specificity.

The major requirements for the PMT are extended response into the infrared combined with low dark current and noise.

The major requirements for the PMT are extended response into the infrared combined with low dark current and noise. To a certain extent, these requirements are contradictory, since photocathodes that have good infrared response tend to have a low work function, leading to increased dark current. Dark current is the photomultiplier signal obtained in the absence of any light emission from the $NO\text{-}O_3$ reaction. This current has two major components: (1) ohmic leakage resulting from imperfect insulating properties of the phototube structure and contamination, and (2) thermionic emission, the emission of electrons from the photocathode due to the absorption of heat energy. Two methods are used to reduce dark current: first, lower the cathode-to-ground voltage (reduce leakage), and second, reduce the temperature of the photocathode (cool the tube).

Depending on the levels of nitric oxide that are to be detected, commercial analyzers may or may not cool the detector. Although detectors have been cooled using water baths, dry ice, or liquid nitrogen, by far the most common approach is to use a thermoelectric module (TEM) utilizing the Peltier effect. The advantages of cooling the detector are reduced background signal (dark current) and noise, and, if the temperature to which the tube is cooled is also controlled, increased stability. The decrease in noise improves the detection limit of the analyzer.

The disadvantages of cooling the PMT are added complexity in detector design and circuitry, increased cost, and increased possibility for analyzer malfunction and service requirements.

Photodiodes have also been used as detectors in commercial nitric oxide analyzers. These types of detectors operate on the same principle of photoemission of electrons as PMTs, but do not use the multiplying secondary stages. Photodiodes can have a much better overlap of detector response with the NO-O_3 emission; however, they generate much larger relative background signals due to the reduced work function of the electron-emitting material. Also, because of the lack of secondary multiplying stages, the signal levels generated for a given concentration of NO or NO_2 (sensitivity) are much lower. These factors have limited the use of photodiodes to nitric oxide analyzers used only at source-level concentrations. As detector technology improves, it remains to be seen whether the use of photodiodes in ambient or subambient concentration detection will become practical.

These factors have limited the use of photodiodes to nitric oxide analyzers used only at source-level concentrations.

Electronics

The detector signal, whether from a PMT or a photodiode, requires further amplification and conditioning to be useful. Although no great detail will be given here, there are several common characteristics of commercial instrumentation.

The optical signal from the NO-O_3 reaction is converted to an electrical current signal by the detector. This electrical signal is converted to a voltage (VDC), then filtered and amplified. The signal can be used in an analog mode and sent to a panel meter (digital or analog), or recorder output, or it can be converted to a frequency and sent to a microprocessor.

Both analog and microprocessor-based (digital) types of instrumentation are available commercially. Analog instrumentation theoretically has the advantage of reduced complexity in circuitry and reduced cost; however, the use of microprocessors has become so commonplace that this advantage has rapidly eroded. Microprocessor-based analyzers offer a variety of parameters and "bells and whistles." As the user has become more sophisticated, the microprocessor has gone from a marketing gimmick to actually improving the operation, serviceability, and performance. The use of a computer interface to collect data, change instrument parameters, and calibrate the analyzer are all advantages gained by proper implementation of microprocessor-based instrumentation.

The major disadvantage of microprocessors is, if improperly utilized, the analyzer becomes only a black box. The user relies on the digital readout or computer interface solely for troubleshooting and serviceability. The fundamental operating principles upon which the analyzer is based become less understood, and ultimately the user can become incapable of operating the instrument properly. The analyzer manufacturer needs to provide a microprocessor-based unit that truly takes advantage of the internal computer and does not isolate the user or let the analyzer "take over."

Converter

To measure NO_x (NO + NO_2), nitrogen dioxide (NO_2) must be changed to NO. This is accomplished by the converter unit of the analyzer.

As described earlier, the nitrogen oxide analyzer is sensitive only to NO molecules. To measure NO_x (NO + NO_2), nitrogen dioxide (NO_2) must be changed to NO. This is accomplished by the converter unit of the analyzer. At present, there are three basic NO_2 to NO conversion methods available commercially.

The first approach discovered, still utilized today in virtually all source- or high-level nitrogen dioxide measurements, is the thermodynamic converter. The thermodynamic converter relies on the fact that at temperatures above 1200°F (650°C), the equilibrium of the NO-NO_2 system favors the NO state virtually 100%. The equation describing this process is

$$2NO_2 \overset{T > 1200°F}{\rightleftharpoons} 2NO + O_2 \qquad (32\text{-}5)$$

Converters manufactured from gold, platinum, palladium, and nickel have been used, but most commonly used has been the stainless steel converter heated above 1200°F (650°C).

However, in order to practically utilize this favorable equilibrium, the employed converter must have a catalytic effect. Although the thermodynamic equilibrium strongly favors the right hand side of Equation (32-5) at high enough temperatures, the activation energy of the above reaction is such that without a catalytic surface, the reaction occurs too slowly to be used in a dynamic chemiluminescence analyzer. Converters manufactured from gold, platinum, palladium, and nickel have been used, but most commonly used has been the stainless steel converter heated above 1200°F (650°C). This type of converter was first developed by researchers at the Environmental Protection Agency (EPA) [Ref. 5].

Numerous studies and practical experience have shown that the use of stainless steel as a catalytic converter requires oxidation of the stainless steel surfaces in contact with the sample being analyzed. This is normally accomplished by the passage of air or oxygen at temperatures above 1200°F (650°C) through the converter for a period ranging from 1 h to several days, depending upon the age and condition of the converter. A nitric acid wash has also been used. The stainless steel itself is thus not the catalytic medium, but rather the oxidized surface layer on the steel.

High concentrations of carbon monoxide, ammonia, hydrocarbons, or other reducing compounds have been shown to accelerate the oxide layer depletion in stainless steel converters, rendering them inefficient.

Because an oxidized layer is required, the sample to be analyzed should contain at least a small percentage of oxygen (minimum 1 to 2%) in order to obtain practical conversion of NO_2 to NO. If a reducing sample is being analyzed, the oxide layer can become depleted and the efficiency of the converter will drop. High concentrations of carbon monoxide, ammonia, hydrocarbons, or other reducing compounds have been shown to accelerate the oxide layer depletion in stainless steel converters, rendering them inefficient. To maintain converter efficiency, oxygen or air must be included as part of the sample matrix. This can be done by dilution of the sample prior to analysis. If a stainless steel converter has become inefficient due to overuse in a reducing atmosphere or due to unusually high concentrations of species as given above, it can be regenerated by operating the converter at its normal operating temperature overnight with clean room air or oxygen. This procedure reoxidizes the surface layers of the converter and makes it once again efficient in NO_2 conversion.

The second type of converter is based on chemical conversion rather than thermodynamics. NO_2 is reduced to NO via a chemical reaction. Typical converter materials include molybdenum, carbon, and ferrous sulfate. The general equation describing the conversion process is

$$NO_2 + A \rightarrow NO + AO \qquad (32\text{-}6)$$

With molybdenum, the reaction process is actually as follows:

$$3NO_2 + Mo \rightarrow MoO_3 + 3NO \qquad (32\text{-}7)$$

Since these types of chemical reactor converters are gradually consumed, they eventually require replacement or replenishment. Molybdenum converters theoretically can be restored by converting the molybdenum trioxide back to molybdenum metal. This is based on the following reduction reaction:

$$3H_2 + MoO_3 \rightarrow Mo + 3H_2O \qquad (32\text{-}8)$$

The converter is disconnected from the analyzer and allowed to cool to room temperature (if not already there). The inlet of the converter is connected to a source of hydrogen, and the outlet of the converter is properly vented. The converter temperature is raised to approximately 840°F (450°C). Hydrogen is allowed to flow through the converter for about 3 h. After this time, the temperature of the converter is lowered and H_2 continues to flow until the converter is back to room temperature. The hydrogen is then disconnected. The converter is flushed with nitrogen to remove traces of hydrogen, then reconnected within the analyzer and allowed to run for approximately 1 h at 750°F (400°C). After this operation, the molybdenum converter should be properly reconditioned.

Although in theory this procedure can restore the molybdenum converter to proper operation, in practice the converter never lasts as long or has as stable a converter efficiency as the original converter. The use of hydrogen at elevated temperatures is also dangerous, presenting an explosion hazard if not properly handled. For the majority of users, it is more practical, from both performance and safety standpoints, to replace rather than recondition molybdenum converters.

The third NO_2 conversion method is primarily used only in research or extremely specialized applications. NO_2 is converted to NO by photolysis using a light source (typically a xenon arc) above 320 nm. Although this approach offers the advantage of increased specificity compared to the other conversion techniques, it suffers from low converter efficiency and instability. At present, it cannot be practically used in source or normal ambient air monitoring. Additional information can be found in References 7 to 9.

The major advantage of thermodynamic converters is that, theoretically at least, they are not consumed as long as a sufficient oxidation layer is present or enough oxygen is in the sample being analyzed to replenish the oxide layer if it is consumed by a reducing agent. The major disadvantages are the high temperature requirement (power consumption, heat generation) to obtain greater than 98% converter efficiency and, more importantly, the high conversion level of other nitrogen-containing compounds — of which ammonia is of most concern.

The major advantage of thermodynamic converters is that, theoretically at least, they are not consumed as long as a sufficient oxidation layer is present or enough oxygen is in the sample being analyzed to replenish the oxide layer if it is consumed by a reducing agent.

Ammonia is often present in industrial air samples either as a generated by-product or as an added agent. Thermodynamic converters will convert ammonia to NO at an efficiency from 60 to 100%, depending on the temperature of the converter. This leads to erroneously high NO_x and NO_2 readings. Other nitrogen-containing compounds, such as hydrogen cyanide, nitric acid, and amines, are also converted at a high efficiency.

The major advantage of chemical converters is that they operate at lower temperatures, therefore consuming less power and producing less heat. Also, the conversion is more specific than high-temperature conversion. Ammonia has been shown to convert only about 5% at typical operating temperatures. Other nitrogen-containing compounds, however, are still converted to NO at high efficiencies. Such compounds as nitric and nitrous acids, organic nitrates and nitrites, and specifically peroxy acetyl nitrate (PAN) are converted to NO in chemical converters with relatively high efficiencies. Thus, the specificity of chemical

The major advantage of chemical converters is that they operate at lower temperatures, therefore consuming less power and producing less heat.

converters, although considerably better than thermodynamic converters, is far from ideal. Another disadvantage of chemical converters is that they are consumed and require replacement or replenishment. This gradual efficiency reduction may not be noticed in slowly deteriorating analyzer performance, and a good preventive maintenance program is required.

Another disadvantage of chemical converters is that they are consumed and require replacement or replenishment. This gradual efficiency reduction may not be noticed in slowly deteriorating analyzer performance, and a good preventive maintenance program is required.

High-temperature converters give a good approximation to total nitrogen conversion, N_t. Low-temperature converters give a good approximation to total oxidized nitrogen conversion, NO_y. Depending upon the sample composition, different converters can be utilized to give useful information. For example, if ammonia is present in the sample, the use of a thermodynamic converter in combination with a chemical converter can give a good approximation of the ammonia concentration:

Depending upon the sample composition, different converters can be utilized to give useful information.

$$[N_t] - [NO_y] = [NH_3] \qquad (32\text{-}9)$$

Pumping System

Nitrogen oxide analyzers utilize two basic systems to generate flow. Instrumentation is usually operated either at reduced pressures by use of a vacuum pump (Figure 32-3), or at ambient (or slightly above) pressure by use of a pump pushing the sample through the system. Because industrial applications occur under conditions of virtually unlimited sample, the maximum sensitivity of the analyzer is limited primarily by the pumping speed of the system. The pumping speed effectively determines the number of NO molecules per unit time seen by the detector in a properly-designed instrument. Therefore, the pumping speed is a critical parameter in maintaining analyzer stability and performance. Detailed discussion and mathematical treatment of the effect of pumping speed and flow within chemiluminescence analyzers are presented in References 10 to 12.

The major advantage of operating an analyzer at lower pressures by utilizing a vacuum pump is reduced quenching effects, resulting in increased sensitivity to nitric oxide concentrations. Sample response and transit times can also be reduced. Operating the reaction chamber and converter at low pressures can minimize unwanted secondary reactions that can affect converter lifetime and efficiency, lower ozone concentration, and generate interfering chemiluminescence.

The major advantage of operating an analyzer at lower pressures by utilizing a vacuum pump is reduced quenching effects, resulting in increased sensitivity to nitric oxide concentrations.

Disadvantages of vacuum analyzers include increased instrument complexity, susceptibility to atmospheric leaks, and the necessity for pump compatibility with high concentrations of ozone from the reaction chamber exhaust. Compatibility is obtained either by scrubbing the ozone before it enters the pump or by using a pump manufactured from materials inert to the high ozone levels. Oil vacuum pumps, in general, generate the highest pumping speed, but require ozone scrubbers and frequent oil changes. Diaphragm pumps, utilizing construction materials such as Teflon and stainless steel, are also used.

Disadvantages of vacuum analyzers include increased instrument complexity, susceptibility to atmospheric leaks, and the necessity for pump compatibility with high concentrations of ozone from the reaction chamber exhaust.

The major advantages of operating the nitric oxide analyzer at ambient pressure are minimized susceptibility to leaks, reduced instrument complexity, and potentially lower cost. Disadvantages are reduced sensitivity, greater quenching effects, possible blower incompatibility and influence upon sample composition, and increased probability of secondary or competing reactions within the chemiluminescent system.

Typical Calibration Requirements

The calibration of all chemiluminescence nitrogen oxide analyzers follows the same basic principles and requires the same basic equipment. However, certain unique procedures may be required, depending on the individual analyzer. The user is cautioned to study and follow the specific guidelines given in the manufacturer's instrument instruction manual. Presented here is a general overview of calibration equipment and techniques.

General Equipment

The basic equipment required for calibration includes the following items:

- Zero gas source
- Nitric oxide concentration standard
- Nitrogen dioxide concentration standard

Depending on the concentration of the sample being monitored and the range of the individual chemiluminescence analyzer being used, these items will have different requirements.

ZERO GAS

Zero gas can be defined as a gas which, when introduced into the chemiluminescence analyzer, produces no detectable response in the parameters of interest. This gas should therefore be "free" of nitric oxide, nitrogen dioxide, or other compounds that, by virtue of chemiluminescence reaction or quenching, generate a measurable response from the analyzer. Depending on the levels of nitrogen oxides to be detected, the actual concentration of the zero air contaminants may vary greatly.

For example, if the analyzer is being used to measure NO or NO_2 in the parts per billion or below range, the zero air requirements are much more stringent than if an analyzer is being used to measure concentrations in the hundreds of parts per million. In the first case, typical of clean ambient air or research analyses, the zero gas may be required to contain less than several tens of parts per trillion impurities. In the second case, typical of source applications, a zero gas with NO or NO_2 contaminants of several hundred parts per billion can be perfectly adequate.

Sources of zero gas include cylinder gases such as nitrogen, oxygen, or air; compressed room air that has been dried, scrubbed, and filtered to bring the contaminant levels down to acceptable limits; and room air, which because of the concentration range being measured, can be used directly without further conditioning. Although the instrument manual and manufacturer should offer reasonable guidelines for determining which source of zero gas is acceptable, the user, being most familiar with the particular application, is ultimately responsible for choosing the appropriate zero gas source.

Sources of zero gas include cylinder gases such as nitrogen, oxygen, or air; compressed room air that has been dried, scrubbed, and filtered to bring the contaminant levels down to acceptable limits; and room air, which because of the concentration range being measured, can be used directly without further conditioning.

Cylinder zero gas, depending on the purity level being purchased, will specify the upper levels of certain component contaminants. The levels of NO_x impurities for high-purity or above grades of cylinder gases are more than adequate for source-level applications. The major advantage of using a cylinder gas as the zero gas source is convenience. The user theoretically requires only a compatible regulator for the cylinder, which has been properly flushed and connected to the analyzer. The disadvantages are high cost, continual need for replacement as cylinder contents are consumed, and general mismatch in application requirements. There are many suppliers of zero gases in cylinders, and occasionally quite a variation in the quality of the supplied gas.

There are many suppliers of zero gases in cylinders, and occasionally quite a variation in the quality of the supplied gas.

Room air that has been compressed and conditioned is often used as zero gas. An air compressor is utilized to raise the gas pressure to allow accurate and reproducible flow control, to initiate water dropout, and to aid in subsequent drying and scrubbing operations. As room air is pressurized, its dew point is raised to the point at which water dropout occurs. The water is usually collected in a storage tank for later removal. The compressed air may be further dried, if required, by passing the gas stream through a drying agent such as calcium sulfate or silica gel, or by using a heatless air dryer or a permeation dryer.

Any NO present in the compressed air is usually oxidized to NO_2 to ease its scrubbing. Oxidation is accomplished by either ozonization or chemical contact. During ozonization, the air is passed through an ozone generator. The O_3 that is produced reacts with NO to form NO_2. Care must be taken to allow sufficient residence time for the reaction to go to completion.

Chemical oxidation is accomplished by passing the air stream through a reacting bed. Such agents as CrO_3 on a alumina support or KMnO4 on alumina (commercial Purafil® or Carusorb® 200) efficiently oxidize NO to NO_2. The chemical contact approach has the advantage of needing no electrical power for its application.

In general, the final step in the generation of zero air from a compressed room air source is to scrub the remaining contaminants (e.g., NO_2, O_3, SO_2) by further reaction or absorption. Chemical scrubbers are usually employed. Soda lime and Purafil® are effective for removing NO_2. Activated charcoal is effective for removing both O_3 and SO_2.

The advantages of using a conditioned compressed-air system are that once the system is in place, the zero air source has a very long lifetime. It is convenient to move from location to location, and because the source of air is virtually infinite (room air), only chemical scrubbers need to be replaced as required. Their cost is generally much lower than that of cylinder gas. Also, depending on the application, the user can choose to scrub or not scrub the contaminants as desired, thus better matching the zero gas to the particular application. Most manufacturers of chemiluminescence analyzers offer zero air supplies as well. The disadvantages of a compressed-air system are the initial cost outlay, the consumption of the chemical scrubbers, and difficulty in proving the effectiveness of the scrubbing system.

Use of room air alone as zero gas is in some ways the most convenient source of all. A small diaphragm or bellows pump can be used to draw or deliver room air to the analyzer. Room air, however, can vary greatly in the concentration level of NO or NO_2. Depending on the particular environment (e.g., near a parking lot, traffic location, or smoke source), room air concentrations can reach parts per million levels. In general, if a chemiluminescence analyzer is being used to measure high concentrations of nitrogen oxides (hundreds of parts per million) and the room air has NO_x contamination levels below a part per million, it is an acceptable zero air source.

In general, if a chemiluminescence analyzer is being used to measure high concentrations of nitrogen oxides (hundreds of parts per million) and the room air has NO_x contamination levels below a part per million, it is an acceptable zero air source.

The great advantage of room air is simplicity of equipment and extremely low cost. The great disadvantage is the potential wide variability in NO_x concentration and lack of control of external contamination sources. As source emission requirements become more stringent, using room air alone as a zero air source will no longer be an acceptable alternative. Obviously, for ambient air analysis, unconditioned room air cannot be used as a zero gas source.

NITRIC OXIDE CONCENTRATION STANDARD (SPAN GAS)

Nitric oxide concentration standards used to calibrate chemiluminescence analyzers are exclusively supplied by gas cylinder manufacturers. Gas mixtures of NO in nitrogen are available from subparts per million to percent levels. Gas suppliers offer differing grades of NO cylinders, the grade (and cost) depending on the concentration, accuracy, and impurity level of the particular cylinder. Although NO mixtures are available with balance gases other than nitrogen (e.g., argon or helium), for the majority of source and ambient applications, NO/balance-nitrogen cylinders should be used. Nitric oxide cylinders are not available as mixtures with air or oxygen. NO reacts with oxygen to form NO_2, thereby making such a cylinder worthless for calibration.

Nitric oxide cylinders purchased from a gas supplier can be made traceable to a National Institute of Standards and Technology (NIST) NO in N_2 standard reference material or NO_2 standard reference material. Procedures for certifying the working standard NO cylinder against a NIST traceable NO or NO_2 standard and for determining the amount of NO_2 impurity are given in EPA Publication No. EPA-600/4-75-003, "Technical Assistance Document for the Chemiluminescence Measurement of Nitrogen Dioxide."

Nitric oxide cylinders are not available as mixtures with air or oxygen. NO reacts with oxygen to form NO_2, thereby making such a cylinder worthless for calibration.

The pressure regulator for the NO cylinder must be manufactured from inert materials and have a nonreactive diaphragm and internal parts and suitable delivery pressure. A common source of error when using NO cylinders is to fail to flush out or evacuate the regulator prior to its use. Any oxygen trapped in the regulator will slowly react with NO to form NO_2. The NO concentration will be lower than it should, and the cylinder will appear to have a significant NO_2 impurity. As the oxygen in the regulator is consumed, the NO concentration will go up and the NO_2 concentration will drift down. At some point, the NO concentration will stabilize; however, this may take days to occur. Whenever a regulator is connected to a NO cylinder, it should be repeatedly flushed or evacuated to remove all traces of oxygen before use. After the regulator has been flushed, it should be removed only if the cylinder contents have been consumed or if the cylinder is not to be used for an extended period of time.

A common source of error when using NO cylinders is to fail to flush out or evacuate the regulator prior to its use.

NITROGEN DIOXIDE CONCENTRATION STANDARD (SPAN GAS)

There are three major NO_2 concentration sources. The first is cylinders supplied by gas manufacturers. As with NO cylinders, concentrations from parts per million to percent levels are available. Normally, NO_2/balance-air cylinders are used. Mixtures with nitrogen or other inert gases are also available.

NO_2 cylinders have traditionally been less stable than NO cylinders or gas mixtures of other major pollutants (CO, SO_2). The exact reasons for this are not clear. NO_2 is more reactive than NO, and certainly heterogeneous wall reactions have been a major cause of NO_2 concentration instability. Over the years, gas suppliers have made great improvements in the reactivity of cylinder wall materials, and NO_2 gas sources today are superior to those of the past. Nevertheless, if cylinders of NO_2 are to be used, high-concentration mixtures have been shown to maintain greater stability, and the cylinder concentration should be considered valid only for a limited period of time before requiring recalibration or certification.

Use of a NO_2 cylinder requires a pressure regulator manufactured from appropriate inert materials and which has a suitable delivery pressure. As with NO cylinders (or any other cylinder gas), the regulator should be properly flushed or evacuated.

The major advantage of using NO_2 cylinders for chemiluminescence analyzer calibration is again convenience. The major disadvantages are potential instability in concentration, cost, and continued need for replacement as the cylinder contents are consumed or become unstable.

The second major NO_2 calibration source is based on Equations (32-1) and (32-2). The rapid gas-phase reaction between NO and O_3 is used to generate stoichiometric quantities of NO_2. This technique is termed gas-phase titration (GPT) and was initially developed by the U.S. EPA. Virtually every

Virtually every chemiluminescence analyzer manufacturer offers GPT calibrators for both ambient and source applications.

A detailed description of the GPT method is given in the Code of Federal Regulations, Title 40, Part 50, Appendix F.

chemiluminescence analyzer manufacturer offers GPT calibrators for both ambient and source applications.

A diagram of a typical GPT system to generate known concentrations of NO_2 is shown in Figure 32-5. A basic system consists of a zero gas source, NO/balance-nitrogen gas standard, flow controllers, ozone generator, reaction chamber, and mixing chamber. A detailed description of the GPT method is given in the Code of Federal Regulations, Title 40, Part 50, Appendix F. The analyzer instruction manual usually provides a full description as well.

In general, the NO channel of the chemiluminescence analyzer is used to determine how much NO_2 is being produced by the GPT system. The analyzer, therefore, must first be calibrated for NO response.

This is accomplished, following the manufacturer's instructions, using a zero gas source and a NO/balance-N_2 cylinder. Once the NO channel of the analyzer is calibrated, a NO concentration of approximately 90% of the desired full-scale range is introduced into the analyzer via the GPT system and the response allowed to stabilize. Ozone is then added to the NO stream by turning on and adjusting the O_3 generator in the GPT system to produce a decrease in the NO concentration equivalent to approximately 80% of the desired full-scale NO_2 range. The decrease in NO concentration is equivalent to the NO_2 concentration generated. The amount of NO_2 generated can be changed by adding variable amounts of O_3 from the O_3 generator.

The major advantage of using a GPT system to generate known concentrations of NO_2 is that all three channels (NO, NO_2, and NO_x) of a typical chemiluminescence analyzer can be calibrated with one system.

The major advantage of using a GPT system to generate known concentrations of NO_2 is that all three channels (NO, NO_2, and NO_x) of a typical chemiluminescence analyzer can be calibrated with one system. Only a single cylinder of gas is required, variable concentrations of both NO and NO_2 are produced, and multipoint calibration of the analyzer can be accomplished. The calibrated NO channel of the analyzer itself is used as the measure of the NO_2 concentration produced; therefore the NO_2 channel calibration can be repeatedly and dynamically checked.

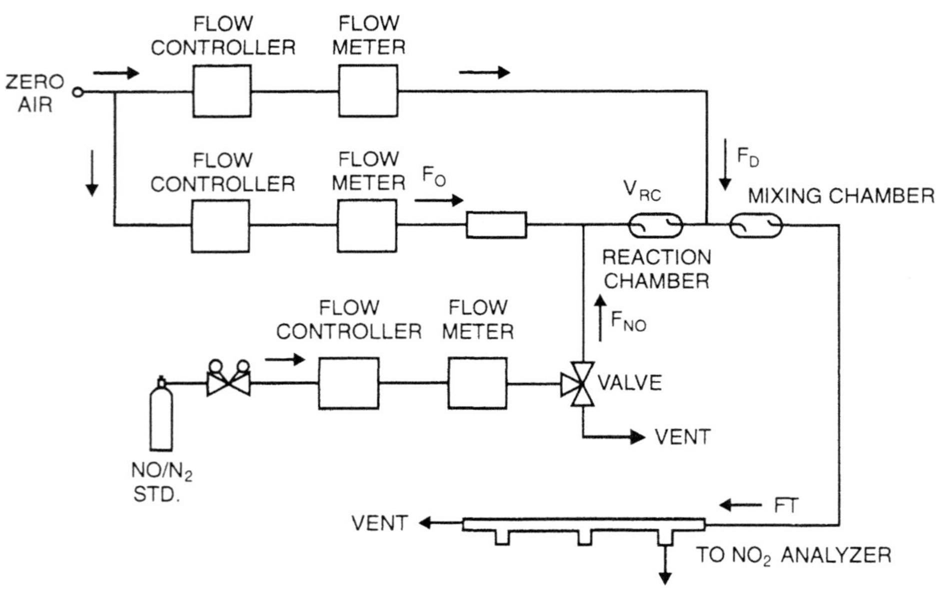

Figure 32-5. Gas-Phase Titration (GPT) System

The major disadvantages of a GPT system are the initial cost outlay, the reliance on the correct calibration of both the NO cylinder and NO channel of the analyzer to obtain accurate NO_2 concentrations, and, finally, the somewhat limited range in NO_2 concentrations available considering the dynamic parameter specifications for the GPT system.

The third major calibration source of NO_2 is the permeation tube. Since permeation tubes are almost solely used to generate ambient concentrations and are not used to calibrate chemiluminescence analyzers at source levels, they will be discussed only briefly.

Permeation tubes are manufactured from either fluorinated ethylene propylene (FEP) or Teflon tubing with variable wall thicknesses and internal diameters. They are partially filled with liquefied samples of pure gas, so as to have a constant vapor pressure of gas at a given temperature within the device. Individual gases permeate at differing rates through the walls of the device and are flushed from the outside surfaces of the tubing into a gas stream of known flow. Depending on the release rate of the tube, and the flow of gas into which the permeating gas is introduced, variable concentrations of the permeating species can be generated.

For NO_2, permeation devices are available as standard reference materials from NIST. There are also a number of commercial sources. The available NO_2 concentrations range from the low parts per billion level to the low parts per million level. The majority of chemiluminescence analyzer manufacturers offer permeation tube systems for instrument calibration.

Calibration (Zero and Span) Procedure

As mentioned before, all chemiluminescence analyzers are similarly calibrated. Detailed calibration procedures for individual commercial analyzers are outlined in the instruction manuals. A general description is presented here.

The analyzer is powered up, the ozonator turned on, and the system allowed to stabilize. Stabilization times for chemiluminescence analyzers with heated converters range from 1/2 to 2 h, depending on whether the analyzer employs a cooled detector. A zero air source, NO standard, and NO_2 standard are required.

A known NO calibration gas of approximately 80% full-scale range is introduced to the instrument and the detector response adjusted (usually by varying the PMT voltage) until the NO channel of the analyzer reads approximately the correct concentration. The analyzer under calibration is then allowed to sample zero gas, stabilize, and zero adjustments are made. Depending on the manufacturer, one, two, or three separate zero adjustments may be required. Note that modern microprocessor-based units can eliminate the need for separate NO, NO_2, or NO_x channel zero adjustments.

Nitric oxide gas is reintroduced at about 80% of the upper limit of the NO range. The instrument is allowed to stabilize, and then the span control is adjusted so that the analyzer response reflects the exact concentration of the introduced gas. If the analyzer has a separate NO_x span control, the instrument NO_x response is adjusted to reflect the exact concentration (which should be the sum of the NO concentration and any NO_2 impurity). Multipoint calibration of the NO and NO_x channels can be done by generating additional concentrations.

Now a known concentration of NO_2 is introduced. Depending on the type of calibration source used and the individual analyzer, different procedures are required. Some analyzers do not have separate NO_2 span adjustments; others allow the microprocessor to calibrate and correct both the NO_2 and NO_x channels for converter inefficiency. The instrument instruction manual should be followed. However, the common parameter for all analyzers is that the NO_2 converter efficiency should be greater than 96% in order to minimize sampling errors. A multipoint calibration should be performed to demonstrate linearity (usually ±1% of full scale) over the desired range.

A multipoint calibration should be performed to demonstrate linearity (usually ±1% of full scale) over the desired range.

Chemiluminescence Sampling Requirements

Chemiluminescence nitrogen oxide analyzers are utilized by numerous industries, federal and state agencies, and university and research institutions. The source (relatively high concentration) areas alone include electric and gas utilities, fossil fuel combustion plants, cogeneration facilities, the petrochemical industry, pulp and paper manufacturers, and the automotive industry.

Each particular industry and/or application has its own unique sampling requirements. It is impossible to detail or even outline fully every individual requirement. Presented here is a brief discussion of those areas that are most critical to the performance of chemiluminescence analyzers in source applications. These include the influence of quenching gases and viscosity changes on the chemiluminescent measurement, the effect and importance of interfering gases, and, finally, the choice and application of the extractive sampling system.

Quenching and Viscosity

Normally, chemiluminescence analyzers are calibrated with standard cylinder mixtures of NO in N_2 (NO/balance-Nitrogen) that may or may not be diluted further with dry zero air. The calibration gases, for all practical purposes, are mixtures of nitrogen oxides in 80% or greater nitrogen with little or no water, carbon dioxide, or other component gases. Actual measurements are made in mixtures having compositions far different than that of the calibration gas. Combustion sources typically have high relative water and CO_2 concentrations, and other sources may have large hydrocarbon, varying oxygen, and/or other component (CO, H_2, Ar) content.

In the section on "Measurement Principles" it is noted that a major limitation to the chemiluminescence reaction is the effect of Equation (32-4) on the fraction of excited NO_2 molecules that deliver their energy in the form of light emission. Equation (32-4) describes the process of third-body-quenching, whereby excited NO_2 molecules are "robbed" of their energy by collisional deactivation. A number of different studies [Ref. 13-15] have shown that component gases such as water vapor, CO_2, and hydrocarbons have different quenching efficiencies relative to nitrogen and thus can cause significant errors when measuring sample gases containing large fractions of these components. For vacuum chemiluminescence instruments, operating at approximately 10 mm Hg pressure, every percent of water in the sample reduces the signal relative to that of nitrogen by about 0.3%. Every percent of CO_2 reduces the signal relative to that of nitrogen by about 0.05%. In combustion samples, where water vapor and CO_2 concentrations are often in the 10 to 20% range, cumulative errors of approximately 10% can occur.

In combustion samples, where water vapor and CO_2 concentrations are often in the 10 to 20% range, cumulative errors of approximately 10% can occur.

It has also been noted in this chapter how critical to analyzer performance is stability of the sample flow in the instrument. Flow stability is generally maintained by using capillaries or critical orifices and by regulating the pressure drop. This approach generally works well only if the sample gas viscosity remains constant. Volumetric flow rate through a capillary with a fixed pressure drop is inversely proportional to sample gas viscosity [Ref. 16]. Sample gases with compositions other than that of the calibration gas will have different viscosities and cause changes in the sample flow rate. The number of NO molecules entering the reaction chamber in time (as well as the reaction chamber pressure) will differ from calibration values, resulting in sample analysis errors.

Several possible approaches can be used to minimize the effects of differing sample composition.

Gases such as water vapor, CO_2, hydrocarbons, H_2, or helium thus alter the response of a chemiluminescence analyzer relative to that of nitrogen (the most common calibration balance gas) in two ways: first, by changing the relative quenching efficiency of the sample; second, by altering the sample flow rate in the analyzer due to viscosity effects. Several possible approaches can be used to minimize the effects of differing sample composition.

The effects of background gas composition on the indicated nitrogen oxide signal can be corrected numerically by utilizing quenching efficiencies and viscosity information for the sample components, either taken from the literature or determined empirically. The advantage of this approach is that the sample is measured directly without changing its composition, and calibration is achieved using the standard approach of NO/balance-nitrogen.

The disadvantages of this approach are many. First, the concentrations of the different component gases must be measured, which requires additional monitoring instrumentation. Second, unless the concentrations of the component gases are invariant, they must be monitored continuously. Third, quenching efficiencies reported in the literature vary considerably, because different instruments, produced by different manufacturers and operating at different pressures, were used to obtain the reported data. Finally, empirical determination of quenching and viscosity data is costly, subject to operator error, and prone to individual instrument variation.

Another approach is to calibrate the chemiluminescence analyzer with NO in a background gas similar to that of the actual sample. Using this method, both quenching and viscosity effects are "canceled out" during calibration. The advantage is one of convenience, and because the actual monitoring analyzer is calibrated directly, individual instrument variations become immaterial. As in the numerical correction approach, concentrations of the component gases within the sampling mixture must be determined, requiring additional monitoring equipment. Another disadvantage is that the component composition must remain the same as used in calibration; otherwise, the "canceling out" assumption is invalid.

A third approach is to dilute the sample with a known amount of nitrogen or air. Variations in sample composition will have less effect on the flow rate and quenching because the majority of the sample is now the balance gas. Also, the absolute concentrations of the quenching gases are reduced and their effect on the analyzer response by Equation (32-4) is minimized. Dilution ratios of 10:1 to 300:1 are commonly used. This method has the added benefit of lowering the dew point of the sample, which reduces the possibility of condensation in the sample line prior to the analyzer as well as in the analyzer itself. The difficulties presented by this technique are that the dilution ratio must be both large and accurately known, additional equipment (flowmeters, capillaries, solenoid valves, pump) is required, resulting in added cost, and the overall sensitivity of the chemiluminescence instrument is reduced by the dilution ratio.

A final approach is to remove the quenching gases from the sample stream, thereby eliminating their influence on the analyzer response and sample flow rate. Removal can be achieved by chemical reaction, permeation, adsorption, or condensation. Water vapor and carbon dioxide are the most common component gases treated this way. The disadvantages of this approach are the added cost of the removal agents, the strong possibility of altering the nitrogen oxide content of the sample, and the change in volume percent of the nitrogen oxides due to the removal of high concentrations of water vapor and/or carbon dioxide from the sample.

Interferences

Although quenching and viscosity influences on a chemiluminescence analyzer may be considered to "interfere" with the instrument measurement, the term "interferant" has been traditionally given to a component in the gas stream which, by virtue of thermodynamic or chemical reaction, causes an increase in chemiluminescence signal. This increase in signal can, in theory, occur in the NO channel, the NO_x channel, or in both channels.

In practice, the NO channel rarely is subject to interference. The detector/filter combination gives a high degree of specificity, as has already been noted. Also, the reaction of NO with O_3 is quite rapid compared to other component reactions.

Therefore, in an instrument designed with an appropriately-sized reaction chamber for the analyzer flow rate, any unwanted chemiluminescence (usually by a slower, but possibly by a faster, reaction with ozone) can be effectively discriminated against. Unless a compound such as ethylene is present at a concentration orders of magnitude greater than NO, the channel is typically free of interference.

The NO_x channel, however, is much more likely to suffer from interfering compounds.

The NO_x channel, however, is much more likely to suffer from interfering compounds. The converter, whether thermodynamic or chemical, is subject to interference from other species such as ammonia or nitric acid. This has already been discussed in some detail.

Dilution of the sample stream does nothing to eliminate this type of interference. Both the nitrogen oxides and the interfering compounds are diluted by the same ratio. Therefore, the error in measurement remains the same.

If the concentration and conversion efficiency to NO of the interfering components are known, a numerical correction to the instrument response can be made. It is extremely difficult in practice, however, to monitor every possible interfering component and obtain an accurate correction. Only in relatively simple sample matrices (backgrounds) are such corrections meaningful.

As an alternative, the interfering components can be removed from the sample stream prior to analysis. Chemical reaction, adsorption, condensation, and permeation are possible approaches. It is very difficult, however, to selectively remove other nitrogen-containing compounds without seriously affecting the NO_2 concentration. Nitrogen dioxide is both water soluble and extremely reactive.

Extractive Sample Conditioning System

An extractive sampling system generally should perform a variety of different functions in order to obtain reliable, reproducible, and accurate data from the analyzer. The system must provide a sample from the source that gives an accurate representation of total emissions. It must not affect the component composition in such a way as to alter the relative concentration ratios or the final reported measurements.

The system must provide a sample from the source that gives an accurate representation of total emissions. It must not affect the component composition in such a way as to alter the relative concentration ratios or the final reported measurements.

The sample must be conditioned to be compatible with the monitoring instrument requirements. An extractive system must also provide a means to calibrate the entire system following a flow path as close as possible to that of the actual sample measurement.

An accurate representation of the total emissions from the source is obtained by optimum positioning of the in-situ sampling probe(s) in the duct (stack or process stream). Details for choosing the appropriate location and/or mounting arrangement are given elsewhere in this handbook. For nitrogen oxides, the probe may or may not be heated; however, it must be constructed of inert materials, such as stainless steel or Inconel, and have suitable filtration to prevent or minimize plugging. A back flush or purge is often incorporated into the probe design to aid in this requirement.

For nitrogen oxides, the transport tubing should be stainless steel or Teflon.

After filtration, the sample may be sent via transport tubing, which can be heated, through a conditioning unit and on to the analyzer. For nitrogen oxides, the transport tubing should be stainless steel or Teflon. Other types of probe systems are also available which do in-situ dilution via critical orifice design, lowering the dew point of the sample to the point at which no further conditioning is required. The appropriateness of a particular system depends on the application.

The major approaches to sample conditioning are chemical drying, condensation, permeation, and dilution. Chemical drying and condensation systems primarily deal with the removal of water vapor from the sample stream to eliminate its effect on the chemiluminescence analyzer. As has been discussed, water vapor is the principal quenching gas in the majority of source sample streams. This type of conditioning is satisfactory if the monitored component does not absorb or react on the chemical dryer or in the water collected during condensation. Nitrogen dioxide is highly reactive, however, and is also water soluble. If accurate NO_2 measurements are required, either of these two drying approaches is questionable. If the nitrogen oxide component consists of NO or a small fraction is NO_2, these approaches can give acceptable results.

Permeation conditioners operate on the principle of selective water vapor permeation through a suitable membrane. This type of device has the theoretical advantages of being nonconsumable, being less prone to corrosion (since water is removed in the gas phase rather than through condensation), and requiring no periodic removal of the condensate. Disadvantages include the necessity for creating the driving force for permeation via back flush or evacuation, cost, and some loss of NO_2 through the permeating membrane.

Chemical drying, condensation, and permeation drying provide a sample whose measurement is on a dry basis. Because the water vapor is removed prior to analysis, the volume percent measurement of the nitrogen oxides does not represent the actual concentration in the source. Assuming no loss of nitrogen oxides during the conditioning portion of the sampling system, this analysis method will lead to a reported concentration that is higher than the true concentration.

Chemical drying, condensation, and permeation drying provide a sample whose measurement is on a dry basis. Because the water vapor is removed prior to analysis, the volume percent measurement of the nitrogen oxides does not represent the actual concentration in the source. Assuming no loss of nitrogen oxides during the conditioning portion of the sampling system, this analysis method will lead to a reported concentration that is higher than the true concentration.

Gas dilution is another major conditioning method. This approach dilutes the source gas sample at known ratios with inert, dry gas. Dilution can be in-situ at the probe, or later, after gas transport. This technique has the advantage of reducing gas-handling problems by decreasing the temperature and dew point of the gas entering the analyzer. The water is not actually removed from the sample, but is diluted at the same ratio as the nitrogen oxide; therefore, the measurement is on a wet basis, giving an accurate representation of the gas composition at the source.

Disadvantages of dilution systems include increased complexity, requirement for precise dilution ratios, initial cost outlay, and reduction in sample sensitivity. Although ambient analyzers can be used to measure nitrogen oxides at adequate sensitivities in the majority of cases, if the sample stream is being analyzed for other components (e.g., carbon monoxide), a dilution approach may be unacceptable because of sensitivity requirements. As modern instrumentation becomes more sensitive and reliable, this will become less of a disadvantage.

Troubleshooting and Maintenance

Individual manufacturers of chemiluminescence analyzers usually provide detailed troubleshooting and maintenance guidelines in the instrument instruction manual. This section offers some general guidelines. In all cases, the manual and the manufacturer of the instrument are the first choices for information regarding problems or questions.

Troubleshooting an analyzer requires certain equipment. This includes a digital multimeter capable of measuring voltages, current, and resistance. An oscilloscope can be helpful to view and analyze waveforms, voltages, and test signals. Calibration and zero gas are necessary to check instrument response to known samples, and a flowmeter in the 20 to 2,000 cm^3/min range is useful. A set of tools containing a variety of wrenches, screwdrivers, and pliers is also helpful.

Electronic problems or checks cannot be adequately outlined here because of the large variation in individual instrumentation. However, a majority of problems can be traced to defective voltage supplies, blown fuses, and unplugged analyzers, pumps, or fans.

In modern instrumentation, the ±15 and ±5 V supplies should be checked as well as any other available voltages. Test points on individual circuit boards often are noted in the instruction manual, which can supply useful troubleshooting information. As discussed in the "Basic Instrumentation" section of this chapter, high voltages are required for proper ozonator and photomultiplier operation. A high-voltage probe may be necessary to check these components. Only a properly trained individual who is thoroughly familiar with the given analyzer should attempt any maintenance or repair.

Because chemiluminescence analyzers are so sensitive to flow and pressure, many sources of problems can be traced to these parameters. Sample leaks, both external and internal to the analyzer, can cause instrument drift, excessive noise, nonlinearity, and sensitivity variation. Plugged capillaries or orifices, blocked plumbing, and/or regulator leaks can also cause flow sensitivity and pressure changes. Changes in sample flow rate brought about by blower or vacuum pump performance cause instrument response variation by altering the reaction chamber pressure and gas flow rates.

Confusing lag time with response time is a common error.

Excessive response time can be caused by a partially blocked sample capillary or orifice. Electronic filtering changes in the signal-processing circuitry may also cause changes in response time. Confusing lag time with response time is a common error. Response time is the interval between the first detectable response of an instrument to an introduced concentration and a response of 90 to 95% of its final value. Lag time is the interval between introduction of source gas to the analyzer and the first response of the instrument to that gas.

Low converter efficiency can be caused by a number of factors. The converter temperature may be set too low, or the line voltage may be below instrument limits. If the converter is a high-temperature, thermodynamic type, it normally requires conditioning with a sample containing oxygen for a period of time before converting NO_2 efficiently. If the converter is of the chemical reaction type, it can be consumed or poisoned. If so, it normally requires replacement. Sometimes "low" efficiency is due not to an improperly operating instrument, but to an improperly calibrated NO_2 source. NO_2 sources are not always reliable or stable with time.

As noted earlier, the analyzer instruction manual and the manufacturer's service department are the user's best sources for troubleshooting and maintenance information. Although chemiluminescence nitrogen oxide instruments can appear to be quite complex and difficult to operate, if properly maintained and understood, they can provide reliable, accurate measurements for many years without need for replacement.

It is important to realize that the analyzer is part of a measuring system, not a standalone unit. What appears to be instrument problems or malfunctions can actually originate in the sampling system, calibration system, or gas supply. The best approach to analyzer maintenance is to understand thoroughly the fundamental principles upon which the process analyzer system is based.

Acknowledgment

The author would like to thank Kathy McElroy for her assistance in preparation of this chapter.

References

1. Clough, P.N., and Thrush, B.A., *Transactions of the Faraday Society*, 63:915, 1967.
2. Fontijn, A., Sabadell, A.J., and Ronco, R.J., *Analytical Chemistry*, 42:575, 1970.
3. Stedman, D.H., Daby, E.E., Stuhl, F., and Niki, H., *Journal of the Air Pollution Control Association*, 22:260, 1972.
4. Niki, H., Warnick, A., and Lord, R.R., Paper 710072, Warrendale, PA: Society of Automotive Engineers, 1971.
5. Sigsby, J.E., Jr., Black, F.M., Bellar, T.A., and Klosterman, D.L., *Environmental Science and Technology*, 7:51, 1973.
6. Breitenbach, L.P., and Shelef, M., *Journal of the Air Pollution Control Association*, 23:128, 1973.
7. Kley, D., and McFarland, M., *Atmospheric Technology*, 12:63, 1980.
8. Fehsenfeld, F.C., et al., *Journal of Geophysical Research*, 92:14710, 1987.
9. Ridley, B.A., et al., *Journal of Geophysical Research*, 93:15803, 1988.
10. Steffenson, D.M., and Stedman, D.H., *Analytical Chemistry*, 46:1704, 1974.
11. Mehrabzadeh, A.A., O'Brien, R.J., and Hood, T.M., *Analytical Chemistry*, 55:1660, 1983.
12. Mehrabzadeh, A.A., O'Brien, R.J., and Hood, T.M., *Review of Scientific Instruments*, 54:1712, 1983.
13. Matthews, R.D., Sawyer, R.F., and Schefer, R.W., *Environmental Science and Technology*, 11:1092, 1977.
14. Zabielski, M.F., Seery, D.J., and Dodge, L.G., *Environmental Science and Technology*, 18:88, 1984.
15. Tidona, R.J., Nizami, A.A., and Cemansky, N.P., *Journal of the Air Pollution Control Association*, 38:806, 1988.
16. Folsom, B.A., and Courtney, C.W., "Fundamental Combustion Research Applied to Pollutant Control, Special Report No. 1: Chemiluminescent Measurement of Nitric Oxide in Combustion Products," Santa Ana, CA: Energy and Environmental Research Corp., 1978.

About the Author

Dieter Kita, Ph.D., M.B.A., received his doctorate in analytical chemistry from the University of Michigan, Ann Arbor, Michigan and his M.B.A. from Worcester Polytechnic Institute, Worcester, Massachusetts. Prior to joining Thermo Environmental Instruments Inc., he worked as a research associate at the Space Physics Research Laboratory at the University of Michigan. He is currently manager of research and development for Thermo Environmental Instruments Inc., Franklin, Massachusetts. He has more than 15 years of experience in the research, design, and development of instrumentation for a variety of analytical and environmental applications including nitric oxide monitoring using chemiluminescence. He has also published a number of papers and spoken at many conferences about instrumental analyses in general, and more specifically about environmental measurement.

Compositional Process Analyzer Systems

Of all the techniques that have been developed to separate and measure components of process streams, the most common types employed are gas chromatographs (GC). Literally thousands of applications for GCs have been developed, enabling this technology to be widely employed in the manufacturing operations of almost every industry.

Four major areas have seen advances in recent years: more substrates available for analytical columns; improvements in sample and column switching valve technology; new and improved detector types and capabilities; and greater utilization of microprocessors (control of the GC hardware, software programming capability and increased data reduction capability). An example is data storage and computations required for Continuous Emission Monitoring System (CEMS) applications.

This has led to new applications of basic hardware allowing sample properties to be measured as well as individual components. For example, Simulated Distillation (SIMDIS) applications have been available for years, but accuracy and reliability problems limited their widespread usage. Recent improvements in hardware and software have largely overcome these limitations and the number of these applications is expected to increase. Manufacturers are also developing applications on the basic GC hardware to measure total sulfur, total (organic) nitrogen, total aromatics, and total olefins in complex hydrocarbon streams. These measurements often require complex correlations (models) developed through comparison of data gathered from thousands of samples analyzed by the new technique results of the same samples whose properties are measured by laboratory (off-line) techniques. The use of data highways to connect large numbers of GCs together has enabled complex models and property databases to reside in personal computers also tied into the data highway. The GC only needs to make the required measurements and transmit the raw data to the personal computer, where the complex modeling and data calculations take place.

Applications of mass spectrometers have been limited in the past by the need to employ high vacuum systems as a part of the sample preparation/introduction system (front end). Recent technological advancements have greatly reduced this front end requirement. Newer techniques hold the promise of measurement at atmospheric pressure. Mass spectrometer applications are expected to greatly increase in the near future.

Although less than 15% of the more than 5 million known organic molecules can be separated by gas chromatographic techniques (vapor pressure, boiling point, and sublimation problems preclude the remaining organic molecules from vapor phase separations) virtually *all* the known organic molecules could be separated by liquid chromatographic (LC) techniques. The potential application of LC has not been realized due to the complexity and reliability of the required hardware. High pressure pumps are often required for adequate liquid carrier flow through the column packing material and are a frequent source of system reliability problems. Unlike their GC counterparts, detectors for LCs are very limited. This further limits potential applications of LC. Ironically, this lack of potential applications limits the development of new and better detectors. These limitations ultimately result in a personnel with skills and experience for development of LC applications. However, the number of LC applications could greatly increase if the hardware reliability problems can be solved.

Chapters 33 through 36 will cover the more common applications of Composition Analyzers as follows:

Chapter 33	Gas Chromatographic (GC) Analyzers
Chapter 34	Liquid Chromatographic (LC) Analyzers
Chapter 35	Distillation (% Off) Analyzers
Chapter 36	Mass Spectrometric (Mass Spec) Analyzers

33

Gas Chromatographic (GC) Analyzers

Chromatography dates back to the mid-1800s, when German chemists separated inorganic cations by passing mixtures of them through capillary porous materials. The term "chromatography" is identified with work published in 1906 by Tswett, a Russian botanist [Ref. 1]. He used packed columns to separate the individual components in mixtures of plant extracts. He found that the column produced a series of brilliantly colored bands (the separated and individually colorful components). The name comes from the Greek words *chromo*, which means color, and *graph*, which means writing.

The development of gas chromatographic (GC) analysis is usually more closely identified with a turn-of-the-century American chemist, D.T. Day [Ref. 2], who experimented with crude petroleum and separated it into many components by passing it through columns of Fuller's earth. In 1941, Martin and Synge [Ref. 3] suggested the use of a gaseous mobile phase instead of a liquid. James and Martin followed up the idea, and the results were presented at the Analytical Chemistry Congress at Oxford in 1952. The potential of the method was immediately recognized, and it developed rapidly. The wide variety of capabilities, generated by a multitude of column, carrier, operating condition, and detector combinations has made GC one of today's most popular analytical tools.

Process gas chromatography quickly followed, and by 1955 [Ref. 4] a small display and control panel manufacturer, Watts Manufacturing of Roncevert, West Virginia, was manufacturing process GCs (on a contract basis at first and later as a commercial product). In 1957, Beckman Instruments acquired the company and helped introduce process GC to its large, existing customer base. Today process gas chromatography is probably the most widely used compositional process analysis technique. Many of the companies that manufacture process GCs are listed in Table 33-1.

Table 33-1. Process GC Vendors

Company	Location
Applied Automation/Hartmann & Braun	Bartlesville, OK
Asea Brown Boveri (ABB)-Process Analytics	Lewisburg, WV
Analytical Instruments Inc.	Houston, TX
Daniel Instruments Inc.	Houston, TX
Foxboro Analytical	Foxboro, MA
Fluid Data Inc.	Angleton, TX
Rosemount Analytical	La Habra, CA
Siemens AG/ES Industries	Karlsruhe, Germany/Marlton, NJ
Wasson-ECE	Fort Collins, CO
Yokogawa Hokushin Electric (YEW)/Johnson Yokogawa Corporation	Tokyo, Japan/Newnan, GA

Measurement Principles

Because of its resolving power, speed, and small sample size requirement, GC is one of the most popular methods applied in analytical chemistry. It is a technique for separating substances in the vapor state and then quantifying the separated substances. It is both a quantitative (how much?) method and a semiqualitative (what? to a limited degree) method. The sample must be vaporizable at the temperature of the GC analyzer oven and in either the liquid or gaseous state when injected into the analyzer. The terms gas solid chromatography (GSC) and gas liquid chromatography (GLC) often are used to describe the different physical separation mechanisms that are employed in most GC separations. The separated components are then passed through a detector and quantitatively measured.

Figure 33-1 depicts a typical basic process GC system. This basic system consists of several components. The process sample tap, sample-conditioning system, sample return, calibration sample, analyzer shelter considerations, and interfacing to other systems are all covered in Part I of this handbook and will be only briefly touched in this section. The carrier gas supply, inject valve, separation column, column oven, detector, detector oven, GC controller, data highway, GC system controller, and special communications features constitute the heart of this chapter.

The purpose of all the sample components prior to the sample inject valve is simply to supply the appropriate homogeneous sample to the sample inject valve. The *carrier gas system* usually consists of a carrier gas supply (often gas cylinders of He, N_2, Ar, or H_2), single or multiple pressure regulators, and flow controllers (needle valves, mechanical flow-controllers, or electronic mass-flow controllers).

The purpose of the *sample inject valve* is to repeatedly inject the same physical volume of sample into the carrier gas stream and onto the column.

The purpose of the *sample inject valve* is to repeatedly inject the same physical volume (usually 0.25 to 5 μL of liquid or 0.1 to 2 cm^3 of vapor) of sample into the carrier gas stream and onto the column. (Some liquid inject valves incorporate a vaporizer in their injection scheme; others rely on the sample vaporizing as it contacts the column. This will be discussed further in the section "Valves.")

The *separation column* is used to separate the desired components that make up the sample.

The *separation column* is used to separate the desired components that make up the sample. Often, the operation of the column is oversimplified and compared to a distillation column. A GC column does not only separate components by their boiling points. With the appropriate column selection, one can impart a variety of separation criteria (polarity, reactivity, absorptivity, boiling point, etc.). These additional separation criteria are especially useful when trying to separate similar boiling compounds and/or trying do a fast analysis.

Figure 33-2 depicts two GC column separation techniques. In Figure 33-2(a) (GLC), component A has a low solubility in the liquid phase and as such moves quickly through the column (spending little time in the stationary liquid phase). Component B has a higher solubility than A, and C has the highest. Thus, C spends the most time in the stationary liquid phase and is the slowest moving through the column. Vapor pressure also influences the ability of a column to retain a component. Essentially, the higher the vapor pressure a component exerts, the more it wants to stay in the vapor phase and move rapidly through the column (all else being equal). These characteristics can be complementary (additive): Low vapor pressure and high solubility will move slowly through the column, whereas the opposites will move rapidly. These characteristics can also be competitive. For example, high-vapor-pressure and high-solubility materials want to be in both phases and as such will move through the column at medium (average) rate.

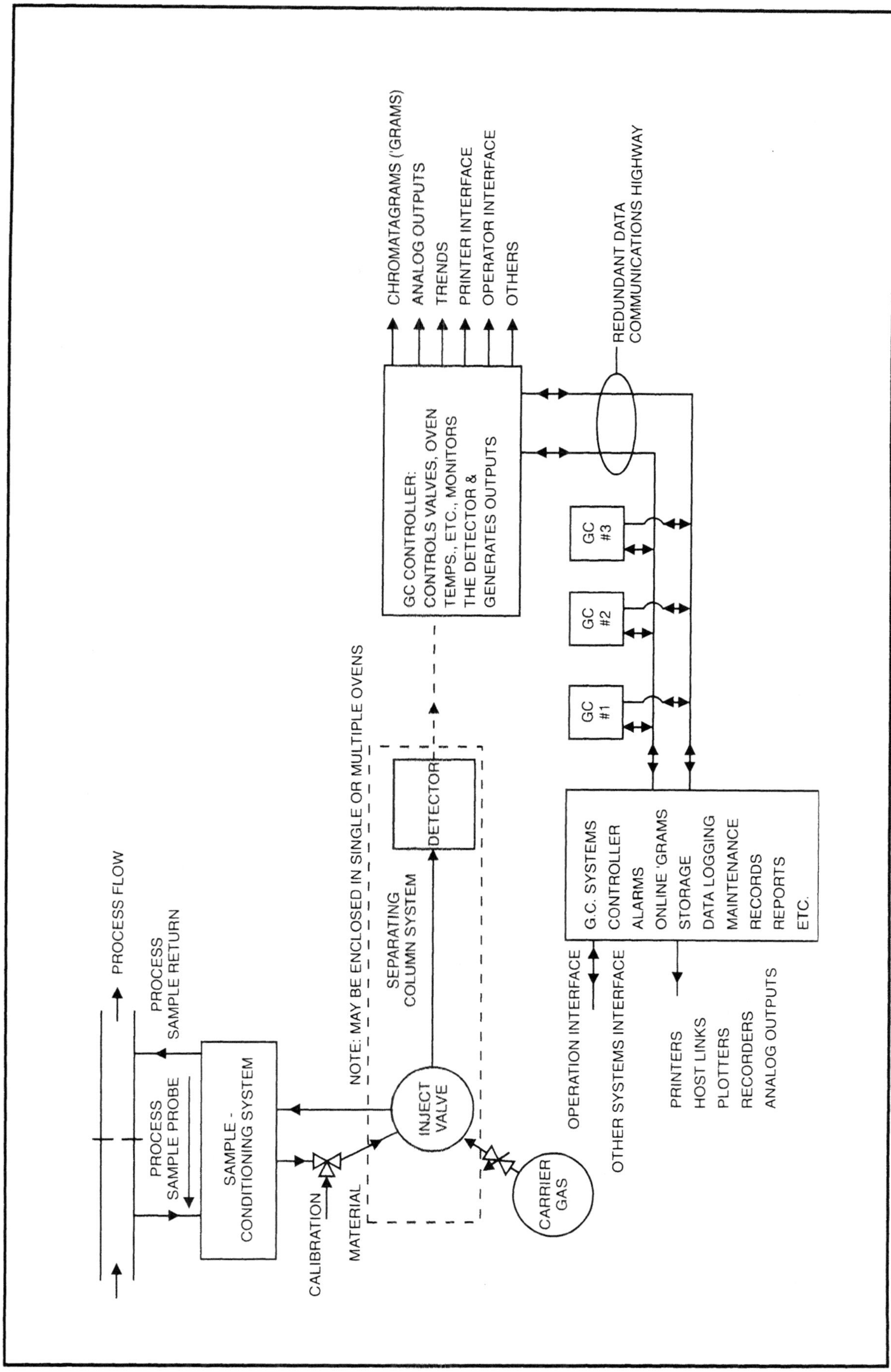

Figure 33-1. Basic Process GC System

Gas solid chromatography (GSC) (Figure 33-2b) is analogous to GLC, except that adsorption on the surface of a solid replaces absorption (solubility) into the liquid phase as a component-separating force. The more strongly a component is absorbed on the surface of the stationary phase, the more time it spends on the stationary phase and the more slowly it moves through the column.

Figure 33-2 (a) and (b) is meant to help an analyzer engineer or technician simply understand what is going on in a GC column. Column selection and design are best left up to the analytical chemist. If you find yourself in a situation where you must design a column system, please refer to the analytical chemistry reference books listed in the bibliography.

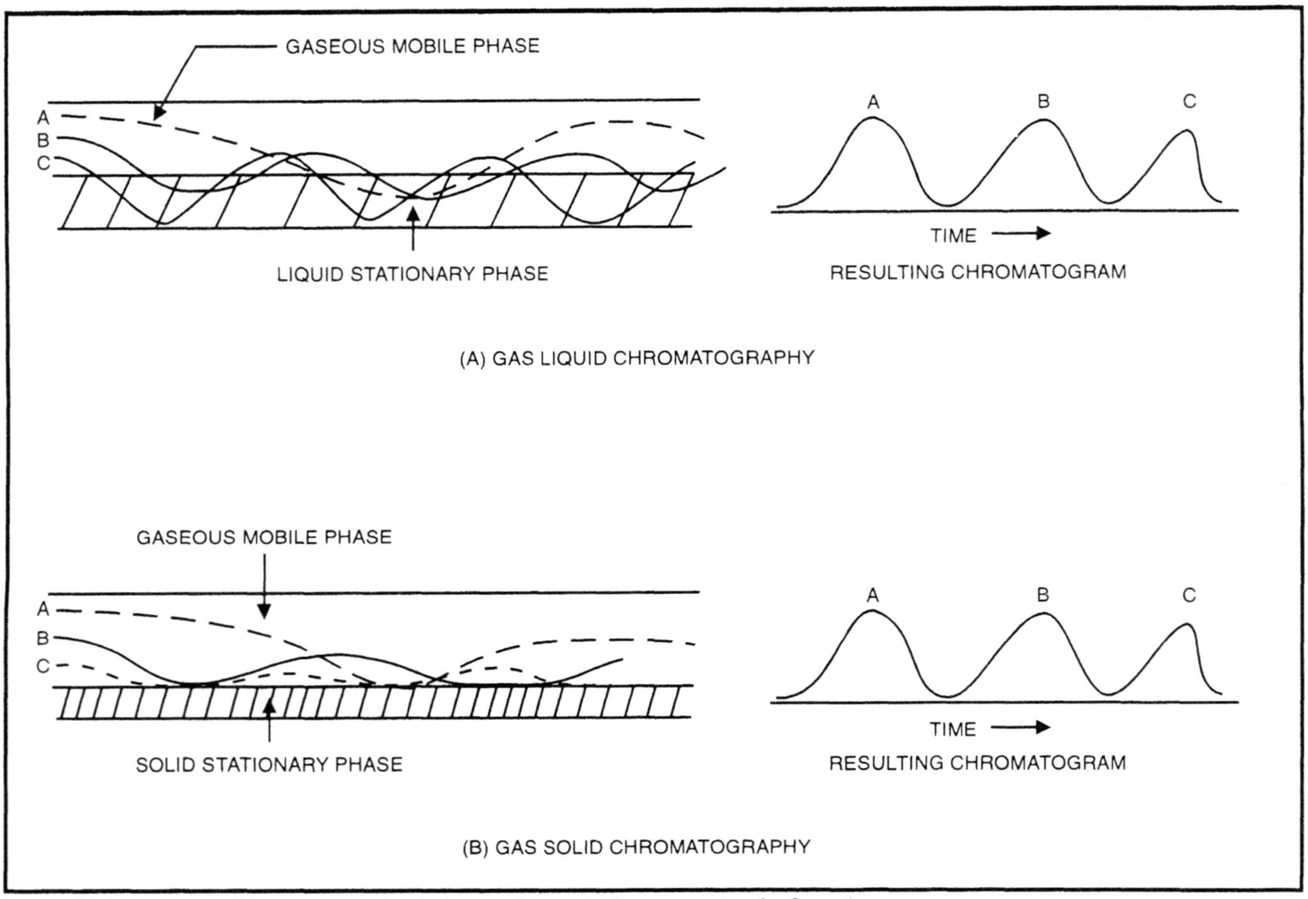

Figure 33-2. How Gas Chromatographic Columns Separate Components of a Sample

The most basic GC separation system consists of a single column between the inject valve and the detector. It is quite common for this single column to be replaced by several columns and valves, and sometimes multiple detectors are used in more complex systems. Some of these optional configurations will be reviewed in the section "Options and Accessories."

The *column oven* is simply an insulated and thermally-heated box that serves two purposes. It assists in providing stability and/or repeatability for the separation mechanism and allows control of the column system at an optimum (isothermal) temperature to effect the desired separation. In some cases, multiple temperature zones and/or temperature programming are utilized. Often, a separate isothermal column oven is used to house and thermally-stabilize the detector.

The *detector* allows the concentration of the component or components of interest to be measured. The separation column has separated the desired components from one another and presents them to the detector as isolated "batches" of the carrier and the component. Several separate types of detectors are available for use in process GC work [e.g., thermal conductivity (TCD), flame ionization (FID), flame photometric (FPD), electron capture (ECD), hall effect (HED), thermionic conductivity (TIC), flame infrared emission (FIE), and others]. Several will be discussed in the section "Options and Accessories." For the moment, a simple thermal conductivity detector (TCD) approach will be used as an example.

The TCD has two sides, a reference and a measure. The reference side normally sees only a regulated flow of pure carrier. The sample side is exposed to the regulated flow of pure carrier and intermittent "batches" (plug flow) of carrier mixed with the separated sample components. Both sides of the detector contain a hot filament that conducts heat through the carrier mixture and to the detector body as the carrier flows through the detector.

The detector is zeroed when both sides are exposed to pure carrier. As the "batches" of carrier mixed with component pass through the measure side, the thermal conductivity of the mixture, being different from the pure carrier, upsets the thermal equilibrium of the detector element; this imbalance is quantified by the TCD control circuitry. In this case, the imbalance is attributable to a comparison of the difference between the thermal conductivity of the sample component plus carrier gas and the concentration of the pure carrier gas. Since the thermal conductivities of materials are a constant (physical property), this measured imbalance can be mathematically solved for the remaining variable, concentration.

The *GC controller* directs the analyzer operation and receives the analog output of the detector. It converts the analog output to a digital value that is mathematically reduced to a sample concentration by one of many mathematical algorithms, and then outputs the result. Several different signal output and mathematical data reduction approaches will be discussed in later sections of this chapter.

Many older (premicroprocessor and centralized-controller) process GC units did not have the data reduction and centralized communication features offered by modern systems. They utilized instead crude cam timers, microswitches, sample-and-hold circuits, RC-integration circuits, and many other circuits and mechanisms. Since these types of units are no longer actively marketed, they will not be discussed in this chapter. Please refer to your vendor's operation and maintenance manuals for assistance if a GC of this type is part of your analyzer inventory. Though quite limited in distribution, they are very reliable and useful in the appropriate application.

One of the many ways that a GC controller can output its results is to a data highway system and then on to a GC system controller that has communication with multiple GC units. Results are output individually or in combination by any of the several methods depicted in Figure 33-1. Figure 33-3 presents a few of the ways that a chromatograph result can be displayed.

Common Applications

Process GC is suitable for any application where the sample components of interest are vaporizable, sufficiently separable on a GC column, and measurable on a compatible detector. The technique is applied in nearly all major industries. Table 33-2 lists several well-known applications.

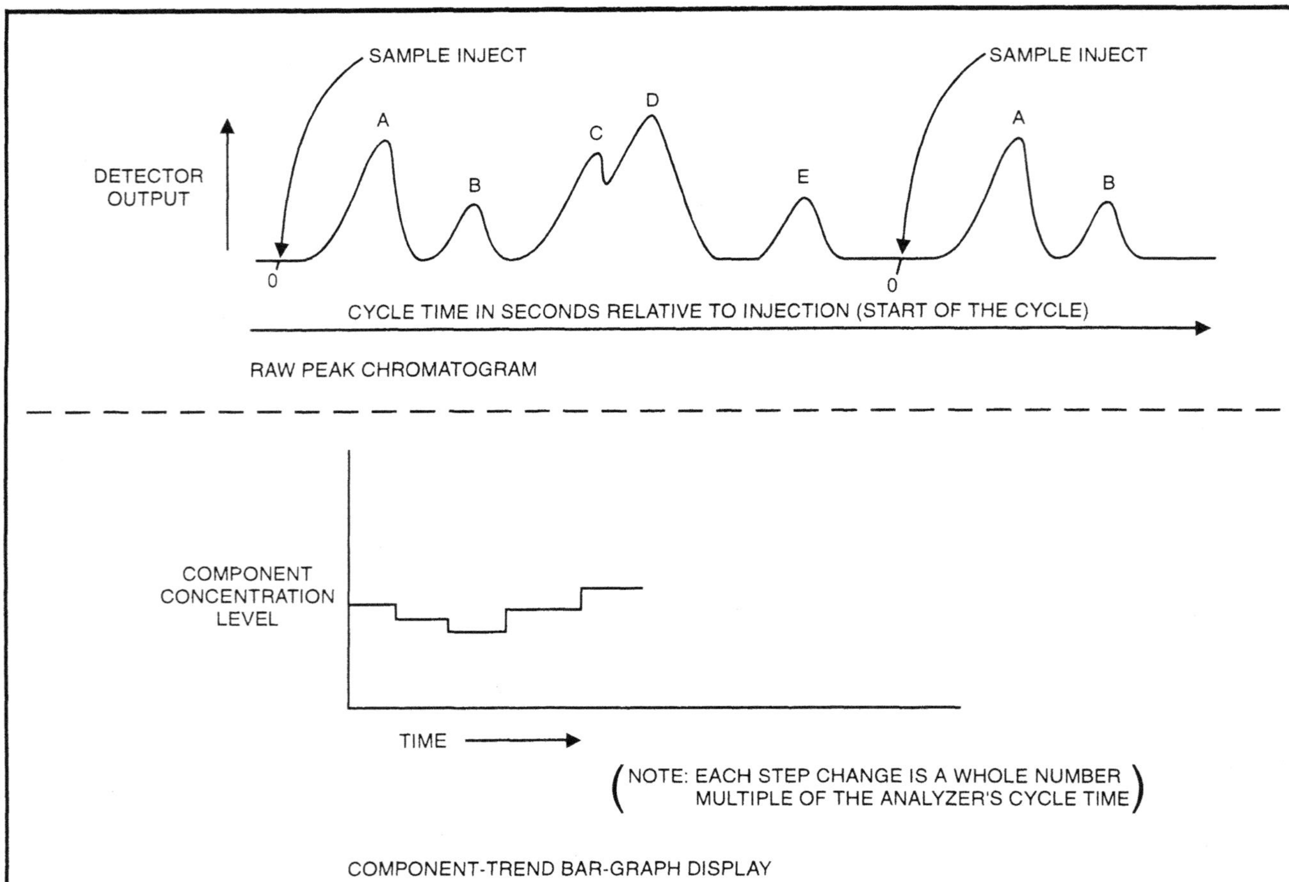

TIME AND DATE:

ANALYZER NAME:
STREAM NUMBER:
ALARM CONDITIONS:

COMPONENT	RETENTION TIME	CONCENTRATION %
A	10.1	20.1
B	22.7	8.3
C	38.2	7.5
D	46.7	48.3
E	65.2	12.7

TYPICAL PRINTED REPORT (SERIAL OUTPUT)

Figure 33-3. Typical Gas Chromatograph Signal Output Examples

Table 33-2. Process GC Applications by Industry Sector

Process GC Applications by Industry Sector
Refining
Gas fractionation
Catalytic cracker
Distillation control
Aromatics BTX unit
Catalytic reforming
Butane and pentane isomerization
Hydrogen plant
Coker VRU
Gasoline blending
Flare gas recovery
Natural gas BTU
Sulfur in tail gas
Chemicals
Ethylene production
Alcohol production
Distillation control
Reactor product analysis
Ammonia production
Vinyl chloride production
Chlorine purity
Polypropylene production
Methyl chloride production
Styrene production
Chlorofluorocarbon production
Silicone production
Safety and Environmental Monitoring
Ambient air (OSHA) monitoring (primarily regulated species)
Emissions vent monitoring (primarily air toxics)
Water pollutants (volatile effluent species)
Fenceline air monitors (air toxics and fugitive emissions)
Biotechnology Industries
Fermenter off-gas monitoring
Pharmaceuticals
Solvent concentration
Ethylene oxide sterilization
Steel Industry
Blast furnace top-gas analysis
Foods and Beverages
Alcohol concentration

Ranges Of Variables

The basic output of a process GC is a concentration. The physical output can be of a current, voltage, or serial format, but it represents a concentration value that can be anything from parts per billion to 100 percent. In addition, recall that a GC can measure multiple components.

The only real limitation on the range is the one imposed by the sample size, the column, and the detector. The detector needs to see a sample concentration in the carrier gas that exceeds its level of detection and is less than the level that results in detector saturation. If you have an appropriate sample and are measuring only one component, you can play with the sample size until you have enough molecules of the desired component to meet the detector's needs. Unfortunately, it is usually not quite that simple. You are often trying to measure multiple components, and the sample size that is ideal for a trace component may be much too large for a major component. Column loading is another concern. Inject too large a sample onto a column can overload it, preventing the normal interaction between the mobile and the stationary phases from taking place. The result is a bad separation.

Options and Accessories

The growth and success of process GC are due in part to the many optional detectors, valves, ovens, column systems, and quantification techniques that are available.

Detectors

Detector options include the thermal conductivity detector (TCD), flame ionization detector (FID), flame photometric detector (FPD), photoionization detector (PID), Hall effect (electrolytic conductivity) detector (HED), electron capture detector (ECD), mass spectrometer detector (MSD), and thermionic conductivity (TIC) detector. The detectors discussed here are of a generic variety. Refer to the GC vendor's operation manual for specific use and maintenance needs.

Each sample component will have its own detector response factor, based on its own thermal conductivity.

THERMAL CONDUCTIVITY DETECTOR

Some of the more popular versions of the TCD are the simple Wheatstone bridge, constant-temperature, and thermistor. All are based on measurement of the amount of heat that a sample-plus-carrier flow (column effluent) conducts relative to a reference flow of carrier gas. The indirect measurement of relative heat loss can then be converted to the sample concentration. This technique works by assuming a constant flow rate and a binary mixture of carrier and sample component. Each sample component will have its own detector response factor, based on its own thermal conductivity.

Figure 33-4 depicts a simple Wheatstone bridge TCD. The filaments (resistance elements) of the bridge are constructed of a material whose electrical resistance varies with temperature. The bridge is balanced when only carrier gas is seen on both sides. This balance is upset when the presence of a sample-component-plus-carrier appears on the measure side and results in a temperature/resistance change. This upset is reflected in the detector's output signal and is proportional to the amount of sample component present.

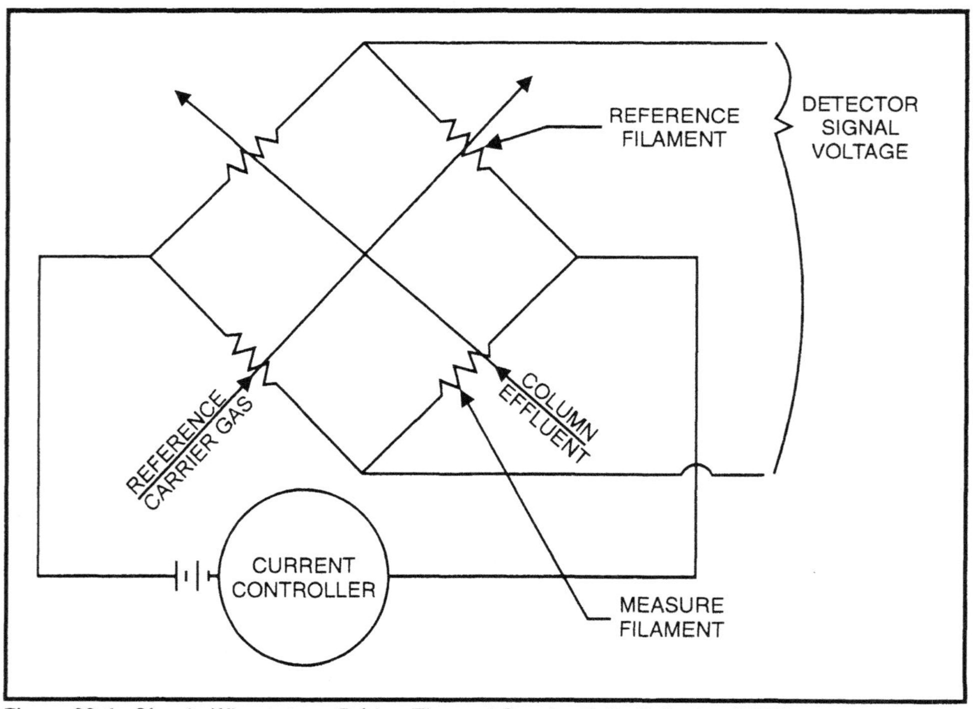

Figure 33-4. Simple Wheatstone Bridge Thermal Conductivity Detector (TCD)

The thermistor version of this detector is similar to the Wheatstone bridge, except that one each of the measure and reference filaments is often replaced with identical fixed resistors and the other filaments with thermistors. Only the thermistors are exposed to column and reference flows.

The constant-temperature version usually consists of two self-balancing Wheatstone bridges. Each bridge has three fixed resistors and a single filament. The filament from one bridge monitors the reference, and the other filament monitors the column effluent. The bridges are automatically balanced, one against the other, and the difference in current that is required to restore the electrical balance is proportional to the sample gas concentration in the column effluent.

The TCD is probably the most popular process GC detector. It is simple, easy to maintain, reliable, and responds to any sample component that is different from the carrier gas. Although not as sensitive as many other detectors, it is more than adequate for most applications. The different TCD designs vary in performance. The thermistor design is generally the most rugged and least sensitive. The constant-temperature version is generally more rugged than the four-filament version and offers about the same sensitivity. The internal volume of the detector itself will influence sensitivity.

The TCD is probably the most popular process GC detector.

The thermistor design is generally the most rugged and least sensitive.

All TCD detectors require very good flow control. Flow fluctuations will change filament temperatures and be interpreted as noise or sample components. These detectors generally lose much of their sensitivity if operated in ovens at temperatures much above 390°F (200°C). At elevated temperatures, there is not a significant temperature differential between the detector body and the filament to allow detection of the small thermal conductivity change generated by small sample component concentrations. To increase detector sensitivity, increase the current on the filaments. This raises the filament temperature and increases detector sensitivity for the same reason that higher oven temperatures lower it. Unfortunately, when increasing the filament current you quickly reach a point of diminishing returns (typically in the 150 to 200 mADC region). The increased current leads to increased noise and shortened filament life. Vibration and exposure to oxygen also can greatly reduce filament life.

Many GC systems come with electronic filament protection, such as switches that remove power from the detector upon carrier gas failure or constant-temperature circuits that reduce the current as a filament begins to heat up. These filaments are very similar to the filaments in a light bulb.

Another way to alter the response of a TCD is to change the carrier gas. Recall that we are measuring the thermal conductivity relative to pure carrier. Consequently, the greater the difference between the sample and the carrier thermal conductivities, the greater the measurement sensitivity. This is one of the reasons that helium is such a popular carrier gas.

Advantages of TCDs:

- High repeatability
- Low maintenance
- Moderate sensitivity
- Broad response
- Low cost

Advantages of TCDs:

- **High repeatability**
- **Low maintenance**
- **Moderate sensitivity**
- **Broad response**
- **Low cost**

FLAME IONIZATION DETECTOR (FID)

The FID is probably the second most popular detector for process GC. It is highly-sensitive and quite versatile. Figure 33-5 depicts a typical FID detector. The GC column effluent is swept into a hydrogen flame, where the combustible components are burned. Aside from water, the products of this combustion process include many ionic species. The flame is positioned in a charged electrical field, and these ionic species provide a source of current flow within this field. This current is directly proportional to the concentration of ions in the flame, which is indirectly proportional to the concentration of appropriate combustible sample components in the column effluent. The detector is essentially nonresponsive to inert gases and most inorganic compounds. The FID is highly sensitive to organic compounds.

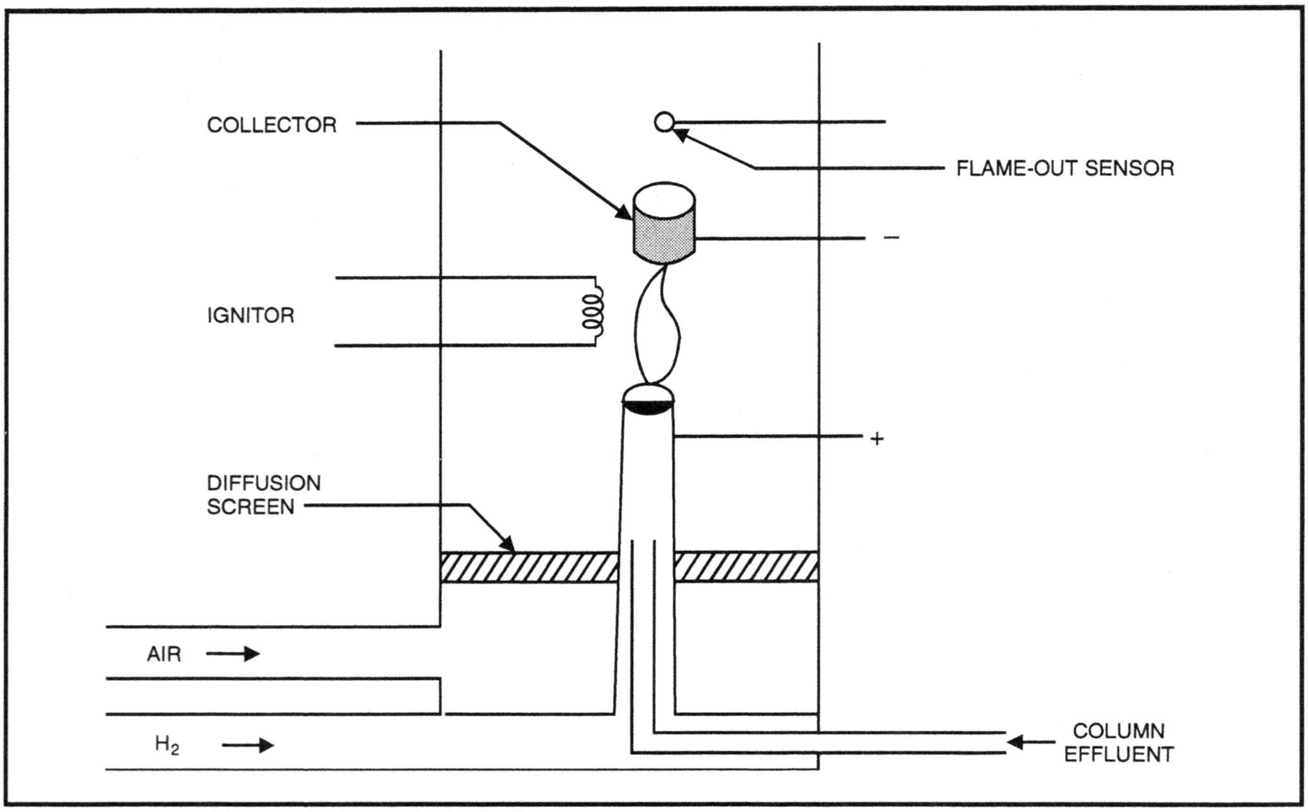

Figure 33-5. Simple Flame Ionization Detector (FID)

For simple organic molecules, the FID detector is often called a carbon counter, because its response is nearly proportional to the number of carbons in the molecule.

The actual ionization process is not well understood, and several theories exist. For simple organic molecules, the FID detector is often called a carbon counter, because its response is nearly proportional to the number of carbons in the molecule. More complex molecules deviate from this principle, and organic molecules containing halides or oxygen have actually been documented as exhibiting a negative contribution from these atoms. Therefore, it is strongly recommended that you calibrate your FID for each component being measured. This is sometimes not done for mixtures of simple, short-chain organic compounds. (You can calibrate a single component and calculate response factors for the others.)

Absolutely clean air is critical for proper FID operation. Any traces of compressor oil or other materials that will burn in the flame will give a noisy background reading. If you are using plant air, be sure to use an air-cleanup system. In addition to filters, coalescers, and absorbers such a system often includes a catalytic combustion unit that precombusts and cleans materials from the air before it reaches the FID.

Another similar catalytic unit is commonly used to modify sample components such as carbon monoxide, carbon dioxide, and a few others into methane for measurement on the FID. This unit is usually called a methanizer or methanator. All major process GC manufacturers supply versions of these devices.

FID maintenance is quite application specific. Systems that include samples that generate ash or soot in the flame (e.g., silanes, siloxanes, and others) will require more frequent detector cleanings. Low oven temperature will tend to generate water condensation problems. The hydrogen flame produces water, which can condense and flood the detector if not maintained as a hot steam and swept from the detector. Several attempts have been made to eliminate this problem, such as inverting the whole detector (flame) or installing a bottom vent (drain). To date, none seems to substitute for a hot system.

FIDs generally come with a flame arrestor in the vent/drain. This should be kept in the hot oven; otherwise, it tends to condense the water out of the vent gas, and the surface tension of the liquid water droplets prevents them from passing effectively through the fine mesh of the flame arrestor. Once the water has been removed from the detector system, a drainable collection container is a good idea. Continuous draining of the detector can be a housekeeping negative; if you attempt to pipe the vent outside, you will find that minor wind variations will affect the pressure and flow balance of the detector. This will result in noise throughout the entire chromatogram.

Automatic flame ignitors have been much improved, but are still a relatively high maintenance item. Either an absolute or a differential temperature measurement is used to sense a flameout and initiate an attempt to autoignite the flame. Column and/or valve switching has been known to be enough disturbance to upset the column effluent flow sufficiently to extinguish the flame. Sometimes a hot oven makes it difficult to sense a flame out or causes false flameouts because of inadequate differential temperatures in the FID. The ignitor is usually a wire filament or a glow plug. Either can require maintenance support.

Advantages of FIDs:

- High repeatability
- Moderate maintenance
- High sensitivity
- Response specific to hydrocarbons
- Moderate cost

Advantages of FIDs:

- **High repeatability**
- **Moderate maintenance**
- **High sensitivity**
- **Response specific to hydrocarbons**
- **Moderate cost**

OTHER PROCESS GC DETECTORS

The *flame photometric detector* (FPD) is primarily used to measure sulfur- and phosphorus-containing compounds. Like the FID, it incorporates a hydrogen flame that burns the sample. However, instead of measuring the ions produced, the FPD measures the emission intensities of selected wavelengths of emitted light that correspond to the characteristic wavelength emissions of molecules energized through the controlled combustion process. These intensities are proportional to the desired sample component concentrations.

The *photoionization detector* (PID) is also similar to, but different from, the FID. It measures the concentration of ionic species in a charged field and converts that to a proportional sample component concentration. The main difference between an FID and a PID is the ionization source. The FID uses a flame and the PID uses a high-energy light source to ionize the species. The key to detecting a molecule with a PID is that the light source's ionization potential in photon-electron-volts must be above that required to ionize the molecule of interest.

The *Hall effect (electrolytic conductivity) detector* (HED) is primarily used for halides and nitrogen compounds. It reacts the sample species in a catalytic reaction chamber and purges the reaction products and carrier through a low-conductivity alcohol solution. The solution absorbs the ionic reaction products, and the change in alcohol solution conductivity is measured. This change in conductivity is then converted to a proportional sample component concentration.

The *electron capture detector* (ECD) is similar to an ionization detector in that it uses a source to generate electrons and ionize the carrier gas molecules, but this source is a low-level radioactive material. Electrophilic molecules (containing oxygen, chlorine, other halogens, organometallic compounds, and others) capture electrons, thus reducing the concentration of ionized carrier-gas molecules. This results in a drop in the measured detector current flow in a way that is proportional to the concentration of the electrophilic species present in the column effluent. Typically, either nitrogen or a mixture of argon and methane is used for a carrier gas with the ECD.

The *mass spectrometer detector* (MSD) is common in the laboratory (referred to as GC-MS), but has only recently been introduced to the process world. It is not a standard option with any of the process GC manufacturers. Wasson will package a Hewlett-Packard laboratory GC-MS suitable for a process environment, but this does not provide all of the standard process output options. Other manufacturers have attached a modified Hewlett-Packard small MS to the column outlet of a process GC. None of these systems could be called "standard," but all are reasonably functional.

These systems are used to primarily monitor process vents. All use an FID or TCD as the primary quantification detector, with the MS used to qualitatively confirm the component separations. The TCD/MS combination offers an advantage in that the TCD will not destroy the sample and as such can measure the same sample before it is delivered to the MS. The FID/MS combination requires switching between detectors, because both destroy the sample.

GC-MS could be helpful with many otherwise unstable, reactive, or difficult-to-quantify samples. Once the sample components have been adequately identified, the GC-MS is often replaced with a simpler GC system for continued quantification. The MS requires a very high vacuum and generally involves more maintenance and technical support than any of the previously discussed detector systems. The MS detector essentially fragments the sample component molecules into charged ions, in a way that creates a reproducible pattern of charge-to-mass ratio fragments. The recognition of these patterns is used to qualitatively identify the sample components.

The details of the workings of this detector and the interpretation of its output are beyond the scope of this chapter. Further information can be found in several of the sources listed in the bibliography and in the last chapter of this book.

Valves

Chromatograph valves are uniquely designed to reliably introduce a fixed sample volume into the carrier gas stream and to switch columns into various positions in the carrier stream, while introducing minimal dead volume into the system. Some of the more popular types of process GC valves include the sliding plate, rotary, linear, diaphragm, and syringe designs. There is also a valveless (pressure-balance) approach to column switching.

SLIDING PLATE, ROTARY, AND LINEAR VALVES

These types of valves have been grouped together because they are similar in concept. Two versions of the sliding plate valve are depicted in Figure 33-6. Both utilize a drilled or grooved surface slider (usually Teflon or some other fluoropolymer), which is quickly moved from one position to another along the finely-machined surface of a metal plate or plates. The grooves in the surface connect various valve ports that have been drilled into the metal plate, and in each position these ports are connected differently. These valves use spring tension to hold the plate(s) and slider against one another.

The rotary valve (not depicted) is essentially the same as the single-sided sliding plate valve, except that its plate and slider, rather than being flat, are formed into a cone. The A and B positions are achieved by rotating the seat (slider) rather than by moving a slider.

The linear valve is depicted in Figure 33-7. This valve is analogous to a sliding plate valve with a cylindrical slider. Grooves in the slider shaft are used to position O-rings, which are used to segment the ports in a switching pattern that supports the application. The elastomeric deformation of the O-rings is used to create the seal. The wide variety of construction material options available for both sliders and plates offers combinations suitable for almost any application. Since new materials are being introduced every year, it is best to review applications and options with your process GC vendor.

All of these valves are multifunction two-position valves. Because they usually direct flows rather than turning them off or on, it is easier to think of them as moving from position A to position B rather than off and on. The off-and-on description in the analyzer controller refers to the air solenoid pilot for the valve actuator position. Since the air can be plumbed to either of two air ports on many of these valves, it is best to identify the on and off positions in the application documentation.

Figure 33-8 depicts several applications for these types of valves. The first (Figure 33-8a) is a simple liquid injection via a sliding plate valve. The sample size can be varied by varying the size of the hole or groove that is machined (or laser cut) into the slider. The second application (Figure 33-8b) is a vapor sample injection via a rotary valve. The sample size on this application is varied by changing the size of the external sample loop. Figures 33-8(c) and (d) depict two vapor injection valve arrangements with column backflush-to-measure valving arrangements.

Figure 33-8 illustrates several popular process GC valve illustration concepts. If you purchase the application from the analyzer vendor, you will be required to use the drawing format on which they have standardized. If you are doing the application engineering, you will have to choose the concept that you prefer. Applications (c) and (d) show that more than one valving approach can accomplish the same goal. Which approach to choose often comes down to a matter of personal preference. Sometimes there is a distinct advantage to one type or arrangement over another (eliminating excessive dead volume, facilitating setup or troubleshooting, using valves that the plant has standardized on and keeps in stock, etc.), but usually there is not. When evaluating various application types or arrangements, visualize yourself as a sample component taking a ride through the system.

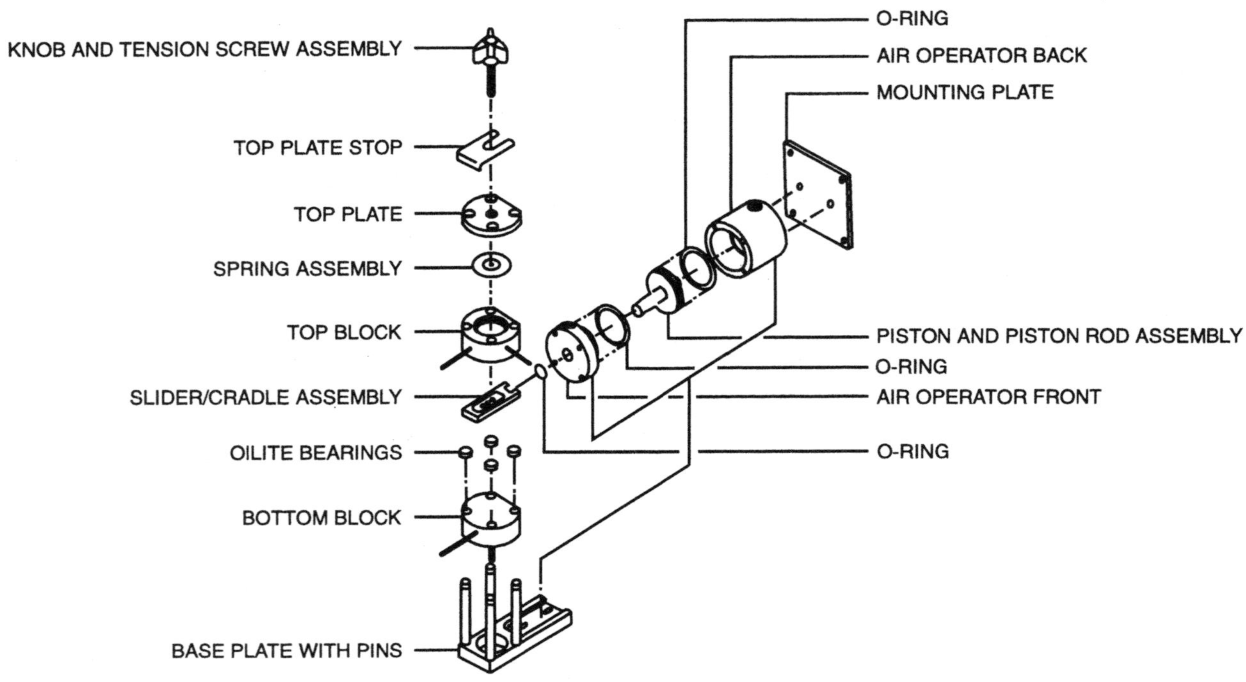

FLUID DATA SERIES 5 VALVE (DUAL-SIDED SLIDING PLATE VALVE)

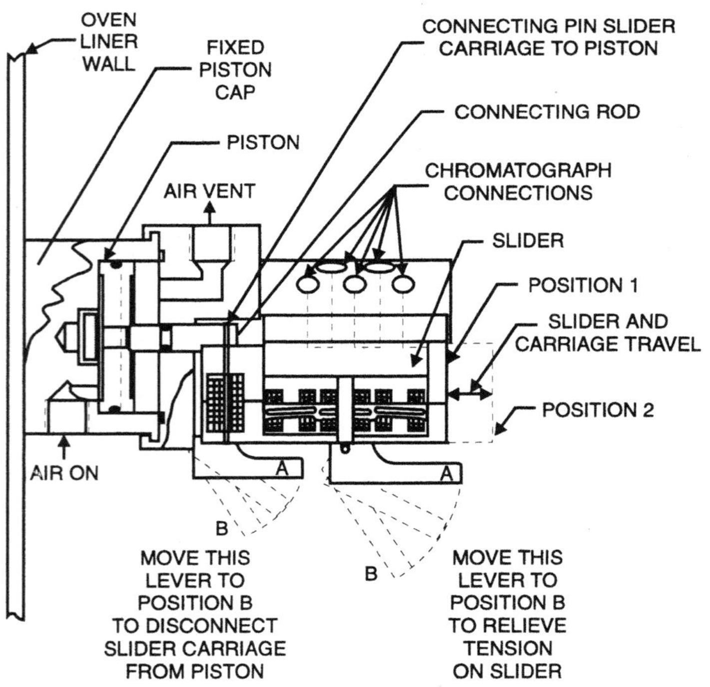

ABB CP VALVE (SINGLE-SIDED SLIDING PLATE VALVE)

Figure 33-6. Sliding Plate Valve Designs for Process Gas Chromatograph
(Courtesy of Fluid Data Inc. and ABB Process Analytics)

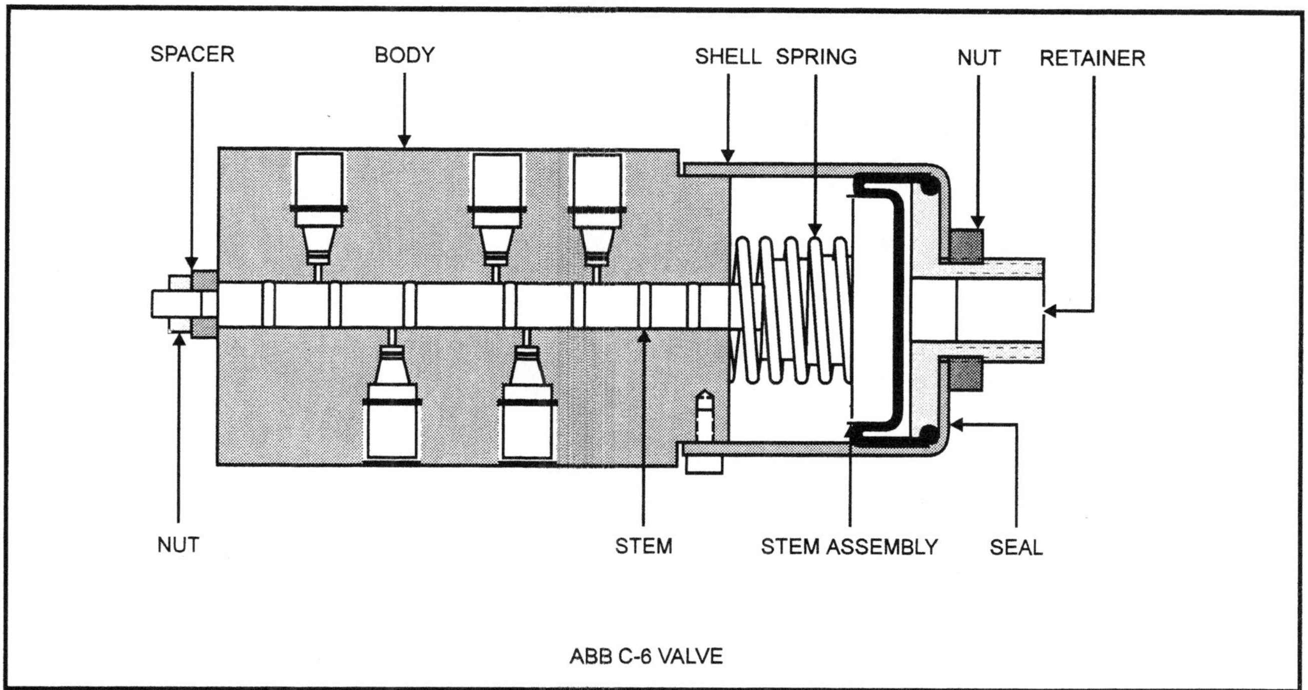

Figure 33-7. Linear Valve Design for Process Gas Chromatograph
(Courtesy of ABB Process Analytics)

DIAPHRAGM VALVES

For vapor-sample injections and column-switching applications, the diaphragm valve can be viewed as a rotary valve. Operationally, it is quite different from a sliding-plate or rotary valve. Instead of one surface rubbing against another to change the port connections, sets of pistons push a Teflon or other flexible diaphragm in a manner that opens and seals alternating sets of ports.

Because of the risk of having all the ports connected during a switch, the valve actually has three positions: (1) switching, (2) all pistons up, and (e) each port isolated from all others. Air-on and air-off positions are analogous to the on-and-off positions used with the rotary valve. The switching position is designed into the valve's operation in a way that is transparent to the user, who controls only the on-and-off positions. Knowing about the switching position can be of assistance in troubleshooting.

Figure 33-9 depicts the operation of a popular analyzer manufacturer's diaphragm process gas chromatography valve. The reduced level of operating friction translates to less valve maintenance than is normally encountered with the higher-friction style of valve (for appropriate applications).

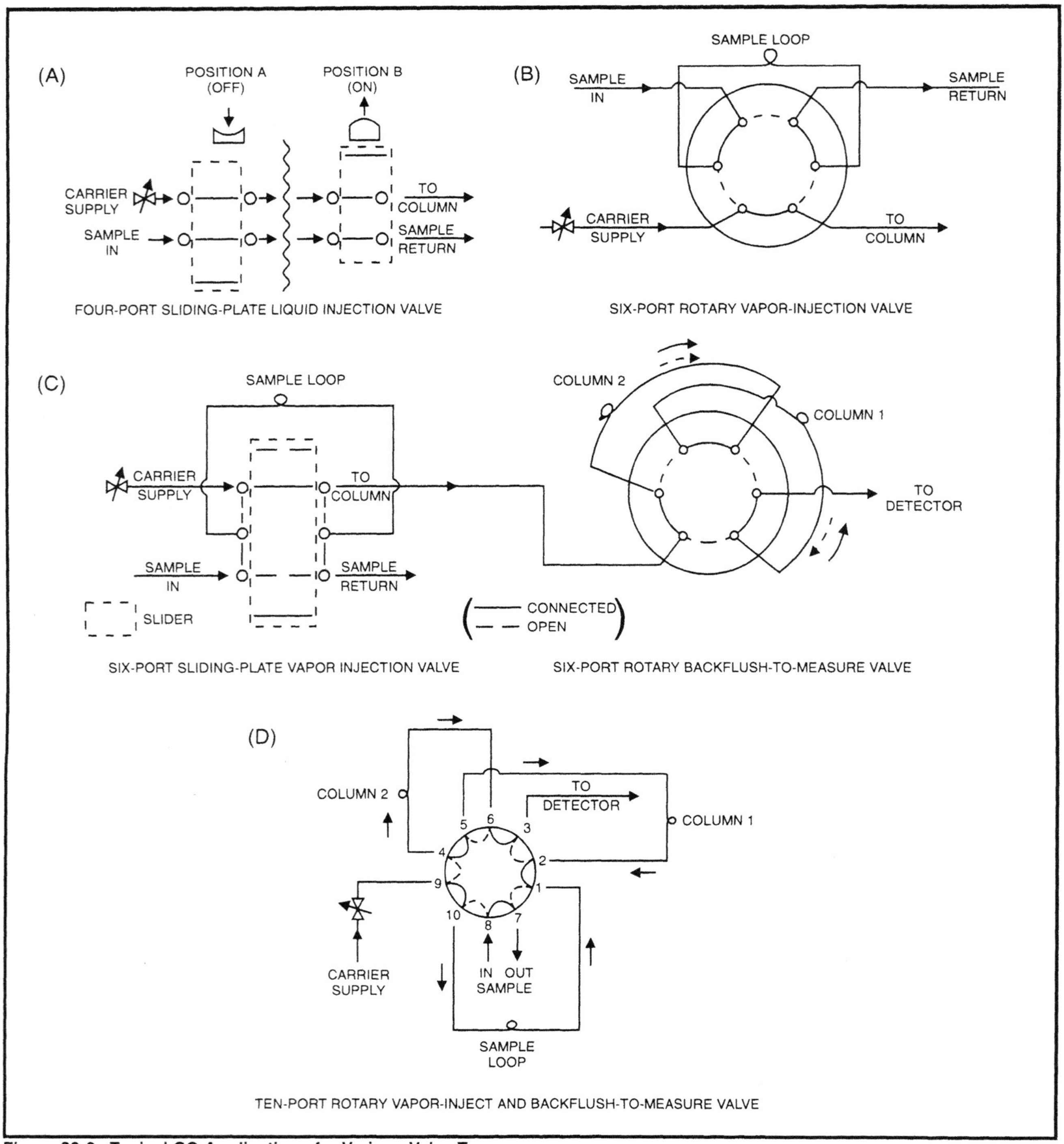

Figure 33-8. Typical GC Applications for Various Valve Types

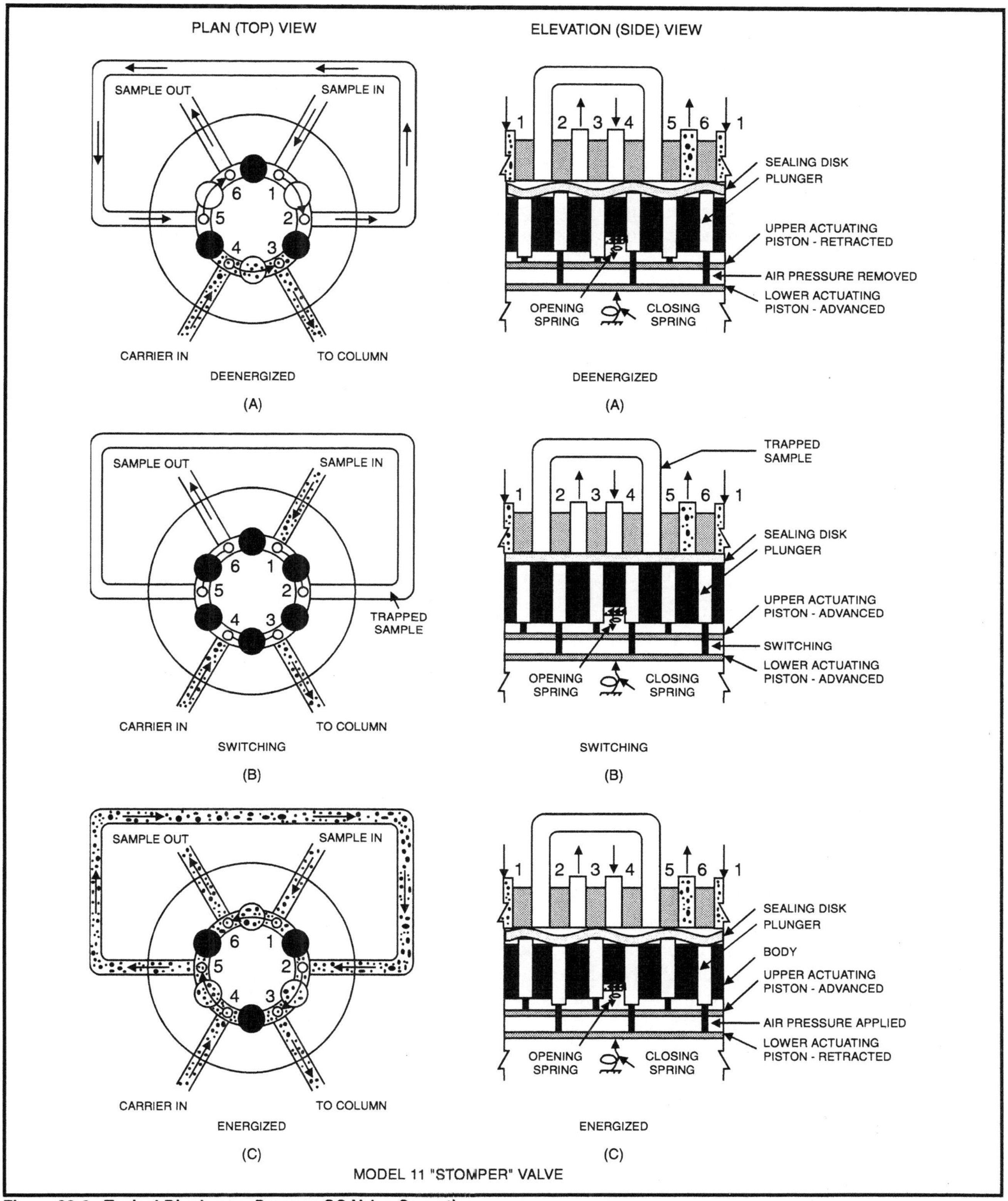

Figure 33-9. Typical Diaphragm Process GC Valve Operation
(Courtesy of Applied Automation/Hartmann & Braun)

Its operation is analogous to a laboratory syringe-type injection, where the tip of a syringe filled with sample is pushed through a septum seal, injected into a vaporizer, and swept onto the column by the carrier gas.

SYRINGE VALVES

These types of valves are called vaporizing valves, liquid injection valves, or syringe valves. The valve (Figure 33-10) essentially collects some sample in a groove or hole and pushes the groove or hole filled with sample through a seal and into a hot vaporizing chamber. This chamber is purged with carrier gas, and the vaporized sample is pushed onto the column. This type of valve is suitable only for injecting liquid samples and is not used in any other aspect of process GC applications. Its operation is analogous to a laboratory syringe-type injection, where the tip of a syringe filled with sample is pushed through a septum seal, injected into a vaporizer, and swept onto the column by the carrier gas.

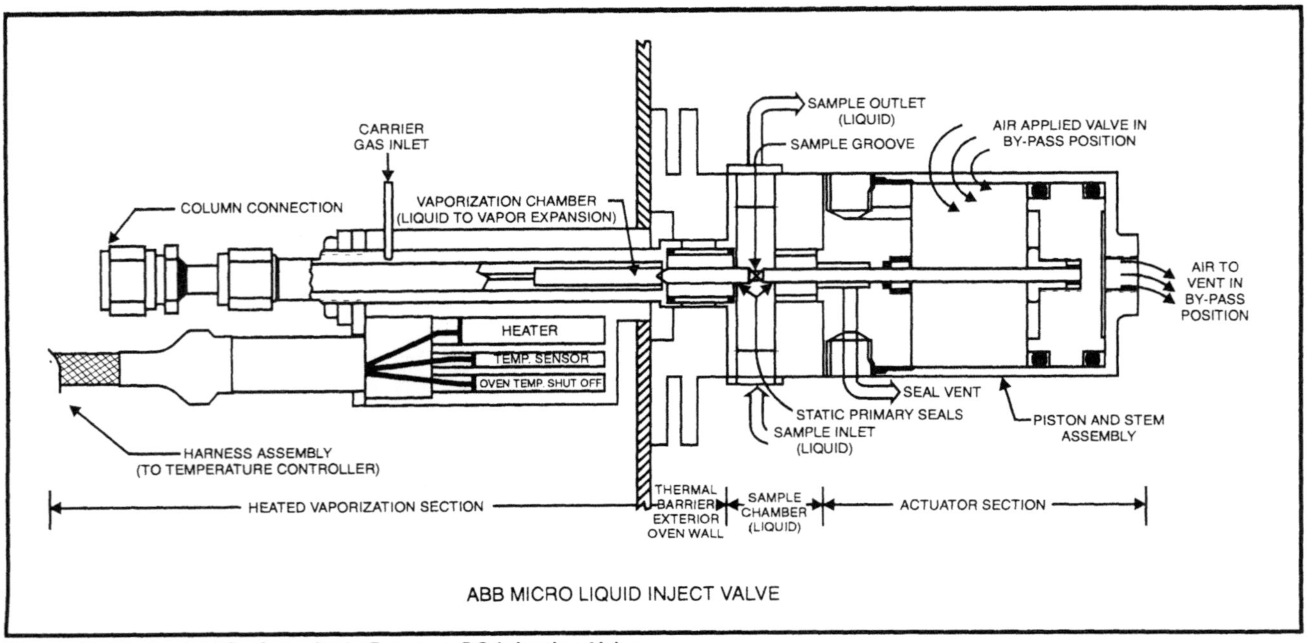

Figure 33-10. Typical Syringe-Style Process GC Injection Valve
(Courtesy of ABB Process Analytics)

Optional Ovens

Certain ovens can only be purchased from certain vendors; others are optional within a vendor's product line. Most are isothermal. Isothermal ovens are used as column, valve, detector, and sample conditioning system ovens on process GC systems. At this time, there are two basic types of isothermal ovens: those with air-bath heaters and those utilizing electrical resistance heaters. Ovens that incorporate air-bath heaters are more popular, offering quicker heat-up, purging of the oven (in case of a leak), easier heater repair, greater oven layout flexibility, and less weight. Resistance-heated units require less or no air and are especially popular for remote locations and plants with limited or poor quality plant-air systems.

Temperature programmed ovens are only supplied with air-bath heaters and are used only as column ovens. Temperature-programming is very helpful when measuring a sample that contains a wide boiling-point mixture of components. This technique is used extensively in lab GC applications. When attempting to duplicate the results of lab methods (ASTM methods, certified test methods, EPA methods, etc.), it is often easiest to use the separation technology (including temperature-programming) of these ovens.

Optional Column Systems

There are many types of column systems, including open-tubular, packed, solid-phase, liquid-phase, boiling-point, and polar. These systems can be mixed and matched in a variety of ways to perform backflushes and heartcuts, and to measure light components and measure heavy components, and so on. Because of the ease with which process GC systems can control the wide variety of valves available, and because of the large isothermal ovens (which can accommodate several valves and columns) that are supplied with most process GC systems, process chromatographers have been using multiple-columns, column-switching, and detector-switching for many years in order to implement high-speed, repeatable, and accurate analyses. Laboratory chromatographers have just rediscovered these techniques, renaming the concepts "multidimensional gas chromatography."

Because column-systems are available in so many application-specific variations, they will not be reviewed in this chapter. If you are trying to design a column system for a specific application, check with your process GC vendor, your lab, and several of the sources listed in the bibliography.

Optional Quantification Techniques

Optional quantification techniques are user-selectable in most process GC software packages. The first step in good quantification occurs during setup, when the GC method is instructed how to collect the relevant data. Should it measure peak height(s) or peak area(s)? How should it do this? How should it turn the measurements into results? Generally speaking, integrating the areas under the peaks leads to more accuracy than measuring peak heights. This is especially true when peak shapes are asymmetrical and/or when peaks are fused (influencing one another's heights). The GC must know how to best integrate the peaks to generate the most reproducible results.

Generally speaking, integrating the areas under the peaks leads to more accuracy than measuring peak heights.

Figure 33-11 shows several approaches to process gas chromatography peak measurement. Different data measurement techniques can generate different results from the same chromatogram. Selecting the correct or best-available approach is an important part of the chromatography method development phase. You want a robust method that generates accurate results over the total anticipated measurement range. Any method will provide an accurate result on a single calibration/measurement sample. Since each software/hardware package combination may execute these options a little differently, it is best to first experiment with the optional approaches on your system.

After the peak measurements have been made, you still have to decide how to best calculate the resulting component concentrations. You can calculate each component concentration by simply multiplying the peak area by the appropriate response factor, or you can ratio (normalize) the peaks, based on either their raw areas or areas that were massaged by appropriate response factors.

After the peak measurements have been made, you still have to decide how to best calculate the resulting component concentrations.

With today's microprocessor-based data systems, this type of calculation can be just the beginning. Many systems allow you to utilize custom-calculation techniques, such as grouping (adding) selected components together by areas, concentrations, or any other relevant and mathematically-achievable result. A simple example of a complex result would be measuring a natural gas sample, generating each component concentration, multiplying each component concentration by its BTU factor, and then combining these results to generate a BTU value for the sampled gas.

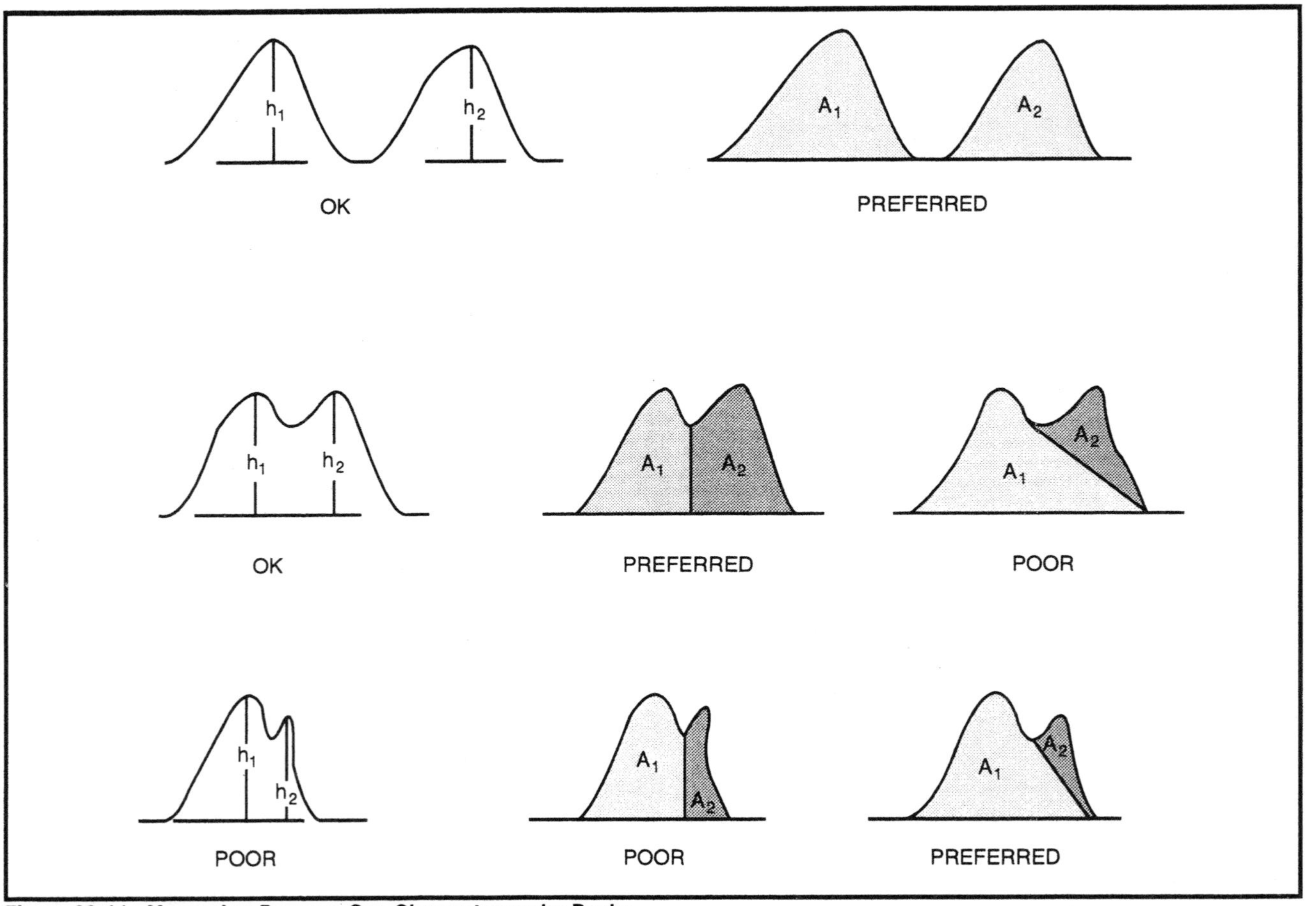

Figure 33-11. Measuring Process Gas Chromatography Peaks

Limitations

The large number of optional detectors and separation-system components available makes process GC a very powerful, broad-based technology for compositional analysis. Each of these options can contribute its own unique limitations. Here are two universal process GC limitations:

1. The analyzer must receive a representative and homogeneous vaporizable sample.
2. The process must be capable of utilizing a cyclical rather than a continuous result.

A process GC system is a batch operation that updates its outputs only once per cycle. This cycle can be anything from seconds to an hour or more in length. The process being controlled or monitored must be able to use information generated in this way; otherwise, the analysis is useless. Many processes are slow to change and have no problem making use of these types of data. Others are very dynamic, requiring continuously-updated inputs of current process conditions.

The trend output shown in Figure 33-3 is the type of discontinuous control input that a system would see from a process GC. The input does not change when the output is changed. It can change only once per analysis cycle, and when it does, the signal represents the process when the sample was injected, not when the signal was updated. This can significantly complicate the control strategy of rapidly changing processes.

For this reason, many processes limit GC results to use in trimming other control inputs, rather than relying on them as the sole control strategy. An example would be a distillation-train control strategy that is primarily controlled from temperature and pressure inputs, but which uses a process GC result to fine-tune the primary control.

Many processes limit GC results to use in trimming other control inputs, rather than relying on them as the sole control strategy.

It is not advisable to drive large or rapid-control actions from a signal that responds slowly and has a significant delay relative to actual process changes.

Process GC is designed for discontinuous analysis using a homogeneous vaporizable sample. This does not mean that the total sample must be vaporizable, only that the components of interest must be and that a sample system can be designed to separate them from nonvaporizable components.

Process GC is designed for discontinuous analysis using a homogeneous vaporizable sample.

Installation Hints

Some installation requirements are unique to certain instrument options and were discussed in the "Options and Accessories" section. The following section will cover general process GC installation requirements.

It is best to install a process GC in a climate-controlled analyzer shelter (the level of control depending on the potential for seasonal ambient temperature range). The acceptable temperature range is somewhat application-specific. Each vendor publishes a range that is generally acceptable for its analyzer, but application-specific characteristics such as low oven temperatures can override these general specifications. A process GC is a complex integrated analytical system; maintenance should not be performed in the falling rain or blowing snow. Maintenance, as well as operating, conditions must be considered when installing a process GC.

The process gas chromatograph system should have clean, dry, well-regulated carrier gas. Most bottled gas and chromatography equipment suppliers offer carrier gas cleanup systems. To avoid problems resulting from loss of carrier, it is recommended that a GC be supplied from a multicylinder manifold system. Most process GC systems utilize microprocessors, which require clean, uninterrupted power. Don't supply the GC from a general process-area lighting panel, or strange and difficult-to-identify problems may arise when large process pumps and heaters turn on-and-off. Even installations using instrument-grade power have required extra isolation transformers and other special devices to get rid of analyzer system electrical "gremlins."

To avoid problems resulting from loss of carrier, it is recommended that a GC be supplied from a multicylinder manifold system.

Vibration induced by large reciprocating process equipment can affect some GC systems. If you are in a high-vibration area and see above-average noise on your baseline, consider vibration-dampening approaches or moving the process GC system. In addition to noise, vibration can cause maintenance problems (everything from broken detector elements and unstable flow regulators to bad electrical connections, fitting leaks, and loose circuit boards).

Purges are usually identified with electrically-hazardous safety requirements, but consider using them in any hostile environment where high humidity, corrosives, or dust is a concern. Purges for these purposes are generally called housekeeping purges. Pressurization and dilution-type purges that are required for safety in an electrically-hazardous area must meet specific installation guidelines (alarms, flow rates, sealing, etc.), which must conform to local, national, and industry-standard electrical safety regulations.

Detector vents generally need to be protected from pressure changes, direct winds, and anything else that could upset them. Factors such as these can affect retention times and detector output signal quality.

Use a good-quality Teflon tape on all threaded carrier, hydrogen, and/or air connections. Solvents and other volatiles from all types of pipe dopes (including Teflon dopes) can bleed into the system and cause detector noise and background problems.

Use a good-quality Teflon tape (such as SWAGELOK® StripTEEZE®) on all threaded carrier, hydrogen, and/or air connections. Solvents and other volatiles from all types of pipe dopes (including Teflon dopes such as SWAGELOK® SWAK® anaerobic pipe thread sealant with TFE) can bleed into the system and cause detector noise and background problems.

Specific Safety Issues

Most safety issues are somewhat application dependent. Is the analyzer located in an electrically-hazardous area? Is the sample toxic or otherwise potentially dangerous? Electrical shock-hazards and the possibility of being burned when working on and inside of a hot oven are safety issues common to all applications. A leak (sample or utility) can be dangerous, depending on the ventilation in the analyzer area. A flammable sample or hydrogen leak could pose an explosion-hazard. A toxic sample leak or carrier gas leak could present an asphyxiation or other inhalation hazard.

Generally, all process GC installations that utilize hydrogen or flammable samples should be reviewed by an electrical-hazard area classification committee. Inadequate ventilation (containment) can result in an electrical-hazard area classification having to be raised from General Purpose to Division 1 (hazard present at all times) or Division 2 (hazard present only under process upset conditions).

Unique Sample Conditioning System Requirements

The sample must be a homogeneous, vaporizable, particle-free fluid by the time it is injected into the chromatograph.

The sample must be a homogeneous, vaporizable, particle-free fluid by the time it is injected into the chromatograph. The original sample can be quite different, but the measured components and sample presented to the analyzer must have been modified to meet these requirements. Sample modification can be accomplished through a variety of sample system components (e.g., filters, coalescers, chillers, heaters, vaporizers, and scrubbers). These are discussed in Chapter 3. Valves, tube fittings, flow sensors, and other GC sample components must meet material compatibility, pressure, and temperature requirements. Refer to your GC vendor's installation and operation manuals and to your company's process construction and operations standards for specific details of field application.

Typical Calibration Requirements

Calibration requirements depend somewhat on the specific application and analyzer options. Analyzers utilizing a response-factor-based normalized-composition method have a built-in theoretical calibration. Most process GC systems have the capability of introducing a known standard sample, measuring that sample, and empirically calculating appropriate response factors based on one known component (the external standard calibration method). Many users choose to do their calibrations off-line for a variety of technical and political reasons. They can take a sample at the same time the analyzer does and send it to the lab. By using the lab result as the standard and ratioing the lab result to the process GC result, they can calculate new response factors for the process GC and enter them manually.

Many users recognize the sampling and analysis variations that can occur in each measurement system and choose to plot the difference between the process GC and the lab GC outputs on a statistical process control (SPC) chart. See Chapter 5 of this volume for SPC/SQC details. The lab sampling and analysis procedures often have more inherent noise in them than the process analysis system. When this is true, it makes no sense to constantly adjust the process analyzer output to conform to a noisy lab output. If a statistical signal from the SPC chart indicates that the difference is beyond control limits, the user can determine whether the problem is in the lab or in the process analysis system and can make the appropriate adjustments.

Some process analyzer systems (like EPA- and OSHA-required applications) have legally-mandated calibration requirements. Some even require the calibration of the total system. This is accomplished by introducing the standard at the process tap and passing it through the total analytical sample conditioning and analyzer system. These requirements can be found in the Federal Register and may be referenced on your plant's operating permits.

These requirements can be found in the Federal Register and may be referenced on your plant's operating permits.

Troubleshooting and Maintenance

A good design and a proper preventive maintenance program will improve a process chromatograph's on-line time. A preventive maintenance program should include checking key voltages and currents, monitoring carrier bottle status, key pressures and flows, and saving copies of results (both printed reports and raw chromatograms). Minor changes in these results can indicate the development of maintenance problems that can be corrected before they cause a larger problem. In the case of a major problem, previous good data can be compared with the current data to help identify differences and their possible causes — in much the same way that a doctor uses a baseline electrocardiogram taken while a patient was healthy for comparison with one taken after the same patient has suffered a heart attack. This type of reference data can be invaluable in assessing problems.

Process GC systems are complex analytical equipment groupings that require trained and skilled maintenance personnel. Plants that have the greatest success with process GC systems are those that create special process analyzer maintenance groups. These groups often service the quality assurance laboratory analyzers in addition to the process analyzers. This helps the process technicians obtain a better feel for the analytical chemistry that is going on inside the lab analyzers and helps them establish a rapport with the lab personnel who provide help with application problems and calibration support.

Most process GC maintenance people will agree that the analyzer itself is often the most useful tool for troubleshooting a process GC problem. Trends and calculated results can provide some information. Shifts in retention time can indicate temperature and/or column problems. Changes in total area counts can indicate detector, sample valve, or sample-conditioning problems. However, the raw detector output — a raw chromatogram — provides the most information. Noisy baselines can indicate detector and electronic problems. Smooth or trending baseline shifts can indicate column bleed, carrier leaks, pressure control problems, or a variety of separation problems. Peak shape changes can indicate a number of things, including actual sample matrix changes, column loading problems, multiple components under one peak, detector saturation, and column deterioration.

Most process GC maintenance people will agree that the analyzer itself is often the most useful tool for troubleshooting a process GC problem.

Your analyzer maintenance procedures should include specific troubleshooting tips for your specific system. Don't ignore this section of the manual. Each vendor knows its instrument and has received feedback from maintenance personnel and customers. Chances are that tips have been included for identifying most known problems or weaknesses in the analyzer's design. Every instrument has a weak component or two, especially a recently introduced model. Minor problems are often identified during a beta test, but not corrected or improved until a future design revision is produced.

If you can't find an answer in the manual, don't hesitate to contact the vendor's service department. They will often provide limited free phone consulting and may be aware of design problems or weaknesses that were unknown when the manual was written.

References

1. Tswett, M., *Berichte Deutsche Botanische Gesellschaft*, 24:316, 384, 1906. The translated excerpt is in T. Robinson, *Journal of Chemistry Education*, 36, 144, 1959.
2. Day, D.T., *Proceedings of the American Philosophical Society*, 36:112, 1897; *Science*, 17:1007, 1903.
3. Martin, A.J.P., and Synge, R.L.M., *Biochemistry Journal (London)*, 35:1358, 1941.
4. Interview by the author with Charles E. Massie, Lewisburg, WV, February 6, 1992.

Bibliography

1. Annino, Raymond, and Villalobos, Richard, *Process Gas Chromatography*, Research Triangle Park, NC: Instrument Society of America, 1992.
2. Clevett, Kenneth J., *Process Analyzer Technology*, New York: Wiley Interscience, 1986.
3. Kolthoff, I.M., Sandell, E.B., Meehan, E.J., and Bruckenstein, Stanley, *Quantitative Chemical Analysis*, New York: MacMillan, 1969.
4. Lee, Milton, Yang, Frank, and Bartle, Keith, *Open Tubular Column Gas Chromatography*, New York: Wiley Interscience, 1984.
5. Poole, Colin, and Poole, Salwa, *Chromatography Today*, New York: Elsevier, 1991.
6. Willard, H.H., Merritt, L.L. Jr., Dean, J.A., and Settle, F.A. Jr., *Instrumental Methods of Analysis*, Belmont, CA: Wadsworth, 1981.

About the Author

James F. (Jim) Tatera is a Senior Analytical Systems Specialist with the Dow Corning Corporation. His experience has included both US and international assignments in analytical research, process engineering, project engineering, production management, and maintenance management. Over 15 of the last 23 years of his career have involved working with process analyzers. He has published several papers and participated in several professional conferences as a paper presenter, session developer, panelist, and workshop presenter.

Mr. Tatera received a B.S. in Chemistry and Math from the University of Wisconsin in 1968, attended a program in Chemical Instrumentation and Analysis at the University of Minnesota, graduated from the US Army Officers Candidate School, and received an MBA from Central Michigan University in 1982. He is one of the original Certified Specialists in Analytical Technology (CSAT) and a member of the ISA, ACS, and AWMA. He is also active in US and international standards activities. He is the ANSI USNC Technical Advisor to IEC SC65D (Analyzing Equipment in Industrial Process Measurement and Control) and has served as the ISA SP76 (Composition Analyzers) committee chairman.

34

Liquid Chromatographic (LC) Analyzers

High-pressure liquid chromatography (HPLC) is a relatively new technique, commercially available in the laboratory since the early 1970s. The principles behind this separation technique are basically the same as those in gas chromatography (GC), the major difference being that in liquid chromatography the analyte is dissolved in the liquid mobile phase, and in gas chromatography the analyte is dissolved in the gaseous mobile phase. In both cases, the separation occurs due to interactions between the analyte and a stationary phase attached to a solid support.

As with any chromatographic technique, HPLC should be used only when one needs to know about the makeup of a process stream in the presence of other components that would interfere with nonchromatographic techniques, such as ultraviolet (UV), Fourier transform infrared (FTIR), and near-infrared (NIR) analysis. The inability to use these techniques usually arises from overlap of absorption bands due to structural similarities present in the materials in the sample stream. Although chromatographic techniques offer the advantage of being able to remove these interferences by separating the components in the process stream before they reach the detector, they have the disadvantage of being discontinuous. In other words, they provide only a "snapshot" of the process at a given time. They are not capable of providing a continuous readout of what is going on in a process stream, as can flow-thru spectroscopic techniques. However, chromatographic techniques may provide the only means for determining the relative concentrations of closely-related substances or analytes that are contained in the stream at very low concentrations.

HPLC is particularly advantageous over GC for the analysis of analytes that have a very high boiling point, since HPLC analysis occurs in the liquid rather than the vapor phase. For this reason, the technology has been widely utilized in the pharmaceutical and polymer industries, where HPLC has replaced GC as the method of choice for analysis of multicomponent systems.

... the technology has been widely utilized in the pharmaceutical and polymer industries, where HPLC has replaced GC as the method of choice for analysis of multicomponent systems.

HPLC also offers a larger number of parameters for controlling the selectivity of an assay than GC (which can also add to the complexity by requiring additional equipment). In GC analysis, the selectivity is adjusted by changing the stationary phase of the column varying the flow rate of the carrier gas, derivatizing the analyte with a reagent, or changing the column temperature. In HPLC, the column stationary phase can be changed and the column temperature adjusted or the analyte derivatized as in GC. In addition, since the analytes are present in the liquid phase, the pH and/or conductivity of the mobile phase can be adjusted. This can have profound effects on how an analyte interacts with the column stationary phase if the analyte molecule contains any ionizable groups. This is particularly true if the stationary phase also is a charged molecule. Charge adjustment can be effected, as well as affected, by the addition of reagents to the mobile phase.

These reagents, or ion pairs, interact with charged groups on the analytes and block their interaction with the stationary phase (see the section "Separation Principles").

The mobile phase can also be utilized as a solvent in which chemical reactions take place. By coupling an automated chemical analyzer to the outlet of the HPLC column, the selectivity of the detection system can be changed. Utilizing this post-column derivitization technique might allow analytes to be measured that otherwise could not be detected. Another application of this technique would be to change the selectivity such that analytes can be detected in the presence of other interfering components.

It is also possible, depending on the choice of detectors, to add an additional dimension to the measurement technique. This can be done through the use of a rapid scanning photometric detector. Examples of detectors that have been used for this purpose in HPLC are UV, NIR, and IR. This technique gives a three-dimensional look at the sample mixture. The first two dimensions are a response of the detector in the time domain, as the peaks elute from the column, and the third dimension is a wavelength scan of each peak. This can be particularly useful for identification or conformation information of the compounds contained in the sample.

The use of HPLC as an on-line analyzer was first reported by Mowery and Roof in 1976 [Ref. 1]. The number of publications on this subject has remained relatively small compared to those that describe on-line gas chromatography. The largest user of on-line analytical methods has been the petrochemical industry. Since this industry deals mainly with lower-boiling-point components, on-line HPLC offers no analytical advantage. Thus on-line GC has been utilized more widely. On the other hand, the pharmaceutical industry, which uses HPLC more frequently than GC due to its analysis of nonvolatile compounds, is just beginning to recognize the advantages of on-line analysis. As a result, on-line HPLC is a technique still in its infancy.

Other contributors to the poor acceptance of on-line liquid chromatography are the notions that the instrument is mechanically too complicated to be dependable and that separations take too long to complete to be useful as a process control methodology. Although HPLC analyzers are mechanically more complex than GCs, attention to proper analyzer design and proper preventive maintenance will allow the analyzer to fulfill most performance expectations. The time to complete an analysis is indeed long compared to nonchromatographic techniques, but the on-line HPLC is able to present data in a time frame that is useful for controlling many chemical processes. One example of an application where HPLC offers a distinct advantage over other techniques is the analysis of biological systems. These systems often contain many closely-related compounds that cannot be differentiated from one another by nonchromatographic techniques such as UV, FTIR, or NIR. These molecules are not volatile and would be charred if injected into a gas chromatograph. In such applications, on-line HPLC may be the only viable technique for obtaining analytical data without performing off-line analysis.

The slow acceptance of on-line HPLC has led to the demise of most commercially-available on-line instruments. Only three companies have been involved in the commercialization of on-line HPLC instruments in the United States: Applied Automation, Waters Associates, and Dionex. At the time of this writing, two of these systems, Waters and Applied Automation, are no longer commercially available. An additional company, Siemens, developed a system which was marketed for a short time in Europe. This system is no longer available. This leaves the potential users of on-line HPLC with few choices. Most applications published in the literature have been performed on laboratory systems adapted for process applications by the user.

Separation Principles

When liquid chromatography was first developed, the most popular method of separation involved the use of nonpolar mobile phases, such as hexane or heptane, with silica gel or alumina as the solid support. The name of this mode, *normal phase chromatography*, arose from the fact that the mobile phase was less polar than the solid phase. The major drawback to this technique was that, because the chromatographic character of silica gel changes drastically with its state of hydration, the mobile phase had to be thoroughly dried before use. If this was not followed, the characteristics of the silica gel would change during the separation, making it very difficult to achieve a reproducible separation. Because of this shortcoming, normal phase chromatography is not widely utilized, although it is still a viable technique for analytes with little or no water solubility.

Reverse Phase Chromatography

To circumvent the problems of normal phase chromatography, *reverse phase chromatography* was developed. In this process, the polarity of the mobile phase and the stationary phase are reversed relative to one another. In other words, the polarity of the mobile phase is greater than the polarity of the stationary phase. This is achieved by utilizing the silanol groups on the silica to bond various functionalities to the silica backbone. Some of the more widely utilized columns and the derivatives attached to the silanol group to prepare them are listed in Table 34-1.

Table 34-1. Sampling of the Most Popular Bonded Phases Used in HPLC Functional group (–Si–O–Si–R) bonded to the stationary phase

Column Type	R Functional Group
Amine	$-NH_2$
Butyl	$-CH_2-(CH_2)_2-CH_3$
Cyano	$-CN$
Diol	$-CH_2-CHOH-CH_2OH$
Octyl (C8)	$-CH_2-(CH2)_6-CH_3$
Octadecyl (C18)	$-CH_2-(CH_2)_{16}-CH_3$
Phenyl	–Phenyl
Trimethylsilyl (TMS)	$-(CH_2)_3$

Reverse phase chromatography is the most widely utilized method in liquid chromatography today. It is useful for the analysis of compounds ranging in size from small organic molecules to large proteins and polymers.

The derivitization of the silica with these functionalities eliminates silica hydration problems, imparts a wide range of available selectivities on the solid phase, and allows aqueous/organic mixtures to be used as the mobile phase. The organic modifiers most commonly used today are low-viscosity, water-soluble solvents (e.g., acetonitrile and acetone) and low-molecular-weight alcohols (e.g., methanol, ethanol, and propanol).

The detection method required to monitor the analyte may dictate the choice of mobile phase. For example, it is extremely difficult to use acetone as a modifier with UV detection since it has a strong UV absorbance. This high background absorbance would render all UV monitors useless.

The detection method required to monitor the analyte may dictate the choice of mobile phase. For example, it is extremely difficult to use acetone as a modifier with UV detection since it has a strong UV absorbance. This high background absorbance would render all UV monitors useless.

The mechanism that produces separations in reverse phase chromatography is the solubility of the analyte in the stationary phase that is bound to the silica gel matrix. As an analyte passes through the column, it will partition between the mobile phase and the stationary phase. A dynamic equilibrium will be established between the stationary phase and the mobile phase of the chromatographic system. this causes the analyte to spend part of its time in the stationary phase and the remainder of its time in the mobile phase. Analytes that are more strongly attracted to the stationary phase than the mobile phase will spend more time in this phase and thus will elute later from the chromatographic column. Inversely, analytes that are less attracted to the stationary phase will spend more time in the mobile phase and will be eluted earlier from the column. This produces a chromatogram with peaks eluting in the order of the affinity that they have for the stationary phase.

Size Exclusion

A second separation mechanism used in liquid chromatography is *size exclusion*. This mode of chromatography is used primarily for characterizing the molecular weight of polymers, either biological or organic. The separation occurs due to differences in molecular size. Porous silica is used to effect the separation, with the pores of the silica acting as a molecular size filter. Analytes with molecular weights below the molecular weight cutoff of the packing material will pass through the pores of the packing material. Analytes with molecular weights above the cutoff will not pass through the pores and will be excluded. These molecules will be forced to proceed around the packing material, remaining in the interstitial space of the chromatography column. The path around the packing material is less tortuous that the path through the particle. Therefore, analytes following this path will travel a shorter distance, and thus will elute first from the column. Analytes with lower molecular weights will pass through the pores, travel a more complex path, and elute later from the column.

Analytes with lower molecular weights will pass through the pores, travel a more complex path, and elute later from the column.

In this mode of chromatography, limiting the chromatographic mechanism to size exclusion is desirable and is affected by two methods. First, the silica is generally derivatized with a diol functionality to prevent interaction with the silanol groups. Secondly, the mobile phase is prepared with a large percentage of organic solvent to minimize any reverse phase interaction with the –OH groups of the diol functionality.

Ion Exchange Chromatography

The last major mechanism available to the chromatographer to achieve separation of the components in a mixture is *ion exchange chromatography*. The separation mechanism is based on the interaction of a charge on the analyte being attracted to an opposing charge on the stationary phase. Elution of analytes in this mode of chromatography is affected by placing an ion in the mobile phase that will compete for sites on the ion exchange resin and thus displace the analyte back into the mobile phase. Another method of eluting the column is to adjust the pH of the mobile phase such that the charge on the analyte is altered in a way that it will no longer be attracted to the charge on the stationary phase.

Ion-pair chromatography is a related mechanism. This technique can be used to achieve a separation of charged molecules on a nonpolar or reverse phase column, making it possible to chromatograph an analyte on a reverse phase column that would normally have minimal affinity for charged analytes. This mode of chromatography is extremely useful for chromatographing mixtures that contain both charged and noncharged analytes. Although there is debate over the exact mechanism of this mode of chromatography, it involves the ionic association of a charged molecule with the opposite charge on the ion-pair reagent. Typical reagents are molecules containing a hydrophobic region and an ionizable functionality, such as a sulfonic acid for cationic analytes and quaternary ammonium compounds for anionic analytes. The nonpolar section of the ion-pair reagent plays an active role in the separation process by interacting with the stationary phase, and thus adds an additional dimension that can be optimized by varying the structure of the hydrophobic region of the ion-pair reagent.

Terms and Equations

Before discussing the factors that affect a chromatographic separation, let's briefly examine some of the terms and equations used by the chromatographer to describe the separation.

First, let's look at the term *retention time* (*RT*). This is the amount of time or volume that has elapsed until a peak elutes from the column (Figure 34-1). This is generally calculated by measuring the time (in seconds) from the point at which the sample was injected to the apex of the peak of interest.

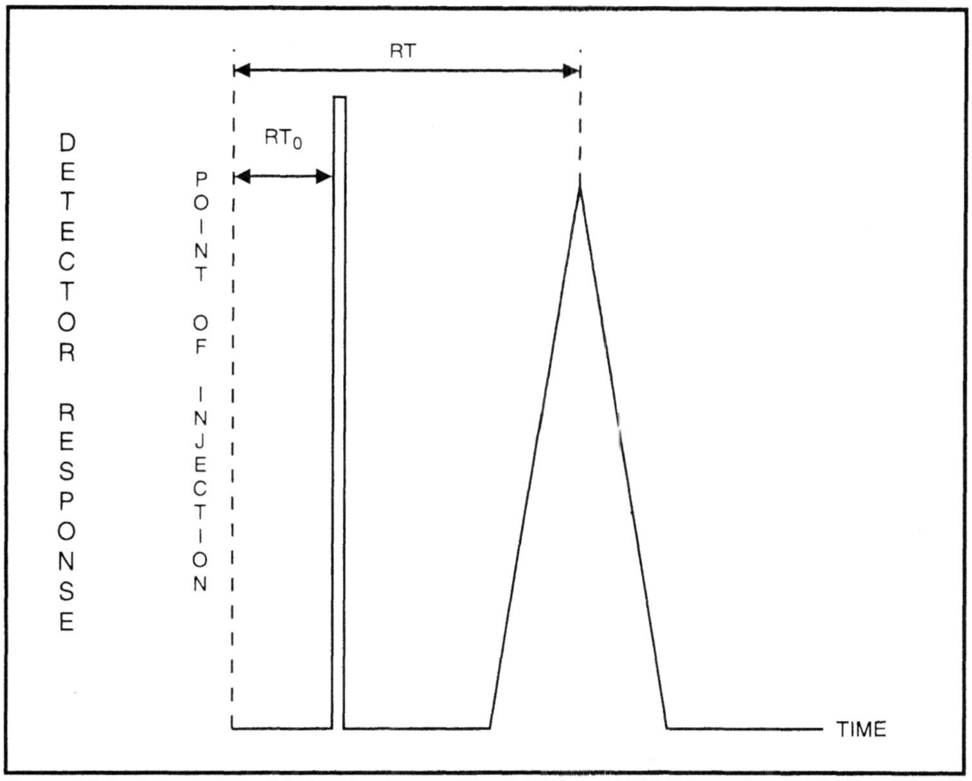

Figure 34-1. Calculation of Retention Time of Peak (*RT*) and Void Volume Peak (RT_0)

The capacity factor (K' value) is a measure of a sample's retention on the column in terms of column volumes. The K' value for a peak is calculated by dividing the difference in retention times of the retained (RT) and nonretained peak (RT_0) by the retention time of the nonretained peak (RT_0) (Figure 34-1). The time it takes for the nonretained peak to elute is referred to as the *void volume* of the column. The void volume is the volume of the column not occupied by the packing material. The K' value for peak 1 (K'_1) would be

$$K'_1 = \frac{RT - RT_0}{RT_0} \tag{34-1}$$

For an optimum separation to occur, the K' value generally should fall in the range of 2 to 10.

Efficiency is used to describe the performance of a column. This term is represented by the plate count N of the column. The larger the N value for a column, the more efficient it is and the better it will be at performing separations. There are two methods for calculating this number. The first (Figure 34-2a) utilizes the width of the peak at the intersection with the baseline of two lines drawn tangentially to the front and back sides of the peak (W):

$$N = 16\left(\frac{RT}{W}\right)^2 \tag{34-2}$$

The second method (Figure 34-2b) uses the width of the peak at one-half the maximum of the peak ($W_{1/2}$):

$$N = 5.54\left(\frac{RT}{W_{\frac{1}{2}}}\right)^2 \tag{34-3}$$

... it is obvious that the narrower the peaks, the larger the plate count obtained or the more efficient the column.

The second method is easier to implement since no tangents have to be drawn to perform the calculation. It can be particularly difficult to decide where to draw these tangents if the peak is not symmetrical or if it exhibits tailing areas on its front or back side. Either method can be used as long as consistency is applied when comparing two chromatographic systems to each other. From these equations it is obvious that the narrower the peaks, the larger the plate count obtained or the more efficient the column.

The relative capacity factor or relative retention factor (α) is a measure of the selectivity of a column for two peaks. It is calculated by ratioing the K' values of the peaks of interest:

$$\alpha = \frac{K'_2}{K'_1} \tag{34-4}$$

If the α value for two peaks is 1, the two peaks exactly correspond and no separation has occurred.

Utilizing the above method of determining whether two peaks are resolved from each other does not take into account the widths of the peaks. If either or both of the peaks are broad, the α value may predict separation when there is actually a fair amount of overlap between the peaks. Another method of determining resolution R takes the peak widths (W_1 and W_2) at their base into account when predicting resolution:

$$R = \frac{K'_2 - K'_1}{0.5\,(W_2 + W_1)} \tag{34-5}$$

Using Equation (34-5), two peaks are baseline resolved when $R = 1.5$.

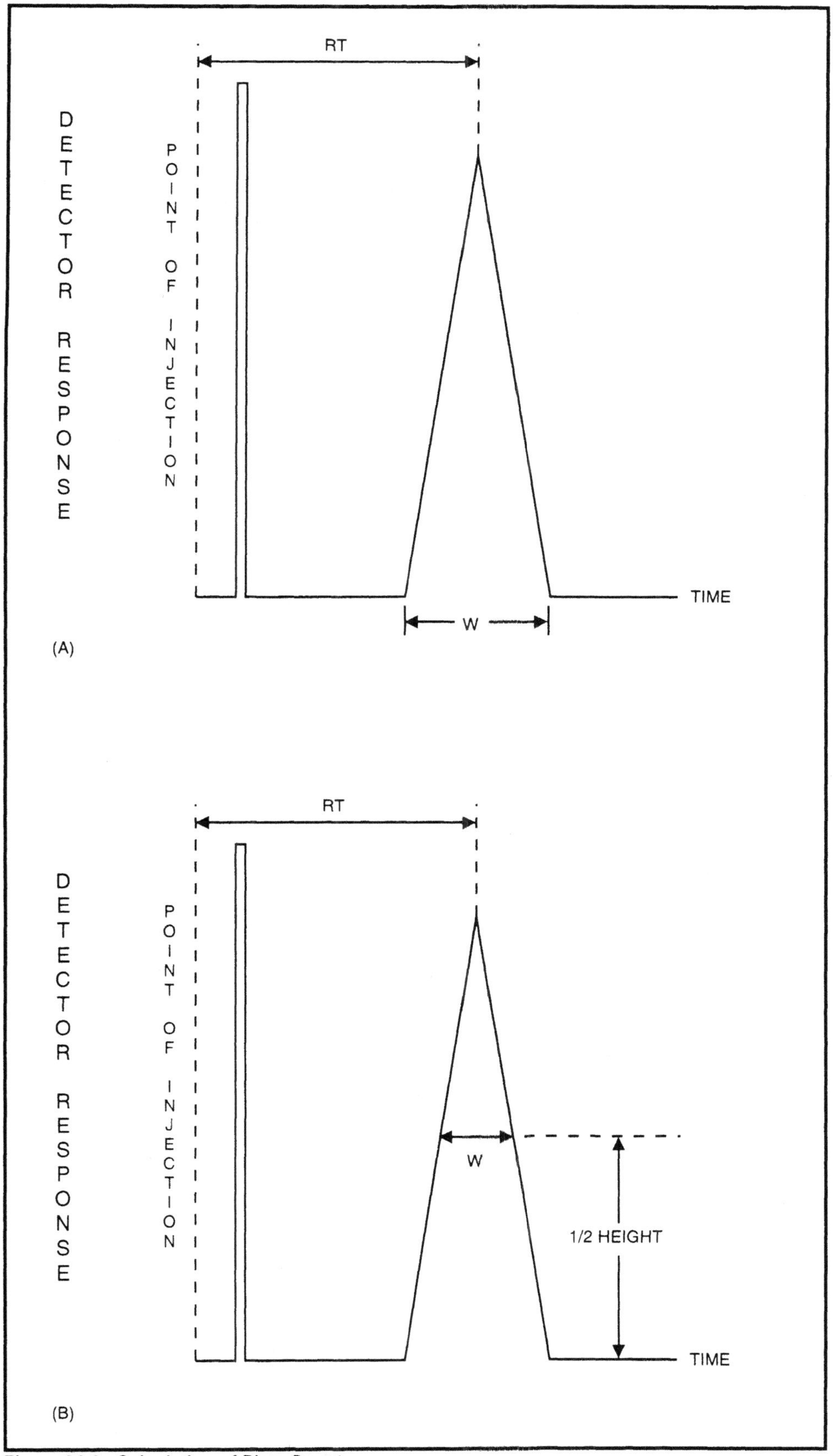

Figure 34-2. Calculation of Plate Count

Column efficiency is one of the most important parameters when optimizing an on-line HPLC separation since it greatly affects how quickly the separation can be achieved. Obviously, if attempting to perform any analytical method on-line, one wants that method to produce data as quickly as possible. In chromatography, this requires familiarity with the factors that affect column efficiency.

Factors that Affect HPLC Separations

The time required to perform a separation can be cut approximately in half by reducing the diameter of the packing material by a factor of two.

The packing material is the main component of the chromatographic system that should be examined when optimizing efficiency. The time required to perform a separation can be cut approximately in half by reducing the diameter of the packing material by a factor of two. The most widely used particle size packing at the current time is 5 µm. Particle size packings of 1 and 3 µm are also commercially available. The 3-µm packings are available in the standard column configurations (4.6-mm ID), whereas the 1-µm packings are available only in columns with 1- or 2-mm ID. These smaller-ID columns, known as "microbore" columns, are operated at low flow rates and give separations in conventional time frames with much lower consumption of solvent. This can be an important factor in on-line HPLC applications.

The major drawback to these microbore columns is their low loading capacity. To get around this requires that extremely small volumes be injected onto the column. Microbore chromatography also requires specialized pumps and detectors to handle the low flow rates and peak volumes, which can also be a drawback. The chromatography community generally believes that particle sizes smaller than 5 µm are difficult to work with due to blocking of the column packing and plugging of the column's end frits. Our group has more than 7 years of experience with 3-µm columns in on-line analyzers for biotech products. The typical column lifetimes span from 3,000 to 5,000 injections. This type of success requires that the solvents used and samples injected onto the column be free of particulate matter.

Our group has more than 10 years of experience with 3-µm columns in on-line analyzers for biotech products. The typical column lifetimes span from 3,000 to 5,000 injections. This type of success requires that the solvents used and samples injected onto the column be free of particulate matter.

When choosing the particle size, consider the pressure limit of the system. The pressure drop across the column increases with the square of the particle size. This means that as the particle size is cut in half, the pressure drop will quadruple. This generally is not a problem since the smaller particle size produces more efficient separations that require shorter columns. Therefore, the increased pressure drop can be compensated for by using a shorter column.

Particle size distribution also affects efficiency. The wider the particle size distribution, the less efficient the packing material. Efficiency is affected by the band-broadening that occurs as the analytes flow around the packing material. This is a negative effect when trying to achieve a separation in the shortest possible time. The narrower the particle diameter distribution, the more similar are the paths followed by the analytes as they traverse the column. This means there will be less band-broadening within the column, resulting in more efficient separations. Look for packings with narrow size-distributions when choosing columns for on-line analysis.

Look for packings with narrow size distributions when choosing columns for on-line analysis.

Packing materials are available in various physical states. The particle can be spherical, irregularly-shaped, porous, nonporous, or any combination of these. Irregularly-shaped particles tend to have a much wider range of particle sizes, which, as just described, leads to more band-broadening and less efficient separations. Spherical packings can be manufactured with much tighter size-distributions, leading to more efficient separations.

Porous and nonporous are terms used to describe the particle surface. Porous packings have many pores on the surface; nonporous particles have smooth surfaces. An analogous comparison might be the "cratered" surface of a golf ball versus the perfectly smooth surface of a ping-pong ball. Nonporous materials will give separations in the shortest time since the analyte, interacting only with the surface of the particle, can achieve equilibrium with the stationary phase in a shorter period of time. Porous packing materials require the analyte to travel into and out of the pores as it traverses the column. This leads to more band-spreading than occurs on a nonporous packing.

The drawback to nonporous packing is that the surface area of such material tends to be smaller than that of porous material, which leads to lower column capacity. This causes the column to be easily "overloaded," limiting the range of concentrations that can be separated on the column before the separating power of the column is lost.

The choice between porous and nonporous particles must be decided for each application. The most widely used silica is the porous, spherical type. Spherical silica's particle size-distribution tends to be much smaller than irregular silica, and it has a much larger surface area than nonporous silica. This gives a material with a capacity that compares with the irregular silicas, but an efficiency that approaches that of nonporous silica. This type of material possesses the most ideal properties for resolution, speed, and a wide range of capacity — necessary characteristics for on-line applications.

Spherical silica's particle size-distribution tends to be much smaller than irregular silica, and it has a much larger surface area than nonporous silica. This gives a material with a capacity that compares with the irregular silicas, but an efficiency that approaches that of nonporous silica.

The amount of time that analytes spend on a reverse phase chromatography column is controlled by many factors. One is the strength of the mobile phase. If a longer retention time is required to achieve separation, the organic content of the mobile phase or mobile phase strength is reduced. Conversely, if the separation can be achieved with less time on the column, the concentration of the organic modifier in the mobile phase or mobile phase strength can be increased. An alternative to this method is to change the organic modifier to one that has a higher solvent strength. The relative eluting powers of the commonly used solvents are as follows:

methanol < acetonitrile < 2-propanol < 1-propanol < acetone

Relative Solvent Eluting Power (most to least):
acetone
1-propanol
2-propanol
acetonitrile
methanol

"Tuning" of the mobile phase to achieve a separation in the shortest possible time is obviously an operation that must be performed on any on-line chromatographic method. Changes in the organic modifier can also impart changes in the selectivity of the assay. Careful control of the mobile phase pH is also important for maintaining proper assay performance. The use of 10 to 100 m*M* buffers to control pH and ionic strength is recommended. This is necessary to minimize ionic interactions between the analytes and the silica of the stationary phase.

The flow rate of the mobile phase through the column is another factor that governs separation speed. As the flow rate is increased, the analytes can be moved more quickly through the column, thus reducing the time for the separation to be completed. There is a limit to how high the flow rate can be increased, however. One limitation is that as the flow rate increases, so does the back pressure of the column. This is a linear relationship that is independent of the analyte and dependent on the viscosity of the mobile phase at a constant particle size and column length.

A second limit on flow rate is the mass transfer rate. This is the time it takes for the analyte to move into and out of the dead-ended pores on the surface of the silica. If the flow rate is increased past the point where the bulk of the molecules can move in and out of the pores, band spreading will occur. Column efficiency has a parabolic relationship with flow rate and depends on the analyte, the packing material, the column temperature, and, to a degree, the viscosity of the solvent. Larger molecules will have much larger mass transfer rates and thus a narrower optimum flow rate range.

Since the interaction of the analyte with the stationary phase is an equilibrium reaction, anything that can increase the rate of this equilibrium will affect separation speed. Obviously, the first thing that comes to mind is temperature. In general, the higher the operating temperature of the column, the faster the separation. Because the viscosity of the solvent is decreased with higher temperature, the mass transfer rate is increased. The exception to this rule occurs when dealing with biomolecules. In these applications, the time required to complete a separation can become longer at higher temperature due to changes that occur in the three-dimensional structure of the protein. At lower temperatures the protein may remain in a coiled state, minimizing the sites that can interact with the stationary phase. At higher temperatures the protein can denature, becoming more linear. This causes an increase in the number of interaction sites and thus in the length of time it interacts with the stationary phase, resulting in longer retention time.

Another advantage of increasing the temperature of the system is the reduction of the viscosity of the solvent which results in a reduction of the system back pressure. This can reduce the back pressure that may be imposed by small-particle-size packing materials to below the pressure limit of the system.

Instrumentation Hardware

Pumps

A liquid chromatographic separation occurs by placing the sample into the liquid mobile phase. The mobile phase and dissolved sample are moved through the column by a high-pressure pump. As the sample moves through the column, it interacts with the stationary phase of the column and is separated into its various components. The more it interacts with the column, the longer it takes for the analyte to exit the column. Separation of the components in the mixture occurs because of the differences in affinities that the components have for the stationary phase. As the components exit the column, their presence is measured by a flow-through detector. The signal from the detector is fed to an integrator or a computer system that runs chromatography software to perform data analysis. The peak areas or peak heights are used to calculate the concentration of components in the mixture. The processed data can then be used by a controller to determine the current state of the process being monitored.

All HPLC systems have components that are common to one another, including a mobile phase reservoir(s), eluent pump, injection valve, analytical column, column heater, detector, and integration device.

All HPLC systems have components that are common to one another, including a mobile phase reservoir(s), eluent pump, injection valve, analytical column, column heater, detector, and integration device. A presaturation column and a guard column are optional. A typical system is diagrammed in Figure 34-3.

Liquid chromatography differs most dramatically from gas chromatography in that the mobile phase that is propelling the sample through the column is liquid rather than gas. This makes HPLC an inherently more complicated technique than GC. In GC, the pressure providing the driving force for the mobile phase is the pressure of the gas cylinder itself after being reduced down by a series of regulators and flow controllers. The earliest HPLC systems utilized a similar methodology to provide the force to propel the mobile phase through the column. This was done by pressurizing the mobile phase reservoir with an inert gas. The disadvantage of such a system is the limited mobile phase volume, which requires that the flow be stopped in order to refill the system. The advantage of this system is that it is extremely pulse-free and has no moving parts to maintain.

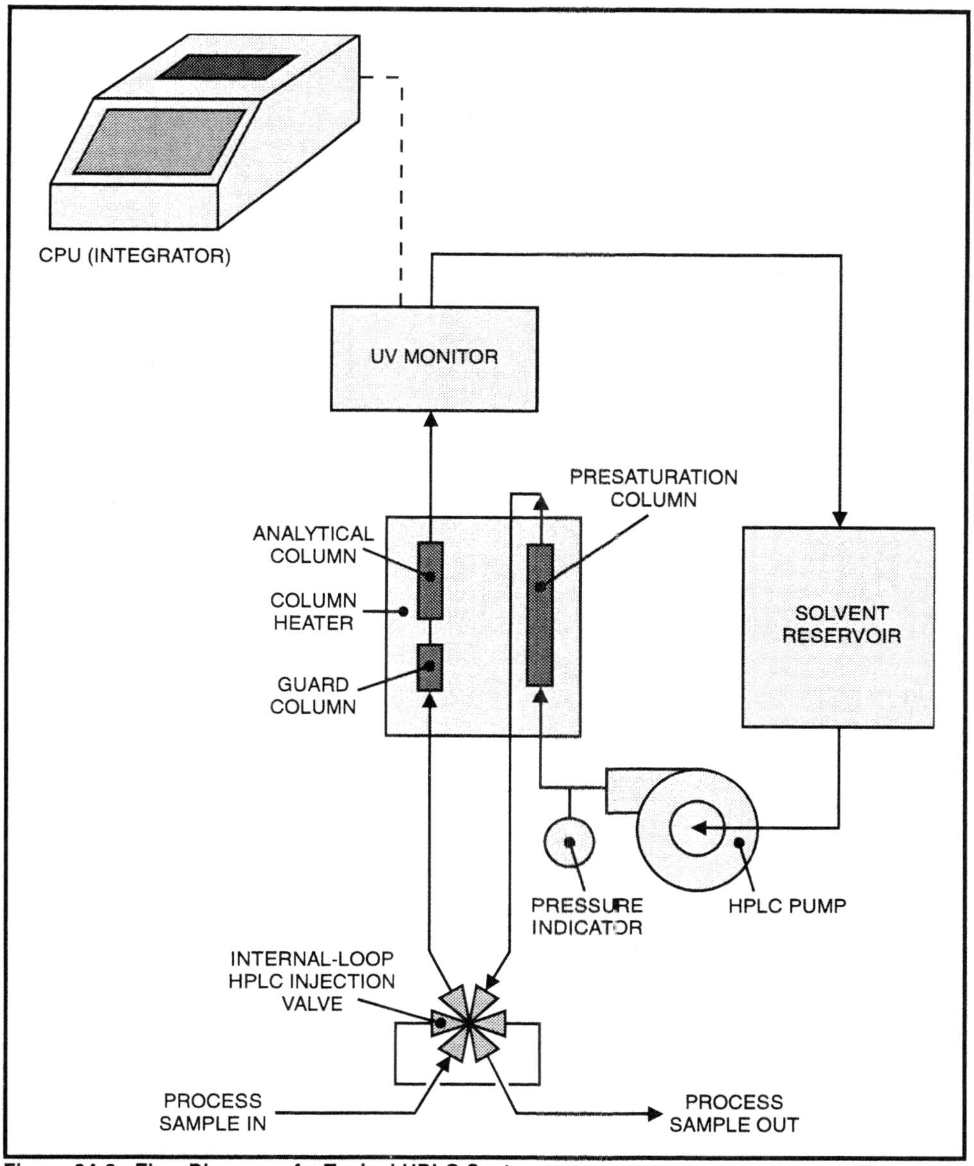

Figure 34-3. Flow Diagram of a Typical HPLC System

SYRINGE-TYPE PUMPS

The next design to evolve utilized a syringe-type pump. In this design, a piston, driven by a stepper motor, continuously compresses the mobile phase contained in a cylinder. The action of the piston provides the force for moving the eluent through the system. This pump design was used in the Waters ProMonix® on-line HPLC system. This type of pump design has no check valves and a minimum number of seals to wear out, which is obviously a big advantage for on-line analysis.

The major drawback is that the cylinder has a finite volume. This could be a limitation in designing the chromatographic method, since the length of the chromatographic separation would be limited by the volume of mobile phase that can be contained in the pump's piston. To refill the pump's cylinder requires that the flow to the column be stopped while the cylinder is refilled with mobile phase. This can be very disruptive to the separation taking place. If the volume of the pump is sufficient to complete a chromatogram, it still requires that the pump be refilled before the next chromatogram can be run. When attempting to get an

One method to avoid this limitation is to have two syringe pumps in the system. This allows one pump to be refilled while the other pump is used for the separation.

assay as quickly as possible, one doesn't want to waste time waiting for the pump to be refilled for the next analysis. One method to avoid this limitation is to have two syringe pumps in the system. This allows one pump to be refilled while the other pump is used for the separation. This limitation led to the replacement of these types of pumps in most systems with pumps capable of continuous compression and refill cycles.

PNEUMATIC AMPLIFIER PUMPS

The pneumatic amplifier pump is a popular design for on-line use. The design has a large-bore piston that drives, under the force of the air supply's pressure, against a small-bore liquid piston, which forces liquid to flow. The piston acts as a hydraulic amplifier to multiply the pressure of the gas supply to the pressure required for performing HPLC analysis. The amplification that is gained is the ratio of the cross-sectional area of the pneumatic piston to the area of the liquid piston.

This type of pump operates in constant pressure mode rather than constant flow, which is viewed as a disadvantage by most chromatographers.

The disadvantage of the design is that there are seals around the pistons and check valves in the pump head. These will wear after a period of time, resulting in a failure to seat and causing fluctuations in the pump's outlet pressure. When this occurs, the valves must be replaced. This type of pump operates in constant pressure mode rather than constant flow, which is viewed as a disadvantage by most chromatographers. The advantages are that the mobile phase volume is not limited, as in the syringe pump, and, since no electricity is required, the pump is intrinsically safe.

DIAPHRAGM PUMPS

The diaphragm pump design eliminates the need for seals. This design uses an electric motor to compress a hydraulic fluid against a flexible diaphragm. The opposite side of the diaphragm is in contact with the mobile phase. As the diaphragm compresses against the mobile phase, the fluid is forced out of the pump and through the column. Since there is no piston moving in and out of the pump head, the seal associated with it is eliminated. The disadvantage is that check valves are required to maintain pressure in the system as the pump goes through its fill-and-compression cycles.

POSITIVE-DISPLACEMENT PISTON PUMP

The positive-displacement piston pump is by far the most popular design.

The positive-displacement piston pump is by far the most popular design. This pump uses a piston moving forward into a cylinder to propel the solvent through the HPLC column. As in the syringe drive pump, the flow is stopped during the pump's refill stroke. This would be very disruptive to a chromatographic separation. To get around this problem, some pumps utilize two cylinders operating approximately 180° out of phase to each other. This allows one pump cylinder to be refilled while the other cylinder is supplying flow to the column. With a slight overlap of the compression cycles of the pumps, no loss of flow to the column ever occurs.

An adaptation of this design allows a single pump to work with minimal disruption of the flow. This is done by replacing the circular cam of the pump with an eccentrically shaped cam to drive the piston (Figure 34-4). This allows the compression stroke to be fairly slow, while the refill stroke occurs quickly. Pulsation is minimized, but not eliminated.

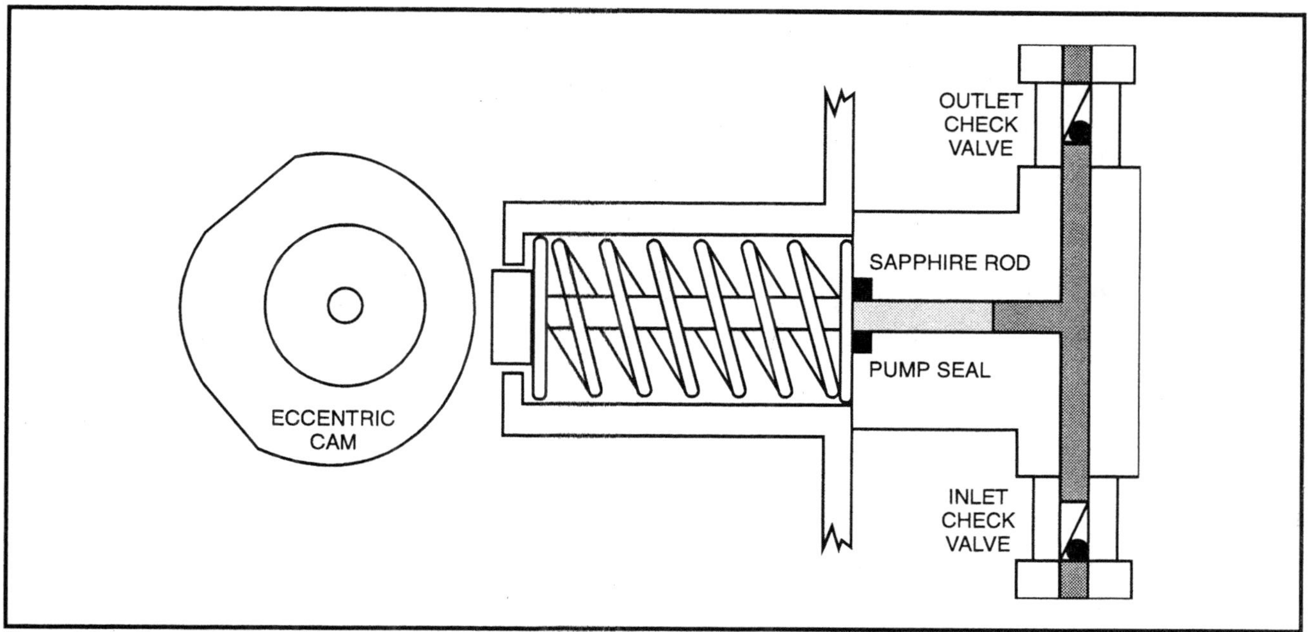

Figure 34-4. Cross-Sectional View of Single-Piston HPLC Pump

A single-piston pump equipped with an eccentric cam still exhibits pressure fluctuations that can be too severe for some detectors. If a detector is sensitive to these fluctuations, baseline noise will be observed. Pulse dampeners, constructed from a coil of small-bore tubing, placed between the pump and column will dampen these pressure pulses to a level that is undetectable by most detectors. Another method for dampening pulses is to install a pressure gage or transmitter between the pump and the rest of the chromatography system. This is quite useful for monitoring system pressure. This is used mainly for troubleshooting purposes, but can also act as a pulse dampener in the system. The pulse dampening occurs due to flexing of the gage element or the diaphragm of the pressure transmitter.

Pressure fluctuations can be caused by faulty check valves or air trapped in the pump head. To check for air, remove the outlet check valve, place your thumb tightly over the opening, then rapidly release your thumb. If air exits from the pump head, repeat the cycle of compressing and releasing your thumb until air no longer exits the pump head. If not done previously, degas the solvent reservoir.

Pressure fluctuations can be caused by faulty check valves or air trapped in the pump head. To check for air, remove the outlet check valve, place your thumb tightly over the opening, then rapidly release your thumb. If air exits from the pump head, repeat the cycle of compressing and releasing your thumb until air no longer exits the pump head. If not done previously, degas the solvent reservoir.

The preferred methods of solvent degassing are to pull a vacuum on the solvent to create boiling or to bubble helium through the solvent. Either method degases most mobile phases after only 2 to 3 minutes of treatment.

A solvent filter should be placed in the inlet line of the HPLC pump. This will minimize damage to the pump's check valves and plugging of the columns due to particulate matter in the mobile phase.

Maintenance associated with this pump design involves the inlet and outlet check valves and the seal located in the cylinder wall where the piston enters. The check valves are not a major maintenance problem, requiring replacement only approximately every 6 months. A large part of the seal wear is due to buffer particles that accumulate on the surfaces of the piston. As the piston reenters the cylinder during its compression stroke, these particles cut into the surface of the seal. These buffer crystals result when the mobile phase evaporates from the surface of the piston, leaving the crystallized buffer behind.

Seal wear can be minimized by the addition of a seal wash, a device that allows a slow flow of water to pass over the piston as it exits the cylinder. This flushes any buffer particles from the piston before it passes back through the seal during the compression stroke, thus reducing wear.

Seal wear can be minimized by the addition of a seal wash, a device that allows a slow flow of water to pass over the piston as it exits the cylinder. This flushes any buffer particles from the piston before it passes back through the seal during the compression stroke, thus reducing wear.

GRADIENT ELUTION

Gradient elution can decrease the amount of time required to elute peaks from the column. In this technique, the eluting strength of the mobile phase is increased versus time by increasing the percentage of organic in the eluent as the chromatography progresses. One method is to use two pumps in the system. The first pump contains the lower-strength buffer and the second pump has the higher-strength buffer. As the chromatogram develops, the percentage of flow from the first pump is decreased as the flow from the second increases.

The disadvantage of gradients is that the column must be reequilibrated with the weak solvent before injecting the next sample. This delays the analysis of subsequent samples, which is generally undesirable in a process analyzer. For this reason and because of the additional complexity of the instrumentation, gradient elution has not been used in on-line analyzers.

An interesting method has been proposed [Ref. 2] that eliminates this delay by having two columns in the system. One column is being reequilibrated while the other is eluted by the gradient. Reference 2 also presents an apparatus to generate an exponentially-shaped gradient without changing either pump's flow rate. This will simplify the pumping system required to generate the gradient.

Gradient elution could become more widely used in the future due to the advent of perfusion chromatography columns.

Gradient elution could become more widely used in the future due to the advent of perfusion chromatography columns (POROS®, PerSeptive Biosystems, Framingham, MA). In some applications, these columns have been able to produce chromatograms orders of magnitude faster than conventional HPLC columns, with many example chromatograms being completed in 1 to 2 min. The reason for this faster chromatography is due to the structure of the resin. Whereas most porous resins have dead-ended pores, the POROS resin has pores that pass completely through the resin, eliminating the stagnant pools of mobile phase that contribute to band spreading. To date, most reported applications of these types of resins have utilized gradient elution. Although gradient elution has been avoided in on-line analyzers, the disadvantages of the technique may be outweighed by the advantages gained by the use of perfusion chromatography.

Valves

Introduction of the sample into the mobile phase is accomplished by an injection valve. The HPLC injection valve is a unique device that allows a fixed volume to be filled with sample while flow to the column is maintained. Most injection valves are two-position devices (Figure 34-5). The first position, or "load" position, allows the fixed-volume sample loop to be filled with a sample while a bypass connection allows flow to continue to the HPLC column. The valve is then turned to the "inject" position, where the sample loop is placed in series with the pump and the chromatography column. In this position, the mobile phase displaces the sample from the loop and into the mobile phase stream, where it is delivered to the analytical column.

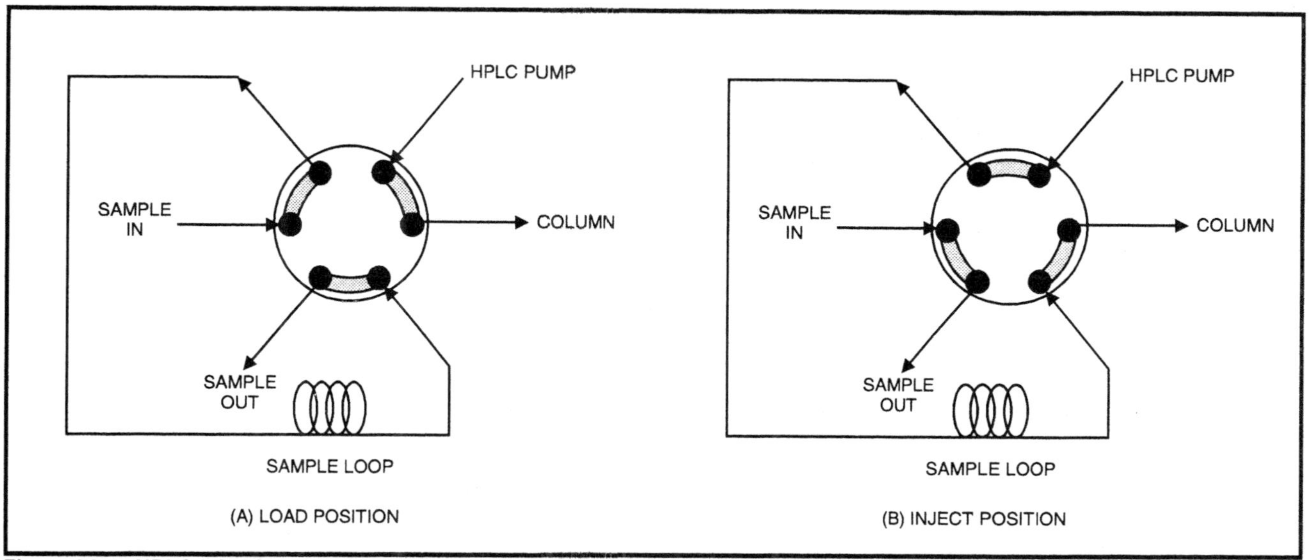

Figure 34-5. HPLC Injection Valve Operation

Changing the length or internal diameter of the sample loop allows the valve to be adapted to various sample volumes. Loop volumes ranging from hundreds of microliters down to approximately 2 µL can be achieved with an "external" loop. This type of loop resides on the outside of the injection valve. For sample sizes from 5 to 0.5 µL, the loop is placed inside the injection valve. If sample sizes less than 0.5 µL are required, the loop is machined into the rotor of the valve. Although some laboratory injectors allow partial filling of the loop, the best precision is obtained by completely filling the injection loop.

The sample loop can be a source of plugging since it can have a small inside diameter. Another maintenance issue is the valve rotor. This section of the valve is usually made of a polymer, which wears over time. This wear keeps the grooves in the rotor seal from properly seating against the stator. If the wear becomes severe enough, it will allow the ports of the valve to leak or to have "crosstalk" between connections. This leakage can be a cause of poor injection precision. Unfortunately, this is the only type of valve available with a pressure rating high enough for use in HPLC.

Since the internal bore of the injection valve is quite small (0.007 in. [0.18 mm]), it is imperative that the sample be free of particulate matter. This is accomplished by filtration of the sample before it enters the injection valve. Compatibility of the sample with the valve's rotor material must be considered in the analyzer's design. The standard rotor material used in many injection valves is attacked by high-pH samples. Manufacturers, on request, can supply an alternate material for the rotor seal that is usable at any pH.

Valve plugging has been observed even with particulate-free samples. The cause was attributed to precipitation of sample components when exposed to the analyzer's mobile phase. The precipitation occurred at the outlet of the injection valve, where the sample is first exposed to the mobile phase. In such cases, the mobile phase makeup may require modification to account for the sample's solubility.

This same type of rotary valve can also be used for column switching [Ref. 3]. This is a useful technique for moving columns into and out of the chromatography eluent stream. Column switching can be used to concentrate analytes from a large volume of sample that contains dilute levels of the analyte of interest. This can be particularly useful for trace analysis in applications such as environmental monitoring. In these types of analyses, the dilute aqueous sample is pumped through the preconcentration column, trapping and concentrating the sample. The preconcentration column is then placed in series with the analytical column and eluted into the analytical column by the mobile phase. In this type of application, the preconcentration column takes the place of the injection loop.

Column switching can also be used to add additional separation mechanisms to the chromatographic process. This can be done by using multiple columns in the analyzer with different selectivity mechanisms — for example, an octyl column and a phenyl column. This might allow a mixture to be partially separated on the first column and then further resolved on the second type of column by diverting a part of the first column's chromatogram onto the second column.

Backflushing is used to speed up the chromatographic process by eluting highly retained peaks back through the inlet of the column.

Switching valves can also be used for reversing the flow through a column. This technique is called "backflushing." Backflushing is used to speed up the chromatographic process by eluting highly retained peaks back through the inlet of the column. This shortens the chromatogram development time by eliminating the wait for the highly retained peaks to traverse the entire length of the column. An alternative to backflushing is the use of gradient elution, described earlier.

If all components of the sample must be quantitated, the backflush valve must be designed to ensure that the backflushed peaks are passed through the detector for quantitation. Otherwise, the purity of the sample would be reported as an artificially high value.

One of the major reasons cited for the lack of use of liquid chromatography as an on-line analyzer is the long analysis time. Switching valves can be used to shorten analysis time by having two identical columns in the analyzer. All samples must be on a column for a finite amount of time for the separation to develop. With two columns and the appropriate switching valves present, the chromatogram from one column can be monitored while the chromatogram on the second column is developing. An example of the valve and column arrangement necessary for performing this task is described in Reference 2.

Temperature Controllers

Temperature control of the column is generally required in any HPLC separation. This is done to minimize environmental temperature influences on the retention time of the peaks. Since all peak identifications are normally based on retention time, it is very important to maintain its reproducibility. Guillemin [Ref. 4] has proposed a technique called "deferred standard" to allow a peak to be identified even though its retention time may change from injection to injection. This is done by injecting a solution either just before or just after the sample so that a reference peak (internal standard) will appear in a noninterfering area of the chromatogram. By ratioing all retention times to the retention time of this reference peak, all peaks in the chromatogram can still be properly identified even though their retention times may be changing. This technique has allowed on-line chromatography to be performed without the need for temperature control.

Temperature control can also be used to increase the efficiency of the chromatographic process. Remember that chromatography is an equilibrium process. Increasing the temperature at which the chromatography is taking place will increase the mass transfer rate of the analyte between the mobile phase and the solid phase of the column. This will increase the efficiency of the column and decrease the time required to complete the separation.

Temperature control is usually achieved through the use of a column heater. The most popular design is a block-type heater, which consists of an aluminum block with grooves machined into it to accept various column dimensions. Heater elements are installed into cavities within the block. These devices are capable of controlling the column temperature to within ±0.2F° (±0.1C°). Temperature-control devices are available that allow subambient control as well as heated operation.

Elevated temperatures can have a deleterious effect on most columns. This occurs by two different mechanisms, depending mainly on the pH of the mobile phase. At pH values above approximately 7.5, the solubility of the silica becomes fairly high. High temperatures increase this solubility, which leads to dissolution of the silica and the formation of voids in the head of the column. At pH values of approximately 2 and lower, hydrolysis of the bonds between the stationary phase and the silica will occur. This allows bare silica to be exposed, which can lead to secondary interactions between the analytes and the solid support. These secondary interactions can be quite strong, resulting in material being irreversibly bound to the silica.

Dissolution of the silica solid support at high pH can be minimized by the use of a presaturation column. This column is placed between the HPLC pump and the injection valve and is packed with underivatized silica (Figure 34-3). The presaturation column is placed in the same column heater as the analytical column to maintain their temperatures at the same value. This column serves to presaturate the mobile phase with silica before the eluent reaches the analytical column, thus effectively eliminating the dissolution of the analytical column.

Detectors

UV MONITOR

The most widely used detector is the UV monitor.

After the analytes exit the analytical column, a means of detecting them is required. The most widely used detector is the UV monitor. This device consists of a source of ultraviolet radiation, a method of selecting the appropriate wavelength, and a detector for measuring the amount of energy transmitted through the sample stream. UV monitors are capable of parts per million detection or even parts per billion analysis in some special cases. The wavelength selection is performed by either a monochromator or an interference filter. Monitors that employ interference filters are more rugged and more attractive for on-line use due to their insensitivity to vibration. Another advantage is that UV sources suited for a narrow wavelength range are usually used in these types of monitors. These sources have lifetimes that are up to an order of magnitude longer than the broadband sources used in an adjustable-wavelength UV monitor. The drawback to this design is its inability to be adapted to a new wavelength without replacing the filter. This generally is not a problem in on-line systems since the instrument is usually dedicated to a particular application.

One interesting possibility provided by monochromator-based instruments is peak identity conformation, which is achieved by utilizing a diode array as a detector rather than a single photocell. In this type of device, the monochromator disperses the source's image onto a diode array after the entire spectrum of the lamp has passed through the sample. The diode array is generally capable of 2 nm resolution. By electronically scanning the output from each of the individual diodes, the entire resulting spectrum can be provided in 2 seconds or less. This process can be performed continuously during sample elution, thereby providing not only a chromatogram of the process sample, but also a spectrum of each peak in the chromatogram. This peak identity conformation can be a powerful method for confirming the peak's identity and validating the data obtained by the on-line HPLC.

REFRACTIVE INDEX MONITOR

The refractive index (RI) monitor is another popular HPLC detector. This device measures the change in refractive index that results from the analytes dissolved in the mobile phase. It is generally less sensitive than the UV monitor, but is more universal in its detectability. This can be an advantage when trying to measure analytes that have no UV chromophore or concentrations that may be too high for the linear range of the UV monitor. It is, however, a drawback if the mobile phase changes in composition during the run, such as occurs in gradient elution. In this case, the monitor may respond to changes in eluent composition as well as to changes in analyte concentration. This can make accurate quantitation difficult due to the baseline's movement in response to the eluent's concentration change.

OTHER MONITORS

Although UV and RI detectors are the only monitors reported in the literature for on-line liquid chromatography applications, several other types of monitors have been used in laboratory HPLCs. These monitors could conceivably be used on-line. The additional detectors used in the lab include fluorescence, electrochemical, conductivity, infrared, molecular weight, flame ionization, and near-infrared. For the most part, these detectors have only been applied in very specialized applications.

Regardless of detector type, the flow cell must have a small internal volume. The use of small-particle-size columns will produce peaks with volumes in the low-microliter range. Use of a flow cell with a volume greater than a few microliters will have a deleterious effect on the separation due to mixing of the peaks in the flow cell.

To test for the source of a noisy chromatogram, stop the HPLC pump to see if the noise disappears. If it does, the noise is likely due to air bubbles in the cell. If it doesn't, it is likely due to electrical noise in the detector.

Air bubbles in the flow cell of the detector can be particularly troublesome, resulting in baseline noise and poor quantitation when the noise spikes are integrated as peaks. Air bubbles are generally due to the release of dissolved gases from the mobile phase caused by the change in pressure experienced by the mobile phase as it travels from the high pressure of the HPLC column to the near-atmospheric pressure of the detector. To eliminate this problem, a small amount of back pressure is added downstream from the detector. The back pressure can be created by use of a commercial back-pressure regulator, by tightly coiling a length of small-bore stainless steel tubing, or by weaving Teflon tubing into a series of knots. Extreme care must be taken to avoid "crimping" the tubing, thus weakening its wall strength and possibly restricting its I.D. and flow capacity. To test for the source of a noisy chromatogram, stop the HPLC pump to see if the noise disappears. If it does, the noise is likely due to air bubbles in the cell. If it doesn't, it is likely due to electrical noise in the detector.

After exiting the detector, the mobile phase can be either sent to a waste collection container for disposal or recycled back into the mobile phase reservoir. Recycling of the solvent seems to work quite well, with little, if any, affect on the chromatographic separation. Mobile phase recycling offers several advantages. Since the mobile phase never leaves the system, solvent consumption and the labor required to prepare it are reduced significantly. Explosion hazards due to vaporization of the solvent from the mobile phase are also reduced since the HPLC analyzer can be assembled as a completely closed system. The reduction of solvent evaporation also increases retention time reproducibility. A possible deleterious effect is the concentration of impurities from the mobile phase onto the analytical column.

The recycled solvent can be used for approximately 6 to 8 weeks (several thousand injections) before it needs to be replaced by fresh solvent. Solvent recycling cannot be used with gradient elution, which is another drawback to this technique.

Tubing

The components of the chromatograph are connected to one another by small-bore, 1/16-in. OD tubing. Most systems use stainless steel, although polymer-based tubing is now available with operating pressures within the range of most assays. Polymer-based tubing must be handled carefully. If it is bent in less than an approximately 1-in. radius, it can develop small stress cracks, which can later develop into leaks. Polymer-based tubing is particularly advantageous in assays of analytes that adsorb to stainless steel. Elimination of stainless steel is also necessary in systems using low-pH buffers that contain halogens.

Polymer-based tubing must be handled carefully. If it is bent in less than an approximately 1-in. radius, it can develop small stress cracks, which can later develop into leaks.

All high-pressure chromatographs use compression-type fittings to make tubing connections. It is very important that no voids are created when tubing connections are made using these types of fittings. When 3-μm columns are used, voids can adversely affect separation efficiency by producing small "mixing" volumes within the chromatograph.

Use 0.007-in. (0.18 mm) diameter tubing between the outlet of the analytical column and the detector to maximize column efficiency. Use 0.010- to 0.020-in. (0.25- to 0.5-mm) ID tubing between the injector and the analytical column. All other sections of tubing should be 0.030 in. ID.

Use 0.007-in. (0.18 mm) diameter tubing between the outlet of the analytical column and the detector to maximize column efficiency. Use 0.010- to 0.020-in. (0.25- to 0.5-mm) ID tubing between the injector and the analytical column. All other sections of tubing should be 0.030 in. ID.

Unique Sampling System Requirements

The sampling systems used in liquid chromatography are very similar to those used in gas chromatography. The major requirement is that the sample contain no particles larger than 0.2 μm when it enters the analytical column in order to prevent plugging of the column's inlet frit. Tangential filters (Collins' "SWIRLKLEAN") have been successfully used to filter the process sample before it enters the chromatograph. These are particularly useful because of their ability to expose the sample to a large surface area with low sample holdup. Their high crossflow velocity also helps to keep the filters clean.

Flow of sample through the small inner passages of the sample injection valve is a major consideration in sampling system design. The considerable pressure drop across these passages requires that the process sample be either pressurized or pumped through the valve.

A peristaltic pump placed after the injection valve allows the sample to be pulled through the valve. This is particularly useful when pressure on the process line is insufficient to push the sample through the injection valve. Such a pump also allows the standard solution to be fed to the valve without having the standard reservoir pressurized.

A peristaltic pump placed after the injection valve allows the sample to be pulled through the valve. This is particularly useful when pressure on the process line is insufficient to push the sample through the injection valve. Such a pump also allows the standard solution to be fed to the valve without having the standard reservoir pressurized.

One of the most important criteria in the sampling system is a method for automatic calibration. Since all peak identities are based on the retention time of a reference peak, it is important to have a method of calibrating the data system before each run. The placement of a three-way valve ahead of the injection valve allows either the process stream or a standard solution to be fed to the analyzer.

A final consideration is sampling process validation. In some sampling systems, this has been achieved through the use of an intravenous drip chamber monitor, which allows the drops of sample emitting from the injection valve to be counted. The total number of drops counted are compared to a set point, and an alarm is sounded if too few drops are observed. Miniature proximity sensors should be utilized to confirm the position of the valves in the sampling system.

Sample dilution and/or conditioning are required in many applications of on-line liquid chromatography. The system shown in Figure 34-6 [Ref. 5] allows the pH of a sample from a biochemical reactor to be adjusted before the sample is fed to an on-line liquid chromatography system. The pH adjustment is necessary to effect dissolution of an insoluble product formed during the initial mixing reaction. Operation of the sample injection system proceeds as follows: Valves 1 (system load valve) and 2 (column inject valve) are placed in the "load" position, which allows solution from the reactor to be withdrawn through the valve 1 sample loop, filling it with the process sample. After proper purging of the sample loop on valve 1, the valve rotor rotates 60° clockwise and the loaded sample loop is placed in the "inject" position. This allows the carrier (diluent) stream to displace valve 1 loop's contents into a mixing coil constructed from 0.050-in. (1.27-mm) ID tubing, where dispersion (mixing) of the sample occurs.

... it is possible to obtain an infinite range of dilution factors ...

The diluent's pH was chosen so that insoluble product would dissolve. Dilution of the product also occurs as a result of the dispersion in the mixing coil. This dispersion produces a near-Gaussian (bell-shaped) distribution of concentrations of the original sample, ranging from infinitely dilute at the tails of the peak to nearly pure sample at the apex of the distribution. This dispersed sample peak flows through the sample loop of valve 2, the on-line liquid chromatograph's injection valve. The volume of the dispersed sample peak is many times larger

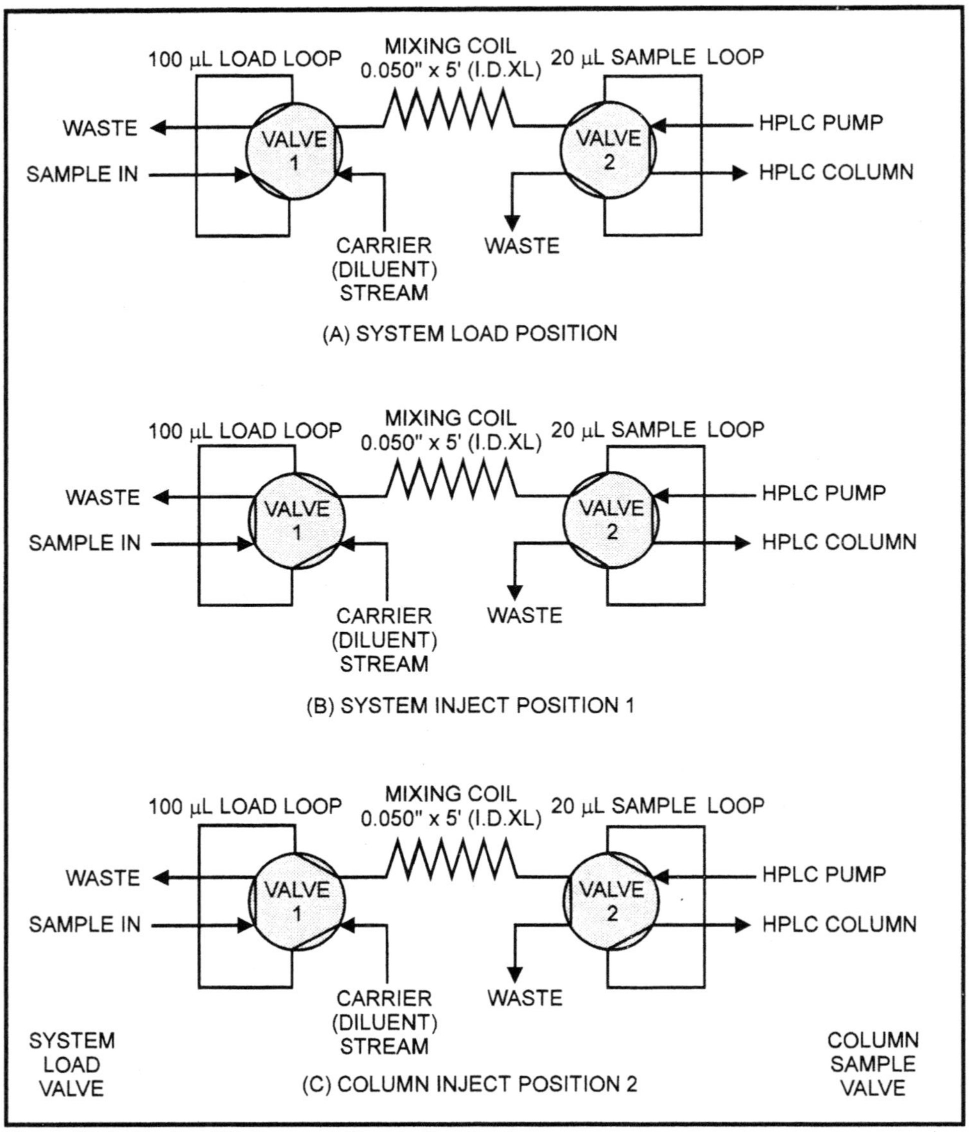

Figure 34-6. On-line Dilution and Dissolution Device

than the volume of the sample loop on valve 2. Because of this, it is possible to obtain an infinite range of dilution factors of the original sample, depending on which part of the dispersed sample peak is in the sample loop at the time the column sample valve (V2) is moved to its "inject" position. The dilution factor is controlled by the amount of delay placed between the movement of valves 1 and 2 to their "inject" positions.

A variation of this dilution system replaces the mixing coil with an 800-µL mixing chamber equipped with a magnetic stirring bar [Ref. 3]. The sample is displaced from the first sample loop (system load valve) into the mixing chamber, where the concentration rapidly rises to a near-steady state. After this rapid rise, there is an exponential decay back to zero concentration. This system produces larger dilution factors than the mixing coil sample inject system.

This system produces larger dilution factors than the mixing coil sample inject system.

The HPLC analyzer contains columns with very small particle-size packing materials and small entrance and exit frits. All of these points can become plugged if the sample stream contains particles larger than 0.5 µm. This requires that the process stream be free of particulate matter. If the possibility exists that the process sample may contain particulates, a filtration device should be installed in the system. Such a device can be installed in the process stream or in the sample line leading to the analyzer.

The best location for the filtration device is in the actual process sample line. This allows use of large-volume devices that do not become plugged as frequently as the small devices used when filtration is performed in the sample stream. A tangential flow filter (Collins' "SWIRLKLEAN") is a small holdup volume device that can be used in the sampling system. It offers high surface area, small internal volume, and high crossflow velocity to help prevent fouling. Whatever type of device is used, it must be installed such that it does not become "air-locked" by air entrained in the sample, which can be particularly troublesome in small, nontangential flow filtration devices.

Installation Hints

Proper installation of the on-line HPLC analyzer is important to the success of an application. Most process environments require that some level of protection be afforded the instrument by an enclosure in order to prevent damage to the analyzer components by water, dust, corrosive fumes, or movable factory equipment, such as forklifts. Protection of this type can be achieved by using a heavy-walled enclosure with gasketed doors. If the area is hosed down frequently, the enclosure might be constructed from stainless steel.

The presence or threat of an explosive (electrical-hazard rated) environment dictates stringent installation requirements. No commercially available unit has an intrinsically safe rating. Therefore, a means of purging the enclosure to meet current electrical codes is required. If the area has a Class I, Division 2 classification (explosive vapors not present during normal operation, but could be during a process upset), this can be handled satisfactorily by the addition of a purge system and a purge pressure indicator. In the case of Class I, Division 1 conditions (explosive vapors present during normal process operation), the power to the analyzer must be disconnected before the enclosure can be opened, which makes it nearly impossible to perform any troubleshooting operations. One solution is to place the analyzer in a safe area and install a sampling system that includes a fast loop to deliver the sample within close proximity of the analyzer. An example would be to place the analyzer in a control room and flow the process along the room's outside wall. The analyzer's small-bore sample line would penetrate the wall to interface with the process stream.

The temperature and humidity of the installation area also must be considered. Unless the column is maintained at a constant temperature, the retention time of the peaks will not be reproducible. This can be achieved by thermostatting the column. If the ambient temperature of the area exceeds the temperature at which the column is controlled, the column will drift up to the ambient temperature. This requires that the column temperature control be capable of cooling as well as heating. An alternative to a combination heating/cooling device would be to operate the column at a temperature above the highest ambient temperature expected. The other possible way around this problem is to use the "deferred standard" technique developed by Guillemin [Ref. 4], which is viable only if the separation required can be achieved over the entire range of temperatures that will be observed.

The solvent's refractive index can also be affected by temperature. Changes in the refractive index can cause baseline excursions that may affect the quantitation. This is a particular problem in chill room installations, where opening the analyzer enclosure allows cold air to enter and contact any exposed capillary tubing of the chromatograph. This can lower the temperature of the mobile phase, change the refractive index, and produce a shift in the baseline. Covering all exposed tubing with foam insulation or using heavy-walled (vacuum) tubing (for its thermal mass) is a successful correction technique.

Humidity generally has less effect on an analyzer than temperature, but can affect the optics of the detector. High humidity can cause condensation to occur, resulting in large shifts in the baseline, an effect most prevalent in chill room installations. The best prevention method is to keep high-humidity air from entering the enclosure. Conduit penetrations into the enclosure should be sealed.

Typical Calibration Requirements

Proper calibration of an on-line HPLC analyzer must be determined for each particular application and is based on the accuracy required, the drift of the detector, and the drift in the retention times of the peaks in the chromatogram. The frequency of instrument failures is also a consideration, since calibration is a good way to assess whether the instrument is operating properly.

Routine calibration can be performed in most cases by using a single-point standard. A multiple-point standard curve is used to determine the linear range of the instrument and must be run before commissioning an analyzer in order to validate its operation and determine the limits of its operating range. With most UV monitors, the linear range will span at least an order of magnitude. The other factor that limits the range of operation is the chromatographic column. Each type of column has a certain range of materials that can be injected onto it before an overload condition occurs. In this overload state, the column begins to lose its resolution power. How much resolution can be lost before the assay is affected depends on how well the peaks in the chromatogram are resolved from one another.

Dual calibration allows the analyzer's data system to be trained to identify a peak in the chromatogram as a particular compound and quantitate its concentration and/or purity.

The calibration operation of any chromatographic analyzer requires that two dimensions be calibrated. One dimension is the time domain (x-axis), the other is the response of the detector (y-axis). Dual calibration allows the analyzer's data system to be trained to identify a peak in the chromatogram as a particular compound and quantitate its concentration and/or purity. Identification of an analyte is based on the peak's retention time.

Peak identity and quantitation are performed by injecting a sample of known content into the chromatograph. The sample may be a primary reference standard solution or a secondary standard prepared from a process sample that has been previously analyzed in the laboratory against a primary reference standard.

Automatic calibration of the system can be performed by inclusion of an algorithm in the data system that utilizes the relative elution order of the peaks in the chromatogram to identify the analytes. An alternative method is to identify the peaks in the chromatogram manually by typing the proper identifiers into the data system after injection of the standard mixture. In either case, a retention time "window" is calculated and stored in the data system. This window is calculated by expanding the actual retention time of the peak by a fixed percentage.

If a peak of interest, Peak A, has a retention time (*RT*) of 200 s, a retention time window of $RT \pm 10\%$ would set the retention time window at 180 to 220 s. Any peak falling within this window would be identified as Peak A.

If a peak of interest, Peak A, has a retention time (*RT*) of 200 s, a retention time window of *RT* ±10% would set the retention time window at 180 to 220 s. Any peak falling within this window would be identified as Peak A.

The width of the retention time window is determined by considering several factors about the chromatogram. If some peaks elute close to the peak of interest, the window may have to be much smaller than ±10% to prevent misidentifying a closely-eluting peak as the peak of interest. On the other hand, if all peaks in the chromatogram are well resolved, the width of the window may be set much wider so that the peak of interest is not likely to be missed any time during the analysis — a critical factor if the peak of interest varies widely in concentration during the analysis. As the concentration goes up, the analytical column may approach an overload condition. When this occurs, the retention time of the peak tends to shorten. If the size of the window is too small, the peak may move out of the window and will no longer be properly identified.

Quantitation of a sample chromatogram can be handled by one of two methods or a combination of both. If concentration units (for example, mg/mL) are to be assigned to peaks in the chromatogram, a relative response factor must be calculated. This is achieved by injecting a reference solution of known concentration. A response factor of peak height or peak area per milligram of analyte is then calculated and stored in the data system. Concentrations are calculated by dividing the peak's response by the appropriate response factor. If all peaks in the chromatogram have approximately the same detector response, the response factor can be used for all peaks in the chromatogram. If this is not the case, a response factor will have to be calculated for each peak that is to be quantitated.

Concentrations are calculated by dividing the peak's response by the appropriate response factor.

Purity values can be calculated without prior measurement of any response factors. This is done by dividing the area of the peak of interest by the area of all peaks in the chromatogram. This calculation method is simple compared to the calculation of concentration, but is accurate only if the response factors of all peaks in the chromatogram are equal. When high accuracy is required, the response factors for each peak will have to be determined and used to correct the purity calculation.

Troubleshooting and Maintenance

As with any on-line analyzer, the quality of data obtained with HPLC is directly related to the quality of the analyzer's maintenance. An on-line HPLC requires more routine attention than, for example, a photometric analyzer. Daily inspection is required in order to assure proper operation. On-line chromatographs also require a more specialized form of maintenance than do most analyzers used in a production setting. The failure point in chromatographs is generally in the area of the chromatographic process itself rather than mechanical or electrical failures of the instrumentation components.

A dedicated team of analyzer maintenance technicians is most important. The analyzer group should be made up of individuals trained in chromatography and, in most cases, experienced in laboratory HPLC. Their knowledge of analytical chromatography can be used not only to assess proper operation of the chromatographic process, but also to troubleshoot instrument components.

A good working knowledge of how the process works can increase the efficiency of the troubleshooting process. Many things that are visible in the chromatogram can indicate which component is failing. Once the defective component is identified, it can be replaced by a backup unit and then the failed unit can be sent to the instrument repair shop.

A constant user concern is that an analyzer will fail during a period when it is unattended. Therefore, it is important that the analyzer be designed with the capability to perform a rudimentary assessment of its operation. Using sensors interfaced to the data system, a program can evaluate whether operating parameters (column pressure, column temperature, detector baseline level, etc.) are within normal ranges. If any of these parameters is out of range, an alert to the process operators is issued.

Most liquid chromatograph (HPLC) failures fall into one of three areas: the columns, the pump, or the injection valve. The major failure in the columns is plugging of column frits, in which case the column usually must be replaced. Although it is possible to replace frits, the cost of the columns does not justify attempting to repair them. HPLC pump failures occur most frequently in the check valves or pump seals. If time permits, these failures can be handled by stopping the analyzer and rebuilding the pump. If time is critical, the pump can be replaced with a backup unit and then rebuilt. Failure of the injection valve is localized to either the rotor seal or plugging of the injection loop. Both of these failures can usually be corrected with the valve in place after temporarily suspending operation of the analyzer.

Most problems in the HPLC can be identified before they become actual failures. This is achieved by a daily inspection that includes examination both of the actual equipment and of chromatograms from the most recent run on the analyzer.

Most problems in the HPLC can be identified before they become actual failures. This is achieved by a daily inspection that includes examination both of the actual equipment and of chromatograms from the most recent run on the analyzer. The total time required to perform routine maintenance and actual repairs to a system averages out to approximately 15 to 30 min per analyzer per day. This is not much different from the amount of time generally dedicated to a laboratory chromatograph and is actually much less if one includes the time it takes to prepare samples and place them on an autoinjector for analysis by a laboratory instrument.

Specific Safety Issues

The major safety issue involves the electrical-hazard classification of the area in which the analyzer is installed; this has already been discussed. Other safety concerns arise from possible leaks of the mobile phase within the HPLC enclosure. Hazardous conditions could be generated if a large amount of the mobile phase were released. This could occur through leaks developed in the HPLC pump or the analyzer tubing. One way to monitor this is to install a pressure transmitter in the mobile phase feed line after the HPLC pump. Any significant drop in system pressure would be indicative of a leak in the solvent delivery system.

Failure of the mobile phase reservoir is a problem that would not be detected by a pressure monitor. An error such as this could be detected by a level sensor installed in the bottom of the enclosure. Examples of such devices include conductivity-type switches, capacitive proximity switches, and fiber-optic level sensors.

The presence of organic vapors within the analyzer could be detected by a monitor mounted inside the enclosure. This device can be wired to an audible alarm, the data system, or a power disconnect. This type of detection system is capable of protection against any type of leak within the enclosure, no matter the source, as long as a detectable level of vapor is produced.

Although the HPLC operates at high pressure (sometimes as high as 5,000 psig), it poses little safety hazard because the mobile phase is in the form of a liquid rather than a gas. Therefore, the pressure can be released without any large changes in volume. Spraying of the mobile phase into the eyes or on the skin could be a concern, depending on the phase makeup.

Examples and Case Histories

Few on-line liquid chromatography applications have been published in the literature. Either there are not many systems on-line or users are reluctant to publish their applications due to the confidentiality of their processes. If one believes the number of working applications cited by the manufacturers of on-line HPLC systems, one would have to assume that the reason for the lack of published applications is confidentiality. Several examples of process applications taken from manufacturer literature [Ref. 6] are listed in Table 34-2.

Table 34-2. Typical Commercial On-Line HPLC Analyzer Applications [Ref.6]

Analyte	Application
2-Naphthaline sulfonacide	Control of a bioreactor for treatment of waste water from a color manufacturing process
Lipophilic by-products, e.g., C10, C12, C14	Process control of a detergent manufacturing process
Caffeine	Continuous determination and documentation of caffeine level in coffee extract
Saccharine, benzamide, toluolsulfonacide, *o*-sulfobenzoacide, nickel	Monitoring of the optimum saccharine/nickel ratio in a galvanic nickel process
N-methylpryridinium, β-picolin	Monitoring of polyamide synthesis
β-caprolactam	Monitoring of interferences in concentrated β-caprolactam used in the synthesis of Perlon (Nylon 6)
Palantinose, trichalulose, isomaltose, saccharose, glucose, fructose	Control of fermentation of a diet sugar
Diethyleneglycol, triethyleneglycol	Monitoring of glycols used as softeners in plastics manufacturing
Cephalosporin antibiotic	Monitoring of a preparative column used to recover a cephalosporin antibiotic from fermentation medium

The greatest number of published applications of on-line liquid chromatography have been in the biotech and polymer areas. This is likely due to the nonvolatility of the sample and the suitability of HPLC for these applications. Gressin [Ref. 7] describes the initial use of an on-line HPLC constructed in-house and then, later, a commercial on-line liquid chromatograph to monitor the progress of a fermentation. An analytical method was developed that allowed the substrate, the fermentation product and two impurities to be monitored simultaneously. Here, HPLC has a distinct advantage over other methods because of its selectivity. It also demonstrates that the long assay time is not a problem for biotech applications. Even though a new assay value could be returned every 20 min, the analyzer provided excellent substrate feed rate data with only one assay per hour. The long-term stability of the analyzer was maintained through use of the deferred standard technique [Ref. 4].

Some of the cited advantages of on-line HPLC include:

1. Increased knowledge of the fermentation process
2. Detection in real-time of unwanted metabolic pathways
3. Indirect help in detection of contaminations
4. Adaptive feeding of substrates to the specific needs of the microorganism

Negatives cited were the equipment's high price and the necessity of rigorous maintenance. The experiences reviewed in the publication reported an uptime of greater than 94%. Although this is low by industry standards [Ref. 8], it is not that unusual for an HPLC analyzer.

The application that drove the development of on-line HPLC at Eli Lilly was automation of the control of columns used in the purification of biosynthetic human insulin [Ref. 9]. Prior to the addition of on-line HPLC, the purification column eluent was collected in fractions that were sampled and submitted to the analytical lab for analysis. After the assay, the fractions that exceeded the mainstream purity value had to be recharged onto the second purification column. Since the columns were eluted with a gradient, the fractions had to be charged onto the second column in the same order that they were eluted from the first column in order to avoid autoeluting the material from the second column. This was a very labor-intensive operation since each fraction had to be manipulated manually.

With the installation of an on-line HPLC [Ref. 5] on the process column (Figure 34-7), the column operations could be totally automated. The installation of several automatic valves allowed a single HPLC system to be used to monitor either of the two process columns. These automatic valves also allowed the two columns to be operated independently or to be connected in series so the mainstream, identified by the on-line HPLC, could flow directly from column 1 onto column 2. On-line HPLC eliminated the delays associated with off-line analysis and the time required to identify the mainstream fractions and charge them onto the next purification column. Throughput for the process step increased approximately tenfold with the addition of the on-line HPLC analyzer to this column operation.

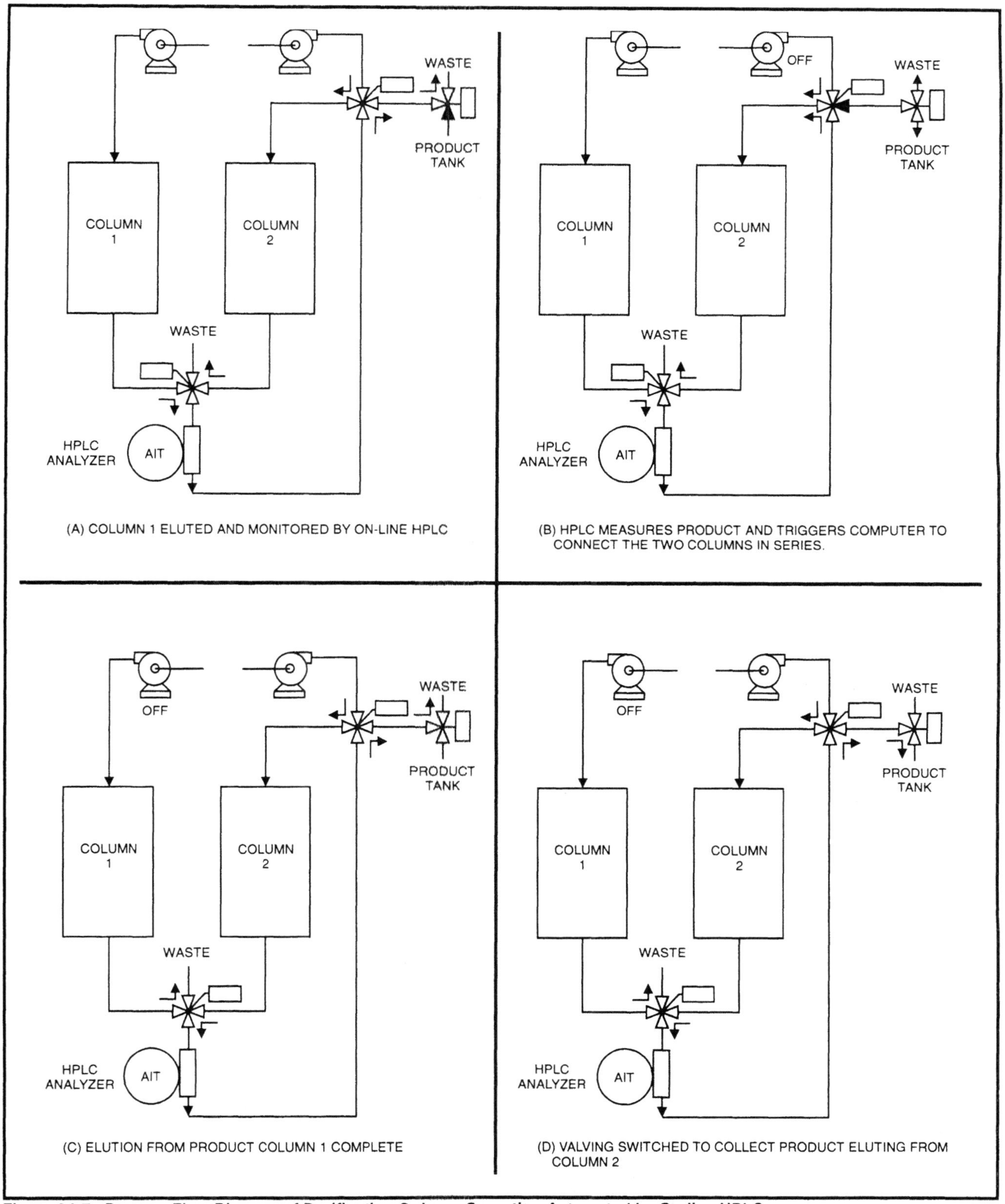

Figure 34-7. Process Flow Diagram of Purification Column Operation Automated by On-line HPLC

References

1. Mowery, R.A., Jr., and Roof, L.B., "On-Stream Analysis by Liquid Chromatography," *Analysis Instrumentation*, 14 (19): 19-24, 1976.

2. Thevenon-Emeric, G., and Regnier, F.E., "Process Monitoring by Parallel Column Gradient Elution Chromatography," *Analytical Chemistry*, 63(11): 1114-1118, 1991.

3. Mowery, R.A., Jr., "Process Liquid Chromatography," *Proceedings from the 35th Instrumentation Symposium for the Process Industries*, College Station, TX: Texas A&M University, 1980, p. 7-17.

4. Guillemin, C.L., "Software Driven Instrumentation: Application to On-line Process Liquid Chromatography," *Trends in Analytical Chemistry*, 8(7): 273-276, 1989.

5. Cooley, R.E., and Stevenson, C.E., "On-line HPLC as a Process Monitor in Biotechnology," *Process Control and Quality*, 2(1): 43-53, 1992.

6. "PLC-502 Process Liquid Chromatograph Applications," Berlin, NJ: ES Industries, Division of Siemens Chromatographs, Siemens AG.

7. Gressin, J.C., "On-line Process Liquid Chromatography: A Useful Tool in Biochemical Processes," *BTF — Biotech-Forum*, 5(1): 38-43, 1988.

8. Clevett, Kenneth J., *Process Analyzer Technology*, New York: John Wiley & Sons, 1986.

9. Cooley, R.E., and Stevenson, C.E., paper presented at International Forum on Process Analytical Chemistry (IFPAC), in *Process Control and Quality Proceedings*, Houston, TX, 1991.

About the Author

Rick E. Cooley is an Associate Engineering Consultant in the Measurement Technology Development group of Eli Lilly and Company where he has been employed for 24 years. He holds a B.S. degree in Chemistry from Purdue University. His initial assignment at Lilly was in the analytical development laboratory where he worked on the development of automated laboratory methods of analysis and interfacing laboratory instruments to the lab's central computer system.

Approximately 15 years ago, Mr. Cooley moved from the traditional analytical laboratory and began working in the process analytics area. His initial assignment was the design and implementation of on-line HPLC analyzers to provide closed loop control of the chromatographic purification processes utilized in the manufacture of biosynthetic human insulin (Humulin®). A total of 26 of these systems have been implemented in Lilly plants worldwide. Mr. Cooley has continued his involvement in the process analytics field utilizing various chromatographic and spectroscopic techniques to monitor and control pharmaceutical processes.

Mr. Cooley has published and presented several papers on automated laboratory instrumentation, on-line HPLC, and process automation utilizing process analytics. He is a Senior Member of the ISA.

35

Distillation (% off) Analyzers

Since the 1960s, chromatographs have been used in laboratories to simulate distillation test methods. Subsequently, on-line chromatographs were developed, and in 1974 the American Society for Testing and Materials (ASTM) published a method to measure the boiling range of petroleum products with gas chromatographs.

ASTM Standard Test Method D-2887 for Boiling Range Distribution of Petroleum Fractions by Gas Chromatography. This test handles samples having final boiling points as high as 1000°F (540°C). It is limited, however, to material with initial boiling point (IBP) above 100°F (40°C) and with vapor pressure low enough to permit sampling at ambient temperatures without losing light ends. Therefore, the method is *not suitable for analyzing gasolines.*

ASTM Standard Test Method D-3710. In 1978, a modified version of the test was developed. This test is designed to measure the entire boiling range of gasolines and gasoline components with either low or high Reid vapor pressure (RVP). It is limited to samples with end points below 500°F (260°C).

Both methods use a gas chromatograph with a column that separates hydrocarbons in order of their increasing boiling points. The column temperature is raised at a controlled rate, and the area under the chromatogram is recorded throughout the analysis. Boiling points are assigned from a calibration curve obtained on a known mixture of hydrocarbons boiling within the expected range of the sample and run under the same temperature and flow conditions. Boiling point distributions obtained by these methods are essentially equivalent to those obtained by ASTM D-2172 (true boiling point [TBP] method).

A process analyzer utilizing these principles, the Monirex Series 404, was developed by Universal Oil Products, Inc. (UOP) (Chicago, IL). Another was developed by Applied Automation (Bartlesville, OK).

On-line Process Analyzers

UOP Monirex Series 404 Boiling Point Monitor

The Monirex analyzer is a microprocessor-controlled, temperature-programmed chromatograph similar in principle to that of ASTM D-2887. It is limited to petroleum products with boiling ranges of 100 to 1000°F (40 to 540°C). A minute sample is injected into a packed tubular column at a fixed low temperature. The column is heated rapidly to raise the temperature at a constant rate, and a thermal conductivity detector (TCD) senses the effluent. Under these conditions, the effluent appearing at a fixed time after sample injection will always have the same boiling point temperature.

With a long column, a separate peak could be produced for each component present, but the analysis would require 20 to 40 min to complete (Figure 35-1). The Monirex analyzer uses a short, packed column, which operates much faster and completes a cycle in 4 min. Individual components are not separated completely (Figure 35-2). The output signal from the thermal conductivity detector is fed to an integrator, the signals are continuously summed and stored. The total integral is used to calculate the percent-off at each selected boiling point, the results are stored and displayed in digital registers. The system can be programmed to record as many as 18 points in the distillation curve. The 0.5% point is recorded as the initial boiling point (IBP) and the 99.5% point as the end point (EP). Other points, for instance, could include the 5, 10, 50, 90, and 95% points. It can also calculate the Reid vapor pressure (RVP) from the concentrations of the lighter components. A calibration sample with a known boiling component at each desired point distilled is used for analyzer calibration.

The analyzer is packaged in two separate units. The analyzer unit is field mounted and suitable for installation in Class I, Group D, Division 1 (ID1) hazardous areas. The readout unit is designed for remote mounting in a nonhazardous area, such as a control room. The field unit contains a sample conditioning system and a calibration introduction system.

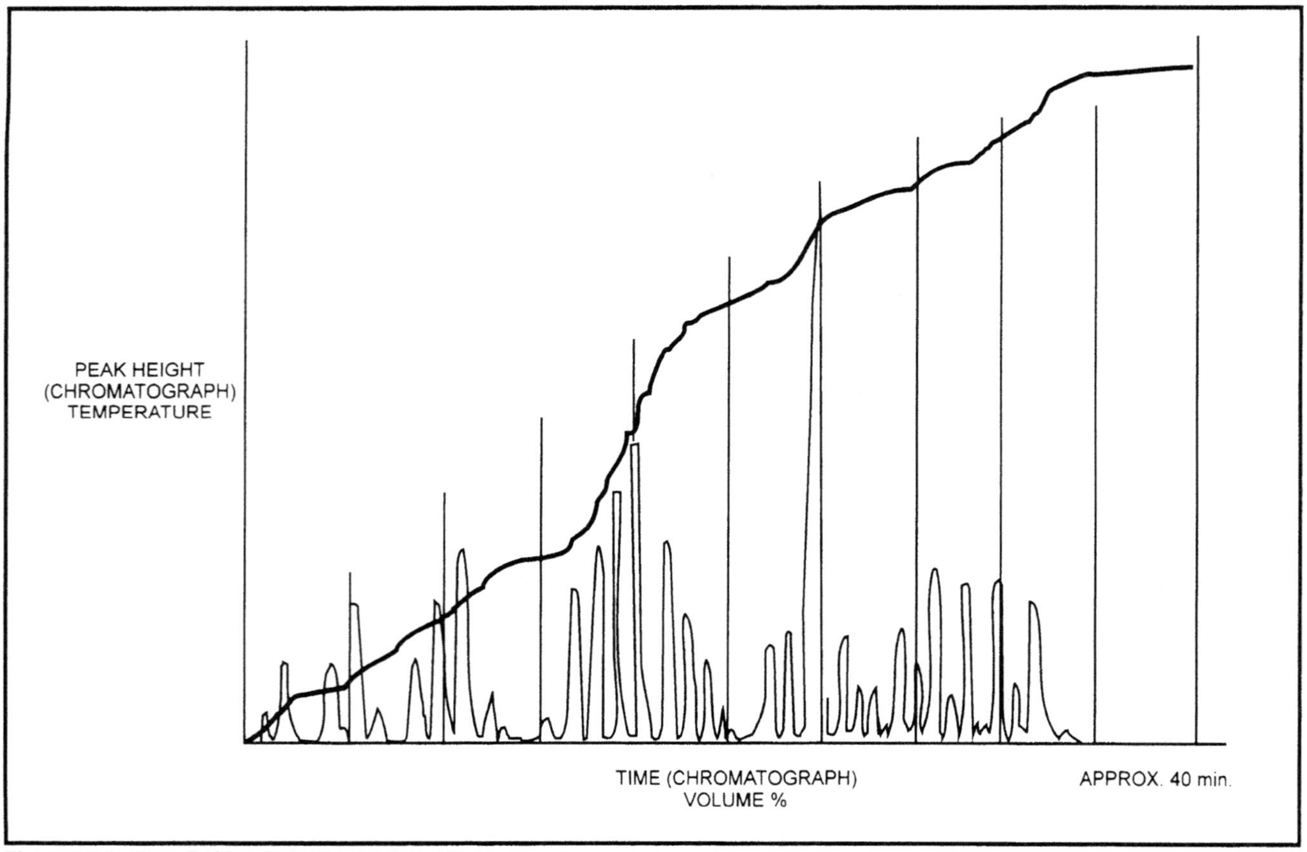

Figure 35-1. True Boiling Point Curve from Chromatogram
(Courtesy of Universal Oil Products, Inc.)

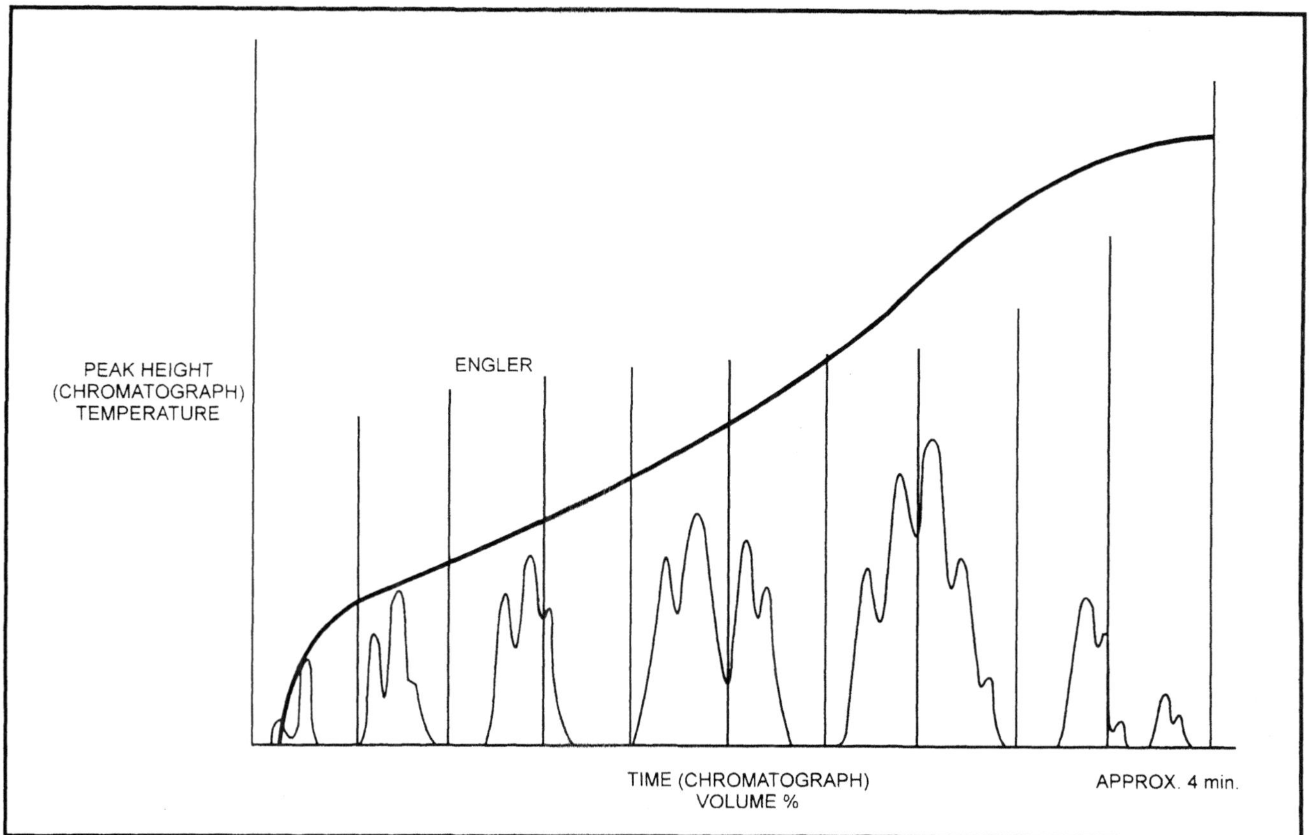

Figure 35-2. Simulated Engler Distillation
(Courtesy of Universal Oil Products, Inc.)

Applied Automation Optichrom Advance Gas Chromatograph

The Optichrom simulated-distillation (SIMDIS) process gas chromatograph is a temperature-programmed instrument that uses a flame ionization detector (FID) and a fused-silica open-tubular (FSOT) capillary column instead of a packed column. The lower enclosure contains the sampling system, which is designed to withdraw a representative sample from the process, transport it to the analyzer, and condition it for the analysis. Since the process sample is never exposed to the atmosphere, the possibility of losing light ends is minimized and the initial boiling point (IBP) data are reproducible.

The center enclosure is a temperature-controlled oven in which are located the sample injection valve, the chromatographic column, and the detector. Within the oven, the outer enclosed oven area is maintained continuously at a temperature close to the maximum to which the column will be programmed. The column, which is in the inner enclosed oven area, is cycled through the sequence illustrated in Figure 35-3.

The upper enclosure contains the controllers which operate the sampling system, the analytical oven, and monitor the operation of the analyzer for possible faults. Electronic components detect and store the analytical results and calculate boiling point(s) at specific percent(s) distilled [%(s) OFF]. Other data can also be calculated, such as Reid vapor pressure (RVP) and ASTM D-86 distillation values. Data are transmitted to printers, recorders, and process control instruments.

The results from this analyzer are compatible with ASTM D-2887 and D-3710 laboratory tests. The capillary column was chosen because it can handle samples having higher boiling ranges than packed columns and can comply more readily with safe temperature limits as specified by National Fire Protective Association (NFPA) codes.

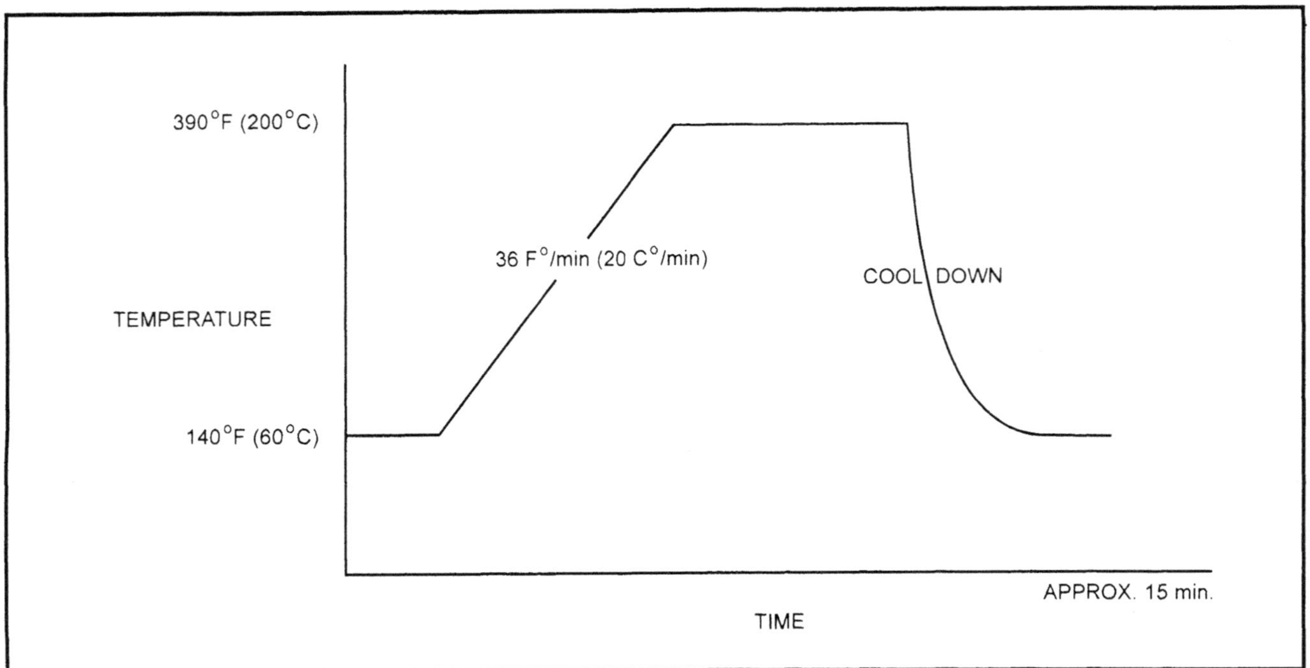

Figure 35-3. Optichrom Gas Chromatograph: Typical Column Oven Temperature Profile
(Courtesy of Applied Automation, Inc.)

Common Applications

Major applications for simulated distillation (SIMDIS) analyzers are in the petroleum and petrochemical industries. For example, such analyzers are being used in on-line gasoline blending systems. In addition to helping to maintain the proper boiling range, the chromatographs can calculate the Reid vapor pressure (RVP) of the finished gasolines in order to meet volatility requirements, which are becoming more and more strict. Such applications can pay for themselves in a very short time.

Range of Variables

The standard Monirex analyzer can handle gasolines with end points up to 400° F (200°C). with short cycle times and up to 750° F (400°C) with longer cycles. Other models are available that can handle end points to 1000°F (540°C). The Applied Automation instrument can monitor end points to 750°F (400°C).

Options and Accessories

Both the Applied Automation and the Monirex units can be programmed to analyze more than one stream in sequence and can record percent distilled at specific temperatures.

Limitations

Excellent correlations have been obtained with the ASTM D-3710 laboratory results. However, correlations with the ASTM D-86 distillation test method may vary considerably. For example, the chromatographic method includes the butanes in gasoline, but they cannot be condensed and recovered in the ASTM D-86 test. Therefore, front-end results will vary. Correlations with the ASTM D-86 distillation test also may be quite different for different types of products, and individual correlation factors could be required for each product.

Installation Hints

These analyzers should be installed indoors for protection against the weather and for convenience in maintenance. Ambient temperature should be within 0 to 110°F (–20 to 45°C) with no abrupt fluctuations. Clean, dry air is required to purge the analyzer housings.

Specific Safety Issues

The analyzers are designed for operation in hazardous areas classified as NEC Class I, Group D, Division I (ID1), and as CENELEC Zone I areas. The analyzer should be installed in a climate-controlled building, vented with a positive pressure fan exhausting out of the upper wall area of the building and have hydrocarbon detectors/flame detectors monitoring the analyzer area.

Unique Sampling System Requirements

Sample-conditioning systems for the analyzers are available from their manufacturers.

Calibration Requirements

Calibration with a known sample can be performed at timed intervals or on demand.

Troubleshooting and Maintenance

Like other liquid chromatographs, simulated distillation analyzers require routine maintenance. See Chapter 34 for additional information.

About the Author

Conard Donnell received a Bachelor's and a Masters degree in Chemistry and Physics from Temple University. His career with automatic process analyzers began and ended at Sunoco Refinery where he developed on-line instrumentation to measure the physical properties of process streams automatically and much faster than laboratory analysis. Since retirement, Mr. Donnell has worked as a consultant for Sunoco, Fluid Data, Precision Scientific Co., Lockwood and McLorie, and DuPont working on projects that included design and installation of process analyzer systems in refineries worldwide.

36

Mass Spectrometric (Mass Spec) Analyzers

Considered one of the most powerful and versatile analytical methods available, mass spectrometry is being applied to a multitude of applications [Ref. 1]. One such application is on-line monitoring of chemical processes, including environmental measurements. This chapter will address various aspects of this application.

Mass spectrometry had its beginnings in the late 19th century when scientists studied positive rays in low-pressure electrical discharge tubes. J.J. Thompson, the discoverer of the electron, built an apparatus and in 1913 demonstrated the existence of two separate isotopes of neon. The first instrument to be used as a mass spectrograph was constructed in 1919 by F.W. Aston [Ref. 2]. About the same time, A.J. Dempster [Ref. 3] built the first mass spectrometer suitable for isotopic abundance measurements.

For about 20 years mass spectrometers were built mainly by universities and used exclusively for research — primarily the precise measurement of atomic masses and isotope abundance. In the late 1930s, Herbert Hoover, Jr., of the United Geophysical Company, developed a mass spectrometer for oil exploration. The instrument proved to be too large to be portable, but turned out to be a good laboratory unit. Spurred by World War II and the need to improve the refining process, the petroleum companies were first to apply mass spectrometry industrially. The first commercial unit was delivered to Atlantic Refinery in 1942.

Mass spectrometers stayed in the lab for many years. A few were adapted for on-line use from the late 1950s to the early 1970s, but none was too successful due to the complexity of operation and data processing. An exception was the use of a modified lab unit by Dunn [Ref. 4] on an acrylonitrile process.

Until the mid-1960s, the majority of commercially-available mass spectrometers were magnetic analyzers. A few time-of-flight units were used, but mainly in research. Paul [Ref. 5] had invented the quadrupole analyzer in 1953. It was first used for research, but quickly was developed for laboratory qualitative use due to its inherent lower cost and simplicity. By the early 1970s, magnetic analyzers and quadrupoles were used widely for quantitative gas chromatography/mass spectrometry (GC/MS) and structural analysis. Little attention had been paid to on-line process analysis.

Then, during the mid-1970s, several companies entered the process market with units specifically designed for this application. Since that time, units have been refined and several additional process mass spectrometers also have been introduced. Today at least ten companies offer mass spectrometers for on-line continuous process monitoring applications.

Measurement Principles

A mass spectrometer can be thought of as an electronic chemist. A multicomponent gas mixture is introduced, the gas molecules ionized, the ions separated according to their mass-to-charge ratio, and the separated ions measured, giving the user a quantitative analysis of the original mixture.

A mass spectrometer can be thought of as an electronic chemist. A multicomponent gas mixture is introduced, the gas molecules ionized, the ions separated according to their mass-to-charge ratio, the separated ions measured, giving the user a quantitative analysis of the original mixture.

The major components of a mass spectrometer are (Figure 36-1):

- Inlet system
- Ion source
- Analyzer
- Detector
- Vacuum system
- Electronics/controller/data system

There are many different types of these component which are applicable to a wide variety of applications. This discussion will deal only with those currently used for the majority of on-line process mass spectrometers.

Inlet System

1 torr = 1 mm Hg

The function of the mass spectrometer inlet system is to accept a representative, conditioned sample from a sample-handling/conditioning system at a pressure of about 1 atm and introduce part of it into the ion source at a pressure of 10^{-4} to 10^{-6} torr. The typical process mass spectrometer inlet uses a flow-by capillary design (Figure 36-2) where 0.5 to 2 L/m of sample passes through the flow-by line, returning to the process or vent. A small-diameter (typically 0.07 in. [1.8 mm] ID) capillary conducts 1 to 2 mL/s from the flow-by line to a "leak," reducing the pressure to about 1 torr in the process.

The leak (a frit or small orifice) provides a second stage of pressure reduction from 1 torr to the source vacuum. Very little gas is admitted through the leak, the majority of the sample being bypassed through a mechanical (roughing) pump. Particular care must be taken in the design of the inlet system to avoid fractionating the sample.

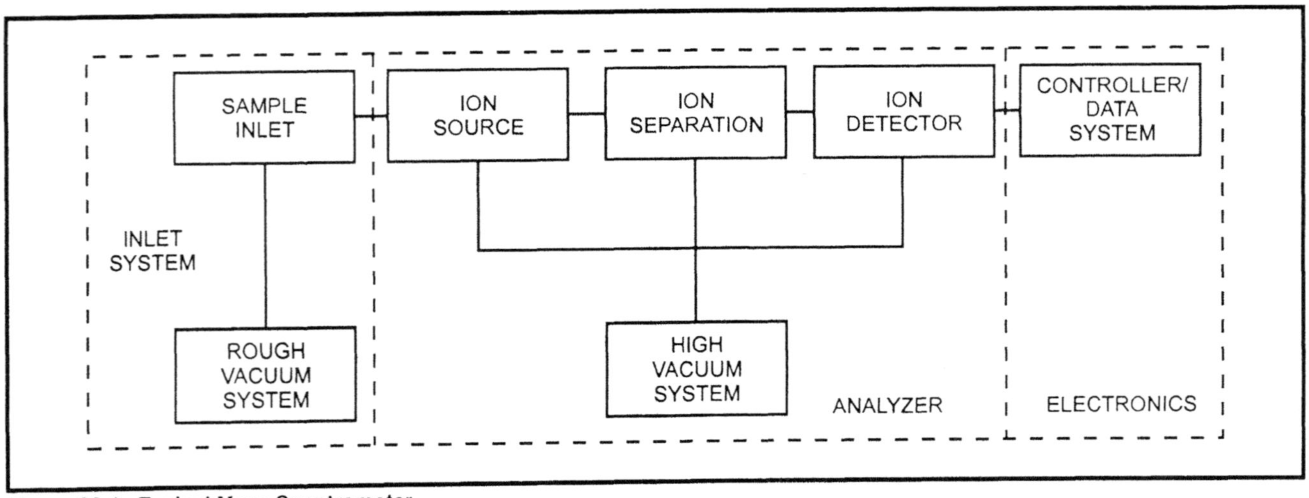

Figure 36-1. Typical Mass Spectrometer

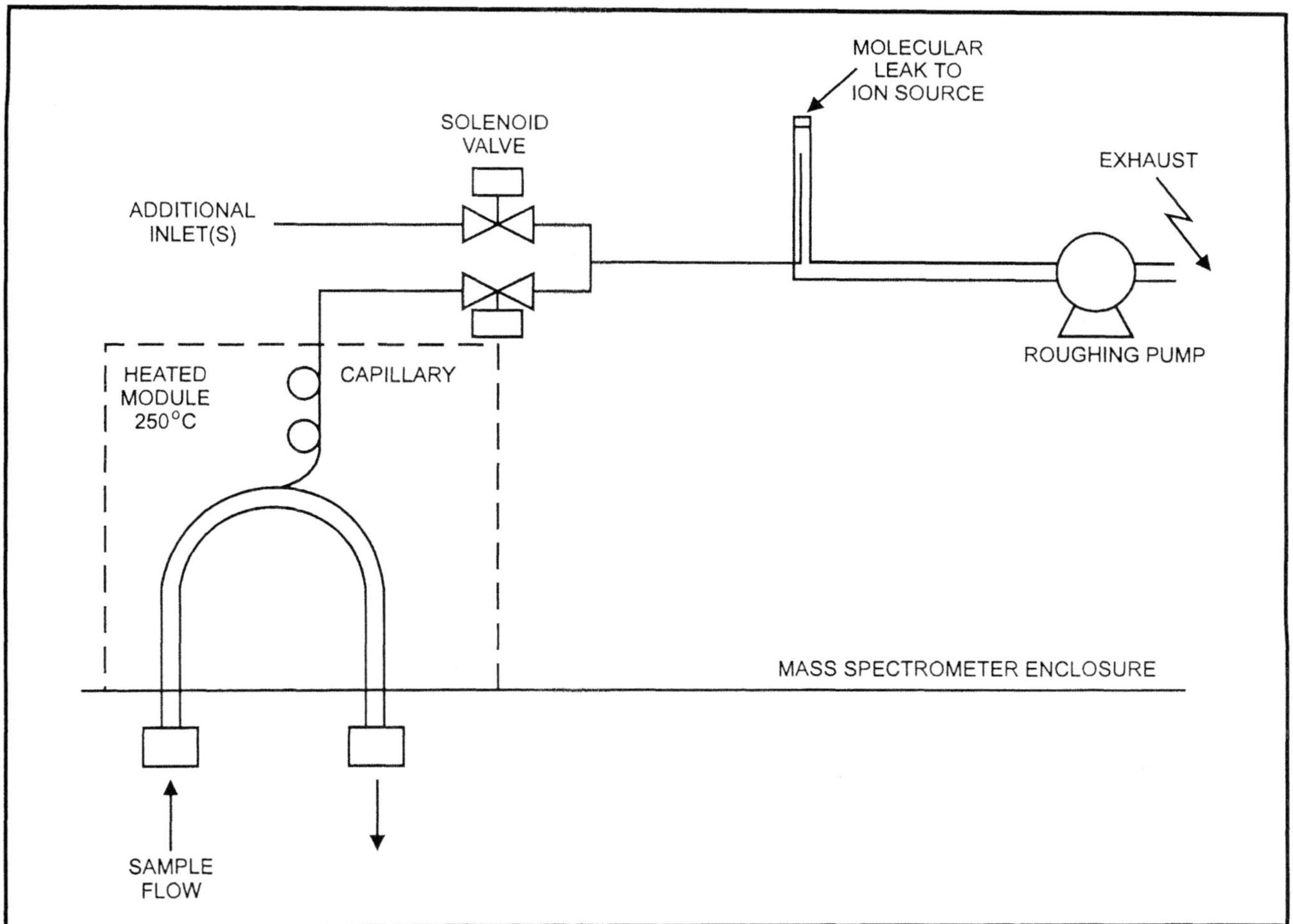

Figure 36-2. Flow-by Capillary Inlet (Molecular Leak) System

Ion Source

Gas molecules that enter the source are bombarded with electrons emitted from an incandescent filament. When these electrons collide with or pass close enough to transfer energy to the molecule, a valence electron is lost, forming a positive ion. Also, sufficient energy is imparted to many molecules, causing them to fragment and form positive ion fragments.

It is these whole and fragmented positive ions that are measured. Negative ions are not analyzed, and neutral fragments are pumped away. The positive ions are injected into the analyzer by the proper voltages applied to elements of the ion source.

Analyzer

The analyzer separates the ions according to their mass (more precisely, their mass-to-charge ratio, m/z). Two types of analyzers are in common use: **quadrupole** (RF/DC focus) and **magnetic sector**. The quadrupole (Figure 36-3) consists of four symmetrical rods in two parallel planes with RF and DC voltages applied to opposing rods. For a given value of these voltages, only ions of one specific mass, injected into the annular space between the rods, will pass all the way through the rod assembly and strike the detector. All other ions of lesser and greater mass will oscillate in ever-increasing amplitudes until they strike the rods, are neutralized, and are pumped away. In the strictest sense, a quadrupole is not an analyzer but a filter, analogous to a narrow-band electronic filter. By programming the rod voltages, each ion used in an analysis is sequentially allowed to pass through the filter.

The second common type of analyzer found in process mass spectrometers is the magnetic sector (Figure 36-4). Here the ions exiting the source encounter a magnetic field and travel at right angles to the lines of magnetic flux. There is a constant force on the ions that makes them travel in curved paths, the radius of which is proportional to the mass (*m*/*z*). Thus, there is mass separation or dispersion in this analyzer, and each separate ion beam can be measured by sweeping it across a collector (by varying the ion source accelerating voltage or the magnetic field strength) or by placing discrete collectors at the focal point of each ion beam.

Detector

Each ion beam separated by the analyzer strikes the detector. Two types of detectors are available: the **Faraday cup** and the **electron multiplier**. The Faraday cup, as the name implies, is a cup into which the ions fall, picking up an electron. The resulting minute current flow to the cup is passed through a high-megohm resistor and the developed voltage is amplified. The Faraday cup provides a one-for-one conversion and is typically used for relatively high-concentration analysis.

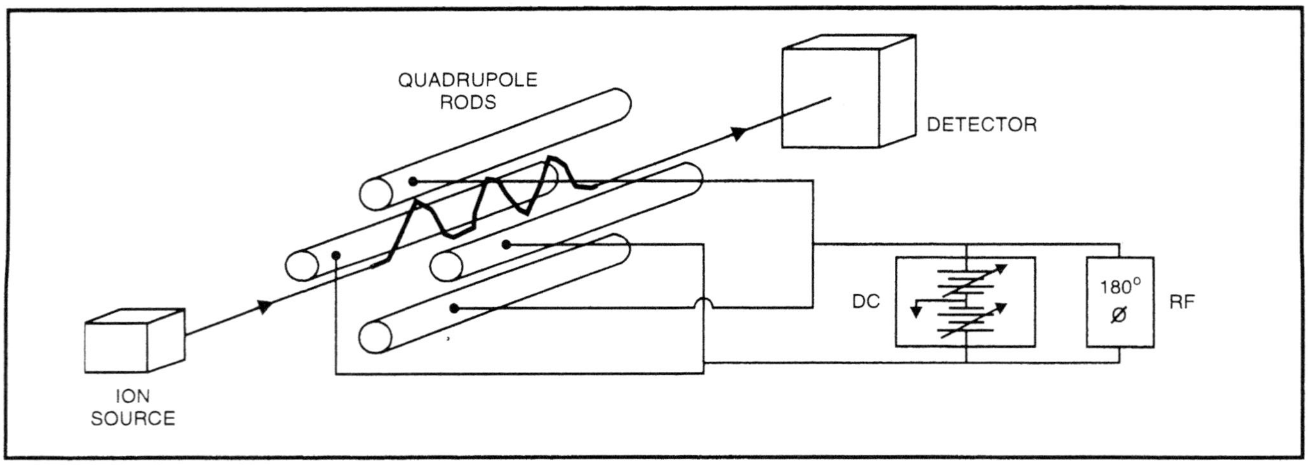

Figure 36-3. Quadrupole Mass Spectrometer

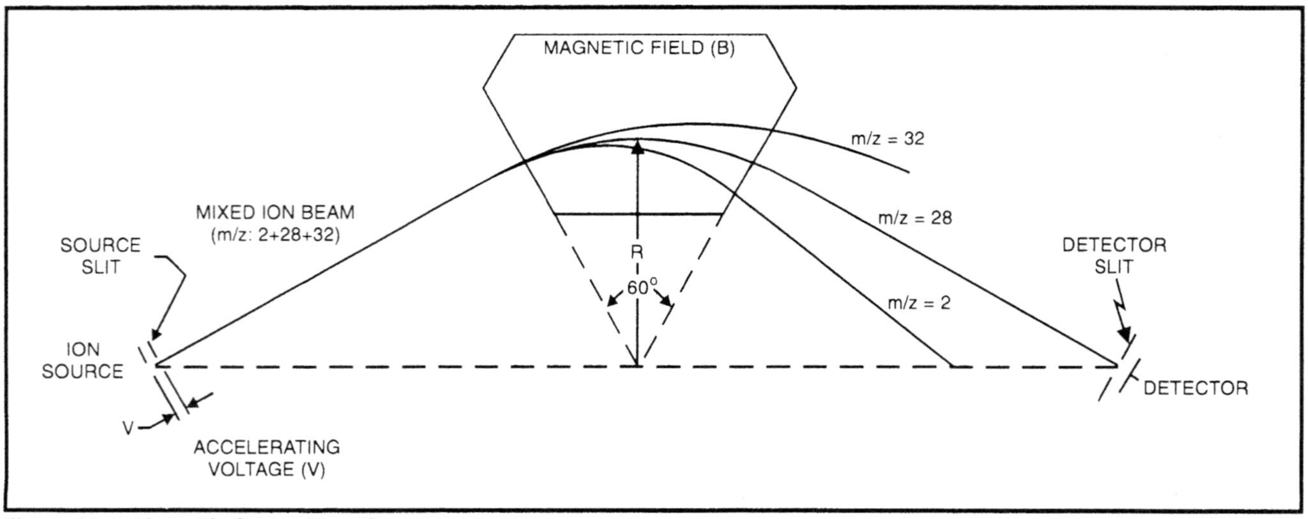

Figure 36-4. Magnetic Sector Mass Spectrometer

For very high-sensitivity requirements, an electron multiplier is used. Most multipliers today are continuous dynode designs. A single ion striking the surface at the entrance of the multiplier causes several secondary electrons to be ejected, which in turn strike the surface farther down the dynode, causing more electrons to be emitted. Commercially available multipliers are capable of producing up to 10^6 electrons for each ion measured. This electron current is then passed through a resistor and the resultant voltage amplified, as with the Faraday cup.

In either case, the developed voltage is directly proportional to the abundance of the particular component in the mixture. The voltage usually is digitized, and this digital value is used in computation. For qualitative work, the voltage can be displayed as an analog peak scan where area (or height) is proportional to the abundance and horizontal position is proportional to mass-to-charge ratio (m/z).

Vacuum System

The ion source/analyzer/detector portions of a mass spectrometer must be at a very low pressure to avoid collisions between gas molecules and/or ions. This vacuum is provided by a two-stage vacuum system. The oil seal mechanical (roughing) pump is used to pull the system down from atmospheric pressure to about 10^{-2} torr. A high-vacuum pump (turbomolecular, oil diffusion, or ion pump) reduces the pressure to instrument operating vacuum, approximately 10^{-6} torr. The mechanical pump continues to back the high-vacuum pump (in the case of the turbomolecular and diffusion pumps) and also pulls the sample through the inlet system, past the molecular leak.

Control and Computation

Most process mass spectrometers use a personal computer (PC) to control and tune the system and to do the actual calculation of component concentrations. The concentration ratios of all fragment and molecular (parent) ions formed to the most abundant ion species (often called the "base" peak) are referred to as fragmentation patterns (Figures 36-5 and 36-6). These patterns are unique and reproducible and can be thought of as a "fingerprint" for identifying a specific component.

A library of spectra and fragmentation patterns is available in printed form and on computer disks from the National Institutes of Health (NIH). If your company's laboratory uses mass spectrometry, you can probably find this library there. Vendors also can usually supply selected spectra.

A library of spectra and fragmentation patterns is available in printed form and on computer disks from the National Institutes of Health (NIH). If your company's laboratory uses mass spectrometry, you can probably find this library there. Vendors also can usually supply selected spectra.

If a peak can be found that is unique to only one component in the mixture (a unicomponent peak), then analysis is simple. One merely divides the value of the peak by a predetermined calibration factor for that component (e.g., peak units divided by peak units/percent concentration). In most cases, however, the fragmentation patterns overlap, so computation must be accomplished by setting up a matrix or set of simultaneous equations and solving for component concentrations. With today's high-speed computers, it is not uncommon for a system to be able to solve a 32 × 32 matrix each time an analysis is made (every few seconds).

ELECTRON IMPACT IONIZATION

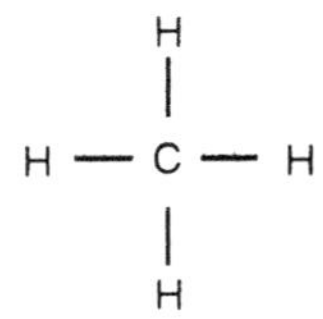

METHANE

MW-16

M/Z	%
.16	100
15	90
14	20
13	10
12	3

Figure 36-5. Fragmentation Pattern of Methane

ELECTRON IMPACT IONIZATION

C≡O

CARBON MONOXIDE

(MW 28)

M/Z	%
28	100
12	5
16	1
14	<1

N≡N

NITROGEN

(MW 28)

M/Z	%
28	100
14	5

Figure 36-6. Fragmentation Pattern of Carbon Monoxide and Nitrogen

To evaluate a new application or the feasibility of adding a new compound to those being monitored, the fragmentation patterns of all gases in the stream must be reviewed. The user (often with the help of the supplier) must determine if interferences are present and whether they are correctable. Suppose that a mixture of ethane and propane is being monitored. The stream composition changes such that propylene is now present (new composition: ethane, 60%; propane, 30%; propylene, 10%). The fragmentation patterns for these gases are as follows:

FRAGMENTATION PERCENTAGES

m/z	Ethane	Propane	Propylene
15	3	11	8
26	22	12	14
27	36	47	42
28	100	59	2
29	21	100	
30	30	2	
39		21	77
40		3	29
41		14	100
42		6	66
43		23	23
44		27	

In the original analysis of ethane and propane, only masses (*m/z*) 44 and 30 needed to be measured in the calibration gases and the process stream. Propane concentration would be determined on *m/z* 44, the contribution at *m/z* 30 subtracted from the total intensity of *m/z* 30, and the remainder would represent ethane.

With the addition of propylene, examination of the spectrum reveals that adding *m/z* 42 to the calibration and analysis will allow measurement of this new component. From the propane calibration, correction is made for the contribution to *m/z* 42 in the mixture, and the remainder represents propylene.

Note that with the normal relative concentrations of each component, the base (100%) peaks or masses, are not used in this analysis. This is because there is better mathematical "leverage" using some of the smaller peaks.

To compensate for any common mode errors, such as drift in the electronics, variations in inlet pressure, and so on, most analyses are normalized to the sum of all individual analyses.

If the analysis is on a normalized basis, then all components must be measured. If, for example, a 5% component is ignored or unknown, then this 5% error will be distributed to all the other components in proportion to their concentrations.

Mass spectrometers have certain advantages over other process analyzers:

- Linear response from parts per million to percent concentrations
- Approximately equal sensitivity to all species
- Ability to analyze inorganic as well as organic species
- Speed of analysis

What may take 5 to 15 min to analyze by other methods can often be accomplished with a mass spectrometer in 20 to 30 s.

The major advantage of the process mass spectrometer over other multicomponent analyzers is speed of analysis.

The major advantage of the process mass spectrometer over other multicomponent analyzers is speed of analysis.

Common Applications

Before discussing applications, it might be well to define a process mass spectrometer now that the basic principles have been covered. Characteristics of today's process instruments include:

- Multicomponent, quantitative analysis
- Continuous sampling (as opposed to batch sampling)
- Microprocessor-controlled
- Automated data acquisition/reduction
- Limited operator interaction
- Automated stream selection
- Packaged for compatibility with surrounding environment

Common applications can be categorized as process monitoring or environmental monitoring.

Process Monitoring

Process monitoring applications using mass spectrometry include hydrocarbon processing where the user needs to monitor the concentrations of feedstock and by-products in the production of a product. Examples of such processes include:

- Ammonia
- Methanol
- Ethylene
- Ethylene oxide
- Vinyl chloride
- Coal gasification
- Acrylonitrile
- Refinery fuel gas
- Hydrogen plant
- Polyvinyl chloride

Another application is monitoring liquid-phase reactions (by inference) from the composition of the gas phase above the liquid. Two common examples are fermentation processes in the biotechnology and pharmaceutical industries and steelmaking (basic oxygen furnaces [BOF] and blast furnaces[BF]).

Although the majority of on-line mass spectrometers are found in processes involving organic gases, instruments are also used in some inorganic applications, as the analyzer is not limited to any particular type of gas. An example of such an application is in the production or use of chlorine; another is monitoring gases in an operating fuel cell.

In process monitoring applications, the concentration of most gases is in the 0.1 to 100% range.

Environmental Monitoring

10,000 PPM = 1 PCT

A second major application area is monitoring low-level (parts per million to 0.1%) gases in ambient air. An example would be area monitoring around a plant producing a substance whose release to the atmosphere is controlled by legislation. Not only does the company need to monitor what is being emitted from its plant, but it also wants to be aware of what may be coming from a neighboring plant.

Worker safety in a manufacturing area is a good application for a process mass spectrometer. Solvents and other volatile compounds used in manufacturing need to be monitored for both time weighted averages (TWAs) and lower explosive limits (LELs). By use of multilevel alarms, corrective action can be taken quickly in case of a spill or other abnormal conditions.

Monitoring of low-level components in vents can be used to show compliance with local environmental regulations. Combined with a flowmeter signal or a fixed, known flow rate, the mass spectrometer computer can report not only percent or ppm levels but also the mass flow rate (pounds per hour, etc.).

A potential low-level environmental application for the process mass spectrometer is the monitoring of volatile organic compounds (VOCs) in water. A proper sample system (e.g., continuous sparging) is required, but once the VOCs are extracted and enriched, the mass spectrometer can be used just as in an air-monitoring application.

Data Obtained

What types of data can a user obtain from a process-monitoring mass spectrometer? The most basic is the concentration of one or all the components in the gas stream being monitored. This information is reported as mole or volume percent (parts per million for low-level applications).

Because most on-line mass spectrometers use a PC for control and data processing, the user can obtain numerous calculations and reports. For example, using user-defined factors, the system can easily calculate and display BTU and specific gravity, or an important ratio such as the H2/N2 ratio in the ammonia process can be calculated continuously. Such calculations can be displayed and trended along with the conventional gas concentration data.

Special reports are available for specific applications. Environmental monitoring usually requires that some type of permanent record be kept and, in many cases, a report sent to the state or federal Environmental Protection Agency (EPA). Such reports can also be generated by the system giving information on TWA, minimum and maximum levels, number of alarm levels exceeded, and so on.

Economic Impact

The economic impact of using a process mass spectrometer depends, of course, on the application and process. In the case of environmental monitoring, the savings that can be achieved are measured in terms of the cost to clean up a spill, contamination of ground water, fines by the local regulatory agency, or lawsuits by workers exposed to toxic or hazardous chemicals.

In process applications, the value of process monitoring can be calculated more precisely, based on measured or anticipated savings. For example, a world-class ethylene plant may produce 3 million lbs of ethylene per day. An increase in efficiency of only 0.1% translates into a yearly gain of more than $210,000 (at an ethylene price of $0.20/lb). For a cost of $100,000 (instrument and installation), the payout on this investment is about 6 months.

Range of Variables

Most components that can be vaporized and are compatible with the inlet system temperature can be analyzed in a mass spectrometer. Spectral interferences will sometimes limit an analysis or the concentration level that can be measured. Special inlets that use membranes to exclude water vapor or air components also will exclude inorganic gases (e.g., HCl, Cl_2, etc.). Thus, in low-level ambient-air applications the use of such an inlet may prohibit complete analysis with the mass spectrometer.

Most available process mass spectrometers have a mass range that is compatible with compounds that can be vaporized at the temperature of the mass spectrometer inlet system. Typically, this mass range is $m/z = 2$ to 200. In a few cases, such as with some fluorochlorocarbons, one may need to look at masses greater than 200. In these cases, a typical mass range would be $m/z = 2$ to 400.

The resolution of a mass spectrometer is its ability to separate adjacent masses, and the degree of separation affects the accuracy of measuring the quantity of each gas present. It does no good to have a mass range to m/z 200 if you can only cleanly separate masses up to m/z 60. (That is, above 60 there is so much overlap between adjacent masses that it is impossible to accurately measure the abundance of one peak when an adjacent peak due to another compound is present.)

It is important for a prospective purchaser to know that the instrument has resolution that is compatible with the mass range.

Dynamic range is another important variable. Most electrometer amplifiers have dynamic ranges of 10^4 to 10^5. Thus, a mass spectrometer equipped only with a Faraday cup collector is usually limited to analyses between 100% and 10 to 100 ppm. Electron multiplier detectors used with a Faraday cup (dual or combination detector) can extend the dynamic range of 10^6 or better and push the detection threshold to 1 ppm or less (usually limited by system noise).

Options and Accessories

The most common options available for process mass spectrometers deal with sampling and data output.The typical sample inlet for the mass spectrometer is a heated flow-by capillary, described earlier in this chapter. Basic instruments usually are supplied with a single inlet, installed inside the analyzer case. Additional inlets can be purchased; the number depends on available room in the enclosure, but four to six is typical. If more streams or calibration gases must be monitored, then an external sampling system with a single internal inlet is used.

Although most mass spectrometer inlets operate at 212 to 260°F (100 to 125°C), sufficient to keep water in the vapor phase, occasionally higher temperatures are required to avoid chemical reactions (e.g., with CO_2 and NH_3). In these cases, a special high-temperature inlet must be used, capable of maintaining the sample at 480 to 530°F (250 to 275°C).

Other options available include special manifolds or valves for automatic calibration. These systems are usually under the control of the mass spectrometer computer through some type of stream selection module. The operator can program the system such that each calibration gas is sequentially admitted to the analyzer, and analysis methodologies are automatically updated with the new calibration factors.

Some commercial process mass spectrometers are available in different optional enclosures. These range from general purpose to electrical hazard rating of Class I, Division 1, typically Groups C and D.

Results of analyses are displayed on either a monitor for computer-based systems or a digital display for analog-based systems. Many optional display modes are available with most commercial systems. The most common is an isolated 4-20 mADC signal for each gas concentration or calculated variable.

Most process control computers can accept these signals directly. The same information can be sent to a host computer via an RS-232 line.

Another display mode is via a printer. Such reports provide the user with a hard copy permanent record.

Many users also want their process-monitoring mass spectrometer to alarm if predetermined high- or low-level limits are exceeded. This is usually accomplished using optional contact closures. Such closures can also be used for stream identification or to signal malfunctions such as loss of system purge, loss of sample flow, and so forth.

Limitations

The process mass spectrometer is a fairly universal analyzer, but it does have a few limitations. One of these is uncorrectable spectral interference. As pointed out in the section "Measurement Principles," gases undergoing electron bombardment ionization will fragment into a distinct pattern. In some cases these patterns are similar, but using matrix calculations the concentrations of each species can usually be calculated.

An example is CO and N_2. Both form ions at *m/z* 28 and 14 (*m/z* 12 or 16 could be used for CO, but the presence of other gases usually contributes to these peaks). The ratio of the *m/z* 14 peak is about five to one in favor of N_2, so in a mathematical solution one has about a fivefold leverage. If, however, the concentration ratios are such that one gas is present at a level of one-tenth (or less) of the second, then this leverage is lost, and it is impossible to accurately measure the two species. The same problems can occur with isomers of various compounds.

Analyses can be limited by signal-to-background or signal-to-noise ratios. In the former case, a good example is trying to analyze ppm amounts of N_2 in a sample. Most vacuum systems have background gases from air (due to residual molecules or minute in-leakage). Therefore, it is not possible to measure these air gases as impurities in a sample when the signal-to-background ratio becomes unfavorable. Even subtracting the background signal often does not yield good precision due to signal size or variability.

In the case of signal-to-noise, the minimum detectable signal will be limited by the overall system noise level.

As pointed out earlier, a mass spectrometer requires a gaseous or vaporized sample. Therefore, volatility can be a limitation if the sample requires a vaporization temperature higher than is compatible with the analyzer.

Resolving power can be a limitation with some analyzers. As described in the section "Range of Variables," the resolving power of a mass spectrometer is its ability to separate adjacent mass peaks. If two compounds in the stream have adjacent peaks that must be used for computation but the peaks cannot be cleanly separated, then the analysis cannot be made. It is important for a prospective user to be aware of this limitation.

Corrosive or reactive samples can be considered a limitation as they either preclude the desired analysis or result in abnormal downtime and maintenance. Such problems often can be minimized by the use of special materials.

In very few cases, samples that exist in one form at atmospheric pressure will change forms when introduced into the vacuum system of a mass spectrometer. Two examples come to mind. One is ozone, which will decompose to 0_2. The other is high-level acetic acid, which forms dimers and trimers at atmospheric pressure but reverts back to the monomer in the mass spectrometer. In such cases, the mass spectrometer analysis does not represent the true sample present in the process.

Examples and Case Histories

Two actual examples of process mass spectrometry in use today are presented to illustrate the utility of this technology.

Ammonia Synthesis

The modern ammonia plant (Figure 36-7) consists of a series of operations whereby hydrocarbon feed, generally natural gas, is steam-reformed to produce a hydrogen-rich stream that is fed to a secondary reformer. Air is introduced at this point in the process to provide nitrogen and complete the reformer operation. Reformer exit gas is delivered to a two-stage shift-converter system, where CO is converted to CO_2. Carbon dioxide is removed in a purification system, and a final purification is accomplished by methanation of residual carbon oxides and drying. The pure synthesis gas, with a H:N ratio of 3:1 is mixed with recycle gas, compressed, and fed to the ammonia converter. The converter effluent gas, containing 10 to 15% NH_3, is cooled for ammonia removal as product, and the gases are recycled to the converter. Some plants conduct a portion of the recycle gas to a hydrogen-recovery section where inerts (argon and methane) are removed, with a relatively pure hydrogen stream being returned to the process.

Composition control has come to be accepted as the primary mode to efficient operation of a modern ammonia plant. Energy-efficient and cost-effective designs require extensive use of composition information, including feed-gas analysis, synthesis-gas purity, H/N ratio control, converter inlet and outlet composition, and, in some cases, hydrogen-recovery systems. Typical streams that are analyzed are listed in Table 36-1.

The H/N ratio in makeup synthesis gas is critical to converter performance. Deviation from the stoichiometric 3:1 ratio of 0.1 ratio units can result in a production loss of up to 1% in a 1,000 ton/day ammonia plant.

The mass spectrometer offers a great advantage in that the full multicomponent analysis is performed in one analyzer without discrimination on the basis of polarity, molecular weight, and so forth, as is the case with chromatographic and infrared analyzers. The high-speed capability of this analyzer permits essentially the same real-time analysis for all streams, resulting in a much higher degree of control. The optimized control strategy objectives are:

- **stabilize the process**
- **maximize ammonia production**
- **minimize energy consumption**

Table 36-1. Ammonia Plant Applications

Typical Streams	Components	Ranges, %
Primary reformer feed	H_2	0-0.1
Secondary reformer effluent	H_2	0-50
High- and low-temperature shift converters	CH_4	0-1
Makeup gas	N_2	0-20
Converter inlet	N_2	0-20
Converter effluent	CO	0-10
Hydrogen-recovery streams	CO_2	0-10

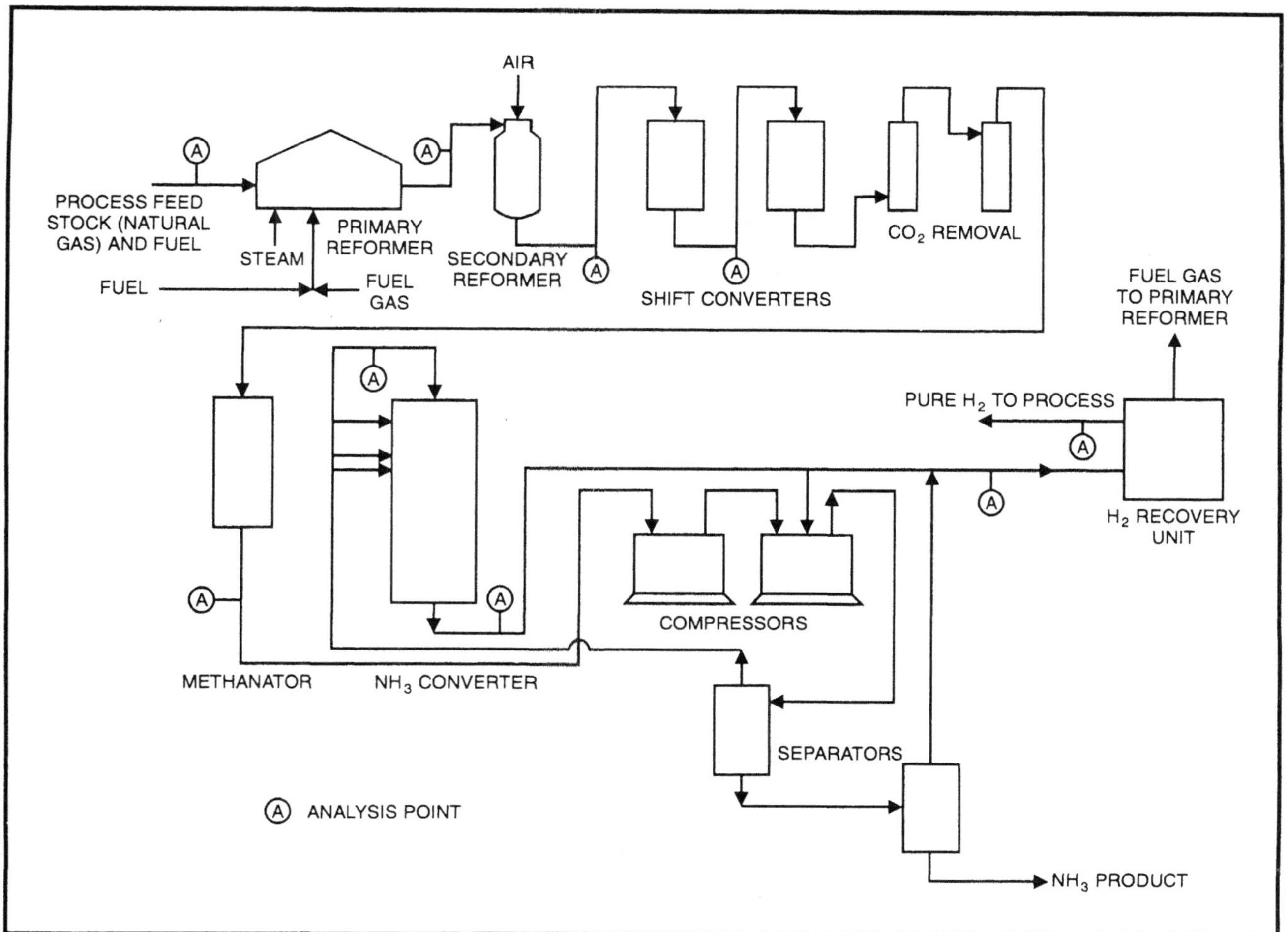

Figure 36-7. Process Flow Diagram: Ammonia Synthesis

Air Monitoring

Facilities that produce or use toxic or hazardous chemicals must monitor the ambient air for these materials in order to provide a safe environment for workers and the surrounding community, and to meet current and pending regulatory legislation (e.g., the Clean Air Act Amendment 1990, title 40, part 53 - Air Programs).

VINYL CHLORIDE MONOMER PRODUCTION

A plant is monitoring ambient air from 30 sample points in the facility, ranging from the process control room to the storage area. The sampling time is approximately 15 s per point, providing a complete plant-wide analysis cycle in as little as 7-1/2 min. Vinyl chloride is monitored with alarm levels set at warn = 2, alarm = 30, and evacuate = 100 ppm. Specific action steps are dictated at each alarm level.

AREA MONITORING

A company is monitoring ambient air for the presence of nine possible components used in the company's proprietary process. Alarm levels, typically in the 1- to 5-ppm range, are set for specific components, and these alarm signals are tied into the plant's host computer. The materials monitored and the mass spectrometer's full-scale ranges are shown in the following table.

Compound	Full-Scale Range, ppm
Vinyl chloride	0-50
Acrylonitrile	0-50
1,3-Butadiene	0-50
Methane	0-100
Styrene	0-50
Ethyl acrylate	0-50
Methyl acrylate	0-50
Methanol	0-50
Vinyl acetate	0-50

Installation Hints

All process mass spectrometers include or can be purchased with factory start-up assistance. To assure a trouble-free installation, it is important that the instruments be started by trained service people. The user, however, should make every effort to have personnel present to work with the supplier's service engineer during start-up for training and for a better understanding of system elements and operation.

Before the mass spectrometer is delivered, there are several things the user should do.

Location

Be sure the proposed location for the analyzer is appropriate. Most process mass spectrometers have relatively wide ambient-temperature specifications, but like other multicomponent analyzers (process gas chromatographs, etc.) they should be installed in an analyzer shelter where the environment is semicontrolled. In other words, don't install the unit outdoors where it is subjected to sun, rain, and snow.

Utilities

Be aware of the power requirements of the analyzer. Note that starting power is significantly higher than running power due to the mechanical pumps. Also, many plants have fluctuating voltage and/or intermittent power dropouts. If this is the case, you may want to install an uninterruptable power supply (UPS).

If the instrument you purchase requires instrument air for purge or operating pneumatic valves, be sure your supply has adequate pressure and volume. Instrument air should be dry (–40°F [–40°C] dew point). On more than one application, wet compressed air has resulted in moisture being sprayed into the electronics modules.

Vibration

Because so much depends on the individual unit and the mounting mechanism, it is difficult to specify a vibration frequency range that would interfere with instrument operation. In general, however, try to avoid installing an analyzer where it is subjected to severe vibration (such as next to a large reciprocating compressor). If this situation cannot be avoided, check with the supplier.

Sampling

The customer is usually responsible for mounting the sample-handling system (plus any sample-conditioning components), installing the sample probe, and running sample transport lines. The mass spectrometer supplier normally provides the probes, sample-handling/calibration systems, sample-conditioning systems, and the user normally provides sample transfer lines (plus heat tracing, if required).

The sample/calibration systems should be installed close to the analyzer to minimize transport time. These systems are typically mounted on a plate installed on a wall next to the analyzer.

Calibration gas cylinders also should be properly restrained close to the calibration manifold to minimize calibration material transport time (and thus calibration time). Typically, these cylinders are mounted on an outside wall of the instrument house or inside a temperature-controlled enclosure, if required.

Sample Disposal

Most surplus or analyzed sample gas must be properly disposed of rather than vented to the atmosphere. Such disposal can consist of putting the gas back in the process stream, venting to a flare, or other measures. Whatever the method required by law or company policy, the user must be aware of pressure balance considerations.

One or two mechanical pumps are used before the high-vacuum (turbomolecular or diffusion) pump and pull the sample through the capillary inlet system. These commercially-available pumps are designed to exhaust to atmospheric or a slightly positive pressure. Since many disposal systems operate at pressures above atmospheric, it is important that the user be familiar both with plant requirements and with system requirements. It may be necessary to design a disposal/sample-recovery system as well as a sample-conditioning system.

Not only must the exhaust of the mass spectrometer's mechanical pumps be vented, but sample from the flow-by inlet system plus any sample system bypass flow, must also be disposed of. The mechanism for putting these gases into the disposal system will depend on the design of the sample-handling system and the plant's disposal system.

Be sure to discuss your requirements for venting excess or analyzed sample with the mass spectrometer supplier before you purchase a system. High-pressure vent lines usually require special design considerations.

Be sure to discuss your requirements for venting excess or analyzed sample with the mass spectrometer supplier before you purchase a system. High-pressure vent lines usually require special design consideration.

Specific Safety Issues

The most obvious initial safety issue is to ascertain that the area classification of the chosen mass spectrometer model is compatible with the proposed location of the analyzer. Some models are designed only for general-purpose locations, while others can be located in Division 2 and/or Division 1 areas. **Group B areas require very special consideration.**

The factors involved with proper disposal of bypassed or analyzed sample gas were discussed in the section "Installation Hints." This is necessary, of course, to protect people or the environment from exposure to corrosive, toxic, carcinogenic, explosive, or otherwise harmful gases being analyzed.

When maintenance or repair is required on the mass spectrometer or sample system, all lines must be flushed with a nonreactive gas before breaking connections to avoid accidental exposure or buildup of dangerous gas. This is particularly essential if the line being opened is inside an enclosure where escaped sample gas can be trapped.

Work on the electronics of the process mass spectrometer is always dangerous, as high voltage is used in several areas (e.g., electron multiplier supplies, accelerating voltage or RF generators). Not only should all power be removed from the system before work is started, but the main power also should be "locked out" so that it cannot be accidentally turned on. If a circuit must be energized to troubleshoot, then the person doing the work should be thoroughly trained and should work with a partner.

The oil in the mechanical pumps of process mass spectrometers must be changed periodically and proper disposal precautions should be taken. This is particularly true if the system has been pumping gases such as benzene or HCl.

Unique Sampling System Requirements

The sample-conditioning requirements for the process mass spectrometer are very similar to those for other commonly used process analyzers, such as the process gas chromatograph. Sample delivered to the flow-by line should be at a constant pressure. If the process is operating at a positive pressure, the sample system will reduce the pressure so that it is only slightly positive (3 to 5 psig), and flow will occur. It is good practice to reduce sample pressure as close to the sample tap as possible to optimize response time.

If the sample is at atmospheric pressure, then a secondary sample pump and vacuum regulator are usually used downstream of the sample system and mass spectrometer. This pump then pulls a continuous flow of sample through all lines of the sample system as well as the bypass line of the mass spectrometer to achieve an acceptable response (transport) time. For samples at subatmospheric pressure, special sample systems must be designed for each individual situation, but this is normally not a problem.

As mentioned earlier, most process-monitoring mass spectrometer inlet systems operate at a temperature of 212 to 260°F (100 to 125°C). (Special high-temperature inlet systems are also available.) The temperature of the sample delivered must be compatible with the inlet system temperature; otherwise, condensation or reactions could occur. It is the function of the sample system to reduce high temperatures to an acceptable level while dealing with any liquid fallout before the sample reaches the mass spectrometer. Conversely, if the sample is liquid in its normal process state, then the sample system must vaporize it and keep it in vapor form while transporting it to the mass spectrometer. This function usually requires a vaporizing pressure regulator and integrally heat-traced-and-insulated sample transfer lines.

Liquids should never be introduced into a mass spectrometer as this can result in plugged inlet lines and total system contamination.

Liquids should never be introduced into a mass spectrometer as this can result in plugged inlet lines and system contamination.

The sample system must also remove any particulate material. Two stages of filtration often are used. A coarse filter is installed at or close to the sample tap. In the case of a high-pressure process sample source, this filter will remove any scale or rust extracted with the sample from the process. In the case of air monitoring, where sample is being pulled through lines by a vacuum pump or blower, the coarse end-of-the-line filter will prevent contaminants (such as insects) from being drawn into the line. Just before sample is introduced to the flow-by line, a fine filter (typically 0.5 µm) should be used.

A word about sample probes: Although not unique to the process mass spectrometer sampling system, this method of extracting a sample from a high-pressure, flowing sample is highly recommended. A properly designed probe in a pipe or vessel will maximize attainment of a representative sample and minimize extraction of pipe scale by pulling sample from the center of the flowing stream or out from the wall of a process vessel.

Figure 36-8 shows a composite sample system that deals with low- and high-pressure gas samples and a liquid stream. A calibration gas manifold is also depicted.

The following sample conditions must be avoided if a satisfactory analysis is to be obtained:

- Polymerization (phase change and composition change)
- Changes in sample state (phase changes)
- Corrosive material
- Particulates (especially material ≥ 1.0 or 0.5 microns)
- Multiple phases
- Condensation (phase change and composition change)
- Pressure/temperature changes (phase change)

Typical Calibration Requirements

When calibrating a process mass spectrometer, two pieces of information must be determined: (1) the fragmentation pattern and (2) the response (peak area or peak height proportional to percent concentration) of the instrument to each gas to be analyzed. Since many gases have similar or overlapping spectra, binary gas mixtures are normally used for a full calibration. The primary gas in the mix is the gas to be calibrated (at a concentration close to that in the stream to be monitored). The balance gas is chosen so that it has no fragments (peaks) that overlap the primary gas. In this way, both the fragmentation pattern and the response factor can be determined.

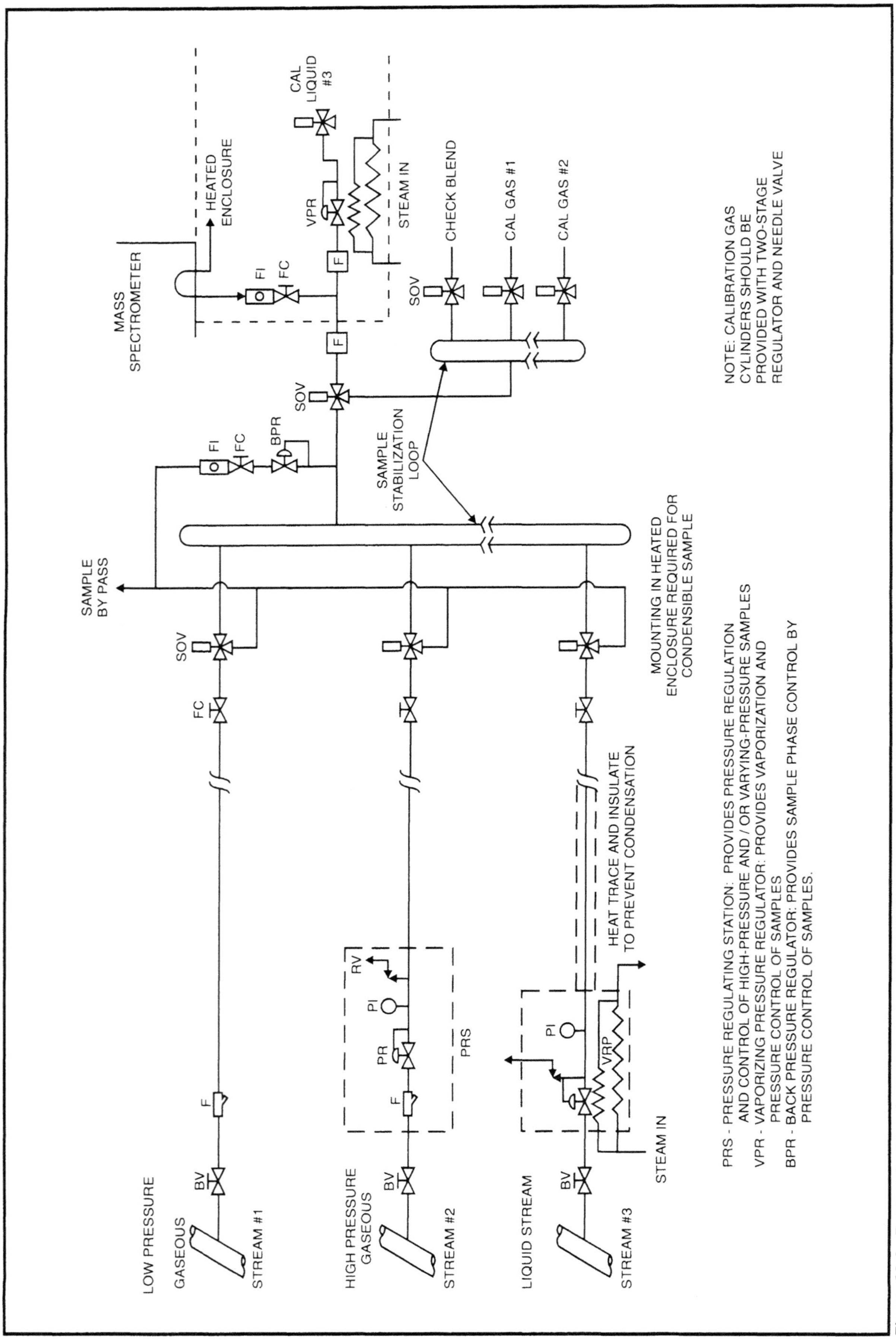

Figure 36-8. Typical Sample System for Low- and High-Pressure Gas and Liquid Streams

With most process mass spectrometers, the sequential introduction of binary mixes, determination of fragmentation patterns, response factors and adjustment of analysis methods for a new calibration are done automatically under computer control, with no intervention by the operator required.

In order to check the validity of the current calibration, a check blend is often analyzed. This gas is a known mixture approximating the sample being analyzed. The operator can set the high and low limits for each gas that are the maximum accepted before recalibration is required. Instruments that have alarm capabilities can be programmed to automatically run the check blend on a predetermined schedule and alarm if the limits are exceeded. The frequency with which any mass spectrometer must be calibrated is determined by the user's analytical requirements.

When analyzing streams with percentage levels of each component, the typical method normalizes the sum of all calculated concentrations to 100%. (This compensates for any common mode errors, such as electronic drift, changing sample pressure, etc.) In the case of analyzing for low-level (ppm) components, such as air monitoring, normalization to the total sample is impractical. Also, one is rarely interested in analyzing the major constituents (N_2 and O_2 in the case of air). In these applications, one calibrates on binary mixtures for only the low-level components to be measured and relates (normalizes) them to a constant peak from one of the major components. For air monitoring, a typical reference peak is the *m/z* 7 peak from nitrogen (from N^{++}). Thus, the requirements for calibration standards are the same for both low- and high-level applications, but the information is used differently.

Sometimes only one component of a mixture is of interest. If a single peak can be found that is due only to the component to be measured, then an absolute calibration can be used. In this case, a sample of pure gas (or a known blend with no interfering peaks) is run, and the response of the mass spectrometer to that gas is determined. When the process stream is analyzed, only the unique peak is monitored and the concentration is determined using the calibration response factor. Absolute calibration usually requires more frequent calibration than when analyses are normalized to compensate for common mode changes. When it can be used, however, it simplifies the analysis and calibration requirements.

With most process mass spectrometers, the sequential introduction of binary mixes, determination of fragmentation patterns, response factors and adjustment of analysis methods for a new calibration are done automatically under computer control, with no intervention by the operator required.

Troubleshooting and Maintenance

Preventive maintenance is the best way to minimize the need to troubleshoot problems. There are few moving parts on a process mass spectrometer, but they do need to be periodically serviced. Mechanical pumps must have the oil changed. It is best to follow the recommendation of the manufacturer, but at a minimum the oil level in the sight gage should be checked every week or so and the oil changed every 6 months. If corrosive or other gases that can break down pump oil are being pumped, it may be necessary to change this oil more frequently. The color of the oil is a good indicator of contamination (contaminated oil gets darker in color).

If the gurgling of the coarse (mechanical) pump does not stop in a minute or two after start-up, there is a leak in the vacuum system.

If the gurgling of the coarse (mechanical) pump does not stop in a minute or two after start-up, there is a leak in the vacuum system.

As with the mechanical pumps, follow the recommendation of the manufacturer in servicing the high-vacuum pump. If the system uses a turbomolecular pump, the lubricating system generally needs to be serviced every 5,000 h (6 months). If a diffusion pump or ion pump is being used, deteriorating vacuum (increasing pressure) is usually an indication that service or replacement is due.

An area that is often overlooked is maintenance of system fans, both in the analyzer and the computer. These fans are normally used for cooling or

Do not turn on the instrument filament until you are certain adequate vacuum has been achieved.

distributing the air within an enclosure. They should be checked and cleaned of dust buildup every 6 months — more frequently if in a dusty area.

Analyzer tuning should be checked every 30 to 60 days. Although this is more of an operating adjustment, it can be considered a form of preventive maintenance. If an abnormal change or drift in analysis is noted, check the tuning constants before recalibrating.

The person responsible for maintaining a process mass spectrometer must understand the basic concepts of mass spectrometry and the functions of the various subassemblies and circuits. This will help immensely in isolating and identifying potential sources of a specific problem.

Mass spectrometer ion source filaments have a finite lifetime — 9 to 12 months on average. Most instruments have two filaments, so that when the first burns out, the operator can switch to the second (usually requiring recalibration). This backup design allows operation to continue until a scheduled plant turnaround when the mass spectrometer can be vented and new filaments installed.

Do not turn on the instrument filament until you are certain adequate vacuum has been achieved.

Whenever a mass spectrometer is vented and disassembled, care should be taken to assure cleanliness and good vacuum practice. If the instrument is in an area of high humidity, venting with dry nitrogen or air passed through a dryer will significantly shorten system downtime.

No part of the ion source or other assemblies in the vacuum housing should be touched with bare hands. Always use lint-free silk, cotton, or surgical gloves when changing a filament or cleaning source/analyzer parts. The oil from skin will outgas in a vacuum system, causing a background spectrum and an elevated pressure. This situation will continue for many weeks.

If it becomes necessary to disassemble and clean the ion source or analyzer, follow the manufacturer's recommendations. Realignment of these parts is usually critical. Some sources are self-aligning, while others require an alignment jig available from the manufacturer.

A final word on safety. It was mentioned in the section "Specific Safety Issues" that high voltages are found in almost all process mass spectrometers. Care should be taken in dealing with these voltages, and the unit should be serviced only by qualified technicians — either factory personnel or customers who have been factory trained. Also, be sure to follow your company's regulations when servicing a system in a Division 1 or 2 area.

References

1. White, F.A., and Wood, G.M., *Mass Spectrometry*, New York: John Wiley & Sons, 1986.

2. Aston, F.W., *Philosophical Magazine*, 38:709, 1919.

3. Dempster, A.J., *Physical Review*, 11:316, 1918.

4. Dunn, B.E., *Acrylonitrile Process Analyzer System*, Research Triangle Park, NC: Instrument Society of America, 1972.

5. Paul, W., and Steinwedel, H., *Zeitschrift für Naturforschung*, 8a:448, 1953.

About the Author

Urie "Jerry" McCleary, Jr. received his B.S. and M.S. degrees in Petroleum Engineering from Stanford University in 1953 and 1954, respectively. He worked for Shell Oil Company for seven years before entering the analytical instrument business as an Application Engineer and then as Product Manager, Mass Spectrometers for Consolidated Electrodynamics Corp. (CEC). At CEC he was involved with the design, marketing, and support of laboratory mass spectrometers and some of the first on-line process units.

In 1970, the instrument division of CEC was acquired by DuPont Instruments where Mr. McCleary was Products Manager for mass spectrometers, leak detectors, ESCA (Electron Scattering Chemical Analysis — an x-ray technique), and moisture analyzers (quartz, crystal, diphosphorous pentoxide and NIR). In 1976, he joined Perkin Elmer as Marketing Manager for Industrial and Medical Mass Spectrometers and then became Director of Medical Products. Perkin Elmer designed, manufactured, and marketed continuous-monitoring mass spectrometers for process applications and for patient monitoring in hospitals.

In 1988, Mr. McCleary joined Milton Roy as Marketing Manager for their line of process and environmental-monitoring mass spectrometers. When Milton Roy was acquired and the mass spectrometer product line sold in 1992, Mr. McCleary began a consulting business for companies interested in on-line industrial mass spectrometers.

Appendix 1
ROI — An Economic Evaluation

The following lists an economic evaluation Excel™ spreadsheet that was developed by Thomas Ward of Clarkson University.

```
A:AI:   'INPUT PARAMETERS
A:EI:   '
A:I1:   '
A:B2:   'DISCOUNT RATE
A:C2:   0.1
A:E2:   'PROJECT START:
A:G2:   1993
A:B3:   'DEBT/TOTAL CAPITAL
A:C3:   0
A:E3:   'PROJECT END:
A:G3:   2002
A:B4:   'FED INC TAX RATE
A:C4:   0.34
A:E4:   'DEPRECIATION: STRAIGHT LINE
A:B5:   'INVEST TAX CREDIT
A:C5:   0
A:E5:   'SALVAGE: NEGLIGIBLE
A:B6:   'INTEREST RATE
A:C6:   0
A:E6:   'END-OF-YEAR FLOWS
A:B7:   'PRESENT TIME:
A:C7:   1993
A:E7:   'LINE-ITEM INTEREST NEGLECTED
A:B8:   '
A:C8:   '
A:D8:   '
A:E8:   '
A:F8:   '
A:G8:   '
A:H8:   '
A:I8:   '
A:J8:   '
A:K8:   '
A:L8:   '
A:M8:   '
A:N8:   '
A:A9:   'LINE
A:B9:   'END OF
A:C9:   1993
A:D9:   +$C9+DII
A:E9:   +$C9+EII
A:F9:   +$C9+FII
A:G9:   +$C9+GII
A:H9:   +$C9+HII
A:I9:   +$C9+III
A:J9:   +$C9+JII
A:K9:   +$C9+KII
A:L9:   2002
A:B10:  'OPERATING YEAR
A:C10:  +F10-3
A:D10:  +F10-2
A:E10:  +F10-1
A:F10:  1
A:G10:  +D10+3
A:H10:  +D10+4
A:I10:  +D10+5
A:J10:  +D10+6
A:K10:  +D10+7
A:L10:  +D10+8
A:B11:  'DISCOUNT YEAR n
A:C11:  0
A:D11:  +C11+1
A:E11:  +D11+1
A:F11:  +E11+1
A:G11:  +F11+1
A:H11:  +G11+1
A:I11:  +H11+1
A:J11:  +I11+1
A:K11:  +J11+1
A:L11:  +K11+1
A:A12:  1
A:B12:  'FIXED CAPITAL
A:C12:  '
A:DI2:  -2000
A:EI2:  -2000
A:AI3:  2
A:BI3:  'LAND
A:CI3:  '
A:DI3:  -200
A:KI3:  '
A:L13:  -D13
A:A14:  3
A:BI4:  'WORKING CAPITAL
A:D14:  '
A:E14:  -400
A:K14:  '
A:L14:  -E14
A:A15:  4
A:B15:  'TOTAL CAPITAL
A:C15:  '
```

```
A:DI5:   +D12+D13+D14
A:E15:   +E12+E13+E14
A:A16:   5
A:B16:   'SALES REVENUE
A:E16:   '
A:F16:   4000
A:G16:   5500
A:H16:   6000
A:I16:   6000
A:J16:   6000
A:K16:   6000
A:L16    6000
A:A17:   6
A:B17:   'PROD COST [LESS DEPR & INT]
A:E17:   '
A:F17:   2400
A:G17:   2000
A:H17:   2000
A:I17:   2000
A:J17:   2000
A:K17:   2000
A:L17:   2200
A:A18:   7
A:B18:   'CAPITALIZED FIXED CAPITAL
A:D18:   '
A:E18:   +D12*(1+$C2)^1+E12
A:A19:   8
A:B19:   'CAPITALIZED TOTAL CAPITAL
A:D19:   '
A:E19:   +D15*((1+$C2)^1)+E15
A:A20:   9
A:B20:   'OPERATING INCOME
A:E20:   '
A:F20:   +F16-F17
A:G20:   +G16-GI7
A:H20:   +H16-H17
A:I20:   +I16-I17
A:J20:   +J16-J17
A:K20:   +K16-K17
A:L20:   +L16-L17
A:A21:   10
A:B21:   'INTEREST ON DEBT
A:E21:   '
A:F21:   0
A:G21:   0
A:H21:   0
A:I21:   0
A:J21:   0
A:K21:   0
A:L21:   0
A:A22:   11
A:B22:   'DEPRECIATION [TAX BASIS]
A:E22:   '
A:F22:   -$E18/$C48
A:G22:   -$E18/$C48
A:H22:   -$E18/$C48
A:I22:   -$E18/$C48
A:J22:   -$E18/$C48
A:K22:   -$E18/$C48
A:L22:   -$E18/$C48
A:A23:   12
A:B23:   'NET TAXABLE INCOME
A:E23:   '
A:F23:   +F20-F21-F22
A:G23:   +G20-G21-G22
A:H23:   +H20-H21-H22
A:I23:   +I20-I21-I22
A:J23:   +J20-J21-J22
A:K23    +K20-K21-K22
A:L23:   +L20-L21-L22
A:A24:   13
A:B24:   'FEDERAL INCOME TAX
A:E24:   '
A:F24:   +F23*$C4
A:G24:   +G23*$C4
A:H24:   +H23*$C4
A:I24:   +I23*$C4
A:J24:   +J23*$C4
A:K24:   +K23*$C4
A:L24:   +L23*$C4
A:A25:   14
A:B25:   'INVESTMENT TAX CREDIT
A:E25:   '
A:F25:   0
A:G25:   0
A:H25:   0
A:I25:   0
A:J25:   0
A:K25:   0
A:L25:   0
A:A26:   15
A:B26:   'NET INCOME
A:E26:   '
A:F26:   +F23-F24+F25
A:G26:   +G23-G24+G25
A:H26:   +H23-H24+H25
A:I26:   +I23-I24+I25
A:J26:   +J23-J24+J25
A:K26:   +K23-K24+K25
A:L26:   +L23-L24+L25
A:A27:   16
A:B27:   'CASH FLOW
A:C27:   +C12+C13+CI4+C26+C22
A:D27:   +D12+D13+DI4+D26+D22
A:E27:   +E12+E13+EI4+E26+E22
A:F27:   +F12+F13+FI4+F26+F22
A:G27:   +G12+G13+GI4+G26+G22
A:H27:   +H12+H13+HI4+H26+H22
A:I27:   +I12+I13+II4+I26+I22
A:J27:   +J12+J13+JI4+J26+J22
A:K27:   +K12+K13+KI4+K26+K22
A:L27:   +L12+L13+LI4+L26+L22
A:A28:   17
A:B28:   'CASH FLOW + INTEREST
A:C28:   +C27+C21
A:D28:   +D27+D21
A:E28:   +E27+E21
A:F28:   +F27+F21
A:G28:   +G27+G21
A:H28:   +H27+H21
A:I28:   +I27+I21
A:J28:   +J27+J21
A:K28:   +K27+K21
A:L28:   +L27+L21
A:A29:   18
A:B29:   'DISCOUNTED FLOW
A:C29:   '
A:D29:   '
A:E29:   '
A:F29:   '
A:G29:   '
A:H29:   '
```

```
A:I29:    '
A:J29:    '
A:K29:    '
A:L29:    '
A:M29:    '
A:N29:    '
A:C30:    +C28*(1+$C2)^-C11
A:C31:    +D28*(1+$C2)^-D11
A:C32:    +E28*(1+$C2)^-E11
A:C33:    +F28*(1+$C2)^-F11
A:C34:    +G28*(1+$C2)^-G11
A:C35:    +H28*(1+$C2)^-H11
A:C36:    +I28*(1+$C2)^-I11
A:C37:    +J28*(1+$C2)^-J11
A:C38:    +K28*(1+$C2)^-K11
A:C39:    +L28*(1+$C2)^-L11
A:C40:    '
A:A41:    'PROFITABILITY PARAMETERS
A:B42:    'DISCOUNTED TOTAL CAPITAL
A:C42:    -D15*((1+$C2)^-D11)-(E15*(1+$C2)^-$E11)
A:B43:    'NET PRESENT VALUE
A:C43:    @SUM(C30..C39)
A:B44:    'NET RETURN RATE [%/YEAR]
A:C44:    (100*C43)/((+C42)*C47)
A:B45:    'NET PAYOUT TIME [YEARS]
A:C45:    2.75
A:B46:    'NET PAYOUT FRACTION
A:C46:    0.392
A:B47:    'VENTURE LIFE [YEARS]
A:C47:    9
A:B48:    'OPERATING LIFE [YEARS]
A:C48:    7
A:A50:    '
A:B50:    '
```

Appendix 2

Computer Simulation of Control Charts

In Chapter 5 it was noted that the control charts used throughout the chapter were computer-simulated in order to protect the confidentiality of the data from otherwise-real case studies. However, the key features of the actual control charts were preserved in the simulations, since simulations are sometimes not believed to represent or even resemble the "real world." This appendix is included in order to help the reader mentally connect the control charts with the actual case studies. Furthermore, some readers may be curious about how to simulate a control chart or may want to experiment with various control chart features without going through tedious manual calculations or waiting until a real process goes out of control just to see what happens. Naturally, we hope that real processes will not get out of control!

Computer simulation and modeling of statistical processes is frequently a fast and efficient way to understand actual process performance without having to experiment on the commercial manufacturing equipment.

Computer simulation and modeling of statistical processes is frequently a fast and efficient way to understand actual process performance without having to experiment on the commercial manufacturing equipment.

Control charts can be simulated with any personal computer spreadsheet program that has a uniform distribution random number generator and a graphics function. The uniform distribution, sometimes called the "boxcar" distribution, exhibits values with equal probability regardless of the distance from the mean. In other words, the distribution is rectangular (like a boxcar) rather than bell-shaped (like the normal distribution).

Figure A2-1 is a control chart simulation template for generating a control chart with 30 points. Thirty points is a conveniently small number of points to use without too much risk of making a decision based on insufficient data. The left part of the template has a set of ten uniformly distributed random numbers for calculating each of the 30 normally distributed random numbers. This is an application of the central limit theorem. One could do an acceptable simulation with only five uniformly distributed numbers, but ten numbers gives a better simulation. The "copy" feature on most spreadsheets requires only a few more keystrokes to achieve statistically better results.

POINT	RANDOM NUMBERS 1	2	3	.....	10	RANDOM NUMBERS AVERAGES	SIMULATED PROCESS
1	RANDOM	RANDOM	RANDOM	.....	RANDOM	$\bar{X}$ for ROW 1	ARITHMETIC OPERATION PERFORMED ON RANDOM NUMBER AVERAGES
2	RANDOM	RANDOM	RANDOM	.....	RANDOM	$\bar{X}$ for ROW 2	
3	RANDOM	RANDOM	RANDOM	.....	RANDOM	$\bar{X}$ for ROW 3	
4	RANDOM	RANDOM	RANDOM	.....	RANDOM	$\bar{X}$ for ROW 4	
.	.	.	.		.	.	
.	.	.	.		.	.	
29	RANDOM	RANDOM	RANDOM	.....	RANDOM	$\bar{X}$ for ROW 29	
30	RANDOM	RANDOM	RANDOM	.....	RANDOM	$\bar{X}$ for ROW 30	

CALCULATED STATISTICS ON SIMULATED PROCESS:

AVERAGE
STANDARD DEVIATION
UPPER CONTROL LIMIT
LOWER CONTROL LIMIT

UPPER PROCESS CAPABILITY
LOWER PROCESS CAPABILITY

Figure A2-1. Control Chart Simulation Template for Using a Spreadsheet to Generate a 30-Point Control Chart

The column next to the random numbers contains the averages of the uniform random numbers; these are the 30 normally distributed random numbers that are the "foundation" of the control chart. However, these normally distributed random numbers are in the range 0 to 1, the same range as the uniformly distributed numbers from which they were calculated. The standard deviation of the normally distributed numbers, each calculated from ten uniformly distributed numbers is about 0.091 if the uniform numbers are in the 0 to 1 range. To obtain normally distributed random numbers in other ranges and with other standard deviations, one can perform combinations of the basic arithmetic operations of addition, subtraction, multiplication, and division on the original set.

The original set of normally distributed random numbers or any set derived from them by performing the same operation(s) on each number should produce an "in-control" control chart. Furthermore, an unlimited number of control charts is possible by using the "recalculate" feature of the spreadsheet. One can also set up individual cells off the block of random numbers for calculating the mean, standard deviation, control limits, the number of out-of-control points, and other statistics derived from these.

Although using the spreadsheet functions for counting the out-of-control points is tedious, it is a useful tool for doing a "rule of thumb" test on the validity of the random number generator. Computer random number generators are "algorithms" programmed into the software like other spreadsheet functions. The difference is that they are dependent upon some changing measurement or number not believed to follow a pattern. The algorithm is not usually published in the software documentation because most users are not interested in the "nuts and bolts" of it.

The random number generator in any software is not random in a strict mathematical sense; most are fairly good approximations, however. Therefore, if the 3σ control limits theoretically include 99.7% of the data, or 997 out of 1,000 points, we should ideally find only three out-of-control points in 1,000 generated points.

If the spreadsheet generates 30 normally distributed random numbers at a time, we can generate a succession of 1,020 such random numbers by using the "recalculate" function 34 times. This takes only a few minutes if the spreadsheet program dedicates one key to "recalculate." By watching the cell that displays the "out-of-control" point count, one should find, on average, only three out-of-control points for every 1,000 normally distributed random numbers, or one for every 11 recalculations. Use the "recalculate" function with caution: Once "recalculate" is used, the previous set of data is lost forever unless a hard copy or disk file has been made of the spreadsheet. A hard copy is better, since it cannot be accidentally recalculated and overwritten.

Computer-generated random numbers are lost forever unless a hard copy or disk file is made before using the recalculate function.

In the same way that the basic set of normally distributed random numbers is scaled using combinations of the arithmetic operations, so can the various out-of-control patterns be simulated. For example, two sudden shifts in the data can be simulated by using one multiplier for the first ten random numbers, a somewhat different one for the second ten, and very different one for the last ten. A trend can be simulated by adding higher and higher (or lower and lower) increments to each successive random number starting with the first one.

Further discussion about random number generators is not within the scope of this Appendix or the topic of process control. However, simulations are of general interest in process control work, and random numbers can be useful for simulating realistic situations since processes of the types discussed here possess inherent random variation.

About the Author

Gary D. Nichols received his B. S. in chemistry from California Polytechnic State University at San Luis Obispo and his M. S. in analytical chemistry from the University of Georgia. He has worked in process analyzer development, applications, engineering, and maintenance with PPG Industries, BASF Corporation, Exxon Chemicals, and Sunoco. He is currently analyzer technologist with Sterling Chemicals in Texas City, TX. He is a Senior Member of ISA. In 1988, he published *On-Line Process Analyzers*. He has been active in ISA's Certified Specialist in Analytical Technology (CSAT) program and its SP76 Standards and Practices Committee for Composition Analyzers. He is also an ASQC Certified Quality Engineer (CQE). He is the News Editor for *Process Control and Quality* and is a contributing author to the *Encyclopedia of Analytical Chemistry*.

Appendix 3
Linear Regression and Least Squares

Linear regression is a statistical method for determining the mathematical equation, or general relationship, between a dependent variable and one or more independent variables. Linear regression is not, strictly speaking, an SPC/SQC tool, but it is a statistical tool that is frequently used with SPC/SQC in process analyzer work. Despite the term "linear," the method can be used for finding polynomial relationships that do not necessarily result in a straight line, although the straight line is the most common objective of linear regression. A straight line is a "polynomial of degree one." The discussion in this Appendix is restricted to straight-line relationships, or "simple linear regression." Reference 1 provides a highly readable and easily applied discussion of linear regression, least squares, and other statistical methods.

Linear regression is a broad term. The most popular method for finding straight lines and other polynomial relationships for experimental data is the method of least squares. The equation for a straight line is:

$$y = mx + b \qquad \text{(A3-1)}$$

where:

y = dependent variable

x = independent variable

m = slope of the line

b = y-intercept

The y-intercept is the point where the line intercepts the y-axis (at $x = 0$). The method of least squares gives the values of the slope and the y-intercept. In SPC/SQC we are usually not so much interested in determining the precise values of m and b as we are in finding whether a general straight-line relationship exists, and whether "significant" slope and y-intercept exist. The qualitative use of these pieces of quantitative information is necessary in process analyzer applications work and in diagnosing and solving process analyzer applications problems.

To illustrate the method of least squares, assume we wish to find out whether the response of a process analyzer is linear with respect to the chemical concentration of interest in the process stream. We carefully prepare a series of standards in the laboratory by a reliable method (e.g., weight percent) using an analytical balance and cross-check them by a reliable laboratory analytical method such as titration. These results are given in Table A3-1. Note that for each equal increment of the component of interest in successive standards, the same additional amount of titrant is required to reach the end point.

Figure A3-1 is the least-squares graph of this data; the slight deviation of the data points from the straight line is attributable to random experimental variation or experimental error. Note that the y-intercept is zero and the slope is 1.00. The zero y-intercept means that if none of the components of interest is present, the analytical method so indicates. The slope shows not only that the analytical method responds proportionally to the amount of the component of interest present but also that the response, or sensitivity of the method, is large enough to be useful for measuring concentration differences. This is ideal behavior for any analytical method. We should note that in some analyzer applications, linearity cannot be attained due to technological limitations and that for specific applications, accuracy of the measurement at a single point may be more important than linearity over a wide range.

... for specific applications, accuracy of the measurement at a single point may be more important than linearity over a wide range.

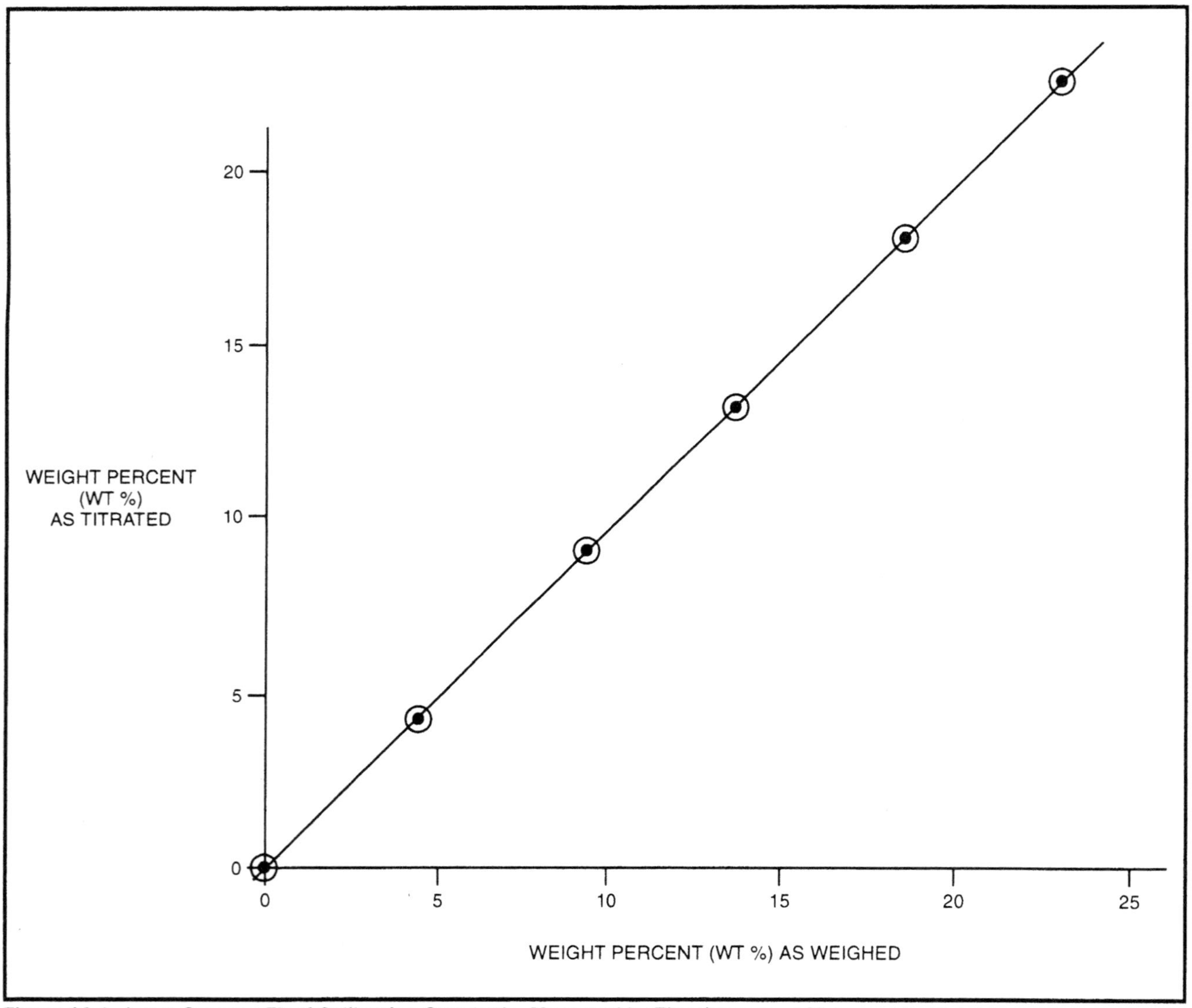

Figure A3-1. Least-Squares Fit of Calibration Standards: Measured by Titration versus Measured by Weight

Table A3-1. Tabular Data of the Concentration of Calibration Standards as Measured by Weight and by Titration

Wt% as Weighed on Balance	Wt% as Measured by Titration
0	0
4.37	4.34
9.28	9.30
13.62	13.63
18.17	18.22
22.93	22.89

We now wish to find out whether the process analyzer behaves as ideally as the laboratory method. At this point, we take the standards to the field, take the process analyzer off-line, run each of the standards through the analyzer system and measure the process analyzer response. These data are shown in Table A3-2, along with the concentrations calculated from the weights when the standards were prepared. Figure A3-2 is the graph of the data and the least-squares fit of a straight line through the points.

Before we discuss the least-squares fit for these data, let us qualitatively find out what "least squares" means. Note that each of the data points is above or below the least-squares straight line. Intuitively, we would expect that the "best straight line" through the points is the one which, overall, is "closest" to the individual points. This is partially correct. Since y is the dependent variable and x is the independent variable, we assume that the x coordinates for the points are correct; therefore, we wish to get the overall closest fit in the y, or vertical, direction. The least-squares method, as the name implies, actually uses the squares of the distances from the points to the fitted line to determine which line is best. If only the distances were used, more than one "best fit" could be calculated; this would be physically meaningless. Using the squares of distances allows the calculation on only one "best fit." The details of the calculations are too lengthy for this discussion, but a one-page worksheet for the calculations is given in Ref. 1.

Table A3-2. Tabular Data of the Concentration Calibration Standards as Measured by Weight and on a Process Analyzer

Wt% as Weighed on Balance	Wt% as Measured on Process Analyzer
0	0.08
4.37	3.75
9.28	8.00
13.62	11.52
18.17	15.76
22.93	19.54

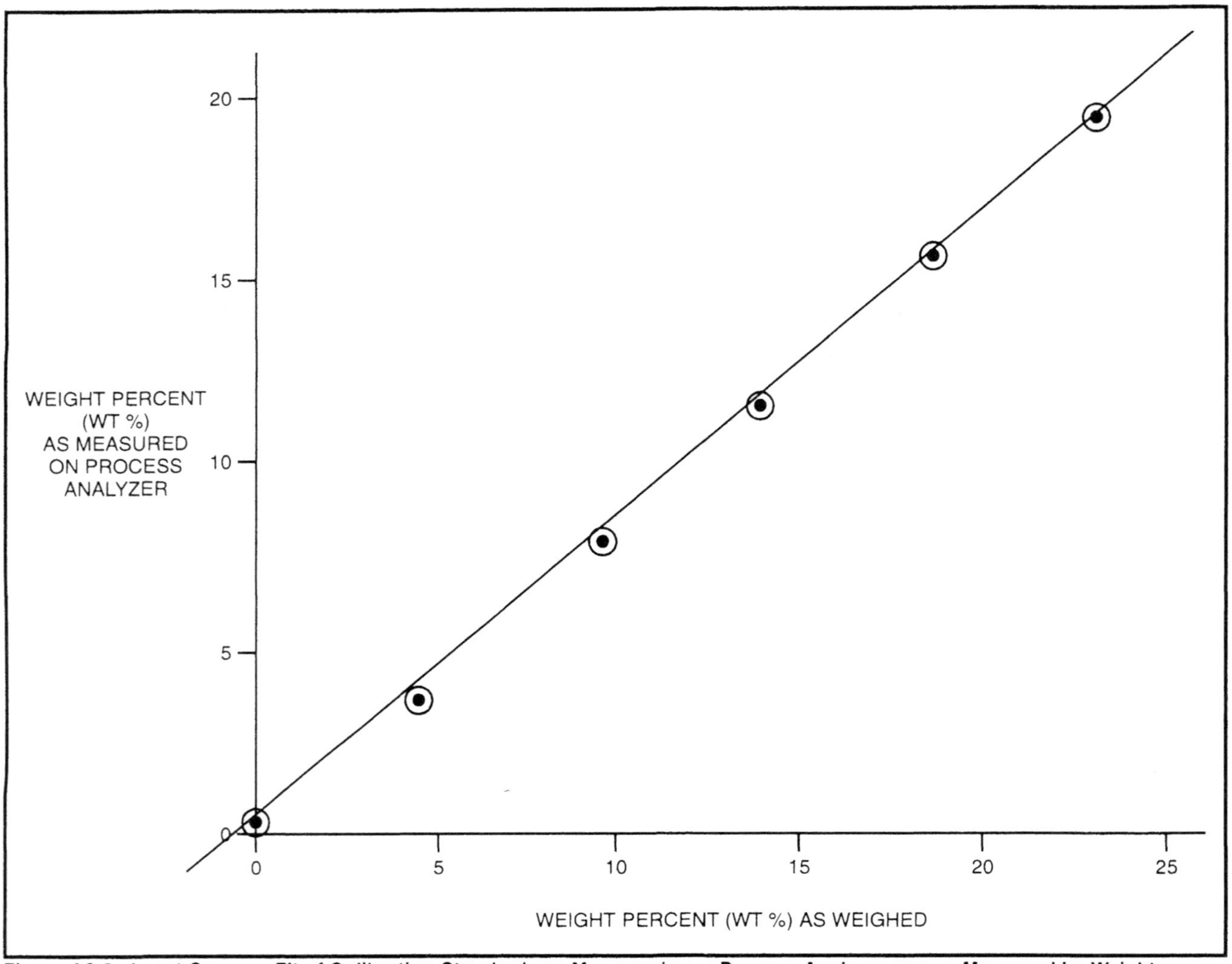

Figure A3-2. Least-Squares Fit of Calibration Standards as Measured on a Process Analyzer versus Measured by Weight

A nonzero *y*-intercept may be acceptable if analyzer precision is more important than absolute accuracy.

The y-intercept for the process analyzer data is 0.08 and the slope is 0.95. The y-intercept does not pass through zero, as we would wish. The cause for this may be that the zero on the analyzer is not set correctly. If this is the case, a "zero" standard should be run through the analyzer system and the zero adjustment changed so that the readout from the instrument is zero under these conditions. In an actual case, there may be other reasons for this, but the point is that the nonzero y-intercept is an indication that the process analyzer may not be functioning as desired. The slope does not appear to be quite as steep as that for the laboratory method, but it indicates adequate sensitivity for the application. If we are using calibration standards within the concentration range of interest and for which the analyzer is designed, we would expect a slope of 1.00, or a 45° angle in the least-squares plot. The gain on the process analyzer can be adjusted to bring the slope to 1.00. After these adjustments on the analyzer are completed, it is a good idea to run the series of standards through the analyzer again to make sure that the adjustments are correct.

Least-squares slope and intercept correspond, respectively, to electronic gain and electronic zero when working with instrumentation.

Least-squares slope and intercept correspond, respectively, to electronic gain and electronic zero when working with instrumentation.

If we compare the least-squares line for the laboratory method to that for the process analyzer, we note that the data points for the former are, in general, much closer to the line than for the latter. The process analyzer data have more "spread" or "scatter." The statistical measure for how well the fitted line describes the data is the correlation coefficient. The correlation coefficient for the laboratory titration method is 0.99999 and for the process analyzer is 0.99986. The lower correlation coefficient for the process analyzer data reflects the greater spread or scatter in the least-squares plot. Both of these correlation coefficients are unusually high, since this is only an illustration. Even "good" fits usually have correlation coefficients much lower; more than two or three decimal places is seldom necessary to judge whether the fit is acceptable.

In this example, we'll use the least-squares method as a statistical tool to find out what effect, if any, ambient temperature has on process analyzer performance. Experimentally, we block a stable sample into the analyzer and sample system and watch the analytical output over several hours as the ambient temperature changes. The following temperature (in °F) and concentration measurement (in %) data pairs are obtained: (60, 41.0); (68, 41.3); (73, 41.6); (79, 42.0); (85, 42.4); and (91, 42.7). When the least-squares function on a pocket scientific calculator is applied to these measurements, the slope, *y*-intercept, and correlation coefficient are found to be 0.057, 37.46, and 0.995, respectively. The correlation coefficient indicates a good first-order fit. In this case, the *y*-intercept is of no particular interest since we do not expect the ambient temperature to be 0°F near the analyzer. But the slope tells us that for every 1F° change in ambient temperature, the analyzer can be expected to read differently by about 0.06%, or greater than 0.57% change for each 10F° ambient temperature change. This may or may not be significant in terms of operating the manufacturing process, but the fact that the effect is measurable using an ordinary weather thermometer and monitoring the analyzer output indicates that there is probably much room for improving the analyzer installation.

The correlation coefficient can be positive, negative, or zero; it must be between −1 and 1. Zero indicates no correlation between the dependent variable and the independent variable. An "inverse" correlation exists if the correlation coefficient is negative. The positive correlations in the example above are more common. The closer the correlation coefficient to 1 or −1, the better the least-squares line is for describing the relationship between the two variables. However, we generally cannot say how close to 1 or −1 the correlation coefficient must be to be "good enough" for the application at hand; that question is answered by the particulars of the application.

In applying least squares, one must be cautious as to whether the dependent variable truly depends on the independent variable. Simply because a mathematical relationship can be demonstrated does not, by itself, indicate that the same is true on the actual manufacturing equipment. A "false correlation" can lead to seriously incorrect decisions.

In most ways, the above example typifies the use of the method of least squares for addressing process analyzer and other control and technical situations. However, it is possible to "force" a least-squares fit onto data for which the method is inappropriate. The calculations can be performed on any set of numbers, irrespective of what the numbers represent. Before applying least-squares calculations, one should graph the data points on paper (or a computer screen) to see their general appearance. Some data will show a distinct curvature or other pattern that differs from a straight line. In these cases, the least-squares procedure is of no advantage and may even be misleading. The shape may, however, indicate that some problem needs to be addressed. In work with analytical instrumentation, a curve may indicate that the equipment is being operated outside the concentration range for which it was intended, that a part is malfunctioning, or that something is occurring in the sample which was not anticipated in the application and design of the analyzer system. It is also important to guard against "extrapolating" the least-squares line outside the range of the experimental data from which it was calculated, because both the manufacturing process and the control instrumentation may behave entirely different and in undesirable ways above and below the original data range.

It cannot be emphasized enough that these and all other statistical methods are simply tools that aid the judgement of experienced technical specialists. These tools cannot take the place of such judgement and, in the absence of that judgement, can be worse than no tools at all. Another caution is in order: People who work in technology are often inclined to "worship at the altar of the straight line." That is, they try to force-fit a straight line on many processes and concepts that do not warrant it by inappropriately modifying manufacturing equipment, making the calculations or control system unnecessarily complicated, or other artificial and arbitrary means. This reaction is only natural since the straight line is one of the simplest and most easily understood mathematical concepts. To avoid falling into this trap, it is sometimes best to apply the method of least squares only after other statistical methods, such a control chart, have been used to find out how the process behaves. If least squares seems appropriate in the early stages of a project, it is okay to go ahead with it, but one must exercise restraint in acting on only the least-squares results before exhausting the power of other SPC/SQC tools. In some cases, analyzer validation at one point is more important than linearity over a wide concentration range.

... statistical methods are simply tools that aid the judgement of experienced technical specialists.

Misapplied statistical tools are usually worse than no statistical tools at all!

The details of the method of least squares are not given in this appendix since the intent is only to make the reader aware of their existence and use. Computational details and additional background are available in many books on statistics. Furthermore, many inexpensive scientific and business pocket calculators have simple linear regression as a built-in function; explanation for its use is usually in the accompanying manual. Least-squares methods are also available in most PC-based statistics and mathematical software.

Reference

1. Natrella, Mary G., *Experimental Statistics*, National Bureau of Standards Handbook 91, Washington, DC: Superintendent of Documents, 1963.

About the Author

Gary D. Nichols received his B. S. in chemistry from California Polytechnic State University at San Luis Obispo and his M. S. in analytical chemistry from the University of Georgia. He has worked in process analyzer development, applications, engineering, and maintenance with PPG Industries, BASF Corporation, Exxon Chemicals, and Sunoco. He is currently analyzer technologist with Sterling Chemicals in Texas City, TX. He is a Senior Member of ISA. In 1988, he published *On-Line Process Analyzers*. He has been active in ISA's Certified Specialist in Analytical Technology (CSAT) program and its SP76 Standards and Practices Committee for Composition Analyzers. He is also an ASQC Certified Quality Engineer (CQE). He is the News Editor for *Process Control and Quality* and is a contributing author to the *Encyclopedia of Analytical Chemistry*.

Appendix 4 Discrimination Ratios

During the normal operation of an infrared-type spectroscopic analyzer, the infrared intensity through the reference beam remains constant and the light energy from the absorption cell is attenuated by the analyte. The amount of infrared attenuation is proportional to the concentration, path length, and infrared absorption properties of the compounds in the cell. This phenomenon has been described (see Chapter 28) by the Beer-Lambert law of light absorption:

$$A = log\ I_0/I = abc$$

Consider a sensitization of an infrared-type analyzer and assume that:

- A' is equivalent to a full-scale signal
- c' concentration of the component of interest
- A'' is equivalent to a partial full-scale signal
- c'' concentration of a background component
- a' and a'' are absorption coefficients (absorptivities) for each component

then,

$$\frac{A''}{A'} = \frac{a''c''b}{a'c'b} = \frac{a''c''}{a'c'} \quad \text{or} \quad \frac{a'}{a''} = \frac{C''A'}{C'A''}$$

Then, for these two compounds and this particular optical system, the ratio of absorptivities, a''/a', can be determined. This ratio is usually referred to as a discrimination ratio (DR) and is usually expressed as:

$$DR = \frac{A}{B \times C} \quad \text{or} \quad B = \frac{A}{DR \times C}$$

Where:

DR = discrimination ratio for the particular background component and specific sensitization

A = average quantity of contaminant expressed in volume or mole percent

B = full-scale reading produced by A quantity and expressed as a decimal value

C = full-scale range of the analyzer expressed in the same terms as A

Therefore, if *DR* values are available from the manufacturer, the feasibility of utilizing a selected optical configuration for a particular compound in the presence of other compounds can be determined. For example, consider the analysis of 0 to 0.1% carbon monoxide in the following stream with the appropriate *DR* values.

Compound	Stream Analysis, %	DR Value	Analysis Error % Full Scale	% CO
Carbon monoxide	0-0.1	1	100	0.1
Carbon dioxide	0-1	5,000	0.2	0.00002
Water	0-2.5	2,500	1.0	0.001
Oxygen	20	00	0	0
Nitrogen	80	00	0	0

The *DR* values for CO_2 and water vapor are obtained from the table, and the *DR* formula is solved for *B*. For example, to calculate the response to 1% CO_2 with the above values:

$$B = \frac{A}{DR \times C} = \frac{1\%}{5,000 \times 0.1\%} = \frac{1}{500} = 0.002$$

$$= \text{equivalent to } 0.00002\% \text{ (0.2 ppm) CO}$$

Index

Numerics

A

B

C

D

E

F

G

H

I

N

O

P

Q

R

T

U